Analysis, Geometry and Topology of Elliptic Operators

Analysis, Geometry and Topology of Elliptic Operators

Editors

Bernhelm Booß-Bavnbek
Roskilde University, Denmark

Slawomir Klimek
IUPUI, USA

Matthias Lesch
Universität Bonn, Germany

Weiping Zhang
Nankai University, China

World Scientific

NEW JERSEY • LONDON • SINGAPORE • BEIJING • SHANGHAI • HONG KONG • TAIPEI • CHENNAI

Published by

World Scientific Publishing Co. Pte. Ltd.

5 Toh Tuck Link, Singapore 596224

USA office: 27 Warren Street, Suite 401-402, Hackensack, NJ 07601

UK office: 57 Shelton Street, Covent Garden, London WC2H 9HE

British Library Cataloguing-in-Publication Data
A catalogue record for this book is available from the British Library.

ANALYSIS, GEOMETRY AND TOPOLOGY OF ELLIPTIC OPERATORS

ISBN 981-256-805-0

Printed in Singapore by World Scientific Printers (S) Pte Ltd

Contents

Preface

On May 20-22, 2005, a workshop was held at Roskilde University in Denmark to honour Krzysztof P. Wojciechowski on his 50th birthday. This volume collects the papers of that workshop.

The purpose of the volume is twofold. The more obvious one is to acknowledge and honour Krzysztof Wojciechowski's contributions over the last 20-25 years to the theory of elliptic operators. Lesch's write up goes over many of Krzysztof's achievements, highlighting those insights that were particularly influential in shaping the direction of the theory. It is supplemented by Park's review of recent work pioneered by Wojciechowski. As our second purpose, we also hope to offer younger researchers and graduate students a snapshot of the current state of affairs. The proceedings contain a mix of review and research papers, both reflecting on the past and looking into the future. We obviously do not attempt to speak for the whole, vast area of the theory of elliptic operators. Most papers in these proceedings are, in one way or another, studying objects and techniques that have interested Krzysztof: spectral invariants, cutting and pasting, boundary value problems, heat kernels, and applications to topology, geometry and physics.

The modern theory of elliptic operators, or simply elliptic theory, has been shaped by the Atiyah-Singer index theorem created some 40 years ago. The Atiyah-Singer index theory expanded the scope of ellipticity to consider relations with and applications to topology. The notion of index acquired a dual personality, both analytical and topological. Consequently, wherever topological invariants appear, one is now tempted to see if the analytical aspects can be developed to interpret the invariant. In other words, analysts are always on the lookout for topological or geometrical invariants hoping to find operators behind them. Developments in topology are therefore of special interest to elliptic theorists. Bleecker's paper revisits some aspects of the so called embedding proof of the Atiyah-Singer index theorem. The contributions of Bunke and Schick on T-duality and Nicolaescu's survey of singularities of complex surfaces detail some topological theories of potential interest to analysts and possible applications of analytical methods.

Heat kernel techniques are at the heart of another one of the several proofs of the Atiyah-Singer index theorem. Different tools and techniques have been developed and are continuing to be developed to understand heat kernels and related spectral functions in a variety of situations. Two problems stand out: to describe and compute variations of heat kernels with respect to parameters and to calculate asymptotics of heat kernels – like functions of operators. These have been the central technical issues for much of Krzysztof Wojciechowski's work. As the scope of elliptic theory increases, so is the variety of contexts for heat kernel calculations which will undoubtedly occupy the interest of people in the future. The papers of Avrimidi on heat kernels of non-Laplace operators, of Furutani on heat kernels on nilpotent Lie groups, of Grubb on expansions of zeta-like functions, and of Paycha and Rosenberg on canonical traces all fall into this category.

Since the original papers on index theory, elliptic theory has continued to develop. More areas of mathematics, other than topology, have started influencing its progress. More and more objects of a similar nature to index have been investigated. For one thing, index is a very simple spectral invariant, and an important branch of elliptic theory looks at other spectral invariants and their geometrical and topological significance. We need to mention here some invariants that have particularly interested Krzysztof: the eta invariant, spectral flow, analytic torsion and infinite dimensional determinants. But there are many other invariants such as Seiberg-Witten invariants and elliptic genus. We expect that this list is not complete and that the future will bring more analytic invariants with topological and geometrical applications.

In the spirit of topological surgery theory, a major effort was undertaken to study elliptic operators and their spectral invariants using "cutting and pasting". This naturally leads to the problem of how to set up an elliptic theory on manifolds with boundary. This is the subject that Krzysztof has devoted most of his mathematical efforts shaping. The papers of Dai on eta invariants, of Ma and Zhang on L^2-torsion, and Park's review of gluing formulas for zeta determinants, as well as the contribution of Lesch, give the state of the art for at least some of the questions in this area.

Beside topology, the operator theory and operator algebras have been and will in the future be a driving force in the development of elliptic theory. What started with the analysis of a single Fredholm operator on a manifold, acquired greater depth and importance by considering whole spaces of operators. With the invention of operator K-theory, elliptic theory is evolving in a more abstract, algebraic fashion. Ellipticity is now defined not just for (pseudo) differential operators and not just on manifolds with or without boundary or even with corners. The proper context for the study of ellipticity is noncommutative differential geometry. Noncommutative geometry aims to consider discrete spaces as well as noncommutative objects on equal footing with topological spaces. Moreover, there is a duality which runs even deeper with the modern interpretation of an elliptic operator as a K-cycle over a C^*-algebra. It seems quite possible, and even likely, that such more algebraic trends will constitute the mainstream of elliptic theory in the future. Operator-theoretic contributions to this volume include papers by Benameur et al. on spectral flow in von Neumann algebras, by Douglas on a new kind of index theorem, by Klimek on a noncommutative disk, by Mickelsson on star products and central extensions and by Wurzbacher on homotopy calculations for some spaces of operators, while Dodziuk explores elliptic theory in a discrete setting.

Theoretical particle, string and membrane physics have and will continue to provide major motivation for elliptic theory. As the world of elementary particles continues to expand, one naturally suspects that the so-called elementary particles are not so elementary any more. Some of the current theories develop the idea that the basic structures of the universe are not point-like but rather stringy– or membrane-like. Such objects would naturally live in dimensions higher than our 4 dimensional world. To write down laws for such objects one is lead to modern global analysis involving arbitrary dimensional manifolds and operators on them. Of course new structures and new ideas also appear, such as supersymmetry, conformal symmetry, mirror symmetry and anomalies. Many exciting new mathematical questions arise. Several papers in this volume follow this line of reserach: Bunke and Schick on very general mirror symmetry, Esposito et al. on quantum gravity, Paycha and Rosenberg on conformal anomalies, Paycha and Scott on superconnections, Zhu on symplectic functional analysis and Hamiltonian dynamics.

With its intricate theory, powerful methods and variety of applications, the theory of elliptic operators should stay in the forefront of mathematics for long years to come. The fact that this has been the case in the recent past, is due in a nontrivial way to the work and insights of Krzysztof Wojciechowski.

Acknowledgements. We thank the authors for their contributions, the Non-commutative Geometry Center and the PDE Network of The Danish Science Research Council for financial support, Roskilde University for hospitality, and Benjamin Himpel (Bonn University) and the Staff of World Scientific Publishing Company for encouragement and help in preparing this volume.

The Editors

Part I

On the Mathematical Work of Krzysztof P. Wojciechowski

Analysis, Geometry and Topology
of Elliptic Operators, pp. 3–22

SELECTED ASPECTS OF THE MATHEMATICAL WORK OF KRZYSZTOF P. WOJCIECHOWSKI

MATTHIAS LESCH
Mathematisches Institut
Beringstr. 1
53115 Bonn, Germany
lesch@math.uni-bonn.de

Dedicated to Krzysztof P. Wojciechowski on his 50th birthday

To honor and to please our friend Krzysztof P. Wojciechowski I will review the milestones of his mathematical work. This will at the same time be a tour of Analysis and Geometry of Boundary Value Problems. Starting in the 80s I will discuss the spectral flow and the general linear conjugation problem, the Calderón projector and the topology of space of elliptic boundary problems. The theme of the 90s is the eta invariant. The paper with Douglas was fundamental for establishing spectral invariants for manifolds *with* boundary and for the investigation of the behavior of spectral invariants under analytic surgery. This was so influential that many different proofs of the gluing formula for the eta-invariant were published. Finally turning to the new millennium we will look at the zeta–determinant. Compared to eta this is a much more rigid spectral invariant which is technically challenging.

2000 *Mathematics Subject Classification.* Primary 58J32; Secondary 58J30, 58J28, 58J52, 35P99

1. Introduction

1.1. *The framework and the problem*

To begin with let us describe in general terms the problems to which Krzysztof P. Wojciechowski has contributed so much in the last 25 years.

Let X be a compact smooth Riemannian manifold with boundary $\Sigma = \partial X$. Furthermore, let E, F be hermitian vector bundles over X and let

$$D : \Gamma^\infty(X, E) \longrightarrow \Gamma^\infty(X, F) \tag{1}$$

Supported partially by Sonderforschungsbereich/Transregio 12 "Symmetries and Universality in Mesoscopic Systems" (Bochum–Duisburg/Essen–Köln–Warszawa)

be an elliptic differential operator: $\Gamma^\infty(X,E)$ denotes the spaces of smooth sections of the bundle E.

In this situation some natural questions occur:

1. *What are appropriate boundary conditions for D on X?*

This question is absolutely fundamental since without imposing boundary conditions we cannot expect D to have any reasonable spectral theory.

A boundary condition is given by a pseudo–differential operator

$$P:\Gamma^\infty(\Sigma,E)\longrightarrow\Gamma^\infty(\Sigma,E) \tag{2}$$

of order 0.[a] The *realization* D_P of the boundary condition given by P is the differential expression D acting on the domain

$$\operatorname{dom}(D_P):=\{u\in L^2_1(X,E)\mid P(u|\Sigma)=0\}. \tag{3}$$

Since D is elliptic what one should expect naturally for P to be "appropriate" is that *elliptic regularity* holds. That is if $Du\in L^2_s(X,E)$ [b] is of Sobolev order $s\geq 0$ and if $P(u|\Sigma)=0$ then $u\in L^2_{s+d}(X,E)$ is already of Sobolev order $s+d$, where d denotes the order of D.

2. *What is the structure of the space of all (nice) boundary conditions and how do spectral invariants of D_P depend on the boundary condition?*

These problems are the Leitfaden of Krzysztof P. Wojciechowski's work. If we are given a realization D_P of a nice boundary value problem we can do spectral theory and study the basic spectral invariants of D_P. We will see that the question in the headline leads to interesting and delicate analytical problems. Let us specify the kind of spectral invariants we mean here.

The most basic spectral invariant of the Fredholm operator D_P is its index

$$\operatorname{ind}D_P=\dim\ker D_P-\dim\operatorname{coker}D_P. \tag{4}$$

More rigid (and analytically more demanding) spectral invariants are derived from the heat trace

$$\operatorname{tr}(e^{-tD_P^2})=\sum_{\lambda\in\operatorname{spec}D_P\setminus\{0\}}e^{-t\lambda^2}, \tag{5}$$

[a]One could think of more general definitely nonlocal boundary operators, but in this paper we will content ourselves to pseudo–differential boundary conditions.
[b]We denote the space of sections of E which are of Sobolev order s by $L^2_s(X,E)$.

where D_P is now assumed to be self–adjoint, via Mellin transforms. The most important examples are the η–invariant

$$\eta(D_P) = \left[\frac{1}{\Gamma(\frac{s+1}{2})}\int_0^\infty t^{(s-1)/2}\operatorname{tr}\left(D_P e^{-tD_P^2}\right)dt\right]_{s=0} \tag{6}$$

and the ζ–determinant

$$\log\det_\zeta(D) = -\frac{d}{ds}\left[\frac{1}{\Gamma(s)}\int_0^\infty t^{s-1}\operatorname{tr}\left(e^{-tD_P^2}\right)dt\right]_{s=0}. \tag{7}$$

The existence of these invariants is highly non–trivial since it depends on the meromorphic continuation of the right hand side of (6) and (7).

In the following sense the index is the least rigid and the ζ–determinant is the most rigid of these three invariants. In order not to get into too much technicalities assume for the moment that $D(s)_{a\le s\le b}$ is a smoothly varying family of elliptic operators on a closed manifold.[c]

The index is insensitive to small perturbations of the operator. Hence $\operatorname{ind} D(s)$ will not depend on s at all. The variation of the η–invariant is easy to understand. First of all the *reduced* η–invariant

$$\widetilde{\eta}(D(s)) = \frac{1}{2}(\dim\ker D(s) + \eta(D(s)) \tag{8}$$

has only integer jumps and the total number of jumps equals the *spectral flow* of the family $D(s)$ over the interval [a,b]. The variation of $\widetilde{\eta}(D(s)) \bmod \mathbb{Z}$ is *local* in the sense that $\frac{d}{ds}(\widetilde{\eta}(D(s) \bmod \mathbb{Z})$ is the integral of a density which is a local expression in terms of the coefficients of the operator and its derivatives, cf. Gilkey [12], Sec. 1.13. The variation of the ζ–determinant is more complicated and depends on global data.

It is therefore most natural that the early work of Krzysztof P. Wojciechowski dealt with problems related to the index. The paper [11] with Douglas is a landmark since it is the starting point of a whole decade seeing a lot of papers focusing on the η–invariant and the ζ–determinant. I was told that it came as an almost unbelievable surprise for the mathematical community when η–function and η–invariant for Dirac operators on compact manifolds *with* boundary were established in [11], since until then the η-invariant was only established for closed manifolds and considered solely as a natural *correction* term associated to index problems on manifolds with boundary and living exclusively on the boundary.

---

[c]Here smoothly varying means that all coefficients depend smoothly on the parameter.

The paper [11] already contained one of the major analytical tools which has been refined and exploited ever since: the adiabatic method (see Section 3.1 below).

There is a variant of the problems mentioned above which I would like to point out. Suppose that M is a closed manifold which is *partitioned* by a separating hypersurface $\Sigma \subset M$. I. e. there are compact manifolds with boundary Y, X such that [d]

$$M = Y \cup_\Sigma X. \tag{9}$$

After having chosen appropriate boundary conditions P^X, P^Y for D on X, Y we have three versions of D: D_{P^X}, D_{P^Y} and the essentially self–adjoint operator D on the closed manifold M. In a sense we have "$D = D_{P^Y} \cup D_{P^X}$" and it is natural to ask how the spectral invariants of D, D_{P^X}, and D_{P^Y} are related. Krzysztof P. Wojciechowski and his collaborators have provided us with spectacular results on this problem.

1.2. *The basic framework of boundary value problems for Dirac type operators*

Let us be a bit more specific now and describe the basic set–up of boundary value problems for Dirac type operators as we understand it today.

Let X and D be as before. We assume that D is an operator of *Dirac type.* That is in local coordinates

$$D^2 = -g^{ij}(x) I_{\operatorname{rank} E} \frac{\partial^2}{\partial x_i \partial x_j} + \text{ lower order terms.} \tag{10}$$

This is the most general notion of Dirac operator. The leading symbol of D

$$\sigma_D(x, df) = i[D, f]_x, \quad f \in C^\infty(X), \tag{11}$$

induces a *Clifford module* structure on E. That is we put for $v \in T_x X$ [e]

$$c(v) := \frac{1}{i} \sigma_D(x, v^\flat). \tag{12}$$

Then $c(v)^2 = -g(v, v)$ and hence by the universal property of the Clifford algebra, c extends to a section of the bundle $\operatorname{Hom}(Cl(TX, g), \operatorname{End} E)$

---

[d]This is a situation which is typical for surgery theory in which we would have $\Sigma = S^k \times S^l$, $Y = S^k \times D^l$, where S^k denotes the unit sphere in $\mathbb{R}^{k+1}$ and D^k denotes the unit disc in $\mathbb{R}^k$.

[e]The Riemannian metric provides us with the "musical" isomorphisms $\flat : T_x M \to T_x^* M$ and $\sharp = \flat^{-1}$.

of algebra–homomorphisms between the bundle of Clifford–algebras $Cl(TX, g)$ and the endomorphism bundle $\operatorname{End} E$. This gives E the structure of a Clifford–module.

If we choose a Riemannian connection ∇ on E we can form the Dirac operator D^∇ on E which is locally given by

$$D^\nabla = \sum g^{ij} c((\frac{\partial}{\partial x_i})^\flat)\nabla_{\frac{\partial}{\partial x_j}}. \tag{13}$$

In the terminology of Booß–Bavnbek and Wojciechowski [7] such operators are called "generalized Dirac operators". The operators D^∇ and D obviously have the same leading symbol, hence

$$D = D^\nabla + V \tag{14}$$

with $V \in \Gamma^\infty(X, \operatorname{End} E)$.

Next we have to take the boundary of X into account. We fix a diffeomorphism from a collar U of the boundary onto $N := [0, \varepsilon) \times \Sigma$. Then we may choose a unitary transformation Φ from $L^2(U, E)$ onto the product Hilbert space $L^2([0, \epsilon), L^2(\Sigma, E))$. The operator $\Phi D \Phi^{-1}$ which, by slight abuse of notation, will again be denoted by D then takes the form

$$D|N = J(\frac{d}{dx} + B(x)) + V(x) \tag{15}$$

where $J \in \Gamma^\infty(\Sigma, \operatorname{End} E)$ is a unitary reflection ($J^2 = -I, J^* = -J$), $V \in C^\infty([0, \varepsilon), \Gamma^\infty(\Sigma, \operatorname{End} E))$ and $(B(x))_{0 \le x \le \varepsilon}$ is a smooth family of first order formally self–adjoint differential operators on the closed manifold Σ (called the tangential operator).

Replacing $B(x)$ by $B(x) + J^{-1}V(x)$ be obtain alternatively

$$D|N = J(\frac{d}{dx} + \widetilde{B}(x)) \tag{16}$$

at the expense that now $\widetilde{B}(x)$ has only self–adjoint leading symbol.

We emphasize that J is independent of x and that (15) holds for all operators of Dirac type (Brüning and Lesch [9], Lemma 1.1). The representation (15) of a generalized Dirac operator is crucial for the geometry of their boundary value problems. In the existing literature, one could sometimes get the impression that for (15) to hold one needs that D is the Dirac operator of a Riemannian connection on E as in (13) or even a compatible Dirac operator.

Furthermore, for many results to be presented below only the following properties of D will be needed:

(1) D is first order formally self–adjoint elliptic,
(2) D has the form (15) near the boundary,
(3) D has the *unique continuation property.*

Properties of Dirac operators which are related to Clifford algebras will more or less play no role.

D is formally self–adjoint. That is for sections $f, g \in \Gamma^\infty(X, E)$ we have Green's formula

$$(Df, g) - (f, Dg) = -\int_\Sigma \langle f, g\rangle_{E_x} \,\mathrm{dvol}(x). \tag{17}$$

In order to obtain an unbounded self–adjoint operator in $L^2(X, E)$ we have to impose appropriate boundary conditions.

For a pseudo–differential orthogonal projection[f]

$$P : L^2(\Sigma, E) \to L^2(\Sigma, E)$$

we define D_P to be the differential expression D acting on the domain (3).

Definition 1.1. 1. In the notation of (15) we abbreviate $B_0 := B(0)$ and denote by $P_+(B_0)$ the orthogonal projection onto the positive spectral subspace of B_0. This is a pseudo–differential operator of order 0. Its principal symbol is denoted by $\sigma_{P_+(B_0)}$.

2. The boundary condition defined by P is called *well–posed* if for each $\xi \in T_x^*\Sigma \setminus \{0\}$ the principal symbol $\sigma_P(\xi)$ of P maps range $\sigma_{P_+(B_0)}(\xi)$ bijectively onto range $\sigma_P(\xi)$.

This is Seeley's definition of well–posedness [22]. If P is well–posed then D_P has nice properties.

Proposition 1.1. *Let P be well–posed. Then D_P is a Fredholm operator with compact resolvent. Moreover it is regular in the sense that if a distributional section u of E satisfies $Du \in L^2_s(X, E)$ and $P(u|\Sigma) = 0$ then $u \in L^2_{s+1}(X, E)$, $s \geq 0$.*

It turns out that for Dirac type operators this notion of regularity already characterizes the class of well–posed boundary conditions as was shown by Brüning and Lesch [9].

So far we have basically presented the status of affairs from the point of view of classical elliptic theory.

---

[f]This is not a big loss of generality. It can be shown that if the boundary operator has closed range then the boundary condition may be represented by an orthogonal projection.

2. The early work on spectral flow and the general linear conjugation problem

[25, 4, 5]

The early papers [25, 4, 5] (in part with Booß) on the general linear conjugation problem are fundamental for our todays understanding of the structure of boundary value problems of Dirac type operators. The linear conjugation problem is the natural generalization of the classical Riemann Hilbert problem to elliptic operators (cf. [7], Sec. 26).

Consider a partitioned manifold $M = Y \cup_\Sigma X$ as in (9) and let

$$D = \begin{bmatrix} 0 & D_- \\ D_+ & 0 \end{bmatrix} \tag{18}$$

be a super-symmetric Dirac operator. That is the bundle $E = E^+ \oplus E^-$ is $\mathbb{Z}_2$-graded and D is odd with respect to this grading.

In a collar $N = (-\varepsilon, \varepsilon) \times \Sigma$ of Σ we write D in the form (16) and hence we get for D_+

$$D_+ = \sigma\big(\frac{d}{dx} + B(x)\big) \tag{19}$$

where $\sigma \in \Gamma^\infty(\Sigma, \mathrm{Hom}(E^+, E^-))$ is unitary (and independent of x) and $(B(x))_{-\varepsilon\leq x\leq\varepsilon}$ is a smooth family of elliptic differential operators with self–adjoint leading symbol.

Furthermore, let $\Phi \in \Gamma^\infty(\Sigma, \mathrm{Aut}(E))$ be a unitary bundle automorphism[g] of E which is even with respect to the grading. Multiplication by Φ is a pseudo–differential operator of order 0 which we denote by the same letter. We assume that Φ commutes with the leading symbol of $B(x)$. As a consequence the operator $\Phi B \Phi^{-1} - B$ is of order 0 and $\Phi P_+(B(x)) - P_+(B(x))\Phi$ is of order -1 and thus acts as a compact operator on $L^2(\Sigma, E^+)$.

We introduce a local boundary value problem by letting the differential expression D_+ act on

$$\mathrm{dom}(D_+^\Phi) := \big\{(u_1, u_2) \in L_1^2(Y, E^+) \oplus L_1^2(X, E^+) \;\big|\; u_1|\Sigma = \Phi u_2|\Sigma\big\}. \tag{20}$$

From Green's formula (17) on derives

$$(D_+^\Phi)^* = D_-^{\sigma\Phi\sigma^*} \tag{21}$$

[g]Krzysztof P. Wojciechowski originally treated more generally Φ's which cover a diffeomorphism of Σ. Then multiplication by Φ is a Fourier integral operator.

and thus

$$D^{\Phi\oplus\sigma\Phi\sigma^*} = \begin{bmatrix} 0 & D_-^{\sigma\Phi\sigma^*} \\ D_+^{\Phi} & 0 \end{bmatrix} = \begin{bmatrix} 0 & (D_+^{\Phi})^* \\ D_+^{\Phi} & 0 \end{bmatrix}. \tag{22}$$

One can show that $D^{\Phi\oplus\sigma\Phi\sigma^*}$ is a realization of a local elliptic boundary value problem. Introducing the *Cauchy data spaces*

$$N(D_+, X) := \{u|\Sigma \mid u \in L^2_{1/2}(\Sigma, E^+), D_+u = 0\} \tag{23}$$

we find

$$\begin{aligned} \operatorname{ind} D_+^{\Phi} &= \dim\big((\Phi N(D_+, X)) \cap N(D_-, Y)\big) \\ &\quad - \dim\big((J\Phi^* J^* N(D_-, X)) \cap N(D_-, Y)\big). \end{aligned} \tag{24}$$

Before we can state the main result on the linear conjugation problem we need to elaborate a bit more on the Cauchy data spaces.

2.1. *Calderón projector and the smooth self–adjoint Grassmannian*

Definition 2.1. The (orthogonalized) Calderón projector $C(D, X)$ is the orthogonal projection onto the Cauchy data space $N(D, X)$.

There is a little subtlety here. The natural construction of the Calderón projector via the invertible double (cf. [7], Sec. 12) gives a pseudo–differential (in general non–orthogonal) projection onto the Cauchy data space. It is an orthogonal projection if D is in product form (cf. (36) below) near the boundary. Of course, for any projection there is an orthogonal projection with the same image and using the results of Seeley [24] it follows that

Proposition 2.1. *The orthogonalized Calderón projector* $C(D, X)$ *is a pseudo–differential operator of order* 0. *Its leading symbol coincides with the leading symbol* $\sigma_{P_+(B_0)}$ *of* $P_+(B_0)$.

The pseudo–differential properties of the Calderón projector had been developed by Calderón [10] and Seeley [23]. In [6] we will show that the orthogonalized Calderón projector can be constructed from a natural boundary value problem on the disconnected double $X \coprod X$. For brevity we will address the orthogonalized Calderón projector just as Calderón projector.

The in my view most important observation of the papers [4, 5] is the fact that the Cauchy data spaces are Lagrangian. To explain this note that on the Hilbert space $L^2(\Sigma, E)$ we have the symplectic form

$$\omega(f, g) := -(Jf, g). \tag{25}$$

This claim may be somewhat bewildering since $L^2(\Sigma, E)$ is firstly a complex vector space and secondly infinite–dimensional. Nevertheless, ω is a non–degenerate skew–adjoint sesqui–linear form and it turns out that it makes perfectly sense to talk about Lagrangians, symplectic reductions, Maslov indices etc. The only difference is that, due to the infinite–dimensionality, Fredholm conditions come into play. This is a fascinating story and an elaboration would definitely need more space. For some basics cf. Kirk and Lesch [14], Sec. 6. We state explicitly what Lagrangians are in $L^2(X, E)$.

Lemma 2.1. *A subspace $L \subset L^2(X, E)$ is Lagrangian if and only if $L^\perp = J(L)$.*

The following is basically a consequence of Green's formula (17).

Proposition 2.2. *A realization D_P of a boundary condition is a symmetric operator if and only if* range P *is an isotropic subspace of $L^2(X, E)$. Moreover, if P is well–posed then D_P is self–adjoint if and only if* range P *is Lagrangian.*

The following Theorem was proved first in [4]:

Theorem 2.1. *Let X be a compact Riemannian manifold with boundary and let D be a Dirac type operator on X. Then the Cauchy data space of $N(D, X)$ is a Lagrangian subspace of $L^2(X, E)$ with respect to the symplectic structure* (25) *induced by Green's form.*

This theorem is not only beautiful. It is of fundamental importance. We are now able to describe spaces of well–posed boundary value problems as Grassmannian spaces:

Definition 2.2. Let $\mathcal{P}$ be the space of all pseudo–differential orthogonal projections acting on $L^2(\Sigma, E)$.

The pseudo–differential Grassmannian $\mathrm{Gr}_1(B_0)$ is the space of $P \in \mathcal{P}$ such that

$$P - P_+(B_0) \text{ is of order } -1. \tag{26}$$

The space of $P \in \mathcal{P}$ such that the difference $P - P_+(B_0)$ is smoothing is denoted by $\mathrm{Gr}_\infty(B_0)$.

Finally the self–adjoint (smooth) pseudo–differential Grassmannian $\mathrm{Gr}_p^*(B_0)$ is the space of $P \in \mathrm{Gr}_p(B_0), p \in \{1, \infty\}$, whose image is additionally Lagrangian.

Since $P_+(B)$ and $C(D,X)$ have the same leading symbol (26) may be replaced by

$$P - C(D,X) \text{ is of order } -1. \tag{27}$$

Hence P and $C(D,X)$ also have the same leading symbol and thus it is obvious from the Definition 1.1 that the boundary condition given by P is well–posed.

Furthermore, since the difference of any two elements $P, Q \in \mathrm{Gr}_1(B_0)$ is compact they form a *Fredholm pair*, that is

$$PQ : \operatorname{range} Q \longrightarrow \operatorname{range} P \tag{28}$$

is a Fredholm operator. The index of this Fredholm operator is denoted by $\operatorname{ind}(P,Q)$. We have

$$\operatorname{ind}(P,Q) = \dim(\ker P \cap \operatorname{range} Q) - \dim(\operatorname{range} P \cap \ker Q). \tag{29}$$

2.2. *The main theorem on the general linear conjugation problem*

We are now in a position to state the main result on the general linear conjugation problem.

Theorem 2.2. *The index of the linear conjugation problem* (20) *is given by*

$$\begin{aligned}\operatorname{ind} D^\Phi &= \operatorname{ind}\big(I - C(D_+,Y), \Phi C(D_+,X)\Phi^{-1}\big)\\ &= \operatorname{ind} D + \operatorname{ind}\big(C(D_+,X) - \Phi C(D_+,Y)\big)\\ &= \operatorname{ind} D + \operatorname{ind}\big(P_+(B_0) - \Phi P_-(B_0)\big).\end{aligned}$$

There would be much more to say. This index theorem is related to a lot. It is a generalization of the classical Riemann Hilbert problem on the complex projective line. It is related to the spectral flow and to the index of generalized Toeplitz operators.

I will not go into that. But let me say that the papers [25, 4, 5] contain much more. They provide a comprehensive presentation of the spectral flow and its topological meaning, Fredholm pairs, and the construction of the Calderón projector. Also it is proved that $P_+(B_0)$ is a pseudo–differential operator.

3. The η–invariant

[11, 26, 27, 16, 28]

Let us start with some general remarks on η– and ζ–functions. Let T be an unbounded self–adjoint operator in the Hilbert space H. Assume that T has compact resolvent such that the spectrum of T consists of a sequence of eigenvalues

$$|\lambda_1| \leq |\lambda_2| \leq \dots \text{ (repeated according to their finite multiplicity)}$$

with $|\lambda_n| \to \infty$. If λ_n satisfies a growth condition

$$|\lambda_n| \geq C n^{\alpha}, \tag{30}$$

for some $\alpha > 0$ then we can form the holomorphic functions

$$\begin{aligned} \eta(T;s) &:= \frac{1}{\Gamma(\frac{s+1}{2})} \int_0^\infty t^{(s-1)/2} \operatorname{tr}\big(Te^{-tT^2}\big) dt \\ &= \sum_{\lambda \in \operatorname{spec} T \setminus \{0\}} |\lambda|^{-s} \operatorname{sign} \lambda \\ &= \operatorname{tr}\big(T|T|^{-s-1}\big), \quad \operatorname{Re} s > \frac{1}{\alpha}, \end{aligned} \tag{31}$$

and

$$\begin{aligned} \zeta(T;s) &:= \sum_{\lambda \in \operatorname{spec} T \setminus \{0\}} \lambda^{-s} \\ &= \operatorname{tr}\big(T^{-s}\big), \quad \operatorname{Re} s > \frac{1}{\alpha}. \end{aligned} \tag{32}$$

If T is non-negative then $\zeta(T;s)$ is also a Mellin transform similar to the first equality in (31)

$$\zeta(T;s) = \frac{1}{\Gamma(s)} \int_0^\infty t^{s-1} \operatorname{tr}\big(e^{-tT^2}\big) dt. \tag{33}$$

For general T the function $\zeta(T;s)$ can still be expressed in terms of Mellin transforms using the formula

$$\zeta(T;s) = \frac{1}{2}\big(\zeta(T^2;s/2) + \eta(T;s)\big) + e^{-i\pi s}\frac{1}{2}\big(\zeta(T^2;s/2) - \eta(T;s)\big). \tag{34}$$

Up to a technical point the existence of a short time asymptotic expansion of $\operatorname{tr}\big(Te^{-tT^2}\big), \operatorname{tr}\big(e^{-tT^2}\big)$ and the meromorphic continuation of the functions $\zeta(T;s), \eta(T;s)$ is equivalent (cf. Brüning and Lesch [8], Lemma 2.2, for the precise statement).

If T is an elliptic operator on a closed manifold then it follows from the celebrated work of Seeley [24] that $\eta(T;s), \zeta(T;s)$ extend meromorphically to $\mathbb{C}$ with a precise description of the location of the poles and their residues.

If $\eta(T;s)$ is meromorphic at least in a half plane containing 0 one defines the *η–invariant* of T as

$$\begin{aligned}\eta(T) &:= \frac{1}{2\pi i}\oint_{|s|=\varepsilon} \frac{\eta(T;s)}{s}ds \\ &= \text{constant term in the Laurent expansion at } 0 \\ &=: \eta(T;0).\end{aligned} \tag{35}$$

In many situations one can even show that $\eta(T;s)$ is regular at 0. The η–invariant was introduced in the celebrated work of Atiyah, Patodi and Singer [1, 2, 3] as a boundary correction term in an index formula for manifolds with boundary.

We return to manifolds with boundary and consider again a compact Riemannian manifold X with boundary $\partial X = \Sigma$ and a formally self–adjoint operator of Dirac type acting on the hermitian vector bundle E.

From now on we assume that D is in *product form* near the boundary. That is in the collar $N = [0,\varepsilon) \times \Sigma$ of the boundary D takes the form

$$D|N = J(\frac{d}{dx} + B) \tag{36}$$

with J, B as in (15) and such that B is independent of x. The formal self–adjointness of D and B then implies

$$JB + BJ = 0. \tag{37}$$

The next Theorem guarantees the existence of the η–invariant and the ζ–determinant on the smooth self–adjoint Grassmannian.

Theorem 3.1. [28] *For $P \in \mathrm{Gr}^*_\infty(B)$ the functions $\eta(D_P;s), \zeta(D_P;s)$ extend meromorphically to a half–plane containing 0 with poles of order at most 1. Furthermore, 0 is not a pole and $\zeta(D_P;0)$ is independent of P.*

Let me say a few words about the strategy of proof. As pointed out before we have to prove short time asymptotic expansions for $\mathrm{tr}\big(D_P e^{-tD_P^2}\big)$ and $\mathrm{tr}\big(e^{-tD_P^2}\big)$. Duhamel's principle[h] allows to separate the interior contributions and the contributions coming from the boundary. Namely, let

[h]A big word for something very simple: the method of variation of the constant for first order inhomogeneous ordinary differential equations.

$\varphi \in C_0^\infty([0,\varepsilon))$ be a cut–off function with $\varphi \equiv 1$ near 0. Extend φ by 0 to a smooth function on X.

Let $\widetilde{D}$ be any elliptic extension of D to a closed manifold[i] and let $D_{P,0}$ be the model operator $J\big(\frac{d}{dx}+B\big)$ on the cylinder $[0,\infty)\times\Sigma$ with boundary condition P at $\{0\}\times\Sigma$. Then

$$\begin{aligned}\operatorname{tr}\big(D_P e^{-tD_P^2}\big) = \operatorname{tr}\big(&\varphi D_{P,0} e^{-tD_{P,0}^2}\big) + \\ &\operatorname{tr}\big((1-\varphi)\widetilde{D}e^{-t\widetilde{D}^2}\big) + O(t^K), \quad t \to 0+\end{aligned} \tag{38}$$

for any $K > 0$.

By local elliptic analysis the second term in (38) has a short time asymptotic expansion [12], Lemma 1.9.1. So one is reduced to the treatment of the model operator $D_{P,0}$. For the Atiyah–Patodi–Singer problem $P = P_+(B)$ there are explicit formulas for $e^{-tD^2_{P_+,0}}$ from which the asymptotic expansion can be derived using classical results on special functions. Finally, for $P \in \mathrm{Gr}^*_\infty(B)$ the operator $D_{P,0}$ can be treated as a perturbation of the APS operator $D_{P_+,0}$ [28].

A completely different approach by Grubb [13] leads to the generalization of Theorem 3.1 to all well–posed boundary value problems.

3.1. *The adiabatic limit*

Let us explain the result of [11, 26, 27] on the adiabatic limit of the η–invariant. We start with a partitioned manifold $M = Y \cup_\Sigma X$. Then we stretch the neck by putting

$$\begin{aligned} X_R &= [0,R]\times\Sigma \cup_{\{R\}\times\Sigma} X, \\ Y_R &= [-R,0]\times\Sigma \cup_{\{-R\}\times\Sigma} Y, \\ M_R &= Y_R \cup_{\{0\}\times\Sigma} X_R. \end{aligned}$$

Denote by $\tilde{\eta}(D, M_R)$ the reduced η–invariant of D on M_R and by $\tilde{\eta}(D_P, X_R)$ the reduced η–invariant of D_P on X_R.

Theorem 3.2. *We have*

$$\begin{aligned}\lim_{R\to\infty} \tilde{\eta}(D, M_R) \equiv &\lim_{R\to\infty} \tilde{\eta}(D_{I-P_+(B)}, Y_R) \\ &+ \lim_{R\to\infty} \tilde{\eta}(D_{P_+(B)}, X_R) \mod \mathbb{Z}.\end{aligned} \tag{39}$$

[i]The existence of such a $\widetilde{D}$ is not essential for the following result but it simplifies the exposition. For Dirac type operators we can choose $\widetilde{D}$ to be the invertible double.

We should be a bit more specific about the meaning of $P_+(B)$ here. The positive spectral projection of B is Lagrangian if and only if B is invertible. If B is not invertible then one has to fix a Lagrangian subspace of the null space of B. So whenever a Lagrangian is needed we choose $P_+(B)$ such that

$$1_{(0,\infty)}(B) \le P_+(B) \le 1_{[0,\infty)}(B).$$

That this is possible follows from the Cobordism Theorem (cf. [11] or Lesch and Wojciechowski [16]).

In [11] it was shown that the η–invariant makes sense for generalized Atiyah–Patodi–Singer boundary conditions, i.e. for $D_{P_+(B)}$. Moreover, it was shown that $\lim_{R\to\infty} \tilde{\eta}(D_{P_+(B)}, X_R)$ exists. The limit can be interpreted as the η–invariant of the operator D on the manifold with cylindrical ends X_∞. The full strength of Theorem 3.2 was proved in [26, 27]. In fact the (mod $\mathbb{Z}$ reductions) of the ingredients of formula (39) do not depend on $\mathbb{Z}$ as was observed by W. Müller [18]. In this way we obtain the gluing formula for the η–invariant for the boundary condition $P_+(B)$. The following generalization to all $P \in \mathrm{Gr}^*_\infty(B)$ is worked out in [28].

Theorem 3.3. *Let $M = Y \cup_\Sigma X$ be a partitioned manifold and let D be a Dirac type operator which is in product form in a collar of Σ. Then for $P \in \mathrm{Gr}^*_\infty(B)$*

$$\tilde{\eta}(D, M) \equiv \tilde{\eta}(D_P, X) + \tilde{\eta}(D_{I-P}, Y) \mod \mathbb{Z}. \tag{40}$$

There is even a formula if $I - P$ is replaced by a general $Q \in \mathrm{Gr}^*_\infty(-B)$. This is an extension of a formula for the variation of the η–invariant under a change of boundary condition from [16], cf. also Theorem 4.1 below.

Because of its importance let us look briefly at the method of proof.

The first observation is that the heat kernel of the model operator $D = J\big(\frac{d}{dx} + B\big)$ on the cylinder $\mathbb{R} \times \Sigma$ is explicitly known since D^2 is just a direct sum of one–dimensional Laplacians $-\frac{d^2}{dx^2} + b^2$. Let $\mathcal{E}_{\rm cyl}(t; x, y)$ be this cylinder heat kernel. Furthermore, denote by $\mathcal{E}_R(t; x, y)$ the heat kernel of D on the stretched manifold M_R.

Next one chooses R–dependent cut–off functions $\phi_{j,R}, \psi_{j,R}, j = 1, 2$, as follows:

$$\psi_{2,R}(x) = \begin{cases} 0 & \text{if } |x| \le 3R/7, \\ 1 & \text{if } |x| \ge 4R/7, \end{cases}$$

$$\psi_{1,R} = 1 - \psi_{2,R}.$$

Finally, choose $\phi_{j,R}$ such that $\phi_{j,R}\psi_{j,R} = \psi_{j,R}$. Then paste the heat kernel $\mathcal{E}_R$ on M_R and the cylinder heat kernel to obtain the kernel

$$Q_R(t;x,y) = \phi_{1,R}(x)\mathcal{E}_{\text{cyl}}(t;x,y)\psi_{1,R}(y) + \phi_{2,R}(x)\mathcal{E}_R(t;x,y)\psi_{2,R}(y). \quad (41)$$

Then Duhamel's principle yields

$$\mathcal{E}_R(t) = Q_R(t) + \mathcal{E}_R\#C_R(t), \quad (42)$$

where # is a convolution and C_R is an error term.

It seems that not much is gained yet. The point is that Douglas and Wojciechowski [11] could show that in the adiabatic limit the error term is negligible in the following sense:

Theorem 3.4. *There are estimates*

$$\|\mathcal{E}_R(t;x,y)\| \le c_1 t^{-\dim X/2} e^{c_2 t} e^{-c_3 d^2(x,y)/t},$$
$$\|(\mathcal{E}_R\#C_R)(t;x,x)\| \le c_1 e^{c_2 t} e^{-c_3 R^2/t}$$

with c_1, c_2, c_3 independent of R.

Note that this result is much more than e.g. (38). For the η– and ζ–determinant the full heat semigroup contributes. It is astonishing that nevertheless in the adiabatic limit the full integrals from 0 to ∞ in (6) and (7) split into contributions coming from the cylinder and from the interior of the manifold.

4. The relative η–invariant and the relative ζ–determinant

[16, 21]

Recall from Theorem 3.1 that for $P \in \mathrm{Gr}^*_\infty(B)$ the ζ–function $\zeta(D_P;s)$ is regular at 0. One puts

$$\det\nolimits_\zeta D_P := \begin{cases} \exp\big(-\zeta'(D_P;0)\big), & 0 \notin \operatorname{spec} D_P, \\ 0, & 0 \in \operatorname{spec} D_P. \end{cases} \quad (43)$$

In view of (34) and Theorem 3.1 a straightforward calculation shows for D_P invertible

$$\det\nolimits_\zeta D_P = \exp\Big(i\frac{\pi}{2}(\zeta(D_P^2;0) - \eta(D_P)) - \frac{1}{2}\zeta'(D_P^2;0)\Big). \quad (44)$$

We emphasize that the regularity of $\eta(D_P;s)$ and $\zeta(D_P;s)$ at $s = 0$ is essential for (44) to hold. (44) shows that the η–invariant is related to the phase of the ζ–determinant and that in general

$$(\det\nolimits_\zeta D)^2 \ne \det\nolimits_\zeta(D^2).$$

The natural question which arises at this point is

Problem 4.1. How does $\det_\zeta(D_P)$ depend on $P \in \mathrm{Gr}^*_\infty(B)$?

The answer to this problem has a long history. Since the only joint paper of Wojciechowski and myself deals with an aspect of the problem I take the liberty to add a few personal comments. In 1992 I was a Postdoc at University Augsburg. At that time the paper [11] had just appeared and the gluing formula for the η–invariant was in the air. Still much of our todays understanding of spectral invariants for Dirac type operators on manifolds with boundary was still in its infancy. When Gilkey visited he posed a special case of the Problem 4.1. If the tangential operator is not invertible there is no canonical Atiyah–Patodi–Singer boundary condition for D. The positive spectral projection of B is not in $\mathrm{Gr}^*_\infty(B)$. Rather one has to choose a Lagrangian subspace $V \subset \ker B$ and put

$$P_V := 1_{(0,\infty)}(B) + \Pi_V,$$

where Π_V denotes the orthogonal projection onto V. Then $P_V \in \mathrm{Gr}^*_\infty(B)$. The boundary condition given by P_V is called a generalized Atiyah–Patodi–Singer boundary condition. Gilkey asked how the eta–invariant depends on V.

I did some explicit calculations on a cylinder which let me guess the correct formula. However, I did not know how to prove it in general. Somewhat later Gilkey sent me a little note of Krzysztof dealing with the same problem. He urged us to work together. I was just a young postdoc and I felt honored that Krzysztof, whose papers I already admired, quickly agreed. Except writing papers with my supervisor this was my first mathematical collaboration. It was done completely by fax and email; Krzysztof and I met for the first time more than a year after the paper had been finished.

In [16] Krzystof and I proved a special case of the following result. The result as stated is a consequence of the Scott–Wojciechowski Theorem as was shown in [14], Sec. 4. The Scott–Wojciechowski Theorem will be explained below.

Theorem 4.1. *Let $P, Q \in \mathrm{Gr}^*_\infty(B)$. Then*

$$\widetilde{\eta}(D_P) - \widetilde{\eta}(D_Q) \equiv \log \det{}_{\mathrm{F}}(\Phi(P)\Phi(Q)^*) \bmod \mathbb{Z}. \tag{45}$$

If P or Q is the Calderón projector then (45) is even an equality [14].

The general answer to Problem 4.1 given by Scott and Wojciechowski [21] is just beautiful. To explain their result we need another bit of notation. Recall that J defines the symplectic form on $L^2(\Sigma, E)$ (25). Let

$$E = E_i \oplus E_{-i}$$

be the decomposition of E into the eigenbundles of J. If $P \in \mathrm{Gr}^*_\infty(B)$ then

$$L = \operatorname{range} P \subset L^2(\Sigma, E)$$

is Lagrangian and from Lemma 2.1 one easily infers that the restrictions of the orthogonal projections $\Pi_{\pm i} = \frac{1}{2i}(i \pm J)$ onto $E_{\pm i}$ map L bijectively onto $L^2(\Sigma, E_{\pm i})$ and

$$\Phi(P) := \Pi_{-i} \circ (\Pi_i | E_i)^{-1}$$

is a unitary operator from $L^2(\Sigma, E_i)$ onto $L^2(\Sigma, E_{-i})$. For P we then have the formula

$$P = \frac{1}{2}\begin{pmatrix} I & \Phi(P)^* \\ \Phi(P) & I \end{pmatrix}. \tag{46}$$

For $P, Q \in \mathrm{Gr}^*_\infty(B)$ the operator $\Phi(P)^*\Phi(Q) - I$ is smoothing and hence $\Phi(P)^*\Phi(Q)$ is of determinant class.

With these preparations, the Scott–Wojciechowski theorem reads as follows.

Theorem 4.2. *Let $P \in \mathrm{Gr}^*_\infty(B)$ and let $C(D, X)$ be the orthogonalized Calderón projector. Then*

$$\det_\zeta(D_P) = \det_\zeta(D_{C(D,X)}) \det_{\mathrm{F}}\Big(\frac{I + \Phi(C(D,X))\Phi(P)^*}{2}\Big). \tag{47}$$

5. Adiabatic decomposition of the ζ–determinant

[15, 19, 20]

When the gluing formula for the η–invariant had been established it was Krzysztof's optimism that eventually lead to a similar result for the ζ–determinant. The author has to admit that he was an unbeliever: I could not see why a reasonable analytic surgery formula for the ζ–determinant should exist. Well, I was wrong. A fruitful collaboration of J. Park and Krzysztof P. Wojciechowski eventually proved that the adiabatic method, which originally had been developed in the paper [11], was even strong enough to prove an adiabatic surgery formula for the ζ–determinant.

Consider again the adiabatic setting M_R, X_R, Y_R as in (3.1). In order not to blow up the exposition too much I will not present the result in its most general form. Rather I will make the following technical assumptions:

(1) The tangential operator B is invertible.
(2) The L^2–kernel of D on $X \cup [0,\infty) \times \Sigma$ and $Y \cup [0,\infty) \times \Sigma$ vanishes.

Then the adiabatic surgery theorems for the Laplacians read as follows:

Theorem 5.1. *Let $\Delta_{\pm,R,d}$ be the Dirichlet extension of D^2 on X_R, Y_R resp.; D_R denotes the operator D on X_R. Then*

$$\lim_{R\to\infty} \frac{\det_\zeta D_R^2}{\det_\zeta \Delta_{+,R,d} \det_\zeta \Delta_{-,R,d}} = \sqrt{\det_\zeta B^2}.$$

Theorem 5.2. *Let D_{+,R,P_+}, D_{-,R,P_-} be the operator D with Atiyah–Patodi–Singer boundary conditions on X_R, Y_R resp. Then*

$$\lim_{R\to\infty} \frac{\det_\zeta D_R^2}{\det_\zeta D^2_{+,R,\Pi_>} \det_\zeta D^2_{-,R,\Pi_<}} = 2^{-\zeta'(B^2,0)}.$$

These technical assumptions mentioned above were removed in Park and Wojciechowski [20]. For details the reader should consult loc. cit.

Finally, the "adiabatic" results on the zeta–determinants obtained by Park and Wojciechowski are not adiabatic any more. Loya and Park [17] showed that most of those results (and more) are true without stretching. Krzysztof P. Wojciechowski did have different (and charming) ideas how to remove stretching of the cylinders. Unfortunately, his serious illness did not allow him to fill all the details and finish the paper.

References .

1. M. F. Atiyah, V. K. Patodi, and I. M. Singer, *Spectral asymmetry and Riemannian geometry I*, Math. Proc. Cambridge Philos. Soc. **77** (1975), 43–69.
2. M. F. Atiyah, V. K. Patodi, and I. M. Singer, *Spectral asymmetry and Riemannian geometry II*, Math. Proc. Cambridge Philos. Soc. **78** (1975), 405–432.
3. ______, *Spectral asymmetry and Riemannian geometry III*, Math. Proc. Cambridge Philos. Soc. **79** (1976), 71–99.
4. B. Booß and K. P. Wojciechowski, *Desuspension and splitting elliptic symbols I*, Ann. Global Anal. Geom. **3** (1985), 337–383.
5. ______, *Desuspension and splitting elliptic symbols II*, Ann. Global Anal. Geom. **4** (1986), 349–400.
6. B. Booß-Bavnbek, M. Lesch, and C. Zhu, *In preparation.*

7. B. Booß-Bavnbek and K. P. Wojciechowski, *Elliptic boundary problems for Dirac operators*, Birkhäuser, Basel, 1993.
8. J. Brüning and M. Lesch, *On the eta–invariant of certain non–local boundary value problems*, Duke Math. J. **96** (1999), 425–468, dg-ga/9609001.
9. ______, *On boundary value problems for Dirac type operators: I. Regularity and self-adjointness*, J. Funct. Anal. **185** (2001), 1–62, math.FA/9905181.
10. A. Calderón, *Boundary value problems for elliptic equations*, Outlines of the joint Soviet–American symposium on partial differential equations (Novosibirsk), 1963, pp. 303–304.
11. R. G. Douglas and K. P. Wojciechowski, *Adiabatic limits of the η-invariants the odd–dimensional Atiyah–Patodi–Singer problem*, Comm. Math. Phys. **142** (1991), 139–168.
12. P. Gilkey, *Invariance theory, the heat equation, and the Atiyah–Singer index theorem*, 2. ed., CRC Press, Boca Raton, 1995.
13. G. Grubb, *Trace expansions for pseudodifferential boundary problems for Dirac–type operators and more general systems*, Ark. Mat. **37** (1999), 45–86.
14. P. Kirk and M. Lesch, *The η–invariant, Maslov index, and spectral flow for Dirac–type operators on manifolds with boundary*, Forum Math. **16** (2004), 553–629, math.DG/0012123.
15. S. Klimek and K. P. Wojciechowski, *Adiabatic cobordism theorems for analytic torsion and η–invariant*, J. Funct. Anal. **136** (1996), 269–293.
16. M. Lesch and K. P. Wojciechowski, *On the η–invariant of generalized Atiyah–Patodi–Singer boundary value problems*, Illinois J. Math. **40** (1996), 30–46.
17. P. Loya and J. Park, *On the gluing problem for the spectral invariants of dirac operators*, To appear.
18. W. Müller, *Eta invariants and manifolds with boundary*, J. Differential Geom. **40** (1994), 311–377.
19. J. Park and K. P. Wojciechowski, *Adiabatic decomposition of the ζ–determinant of the Dirac Laplacian I. The case of invertible tangential operator*, Comm. Partial Differential Equations **27** (2002), 1407–1435, (with an Appendix by Y. Lee).
20. ______, *Scattering theory and adiabatic decomposition of the ζ–determinant of the Dirac Laplacian*, Math. Res. Lett. **9** (2002), 17–25.
21. S. G. Scott and K. P. Wojciechowski, *The ζ–determinant and Quillen determinant for a Dirac operator on a manifold with boundary*, Geom. Funct. Anal. **10** (1999), 1202–1236.
22. R. T. Seeley, *Topics in pseudo–differential operators*, C.I.M.E., Conference on pseudo–differential operators 1968, Edizioni Cremonese, Roma, 1969, pp. 169–305.
23. ______, *Singular integrals and boundary value problems*, Amer. J. Math. **88** (1966), 781–809.
24. ______, *Complex powers of an elliptic operator*, Proc. Sympos. Pure Math. **10** (1967), 288–307.
25. K. P. Wojciechowski, *Spectral flow and the general linear conjugation problem*, Simon Stevin **59** (1985), 59–91.

26. ———, *The additivity of the η–invariant. The case of an invertible tangential operator*, Houston J. Math. **20** (1994), 603–621.
27. ———, *The additivity of the η–invariant. The case of a singular tangential operator*, Comm. Math. Phys. **109** (1995), 315–327.
28. ———, *The ζ-determinant and the additivity of the η-invariant on the smooth, self-adjoint Grassmannian*, Comm. Math. Phys. **201** (1999), 423–444.

Received by the editors September 14, 2005; revised January 5, 2006

Analysis, Geometry and Topology
of Elliptic Operators, pp. 23–38

GLUING FORMULAE OF SPECTRAL INVARIANTS AND CAUCHY DATA SPACES

JINSUNG PARK
School of Mathematics, Korea Institute for Advanced Study
207-43, Cheongnyangni 2-dong, Dongdaemun-gu
Seoul 130-722, Korea
jinsung@kias.re.kr

Dedicated to Krzysztof P. Wojciechowski on his 50th birthday

We review the gluing formulae of the spectral invariants - the ζ-regularized determinant of a Laplace type operator and the eta invariant of a Dirac type operator. In particular, we explain the crucial role of the Cauchy data spaces in these gluing formulae.

2000 *Mathematics Subject Classification.* Primary 58J28; Secondary 58J52

1. Introduction

In this article, we survey the gluing formulae of the spectral invariants - the ζ-regularized determinant of a Laplace type operator and the eta invariant of a Dirac type operator. After these spectral invariants had been originally introduced by Ray and Singer [30] and Atiyah, Patodi, and Singer [1] respectively, these invariants have been studied by many people in many different parts of mathematics and physics. Here we discuss the gluing formulae of these spectral invariants. These formulae have been proved independently by several authors using different techniques. For nice introductions to this subject, we refer to Bleecker and Booß-Bavnbek [3] and Mazzeo and Piazza [21] where the reader can find many technical details and ideas of proofs. Therefore, instead of repeating the details of these introductions, we explain one principle which holds for all the known gluing formulae of the spectral invariants. This principle also enabled us to get a new proof of the gluing formulae of the eta invariant of a Dirac type operator and simultaneously to prove the gluing formula of the ζ-regularized determinant of a Dirac Laplacian [17], [18]. We hope that this article would be helpful in the understanding of gluing formulae of the spectral invariants and other related gluing problems in similar situations.

Now let us review briefly the history of this subject: gluing problems of the spectral invariants. This will help the reader to understand the motivation of this article.

First, it is appropriate to begin with mentioning the pioneering work of Wojciechowski. In his paper with Douglas [10], they found a striking formula, which states that the eta invariant of a Dirac type operator over a manifold with boundary converges to a local expression as the cylindrical length near the boundary is getting longer and longer (the *adiabatic limit*). Although they did not formulate the gluing formula of the eta invariant in their paper [10] explicitly, this result suggested the existence of the gluing formula of the eta invariant.

The work of Douglas and Wojciechowski [10] stimulated many mathematicians working around the eta invariant, so after their paper appeared, during the last 15 years the gluing formula of the eta invariant has been proved independently and using different techniques by Hassell, Mazzeo, and Melrose [12], Wojciechowski [34], Bunke [6], Müller [23], Brüning and Lesch [5], Kirk and Lesch [14], Park and Wojciechowski [27] and many others. Although their proofs are different from each other, they altogether used the generalized Atiyah-Patodi-Singer spectral projection to impose the boundary conditions for the Dirac operator over manifold with boundary. Among the aforementioned works, the formula of Kirk and Lesch is the most complete in the sense that their formula has no integer ambiguity (Bunke's formula also holds without the integer ambiguity) and they show the origin of the integer contribution in their proof. In fact, they needed to use the Calderón projector for the boundary condition to formulate their formula. Hence, this seems to suggest that the Calderón projector might be the natural projection in the gluing formula instead of the generalized Atiyah-Patodi-Singer spectral projection.

We can also see such a suggestion from the adiabatic decomposition formula of the ζ-regularized determinant of the Dirac Laplacian proved by Park and Wojciechowski [25], [26], [27]. In their formula, the adiabatic limit of the ratio of the ζ-regularized determinants of the Dirac Laplacians over the original manifold and decomposed submanifolds is mainly described in terms of the scattering matrices of the corresponding Dirac operators over manifolds with cylindrical ends, which are obtained by attaching half infinite cylinders to the decomposed manifolds with boundaries. Here, we can regard the scattering matrix for a noncompact manifold with cylindrical end as corresponding to the Calderón projector for a manifold with boundary.

Following this suggestion - the use of the Calderón projector, Loya and Park [17] could find a new proof of the gluing formula of the eta invariant of a Dirac type operator, which also provides us, simultaneously, with the gluing formula of the ζ-regularized determinant of a Dirac Laplacian. Actually, these formulas are not two different formulas, but just two aspects - phase and modulus - of one unified formula. To state their formulae, they needed to introduce an operator U which is defined by the Cauchy data spaces of the restricted Dirac operators over the decomposed manifolds. The proof in [17] can be easily employed for more general situations, for instance, for noncompact manifolds [18]. (Technically, the proof in [17] does not use the fact the variation of the eta invariant is locally computable, which holds only for compact manifolds.)

We can also see such a suggestion in the gluing formula of the ζ-regularized determinant proved in the end of 80's by Burghelea, Friedlander, and Kappeler (BFK) [7]. In (a special case of) their formula, the ratio of the ζ-regularized determinants of Laplace type operators over the original manifold and decomposed submanifolds is mainly described by the ζ-regularized determinant of a certain operator $\mathcal{R}$ over the cutting hypersurface, and this operator $\mathcal{R}$ has an expression in terms of the Cauchy data spaces of the restricted Laplace operators over decomposed manifolds. As we will explain, the operators U and $\mathcal{R}$ can be understood under the following principle: *The gluing formulae of the spectral invariants are mainly described by the difference of the Cauchy data spaces.*

Now let us explain the structure of this article. In Section 2, we first review the gluing formula of BFK [7] and one of its generalization [16]. In Section 3, we explain how the operator $\mathcal{R}$ can be understood in terms of the Cauchy data spaces. In Section 4, we review the gluing formulae of the spectral invariants of a Dirac type operator proved by Loya and Park [17]. Finally, in Section 5, we explain the operator U, which plays the crucial role in the gluing formulae of the spectral invariants of a Dirac type operator, and its underlying meaning compared with $\mathcal{R}$.

The author gives his sincere thanks to Paul Loya for his helpful comments and the referee for corrections and suggestions, all of which considerably improved this article.

2. Gluing formula of the ζ-regularized determinant of a Laplace type operator

In this section, we review the gluing formula of the ζ-regularized determinant of a Laplace type operator. This type of formula was first proved by Burghelea, Friedlander and Kappeler (BFK) in [7] where they call this a Mayer-Vietoris type formula. Although their formula holds for a more general situation, that is, more general differential operators and general local elliptic boundary conditions, here we just restrict our discussion to a Laplace type operator and the Dirichlet boundary condition.

Now let us explain the BFK formula in more detail. Let M is a compact manifold and Y is a hypersurface in M, which decomposes M into two submanifolds M_- and M_+ (here we assume that M_- is the left side manifold and M_+ is the right side manifold). Hence we have

$$M = M_- \cup M_+, \qquad Y = M_- \cap M_+.$$

Let us consider a Laplace type operator over M,

$$\Delta_M : H^2(M,E) \longrightarrow L^2(M,E),$$

where E is a Hermitian vector bundle over M. For the restrictions of Δ_M to M_- and M_+, we impose the Dirichlet boundary conditions so that we obtain

$$\Delta_{M_\pm} := \Delta_M|_{M_\pm} : \operatorname{dom}(\Delta_{M_\pm}) := \{\phi \in H^2(M_\pm, E) \,|\, \gamma_0(\phi) = 0\} \longrightarrow L^2(M_\pm, E)$$

where $\gamma_0 : H^2(M_\pm, E) \to H^{\frac{3}{2}}(Y, E_0)$ $(E_0 := E|_Y)$ denotes the restriction map to Y. Now let us recall that the ζ-function of Δ_M is defined by

$$\zeta(s, \Delta_M) := \frac{1}{\Gamma(s)} \int_0^\infty t^{s-1} \big[\operatorname{Tr}(e^{-t\Delta_M}) - \dim\ker \Delta_M \big]\, dt \tag{1}$$

for $\Re s \gg 0$ and this has the meromorphic extension over $\mathbb{C}$ with $s = 0$ as a regular value. Then the ζ-regularized determinant of Δ_M is defined by

$$\det_\zeta \Delta_M := \exp(-\zeta'(0, \Delta_M)). \tag{2}$$

The ζ-regularized determinant of $\Delta_{M_\pm}$, $\det_\zeta \Delta_{M_\pm}$ is defined in a similar way as $\det_\zeta \Delta_M$. Now a natural question in this circumstance is the relation of $\det_\zeta \Delta_M$ with $\det_\zeta \Delta_{M_-}$, $\det_\zeta \Delta_{M_+}$, and the BFK formula gives us the answer to this question.

To explain their formula, we need to introduce an operator $\mathcal{R}$ acting on $C^\infty(Y, E_0)$, which is defined as follows: First, given $f \in C^\infty(Y, E_0)$, let us denote by $\phi_\pm$ the (assumed to be unique) solutions of the Dirichlet problems for the restrictions of Δ_M to $M_\pm$:

$$\Delta_M \phi_i = 0 \qquad \text{over} \quad M_\pm \setminus Y, \qquad \phi_\pm|_Y = f.$$

Then, the operator $\mathcal{R}$ is defined by

$$\mathcal{R}f := (\partial_u \phi_-)\big|_Y - (\partial_u \phi_+)\big|_Y$$

where u is the normal variable to Y such that $\pm\partial_u$ is the inward directional derivative for $M_\pm$.

Remark 2.1. For $\mathcal{R}$ to be well defined, we need the condition that the Dirichlet problem of $\Delta_M|M_\pm$ has a unique solution. This is the case for the Laplace-Beltrami operator acting on the space of k-forms (see Remark 3.1). We always assume that the Laplace type operator Δ_M satisfies this condition in this article.

It is known that $\mathcal{R}$ is a nonnegative pseudodifferential operator over Y of order 1, hence its ζ-regularized determinant is well defined. Now we are ready to state the BFK formula.

Theorem 2.1. [7] *When* $\ker \Delta_M = \{0\}$, *we have*

$$\frac{\det_\zeta \Delta_M}{\det_\zeta \Delta_{M_+} \cdot \det_\zeta \Delta_{M_-}} = C(Y) \cdot \det_\zeta \mathcal{R} \tag{3}$$

where $C(Y)$ *is a constant depending only on the symbols of* $\Delta_M, \Delta_{M_\pm}, \mathcal{R}$ *over* Y.

The BFK formula in (3) describes the ratio of $\det_\zeta \Delta_M$ and $\det_\zeta \Delta_{M_+} \cdot \det_\zeta \Delta_{M_-}$ in terms of $\det_\zeta \mathcal{R}$ modulo the constant $C(Y)$, which can be considered as data near Y. Note that although the operator $\mathcal{R}$ is defined over Y, $\mathcal{R}$ contains global information over M via the null solutions of the restrictions of Δ_M to $M_\pm$.

Remark 2.2. By definition of $\mathcal{R}$, $\ker \Delta_M = \{0\}$ implies that $\ker \mathcal{R} = \{0\}$. Hence, under this condition, all the operators occurring in (3) have trivial kernels. Without this condition, we have an additional term on the right side of (3).

Remark 2.3. When we assume that Δ_M has the following product form over a collar neighborhood $\mathcal{U} \cong Y \times [-1,1]_u$ of Y,

$$\Delta_M|\mathcal{U} = -\frac{d^2}{du^2} + \Delta_Y$$

where u denotes the variable of the normal direction to Y and Δ_Y is a Laplace type operator over Y, we can obtain the exact value of $C(Y)$ as in [9], [15], [28],

$$C(Y) = 2^{-\zeta(0,\Delta_Y)-h_Y}. \tag{4}$$

Here $\zeta(s, \Delta_Y)$ is the ζ-function of Δ_Y and $h_Y := \dim\ker \Delta_Y$.

For a noncompact manifold M, the operator $e^{-t\Delta_M}$ is not of trace class. Hence the ζ-regularized determinant cannot be defined as in compact case. But, in this case, one can use the b-trace of Melrose [22] or the relative trace of Müller [24] instead of using the ordinary trace of $e^{-t\Delta_M}$. For the ζ-regularized determinant defined by the b-trace or relative trace over a noncompact manifold, its gluing formulae have been proved by Hassell and Zelditch [13] for the decomposition of $M = \mathbb{R}^2$ into a compact smooth domain and its complement, and by Carron [9] for the general noncompact case. Here we just explain one generalization of the BFK formula in (3) for a noncompact manifold X with cylindrical end. The manifold X with cylindrical end has the following decomposition,

$$X = N \cup_Y Z$$

where N is a manifold with boundary Y and $Z \cong Y \times [0, \infty)_u$. We may assume there is a collar neighborhood $W \cong Y \times [-1, 0]_u$ of Y within N. We consider a Laplace type operator Δ_X acting on $C^\infty(X, E)$ where E is a Hermitian vector bundle over X. We also assume product structures of the Riemannian metric of X and the Hermitian metric of E over

$$W \cup_Y Z \cong Y \times [-1, \infty)_u .$$

Finally we assume the following expression of Δ_X over $W \cup_Y Z$,

$$\Delta_X|_{W\cup_Y Z} = -\frac{d^2}{du^2} + \Delta_Y \tag{5}$$

where Δ_Y is a Laplace type operator over Y. As before, we impose the Dirichlet boundary conditions for the restrictions of Δ_X to N and Z and denote by Δ_N, Δ_Z the resulting operators. Then the relative ζ-function for (Δ_X, Δ_Z) is defined by

$$\zeta(s, \Delta_X, \Delta_Z) := \frac{1}{\Gamma(s)}\Big(\int_0^1 + \int_1^\infty\Big) t^{s-1}\,\mathrm{Tr}(e^{-t\Delta_X} - e^{-t\Delta_Z})\, dt.$$

Here the integral $\int_0^1 \cdot\, dt$ has a meromorphic extension from $\Re s \gg 0$ to $\mathbb{C}$ and the integral $\int_1^\infty \cdot\, dt$ has a meromorphic extension from $\Re s \ll 0$ to $\mathbb{C}$. The resulting meromorphic extension of $\zeta(s, \Delta_X, \Delta_Z)$ is regular at $s = 0$. (In the above definition of $\zeta(s, \Delta_X, \Delta_Z)$, the relative trace $\mathrm{Tr}(e^{-t\Delta_X} - e^{-t\Delta_Z})$ contains the zero eigenvalues of Δ_X, but these are cancelled out after taking the sum of $\int_0^1 \cdot\, dt$ and $\int_1^\infty \cdot\, dt$.) Then the relative ζ-regularized determinant is defined by

$$\det{}_\zeta(\Delta_X, \Delta_Z) := \exp\big(-\zeta'(0, \Delta_X, \Delta_Z)\big).$$

To explain the generalization of the BFK formula to manifolds with cylindrical end, we need to introduce some more notations. Let $\{u_j\}$ be an orthonormal basis for the kernel of Δ_X on $L^2(X, E)$ and let $\{U_j\}$ be a basis of the 'extended L^2-solutions' (bounded solutions of $\Delta_X U_j = 0$) such that at ∞ on the cylinder, $\{U_j(\infty)\}$ are orthonormal in $L^2(Y, E_0)$ where $E_0 := E|_Y$. Let $v_j = u_j|_Y$ and $V_j = U_j|_Y$ be the restrictions of u_j and U_j, respectively, to the hypersurface $\{0\} \times Y$. It can be shown that the sections $\{v_j, V_j\}$ are linearly independent in $L^2(Y, E_0)$, therefore both operators

$$L = \sum_j v_j \otimes v_j^*, \qquad \widetilde{L} = \sum_j V_j \otimes V_j^*$$

are nonnegative linear operators on the finite-dimensional vector space $V = \mathrm{span}\{v_j,\ V_j\} \subset L^2(Y, E_0)$. Since the set $\{v_j,\ V_j\}$ is a linearly independent set spanning V, the operator

$$L + \widetilde{L} : V \longrightarrow V$$

is positive. In particular, $\det(L + \widetilde{L})$ is nonzero. Now we can state the main result of [16].

Theorem 2.2. [16] *The following equality holds,*

$$\frac{\det_\zeta(\Delta_X, \Delta_Z)}{\det_\zeta \Delta_N} = 2^{-\zeta(0,\Delta_Y) - h_Y} \frac{\det_\zeta \mathcal{R}}{\det(L + \widetilde{L})} \tag{6}$$

where $\zeta(s, \Delta_Y)$ is the ζ-function of Δ_Y and $h_Y = \dim\ker \Delta_Y$.[a]

Remark 2.4. Originally the formula (6) was given in terms of the b-trace in [16]. We refer to [16] for this and an elementary introduction to the b-trace. Since we do not assume any condition on the kernel of Δ_X, we have the additional term $\det(L + \widetilde{L})$ on the right side of (6) (see Remark

[a]After [16] was completed, we learned that this result was also proved in "Regularized determinants of Laplace type operators, analytic surgery and relative determinants" by J. Müller and W. Müller.

2.2). By the product structure in (5), we could obtain the explicit form of the constant $C(Y) = 2^{-\varsigma(0,\Delta_Y)-h_Y}$ as in (4).

3. The operator $\mathcal{R}$ and the Cauchy data spaces of Laplace type operator

The discussion in this section holds for a more general situation, but we just restrict our concern to the closed manifold M. The main purpose of this section is to investigate the operator $\mathcal{R}$ in terms of the Cauchy data spaces of $\Delta_M|_{M_\pm}$.

Now we recall the operator Δ_M over M which is decomposed into M_-, M_+ by the hypersurface Y. The trace map γ is defined by

$$\gamma(\phi) = (\phi|_Y, (\partial_u\phi)|_Y) : H^2(M,E) \longrightarrow H^{\frac{3}{2}}(Y,E_0) \oplus H^{\frac{1}{2}}(Y,E_0)$$

where $E_0 := E|_Y$. Here u denotes the normal variable for the collar neighborhood

$$\mathcal{U} \cong Y \times [-1,1]_u$$

of Y and $\pm\partial_u$ is the inward directional normal derivative to $M_\pm$. Then the restriction $\Delta_\pm := \Delta_M|_{M_\pm}$ determines the Cauchy data space $\mathcal{H}(\Delta_\pm)$ defined by

$$\mathcal{H}(\Delta_\pm) := \{\, (f,g) \in C^\infty(Y,E_0) \oplus C^\infty(Y,E_0) \,|\, \exists\phi \in C^\infty(M_\pm,E) \text{ such that } \Delta_M\phi = 0 \text{ on } M_\pm \setminus Y \text{ and } \gamma(\phi) = (f,g)\,\}.$$

Hence, $\mathcal{H}(\Delta_\pm)$ consists of the pair of the Dirichlet and Neumann data of $\Delta_\pm$ over $M_\pm$.

The Dirichlet-to-Neumann operator $\mathcal{N}_\pm$ over $C^\infty(Y,E_0)$ is defined to be the map sending $f \in C^\infty(Y,E_0)$ to the corresponding Neumann data $g \in C^\infty(Y,E_0)$ such that $(f,g) \in \mathcal{H}(\Delta_\pm)$. Note that the well-definedness of $\mathcal{N}_\pm$ ($\ker\mathcal{N}_\pm = \{0\}$) is equivalent to the condition that the operator $\Delta_\pm$ with the Dirichlet (Neumann) boundary condition has no null solution.

Remark 3.1. Let us consider the case of the Laplace-Beltrami operator $(d+d^*)^2$ acting on the space of k-forms over a manifold with boundary. The operator $(d+d^*)^2$ with the Dirichlet boundary condition has no null solution. Indeed, a null solution of $(d+d^*)^2$ would also be a null solution of $d+d^*$ by the Green formula. But, this is impossible by the unique continuation theorem for $d+d^*$ [2]. Hence, the Dirichlet-to-Neumann operator for the Laplace-Beltrami operator is well defined.

Under the condition assumed in Remark 2.1, the operator $\mathcal{N}_\pm$ defines its graph which is exactly $\mathcal{H}(\Delta_\pm)$. We refer to [29], [20] for some more detailed explanations about the Dirichlet-to-Neumann operator and its application to the relative formula of the Dirichlet/Neumann Laplacians.

Now, recalling the definition of $\mathcal{R}$ and $\mathcal{N}_\pm$, we can easily see that

$$\mathcal{R} = \mathcal{N}_- - \mathcal{N}_+ : C^\infty(Y, E_0) \longrightarrow C^\infty(Y, E_0). \tag{7}$$

Hence, we can see that the operator $\mathcal{R}$ describes the *difference of two Cauchy data spaces* $\mathcal{H}(\Delta_\pm)$ in the sense that

$$(f, \mathcal{R}f) = (f, \mathcal{N}_- f) - (f, \mathcal{N}_+ f) \qquad \text{for} \quad (f, \mathcal{N}_\pm f) \in \mathcal{H}(\Delta_\pm).$$

In conclusion, the gluing formula of $\det_\zeta \Delta_M$ is mainly described by the difference of the Cauchy data spaces $\mathcal{H}(\Delta_\pm)$.

4. Gluing formula of the spectral invariants of a Dirac type operator

In this section, we discuss the gluing formulae [17] of the spectral invariants of Dirac type operators, that is, the eta invariant of a Dirac type operator and the ζ-regularized determinant of a Dirac Laplacian.

Let $\mathcal{D}$ be a Dirac type operator acting on $C^\infty(M, S)$ where M is a closed compact Riemannian manifold of arbitrary dimension and S is a Clifford bundle over M. Let Y be an embedded hypersurface in M which decomposes M into two submanifolds M_- and M_+. Hence we have

$$M = M_- \cup M_+, \qquad Y = M_- \cap M_+.$$

We assume all geometric structures are of product type over a tubular neighborhood $\mathcal{U} \cong [-1,1]_u \times Y$ of Y where the Dirac operator takes the product form over $\mathcal{U}$,

$$\mathcal{D}|_\mathcal{U} = G(\partial_u + D_Y),$$

where G is a unitary operator on $S_0 := S|_Y$ and D_Y is a Dirac type operator over Y satisfying

$$G^2 = -\mathrm{Id} \quad \text{and} \quad D_Y G = -G D_Y\,.$$

Recall that the eta function of $\mathcal{D}$ and the zeta function of $\mathcal{D}^2$ are defined through the heat operator $e^{-t\mathcal{D}^2}$ by

$$\eta(s, \mathcal{D}) := \frac{1}{\Gamma(\frac{s+1}{2})} \int_0^\infty t^{\frac{s-1}{2}} \operatorname{Tr}(\mathcal{D} e^{-t\mathcal{D}^2})\, dt,$$

$$\zeta(s, \mathcal{D}^2) := \frac{1}{\Gamma(s)} \int_0^\infty t^{s-1} \big[\operatorname{Tr}(e^{-t\mathcal{D}^2}) - \dim\ker \mathcal{D} \big]\, dt,$$

which are defined a priori for $\Re s \gg 0$ and extend to be meromorphic functions on $\mathbb{C}$ that are regular at $s = 0$. The eta invariant and the reduced eta invariant of $\mathcal{D}$ are defined by

$$\eta(\mathcal{D}) := \eta(0, \mathcal{D}), \qquad \tilde{\eta}(\mathcal{D}) := \frac{1}{2}(\eta(0, \mathcal{D}) + \dim \ker \mathcal{D}),$$

and the ζ-determinant of $\mathcal{D}^2$, $\det_\zeta \mathcal{D}^2$ is defined as in (2) using $\zeta(s, \mathcal{D}^2)$.

By restriction, $\mathcal{D}$ induces Dirac type operators $\mathcal{D}_\pm$ over $M_\pm$. For these operators, we choose orthogonal projections $\mathcal{P}_\pm$ over $L^2(Y, S_0)$ that provide us with *well-posed boundary conditions* for $\mathcal{D}_\pm$ in the sense of Seeley [33]. Then the following operators

$$\mathcal{D}_{\mathcal{P}_\pm} : \mathrm{dom}(\mathcal{D}_{\mathcal{P}_\pm}) \to L^2(M_\pm, S) \tag{8}$$

where

$$\mathrm{dom}(\mathcal{D}_{\mathcal{P}_\pm}) := \{ \phi \in H^1(M_\pm, S) \mid \mathcal{P}_\pm(\phi|_Y) = 0 \},$$

share many of the analytic properties of $\mathcal{D}$; in particular, they are Fredholm and have discrete spectra, but are not necessarily self-adjoint. Amongst such projectors are the (orthogonalized) Calderón projectors $\mathcal{C}_\pm$ [8], [32] which are projectors defined intrinsically as the unique orthogonal projectors onto the closures in $L^2(Y, S_0)$ of the infinite-dimensional *Cauchy data spaces* of $\mathcal{D}_\pm$:

$$\begin{aligned} \mathcal{H}(\mathcal{D}_\pm) := \{ f \in C^\infty(Y, S_0) \mid \exists \phi \in C^\infty(M_\pm, S) \quad &\text{such that} \\ \mathcal{D}\phi = 0 \ \text{ on } \ M_\pm \setminus Y \ &\text{ and } \ \phi|_Y = f \}. \end{aligned} \tag{9}$$

To state our main theorem, we recall that the Calderón projectors $\mathcal{C}_\pm$ have the matrix forms

$$\mathcal{C}_\pm = \frac{1}{2} \begin{pmatrix} \mathrm{Id} & \kappa_\pm^{-1} \\ \kappa_\pm & \mathrm{Id} \end{pmatrix} \tag{10}$$

with respect to the decomposition

$$L^2(Y, S_0) = L^2(Y, S^+) \oplus L^2(Y, S^-),$$

where $S^\pm \subset S_0$ are the $(\pm i)$-eigenspaces of G. The maps

$$\kappa_\pm : L^2(Y, S^+) \to L^2(Y, S^-) \tag{11}$$

are isometries, so that

$$U := -\kappa_- \kappa_+^{-1}$$

is a unitary operator over $L^2(Y, S^-)$, which is moreover of Fredholm determinant class. The last property follows easily from the fact that the differences of $\mathcal{C}_\pm$ and the Atiyah-Patodi-Singer spectral projections are smoothing operators; we refer to [31] for the details. We denote by $\widehat{U}$ the restriction of U to the orthogonal complement of its (-1)-eigenspace. We also put

$$L := \sum_{j=1}^{h_M} v_j \otimes v_j^* \tag{12}$$

where $h_M = \dim \ker \mathcal{D}$ and $v_j = u_j|_Y$ with the orthonormal basis $\{u_j\}$ of $\ker \mathcal{D}$. Then L is a positive operator on the finite-dimensional vector space $(\ker \mathcal{D})|_Y$. We are now ready to introduce the main result of [17].

Theorem 4.1. [17] *The following gluing formulae hold:*

$$\begin{aligned} \frac{\det_\zeta \mathcal{D}^2}{\det_\zeta \mathcal{D}^2_{\mathcal{C}_-} \cdot \det_\zeta \mathcal{D}^2_{\mathcal{C}_+}} &= 2^{-\zeta(0, D_Y^2) - h_Y} (\det L)^{-2} \det_F \Big(\frac{2\mathrm{Id} + \widehat{U} + \widehat{U}^{-1}}{4} \Big), \\ \tilde{\eta}(\mathcal{D}) - \tilde{\eta}(\mathcal{D}_{\mathcal{C}_-}) - \tilde{\eta}(\mathcal{D}_{\mathcal{C}_+}) &= \frac{1}{2\pi i} \operatorname{Log} \det_F U \mod \mathbb{Z}, \end{aligned}$$

where $\zeta(s, D_Y^2)$ is the ζ-function of D_Y^2, $h_Y = \dim \ker D_Y$, $\det_F$ denotes the Fredholm determinant, and Log *the principal value logarithm.*

We can replace the Calderón projector $\mathcal{C}_\pm$ in the gluing formulae in Theorem 4.1 by other orthogonal projection in the *smooth, self-adjoint Grassmannian* $Gr^*_\infty(\mathcal{D}_\pm)$, which consists of orthogonal projections $\mathcal{P}_\pm$ such that $\mathcal{P}_\pm - \mathcal{C}_\pm$ are smoothing operators and

$$G\mathcal{P}_\pm = (\mathrm{Id} - \mathcal{P}_\pm) G.$$

Let $\mathcal{P}_1 \in Gr^*_\infty(\mathcal{D}_-)$ and $\mathcal{P}_2 \in Gr^*_\infty(\mathcal{D}_+)$. Then the eta invariant of $\mathcal{D}_{\mathcal{P}_i}$ and the ζ-regularized determinant of $\mathcal{D}^2_{\mathcal{P}_i}$ are well defined by the results of Grubb [11] and Wojciechowski [35]. The orthogonal projections $\mathcal{P}_1$ and $\mathcal{P}_2$ determine maps κ_1 and κ_2 as in (10), and we define

$$U_1 := \kappa_- \kappa_1^{-1}, \qquad U_2 := \kappa_2 \kappa_+^{-1}, \qquad U_{12} := -\kappa_1 \kappa_2^{-1} \qquad \text{over } L^2(Y, S^-).$$

As before, let us denote by $\widehat{U}_i$ the restriction of U_i to the orthogonal complement of its (-1)-eigenspace. We define the operator L_1 over the finite-dimensional vector space $\operatorname{ran}(\mathcal{C}_-) \cap \operatorname{ran}(\mathrm{Id} - \mathcal{P}_1)$ by

$$L_1 = -P_1\, G \mathcal{R}_-^{-1}\, G\, P_1 \tag{13}$$

where $\mathcal{R}_-$ is the BFK operator for the double of $(\mathcal{D}_-)^2$ (see chapter 9 of [4] for the double construction), and P_1 is the orthogonal projection onto

$\mathrm{ran}(\mathcal{C}_-) \cap \mathrm{ran}(\mathrm{Id} - \mathcal{P}_1)$. Then L_1 is a positive operator [19]. We define L_2 in a similar way. We can now state the general gluing formulae for the spectral invariants of Dirac type operators.

Theorem 4.2. [17] *The following general gluing formulae hold:*

$$\frac{\det_\zeta \mathcal{D}^2}{\det_\zeta \mathcal{D}^2_{\mathcal{P}_1} \cdot \det_\zeta \mathcal{D}^2_{\mathcal{P}_2}} = 2^{-\zeta(0,D_Y^2)-h_Y} (\det L)^{-2} \det{}_F \Big(\frac{2\mathrm{Id} + \widehat{U} + \widehat{U}^{-1}}{4} \Big)$$
$$\cdot \prod_{i=1}^{2} (\det L_i)^{-2} \det{}_F \Big(\frac{2\mathrm{Id} + \widehat{U}_i + \widehat{U}_i^{-1}}{4} \Big)^{-1},$$
$$\tilde{\eta}(\mathcal{D}) - \tilde{\eta}(\mathcal{D}_{\mathcal{P}_1}) - \tilde{\eta}(\mathcal{D}_{\mathcal{P}_2}) = \frac{1}{2\pi i} \operatorname{Log} \det{}_F U_{12} \mod \mathbb{Z}.$$

Remark 4.1. As in Theorem 2.2, we can generalize Theorem 4.2 to non-compact manifolds with cylindrical end. We refer to [18] for this result and its proof.

The gluing formula of the eta invariant in Theorem 4.2 (when $\mathcal{P}_i$ are the generalized Atiyah-Patodi-Singer spectral projections) has the same form as (or its reduced form modulo $\mathbb{Z}$ of) the gluing formulae proved by Hassell, Mazzeo, and Melrose [12], Wojciechowski [34], Bunke [6], Müller [23], Brüning and Lesch [5], Kirk and Lesch [14], Park and Wojciechowski [27]. In this formula, the data given by the Calderón projector or the Cauchy data spaces are cancelled so these data do not appear in the gluing formula. This is the main reason that the important role of the Cauchy data spaces in the gluing formula of the eta invariant has not been noticed before the work of Kirk and Lesch [14]. But, in the gluing formula of the ζ-regularized determinant in Theorem 4.2, these terms appear via $\widehat{U}$ with the additional terms $\widehat{U}_i$ when we impose boundary conditions given by orthogonal projections other than the Calderón projectors $\mathcal{C}_\pm$. Comparing the gluing formulae in Theorems 4.1 and 4.2, we can see that the Calderón projectors are the most natural projections for the gluing formulae of the spectral invariants since imposing the boundary conditions by these projections makes the formulae the simplest possible.

5. The operator U and the Cauchy data spaces of a Dirac type operator

In this section, we investigate the operator U, which appears both gluing formulae of the ζ-regularized determinant and the eta invariant in Theorem

4.1. The purpose of this section is to explain that the unitary operator U also describes the difference of the Cauchy data spaces $\mathcal{H}(\mathcal{D}_\pm)$ as $\mathcal{R}$ did the difference of $\mathcal{H}(\Delta_\pm)$.

As before, let us consider the operator $\mathcal{D}$ over the closed manifold M which is decomposed into M_-, M_+ by the hypersurface Y. (The following discussion holds for more general situations, but we just restrict our concern to the closed manifold M.) Now let us recall that $\mathcal{H}(\mathcal{D}_\pm)$ consists of the boundary values of the null-solutions of $\mathcal{D}_\pm$ over $M_\pm$ (see the definition (9)) and the (orthogonalized) Calderón projectors $\mathcal{C}_\pm$ are the unique orthogonal projectors onto the closures of $\mathcal{H}(\mathcal{D}_\pm)$, $\overline{\mathcal{H}(\mathcal{D}_\pm)}$ in $L^2(Y, S_0)$.

To explain the underlying meaning of U, let us explain how we could derive the definition of the operator U by modelling on the operator $\mathcal{R}$.

First, let us recall the expression of $\mathcal{R}$ in (7) with $\mathcal{N}_\pm$ and that $\mathcal{N}_\pm$ determines $\mathcal{H}(\Delta_\pm)$ as its graph. So, our task is to find the operator whose graph is $\mathcal{H}(\mathcal{D}_\pm)$. Noting that the Calderón projector $\mathcal{C}_\pm$ has the matrix form in (10), we can see that $\kappa_\pm$ determines (the closure in $L^2(Y, S_0)$ of) $\mathcal{H}(\mathcal{D}_\pm)$ as its graph.

Now the question is how we define U using $\kappa_\pm$. We here observe that $\overline{\mathcal{H}(\mathcal{D}_\pm)}$ is a Lagrangian subspace in $L^2(Y, S_0)$ with respect to the symplectic form $\langle G\ ,\ \rangle$. Hence, there is the unitary operator over $L^2(Y, S^-)$ which transforms $\mathcal{H}(\mathcal{D}_+)$ to $\mathcal{H}(\mathcal{D}_-)$, that is, describes the difference of them. From this reasoning, we can see that the operator $\kappa_-\kappa_+^{-1}$ does this since

$$(x, \kappa_- x) = (x, (\kappa_-\kappa_+^{-1})\kappa_+ x) \qquad \text{for} \qquad (x, \kappa_\pm x) \in \mathcal{H}(\mathcal{D}_\pm).$$

But, we actually need to find the unitary operator which transforms the Cauchy data space of $\mathcal{D}_+^*$ to $\mathcal{H}(\mathcal{D}_-)$, where $\mathcal{D}_+^*$ is the reflection of the Dirac type operator $\mathcal{D}_+$ to the manifold M_+^*, which is the left side manifold on the double of M_+. This is because $M_+(M_-)$ is a right (left) side manifold and we have to compare the Cauchy data spaces over the left side manifolds to measure the true difference of the Cauchy data spaces. Recalling the double construction in chapter 9 of [4], we can see that the corresponding Calderón projector $\mathcal{C}_+^*$ on the left side manifold M_+^* to $\mathcal{C}_+$ on M_+ is given by

$$\mathrm{Id} - \mathcal{C}_+ = \frac{1}{2}\begin{pmatrix} \mathrm{Id} & -\kappa_+^{-1} \\ -\kappa_+ & \mathrm{Id} \end{pmatrix}.$$

Therefore, the operator $-\kappa_+$ determines the Cauchy data space $\mathcal{H}(\mathcal{D}_+^*)$ as its graph. In conclusion, we can see that $U = -\kappa_-\kappa_+^{-1}$ is the correct operator measuring the true difference of the Cauchy data spaces.

The interesting point is that we can obtain the modulus and the phase data from U via the following equality:

$$\left(\frac{\mathrm{Id} + U}{2}\right)^2 = U\left(\frac{2\mathrm{Id} + U + U^{-1}}{4}\right),$$

where the principal logarithm of the first factor U, $\operatorname{Log} U$ describes the gluing formula of the eta invariant and the second modulus part describes the gluing formula of the ζ-regularized determinant in Theorem 4.1. In fact, this is not a simple coincidence but follows from the deep relation between the eta invariant of the Dirac operator and the ζ-regularized determinant of the Dirac Laplacian as the proof of Theorem 4.1 shows. We refer to [17] for the details of its proof.

References

1. M.F. Atiyah, V.K. Patodi, and I.M. Singer, *Spectral asymmetry and Riemannian geometry. I*, Math. Proc. Cambridge Philos. Soc. **77** (1975), 43–69.
2. C. Bär, *Zero sets of solutions to semilinear elliptic systems of first order*, Invent. Math. **138** (1999), 183–202.
3. D. Bleecker and B. Booß-Bavnbek, *Spectral invariants of operators of Dirac type on partitioned manifolds*, Aspects of Boundary Problems in Analysis and Geometry, 1–130, Oper. Theory Adv. Appl., 151, Birkhauser, Basel, 2004.
4. B. Booß-Bavnbek and K. P. Wojciechowski, *Elliptic Boundary Problems for Dirac Operators*, Birkhäuser Boston Inc., Boston, MA, 1993.
5. J. Brüning and M. Lesch, *On the η-invariant of certain nonlocal boundary value problems*, Duke Math. J. **96** (1999), 425–468.
6. U. Bunke, *On the gluing formula for the η-invariant*, J. Differential Geom. **41** (1995), 397–448.
7. D. Burghelea, L. Friedlander, and T. Kappeler, *Mayer-Vietoris type formula for determinants of differential operators*, J. Funct. Anal. **107** (1992), 34–65.
8. A.-P. Calderón, *Boundary value problems for elliptic equations*, 1963 Outlines Joint Sympos. Partial Differential Equations (Novosibirsk, 1963) pp. 303–304 Acad. Sci. USSR Siberian Branch, Moscow
9. G. Carron, *Déterminant relatif et la fonction Xi*, Amer. J. Math. **124**, no. 2 (2002), 307–352.
10. R. G. Douglas and K. P. Wojciechowski: *Adiabatic limits of the η–invariants. The odd–dimensional Atiyah–Patodi–Singer problem*, Comm. Math. Phys. **142** (1991), 139–168.
11. G. Grubb, *Poles of zeta and eta functions for perturbations of the Atiyah-Patodi-Singer problem*, Comm. Math. Phys. **215** (2001), 583–589.
12. A. Hassell, R. R. Mazzeo, and R. B. Melrose, *Analytic surgery and the accumulation of eigenvalues*, Comm. Anal. Geom. **3** (1995), 115–222.
13. A. Hassell and S. Zelditch, *Determinants of Laplacians in exterior domains*, IMRN **18** (1999), 971–1004.

14. P. Kirk and M. Lesch, *The eta invariant, Maslov index, and spectral flow for Dirac-type operators on manifolds with boundary*, Forum Math. **16** (2004), 553–629.
15. Y. Lee, *Burghelea-Friedlander-Kappeler's gluing formula for the zeta-determinant and its applications to the adiabatic decompositions of the zeta-determinant and the analytic torsion*, Trans. Amer. Math. Soc. **355**, no. 10 (2003), 4093–4110.
16. P. Loya and J. Park, *Decomposition of the zeta-determinant for the Laplacian on manifolds with cylindrical end*, Illinois J. Math. **48**, no. 4 (2004), 1279–1303.
17. P. Loya and J. Park, *On the gluing problem for the spectral invariants of Dirac operators*, Adv. Math., to appear.
18. P. Loya and J. Park, *On the gluing problem for Dirac operators on manifolds with cylindrical ends*, J. Geom. Anal. **15** (2005), 285–319.
19. P. Loya and J. Park, *The comparison problem for the spectral invariants of Dirac type operators*, Preprint, 2004.
20. P. Loya and J. Park, *ζ-determinants of Laplacians with Neumann and Dirichlet boundary conditions*, J. Phys. A, **38** (2005), 8967–8977.
21. R. Mazzeo and P. Piazza, *Dirac operators, heat kernels and microlocal analysis. II. Analytic surgery*, Rend. Mat. Appl. (7) **18**, no. 2 (1998), 221–288.
22. R. B. Melrose, *The Atiyah-Patodi-Singer Index Theorem*, A.K. Peters, Wellesley, 1993.
23. W. Müller, *On the L^2-index of Dirac operators on manifolds with corners of codimension two. I*, J. Differential Geom. **44** (1996), 97–177.
24. W. Müller, *Relative zeta functions, relative determinants and scattering theory*, Comm. Math. Phys. **192** (1998), 309–347.
25. J. Park and K. P. Wojciechowski, *Adiabatic decomposition of the ζ-determinant of the Dirac Laplacian. I. The case of an invertible tangential operator. With an appendix by Yoonweon Lee*, Comm. Partial Differential Equations **27** (2002), no. 7-8, 1407–1435.
26. J. Park and K. P. Wojciechowski, *Scattering theory and adiabatic decomposition of the ζ-determinant of the Dirac Laplacian*, Math. Res. Lett. **9** (2002), no. 1, 17–25.
27. J. Park and K. P. Wojciechowski, *Adiabatic decomposition of the ζ-determinant and Scattering theory*, Michigan Math. J., to appear, math.DG/0111046 .
28. J. Park and K. P. Wojciechowski, *Adiabatic decomposition of the zeta-determinant and the Dirichlet to Neumann operator*, J. Geom. Phys. **55** (2005), 241–266.
29. J. Park and K. P. Wojciechowski, *Agranovich-Dynin formula for the zeta-determinants of the Neumann and Dirichlet problems*, Spectral geometry of manifolds with boundary and decomposition of manifolds, 109–121, Contemp. Math., 366, Amer. Math. Soc., Providence, RI, 2005.
30. D. B. Ray and I. M. Singer, *R-torsion and the Laplacian on Riemannian manifolds*, Adv. Math. **7** (1971), 145–210.
31. S. G. Scott and K. P. Wojciechowski, *The ζ-determinant and Quillen determinant for a Dirac operator on a manifold with boundary*, Geom. Funct.

Anal. **10**, no. 5 (1999), 1202–1236.
32. R. T. Seeley, *Singular integrals and boundary value problems*, Amer. J. Math. **88** (1966), 781–809.
33. R. T. Seeley, *Topics in pseudo-differential operators*, Pseudo-Diff. Operators (C.I.M.E., Stresa, 1968), 1969, pp. 167–305.
34. K. P. Wojciechowski, *The additivity of the* η*-invariant. The case of a singular tangential operator*, Comm. Math. Phys. **169** (1995), 315–327.
35. K. P. Wojciechowski, *The* ζ*-determinant and the additivity of the* η*-invariant on the smooth, self-adjoint Grassmannian*, Comm. Math. Phys. **201**, no. 2 (1999), 423–444.

Received by the editors September 15, 2005 ; Revised January 4, 2006

Part II

Topological Theories

Analysis, Geometry and Topology
of Elliptic Operators, pp. 41–62

THE BEHAVIOR OF THE ANALYTIC INDEX UNDER NONTRIVIAL EMBEDDING

DAVID BLEECKER
Department of Mathematics
University of Hawaii
Honolulu, HI 96822, USA
bleecker@math.hawaii.edu

Dedicated to Krzysztof P. Wojciechowski on his 50th birthday

The Atiyah-Singer index formula states that the analytic index of an elliptic pseudo-differential operator equals the topological index of the K-theoretic class of its asymptotic symbol. In the embedding proof of the index formula, it is shown that both indices obey certain axioms and which guarantee that they are the same. For the most part, the verification that the topological index obeys the axioms is fairly direct, once one has the relevant background in K-theory. However, the verification is not as straightforward for the analytic index. This is particularly true for the multiplication axiom which is used to show how the index behaves under embedding. Complications arise when the normal bundle of the embedding is nontrivial. Our aim is to clarify the verification of the multiplication axiom for the analytic index by defining the participating elliptic pseudo-differential operators in terms of *global* symbols invariantly defined on the cotangent bundles of the relevant base spaces.

2000 *Mathematics Subject Classification.* Primary 58J20; Secondary 58J40

1. Introduction

A major part of the embedding proof, carried out by Atiyah and Singer [1], of the index formula amounts to verifying a multiplication axiom which allows one to show that both the topological and analytic index are preserved under the Thom isomorphism induced by the normal bundle of the tangent bundle of an embedding. As mentioned in the abstract, while the verification of this multiplication axiom is fairly clean for the topological index, it is not so easy for the analytic index, particularly when the normal bundle of the embedding is nontrivial. Our aim is to provide a framework for a proof of this result, which, when compared to existing approaches, we believe is somewhat more streamlined and globally expressed (i.e., free of local coordinates). This method is based on defining pseudo-differential operators from sections of a vector bundle $E \to X$ to sections of a bundle $F \to X$ in terms of a *globally* defined symbol which is a section $p \in C^\infty(\mathrm{Hom}(\pi^*E, \pi^*F))$ of the bundle of homomorphisms between the lifts π^*E and π^*F to the cotangent bundle T^*X, where $\pi : T^*X \to X$. The definition of the operator, say $\mathrm{Op}(p) \in C^\infty(E, F)$, associated with p entails the introduction of a metric on X and connections on E and F. It could be argued that (without a considerable background in modern differential geometry) this is not easier than using local coordinates and framings, but there are advantages. This global approach to pseudo-differential operators is not new. It seems to have first appeared in the paper of Juliane Bokobza-Haggiag [2]. Many subsequent developments and applications have appeared steadily since then, in the work of Harold Widom [3], Ezra Getzler [4], and Theodore Voronov [5], just to mention a few. The application of the global approach to the index theorem, has been mostly in the context of the heat equation proof, rather than in the embedding proof. Nevertheless, in preliminaries leading up to their treatment of the embedding proof in their enlightening book, H. Blaine Lawson and Marie-Louise Michelsohn [6, p. 188] point out the possible desirability of defining pseudo-differential operators with a global symbol. In essence, here we are exploring this possibility. We find that some of the difficulties are softened. In particular, the lifting a pseudo-differential operator to an invariant one in the proof of the twisted multiplication formula is made easier, since the global symbol can be lifted by means of a connection. Moreover the thorny problem of forming suitable products of individual pseudo-differential operators with identity operators over product manifolds (see [6, p. 250f], or [1, p. 514f]) is alleviated by performing operations on the globally-defined total product symbols. This will be made clearer below. One fundamental challenge inspired by this program is the task of constructing a global symbol whose associated pseudo-differential operator is exactly the operator that one may want, as opposed to an approximate operator with essentially the same asymptotic principal symbol.

2. Pseudo-differential operators and symbols

We work within the C^∞ category unless stated otherwise. Let ρ be the injectivity radius of the compact manifold X with Riemannian metric g and Levi-Civita connection ∇; i.e., for all $x \in X$, the exponential map $\exp_x : T_xX \to X$ relative to g is injective on the disk of radius ρ about $0_x \in T_xX$. Let $\pi_E : E \to X$ and $\pi_F : F \to X$ be complex hermitian vector bundles equipped with hermitian connections $\nabla^E : C^\infty(E) \to C^\infty(T^*X \otimes E)$ and $\nabla^F : C^\infty(F) \to C^\infty(T^*X \otimes F)$, where $C^\infty(E)$ denotes the space of (smooth) sections of $\pi_E : E \to X$. For $x, y \in X$, with $d(x,y) < \rho$, let $\tau^E_{x,y} : E_y \to E_x$ denote parallel translation relative to ∇^E along the unique geodesic from y to x with minimal length $d(x,y)$. Let $\psi : [0,\infty) \to [0,1]$ be smooth, with $\psi(r) = 1$ for $r \in [0,\rho/3]$ and $\psi(r) = 0$ for $r \in [2\rho/3, \infty]$. For $\pi : T^*X \to X$ and $u \in C^\infty(E)$, let $u^\wedge \in C^\infty(\pi^*E)$ be defined (where $x = \pi(\xi)$ and $\xi \in T^*X$) by

$$u^\wedge(\xi) := \int_{T_xX} e^{-i\xi(v)}\psi(|v|)\tau^E_{x,\exp_x v}\,[u(\exp_x v)]\,d'v \in E_x \text{ for } \xi \in T^*_xX.$$

where $d'v = (2\pi)^{-n/2}dv$ and dv is the volume element on T_xX associated with g_x. For $x, y \in X$ with $d(x,y) < \rho$, we have $y = \exp_x v$ for a unique $v \in T_xX$ with $|v| = d(x,y)$, and we may define $\alpha \in C^\infty(X \times X, [0,1])$ by

$$\alpha(x,y) := \begin{cases} \psi\,(d(x,y)) = \psi(|v|) & \text{for } d(x,y) < \rho \\ 0 & \text{for } d(x,y) \geq \rho. \end{cases}$$

Note that we can think of the function (in $C^\infty(T_xX, E_x)$)

$$v \mapsto \psi(|v|)\tau^E_{x,\exp_x v}\,[u(\exp_x v)] = \tau^E_{x,\exp_x v}\,[\alpha(x,\exp_x v)u(\exp_x v)] \quad (v \in T_xX)$$

as a "pull-back" (of sorts), using τ^E and $\exp_x : T_xX \to X$, of the bump function $\alpha(x,\cdot)$ times $u(\cdot)$ in a neighborhood of x, and $u^\wedge|_{T^*_xX}$ is the Fourier transform of this "pull-back" of $\alpha(x,\cdot)u(\cdot)$. The "inverse Fourier transform" $(u^\wedge)^\vee : TX \to E$ of $u^\wedge$ is given by

$$(u^\wedge)^\vee(v) := \int_{T^*_xX} e^{i\xi(v)}u^\wedge(\xi)\,d'\xi = \psi(|v|)\tau_{x,\exp_x v}\,[u(\exp_x v)],$$

where $d'\xi = (2\pi)^{-n/2}d\xi$. Since $(u^\wedge)^\vee(v) \in E_x$, $(u^\wedge)^\vee$ is a section of the pull-back of E to TX via $\pi : T^*X \to X$. Moreover, we can recover u locally about x from $u^\wedge|_{T^*_xX}$. In particular, for $v = 0_x \in T_xX$, we have $(u^\wedge)^\vee(0_x) = u(x)$. For $\pi : T^*X \to X$ and a section $p \in C^\infty(\mathrm{Hom}(\pi^*E, \pi^*F))$, (of $\mathrm{Hom}(\pi^*E, \pi^*F) \to T^*X$) we define an operator

$\mathrm{Op}(p) : C^\infty(E) \to C^\infty(F)$ via

$$\mathrm{Op}(p)(u)_x := \int_{T_x^*X} e^{i\xi(v)} p(\xi) \left(u^\wedge(\xi)\right) d'\xi \Big|_{v=0} = \int_{T_x^*X} p(\xi) \left(u^\wedge(\xi)\right) d'\xi$$

$$= \int_{T_x^*X} p(\xi) \left(\int_{T_xX} e^{-i\xi(v)} \psi(|v|) \tau^E_{x,\exp_x v} \left[u(\exp_x v)\right] d'v \right) d'\xi$$

$$= \int_{T_xX \times T_x^*X} p(\xi) \left(e^{-i\xi(v)} \psi(|v|) \tau^E_{x,\exp_x v} \left[u(\exp_x v)\right] \right) d'v d'\xi$$

$$= \int_{T_xX \times T_x^*X} e^{-i\xi(v)} p(\xi) \left(\tau^E_{x,\exp_x v} \left[\alpha(x, \exp_x v) u(\exp_x v)\right] \right) d'v d'\xi. \quad (1)$$

Recall roughly that quantization in quantum mechanics attempts to convert functions of position and momentum (i.e., functions on T_x^*X) into operators. One may think of $\mathrm{Op}(p)$ as a quantization of p, but $\mathrm{Op}(p)$ depends on many choices (e.g., the choice of metric, connections, and $\alpha : X \times X \to [0,1]$). Apart from these choices, there are other choices one can make, as is discussed in [5]. For example, if $s \in [0,1]$, let

$$T_{x,\exp_x sv} : T^*_{\exp_x sv} X \to T_x^* X$$

denote parallel translation (with respect to the Levi-Civita connection) for T^*X along the geodesic $t \mapsto \exp_x tv$ in the reverse direction from $\exp_x sv$ to x. In [5] (but with notation that differs from ours), an operator $\mathrm{Op}(p; s)$ (depending on s) is associated to p via

$$\mathrm{Op}(p;s)(u)_x = \int_{T_xX \times T_x^*X} d'v d'\xi \; e^{-i\xi(v)}$$
$$\alpha(x, \exp_x v) \tau^F_{x,\exp_x sv} p(T_{x,\exp_x sv}(\xi)) \tau^E_{\exp_x sv,\, \exp_x v} u\left(\exp_x v\right).$$

When $s = 0$, we get

$$\mathrm{Op}(p;0)(u)_x = \int_{T_xX \times T_x^*X} d'v d'\xi \; e^{-i\xi(v)} \alpha(x, \exp_x v) p(\xi) \tau^E_{x,\, \exp_x v} \left(u\left(\exp_x v\right)\right)$$

which is precisely our $\mathrm{Op}(p)$. In cases of interest, the operators $\mathrm{Op}(p; s)$ for different s differ by "lower order" operators which do not affect the index (if defined). Hence, for simplicity, we only use $s = 0$. As stated in [5] the choice of s is related to the choice of operator ordering of monomials in position and momentum variables under quantization.

The connections ∇^E and ∇^F pull back via $\pi : T^*X \to X$ to connections on the bundles $\pi^*E \to T^*X$ and $\pi^*F \to T^*X$, which we continue to denote by ∇^E and ∇^F. The Levi-Civita connection for (X, g) determines a subbundle H of $T(T^*X)$ consisting of horizontal subspaces of $T(T^*X)$, which is complementary to the subbundle V of $T(T^*X)$ consisting of vectors which are tangent to the fibers of $T^*X \to X$. There is a natural Riemannian metric, say g^*, on T^*X such that V and H are orthogonal and g^* equals g on V and π^*g on H. Using ∇^E and ∇^F, along with the Levi-Civita connection for g^*, say ∇^*, we may construct a covariant derivative

$$\widetilde{\nabla} : C^\infty (\mathrm{Hom}(\pi^*E, \pi^*F)) \to C^\infty (T^* (T^*X) \otimes \mathrm{Hom}(\pi^*E, \pi^*F)) .$$

Since ∇^* extends to $\otimes^k T^* (T^*X)$, we may "iterate" $\widetilde{\nabla}$ to obtain

$$\widetilde{\nabla}^k : C^\infty (\mathrm{Hom}(\pi^*E, \pi^*F)) \to C^\infty \left(\otimes^k T^* (T^*X) \otimes \mathrm{Hom}(\pi^*E, \pi^*F)\right) .$$

Definition 1. We say that $p \in C^\infty(\mathrm{Hom}(\pi^*E,\pi^*F)$ is a **symbol of order** $m \in \mathbb{R}$ if for any $H_1, \ldots, H_I \in C^\infty(H)$ with $|H_1|, \ldots, |H_I| \leq 1$ and $V_1, \ldots, V_J \in C^\infty(V)$, there are constants C_{IJ} (depending only on I, J and p), such that

$$\left|\left(\widetilde{\nabla}^{I+J} p\right) (H_1, \ldots, H_I, V_1, \ldots, V_J)\right| \leq C_{IJ} \left(1 + \sum_{j=1}^{J} |V_j|\right)^{m-J} .$$

Moreover, we require that the m-th order asymptotic symbol of p, namely

$$\sigma_m(p) (\xi) := \lim_{t\to\infty} \frac{p (t\xi)}{t^m} \quad (\text{for } \xi \neq 0) \tag{2}$$

exist, where the convergence is uniform on $S(T^*X)$. Then we call $\mathrm{Op}(p)$ a pseudo-differential **operator of order** m. We denote the set of symbols of order m by $\mathrm{Symb}_m(E, F)$.

Clearly, for $m' > m$,

$$\mathrm{Symb}_{m'}(E, F) \supset \mathrm{Symb}_m(E, F) \supset \mathrm{Symb}_{-\infty}(E, F) := \bigcap_{m=0}^{-\infty} \mathrm{Symb}_m(E, F).$$

For $p \in \mathrm{Symb}_m(E, F)$, we then have the operator, say $\mathrm{Op}(p) : C^\infty(E) \to C^\infty(F)$, given by (1), which extends to a bounded operator $\mathrm{Op}_s(p) : L^2_s(E) \to L^2_{s-m}(F)$, where for any $s \in \mathbb{R}$, $L^2_s(E)$ is the s-th Sobolev space of sections of E, namely the completion of $C^\infty(E)$ with respect to the norm $\|\cdot\|_s$ defined by

$$\|u\|_s^2 := \int_{T^*X} \left(1 + |\xi|^2\right)^s |u^\wedge(\xi)|^2 \, d\xi.$$

Recall that for $k \in \mathbb{Z}^+$, and $s > n/2 + k$, there is a compact inclusion $L^2_s(E) \subset C^k(E)$. For each s, the linear map

$$\mathrm{Op}_s : \mathrm{Symb}_m(E, F) \to \mathrm{B}(L^2_s(E), L^2_{s-m}(F))$$

into the Banach space $\mathrm{B}(L^2_s(E), L^2_{s-m}(F))$ of bounded linear transformations is continuous (see [6, p. 177f]). Moreover, for $\varphi \in \mathrm{Symb}_{-\infty}(E,F)$, $\mathrm{Op}_s(\varphi)$ is a compact operator for any $s \in \mathbb{R}$, and $\mathrm{Op}_s(\varphi)\left(L^2_s(E)\right) \subset C^\infty(F)$; i.e., $\mathrm{Op}_s(\varphi)$ is a smoothing operator.

Definition 2. We say that $p \in \mathrm{Symb}_m(E,F)$, and the corresponding operator $\mathrm{Op}(p)$, are **elliptic** if for some constant $c > 0$, $p(\xi)^{-1}$ exists for $|\xi| > c$, and for some constant $K > 0$

$$\left|p(\xi)^{-1}\right| \le K\left(1+|\xi|\right)^{-m} \text{ for all } \xi \in T^*X \text{ with } |\xi| > c.$$

We set $\mathrm{Ell}_m(E,F) := \{p \in \mathrm{Symb}_m(E,F) : p \text{ is elliptic}\}$.

For $p \in \mathrm{Ell}_m(E,F)$, there are $q \in \mathrm{Symb}_{-m}(E,F)$, $\varphi_E \in \mathrm{Symb}_{-\infty}(E,E)$ and $\varphi_F \in \mathrm{Symb}_{-\infty}(F,F)$, such that

$$\mathrm{Op}_{s-m}(q) \circ \mathrm{Op}_s(p) = \mathrm{Id}_{L^2_s(E)} + \mathrm{Op}_s(\varphi_E) \text{ and}$$
$$\mathrm{Op}_s(p) \circ \mathrm{Op}_{s-m}(q) = \mathrm{Id}_{L^2_{s-m}(F)} + \mathrm{Op}_{s-m}(\varphi_F).$$

Since $\mathrm{Op}_s(\varphi_E)$ and $\mathrm{Op}_{s-m}(\varphi_F)$ are compact operators, it follows that $\mathrm{Op}_s(p)$ is Fredholm, and hence we may define

$$\mathrm{index}(\mathrm{Op}_s(p)) := \dim\ker(\mathrm{Op}_s(p)) - \dim\mathrm{coker}(\mathrm{Op}_s(p)).$$

Note also that if $\mathrm{Op}_s(p)u \in C^\infty(F)$, then

$$u = \mathrm{Op}_{s-m}(q)\left(\mathrm{Op}_s(p)u\right) - \mathrm{Op}_s(\varphi_E)u \in C^\infty(E).$$

Thus, $\dim\ker\left(\mathrm{Op}_s(p)\right) < \infty$, $\ker\left(\mathrm{Op}_s(p)\right) \subset C^\infty(E)$, and $\ker\left(\mathrm{Op}_s(p)\right)$ is independent of s. As a consequence,

$$\mathrm{index}(\mathrm{Op}_s(p)) = \dim\ker(\mathrm{Op}(p)) - \dim\mathrm{coker}(\mathrm{Op}(p))$$

is independent of s.

3. Definition of the analytic index

Before proceeding with the definition of the analytic index of an element of $K(T^*X)$, we consider the more familiar case of the (analytic) index of an elliptic, linear differential operator $D : C^\infty(E) \to C^\infty(F)$ of given order m. Associated with D is its principal symbol $\sigma_m(D) \in C^\infty\left(\mathrm{Hom}(\pi^*E, \pi^*F)\right)$ which is defined as follows. If in local coordinates $(x^1, \ldots, x^n)$ about a point $x \in X$,

$$D = \sum\nolimits_{k=0}^{m} \sum\nolimits_{j_1,\ldots,j_k=1}^{n} i^{-k} A_{j_1\ldots j_k}(x) \frac{\partial^k}{\partial x_{j_1} \cdots \partial x_{j_k}},$$

where $A_{j_1\dots j_k}(x) \in \operatorname{Hom}(E_x, F_x)$, then for $\xi_x = \xi_1 dx^1 + \cdots + \xi_n dx^n \in T_x^*X$,

$$\sigma_m(D)(\xi) := \sum\nolimits_{j_1,\dots,j_m=1}^{n} A_{j_1\dots j_m}(x)\xi_{j_1}\cdots\xi_{j_m}.$$

One can check that $\sigma_m(D)$ is independent of the choice of local coordinates, although this would not be the case if lower-order terms were included. If $\sigma_m(D)$ is invertible outside of the zero section of T^*M, then D is said to be elliptic, which we assume. If lower order terms were included and if we denoted this coordinate-dependent, locally-defined "full symbol" by $p_{\mathrm{loc}}(D)(\xi)$, then $\sigma_m(D)(\xi)$ at $\xi \in T_x^*X$ would be given by

$$\lim_{t\to\infty} \frac{p_{\mathrm{loc}}(D)(t\xi)}{t^m}$$

in comparison with (2). However, it is not clear that D is precisely $\mathrm{Op}(p)$ for some globally defined $p \in \mathrm{Symb}_m(E, F)$. In the language of physicists, it is not clear that D can be precisely dequantized. If such p exists, it would clearly depend on choices of a Riemannian metric on M, connections for E and F and on the function $\alpha : X \times X \to [0, 1]$ supported near the diagonal. However, in [2] and [3], it is shown that given such choices, p can be found so that $\mathrm{Op}(p)$ and D differ by an operator which is infinitely smoothing (and hence compact); i.e.,

$$D - \mathrm{Op}(p) = \mathrm{Op}(\sigma) \text{ for } \sigma \in \mathrm{Symb}_{-\infty}(E, F).$$

By methods that are standard by now, it follows that D has Fredholm Sobolev extensions $D_s : L_s^2(E) \to L_{s-m}^2(F)$ for all s, with a common index, which is sometimes called the *analytic* index of D; it is just the usual operator-theoretic index. It is simply denoted by $\mathrm{index}(D)$ and if $D^* : C^\infty(F) \to C^\infty(E)$ is the formal L^2-adjoint of D, then

$$\begin{aligned}\dim\ker(D) - \dim\ker(D^*) &= \mathrm{index}(D)\\ &= \mathrm{index}(\mathrm{Op}(p) + \mathrm{Op}(\sigma)) = \mathrm{index}(\mathrm{Op}(p)).\end{aligned}$$

Thus, readers (including the author) who are bothered by the fact that differential operators may not be precisely dequantized, may take some solace in the fact that elliptic differential operators may be approximated by a pseudo-differential operator of the form $\mathrm{Op}(p)$, modulo smoothing operators which preserve the index.

The analytic index of $A \in K(T^*X)$ is defined as follows. Recall that A can be regarded as an equivalence class of bundle maps $V_0 \xrightarrow{\alpha} V_1$ for complex vector bundles V_0 and V_1 over T^*X. Moreover, it is required that the support $\mathrm{supp}(\alpha) := \left\{\xi \in T^*X \mid \alpha(\xi) \notin \mathrm{Iso}((V_0)_\xi, (V_1)_\xi)\right\}$ be a compact subset of T^*X. Using the compactness of X and the fact that the zero section of T^*X is clearly retract of the unit ball bundle $B(T^*X)$, it is not difficult to show (e.g., see [1], [6] and [7]) that we can represent any $A \in K(T^*X)$ by some $p \in \mathrm{Ell}_m(E,F)$ for some complex vector bundles E and F over X, where $p(\xi)$ is an isomorphism for $|\xi| \geq c > 0$ and $m \in \mathbb{R}$ can be chosen arbitrarily. The analytic index of $A \in K(T^*X)$ is defined by $\mathrm{index}_a(A) := \mathrm{index}(\mathrm{Op}(p))$. Of course one needs to check that $\mathrm{index}(\mathrm{Op}(p))$ is independent of the choice of m and $p \in \mathrm{Ell}_m(E,F)$ representing A. It can be shown that $p_0 \in \mathrm{Ell}_0(E_0,F_0)$ and $p_1 \in \mathrm{Ell}_0(E_1,F_1)$ both represent A precisely when there are vector bundles $\widetilde{E}$ and $\widetilde{F}$ over $X \times I$, and for $\pi \times \mathrm{Id} : (T^*X) \times I \to X \times I$, a bundle map $P : (\pi \times \mathrm{Id})^* \widetilde{E} \to (\pi \times \mathrm{Id})^* \widetilde{F}$ such that $P_{|(T^*X)\times\{t\}} \in \mathrm{Ell}_0(\widetilde{E}|_{X\times\{t\}}, \widetilde{F}|_{X\times\{t\}})$, and (for $k = 0,1$) isomorphisms η_k and φ_k, such that we have a commutative diagram of bundle maps

$$\begin{array}{ccc} (\pi \times \mathrm{Id})^* \widetilde{E}|_{(T^*X)\times\{k\}} & \overset{P|_{(T^*X)\times\{k\}}}{\rightarrow} & (\pi \times \mathrm{Id})^* \widetilde{F}|_{(T^*X)\times\{k\}} \\ \downarrow \eta_k & & \downarrow \varphi_k \\ \pi^* E_k \oplus \mathbb{C}^{n_k} & \overset{p_k \oplus I_{n_k}}{\rightarrow} & \pi^* F_k \oplus \mathbb{C}^{n_k}, \end{array}$$

where $\mathbb{C}^{n_k}$ denotes a bundle over T^*X of dimension n_k (see [6, p. 247]). Using the invariance of the index under continuous deformation, we have

$$\mathrm{index}\left(\mathrm{Op}\left(P|_{(T^*X)\times\{0\}}\right)\right) = \mathrm{index}\left(\mathrm{Op}\left(P|_{(T^*X)\times\{1\}}\right)\right).$$

Then using other standard properties of the index, we obtain

$$\begin{aligned} \mathrm{index}\left(\mathrm{Op}\left(p_0\right)\right) &= \mathrm{index}\left(\mathrm{Op}\left(p_0 \oplus I_{n_0}\right)\right) \\ &= \mathrm{index}\left(\mathrm{Op}\left(\varphi_0 \circ \left(p_0 \oplus I_{n_0}\right) \circ \eta_0^{-1}\right)\right) \\ &= \mathrm{index}\left(\mathrm{Op}\left(P|_{(T^*X)\times\{0\}}\right)\right) = \mathrm{index}\left(\mathrm{Op}\left(P|_{(T^*X)\times\{1\}}\right)\right) \\ &= \mathrm{index}\left(\mathrm{Op}\left(\varphi_1 \circ \left(p_1 \oplus I_{n_1}\right) \circ \eta_1^{-1}\right)\right) = \mathrm{index}\left(\mathrm{Op}\left(p_1 \oplus I_{n_1}\right)\right) \\ &= \mathrm{index}\left(\mathrm{Op}\left(p_1\right)\right). \end{aligned}$$

For $q(\xi) := (1 + |\xi|^2)^{1/2} \, \mathrm{Id}_E \in \mathrm{Ell}_1(E,E)$, we have $M_{q^{-m}} : \mathrm{Ell}_m(E,F) \to \mathrm{Ell}_0(E,F)$ given by

$$M_{q^{-m}}(p) = p \circ q^{-m} \in \mathrm{Ell}_0(E,F) \text{ for } p \in \mathrm{Ell}_m(E,F).$$

Moreover, since the operator $\mathrm{Op}(q^{-m})$ is invertible,

$$\mathrm{index}\left(p \circ q^{-m}\right) = \mathrm{index}\left(p\right),$$

showing that the definition of $\mathrm{index}_a(A)$ is independent of the choice of m.

4. The formulation of the multiplicative property

In the embedding proof of the Atiyah-Singer Index Theorem, the multiplicative property is used to reduce the index formula for an elliptic pseudo-differential operator over a topologically complicated compact manifold X to the case of a related pseudo-differential operator over an ordinary sphere in which X can be embedded. For operators over spheres, the index formula can then either be checked explicitly or further reduced (by means of Bott periodicity) to the case of an operator over S^2 or S^1. At the end of this section, we will explain more precisely how the multiplicative property fits into the general scheme of the embedding proof.

Consider an embedding $f : X \to Y$ of the compact manifold X into some manifold Y (say $\mathbb{R}^{n'}$ or $S^{n'}$). From an elliptic pseudo-differential operator on X, we will construct an appropriate elliptic pseudo-differential operator, with the same index, on a suitably compactified tubular neighborhood, say S, of $f(X)$ in Y. In other words, from a symbol

$$a \in \mathrm{Ell}_m(E, F) \subset C^\infty(T^*X, \mathrm{Hom}(\pi_X^*E, \pi_X^*F)), \text{ where } \pi_X : T^*X \to X$$

with associated operator $\mathrm{Op}(a) : C^\infty(E) \to C^\infty(F)$, one needs to construct suitable complex vector bundles $\widetilde{E} \to S$ and $\widetilde{F} \to S$ and a symbol

$$c \in \mathrm{Ell}_m(\widetilde{E}, \widetilde{F}) \subset C^\infty(T^*S, \mathrm{Hom}(\pi_S^*\widetilde{E}, \pi_S^*\widetilde{F})), \tag{3}$$

with associated operator $\mathrm{Op}(c) : C^\infty(\widetilde{E}) \to C^\infty(\widetilde{F})$; here $\pi_S : T^*S \to S$. The essential ingredient which is needed to produce c is an equivariant K-theory element $b \in K_{\mathrm{O}(m)}(T^*S^m)$, where $m = n' - n$ and S^m is the unit m-sphere. The choice of $b \in K_{\mathrm{O}(m)}(T^*S^m)$ which yields $\mathrm{index}\,\mathrm{Op}(c) = \mathrm{index}\,\mathrm{Op}(a)$ is essentially the famous generating Bott element, but b will be arbitrary here. We begin with a short review of relevant equivariant K-theory for those who desire it. The work of Graeme Segal [8] is an excellent, authoritative exposition of the foundations of equivariant K-theory.

Let G be a group which acts to the left on X, via a $L : G \times X \to X$. We write $g \cdot x = L_g(x) = L(g, x)$. Let $\pi : E \to X$ be a complex vector bundle over X and suppose that there is a left action of G on E such that $\pi(g \cdot e) = g \cdot \pi(e)$ and $e \mapsto g \cdot e$ is linear on each fiber E_x. Then $\pi : E \to X$ is called a G-vector bundle. As an example, if X is a manifold and G acts on X smoothly, then the action on $T_{\mathbb{C}}X := \mathbb{C} \otimes TX$ given by $v \mapsto d(L_g)(v)$ for $v \in T_{\mathbb{C}}X$ makes $T_{\mathbb{C}}X \to X$ a G-vector bundle. More generally, $\Lambda^k(T_{\mathbb{C}}X) \to X$ is a G-vector bundle. A morphism from G-vector bundle $\pi_1 : E_1 \to X$ to G-vector bundle $\pi_2 : E_2 \to X$ is a vector bundle morphism (linear on fibers) $\varphi : E_1 \to E_2$ such that $\varphi(g \cdot e) = g \cdot \varphi(e)$. An isomorphism of G-vector bundles is a morphism which is bijective. The direct sum of G-vector bundles is clearly a G-vector bundle and this operation induces an abelian semi-group structure on the set of isomorphism classes of G-vector bundles. We can then form the associated abelian group $K_G(X)$ via the Grothendieck construction. Moreover, the tensor product of G-vector bundles yields a G-vector bundle, and this induces a ring structure on $K_G(X)$. For a homogeneous space G/H where H is a closed subgroup of G, there is a ring isomorphism $K_G(G/H) \cong R(H) :=$ the representation ring of H. Recall that $R(H)$ is the Grothendieck ring obtained from the abelian semi-group of equivalence classes of representations of H with addition induced by the direct sum. Tensor product of representations induces a multiplication on $R(H)$ making it a ring. More concisely, $R(H) = K_H(\{\text{point}\})$. As with ordinary K-theory, an element of $K_G(X)$ can also be described as equivalence classes of G-equivariant morphisms $E \to F$ of G-bundles which are isomorphisms outside of a compact support (i.e., morphisms with compact support).

We proceed with the construction of $c \in \mathrm{Ell}_m(\widetilde{E}, \widetilde{F})$ in (3). Let $\pi_P : P \to X$ be the principal $\mathrm{O}(m)$-bundle of orthonormal frames of the normal bundle $N \to X$ for the embedding $f : X \to Y$, where $\dim X = n$ and $\dim Y = n'$. We regard a frame $p \in P_x$ as a linear isometry $p : \mathbb{R}^m \to N_x$, where $m = n' - n$ and N_x is the fiber of the normal bundle at $x \in X$. In terms of associated bundles, we have

$$N = P \times_{\mathrm{O}(m)} \mathbb{R}^m = (P \times \mathbb{R}^m) \,/\, \mathrm{O}(m),$$

where $\mathrm{O}(m)$ acts on $P \times \mathbb{R}^m$ via $(p, v) \cdot A := (p \circ A, A^{-1}v)$. Note that $\mathrm{O}(m)$ also acts on $\mathbb{R}^{m+1} = \mathbb{R}^m \times \mathbb{R}$ via $A \cdot (v, a) = (A(v), a)$, and the m-sphere $S^m \subset \mathbb{R}^{m+1}$ is invariant under this action with two fixed points, the poles $(\mathbf{0}, \pm 1) \in S^m$. Let

$$S := P \times_{\mathrm{O}(m)} S^m \text{ and let } Q : P \times S^m \to P \times_{\mathrm{O}(m)} S^m = (P \times S^m) \,/\, \mathrm{O}(m)$$

be the quotient map. We may regard $\pi_S : S \to X$ as the m-sphere bundle over X obtained by compactification of the normal bundle N via adjoining the section at infinity. Choose a so(m)-valued connection 1-form ω on P; there is actually a natural ω induced by $f : X \to Y$ and a given Riemannian metric on Y. Then we have an $\mathrm{O}(m)$-invariant distribution H of horizontal subspaces (i.e., $H_p = \mathrm{Ker}\, \omega_p$) on P and hence on $P \times S^m$. By the $\mathrm{O}(m)$-invariance of H, $Q_*(H)$ is a well defined distribution on S. Moreover, since $\pi_{S*} Q_*(H_p) = \pi_{P*}(H_p) = T_{\pi(p)}X$, $Q_*(H)$ is complementary to the vertical distribution V_S of tangent spaces of the fibers of $\pi_S : S \to X$. We denote

$Q_*(H)$ by H_S. Thus, we have a splitting

$$TS = V_S \oplus H_S = V_S \oplus Q_*(H). \tag{4}$$

We also have $T^*S = \widetilde{V}_S^* \oplus \widetilde{H}_S^*$, where

$$\widetilde{H}_S^* := \{\alpha \in T^*S : \alpha(V_S) = 0\} \text{ and}$$
$$\widetilde{V}_S^* := \{\beta \in T^*S : \beta(H_S) = 0\} \cong P \times_{\mathrm{O}(m)} T^*S^m.$$

In view of the splitting (4), there are identifications $\widetilde{V}_S^* \cong V_S^* := (V_S)^*$ and $\widetilde{H}_S^* \cong H_S^* := (H_S)^*$. Note that $\mathrm{O}(m)$ acts on the sphere S^m, and hence on T^*S^m via pull-back of covectors. Thus, we may consider $K_{\mathrm{O}(m)}(T^*S^m)$. The projection $P \times T^*S^m \to T^*S^m$ induces a map $K_{\mathrm{O}(m)}(T^*S^m) \to K_{\mathrm{O}(m)}(P \times T^*S^m)$. Moreover, there is the general fact that if G acts freely on X, then the projection $Q : X \to X/G$ induces an isomorphism $Q^* : K(X/G) \cong K_G(X)$ (see [8, p. 133]). Thus, we have

$$K_{\mathrm{O}(m)}(T^*S^m) \to K_{\mathrm{O}(m)}(P \times T^*S^m) \overset{(Q^*)^{-1}}{\underset{\to}{\cong}} K(P \times_{\mathrm{O}(m)} T^*S^m) = K(V_S^*). \tag{5}$$

We define

$$K(T^*X) \otimes K(V_S^*) \to K(T^*S), \tag{6}$$

as follows. If $E \to T^*X$ and $F \to V_S^*$ are complex vector bundles, then for $\alpha' \in T^*X$ and $\beta' \in V_S^*$, we have unique $\alpha \in \widetilde{H}_S^*$ and $\beta \in \widetilde{V}_S^*$ such that $\alpha(v) = \alpha'\left((\pi_S)_*(v)\right)$ for v in TS, and $\beta|_{V_S} = \beta'$ and $\beta(H_S) = 0$. Then $E_{\alpha'} \otimes F_{\beta'}$ is the fiber of a bundle over T^*S at the point $\alpha + \beta$. Thus, we have $K(T^*X) \otimes K(V_S^*) \to K(T^*S)$ induced by $[E] \otimes [F] \mapsto [E \otimes F]$. Using the homomorphisms (5) and (6), we then have

$$K(T^*X) \otimes K_{\mathrm{O}(m)}(T^*S^m) \to K(T^*X) \otimes K(V_S^*) \to K(T^*S). \tag{7}$$

For any representation $\rho : \mathrm{O}(m) \to \mathrm{GL}(\mathbb{C}^q)$, we have the associated vector bundle $P \times_\rho \mathbb{C}^q \to X$. Let $R(\mathrm{O}(m))$ be the representation ring of $\mathrm{O}(m)$. The assignment $\rho \mapsto P \times_\rho \mathbb{C}^q$ extends to a ring homomorphism

$$R(\mathrm{O}(m)) \to K(X),$$

which is to say that $K(X)$ is a $R(\mathrm{O}(m))$-module. Moreover, recall that $K(T^*X)$ is a $K(X)$-module via $u \cdot v = (\pi^* u)\, v$. Thus, ultimately $K(T^*X)$ is an $R(\mathrm{O}(m))$-module. We are now in a position to state

The Multiplicative Property. For $v \in K_{\mathrm{O}(m)}(T^*S^m)$ and $u \in K(T^*X)$, we have $u \cdot v \in K(T^*S)$, via (7). Moreover,

$$\mathrm{index}_a(u \cdot v) = \mathrm{index}_a\left(\left(\mathrm{index}_{\mathrm{O}(m)}\, v\right) \cdot u\right),$$

where $\left(\mathrm{index}_{\mathrm{O}(m)} v\right) \cdot u \in K(T^*X)$ makes sense since $\mathrm{index}_{\mathrm{O}(m)} v \in R(\mathrm{O}(m))$, and as we have just noted, $K(T^*X)$ is an $R(\mathrm{O}(m))$-module. In particular, if $\mathrm{index}_{\mathrm{O}(m)} v = 1 \in R(\mathrm{O}(m))$, then $\mathrm{index}_a (u \cdot v) = \mathrm{index}_a u$.

So far, we have not indicated how the multiplicative property fits into the embedding proof of the index formula, namely $\mathrm{index}_a(u) = \mathrm{index}_t(u)$, where $u \in K(T^*X)$, nor have we defined the topological index $\mathrm{index}_t(u)$. Since we suspect that most readers would like to see this, we close this subsection with a necessarily sketchy outline of the argument.

Let $\pi : V \to X$ be a complex vector bundle, where X is compact. Let $\Lambda^i(V)$ be the i-th exterior bundle of V over X. The pull-backs $\pi^*\Lambda^i(V)$ are then bundles over V, say $\pi^i : \pi^*\Lambda^i(V) \to V$. At each $v \in V$, we have a linear map $\alpha_v^i : \left(\pi^*\Lambda^i(V)\right)_v \to \left(\pi^*\Lambda^{i+1}(V)\right)_v$, given by $\alpha_v^i(w) = v \wedge w$. Since $\alpha_v^{i+1} \circ \alpha_v^i = 0$, we have a complex over V, namely

$$0 \to \pi^*\Lambda^0(V) \stackrel{\alpha^0}{\to} \pi^*\Lambda^1(V) \stackrel{\alpha^1}{\to} \cdots \stackrel{\alpha^{n-1}}{\to} \pi^*\Lambda^n(V) \to 0,$$

where n is the fiber dimension of V. If $v \neq 0$, we have $\mathrm{Im}\left(\alpha_v^i\right) = \mathrm{Ker}\left(\alpha_v^{i+1}\right)$, so that the complex is exact over V minus the zero section. Thus, the complex defines an element $\lambda_V \in K(V)$. The standard Thom Isomorphism Theorem states that

$$\varphi : K(X) \to K(V), \text{ given by } \varphi(a) = (\pi^*a)\,\lambda_V,$$

is an isomorphism. To indicate the dependence of φ on $\pi : V \to X$, we use the notation $\varphi_{V \to X} : K(X) \to K(V)$. A special case of this isomorphism arises as follows. Let X and Y be manifolds and $f : X \to Y$ a smooth, proper embedding. We have $f_* : TX \to TY$. While the normal bundle N of X in Y does not have a complex structure, the normal bundle $\widetilde{N}$ of TX in TY does. Thus, we have

$$\varphi_{\widetilde{N} \to TX} : K(TX) \to K(\widetilde{N}).$$

Note that $\widetilde{N}$ can be embedded into TY as an open subset, and this embedding induces a homomorphism $h : K(\widetilde{N}) \to K(TY)$. The composition $h \circ \varphi_{\widetilde{N} \to TX}$ gives us a homomorphism

$$f_! := h \circ \varphi_{\widetilde{N} \to TX} : K(TX) \to K(TY).$$

In the case where $Y = \mathbb{R}^{n+m}$, we have $TY = \mathbb{R}^{2(n+m)}$. If $i : \{\mathbf{0}\} \to \mathbb{R}^{n+m}$ is the inclusion of the origin, then $i_! : K(T\{\mathbf{0}\}) \cong K(\mathbb{R}^{2(n+m)})$, and plainly

$K(T\{0\}) \cong \mathbb{Z}$, since $T\{0\}$ is just a point. By definition, the composition $i_!^{-1} \circ f_!$ is the topological index, namely

$$\text{index}_t : K(TX) \xrightarrow{f_!} K(\mathbb{R}^{2(n+m)}) \overset{i_!^{-1}}{\underset{\rightarrow}{\cong}} \mathbb{Z}.$$

Of course, some work is needed to show that this is well defined (e.g., independent of the choice of f). To prove the index formula, one needs to show that $\text{index}_a(u) = \text{index}_a(f_! u)$. Then

$$\begin{aligned}\text{index}_a(u) &= \text{index}_a(f_! u) = \text{index}_a((i_! i_!^{-1})(f_! u)) = \text{index}_a(i_!(i_!^{-1} f_! u)) \\ &= \text{index}_a(i_!^{-1} f_! u) = i_!^{-1} f_! u = \text{index}_t(u).\end{aligned}$$

Since $f_! = h \circ \varphi_{\widetilde{N} \to TX}$ is a composition of two maps, the proof that $\text{index}_a(u) = \text{index}_a(f_! u)$ has two parts, namely

1. $\text{index}_a(\varphi_{\widetilde{N} \to TX}(u)) = \text{index}_a(u)$ and
2. $\text{index}_a(\varphi_{\widetilde{N} \to TX}(u)) = \text{index}_a(h(\varphi_{\widetilde{N} \to TX}(u)))$.

Part 2 follows from the Excision Property and its proof is easier than part 1 (e.g., see [6, p. 248 and p. 254]). Part 1 is a consequence of the multiplicative property. Indeed, for $\pi_{\widetilde{N}}^* : \widetilde{N} \to TX$,

$$\varphi_{\widetilde{N} \to TX}(u) = (\pi_{\widetilde{N}}^* u)\lambda_{\widetilde{N}} = u \cdot i_! 1,$$

where the last equality follows (in part) from the fact that the associated bundle $P \times_{\mathrm{O}(m)} T^* S^m \cong V_S^*$ is isomorphic to $\widetilde{N}$ with one of its two summands compactified; note that $\widetilde{N} \cong \pi^* N \oplus \pi^* N$ where $\pi : TX \to X$. By various means (none very easy) it is known that $\text{index}_{\mathrm{O}(m)} i_! 1 = 1 \in R(\mathrm{O}(m))$; see [1, Proposition (4.4), p. 505] or incompletely in [6, p. 253]. Thus,

$$\begin{aligned}\text{index}_a(\varphi_{\widetilde{N} \to TX}(u)) &= \text{index}_a\left((\pi_{\widetilde{N}}^* u)\lambda_{\widetilde{N}}\right) = \text{index}_a(u \cdot i_! 1) \\ &= \text{index}_a((\text{index}_{\mathrm{O}(m)} i_! 1) \cdot u) = \text{index}_a(u).\end{aligned}$$

5. Proving the multiplicative property

Let $u = [a] \in K(T^* X)$ and $v = [b] \in K_{\mathrm{O}(m)}(T^* S^m)$ for first-order elliptic symbols $a \in \mathrm{Ell}_1(E, F)$ and $b \in \mathrm{Ell}_{\mathrm{O}(m)1}(E', F')$ which means the following. For $g \in \mathrm{O}(m)$, let $L_g : S^m \to S^m$ be given by $L_g x = gx$. The differential $L_{g*} : T_x S^m \to T_{gx} S^m$ induces $L_g^* : T_{gx}^* S^m \to T_x^* S^m$ given by $L_g^*(\xi_{gx})(Y_x) = \xi_{gx}(L_{g*}(Y_x))$ for $Y_x \in T_x S^m$. Then $b \in \mathrm{Ell}_{\mathrm{O}(m)1}(E', F')$ means that, for

$\pi : T^*S^m \to S^m$, $g \in \mathrm{O}(m)$, $e' \in E'_x$ and $\xi \in T^*_x S^m$, we require that $b \in C^\infty(T^*S^m, \mathrm{Hom}(\pi^*E', \pi^*F'))$ satisfy

$$\rho_{F'}(g)\left(b(L_g^*\xi_{gx})(e')\right) = b(\xi_{gx})\left(\rho_{E'}(g)(e')\right) \in F'_{gx},$$

where $\rho_{E'}$ and $\rho_{F'}$ are the given actions of $\mathrm{O}(m)$ on E' and F'. Note that $\rho_{E'}(g)e' \in E'_{gx}$, and $L_g^*\xi_{gx} \in T_x^*S^m$, since $\left(L_g^*\xi_{gx}\right)(Y_x) = \xi_{gx}(L_{g*}Y_x)$. Associated with a and b, there are pseudo-differential operators $\mathrm{Op}(a) : C^\infty(E) \to C^\infty(F)$ on X, and $\mathrm{Op}(b) : C^\infty(E') \to C^\infty(F')$ on S^m. We show that the assumption that $b \in \mathrm{Ell}_{\mathrm{O}(m)1}(E', F')$ together with appropriate choice of connections for E' and F' implies that $\mathrm{Op}(b)$ is $\mathrm{O}(m)$-invariant, in the sense that

$$\mathrm{Op}(b)\left(\rho_{E'}(g)\phi\right) = \rho_{F'}(g)\,\mathrm{Op}(b)(\phi), \text{ for all } \phi \in C^\infty(E'). \tag{8}$$

For this, we assume that $\nabla^{E'}$ and $\nabla^{F'}$ are compatible with the $\mathrm{O}(m)$-actions in the sense that for any curve $\gamma : [c,d] \to S^m$, and parallel translations $\tau_\gamma^{E'} : E'_{\gamma(c)} \to E'_{\gamma(d)}$ and $\tau_\gamma^{F'} : F'_{\gamma(c)} \to F'_{\gamma(d)}$, we have

$$\rho_{E'}(g) \circ \tau_\gamma^{E'} = \tau_{L_g\gamma}^{E'} \circ \rho_{E'}(g) \text{ and } \rho_{F'}(g) \circ \tau_\gamma^{F'} = \tau_{L_g\gamma}^{F'} \circ \rho_{F'}(g);$$

i.e., there are commutative diagrams

$$\begin{array}{ccc} E'_{\gamma(c)} & \to & E'_{\gamma(d)} \\ \downarrow \rho_{E'}(g) & & \downarrow \rho_{E'}(g) \\ E'_{L_g\gamma(c)} & \to & E'_{L_g\gamma(d)} \end{array} \quad \text{and} \quad \begin{array}{ccc} F'_{\gamma(c)} & \to & F'_{\gamma(d)} \\ \downarrow \rho_{F'}(g) & & \downarrow \rho_{F'}(g) \\ F'_{L_g\gamma(c)} & \to & F'_{L_g\gamma(d)} \end{array}$$

Then the invariance (8) of $\mathrm{Op}(b)$ is shown as follows

$$\begin{aligned}
&\mathrm{Op}(b)\left(\rho_{E'}(g)(\phi)\right)_x \\
&= \int_{T_xX \times T_x^*X} d'v d'\xi\; e^{-i\xi(v)}\psi(|v|)b(\xi)\left(\tau^{E'}_{x,\exp_x v}\left[\rho_{E'}(g)(\phi)(\exp_x v)\right]\right) \\
&= \int_{T_xX \times T_x^*X} d'v d'\xi\; e^{-i\xi(v)}\psi(|v|)b(\xi)\left(\rho_{E'}(g)\tau^{E'}_{g^{-1}x, g^{-1}\exp_x v}\left[\phi(g^{-1}\exp_x v)\right]\right) \\
&= \int_{T_xX \times T_x^*X} d'v d'\xi\; e^{-i\xi(v)}\psi(|v|)\rho_{F'}(g)b(L_g^*\xi)\left(\tau^{E'}_{g^{-1}x, g^{-1}\exp_x v}\left[\phi(g^{-1}\exp_x v)\right]\right) \\
&= \rho_{F'}(g)\int_{T_xX \times T_x^*X} d'v d'\xi\; e^{-i\xi(v)}\psi(|v|)b(L_g^*\xi)\left(\tau^{E'}_{g^{-1}x, g^{-1}\exp_x v}\left[\phi(g^{-1}\exp_x v)\right]\right) \\
&= \rho_{F'}(g)\int_{(T_{g^{-1}x}X) \times (T^*_{g^{-1}x}X)} d'\left(L_{g^{-1}*}v\right) d'\left(L_g^*\xi\right) e^{-iL_g^*\xi(L_{g^{-1}*}v)} \\
&\qquad \psi(\left|L_{g^{-1}*}v\right|)b(L_g^*\xi)\left(\tau^{E'}_{g^{-1}x, \exp_{g^{-1}x} L_{g^{-1}*}v}\left[\phi(\exp_{g^{-1}x} L_{g^{-1}*}v)\right]\right)
\end{aligned}$$

$$= \rho_{F'}(g) \int_{(T_{g^{-1}x}X)\times(T^*_{g^{-1}x}X)} d'\widetilde{v}d'\widetilde{\xi}\; e^{-i\widetilde{\xi}(\widetilde{v})}\psi(|\widetilde{v}|)b(\widetilde{\xi}) \left(\tau^{E'}_{g^{-1}x,\exp_{g^{-1}x}\widetilde{v}}\left[\phi(\exp_{g^{-1}x}\widetilde{v})\right]\right)$$

$$= \rho_{F'}(g)\left(\mathrm{Op}(b)\,(\phi)_{g^{-1}x}\right).$$

Recall that $\pi_P : P \to X$ is a principal $\mathrm{O}(m)$-bundle over X, the bundle of orthonormal frames of the normal bundle for the embedding $f : X \to Y$. There is a natural connection, say ω, on P which is inherited from the Levi-Civita connection on the orthonormal frame bundle for Y. We have $a \in \mathrm{Ell}_1(E,F) \subset C^\infty(T^*X, \mathrm{Hom}\,(\pi^*E, \pi^*F))$. For $\pi_{T^*P} : T^*P \to X$, we wish to obtain a lift of a, namely

$$\widetilde{a} \in C^\infty(T^*P, \mathrm{Hom}\,(\pi^*_{T^*P}E, \pi^*_{T^*P}F)),$$

which is $\mathrm{O}(m)$-invariant in the sense that $\widetilde{a}\left(R^*_g\xi_p\right) = \widetilde{a}\left(\xi_p\right)$. Note that ω gives us a splitting $T_pP = H_p \oplus V_p$ and a corresponding splitting $T^*_pP = \widetilde{H}^*_p \oplus \widetilde{V}^*_p$, where

$$\widetilde{H}^*_p := \left\{\xi \in T^*_pP : \xi(V_p) = 0\right\} \text{ and } \widetilde{V}^*_p := \left\{\xi \in T^*_pP : \xi(H_p) = 0\right\}$$

We have a pull-back $\pi^*_P : T^*X \to T^*P$ and note that $\pi^*_P\xi_x \in \widetilde{H}^*_p$ for $\xi_x \in T^*_xX$ and $x = \pi_P(p)$. Indeed, $\pi^*_P : T^*_xX \cong \widetilde{H}^*_p$. Any $\xi_p \in T^*_pP$ decomposes uniquely as $\xi_p := \eta_p + \pi^*_P\xi_x$ for some $\eta_p \in \widetilde{V}^*_p$ and some $\xi_x \in T^*_xX$. We simply define

$$\widetilde{a}\left(\xi_p\right) := a\left(\xi_x\right) = a\left(\pi^{*-1}_P\pi_{\widetilde{H}^*_p}\left(\xi_p\right)\right).$$

Actually, for $\xi_p \in \widetilde{H}^*_p$, $\widetilde{a}\left(\xi_p\right)$ is well-defined without the use of the connection, since

$$\xi_p \in \widetilde{H}^*_p \Rightarrow \xi_p = \pi^*_P\xi_x \text{ for a unique } \xi_x \Rightarrow \widetilde{a}\left(\xi_p\right) = a\left(\xi_x\right).$$

Note that $\widetilde{a}$ is $\mathrm{O}(m)$-invariant, since the decomposition $T^*_pP = \widetilde{H}^*_p \oplus \widetilde{V}^*_p$ is invariant, i.e.,

$$R^*_g\left(T^*_pP\right) = R^*_g(\widetilde{H}^*_p) \oplus R^*_g(\widetilde{V}^*_p) = \widetilde{H}^*_{pg^{-1}} \oplus \widetilde{V}^*_{pg^{-1}}$$

by the R_{g*}-invariance of H_p and π_P.

By means of the projection $\pi_1 : T^*(P \times S^m) \to X$, we may pull back $E \to X$ and $F \to X$ to bundles $\pi^*_1E \to T^*(P\times S^m)$ and $\pi^*_1F \to T^*(P\times S^m)$. Similarly, we simply write π^*_2E' and π^*_2F' for the pull-backs of $E' \to S^m$ and $F' \to S^m$ to $T^*(P \times S^m)$ via $T^*(P \times S^m) \to S^m$. Let $1_{\pi^*_2E'}$ denote the identity automorphism of $\pi_2^* E'$ and let the trivial extension of $\widetilde{a}$ on T^*P to a function on $T^*(P \times S^m)$ be denoted by $\widetilde{a}$ as well. We then obtain

$$\widetilde{a} \otimes 1_{\pi_2^* E'} \in C^\infty(T^*(P \times S^m), \operatorname{Hom}(\pi_1^* E \otimes \pi_2^* E', \pi_1^* F \otimes \pi_2^* E')\,.$$

This is but one of the four blocks in the matrix which will yield a representative of $[a] \cdot [b] \in K(T^*S)$; see (9) below. However, there are difficulties with the required uniform convergence near $\xi = 0$ on the sphere bundle $|\xi|^2 + |\eta|^2 = 1$ in the limit defining the asymptotic symbol (see (2))

$$\begin{aligned}
\sigma_1(\widetilde{a} \otimes 1_{\pi_2^* E'})(\xi, \eta) &= \lim_{t \to \infty} \frac{\left(\widetilde{a} \otimes 1_{\pi_2^* E'}\right)(t\xi, t\eta)}{t} \\
&= \lim_{t \to \infty} \frac{\widetilde{a}(t\xi) \otimes 1_{\pi_2^* E'}}{t} = \begin{cases} \sigma_1(\widetilde{a})(\xi) \otimes 1_{\pi_2^* E'}, & \xi \neq 0 \\ \lim_{t \to \infty} \frac{\widetilde{a}(0)}{t} \otimes 1_{\pi_2^* E'}, & \xi = 0, \eta \neq 0, \end{cases} \\
&= \begin{cases} \sigma_1(\widetilde{a})(\xi) \otimes 1_{\pi_2^* E'}, \; \xi \neq 0 \\ 0 \otimes 1_{\pi_2^* E'} = 0, \quad \xi = 0, \eta \neq 0. \end{cases}
\end{aligned}$$

This can be remedied by multiplying $\left(\widetilde{a} \otimes 1_{\pi_2^* E'}\right)(\xi, \eta)$ by $\varphi_{r_0}(|\xi|, |\eta|)$, where the C^∞ function $\varphi_{r_0} : [0, \infty)^2 \to [0, 1]$ is chosen so that

$$\varphi_{r_0}(r \cos\theta, r \sin\theta) = \begin{cases} 1 & \text{for } r \leq r_0 \text{ or } \frac{\tan\theta}{r_0} \leq 1 \\ h(\frac{\tan\theta}{r_0}) & \text{for } r \geq 2r_0, \end{cases}$$

where the C^∞ function $h : [0, \infty) \to [0, 1]$ is chosen so that

$$h(s) = \begin{cases} 1 \text{ for } s \leq 1 \\ 0 \text{ for } s \geq 2. \end{cases}$$

Then

$$\begin{aligned}
&\varphi_{r_0}(|\xi|, |\eta|) \left(\widetilde{a} \otimes 1_{\pi_2^* E'}\right)(\xi, \eta) \\
&= \begin{cases} \left(\widetilde{a} \otimes 1_{\pi_2^* E'}\right)(\xi, \eta), & \text{for } \sqrt{|\xi|^2 + |\eta|^2} \leq r_0 \text{ or } \frac{|\eta|}{|\xi|} \leq r_0 \\ h(\frac{|\eta|}{r_0 |\xi|}) \left(\widetilde{a} \otimes 1_{\pi_2^* E'}\right)(\xi, \eta), & \text{for } \sqrt{|\xi|^2 + |\eta|^2} \geq 2r_0. \end{cases}
\end{aligned}$$

Note that the two formulas agree on the overlap region

$$\left\{(\xi, \eta) : \frac{|\eta|}{|\xi|} \leq r_0 \text{ and } \sqrt{|\xi|^2 + |\eta|^2} \geq 2r_0\right\},$$

since $h(\frac{|\eta|}{r_0|\xi|}) = 1$ for $\frac{|\eta|}{|\xi|} \leq r_0$. Then

$$\begin{aligned}
\sigma_1(\varphi_{r_0}\widetilde{a} \otimes 1_{\pi_2^* E'})(\xi, \eta) &= \lim_{t\to\infty} \frac{\varphi_{r_0}(|t\xi|, |t\eta|)\left(\widetilde{a} \otimes 1_{\pi_2^* E'}\right)(t\xi, t\eta)}{t} \\
&= \lim_{t\to\infty} \frac{\varphi_{r_0}(|t\xi|, |t\eta|)\widetilde{a}(t\xi) \otimes 1_{\pi_2^* E'}}{t} = \lim_{t\to\infty} h\left(\frac{|t\eta|}{r_0\,|t\xi|}\right)\frac{\widetilde{a}(t\xi)}{t} \otimes 1_{\pi_2^* E'} \\
&= \lim_{t\to\infty} h\left(\frac{|\eta|}{r_0\,|\xi|}\right)\frac{\widetilde{a}(t\xi)}{t} \otimes 1_{\pi_2^* E'} = h\left(\frac{|\eta|}{r_0\,|\xi|}\right)\sigma_1(\widetilde{a})(\xi) \otimes 1_{\pi_2^* E'}.
\end{aligned}$$

The factor $h\left(\frac{|\eta|}{r_0|\xi|}\right)$ ensures uniform convergence on the sphere bundle $|\xi|^2 + |\eta|^2 = 1$ as $t \to \infty$. However, $\varphi_{r_0}\widetilde{a} \otimes 1_{\pi_2^* E'}$ is not an isomorphism for $|\xi|^2 + |\eta|^2$ sufficiently large because $\varphi_{r_0}(0, |\eta|) = 0$ if $|\eta| > 2r_0$. Thus, $\varphi_{r_0}\widetilde{a} \otimes 1_{\pi_2^* E'}$ is not an elliptic symbol even if restricted to the subbundle $\widetilde{H}^* \oplus T^*S^m \subset T^*(P \times S^m)$. This will be remedied when we consider the full symbol $\widetilde{c}_{r_0}(\xi, \eta)$ in (9), which is elliptic on $\widetilde{H}^* \oplus T^*S^m$. Due to the $\mathrm{O}(m)$-equivariance of

$$\varphi_{r_0}\widetilde{a} \otimes 1_{\pi_2^* E'} \in C^\infty(T^*(P \times S^m), \mathrm{Hom}(\pi_1^* E \otimes \pi_2^* E', \pi_1^* F \otimes \pi_2^* E')),$$

we can push $\varphi_{r_0}\widetilde{a} \otimes 1_{\pi_2^* E'}$ down to some $\left(\varphi_{r_0}\widetilde{a} \otimes 1_{\pi_2^* E'}\right)_Q$ in

$$C^\infty(T^*S, \mathrm{Hom}((q^*)^{-1}(\pi_1^* E \otimes \pi_2^* E'), (q^*)^{-1}(\pi_1^* F \otimes \pi_2^* E'))),$$

where q^* denotes the the precursor of

$$Q^* : K(P \times_{\mathrm{O}(m)} T^*S^m) \underset{\to}{\cong} K_{\mathrm{O}(m)}(P \times T^*S^m)$$

on the level of representative vector bundles. However, rather than consider $\mathrm{Op}\left(\left(\varphi_{r_0}\widetilde{a} \otimes 1_{\pi_2^* E'}\right)_Q\right)$, it is easier to work with $\mathrm{Op}\left(\varphi_{r_0}\widetilde{a} \otimes 1_{\pi_2^* E'}\right)$ acting on the *equivariant* sections of $\pi_1^* E \otimes \pi_2^* E' \to P \times S^m$ which correspond to the sections of $(q^*)^{-1}(\pi_1^* E \otimes \pi_2^* E') \to S$. One adjustment that must be made when working over $P \times S^m$ is that when defining $\mathrm{Op}\left(\varphi_{r_0}\widetilde{a} \otimes 1_{\pi_2^* E'}\right)$ (or ultimately $\mathrm{Op}(\widetilde{c}_{r_0}(\xi, \eta))$) via a double integral as in (1), the integration is restricted to the product

$$(H_p \oplus T_f S^m) \times (\widetilde{H}_p^* \oplus T_f^* S^m),$$

as opposed to integrating over all of $T_{(p,f)}(P \times S^m) \times T^*_{(p,f)}(P \times S^m)$, where

$$T^*_{(p,f)}(P \times S^m) = T_p^* P \oplus T_f^* S^m = \widetilde{V}_p \oplus \widetilde{H}_p \oplus T_f S^m.$$

Note that $Q_*(\widetilde{H}_p \oplus T_f S^m) = T_{Q(p,f)}S$, and $\mathrm{Ker}\, Q_*$ consists of tangent vectors to orbits of the $\mathrm{O}(m)$-action on $P \times S^m$. Also,

$$\mathrm{Ker}\, Q_{*(p,f)} \oplus \widetilde{H}_p \oplus T_f S^m = T_{(p,f)}(P \times S^m),$$

but generally $\operatorname{Ker} Q_{*(p,f)} \nsubseteq \widetilde{V}_p$.

Repeating the analogous construction (that we did for $a \in \operatorname{Ell}_1(E,F)$) in the case of the (pointwise) adjoint

$$a^* \in \operatorname{Ell}_1(F,E) \subset C^\infty(T^*X, \operatorname{Hom}(\pi^*F, \pi^*E),$$

we obtain

$$\varphi_{r_0}\widetilde{a}^* \otimes 1_{\pi_2^*F'} \in C^\infty(T^*(P \times S^m), \operatorname{Hom}(\pi_1^*F \otimes \pi_2^*F', \pi_1^*E \otimes \pi_2^*F')).$$

In a straightforward way, we also obtain lifts of

$$b \in \operatorname{Ell}_{O(m)1}(E',F') \subset C^\infty\left(T^*S^m, \operatorname{Hom}_{O(m)}(E',F')\right) \text{ and}$$
$$b^* \in \operatorname{Ell}_{O(m)1}(F',E') \subset C^\infty\left(T^*S^m, \operatorname{Hom}_{O(m)}(F',E')\right).$$

to $T^*(P \times S^m)$ and form

$$\varphi_{r_0}1_{\pi_1^*E} \otimes \widetilde{b} \in C^\infty(T^*(P \times S^m), \operatorname{Hom}(\pi_1^*E \otimes \pi_2^*E', \pi_1^*E \otimes \pi_2^*F')) \text{ and}$$
$$\varphi_{r_0}1_{\pi_1^*F} \otimes \widetilde{b}^* \in C^\infty(T^*(P \times S^m), \operatorname{Hom}(\pi_1^*F \otimes \pi_2^*F', \pi_1^*F \otimes \pi_2^*E').$$

We now define (note the switch from $(|\xi|, |\eta|)$ to $(|\eta|, |\xi|)$)

$$\left(\varphi_{r_0}1_{\pi_1^*E} \otimes \widetilde{b}\right)(\xi,\eta) := \varphi_{r_0}(|\eta|, |\xi|)1_{\pi_1^*E} \otimes \widetilde{b}(\eta) \neq \varphi_{r_0}(|\xi|, |\eta|)1_{\pi_1^*E} \otimes \widetilde{b}(\eta),$$

since there is now a non-uniformity of convergence of the asymptotic symbol for small $|\eta|$, as opposed to small $|\xi|$. For $(\xi,\eta) \in \widetilde{H}^* \oplus T^*S^m \subset T^*(P \times S^m)$ and for $r_0 > 0$, we define

$$\widetilde{c}_{r_0}(\xi,\eta) := \begin{bmatrix} \varphi_{r_0}(|\xi|, |\eta|)\widetilde{a} \otimes 1_{\pi_2^*E'} & -\varphi_{r_0}(|\eta|, |\xi|)1_{\pi_1^*F} \otimes \widetilde{b}^* \\ \varphi_{r_0}(|\eta|, |\xi|)1_{\pi_1^*E} \otimes \widetilde{b} & \varphi_{r_0}(|\xi|, |\eta|)\widetilde{a}^* \otimes 1_{\pi_2^*F'} \end{bmatrix}. \tag{9}$$

Note that $\widetilde{c}_{r_0}(\xi,\eta)$ is homogeneous outside a ball bundle of fixed positive radius about the zero section of $\widetilde{H}^* \oplus T^*S^m$. Although we have noted above that the individual entries, such as $\varphi_{r_0}(|\xi|, |\eta|)\widetilde{a} \otimes 1_{\pi_2^*E'}$, are not isomorphisms for large $|\eta|$ when $\xi = 0$ (or in other cases, for large $|\xi|$ when $\eta = 0$), we will show that the *entire* transformation $\widetilde{c}_{r_0}(\xi,\eta)$ is an isomorphism for $|\xi|^2 + |\eta|^2$ large, as follows. Note that

$$(\widetilde{c}_{r_0}(\xi,\eta))^* := \begin{bmatrix} \varphi_{r_0}(|\xi|, |\eta|)\widetilde{a}^* \otimes 1_{\pi_2^*E'} & \varphi_{r_0}(|\eta|, |\xi|)\left(1_{\pi_1^*E} \otimes \widetilde{b}^*\right) \\ -\varphi_{r_0}(|\eta|, |\xi|)\left(1_{\pi_1^*F} \otimes \widetilde{b}\right) & \varphi_{r_0}(|\xi|, |\eta|)\widetilde{a} \otimes 1_{\pi_2^*F'} \end{bmatrix},$$

and $(\widetilde{c}_{r_0}(\xi,\eta))^* \widetilde{c}_{r_0}(\xi,\eta)$ is block diagonal with entries

$$\varphi_{r_0}(|\xi|, |\eta|)^2 \left(\widetilde{a}^*\widetilde{a} \otimes 1_{\pi_2^*E'}\right) + \varphi_{r_0}(|\eta|, |\xi|)^2 \left(1_{\pi_1^*E} \otimes \widetilde{b^*b}\right) \text{ and}$$
$$\varphi_{r_0}(|\eta|, |\xi|)^2 \left(1_{\pi_1^*F} \otimes \widetilde{bb^*}\right) + \varphi_{r_0}(|\xi|, |\eta|)^2 \left(\widetilde{a}\widetilde{a}^* \otimes 1_{\pi_2^*F'}\right). \tag{10}$$

Note that for r_0 sufficiently large, $\varphi_{r_0}(|\xi|, |\eta|)^2$ and $\varphi_{r_0}(|\eta|, |\xi|)^2$ are not simultaneously 0, since $\varphi_{r_0}(|\xi|, |\eta|)^2 = 0$ only in a narrow cone-like wedge about the subspace $\xi = 0$, truncated by removing a ball of radius r_0, and $\varphi_{r_0}(|\eta|, |\xi|)^2 = 0$ only in a similar region about the subspace $\eta = 0$. Thus, each of the entries in (10) are invertible (indeed, positive) operators on $\pi_1^*E \otimes \pi_2^*E'$ and $\pi_1^*F \otimes \pi_2^*F'$ respectively for $|\xi|^2 + |\eta|^2$ sufficiently large, and then $\widetilde{c}_{r_0}(\xi, \eta)$ is also invertible for $|\xi|^2 + |\eta|^2$ sufficiently large. Since $\varphi_{r_0}(|\xi|, |\eta|) = \varphi_{r_0}(|\eta|, |\xi|) = 1$ for $|\xi|^2 + |\eta|^2 \le r_0^2$, we know that for $|\xi|^2 + |\eta|^2 \le r_0^2$, $\widetilde{c}_{r_0}(\xi, \eta) = \widetilde{c}(\xi, \eta)$ which (by definition) is the transformation $\widetilde{c}_{r_0}(\xi, \eta)$ without the φ_{r_0} factors. Thus, for r_0 sufficiently large, the support of $\widetilde{c}_{r_0}$ is the same as that for $\widetilde{c}$, and the push down of $\widetilde{c}_{r_0}$ to a function on $T^*(P \times_{O(m)} S^m)$ is elliptic; i.e.,

$$(\widetilde{c}_{r_0})_Q \in \mathrm{Ell}_1((q^*)^{-1}((\pi_1^*E \otimes \pi_2^*E') \oplus (\pi_1^*F \otimes \pi_2^*F')), \\ (q^*)^{-1}((\pi_1^*F \otimes \pi_2^*E') \oplus (\pi_1^*E \otimes \pi_2^*F')).$$

To see that $(\widetilde{c}_{r_0})_Q$ represents $[a] \cdot [b]$ for $a \in \mathrm{Ell}_1(E, F)$ and $b \in \mathrm{Ell}_{O(m)1}(E', F')$, one goes through the steps leading to the definition (7), bearing in mind that when K-theory elements are defined in terms of compactly supported length-one complexes, products formed from them (such as the one in (6)) are defined in terms of a length-one complex between sums of tensor products; see [1, p. 490 and p. 528]. Thus,

$$\mathrm{index}_a([a] \cdot [b]) = \mathrm{index}\,\mathrm{Op}\left((\widetilde{c}_{r_0})_Q\right).$$

Although we cannot go into the details here (however, see [1, p. 513f]), even though $\widetilde{c}_Q$ is not elliptic, it is a limit of the elliptic symbols $(\widetilde{c}_{r_0})_Q$ as $r_0 \to \infty$ in a strong enough sense that $\mathrm{Op}_s(\widetilde{c}_Q)$ (for any $s \in \mathbb{R}$) is Fredholm and

$$\mathrm{index}\,\mathrm{Op}(\widetilde{c}_Q) = \mathrm{index}\,\mathrm{Op}_s(\widetilde{c}_Q) = \mathrm{index}\,\mathrm{Op}\left((\widetilde{c}_{r_0})_Q\right) = \mathrm{index}_a([a] \cdot [b]).$$

We compute $\mathrm{index}\,\mathrm{Op}_s(\widetilde{c}_Q)$ as follows. Let $\bar{\pi}_1 : P \times S^m \to X$ and $\bar{\pi}_2 : P \times S^m \to S^m$ be the obvious projections. As we have observed, instead of directly computing the index of $\mathrm{Op}(\widetilde{c}_Q)$, we can instead compute the index of the equivalent operator

$$\mathrm{Op}(\widetilde{c}) : C^\infty_{O(m)}((\bar{\pi}_1^*E \otimes \bar{\pi}_2^*E') \oplus (\bar{\pi}_1^*F \otimes \bar{\pi}_2^*F')) \\ \to C^\infty_{O(m)}((\bar{\pi}_1^*F \otimes \bar{\pi}_2^*E') \oplus (\bar{\pi}_1^*E \otimes \bar{\pi}_2^*F'))$$

acting on equivariant (indicated by the subscript $\mathrm{O}(m)$) sections defined on $P \times S^m$. Using

$$\mathrm{Op}(\widetilde{c}) = \begin{bmatrix} \mathrm{Op}\,(\widetilde{a}) \otimes 1_{\pi_2^* E'} & -1_{\pi_1^* F} \otimes \mathrm{Op}(\widetilde{b}^*) \\ 1_{\pi_1^* E} \otimes \mathrm{Op}(\widetilde{b}) & \mathrm{Op}(\widetilde{a}^*) \otimes 1_{\pi_2^* F'} \end{bmatrix}$$

and

$$\mathrm{Op}(\widetilde{c}^*) = \begin{bmatrix} \mathrm{Op}\,(\widetilde{a}^*) \otimes 1_{\pi_2^* E'} & 1_{\pi_1^* E} \otimes \mathrm{Op}(\widetilde{b}^*) \\ -1_{\pi_1^* F} \otimes \mathrm{Op}(\widetilde{b}) & \mathrm{Op}(\widetilde{a}) \otimes 1_{\pi_2^* F'} \end{bmatrix},$$

we get that $\mathrm{Op}(\widetilde{c}^*)\,\mathrm{Op}(\widetilde{c})$ is block diagonal with entries

$$\mathrm{Op}\,(\widetilde{a}^*)\,\mathrm{Op}\,(\widetilde{a}) \otimes 1_{\pi_2^* E'} + 1_{\pi_1^* E} \otimes \mathrm{Op}(\widetilde{b}^*)\,\mathrm{Op}(\widetilde{b}) \text{ and}$$
$$1_{\pi_1^* F} \otimes \mathrm{Op}(\widetilde{b})\,\mathrm{Op}(\widetilde{b}^*) + \mathrm{Op}(\widetilde{a})\,\mathrm{Op}(\widetilde{a}^*) \otimes 1_{\pi_2^* F'}.$$

Thus,

$$\begin{aligned} \mathrm{Ker}\,\mathrm{Op}(\widetilde{c}) &= \mathrm{Ker}\,(\mathrm{Op}(\widetilde{c}^*)\,\mathrm{Op}(\widetilde{c})) \\ &= \Big((\mathrm{Ker}\,((\mathrm{Op}(\widetilde{a}^*)\,\mathrm{Op}(\widetilde{a})) \otimes 1_{\pi_2^* E'})) \cap \mathrm{Ker}(1_{\pi_1^* E} \otimes (\mathrm{Op}(\widetilde{b}^*)\,\mathrm{Op}(\widetilde{b})))\Big) \\ &\quad \oplus \Big(\mathrm{Ker}\,(\mathrm{Op}(\widetilde{a})\,\mathrm{Op}(\widetilde{a}^*) \otimes 1_{\pi_2^* F'}) \cap \mathrm{Ker}(1_{\pi_1^* F} \otimes \mathrm{Op}(\widetilde{b})\,\mathrm{Op}(\widetilde{b}^*))\Big) \\ &= \Big(\mathrm{Ker}\,(\mathrm{Op}(\widetilde{a}) \otimes 1_{\pi_2^* E'}) \cap \mathrm{Ker}(1_{\pi_1^* E} \otimes \mathrm{Op}(\widetilde{b}))\Big) \\ &\quad \oplus \Big(\mathrm{Ker}(\mathrm{Op}(\widetilde{a}^*) \otimes 1_{\pi_2^* F'}) \cap \mathrm{Ker}(1_{\pi_1^* F} \otimes \mathrm{Op}(\widetilde{b}^*))\Big), \end{aligned}$$

and

$$\begin{aligned} \mathrm{Ker}\,\mathrm{Op}(\widetilde{c}^*) &= \mathrm{Ker}\,(\mathrm{Op}(\widetilde{c})\,\mathrm{Op}(\widetilde{c}^*)) \\ &= \Big(\mathrm{Ker}\,(\mathrm{Op}(\widetilde{a})\,\mathrm{Op}(\widetilde{a}^*) \otimes 1_{\pi_2^* E'}) \cap \mathrm{Ker}(1_{\pi_1^* F} \otimes \mathrm{Op}(\widetilde{b}^*)\,\mathrm{Op}(\widetilde{b}))\Big) \\ &\quad \oplus \Big(\mathrm{Ker}\,(\mathrm{Op}(\widetilde{a}^*)\,\mathrm{Op}(\widetilde{a}) \otimes 1_{\pi_2^* F'}) \cap \mathrm{Ker}(1_{\pi_1^* E} \otimes \mathrm{Op}(\widetilde{b})\,\mathrm{Op}(\widetilde{b}^*))\Big) \\ &= \Big(\mathrm{Ker}\,(\mathrm{Op}(\widetilde{a}^*) \otimes 1_{\pi_2^* E'}) \cap \mathrm{Ker}(1_{\pi_1^* F} \otimes \mathrm{Op}(\widetilde{b}))\Big) \\ &\quad \oplus \Big(\mathrm{Ker}\,(\mathrm{Op}(\widetilde{a}) \otimes 1_{\pi_2^* F'}) \cap \mathrm{Ker}(1_{\pi_1^* E} \otimes \mathrm{Op}(\widetilde{b}^*))\Big). \end{aligned}$$

Since $\mathrm{Op}(\widetilde{a}) \otimes 1_{\pi_2^* E'}$ commutes with $1_{\pi_1^* E} \otimes \mathrm{Op}(\widetilde{b})$, we have that $\mathrm{Op}(\widetilde{a}) \otimes 1_{\pi_2^* E'}$ preserves $\mathrm{Ker}\left(1_{\pi_1^* E} \otimes \mathrm{Op}(\widetilde{b})\right)$, and

$$\begin{aligned} &\mathrm{Ker}\,(\mathrm{Op}(\widetilde{a}) \otimes 1_{\pi_2^* E'}) \cap \mathrm{Ker}\left(1_{\pi_1^* E} \otimes \mathrm{Op}(\widetilde{b})\right) \\ &\qquad\qquad = \mathrm{Ker}\left(\left(\mathrm{Op}(\widetilde{a}) \otimes 1_{\pi_2^* E'}\right)|_{\mathrm{Ker}\left(1_{\pi_1^* E} \otimes \mathrm{Op}(\widetilde{b})\right)}\right). \end{aligned}$$

Similarly,

$$\operatorname{Ker}\left(\operatorname{Op}(\widetilde{a}^*)\otimes 1_{\bar{\pi}_2^* F'}\right)\cap\operatorname{Ker}\left(1_{\bar{\pi}_1^* F}\otimes\operatorname{Op}(\widetilde{b}^*)\right)$$
$$=\operatorname{Ker}\left((\operatorname{Op}(\widetilde{a}^*)\otimes 1_{\bar{\pi}_2^* F'})|_{\operatorname{Ker}\left(1_{\bar{\pi}_1^* F}\otimes\operatorname{Op}(\widetilde{b}^*)\right)}\right).$$

Thus,

$$\operatorname{Ker}\operatorname{Op}(\widetilde{c})=\operatorname{Ker}\left(\left(\operatorname{Op}(\widetilde{a})\otimes 1_{\bar{\pi}_2^* E'}\right)|_{\operatorname{Ker}\left(1_{\bar{\pi}_1^* E}\otimes\operatorname{Op}(\widetilde{b})\right)}\right)$$
$$\oplus\operatorname{Ker}\left((\operatorname{Op}(\widetilde{a}^*)\otimes 1_{\bar{\pi}_2^* F'})|_{\operatorname{Ker}\left(1_{\bar{\pi}_1^* F}\otimes\operatorname{Op}(\widetilde{b}^*)\right)}\right),$$

and similarly

$$\operatorname{Ker}\operatorname{Op}(\widetilde{c}^*)=\operatorname{Ker}\left((\operatorname{Op}(\widetilde{a}^*)\otimes 1_{\bar{\pi}_2^* E'})_{|\operatorname{Ker}\left(1_{\bar{\pi}_1^* F}\otimes\operatorname{Op}(\widetilde{b})\right)}\right)$$
$$\oplus\operatorname{Ker}\left(\left(\operatorname{Op}(\widetilde{a})\otimes 1_{\bar{\pi}_2^* F'}\right)|_{\operatorname{Ker}\left(1_{\bar{\pi}_1^* E}\otimes\operatorname{Op}(\widetilde{b}^*)\right)}\right).$$

We note that

$$\operatorname{Ker}\left(1_{\bar{\pi}_1^* E}\otimes\operatorname{Op}(\widetilde{b})\right)=C^\infty\left(\bar{\pi}_1^*(E)\otimes\operatorname{Ker}\operatorname{Op}(\widetilde{b})\right)\text{ and}$$
$$\operatorname{Ker}\left(1_{\bar{\pi}_1^* F}\otimes\operatorname{Op}(\widetilde{b}^*)\right)=C^\infty\left(\bar{\pi}_1^*(F)\otimes\operatorname{Ker}\operatorname{Op}(\widetilde{b}^*)\right).$$

Thus, $\left(\operatorname{Op}(\widetilde{a})\otimes 1_{\bar{\pi}_2^* E'}\right)|_{\operatorname{Ker}\left(1_{\bar{\pi}_1^* E}\otimes\operatorname{Op}(\widetilde{b})\right)}$ is a differential operator on $C^\infty_{\mathrm{O}(m)}\left(\bar{\pi}_1^*(E)\otimes\operatorname{Ker}\operatorname{Op}(\widetilde{b})\right)$; i.e., on the $\mathrm{O}(m)$-invariant sections of $\bar{\pi}_1^*(E)\otimes\operatorname{Ker}\operatorname{Op}(\widetilde{b})$, where $\operatorname{Ker}\operatorname{Op}(\widetilde{b})$ is a finite-dimensional $\mathrm{O}(m)$-module, and similarly for $(\operatorname{Op}(\widetilde{a}^*)\otimes 1_{\bar{\pi}_2^* F'})|_{\operatorname{Ker}\left(1_{\bar{\pi}_1^* F}\otimes\operatorname{Op}(\widetilde{b}^*)\right)}$. Since $\operatorname{Op}(\widetilde{a})$ is an $\mathrm{O}(m)$-invariant lift of $\operatorname{Op}(a)$, we have an isomorphism of $\mathrm{O}(m)$-modules,

$$\operatorname{Ker}\left(\left(\operatorname{Op}(\widetilde{a})\otimes 1_{\bar{\pi}_2^* E'}\right)|_{\operatorname{Ker}\left(1_{\bar{\pi}_1^* E}\otimes\operatorname{Op}(\widetilde{b})\right)}\right)\cong\operatorname{Ker}\left(\operatorname{Op}(a)\right)\otimes\operatorname{Ker}_{\mathrm{O}(m)}\operatorname{Op}(b),$$

where the action is trivial on the $\operatorname{Ker}(\operatorname{Op}(a))$ factor. Similarly,

$$\operatorname{Ker}((\operatorname{Op}(\widetilde{a}^*)\otimes 1_{\bar{\pi}_2^* F'})|_{\operatorname{Ker}(1_{\bar{\pi}_1^* F}\otimes\operatorname{Op}(\widetilde{b}^*))})\cong\operatorname{Ker}\left(\operatorname{Op}(a^*)\right)\otimes\operatorname{Ker}_{\mathrm{O}(m)}\operatorname{Op}(b^*),$$
$$\operatorname{Ker}((\operatorname{Op}(\widetilde{a}^*)\otimes 1_{\bar{\pi}_2^* E'})_{|\operatorname{Ker}(1_{\bar{\pi}_1^* F}\otimes\operatorname{Op}(\widetilde{b}))})\cong\operatorname{Ker}\left(\operatorname{Op}(a^*)\right)\otimes\operatorname{Ker}_{\mathrm{O}(m)}\operatorname{Op}(b),$$
$$\operatorname{Ker}(\left(\operatorname{Op}(\widetilde{a})\otimes 1_{\bar{\pi}_2^* F'}\right)|_{\operatorname{Ker}(1_{\bar{\pi}_1^* E}\otimes\operatorname{Op}(\widetilde{b}^*))})\cong\operatorname{Ker}\left(\operatorname{Op}(a)\right)\otimes\operatorname{Ker}_{\mathrm{O}(m)}\operatorname{Op}(b^*).$$

Hence, as required,

$$
\begin{aligned}
\operatorname{index}\operatorname{Op}(\widetilde{c}) &= \dim\left(\operatorname{Ker}\operatorname{Op}(\widetilde{c})\right) - \dim(\operatorname{Ker}\operatorname{Op}(\widetilde{c}^*)) \\
&= \dim\begin{bmatrix} \operatorname{Ker}\left((\operatorname{Op}(\widetilde{a})\otimes 1_{\bar{\pi}_2^* E'})\,|_{\operatorname{Ker}\left(1_{\bar{\pi}_1^* E}\otimes\operatorname{Op}(\widetilde{b})\right)}\right) \\ \oplus\operatorname{Ker}\left((\operatorname{Op}(\widetilde{a}^*)\otimes 1_{\bar{\pi}_2^* F'})|_{\operatorname{Ker}\left(1_{\bar{\pi}_1^* F}\otimes\operatorname{Op}(\widetilde{b}^*)\right)}\right)\end{bmatrix} \\
&\quad - \dim\begin{bmatrix} \operatorname{Ker}\left((\operatorname{Op}(\widetilde{a}^*)\otimes 1_{\bar{\pi}_2^* E'})_{|\operatorname{Ker}\left(1_{\bar{\pi}_1^* F}\otimes\operatorname{Op}(\widetilde{b})\right)}\right) \\ \oplus\operatorname{Ker}\left((\operatorname{Op}(\widetilde{a})\otimes 1_{\bar{\pi}_2^* F'})\,|_{\operatorname{Ker}\left(1_{\bar{\pi}_1^* E}\otimes\operatorname{Op}(\widetilde{b}^*)\right)}\right)\end{bmatrix} \\
&= \dim\left(\operatorname{Ker}(\operatorname{Op}(a))\otimes\operatorname{Ker}_{O(m)}\operatorname{Op}(b)\right) \\
&\qquad - \dim(\operatorname{Ker}(\operatorname{Op}(a^*))\otimes\operatorname{Ker}_{O(m)}\operatorname{Op}(b)) \\
&\qquad\qquad + \dim(\operatorname{Ker}(\operatorname{Op}(a^*))\otimes\operatorname{Ker}_{O(m)}\operatorname{Op}(b^*)) \\
&\qquad\qquad\qquad - \dim(\operatorname{Ker}(\operatorname{Op}(a))\otimes\operatorname{Ker}_{O(m)}\operatorname{Op}(b^*)) \\
&= \operatorname{index}\left([a]\cdot\left(\operatorname{Ker}_{O(m)}\operatorname{Op}(b) - \operatorname{Ker}_{O(m)}\operatorname{Op}(b^*)\right)\right) \\
&= \operatorname{index}\left([a]\cdot\operatorname{index}_{O(m)}[b]\right) = \operatorname{index}\left(u\cdot\operatorname{index}_{O(m)} v\right).
\end{aligned}
$$

References

1. Atiyah, M. F. and I. M. Singer, The index of elliptic operators I, *Ann. of Math.* **87** (1968), 484-530.
2. Bokobza-Haggiag, J., Opérateurs pseudo-différentiels sur une variété différentiable, *Ann. Inst. Fourier* (Genoble) **19** (1969), 125-177.
3. Widom, H., A complete symbolic calculus for pseudodifferential operators, *Bull. Sci. Math.* (2) **104** (1980), 19-63.
4. Getzler, E., Pseudodifferential operators on supermanifolds and the Atiyah-Singer index theorem, *Comm. Math. Phys.* **92** (1983), 163-178.
5. Voronov, T., Quantization of forms on the cotangent bundle, *Comm. Math. Phys.* **205** (1999), 315–336.
6. Lawson, H. B. and M. Michelsohn, *Spin Geometry*, Princeton Univ. Press, Princeton, 1989.
7. Booss–Bavnbek, B. and D. Bleecker, *Topology and Analysis*, Springer-Verlag, Berlin-Heidelberg-New York, 1985.
8. Segal, G. B., Equivariant K-theory, *Inst. Hautes Études Sci. Publ. Math.* **34** (1968), 129–151.

Received by the editors September 14, 2005; Revised January 10, 2006

Analysis, Geometry and Topology
of Elliptic Operators, pp. 63–78

CRITICAL POINTS OF POLYNOMIALS IN THREE COMPLEX VARIABLES

LIVIU I. NICOLAESCU

Department of Mathematics, University of Notre Dame
Notre Dame, IN 46556, USA
nicolaescu.1@nd.edu

Dedicated to Krzysztof P. Wojciechowski on his 50th birthday

In this survey we describe situations when *analytic* invariants of an isolated complex surface singularity can be computed by *topological* methods.

2000 *Mathematics Subject Classification.* Primary 14B05; Secondary 14J17, 14J70, 14J80, 57M27

1. The Milnor number

Denote by $\mathcal{O} = \mathcal{O}_{N+1}$ the ring of germs of holomorphic functions $f = f(z_0, \dots, z_N)$ defined in a neighborhood of $0 \in \mathbb{C}^{N+1}$. We denote by $\mathfrak{m} \subset \mathcal{O}$ the maximal ideal of $\mathcal{O}$,

$$f \in \mathfrak{m} \Longleftrightarrow f(0) = 0.$$

Let $f \in \mathfrak{m}$. Assume 0 is an isolated critical point of f, i.e. 0 is an isolated point of the variety

$$\partial_{z_i} f = 0, \quad \forall i = 0, \dots, N.$$

We define the Jacobian ideal of f to be the ideal $J_f \subset \mathcal{O}$ generated by $\partial_{z_i} f$, $i = 0, \dots, N$. From the analytical Nullstellensatz we deduce

$$\sqrt{J_f} = \mathfrak{m} \Longleftrightarrow \exists k > 0 : \mathfrak{m}^k \subset J_f \Longleftrightarrow \dim_{\mathbb{C}} \mathcal{O}/J_f < \infty.$$

The $\dim_{\mathbb{C}} \mathcal{O}/J_f$ number is called the *Milnor number* of f at 0 and it is denoted by $\mu = \mu(f, 0)$.

For every positive integer N we denote by $j_N(f)$ the N-th jet of f. It can be identified with a polynomial of degree N in $n+1$ complex variables.

Two germs $f, g \in \mathfrak{m}$ are called *right-equivalent* and we write this $f \sim_r g$ if g is obtained from f by a change in variables fixing the origin. The next result due to Tougeron [27] can be viewed as a generalization of Morse lemma.

Theorem 1.1. *Let $f \in \mathfrak{m}$ have an isolated singularity at* 0. *Then*

$$f \sim_r j_{\mu+1}(f).$$

Thus, when studying local properties of isolated critical points of a holomorphic function we may asume it is a polynomial.

Example 1.1. (a) Consider three integers $p, q, r \geq 2$ and consider the function

$$f = f_{p,q,r}(x, y, z) = az^p + by^q + cz^r.$$

Then $\mu = (p-1)(q-1)(r-1)$. The singularity described by $f_{2,2,n+1}$ is called the A_n singularity. It has Milnor number n.

(b) Consider the polynomial

$$D_4 = D_4(x, y, z) = x^2y - y^3 + z^2.$$

0 is an isolated critical point of D_4 with Milnor number 4.

Note that the D_4-singularity is *weighted homogeneous*. We recall that a function $f = f(z_1, \ldots, z_N)$ is called weighted homogeneous if there exist nonnegative integers $m_1, \cdots, m_N, m$ such that

$$f(t^{m_1}z_1, \ldots, t^{m_N}z_N) = t^m f(z_1, \ldots, z_N), \quad \forall t \in \mathbb{C}^*.$$

The rational numbers $w_i = m_i/m$ are called the *weights*. The weights of the D_4 singularity are $w_1 = w_2 = \frac{1}{3}, \ \ w_3 = \frac{1}{2}$.

2. The Milnor fibration and its monodromy

Let $f \in \mathfrak{m}$ have an isolated singularity at 0. Set $\mu = \mu(f, 0)$. For $r, \varepsilon > 0$ sufficiently small we can find a close ball $B_\varepsilon \subset \mathbb{C}^{N+1}$ centered at $0 \in \mathbb{C}^{N+1}$ so that

$$f(B_\varepsilon) \supset \mathbb{D}_r = \{u \in \mathbb{C}; \ \ |u| \leq r\}.$$

Set $Z = Z(\varepsilon, r) = B_\varepsilon \cap f^{-1}(\mathbb{D}_r)$. The induced map

$$f : Z \setminus f^{-1}(0) \to \mathbb{D}_r \setminus \{0\}$$

is a locally trivial fibration called the *Milnor fibration* (see Looijenga [12, Sec. 2.B]). Its typical fiber Z_f, called the *Milnor fiber*, is a smooth $2N$-dimensional manifold with boundary. Its boundary is a $(2N-1)$-manifold called the *link of the singularity* and denoted by Lk_f. It is diffeomorphic to $\partial B_\varepsilon \cap \{f^{-1}(0)\}$.

Its typical fiber Z_f has the homotopy type of a wedge of μ spheres of dimension N (see Milnor [15], Thm. 5.1) and thus the Milnor number completely determines the homotopy type of the Milnor fiber. Moreover, the restriction of the Milnor fibration to $\partial \mathbb{D}_r$ is isomorphic to the locally trivial fibration (see Milnor [15, §5])

$$S^{2N+1} \setminus \mathrm{Lk}_f \cong \partial B_\varepsilon \setminus f^{-1}(0) \to \partial \mathbb{D}_r, \quad z \longmapsto r\frac{f(z)}{|f(z)|}. \tag{2.1}$$

When N is even, the intersection pairing q_f on $H_N(Z_f, \mathbb{R})$ is bilinear and symmetric and its isomorphism type is characterized by the integers μ_0, μ_-, μ_+ describing the number of zero (resp. negative, positive) eigenvalues of a symmetric matrix representing this intersection form. In particular we define the *signature of the Milnor fiber*

$$\tau = \tau(f, 0) = \mu_+ - \mu_-.$$

Using the long exact sequence of the pair $(Z_f, \partial Z_f) = (Z_f, \mathrm{Lk}_f)$ and the Poincaré duality isomorphism $H_N(Z_f, \partial Z_f; \mathbb{R}) \cong \mathrm{Hom}\big(H_N(Z_f, \mathbb{R}), \mathbb{R}\big)$ we deduce that

$$\mu_0(f, 0) = b_{N-1}(\mathrm{Lk}_f).$$

The Milnor fibration defines a monodromy map

$$\mathcal{M}_f : \pi_1(\mathbb{D}_r^*) \to \mathrm{Aut}_{\mathbb{Z}}(\tilde{H}_N(Z_f, \mathbb{Z})),$$

where $\tilde{H}_\bullet$ denotes *reduced homology*. We identify $\mathcal{M}_f$ with $\mathcal{M}_f(1) \in \mathrm{Aut}_{\mathbb{Z}}(\tilde{H}_N(Z_f, \mathbb{Z}))$, we denote by $[\mathcal{M}_f]_{\mathbb{Z}}$ its $\mathbb{Z}$-conjugacy class and by $[\mathcal{M}_f]_{\mathbb{C}}$ its $\mathbb{C}$-conjugacy class. The Wang long exact sequence (see Hatcher [7, Ex. 2.48]) of the fibration (2.1) together with the Alexander duality applied to the embedding $\mathrm{Lk}_f \hookrightarrow S^{2N+1}$ imply that

$$\mu_0(f, 0) = b_{N-1}(\mathrm{Lk}_f) = \dim \ker(\mathbb{1} - \mathcal{M}_f).$$

When $N = 1$ the link Lk_f is a 3-manifold and we deduce that it is a rational homology 3-sphere iff $\mu_0 = 0$, iff 1 is not in the spectrum of $\mathcal{M}_f$.

The complex conjugacy class of $\mathcal{M}_f$ is completely determined by the complex Jordan normal form of $\mathcal{M}_f$. In particular, if $\mathcal{M}_f$ is semi-simple (i.e. diagonalizable) then the complex conjugacy class is completely determined by the characteristic polynomial $\det(t\mathbb{1} - \mathcal{M}_f)$. This happens for example when f is weighted homogeneous.

The celebrated monodromy theorem Brieskorn [2], Clemens [3], Deligne [4], Landman [8] describes some constraints on the $\mathbb{C}$-conjugacy class of the monodromy.

Theorem 2.1. *All the eigenvalues of $\mathcal{M}_f$ are roots of 1 and its Jordan cells have dimension $\leq (N+1)$.*

Example 2.1. (a) Consider the germ $f = f_n : (\mathbb{C}, 0) \to (\mathbb{C}, 0)$, $f_n(z) = z^n$. Then the Milnor fiber f_n^{-1} can be identified with the group $\mathcal{R}_n$ of n-th roots of 1,

$$\mathcal{R}_n = \{1, \rho, \dots, \rho^{n-1};\ \ \rho = e^{2\pi \mathbf{i}/n}\}.$$

The Milnor number is $(n-1)$. This is equal to the rank of the reduced homology $\tilde{H}_0(f_n^{-1}(1), \mathbb{Z})$ which can be identified with an additive subgroup of the group algebra $\mathbb{Z}[\mathcal{R}_n]$

$$\tilde{H}_0(f_n^{-1}(1), \mathbb{Z}) \cong \Big\{ \sum_{k=0}^{n-1} a_k \rho^k \in \mathbb{Z}[\mathcal{R}_n];\ \sum_{k=0}^{n-1} a_k = 0 \Big\}.$$

As basis in this group we can choose the "polynomials"

$$e_k := \rho^k - \rho^{k-1},;\ \ k = 1, \dots, n-1.$$

Then

$$\mathcal{M}_{f_n}(e_k) = \begin{cases} e_{k+1} \text{ if } k < n-1 \\ -(e_1 + \dots + e_{n-1}) \text{ if } k = n-1 \end{cases}.$$

We deduce $\mathcal{M}_{f_n}^n = \mathbb{1}$, i.e. all the eigenvalues of the monodromy are n-th roots of 1.

(b) (Sebastiani-Thom, [26]) If $f = f(x_1, \dots, x_p) \in \mathcal{O}_p$ and $g = g(y_1, \dots, y_q) \in \mathcal{O}_q$ have isolated singularities at the origin, then so does $f * g \in \mathcal{O}_{p+q}$ defined by

$$f * g(x, y) = f(x_1, \dots, x_p) + g(y_1, \dots, y_q).$$

Moreover

$$Z_{f*g} \simeq Z_f * Z_f := \text{the join of the Milnor fibers } Z_f \text{ and } Z_f$$

("$\simeq$" denotes homotopy equivalence) and

$$\mu(f * g, 0) = \mu(f, 0) \cdot \mu(g, 0),\ \ [\mathcal{M}_{f*g}]_{\mathbb{C}} = [\mathcal{M}_f]_{\mathbb{C}} \otimes [\mathcal{M}_g]_{\mathbb{C}}.$$

In particular

$$[\mathcal{M}_{f_{p,q,r}}]_{\mathbb{C}} = [\mathcal{M}_{f_p}]_{\mathbb{C}} \otimes [\mathcal{M}_{f_q}]_{\mathbb{C}} \otimes [\mathcal{M}_{f_r}]_{\mathbb{C}}.$$

3. The geometric genus of an isolated surface singularity

Suppose $(X,0) \subset (\mathbb{C}^m,0)$ is a germ of an isolated surface (i.e. complex dimension 2) singularity. Assume X is Stein and set

$$M = X \cap S_\varepsilon^{2m-1}(0).$$

M is an oriented 3-manifold independent of the embedding and the choice of $\varepsilon \ll 1$. We denote its diffeomorphism type by $\mathrm{Lk}(X,0)$. When $(X,0)$ is a hypersurface singularity, i.e. $m = 3$, and X is defined as the zero set of a holomorphic function in three variables the link $\mathrm{Lk}(X,0)$ coincides with the previously introduced Lk_f.

A *resolution* of $(X,0)$ is a pair $(\tilde{X},\pi)$ where $\tilde{X}$ is a smooth complex surface, $\tilde{X} \xrightarrow{\pi} X$ is holomorphic and proper and $\tilde{X} \setminus \pi^{-1}(0) \to X \setminus 0$ is biholomorphic. The resolution is called *good* if the exceptional divisor $E := \pi^{-1}(p)$ is a normal crossing divisor, i.e its irreducible components $(E_i)_{1\leq i\leq s}$ are smooth curves intersecting transversally. We have the following result, Laufer [9].

Theorem 3.1. (a) *Resolutions exist but are not unique.*
(b) *There exists a unique minimal resolution $\tilde{X}$, i.e. a resolution containing no (-1)-spheres. Any resolution is obtained from the minimal one by blowing-up/down (-1) spheres.*
(c) *There exists a unique minimal good resolution which may have (-1)-spheres, but when blown down the exceptional divisor will no longer be a normal crossing divisor.*

If $\tilde{X}$ is a resolution of X then $\tilde{X}$ is Levi pseudoconvex and we deduce $\dim H^1(\tilde{X},\mathcal{O}_{\tilde{X}}) < \infty$. The integer $\dim H^1(\tilde{X},\mathcal{O}_{\tilde{X}})$ is independent of the resolution, and thus it is an *analytic* invariant of $(X,0)$. It is called the *geometric genus* and is denoted by $p_g(X,0)$.

Suppose $\tilde{X}$ is a good resolution of X. The exceptional divisor $E = \pi^{-1}(0)$ is a normal crossings divisor with components $(E_i)_{1\leq i\leq s}$. Each component E_i is a smooth complex curve of genus g_i and self intersection number e_i. They form an integral basis of $\Lambda = H_2(\tilde{X},\mathbb{Z})$ so that $s = b_2(\tilde{X})$. Set

$$\Lambda_+ := \Big\{ \sum_i n_i E_i \in \Lambda;\ \ n_i \geq 0 \Big\}.$$

The topology of $\tilde{X}$ is completely determined by the dual resolution graph. This is a decorated graph with one vertex for each component E_i. The decoration of a vertex is the pair (g_i, e_i). Two different vertices E_i, E_j are connected by $E_i \cdot E_j$ edges. The manifold $\tilde{X}$ is obtained by plumbing disk bundles of degree e_i over E_i using the adjacency relation described by the edges of the resolution graph.

Theorem 3.2. *The intersection form $\mathcal{I}$ on Λ is negative definite.*

Remark 3.1. The above result has a sort of converse. More precisely a result of H. Grauert (see Laufer [9, Thm. 4.8, 4.9]) states that if $\tilde{X}$ is a smooth complex surface and $E \hookrightarrow \tilde{X}$ is a compact normal crossings divisor with components $(E_i)_{1 \leq i \leq s}$ such that the matrix $(E_i \cdot E_j)_{1 \leq i,j \leq s}$ is negative definite then there exists a complex surface X with only isolated singularities and a proper holomorphic map $\pi : \tilde{X} \to X$ which is bi-holomorphic off E, i.e. $\tilde{X}$ is a resolution of the singularities of X.

For example, consider a holomorphic line bundle L over a compact Riemann surface Σ of genus g such that $d = \deg L < 0$. Then Σ can be viewed as a smooth divisor in the total space of L with negative self-intersection. According to Grauert's Theorem mentioned above, this divisor can be blown-down to obtain a (non-compact) surface $X(\Sigma, L)$ with an isolated singularity. The total space of L is a good resolution of this singularity with dual resolution graph consisting of a single vertex decorated by the pair (g, d)

$$\bullet \;\; (g, d).$$

The geometric genus of this singularity is

$$p_g(X(\Sigma, L))) = \sum_{n \geq 0} \dim H^1(\Sigma, \mathcal{O}(-nL)). \tag{3.1}$$

This formula shows that the geometric genus depends on the choice of complex structures on Σ and the line bundle L.

We identify the dual lattice $\Lambda^* = \mathrm{Hom}(\Lambda, \mathbb{Z})$ with $H^2(\tilde{X}, \mathbb{Z})$. The intersection pairing on Λ defines a map $I : \Lambda \to \Lambda^*$ which is $1 - 1$ since the intersection form is negative definite. In particular, we can identify Λ^* with the lattice $I^{-1}\Lambda^* \subset \Lambda_{\mathbb{Q}} := \Lambda \otimes \mathbb{Q}$.

Denote by $K = c_1(\tilde{X}) \in H^2(\tilde{X}, \mathbb{Z}) = \Lambda^*$ the dual canonical class of the resolution. This element is uniquely determined by the adjunction formulæ

$$\langle K, E_i \rangle = e_i + 2 - 2g_i, \quad \forall i = 1, \dots, s = b_2(\tilde{X}).$$

In particular we can identify K with an element $Z \in \Lambda_{\mathbb{Q}}$ and thus we obtain a (negative) rational number Z^2. We set

$$\gamma(X, 0) := Z^2 + s = Z^2 + b_2(\tilde{X}).$$

The rational number $\gamma(X,0))$ is independent of the choice of the resolution $\tilde{X}$. As we will see in the next section, γ is in fact a *topological invariant* of the link of the singularity $(X,0)$.

When $(X,0)$ is a hypersurface singularity $X=\{f=0\}$, $f\in\mathcal{O}_3$, then we set $p_g(f,0)=p_g(X,0)$. In this case the geometric genus of $(X,0)$ can be expressed in terms of the previously introduced invariants $\mu_\pm$, μ_0 and τ. More precisely we have the Laufer, [10] equality

$$\mu = 12p_g(f,0)+\gamma-\mu_0, \tag{3.2}$$

and the Durfee, [5] identity

$$\tau = -\frac{1}{3}(2\mu+\gamma+2\mu_0). \tag{3.3}$$

We deduce

$$\gamma = \mu_- - 4\mu_0 - 5\mu_+,\ \ p_g = \frac{1}{2}(\mu_0+\mu_+). \tag{3.4}$$

The last identities imply

$$\tau = -\gamma - 8p_g. \tag{3.5}$$

The right-hand side of the above equality makes sense for any isolated surface singularity and thus we can refer to the expression $\tau(X,0) = -\gamma(X,0)-8p_g(X,0)$ as the *virtual signature* of the (possibly nonexistent) Milnor fiber of the singularity.

For a generic polynomial f in three variables the geometric genus $p_g(f,0)$ can be described in terms of the arithmetic of its Newton diagram. Consider a polynomial

$$f(z_0,\dots,z_N)=\sum_\nu a_\nu\vec{z}^\nu,\ \ \nu\in\mathbb{Z}^{N+1}_{\geq 0},\ \vec{z}^\nu := z_1^{\nu_1}\cdots z_N^{\nu_N}.$$

Denote by $\mathcal{C}$ the positive octant $\mathcal{C}=\mathbb{R}^{N+1}_{\geq 0}$ and by $\operatorname{supp} f$ the set of multi-indices ν such that $a_\nu\neq 0$. The *Newton polyhedron* of f is the convex hull of $\operatorname{supp}(f)+\mathcal{C}$ in $\mathbb{R}^{N+1}$. We denote it by $\Gamma(f)$. The polynomial f is called *convenient* if its support intersects all the coordinate axes of $\mathcal{C}$, or equivalently, if its Newton polyhedron intersects all the coordinate axes. Assume f is a convenient polynomial.

Let $c=(1,1,\dots,1)$. A lattice vector ν is called *subdiagrammatic* if

$$\nu + c\in\mathcal{C}\setminus\operatorname{Int}\Gamma(f).$$

Denote by $\mathcal{P}_\Gamma[z_0,\dots,z_n]$ the space of convenient polynomials in the variables $z_0,\dots,z_N$ with Newton polyhedron $\Gamma(f)=\Gamma$. We have the following remarkable result of Varchenko and Khovanski, [28].

Theorem 3.3. *For a generic*[a] $f\in\mathcal{P}_\Gamma[z_0,z_1,z_2]$ *the number of subdiagramatic lattice points is equal to the geometric genus* $p_g(f,0)$.

Example 3.1. Denote by $X_{p,q,r}$ the Brieskorn singularity given by

$$ax^p+by^q+cz^r=0.$$

The dual resolution graph of the Brieskorn singularity $X_{2,3,5}$ is given by the E_8 Dynkin diagram in Figure 1, where the genus of each vertex is zero. This resolution is *diffeomorphic* but *not biholomorphic* to the Milnor fiber of this singularity.

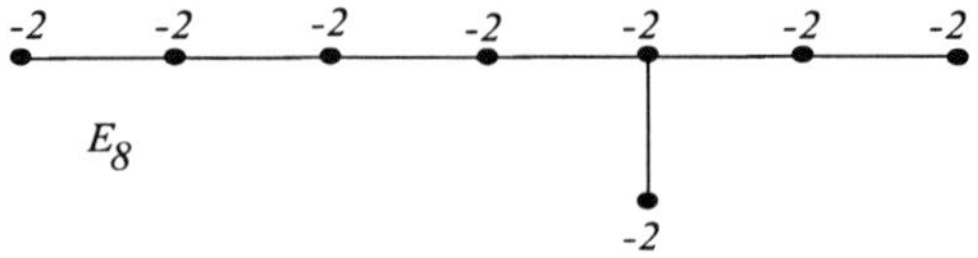

Figure 1. *The E_8 plumbing.*

The singularities $X_{3,7,21}$ and $X_{4,5,20}$ have the same dual resolution graph with a single vertex decorated by $(g,d)=(6,-1)$. Note however that they have different Milnor numbers

$$\mu(X_{3,7,21})=2\cdot 6\cdot 20,\quad \mu(X_{4,5,20})=3\cdot 4\cdot 19.$$

The singularities $X_1=\{x^2+y^3+z^{18}=0\}$ and $X_2=\{z^2+y(x^4+z^6)\}$

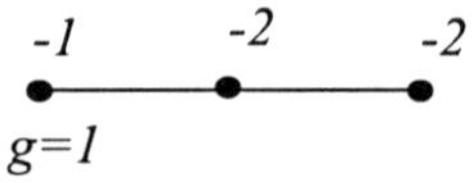

Figure 2. *An elliptic plumbing.*

have the same resolution graph depicted in Figure 2 but different geometric genera and Milnor numbers

$$p_g(X_1,0)=3,\quad \mu(X_1,0)=34,\quad p_g(X_2,0)=2,\quad \mu(X_2,0)=22.$$

[a]The paper [28] describes explicitly the generic polynomials.

4. Topological invariants of the link of an isolated surface singularity

The key to our construction of topological invariants of the link of an isolated surface singularity is the following result of W. Neumann, [22]

Theorem 4.1. *Suppose (X_i, p), $i = 0, 1$ are two isolated surface singularities. Denote by M_i their links, and by $\tilde{X}_i$ their minimal good resolutions. The following statements are equivalent.*

(a) *The dual resolution graphs $\Gamma_{\tilde{X}_i}$ are isomorphic (as decorated graphs).*
(b) *The links M_i are homeomorphic as oriented 3-manifolds.*

In particular, this theorem shows that any combinatorial invariant of the dual resolution graph of an isolated surface singularity is a topological invariant of the link $\mathrm{Lk}(X, 0)$. In particular the invariant $\gamma(X, 0)$ is topological.

Another topological invariant of the link is the *arithmetic genus* $p_a(X, 0)$ defined by

$$p_a(\tilde{X}) := \max\Big\{ \frac{1}{2}Z^2 - \frac{1}{2}\langle c_1(\tilde{X}), Z\rangle + 1;\ \ Z \in \Lambda_+ \setminus 0 \Big\}.$$

This maximum exists since the intersection pairing is negative definite. It is a nonnegative integer independent of the resolution and thus it is a *topological* invariant of (X, p). We will denote it by $p_a(X, p)$, and we will refer to it as the *arithmetic genus* of the singularity.

We can now state the main problem discussed in this survey.

Main Problem *Describe topological and geometric constraints on an isolated surface singularity $(X, 0)$ which will ensure that the geometric genus coincides with a* topological *invariant of the link.*

The computations in Example 3.1 may suggest that this is an ill conceived task. The next list of example will hopefully convince the reader that there is nontrivial content hiding in this question.

Example 4.1. (a) Neumann's Theorem implies that $\mathrm{Lk}(X, 0)$ is homeomorphic to S^3 if and only if 0 is a *smooth* point of X, a result established by D. Mumford in the early 60s. This result is certainly not true in other complex dimensions. For example, the link of the singularity

$$Z_k = \{z_0^2 + z_1^2 + z_3^2 + z_4^3 + z_4^{6k-1} = 0\},$$

is a 7-dimensional manifold M_k homeomorphic to S^7 but the point 0 is not a smooth point of Z_k. Moreover, M_k is diffeomorphic to M_ℓ if and only if $k \equiv \ell \mod 28$ and the collection $(M_k)_{k\geq 1}$ produces all the 28 exotic 7-spheres.
(b) (Artin, [1])

$$p_g(X,0) = 0 \Longleftrightarrow p_a(X,0) = 0.$$

The singularities satisfying $p_g = 0$ are called *rational* and one can show that their links are rational homology spheres.
(c) (Fintushel-Stern, [6]) If the link of $X_{p,q,r}$ is an integral homology 3-sphere then its Casson invariant is equal to $-\frac{1}{8}\tau(X_{p,q,r})$.
(d) (Nemethi, [16]) If $(X,0)$ is an isolated surface singularity whose link is a rational homology sphere, $p_a(X,0) = 1$ and its complex structures satisfies a certain condition then the geometric genus is described by a topological invariant of the link.

The above examples suggest that we should restrict our attention to surface singularities whose links are rational homology spheres and that the Casson-Walker invariant should play a role in elucidating the Main Problem. Let us observe that the link is a rational homology sphere if and only if the dual resolution graph is a tree and the genera of all the vertices are all zero. The results of C. Lescop in [11] lead to a description of the Casson-Walker invariant in terms of the dual resolution graph.

Theorem 4.2. *Suppose $(X,0)$ is an isolated surface singularity whose link M is a rational homology 3-sphere. Denote CW_M its Casson-Walker invariant, by $\det \mathcal{I}$ the determinant of the matrix of the intersection of form $\mathcal{I}$ with respect to the basis $(E_i)_{1\leq i\leq s}$, by δ_i the degree of the vertex E_i and by $\mathcal{I}^{-1}_{ij}$ the (i,j) entry in the matrix $\mathcal{I}^{-1}$. Then*

$$-\frac{24}{|\det \mathcal{I}|}CW_M = 3s + \sum_i e_i + \sum_i (2-\delta_i)\mathcal{I}^{-1}_{ii}.$$

Example 4.2. The dual resolution graph of D_4 is depicted in Figure 3, where all the genera g_i are equal to zero. The intersection form is described by the matrix

$$\mathcal{I} = \begin{bmatrix} -2 & 1 & 1 & 1 \\ 1 & -2 & 0 & 0 \\ 1 & 0 & -2 & 0 \\ 1 & 0 & 0 & -2 \end{bmatrix} \quad \text{with inverse } \mathcal{I}^{-1} = \begin{bmatrix} -2 & -1 & -1 & -1 \\ -1 & -1 & -1/2 & -1/2 \\ -1 & -1/2 & -1 & -1/2 \\ -1 & -1/2 & -1/2 & -1 \end{bmatrix}.$$

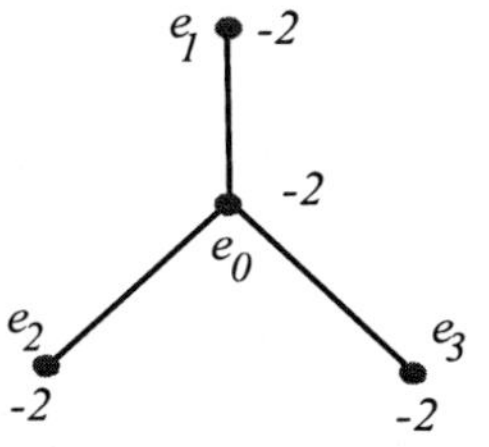

Figure 3. *The Dynkin diagram* D_4.

We deduce

$$-6CW_{\mathrm{Lk}_{D_4}} = 12 - 8 + 2 - 3 \Longrightarrow CW_{\mathrm{Lk}_{D_4}} = -\frac{1}{2}.$$

From the equalities

$$\langle K, E_i \rangle = 0, \quad \forall i = 0, 1, 2, 3$$

we deduce $Z = 0$ so that $Z^2 = 0$ and $\gamma(D_4) = 4$. Using Laufer's identity (3.2) and the equality $\mu(D_4) = 4$ we deduce

$$p_g(D_4) = 0, \quad \tau(D_4) = -4 \Longrightarrow -\frac{1}{8}\tau(D_4) \neq CW_{\mathrm{Lk}_{D_4}}.$$

This shows that the Casson-Walker invariant does not contain all the information required to solve the Main Problem. The correct invariant is a bit more involved and is described in the next section.

5. The Seiberg-Witten invariant

Suppose M is a rational homology 3-sphere. We set $H = H_1(M, \mathbb{Z}) \cong H^2(M, \mathbb{Z})$ and we denote the group operation on H multiplicatively. We denote by CW_M the Casson-Walker invariant of M and by $Spin^c(M)$ the set of *spin*^{c} structures on M. $Spin^c(M)$ is naturally an H-torsor, i.e. H acts freely and transitively on $Spin^c(M)$

$$H \times Spin^c(M) \ni (h, \sigma) \mapsto h \cdot \sigma \in Spin^c(M).$$

The Seiberg-Witten invariant is a function (see Nicolaescu [24] for more details)

$$\boldsymbol{sw}_M : Spin^c(M) \to \mathbb{Q}, \quad \sigma \mapsto \boldsymbol{sw}_M(\sigma).$$

For every $\sigma \in Spin^c(M)$ we define

$$\boldsymbol{SW}_{M,\sigma} : H \to \mathbb{Q}, \quad \boldsymbol{SW}_{M,\sigma}(h) = \boldsymbol{sw}_M(h^{-1} \cdot \sigma).$$

One can give a combinatorial description of this invariant. As explained in Nicolaescu [23] for each *spin*^{c} structure σ the Reidemeister-Turaev torsion of (M, σ) can be viewed as a function $\mathcal{T}_{M,\sigma} : H \to \mathbb{Q}$ satisfying the equivariance condition

$$\mathcal{T}_{M,h\cdot\sigma}(g) = \mathcal{T}_{M,\sigma}(h^{-1}g), \ \ \forall h, g \in H.$$

We have the following result, Nicolaescu [24].

Theorem 5.1.

$$\boldsymbol{SW}_{M,\sigma}(h) = -\frac{1}{|H|}CW_M + \mathcal{T}_{M,\sigma}(h), \ \ \forall h \in H, \ \ \sigma \in Spin^c(M).$$

Remark 5.1. Recently, R. Rustamov [25] showed that the Seiberg-Witten invariant is determined by a renormalized Euler characteristic of the Heegard-Floer homology of M defined by Ozsváth and Szabó.

The link an isolated surface singularity $(X, 0)$ is equipped with a natural *spin*^{c} structure σ_{can}. To define it let us recall that a choice of a *spin*^{c} structure on the link $M = \mathrm{Lk}(X, 0)$ is equivalent to a choice of an almost complex structure on the stable tangent bundle $\underline{\mathbb{R}} \oplus TM$ of M. The stable tangent bundle of M is equipped with a natural complex structure induced by the complex structure on a good resolution $\tilde{X}$. σ_{can} is the *spin*^{c} structure associated to this complex structure. In [19] Nemethi and the present author proved the following result.

Proposition 5.1. *If* $\mathrm{Lk}(X, 0)$ *is a rational homology sphere then* σ_{can} *can be described only in terms of the combinatorics of the dual resolution graph and thus it is a topological invariant of* $\mathrm{Lk}(X, 0)$.

Remark 5.2. σ_{can} first appeared in work of Looijenga-Wahl [13] under a different guise.

Suppose $(X, 0)$ is an isolated surface singularity whose link $M = \mathrm{Lk}(X, 0)$ is a rational homology 3-sphere. A. Nemethi and the present author gave in [19] description of $\boldsymbol{sw}_M(\sigma_{can})$ in terms of the combinatorics of the dual resolution graph. We sketch this description below.

The group $H = H_1(M, \mathbb{Z})$ is a finite Abelian group which admits the presentation

$$0 \to \Lambda \overset{\mathcal{J}}{\hookrightarrow} \Lambda^* \twoheadrightarrow H \to 0.$$

The dual basis $\{E^i\}$ of Λ^* defines generators of H which we denote by the same letters. Denote by $\hat{H} = \mathrm{Hom}(H, \mathbb{C}^*)$ the group of characters of H. Then to every character χ we can associate in a canonical way a weight $\vec{w}_\chi \in \Lambda^*$ (see Nemethi-Nicolaescu [19] for details) and we have

$$\boldsymbol{sw}_M(\sigma_{can}) = -\frac{CW_M}{|H|} + \underbrace{\frac{1}{|H|}\sum_{\chi\neq 1}\lim_{t\to 1}\prod_{i=1}^{s}\Big(t^{-\langle \vec{w}_\chi, E_i\rangle}\chi(E^i) - 1\Big)^{\delta_i - 2}}_{=:\mathfrak{T}_{M,\sigma_{can}}(1)}. \quad (5.1)$$

Example 5.1. Let us explain the above formula in the case of the D_4 singularity. In this case $H \cong \mathbb{Z}_2 \times \mathbb{Z}_2$. For $i = 0, 1, 2, 3$ and $\chi \in \hat{H}$ we set

$$\chi_i = \chi(E_i).$$

Then

$$\hat{H} = \big\{(\chi_0, \chi_1, \chi_2, \chi_3) \in (\mathbb{C}^*)^4;\ \chi_0^2 = \chi_1\chi_2\chi_3,\ \ \chi_0 = \chi_j^2,\ \ \forall j = 1, 2, 3\big\}.$$

We deduce that

$$(\chi_0, \chi_1, \chi_2, \chi_3) \in \hat{H} \iff \chi_0 = 1 = \chi_1\chi_2\chi_3,\ \ \chi_j = \pm 1,\ \ j = 1, 2, 3.$$

In this case the weight $\vec{w}_\chi$ is the same for all characters $\chi \neq 1$ and can be identified with the E_0-row of $\mathfrak{I}^{-1}$,

$$\vec{w} = (-2, -1, -1, -1).$$

We deduce

$$\mathfrak{T}_{D_4} = \frac{1}{4}\sum_{\chi\neq 1}\lim_{t\to 1}\frac{(t^2\chi_0 - 1)}{(t\chi_1 - 1)(t\chi_2 - 1)(t\chi_3 - 1)}$$

$$= \frac{1}{4}\sum_{\chi\neq 1}\lim_{t\to 1}\frac{(t^2 - 1)}{(t\chi_1 - 1)(t\chi_2 - 1)(t\chi_3 - 1)} = \frac{3}{8}.$$

Hence

$$\boldsymbol{sw}_{D_4}(\sigma_{can}) = -\frac{1}{4}CW_{D_4} + \mathfrak{T}_{D_4} = \frac{4}{8} = -\frac{\tau(D_4)}{8}.$$

Definition 5.1. An isolated surface singularity is called *special* if $M = \mathrm{Lk}(X, 0)$ is a rational homology sphere and

$$\boldsymbol{sw}_M(\sigma_{can}) = p_g(X, 0) + \frac{1}{8}\gamma(X, 0) = -\frac{1}{8}\tau(X, 0).$$

The Fintushel-Stern result in Example 4.1(c) shows that the Brieskorn singularities $X_{p,q,r}$ whose links are integral homology spheres are special. The above example shows that the D_4 singularity is special. In fact, the class of special singularities is quite large. The next theorem summarizes recent results by A. Némethi and the author, [17, 18, 19, 20, 21].

Theorem 5.2. *The following classes of singularities are special.*
(a) *rational singularities,*
(b) *singularities with good $\mathbb{C}^*$-action and rational homology sphere link,*
(c) *hypersurface singularities of the form*

$$z^n + P(x, y) = 0$$

with rational homology sphere link.

Remark 5.3. For a while the author strongly believed that any hypersurface singularity whose link is a rational homology 3-sphere must be special. The situation changed after I. Luengo-Velasco, A. Melle-Hernandez and A. Némethi constructed in [14] counterexamples to many long held beliefs in singularity theory. We present below one example from [14].

Consider the polynomial

$$f(x, y, z) = \underbrace{27x^5 + 18x^3yz - 2x^2y^3 - x^2z^3 + 2xy^2z^2 - y^4z}_{f_5} + (x - y + z)^6.$$

The equation $f_5 = 0$ defines a rational cuspidal curve in $\mathbb{P}^2$, i.e. a rational plane curve whose singularities are locally irreducible. It can be given the parametric description

$$t \longmapsto (t : t^3 - 1 : t^5 + 2t).$$

The curve has 4 singularities at ∞, and at the cubic roots of $-\frac{3}{2}$. The line $x - y + z = 0$ does not go through these singular points. The multiplicity sequence of each singularity is 2_3; 2; 2; 2. The corresponding Milnor numbers of these plane curve singularities are $3 \cdot 2(2 - 1)$, 2, 2, 2. These add up to 12 which is the genus of a generic plane quintic. In this case we have

$$\boldsymbol{sw}_{\mathrm{Lk}_f}(\sigma_{can}) - \frac{1}{8}\gamma(f) = 2,$$

while $p_g(f) = 10$ so that the singularity $\{f = 0\}$ is not special.

At the present moment the author is unaware of a *geometric* condition on a polynomial in three variables f which together with the assumption $b_1(\mathrm{Lk}_f) = 0$ will guarantee that the singularity $f = 0$ is special. It is however tempting to conjecture that if f is generic in the sense of Theorem 3.3 then the singularity $\{f = 0\}$ is special. The above example is not generic in the sense of Theorem 3.3 because the corresponding geometric genera are not equal to the values predicted by the theorem.

Acknowledgement. I want to thank Bernhelm Booß-Bavnbek and Krysztof Wojciechowski for their friendship and continued support over the years. I also want to thank the anonymous referee for his/her many useful comments. This work was partially supported by NSF grant DMS 0303601.

References

1. M. Artin, *Some numerical criteria for contractibility on algebraic surfaces*, Amer. J. Math., **84**(1982), 485-496.
2. E. Brieskorn, *Die Monodromie der isolierten Singularitäten von Hyperflächen*, Manuscripta Math., **2**(1970), 103-161.
3. C.H. Clemens, *Picard-Lefschetz theorem for families of nonsingular algebraic varieties acquiring ordinary singularities*, Trans. Amer. Math .Soc., **136**(1969), 93-108.
4. P. Deligne, *Equations Différentieles à Points Singuliers Reguliers*, Lectures Notes in Math., vol. 163, Springer Verlarg 1970.
5. A. Durfee, *The signature of smoothings of complex surface singularities*, Math. Ann., **232**(1978), 85-89.
6. R. Fintushel, R. Stern, *Instanton homology of Seifert fibered homology three spheres*, Proc. London Math. Soc., **61**(1990), 109-137.
7. A. Hatcher: *Algebraic Topology*, Cambridge University Press, 2002.
8. A. Landman, *On the Picard-Lefschetz, transformation for algebraic manifolds aquiring general singularities*, Trans. Amer. Math. Soc., **181**(1973), 89-126.
9. H. Laufer, *Normal Two-Dimensional Singularities*, Ann. of Math. Stud., vol. 71, Princeton University Press, 1971.
10. H. Laufer, *On μ for surface singularities*, Proc. of Symp. in Pure Math., **30**(1977), 45-49.
11. C. Lescop: *Global Surgery Formula For The Casson-Walker Invariant*, Ann. of Math. Stud., vol. 140, Princeton University Press, 1996.
12. E. Looijenga: *Isolated singular points on complete intersections*, London Math. Soc. Lect. Notes Series, vol.77, Cambridge Univ. Press, 1984.
13. E. Looijenga, J. Wahl, *Quadratic functions and smoothings of surface singularities*, Topology, **25**(1986), 261-291.
14. I. Luengo-Velasco, A. Melle-Hernandez, A. Nemethi, *Links and analytic invariants of superisolated singularities*, mathAG/0312416.
15. J.W. Milnor, *Singular Points Of Complex Hypersurfaces*, Ann. of Math. Stud., vol. 61, Princeton University Press, 1968.

16. A. Némethi, *"Weakly" elliptic Gorenstein singularities of surfaces*, Invent. Math., **137**(1999), 145-167.
17. A. Némethi, *Line bundles associated with normal surface singularities*, mathAG/0310084.
18. A. Némethi, *On the Ozsváth-Szabó invariant of negative definite plumbed 3-manifolds*, Geom. Top., **9**(2005) 991-1042
19. A. Némethi, L.I. Nicolaescu, *Seiberg-Witten invariants and surface singularities*, Geom. Top., **6**(2002), 269-328.
20. A. Némethi, L.I. Nicolaescu, *Seiberg-Witten invariants and surface singularities II. Singularities with good* $\mathbb{C}^*$ *action*, J. London Math. Soc., **69**(2004), 593-607, mathAG/0101120.
21. A. Némethi, L.I. Nicolaescu, *Seiberg-Witten invariants and surface singularities III. Splicings and cyclic covers*, mathAG/0207018.
22. W. Neumann, *A calculus for plumbing applied to the topology of complex surface singularities and degenerating complex curves*, Trans. Amer. Math. Soc., **268**(1981), 299-344..
23. L.I. Nicolaescu, *The Reidemeister Torsion Of 3-Manifolds*, Walter de Gruyter, 2003.
24. L.I. Nicolaescu, *Seiberg-Witten invariants of rational homology spheres*, Comm. Contemp. Math., **6**(2004), 833-866, mathGT/0103020.
25. R. Rustamov, *Surgery formula for the renormalized Euler characteristic of Heegard-Floer homology*, mathGT/0409294.
26. M. Sebastiani, R. Thom, *Un resultat sur la monodromie*, Invent. Math, **13**(1971), 90-96.
27. J.C. Tougeron, *Ideaux de fonctions différentiables*, Ann. Inst. Fourier, **18**(1968), 177-240.
28. A. N. Varchenko, A. G. Khovanski, *Asymptotics of integrals over vanishing cycles and the Nedwton polyhedron*, Soviet Math. Dokl. **32**(1985), 122-127.

Received by the editors August 16, 2005; Revised October 11, 2005

Analysis, Geometry and Topology
of Elliptic Operators, pp. 79–104

CHERN-WEIL FORMS ASSOCIATED WITH SUPERCONNECTIONS

SYLVIE PAYCHA

Laboratoire de Mathématiques
Complexe des Cézeaux
Université Blaise Pascal
63 177 Aubière Cedex F
sylvie.paycha@math.univ-bpclermont.fr

SIMON SCOTT

Department of Mathematics,
King's College London
WC2R 2LS
sgs@mth.kcl.ac.uk

Dedicated to our friend Krzysztof Wojciechowski

We define Chern-Weil forms $c_k(\mathbb{A})$ associated to a superconnection $\mathbb{A}$ using ζ-regularisation methods extended to ΨDO valued forms. We show that they are cohomologous in the de Rham cohomology to $\mathrm{tr}(\mathbb{A}^{2k}\,\pi_P)$ involving the projection π_P onto the kernel of the elliptic operator P to which the superconnection $\mathbb{A}$ is associated. A transgression formula shows that the corresponding Chern-Weil cohomology classes are independent of the scaling of the superconnection. When P is a differential operator of order p with scalar leading symbol, the k-th Chern-Weil form corresponds to the regularised k-th derivative at $t = 0$ of the Chern character $\mathrm{ch}(t\mathbb{A})$ and it has a local description

$$c_k(\mathbb{A}) = -\frac{1}{2p}\mathrm{res}\left(\mathbb{A}^{2k}\log(\mathbb{A}^2 + \pi_P)\right)$$

in terms of the Wodzicki residue extended to ΨDO -valued forms.

2000 *Mathematics Subject Classification.* Primary 58J20; Secondary 58J40, 47G99

1. Introduction

Given a superconnection $\mathbb{A}$ adapted to a family of self-adjoint elliptic ΨDOs with odd parity parametrised by a manifold B, the ΨDO valued form

$e^{-\mathbb{A}^2}$ is trace class so that one can define the associated Chern character

$$\mathrm{ch}(\mathbb{A}) := \mathrm{tr}\left(e^{-\mathbb{A}^2}\right) \tag{1}$$

which defines a characteristic class independent of any scaling $t \mapsto \mathbb{A}_t$ of the super connection. Here, tr is the supertrace associated to the family of ΨDOs , which reduces to the usual trace when the grading is trivial. In the case of a family of Dirac operators the limit $\lim_{t\to 0} \mathrm{tr}\left(e^{-\mathbb{A}_t^2}\right)$ of the Chern character built from the rescaled Bismut superconnection exists and the local Atiyah-Singer index formula for families provides a formula for it in terms of certain canonical characteristic forms integrated along the fibres (see the works of Bismut [1], Bismut and Freed [2] and the book by Berline, Getzler and Vergne [3]). When the fibre of the manifold fibration over B reduces to a point, this family setup reduces to an ordinary finite rank vector bundle situation; $\mathbb{A}$ can be replaced by an ordinary connection ∇ on a vector bundle over B with Chern character $\mathrm{ch}(\nabla) := \mathrm{tr}\left(e^{-\nabla^2}\right)$, and the local family index theorem reduces to the usual Atiyah-Singer index formula for a single operator. This can be expressed as a linear combination

$$\mathrm{ch}(\nabla) = \sum_{k=0}^{\mathrm{dim M}} (-1)^k \frac{c_k(\nabla)}{k!}$$

of the associated Chern-Weil forms $c_k(\nabla) := \mathrm{tr}(\nabla^{2k})$ of degree $2k$. Conversely, the Chern-Weil forms can be interpreted as the coefficients of a Taylor expansion of the map $t \mapsto \mathrm{ch}_t(\nabla) := \mathrm{tr}\left(e^{-t\nabla^2}\right)$ at $t = 0$

$$\partial_t^k \mathrm{ch}_t(\nabla)|_{t=0} = (-1)^k\, c_k(\nabla). \tag{2}$$

Replacing traces by appropriate regularised traces gives another insight into the family setup close to this finite dimensional description. Given a superconnection $\mathbb{A}$ adapted to a family of self-adjoint elliptic ΨDOs with odd parity parametrised by a manifold B:

(1) using regularised traces, one can build Chern-Weil type forms that relate to the Chern character (1) as in (2).
(2) If $\mathbb{A}$ is a superconnection associated with a family of *differential operators*, one can show an a priori locality property for these Chern-Weil forms, without having to compute them explicitly. This can be achieved by expressing them in terms of Wodzicki residues.

Let us start with the first of these two issues.

1. Just as ordinary Chern-Weil forms are traces of the k-th power of the curvature, we define k-th Chern-Weil forms associated with a superconnection $\mathbb{A}$ as weighted traces:

$$c_k(\mathbb{A}) := \mathrm{tr}^{\mathbb{A}^2}(\mathbb{A}^{2k})$$

of the k-th power of the curvature $\mathbb{A}^2$. To do this, we extend weighted traces (according to the terminology used in [4] and which were also investigated by Melrose and Nistor [9] and Grubb [5]) to families $\mathbb{A}, \mathbf{Q}$ of classical ΨDO valued forms setting

$$\begin{aligned} \mathrm{tr}^{\mathbf{Q}}(\mathbb{A}) &:= \zeta(\mathbb{A}, \mathbf{Q} + \pi_{\mathbf{Q}_{[0]}}, z)|_{z=0}^{\mathrm{mer}} \\ &= \zeta(\mathbb{A}, \mathbf{Q}, 0)|^{\mathrm{mer}} + \mathrm{tr}(\mathbb{A}\,\pi_{\mathbf{Q}_{[0]}}), \end{aligned} \tag{3}$$

including weights which are themselves ΨDO valued forms. Here, $\pi_{Q_{[0]}}$ is the projection onto the bundle of kernels $\cup_{x\in B}\mathrm{Ker}(Q_x)$ of the zero degree part $Q_{[0]}$ of Q (we assume that $\dim(\mathrm{Ker}(Q_x))$ is constant), while $\zeta(A, P, z)|^{\mathrm{mer}}$ is the mixed degree meromorphic differential form studied by Scott in [14] which extends the holomorphic form $\mathrm{Tr}\,(AP^{-z})$ from a suitable half-plane $\mathrm{Re}(z) >> 0$. We show that the forms $c_k(\mathbb{A})$ are closed and that the associated characteristic classes are independent of the scaling of $\mathbb{A}$. This is equivalent to the fact that the zeta forms $\zeta(\mathbb{A}^2, -k) := \zeta(\mathbb{A}^{2k}, \mathbb{A}^2, 0)|^{\mathrm{mer}}$ are exact and independent of scaling (this fact was proved in [14]); the (closed) Chern-Weil forms $c_k(\mathbb{A})$ differ from these forms by a term $\mathrm{tr}(\pi\,\mathbb{A}^{2k}\,\pi)$ involving the projection π on the kernel of the operator $\mathbb{A}_{[0]}$ to which the superconnection $\mathbb{A}$ is adapted. As a consequence of the exactness of $\zeta(\mathbb{A}^2, -k)$ we obtain that $c_k(\mathbb{A})$ is cohomologous to $\mathrm{tr}(\pi\,\mathbb{A}^{2k}\,\pi)$.

2. The second step can be carried out whenever $\mathbb{A}$ is a superconnection associated with a family of *differential operators* of order p, with scalar leading symbol. In this case, we relate the forms $c_k(\mathbb{A})$ to the Chern character $\mathrm{ch}(\mathbb{A})$ via a formula which mimics equation (2): for $t > 0$,

$$\mathrm{fp}_{t=0}\,\partial_t^k \mathrm{ch}_t(\mathbb{A}) = (-1)^k\,c_k(\mathbb{A}), \tag{4}$$

where fp denotes the finite part (the constant term in the asymptotic expansion as $t \to 0+$). With the same assumptions, we show that the following local formula for these weighted Chern forms (see equation (23)) holds

$$c_k(\mathbb{A}) := -\frac{1}{2p}\,\mathrm{res}(\mathbb{A}^{2k}\,\log(\mathbb{A}^2 + \pi_P)),$$

where the right-side is the residue trace extended these families; this object would not be defined for general families of ΨDOs. This equation generalises formulae obtained by Paycha and Scott in [12] which relate weighted traces of differential operators to Wodzicki residues extended to logarithms (see theorem 6.2) and provides a local expression on the grounds of the locality of the Wodzicki residue.
These results are summarised in Theorem 7.1.

The paper is organised as follows

(1) ΨDO valued forms
(2) Complex powers and logarithms of ΨDO valued forms
(3) The Wodzicki residue and the canonical trace extended to ΨDO valued forms
(4) Holomorphic families of ΨDO valued forms
(5) Weighted traces of differential operator valued forms; locality
(6) Chern-Weil forms associated with a superconnection

2. ΨDO valued forms

In this section we recall the construction of form valued geometric families of ΨDOs from [14]. Consider a smooth fibration $\pi : M \to B$ with closed n-dimensional fibre $M_b := \pi^{-1}(b)$ equipped with a Riemannian metric $g_{M/B}$ on the tangent bundle $T(M/B)$. Let $|\Lambda_\pi| = |\Lambda(T^*(M/B))|$ be the line bundle of vertical densities, restricting on each fibre to the usual bundle of densities $|\Lambda_{M_b}|$ along M_b. Let $\mathcal{E} := \mathcal{E}^+ \oplus \mathcal{E}^-$ be a vertical Hermitian $\mathbb{Z}_2$-graded vector bundle over M and let $\pi_*(\mathcal{E}) := \pi_*(\mathcal{E}^+) \oplus \pi_*(\mathcal{E}^-)$ be the graded infinite dimensional Fréchet bundle with fibre $C^\infty\left(M_b, \mathcal{E}^b \otimes |\Lambda_{M_b}|^{\frac{1}{2}}\right)$ at $b \in B$, where $\mathcal{E}^b$ is the $\mathbb{Z}_2$-graded vector bundle over M_b obtained by restriction of $\mathcal{E}$. By definition, a smooth section ψ of $\pi_*(\mathcal{E})$ over B is a smooth section of $\mathcal{E} \otimes |\Lambda_\pi|^{\frac{1}{2}}$ over M, so that $\psi(b) \in C^\infty(M_b, \mathcal{E}^b \otimes |\Lambda_{M_b}|^{\frac{1}{2}})$ for each $b \in B$. More generally, the de Rham complex of smooth forms on B with values in $\pi_*(\mathcal{E})$ is defined by

$$\mathcal{A}(B, \pi_*(\mathcal{E})) = C^\infty\left(M, \pi^*(\wedge T^*B) \otimes \mathcal{E} \otimes |\Lambda_\pi|^{\frac{1}{2}}\right)$$

with $\otimes$ the $\mathbb{Z}_2$-graded tensor product. Let $C\ell(\mathcal{E})$ denote the infinite-dimensional bundle of algebras with fibre $C\ell(\mathcal{E}^b) = C\ell\left(M_b, \mathcal{E}^b \otimes |\Lambda_{M_b}|^{\frac{1}{2}}\right)$. A section $Q \in \mathcal{A}(B, C\ell(\mathcal{E}))$ defines a smooth family of classical ΨDOs with differential form coefficients parametrized by B.

Such an operator valued form Q is locally described by a *vertical* symbol

$$\mathbf{q}(x,y,\xi) \in C^\infty\left((U_M \times_\pi U_M) \times \mathbb{R}^n, \pi^*(\Lambda T^* U_B) \otimes \mathbb{R}^N \otimes \left(\mathbb{R}^N\right)^*\right),$$

where $\times_\pi$ is the fibre product, ξ may be identified with a vertical vector in $T_b(M/B)$, and U_M is a local coordinate neighbourhood of M over which $\mathcal{E}_{U_M} \simeq U_M \times \mathbb{R}^N$ is trivialized and $\mathbb{R}^N$ inherits the grading of $\mathcal{E}$. With respect to the local trivialisation of $\pi_*(\mathcal{E})$ over $U_B = \pi(U_M)$ one has

$$\mathcal{A}\left(U, \pi_*(\mathcal{E})_{|U_B}\right) \simeq \mathcal{A}(U) \otimes C^\infty\left(M_{b_0}, \mathcal{E}^{b_0}\right)$$

with $M_{b_0} = \pi^{-1}(b_0)$ relative to a base point $b_0 \in U_B$, so that $\mathbf{q}$ can be written locally over U_B as a finite sum of terms of the form $\omega_k \otimes \mathbf{q}_{[k]}$, where $\omega_k \in \mathcal{A}^k(U_B)$ and $\mathbf{q}_{[k]} \in C^\infty\left(U_{b_0} \times \mathbb{R}^n/\{0\}, \mathbb{R}^N \times \left(\mathbb{R}^N\right)^*\right)$ is a symbol (in the single manifold sense) of form degree zero so that for all multi-indices α, β and each compact subset $K \subset U_{b_0}$ the growth estimate holds

$$|\partial_x^\alpha \partial_y^\beta \partial_\xi^\gamma \mathbf{q}_{[k]}(x,y,\xi)| < C_{k,\alpha,\beta,K}(1+|\xi|)^{q_k - |\gamma|}. \tag{5}$$

For clarity we will work only with local symbols which are *simple*, meaning they have the local form $\sum_{k=0}^{\dim B} \omega_k \otimes \mathbf{q}_{[k]}$, with just one term in each form degree, extending by linearity to general sums. The order of a simple symbol is defined to be the $(\dim B + 1)$-tuple $(q_0, \cdots, q_{\dim B})$ with q_k the order of the symbol $\mathbf{q}_{[k]}$; for simplicity we consider the case where q_k is constant on B.

In accordance with the splitting of the local symbol into form degree $\mathbf{q} = \mathbf{q}_{[0]} + \cdots + \mathbf{q}_{[\dim B]}$ the operator

$$(Q\psi)(x) = \frac{1}{(2\pi)^n} \int_{M/B} d\,vol_{M/B} \int_{\mathbb{R}^n} e^{i(x-y)\cdot\xi} \mathbf{q}(x,y,\xi)\psi(y)d\xi,$$

for ψ with compact support in U_M, splits as $Q = Q_{[0]} + Q_{[1]} + \cdots + Q_{[\dim B]}$, where Q is a simple family of ΨDOs in so far as locally there is one component $Q_{[k]} = \omega_k \otimes Q_k \in \mathcal{A}\left(U_B, \pi_*(\mathcal{E})_{|U_B}\right)$ in each form degree. $Q_{[k]}$ raises form degree by k and $Q_k(b) : C^\infty(M_b, \mathcal{E}_b^+) \to C^\infty(M_b, \mathcal{E}_b^-)$ is a pseudodifferential operator in the usual sense acting on sections of the bundle $\mathcal{E}_b := \mathcal{E}_{|M_b}$ over the fibre M_b.

The composition of ordinary symbols naturally extends to a composition of families of vertical symbols defined fibrewise by:

$$\mathbf{q} \circ \mathbf{q}' := \omega \wedge \omega' \otimes q \circ q'$$

where $q \circ q'$ is the ordinary composition of symbols corresponding to the ΨDO algebra multiplication

$$\mathcal{A}^i(B, \Psi^\nu(\mathcal{E})) \times \mathcal{A}^j(B, \Psi^\mu(\mathcal{E})) \longrightarrow \mathcal{A}^{i+j}(B, \Psi^{\nu+\mu}(\mathcal{E})). \tag{6}$$

By a standard method the vertical symbol $\mathbf{q}$ in (x,y)-form can be replaced by an equivalent (modulo $S^{-\infty}$) symbol in x-form. A (simple) family of vertical symbols $\mathbf{q}$ of order $(q_0, \cdots, q_{\dim B})$ is then called *classical* if for each $k \in \{0, \cdots, \dim B\}$ one has $\mathbf{q}_{[k]}(x,\xi) \sim \sum_{j=0}^{\infty} \mathbf{q}_{[k],j}(x,\xi)$ with $\mathbf{q}_{[k],j}(x, t\xi) = t^{q_k - j}\mathbf{q}_{[k],j}(x,\xi)$ for $t \geq 1, |\xi| \geq 1$. A family of vertical ΨDOs is called *classical* if each of its local component simple symbols is classical.

Definition 2.1. A smooth family $Q \in \mathcal{A}(B, C\ell(\mathcal{E}))$ of vertical ΨDOs is elliptic if its form degree zero component $Q_{[0]}$ is pointwise (with respect to the parameter manifold B) elliptic.

In this case Q has spectral cut θ if $Q_{[0]}$ admits a spectral cut θ. Likewise, it is invertible if $Q_{[0]}$ is invertible. Given that Q admits a spectral cut then it has a well-defined resolvent, which is a sum of simple families of ΨDOs. Setting $Q_{[>0]} := Q - Q_{[0]} \in \mathcal{A}^1(B, C\ell(\mathcal{E}))$ then there is an open sector $\Gamma_\theta \in \mathbb{C} - \{0\}$ containing the ray L_θ such that on any compact codimension zero submanifold B_c of B for large $\lambda \in \Gamma_\theta$ one has in $\mathcal{A}(B_c, C\ell(\mathcal{E})))$ using the idempotence of forms on B

$$\begin{aligned}(Q-\lambda)^{-1} &= \left(Q_{[0]} - \lambda\right)^{-1} \\ &\quad + \sum_{k=1}^{\dim B} (-1)^k \left(Q_{[0]} - \lambda\right)^{-1} \left(Q_{[>0]}(Q_{[0]} - \lambda)^{-1}\right)^k. \end{aligned} \tag{7}$$

In particular $\left((Q-\lambda)^{-1}\right)_{[0]} = \left(Q_{[0]} - \lambda\right)^{-1}$.

3. Complex powers and logarithms of ΨDO valued forms

Here we use the complex powers for ΨDO valued forms introduced in [14] to define and investigate the properties of the logarithm of a simple family of invertible admissible elliptic vertical ΨDOs.

Let Q be a smooth family of vertical admissible elliptic invertible ΨDOs, the orders $(q_0, \cdots, q_{\dim B+1})$ of which fulfill the assumption

$$q_0 = \mathrm{ord}(Q_{[0]}) > 0$$

and

$$q_k \leq q_0 \quad \forall k \geq 1. \tag{8}$$

Under these assumptions one obtains an operator norm estimate in $\mathcal{A}(B)$ as $\lambda \to \infty$ in Γ_θ

$$\|(Q-\lambda I)^{-1}\|_{M/B}^{(l)} = O(|\lambda|^{-1})$$

where $\|\cdot\|_{M/B}^{(l)} : \mathcal{A}\left(B, C\ell(\mathcal{E})\right) \to \mathcal{A}(B)$ is the vertical Sobolev endomorphism norm associated to the vertical metric.

Lemma 3.1. *Let Q be an admissible elliptic invertible* ΨDO *valued form on B with spectral cut θ. Then*

$$Q_\theta^{-z} = \frac{i}{2\pi}\int_{C_{\theta,r}} \lambda_\theta^{-z}(Q-\lambda I)^{-1} d\lambda$$

defines a family of ΨDOs *in $\mathcal{A}\left(B, C\ell(\mathcal{E})\right)$ which is a finite sum of simple* ΨDO *families with holomorphic orders*

$$\alpha(z) = -q_0 \cdot z + \alpha(0)$$

where $q_0 = \mathrm{ord} Q_{[0]}$ and the constant term $\alpha(0)$ is determined by the q_i and the form degree. In particular, $\left(Q_\theta^{-z}\right)_{[0]} = \left((Q_\theta)_{[0]}\right)^{-z}$.
Here λ_θ^{-z} is the branch of λ^{-z} defined by $\lambda_\theta^{-z} = |\lambda|^{-z} e^{-iz \quad Arg\lambda}$, $\theta - 2\pi \leq Arg\lambda < \theta$ and r being a sufficiently small positive number, $C_{\theta,r}$ is a contour defined by $C_{\theta,r} = C_{1,\theta,r} \cup C_{2,\theta,r} \cup C_{3,\theta,r}$ with $C_{1,\theta,r} = \{\lambda = |\lambda| e^{i\theta} \mid +\infty > |\lambda| \geq r\}$, $C_{2,\theta,r} = \{\lambda = re^{i\phi} \mid \theta \geq \phi \geq \theta - 2\pi\}$ and $C_{3,\theta,r} = \{\lambda = |\lambda| e^{i(\theta-2\pi)} \mid r \leq |\lambda| < +\infty\}$.
When $Q_{[0]}$ has non negative leading symbol we can choose $\theta = \frac{\pi}{2}$ in which case this complex power is the Mellin transform of the corresponding heat-operator form:

$$Q^{-z} = \frac{1}{\Gamma(z)}\int_0^\infty t^{z-1}\, e^{-tQ}\, dt.$$

Remark 3.1. When Q is not invertible, we can apply the Lemma to $Q + \pi_{Q_{[0]}}$ which is invertible. Here $\pi_{Q_{[0]}}$ denotes the projection onto the kernel of $Q_{[0]}$.

Proof. Let us first check the last formula, which is very straightforward.

As for ordinary pseudodifferential operators, we write

$$\begin{aligned} Q^{-z} &= \frac{i}{2\pi}\int_{C_{R,\theta}} \lambda_\theta^{-z}(Q-\lambda I)^{-1}\, d\lambda \\ &= \frac{i}{2\pi}\int_{C_{R,\theta}} \left(\frac{1}{\Gamma(z)}\int_0^\infty t^{z-1}\, e^{-t\,\lambda}\, dt\right)(Q-\lambda I)^{-1}\, d\lambda \\ &= \frac{1}{\Gamma(z)}\int_0^\infty dt\, t^{z-1}\, \frac{i}{2\pi}\int_{C_{R,\theta}} e^{-t\,\lambda}\,(Q-\lambda I)^{-1}\, d\lambda \\ &= \frac{1}{\Gamma(z)}\int_0^\infty dt\, t^{z-1}\, e^{-t\,Q}. \end{aligned}$$

To prove the first part of the lemma, we follow [14]. Let us first assume (8). Following Seeley's analysis, one can define the complex power $Q_\theta^{-z} = \frac{i}{2\pi}\int_{C_{R,\theta}} \lambda_\theta^{-z}(Q-\lambda I)^{-1} d\lambda$ for Re $(z) > 0$ in the usual way provided Q satisfies assumption (8). Since $\left((Q-\lambda I)^{-1}\right)_{[0]} = \left(Q_{[0]}-\lambda I\right)^{-1}$ it follows that $\left(Q_\theta^{-z}\right)_{[0]} = \left((Q_\theta)_{[0]}\right)^{-z}$. The order of $\left(Q_\theta^{-z}\right)_{[d]}$ is derived from the expression of $\left((Q-\lambda I)^{-1}\right)_{[d]}$, which from (7) can be rewritten as

$$\left((Q-\lambda)^{-1}\right)_{[d]} = \left(Q_{[0]}-\lambda\right)^{-1} + \sum_{k=1}^{\dim B}(-1)^k \sum_{p_1+\cdots+p_k=d} \left(Q_{[0]}-\lambda\right)^{-1} W_{[p_1]}\left(Q_{[0]}-\lambda\right)^{-1}\cdots W_{[p_k]}\left(Q_{[0]}-\lambda\right)^{-1} \tag{10}$$

where $W := Q_{[>0]}$ so that $W_{[p_j]} = (Q - Q_{[>0]})_{[p_j]}$, a simple family of ΨDOs of pure form degree $p_j \in \{1,\ldots,\dim B\}$. Let $\alpha_j = \mathrm{ord}(W_{[p_j]})$. This contributes to $\left(Q_\theta^{-z}\right)_{[d]}$ a (simple) family of ΨDOs order $-q_0\, z - q_0\, k + \alpha_1 + \cdots + \alpha_k$. It follows that the order of each simple component of the complex power is holomorphic with derivative at 0 independent of k.

Complex powers can be extended to operators which do not satisfy assumption (8). By (10)

$$\partial_\lambda^m\,(Q-\lambda)_{[d]}^{-1} = \sum_{k=0}^{\dim B}\ \sum_{\substack{p_1+\ldots+p_k=d\\ m_0+\cdots+m_k=m}} \partial_\lambda^{m_0}\left(Q_{[0]}-\lambda\right)^{-1} W_{[p_1]}\,\partial_\lambda^{m_1}\left(Q_{[0]}-\lambda\right)^{-1}\cdots W_{[p_k]}\partial_\lambda^{m_k}(Q_{[0]}-\lambda)^{-1}. \tag{11}$$

Since $\partial_\lambda^{m_i}(Q_{[0]}-\lambda I)^{-1}$ is of order $-q(m_i+1)$, taking m sufficiently large, we can ensure that

$$\|\partial_\lambda^m(Q_{[0]}-\lambda I)^{-1}\|_{M/B}^{(l)} = O(|\lambda|^{-1})$$

without assumption (8). In this way, using integration by parts we may define

$$Q_\theta^{-z} = \frac{1}{(z-1)\cdots(z-m)} \frac{i}{2\pi} \int_{C_{R,\theta}} \lambda_\theta^{m-z} \partial_\lambda^m (Q - \lambda I)^{-1} d\lambda. \tag{12}$$

The order computation is essentially unchanged. □

Let Q be a smooth family of vertical admissible elliptic invertible ΨDOs. The logarithm is also built as in the single operator case by defining

$$\log_\theta Q := \frac{\partial}{\partial z}_{|z=0} (Q_\theta^z).$$

As in the single operator case the logarithm is *not* quite a family of *classical* ΨDOs, but the non-logarithmic component is located only in the form degree zero component:

Lemma 3.2. *Let $Q \in \mathcal{A}(B, C\ell(\mathcal{E}))$ be a family of differential form valued vertical classical invertible elliptic* ΨDOs *with spectral cut θ. Then*

$$(\log_\theta Q)_{[0]} = \log_\theta Q_{[0]}.$$

If Q moreover satisfies assumption (8), then

$$\log_\theta Q - \log_\theta Q_{[0]} \in \mathcal{A}(B, C\ell(\mathcal{E})).$$

Remark 3.2.

- Since the order of the components of $(Q^{-z})_{[d]}$ are of the form $-qz + \alpha(0)$, from Lemma 3.1, one expects $\log_\theta Q_{[d]}$ to contain $\log|\xi|$ terms, which would contradict the statement of Lemma 3.2. A closer look shows that the terms of the form $|\xi|^{-qz}$ that arise in $Q_{[d]}^{-z}$ when $d > 0$ come with a factor of z and therefore do not yield any $\log|\xi|$ term when differentiated with respect to z at $z = 0$.
- A priori $|\xi|$ is $b \in B$-dependent, this reflecting the fact that the decomposition depends on the choice of metric on the fibre M_b.

Proof. We drop the index θ to simplify notations. First, we show that $(\log Q)_{[0]} = \log Q_{[0]}$. For some integer m chosen large enough, we have for $\mathrm{Re}(z) > 0$:

$$\partial Q^{-z} = \frac{1}{(z-1)\cdots(z-m)} \frac{1}{2i\pi} \int_C \log\lambda\, \lambda^{m-z} \partial_\lambda^m (Q - \lambda I)^{-1} d\lambda$$

and hence

$$\log Q = (-1)^m \frac{1}{m!}\frac{1}{2i\pi}\int_C \log\lambda\, \lambda^m \partial_\lambda^m (Q-\lambda I)^{-1} d\lambda.$$

From (10) it follows that setting $d = k = 0$ yields:

$$(\log Q)_{[0]} = \frac{1}{2i\pi}\int_C \log\lambda\, \partial_\lambda^m (Q_{[0]}-\lambda I)^{-m-1} d\lambda = \log\left(Q_{[0]}\right).$$

It follows that $\log Q = \log\left(Q_{[0]}\right) + (\log Q)_{[>0]}$.

To see that $\log Q - \log Q_{[0]} \in \mathcal{A}(B, C\ell(\mathcal{E}))$, recall that $\log Q = \partial_z Q^z|_{z=0}$ and that for each z the operator Q^z is represented with respect to local trivializations by a local polyhomogeneous symbol of mixed differential-form degree

$$\mathbf{q}^z(b,x,\xi) = \frac{i}{2\pi}\int_{C_0} \lambda_\theta^z\ \mathbf{q}[\lambda](b,x,\xi) \circ \left(\mathbf{w}(b,x,\xi) \circ \mathbf{q}[\lambda](b,x,\xi)\right)^k\, d\lambda \quad (13)$$

where $\circ$ is the vertical form-valued symbol product, C_0 is a finite closed key-hole contour enclosing the spectrum of the leading vertical symbol of $(Q-\lambda)^{-1}_{[0]} = (P-\lambda)^{-1}$, and where $\mathrm{Op}(\mathbf{q}[\lambda]) \sim (Q-\lambda)^{-1}$, $\mathrm{Op}(\mathbf{w}) \sim \mathsf{W} := Q - \mathsf{P}$.

The symbol $\mathbf{q}[\lambda] \circ (\mathbf{w} \circ \mathbf{q}[\lambda])^k$ in (13) is polyhomogeneous, with an asymptotic expansion into terms of decreasing homogeneity, each of mixed form degree, in the usual way. In general this is a complicated expression. Nevertheless, the log-type of $\log Q$ can be inferred just from the leading symbol (top homogeneity). This is a consequence of the following simple lemma.

Lemma 3.3. $\mathbf{q}^z|_{z=0}$ *is the identity vertical symbol* $\mathbf{I}$ *defined by*

$$\mathbf{I}_{[0]} = (I,0,0,\ldots) \quad \textit{and} \quad \mathbf{I}_{[p]} = \mathbf{o} := (0,0,0,\ldots)\,, \qquad p > 0\,,$$

where $\mathbf{I}_{[p]}$ *indicates the component of form degree* p*, and the sequence on the right-side are the homogeneous terms.*

Proof. This is immediate from (13) since $\mathbf{q}[\lambda] \circ (\mathbf{w} \circ \mathbf{q}[\lambda])^k$ is $O(\lambda^{-2})$ for $k > 0$ (general ΨDO s). When $k = 0$ then the integrand is $O(\lambda^{-1})$, and we have the usual situation of form degree zero operators. □

Let $\mathbf{q}^z_w(x,\xi)$ denote the (mixed-form degree) term in the asymptotic expansion of Q^z of homogeneity w, and let $\mathbf{q}^z_{w_{\max}}(x,\xi)$ be the leading symbol (with maximum homogeneity). Then

$$\mathbf{q}^z_{w_{\max}}(x,\xi) = \frac{i}{2\pi}\int_{C_0} \lambda_\theta^z \mathbf{g}(x,\xi,\lambda)\, d\lambda\,, \quad (14)$$

where

$$\mathbf{g}(x,\xi,\lambda) = \mathbf{q}_{-m}[\lambda](x,\xi)\,\mathbf{w}_{\nu}(x,\xi)\,\mathbf{q}_{-m}[\lambda](x,\xi)\ldots\mathbf{w}_{\nu}(x,\xi)\,\mathbf{q}_{-m}[\lambda](x,\xi)$$

is the ordinary form-valued matrix product (i.e. not a symbol product) of leading order symbols $\mathbf{b}_{-m}[\lambda](x,\xi)$ of $\mathsf{P} - \lambda I$ (P of order $m > 0$), and $\mathbf{w}_{\nu}$ the leading ΨDO-order symbol of W which has maximum homogeneity ν.

Each $\mathbf{q}_{-m}$ has the quasi-homogeneity property for $t > 0$

$$\mathbf{q}_{-m}[t^m\lambda](x,t\xi) = t^{-m}\,\mathbf{q}_{-m}[\lambda](x,\xi)$$

and so

$$\mathbf{g}(x,t\xi,t^m\lambda) = t^{-m(k+1)+k\nu}\,\mathbf{g}(x,\xi,\lambda)\ .$$

Hence, making the change of variable $\lambda = t^m\mu$ in (14) we have

$$\mathbf{q}^z_{w_{\max}}(x,t\xi) = t^{mz-mk+k\nu}\,\mathbf{f}^z_{w_{\max}}(x,\xi). \tag{17}$$

It follows that $\mathbf{q}^z$ has an expansion into terms of mixed form degree

$$\mathbf{q}^z(x,\xi) \sim \sum_{j\geq 0}\mathbf{q}^z_{mz-mk+k\nu-j}(x,\xi)\ .$$

From Lemma 3.3 we therefore have

$$\mathbf{q}^z_{mz-mk+k\nu-j}(x,\xi)|_{z=0} = \delta_{j,0}\,\mathbf{I} \tag{19}$$

(since from the lemma terms of positive form degree do not contribute).

The final conclusion for $\log\mathbf{q} = \partial_z\mathbf{q}^z_{z=0}$ now follows in the usual way by differentiating

$$\mathbf{q}^z_{mz-mk+k\nu-j}(x,\xi) = |\xi|^{mz-mk+k\nu-j}\,\mathbf{q}^z_{mz-mk+k\nu-j}\left(x,\frac{\xi}{|\xi|}\right)$$

with respect to z, then evaluating at $z = 0$ and using (19) to get $\log\mathbf{q} \sim \sum_{j\geq 0}\log\mathbf{q}_j$ with

$$\log\mathbf{q}(x,\xi) = m\log|\xi|\,\delta_{j,0}\mathbf{I} + |\xi|^{k(\nu-m)-j}\partial_z|_{z=0}\mathbf{q}^z\left(x,\frac{\xi}{|\xi|}\right)\ .$$

Note, since $\partial_z|_{z=0}\mathbf{q}^z(x,\xi/|\xi|)$ has order zero, that provided $\mathrm{ord}(\mathsf{P}) = m \geq \mathrm{ord}(\mathsf{W})$ (which we assumed) the second term is of order no larger than zero.

Remark 3.3. A formal argument based on the Campbell-Hausdorff formula provides some intuition why the second part of the lemma holds. We show here how the Campbell-Hausdorff formula for ordinary ΨDOs obtained by Okikiolu [10] formally extended to families of vertical

ΨDOs, yields that $(\log Q)_{[0]}$ is a *classical* vertical ΨDO. Indeed, the splitting $Q = Q_{[0]} + Q_{[>0]}$ yields

$$\log Q = \log\left(Q_{[0]} + Q_{[>0]}\right) = \log\left[Q_{[0]}\left(I + Q_{[0]}^{-1} Q_{[>0]}\right)\right]$$

$$\sim \log Q_{[0]} + \log(I + Q_0^{-1} Q_{[>0]}) + \sum_{k=2}^{\infty} C^{(k)}\left(\log Q_{[0]}, \log(I + Q_{[0]}^{-1} Q_{[>0]})\right)$$

where $C^{(k)}(M,N)$ stands for a linear combination of Lie polynomials of degree k in M and N given by:

$$C^{(k)}(M,N) = \sum_{j=1}^{\infty} c_j \sum_{\alpha_l+\beta_l>0,\ \sum_{j=1}^{l}\alpha_j+\beta_j+1=k} (\operatorname{ad} M)^{\alpha_1}(\operatorname{ad} N)^{\alpha_1}\cdots(\operatorname{ad} M)^{\alpha_1}(\operatorname{ad} N)^{\alpha_1} N$$

for some coefficients $c_j \in \mathbb{R}$ and where $(\operatorname{ad} M)M' := [M, M']$. Under assumption (8), the operator $Q_{[0]}^{-1} Q_{[>0]}$ has a vanishing form degree zero part and negative orders $(\beta_{[1]}, \cdots, \beta_{[dim B+1]})$. The logarithm therefore coincides with the logarithm on bounded operators and yields an asymptotic expansion:

$$\log(I + Q_{[0]}^{-1} Q_{[>0]}) \sim \sum_{k=1}^{\infty} \frac{(-1)^{k-1}}{k}\left(Q_{[0]}^{-1} Q_{[>0]}\right)^k,$$

which shows that $\log(I + Q_{[0]}^{-1} Q_{[>0]})$ is a classical vertical ΨDO. It follows that each of the $C^{(k)}(M,N)$'s is also a classical vertical ΨDO. Indeed, these are built up from iterated brackets of an ordinary (corresponding to the 0-form degree part) logarithmic ΨDO $\log Q_{[0]}$ and the classical vertical ΨDO $\log(I + Q_{[0]}^{-1} Q_{[>0]})$. The ordinary symbol analysis shows that such brackets are classical and hence that $\log Q - \log Q_{[0]} \in \mathcal{A}(B, C\ell(\mathcal{E}))$ as claimed in the lemma.

This argument therefore provides a heuristic but maybe more intuitive explanation of the lemma.

4. The Wodzicki residue and the canonical trace extended to geometric families

Both the Wodzicki residue and the cut-off integral defined on ordinary classical symbols extend to smooth families of vertical differential form valued classical symbols.

Let $Q \in \mathcal{A}(B, \Psi(\mathcal{E}))$ be a simple family of ΨDOs of non-integer order, so that in each form degree $\mathrm{ord}(Q_{[k]}) \in \mathbb{R}\backslash\mathbb{Z}$. Then working in local coordinates U_M on M where Q is represented by a smooth family of symbols $\sigma_Q(x,\xi)$, we find that the local matrix valued forms [a]

$$-\!\!\!\!\!\!\int_{T_x^*M} \sigma_Q(x,\xi)\, đ\xi$$

patch together to determine a global section of the bundle $\pi^*(\wedge T^*B) \otimes \mathrm{End}(\mathcal{E}) \otimes |\Lambda_\pi|$ over M; this is proved by an obvious fibrewise version of the usual existence proof of the Kontsevich Vishik canonical trace. Taking the fibrewise trace we consequently have an element

$$\mathrm{TR}_x(Q) := -\!\!\!\!\!\!\int_{T_x^*M} \mathrm{tr}_x(\sigma_Q(x,\xi))\, đ\xi \in C^\infty(M, \pi^*(\wedge T^*B) \otimes |\Lambda_\pi|)$$

which can then be integrated over the fibres to define the canonical trace for families of non-integer order ΨDOs, a differential form on the parameter manifold B, by

$$\mathrm{TR}(Q) := \int_{M/B} \mathrm{TR}_x(Q) \ \in \mathcal{A}(B).$$

In the case when each component of Q has order less than $-n$, then $\mathrm{TR}(Q) = \mathrm{Tr}\,(Q)$, the usual fibrewise trace.

In a similar way, for a simple family $Q \in \mathcal{A}(B, \Psi(\mathcal{E}))$ of ΨDOs of any real (or complex) order one has a residue trace density[b]

$$\mathrm{res}_x(Q) := -\!\!\!\!\!\!\int_{S_x^*M} \mathrm{tr}_x(\sigma_Q(x,\xi)_{-n})\, đ_S(\xi) \in C^\infty(M, \pi^*(\wedge T^*B) \otimes |\Lambda_\pi|)$$

where $\sigma_Q(x,\xi)_{-n} = \sum_{k=0}^{\dim B} (\sigma_Q(x,\xi)_{-n})_{[k]}$ is the homogeneous part of the local symbol of homogeneity $-n$. This can then be integrated over the fibres to define the residue trace for families of arbitrary order ΨDOs, once more defining a differential form on the parameter manifold B, by

$$\mathrm{res}(Q) := \int_{M/B} \mathrm{res}_x(Q)\, dx \ \in \mathcal{A}(B).$$

Notice that if all the components of Q have non-integer ΨDO order then $\mathrm{res}(Q)$ vanishes; as in the case of a single operator, the functionals TR and res are roughly complementary.

[a] $d\bar{\xi} := \frac{1}{(2\pi)^n}\, d\xi$ where $d\xi$ is the ordinary Lebesgue measure on $T_x^*M \simeq \mathbb{R}^n$

[b] In the following formula $đ_S\,\xi := \frac{1}{(2\pi)^n}\, d_S\,\xi$ where $d_S\,\xi$ is the canonical volume measure on the cotangent unit sphere S_x^*M.

As in the case of a single operator, res : $\mathcal{A}(B, \Psi(\mathcal{E})) \longrightarrow \mathcal{A}(B)$ defines a trace, vanishing on (graded) brackets $[Q_1, Q_2]$ of families ΨDOs , while TR vanishes provided $[Q_1, Q_2]$ has non-integer order components.

Let us now extend the Wodzicki residue to forms on B with values in ΨDOs of logarithmic type.

Let $A \in \mathcal{A}(B, \Psi(\mathcal{E}))$ be a family of *differential operators.* Since by Lemma 3.2 the operator valued form $\log_\theta Q - \log_\theta Q_{[0]}$ is classical and since $\mathrm{res}_x(A \log Q_{[0]})dx$ defines a global top degree form on M, as A is a family of differential operators, by the results of [12], so does

$$\mathrm{res}_x(A \log Q) = \mathrm{res}_x(A(\log Q - \log Q_{[0]})) + \mathrm{res}_x(A \log Q_{[0]}) .$$

Hence, in this case, we may define the differential form by

$$\mathrm{res}(A \log Q) = \int_{M/B} \mathrm{res}_x(A \log Q)\, dx \in \mathcal{A}(B).$$

The fact that we restrict to differential operators A ensures the independence of this extended residue on the choice of the metric on the fibre M_b since a change of metric brings in a vertical multiplication operator, which combined with the vertical differential operator A modifies the expression by another differential operator, for which the Wodzicki residue will vanish.

We comment that this is possibly taking place in a $\mathbb{Z}_2$-graded context, where the residue density is the super residue density and so forth.

5. Holomorphic families of ΨDO valued forms

We call a family $Q_z = \sum_{k=0}^{dimB} (Q_z)_{[k]} \in \mathcal{A}(B, C\ell(\mathcal{E}))$ parametrised by $z \in W \subset \mathbb{C}$ *holomorphic* if in each local trivialization of $\pi_*\mathcal{E}$ over a neighbourhood U_B of b, $\left((Q_z)_{|U_B}\right)_{[k]} = \omega_k \otimes Q_{k,z}$ for some $\omega_k \in \mathcal{A}(U_B)$ we have that $z \mapsto Q_{k,z} \in C\ell(M_b, \mathcal{E}_b)$ is holomorphic family of ΨDOs parametrised by W in the usual single operator sense (following Kontsevich and Vishik [7], Lesch [8], see also [12]). In particular, the corresponding symbols $\mathbf{q}_z$ then define a holomorphic family of symbols in the usual sense.

Definition 5.1. We call a holomorphic regularisation procedure a map $\mathcal{R}$ which to any $A \in \mathcal{A}((B, C\ell(\mathcal{E}))$ associates a holomorphic family $A_z \in \mathcal{A}(B, C\ell(\mathcal{E}))$ such that $A_0 = A$ and with order $\alpha(z)$ such that $\alpha'_{[k]}(0) \neq 0$ or any $k \in \{0, \cdots, \mathrm{dim}B + 1\}$. Similarly, one defines holomorphic regularisation procedures on the level of symbols in such a way that a regularisation procedure $\mathcal{R} : A \mapsto A_z$ induces one for the corresponding symbols.

Let us illustrate these definitions with two examples.

Example 5.1.

(1) For any holomorphic map H such that $H(0) = 1$, the map $\mathcal{R}^H : \mathbf{q} \mapsto H(z)\,|\xi|^{-z}\,\mathbf{q}$ defines a holomorphic regularisation procedure on local classical vertical symbols. For a certain choice of H it gives back dimensional regularisation.

(2) Given a family $Q \in \mathcal{A}(B, C\ell(\mathcal{E}))$ of differential form valued vertical classical invertible elliptic ΨDOs with spectral cut θ such that Q moreover satisfies assumption (8), the map

$$\mathcal{R}^Q : A \mapsto A\,Q_\theta^{-z}$$

is a holomorphic regularisation procedure on vertical classical ΨDOs called ζ-regularisation.

Theorem 5.1.

(1). For any family $z \mapsto \mathbf{q}_z := \sum_{k=0}^{\mathrm{dimB}+1} (\mathbf{q}_z)_{[k]}(b, x, \xi)$ of classical symbols locally parametrised by $b \in B$ and holomorphic on an open subset $W \subset \mathbb{C}$ with order $z \mapsto \alpha(z) = (\alpha_{[0]}(z), \cdots, \alpha_{[\mathrm{dimB}+1]}(z))$ such that $z \mapsto \left(\alpha_{[k]}(b)\right)'(z)$ does not vanish for any k, then the functions $-\!\!\!\!\!\!\int_{T_x^* M_b} (\mathbf{q}_z)_{[k]}(b, x, \xi) d\xi$ are meromorphic with simple poles in $\left(\alpha_{[k]}(b)\right)^{-1}(\mathbb{Z}) \cap W$. The pole of the map $z \mapsto -\!\!\!\!\!\!\int_{T_x^* M_b} (\mathbf{q}_z)_{[k]}(b, x, \xi) d\xi$ at a point z_0 in $\left(\alpha_{[k]}(b)\right)^{-1}(\mathbb{Z}) \cap W$ is expressed in terms of a Wodzicki residue:

$$\mathrm{Res}_{z=z_0} -\!\!\!\!\!\!\int_{T_x^* M_b} (\mathbf{q}_z)_{[k]}(b, x, \xi) d\xi = -\frac{1}{\left(\alpha_{[k]}(b)\right)'(z_0)} \mathrm{res}\left(\left(\mathbf{q}_{z_0}\right)_{[k]}(b)\right). \quad (20)$$

(2). As a consequence, given a holomorphic family $z \mapsto Q_z := \sum_{k=0}^{\mathrm{dimB}+1} (Q_z)_{[k]}$ at point b on $W \subset \mathbb{C}$ of differential form valued vertical classical ΨDOs with holomorphic order $z \mapsto \alpha(b) = (\alpha_{[0]}(b)(z), \cdots, \alpha_{[\mathrm{dimB}]}(b)(z))$ such that $z \mapsto \left(\alpha_{[k]}(b)\right)'(z)$ does not vanish for any $k \in \{0, \cdots, \mathrm{dim}B\}$, the map $z \mapsto \mathrm{TR}\left((Q_z(b))_{[k]}\right)$ is meromorphic with simple poles in $\left(\alpha_{[k]}(b)\right)^{-1}(\mathbb{Z}) \cap W$. The pole of $\mathrm{TR}\left((Q_z(b))_{[k]}\right)$ at a point z_0 in $\left(\alpha_{[k]}(b)\right)^{-1}(\mathbb{Z})$ is expressed in terms of the Wodzicki residue of $\left((Q_{z_0})_{[k]}\right)$ at point $b \in B$:

$$\mathrm{Res}_{z=z_0} \mathrm{TR}\left((Q_z(b))_{[k]}\right) = -\frac{1}{\left(\alpha_{[k]}(b)\right)'(z_0)} \mathrm{res}\left((Q_{z_0}(b))_{[k]}\right). \quad (21)$$

Proof. The similar result for ordinary classical ΨDOs [7] applied to each $(q_z(b))_{[k]}$ and each $(Q_z(b))_{[k]}$ yields the result. □

On the grounds of this theorem, the holomorphic regularisation $\mathcal{R}^Q : A \mapsto A\, Q_\theta^{-z}$ on differential form valued vertical classical ΨDOs gives rise to meromorphic maps

$$z \mapsto \zeta_\theta(A, Q, z) := \mathrm{TR}\left(A\, Q_\theta^{-z}\right)$$

with simple poles so that it makes sense to extract the finite part at $z = 0$ denoted by $\zeta(A,\, Q, 0)|^{mer}$. As a consequence of the above theorem, we have as in the case of ordinary ΨDOs, the following formula relating the Wodzicki residue with the complex residue at $z = 0$

$$\mathrm{res}(A) = q_0\, \mathrm{Res}_{z=0}\mathrm{TR}\left(A\, Q^{-z}\right) = q_0\, \mathrm{Res}_{z=0}\mathrm{TR}\left(A\, Q_{[0]}^{-z}\right)$$

since $A\, Q_\theta^{-z}$ has order $\alpha_{[k]}(z) = -q_0\, z + \alpha_{[k]}(0)$ with q_0 the order of $Q_{[0]}$. Indeed, notice this formula is independent of the choice of Q apart from $q_0 = \mathrm{ord}(Q_{[0]})$.

Definition 5.2. Let $Q \in \mathcal{A}(B, C\ell(\mathcal{E}))$ be a family of differential form valued vertical classical invertible elliptic ΨDOs with spectral cut θ such that Q moreover satisfies assumption (8). Provided the dimension of the kernel $\ker(Q(b)_{[0]})$ is independent of b, the map $b \mapsto \left(\Pi_{Q(b)_{[0]}}\right)$ built from the orthogonal projection onto this kernel is smooth and for any $A \in \mathcal{A}(B, C\ell(\mathcal{E}))$,

$$\begin{aligned} &\mathrm{tr}^Q(A)_{[k]} \\ &:= \lim_{z \to 0}\left(\mathrm{TR}\left(A\, \left(Q + \pi_{Q_{[0]}}\right)_\theta^{-z}\right)_{[k]} - \frac{1}{z}\mathrm{Res}_{z=0}\mathrm{TR}\left(A\, \left(Q + \pi_{Q_{[0]}}\right)_\theta^{-z}\right)_{[k]}\right) \\ &:= \zeta(A,\, Q, 0)|^{mer}_{[k]} + \mathrm{tr}\left(A_{[k]}\, \pi_{Q_{[0]}}\right), \end{aligned}$$

defines a differential form $\mathrm{tr}^Q(A)$ on B called the Q-weighted trace of A.

Let us compare these weighted traces to the finite part of heat-operator regularised traces.
When $Q_{[0]}$ has non negative leading symbol the operator $A\, e^{-\epsilon Q}$ is trace-class for positive ϵ and we can write (these formulae are similar to the ones used by Higson [6] to derive the local formula for the Chern character in a

non commutative geometric setup):

$$\begin{aligned}
&\operatorname{tr}\left(A\, e^{-\epsilon Q}\right)\\
&= \sum_{n\geq 0}(-\epsilon)^n \int_{\Delta_n} du \operatorname{tr}\left(A\; e^{-u_0\epsilon Q_{[0]}}\, Q_{[>0]} \cdots e^{-u_{n-1}\epsilon Q_{[0]}}\, Q_{[>0]}\, e^{-u_n\epsilon Q_{[0]}}\right)\\
&= \sum_{n\geq 0}\sum_{|k|\geq 0} \frac{c(k)\,\epsilon^{|k|+2n-1}}{(|k|+n-1)!} \operatorname{tr}\left(A\, Q_{[>0]}^{(k_1)} \cdots Q_{[>0]}^{(k_n)}\, e^{-\epsilon Q_{[0]}}\right),
\end{aligned}$$

where $c(k)$ is defined by induction for any multiindex $k = (k_1, \cdots, k_n)$ by $c(k_1) = 1$ and

$$\begin{aligned}
c(k_1, \cdots, k_n) &= c(k_1, \cdots, k_{n-1})\cdot\\
&\quad \cdot \frac{(k_1 + \cdots k_{n-1} + 1)\cdots(k_1 + \cdots k_{n-1} + n - 1)}{k_n!}.
\end{aligned}$$

For an operator B, the operator $B^{(i)}$ is also defined by induction; $B^{(0)} := B$ and for any non negative integer i, $B^{(i+1)} := [Q_{[0]}, B^{(i)}]$ so that $B^{(i)} = \left(\operatorname{ad}^i Q_{[0]}\right)(B)$.
The sum over n is finite for each fixed form degree d whereas the sum over $k = (k_1, \cdots, k_n) \in \mathbf{N}^n$ is a priori infinite. However, if $Q_{[0]}$ is assumed to have *scalar leading symbol*, then $B^{(k)}$ has order $b + k(q_0 - 1)$ where b is the order of B and q_0 the order of $Q_{[0]}$ and $\left(A\, Q_{[>0]}^{(k_1)} \cdots Q_{[>0]}^{(k_n)}\right)_{[d]}$ has order $a + nq_d + |k|(q_0 - 1)$ where q_d is the order of $Q_{[d]}$. It follows that for each fixed multiindex k, there are coefficients α_{j_k}, $j_k \geq 0$ and β_k such that

$$\begin{aligned}
&\epsilon^{|k|+2n-1} \operatorname{tr}\left(A\, Q_{[>0]}^{(k_1)} \cdots Q_{[>0]}^{(k_n)}\, e^{-\epsilon Q_{[0]}}\right)\\
&\sim_{\epsilon\to 0} \sum_{j_k=0}^{\infty} \alpha_{j_k}\, \epsilon^{\frac{q_0|k|+q_0(2n-1)+j_k-(a+nq_d+|k|\,(q_0-1))-\dim M_b}{q_0}} + \beta_k \log\epsilon\\
&\sim_{\epsilon\to 0} \sum_{j_k=0}^{\infty} \alpha_{j_k}\, \epsilon^{\frac{q_0(2n-1)+j_k-a-nq_d+|k|-\dim M_b}{q_0}} + \beta_k \log\epsilon
\end{aligned}$$

so that the fractional powers of ϵ increase with $|k|$; in the $\epsilon \to 0$ limit, they will not contribute for large enough $|k|$. Extracting a finite part when $\epsilon \to 0$, we can therefore define for any non negative integer d:

$$\begin{aligned}
&\operatorname{fp}_{\epsilon=0}\operatorname{tr}\left(A\, e^{-\epsilon Q}\right)_{[d]}\\
&= \sum_{n\geq 0} \operatorname{fp}_{\epsilon=0}\left[\sum_{|k|\geq 0} \frac{c(k)\,\epsilon^{|k|+2n-1}}{(|k|+n-1)!} \operatorname{tr}\left(A\, Q_{[>0]}^{(k_1)} \cdots Q_{[>0]}^{(k_n)}\, e^{-\epsilon Q_{[0]}}\right)_{[d]}\right].
\end{aligned}$$

Since for ordinary ΨDOs A, Q we have (a folklore result, the proof of which can be found e.g. in a survey by Paycha [11])

$$\mathrm{fp}_{\epsilon=0}\mathrm{tr}(Ae^{-\epsilon Q}) = \mathrm{tr}^Q(A) + \gamma\,\mathrm{res}(A)$$

where γ is the Euler constant, weighted traces coincide with heat-kernel regularised traces for operators with vanishing residue, so this holds in particular for differential operators.
Applying this to each operator $\left(A\,Q_{[>0]}^{(k_1)}\cdots Q_{[>0]}^{(k_n)}\right)_{[d]}$ we get that provided $Q_{[d]}$ and $A_{[d]}$ are differential operators for any non negative integer d, then:

$$\mathrm{fp}_{\epsilon=0}\mathrm{tr}\left(A\,e^{-\epsilon Q}\right) = \mathrm{tr}^Q(A). \tag{22}$$

6. Weighted traces of differential operator valued forms; locality

A connection ∇ on $\mathcal{E}\otimes|\Lambda M_b|^{\frac{1}{2}}$ induces a connection ∇^{Hom} on $C\ell(\mathcal{E})$ which locally reads $\nabla^{\mathrm{Hom}} = d + [\Theta,\cdot]$ if ∇ reads $\nabla = d+\Theta$. Applying Theorem 5.1 to the holomorphic family $Q_z = A\,[\nabla,(Q_\theta+\pi_{Q_{[0]}})^{-z}]$ where $A\in\mathcal{A}(B, C\ell(\mathcal{E}))$ yields:

Theorem 6.1. *Let $Q\in\mathcal{A}(B, C\ell(\mathcal{E}))$ be a differential form on B with values in vertical classical elliptic ΨDOs with spectral cut θ and with kernel* $\mathrm{Ker}\,Q(b)_{[0]}$ *independent of b. Let $A\in\mathcal{A}(B, C\ell(\mathcal{E}))$. Given a connection ∇ on $\mathcal{E}\otimes|\Lambda M_b|^{\frac{1}{2}}$ then we have the equality of forms:*

$$d\,\mathrm{tr}^Q(A) = \mathrm{tr}^Q([\nabla, A]) + \frac{(-1)^{a+1}}{q_0}\mathrm{res}\left(A\,[\nabla, \log_\theta(Q+\pi_{Q_{[0]}})]\right)$$

where q_0 is the order of $Q_{[0]}$ and where a is the degree of A as a form.

Proof. For simplicity we assume Q is invertible, but the proof extends to the non invertible case replacing Q by $Q+\pi_{Q_{[0]}}$ in the complex powers. The proof goes as in [4] where Q was a ΨDO valued 0-form; indeed we have

$$\begin{aligned}
d\,\mathrm{tr}^Q(A) - \mathrm{tr}^Q([\nabla, A]) &= \mathrm{fp}_{z=0}\left(d\mathrm{TR}(AQ_\theta^{-z}) - \mathrm{TR}([\nabla, A]\,Q_\theta^{-z})\right)\\
&= (-1)^a\,\mathrm{fp}_{z=0}\mathrm{TR}\left(A\,[\nabla, Q_\theta^{-z}]\right)\\
&= (-1)^a\,\mathrm{Res}_{z=0}\left(\mathrm{TR}\left(\frac{A\,[\nabla, Q_\theta^{-z}]}{z}\right)\right)\\
&= \frac{(-1)^{a+1}}{q_0}\mathrm{res}\left(A\,[\nabla, \log_\theta Q]\right)
\end{aligned}$$

where we have applied Theorem 5.1 to the holomorphic family $Q_z = A\,[\nabla, Q_\theta^{-z}]$ to get the last identity using the fact that the degree k part of Q_θ^{-z} has order $-q_0\, z + cst$. □

Applying Theorem 5.1 to the holomorphic family $Q_z = A\,(Q_\theta + \pi_{Q_{[0]}})^{-z}$ where $A \in \mathcal{A}(B, C\ell(\mathcal{E}))$ is such that $A_{[i]}$ is a differential operator for any non negative integer i leads to the a description of the weighted trace of a differential operator valued differential form in terms of a Wodzicki residue. In order to make these notes self-contained, we include the full proof for ΨDO valued forms although it mimics the proof derived in [12] in the case of ordinary ΨDOs. As in [12] we use the following preliminary lemma.

Lemma 6.1. *Let $A \in \mathcal{A}(B, C\ell(\mathcal{E}))$ be a family of vertical ΨDOs such that $A_{[i]}(b)$ is a differential operator on M_b at any point $b \in B$ and for any non negative integer i. Then, for any $x \in M_b$, for any positive real number α*

$$z \mapsto -\!\!\!\!\!\!\int_{T_x^* M_b} |\xi|^{-\alpha\, z} \sigma_A(x, \xi)\, đ\xi$$

is meromorphic with simple poles and if $\mathrm{fp}_{z=0}$ *denotes its finite part at $z = 0$ we have:*

$$0 = -\!\!\!\!\!\!\int_{T_x^* M_b} \sigma_A(b, x, \xi)\, đ\xi = \mathrm{fp}_{z=0} -\!\!\!\!\!\!\int_{T_x^* M_b} |\xi|^{-\alpha\, z} \sigma_A(b, x, \xi)\, đ\xi.$$

Proof. The fact that $x \mapsto -\!\!\!\!\!\int_{T_x^* M_b} |\xi|^{-\alpha z} \sigma_A(b, x, \xi)\, đ\xi$ defines a meromorphic function with simple poles follows from Theorem 5.1 applied to $\sigma_z(b, x, \xi) = |\xi|^{-\alpha\, z} \sigma_A(b, x, \xi)$ of order $\alpha(z) = -\alpha\, z + \alpha$ where α is the order of A. let us fix a non negative integer i. The symbol of the differential operator $A_{[i]}$ reads $\sigma_{A_{[i]}}(b, x, \xi) = \sum_{k=0}^{\mathrm{ord} A_{[i]}} \sigma_k(b, x, \xi)$ where for any multiindex $k = (k_1, \cdots, k_{\mathrm{dim} M_b})$, $\sigma_k(b, x, \xi) = a(b, x)\xi^k$ is positively homogeneous. Hence, its cut-off integral on the cotangent space at $x \in M_b$ reads (here $B_{b,x}^*(0, R)$

is the ball of radius R centered at 0 in $T_x^*M_b$):

$$
\begin{aligned}
\int\!\!\!\!\!\!-_{T_x^*M_b} \sigma_{A_{[i]}}(b,x,\xi)\, đ\xi &= \mathrm{fp}_{R\to\infty} \int_{B_{b,x}^*(0,R)} \sigma_{A_{[i]}}(b,x,\xi)\, đ\xi \\
&= \sum_{k=0}^{\mathrm{ord}A_{[i]}} a_k(b,x)\, \mathrm{fp}_{R\to\infty} \int_{B_{b,x}^*(0,R)} \xi^k\, đ\xi \\
&= \sum_{k=0}^{\mathrm{ord}A_{[i]}} a_k(b,x)\, \mathrm{fp}_{R\to\infty} \left(\int_0^R r^{k+n-1} dr \right) \int_{S_x^*M_b} \xi^k d\xi \\
&= \mathrm{fp}_{R\to\infty} \frac{R^{k+n}}{k+n} \int_{S_x^*M_b} \xi^k\, đ\xi = 0.
\end{aligned}
$$

Similarly,

$$
\begin{aligned}
&\mathrm{fp}_{z=0} \int\!\!\!\!\!\!-_{T_x^*M_b} \sigma_{A_{[i]}}(b,x,\xi)|\xi|^{-z}\, đ\xi \\
&= \sum_{k=0}^{\mathrm{ord}A_{[i]}} a_k(b,x)\, \mathrm{fp}_{z=0} \int\!\!\!\!\!\!-_{T_x^*M_b} |\xi|^{-z}\xi^k\, đ\xi \\
&= \sum_{k=0}^{\mathrm{ord}A_{[i]}} a_k(b,x)\, \mathrm{fp}_{z=0} \left(\mathrm{fp}_{R\to\infty} \int_0^R r^{k+n-z-1} dr \right) \int_{S_x^*M_b} \xi^k\, đ\xi \\
&= \sum_{k=0}^{\mathrm{ord}A} a_k(b,x)\, \mathrm{fp}_{z=0} \left(\mathrm{fp}_{R\to\infty} R^{k+n-z}\right) \int_{S_x^*M_b} \xi^k\, đ\xi = 0.
\end{aligned}
$$

The fact that the finite part vanishes in the line before last follows from the fact that $\mathrm{fp}_{R\to\infty} R^{k+n-z}$ vanishes for $\mathrm{Re}(z)$ sufficiently small, as the finite part of a meromorphic extension of a function which vanishes on some half plane. □

We are now ready to prove the main result of this section:

Theorem 6.2. *Let $Q \in \mathcal{A}(B, C\ell(\mathcal{E}))$ be a differential form on B with values in vertical classical elliptic* ΨDOs *with spectral cut θ such that Q moreover satisfies assumption (8) and has kernel* $\mathrm{Ker}\, Q(b)_{[0]}$ *with constant dimension. Let $A \in \mathcal{A}(B, C\ell(\mathcal{E}))$ such that $A_{[i]}$ is a differential operator for any non negative integer i then we have the equality of forms:*

$$
\mathrm{tr}^Q(A) = -\frac{1}{q_0} \mathrm{res}\left(A \log_\theta(Q + \pi_{Q_{[0]}})\right)
$$

where q_0 *is the order of* $Q_{[0]}$ *and* $\pi_{Q_{[0]}}$ *the orthogonal projection onto the* Ker $Q_{[0]}$.

Proof. Here again, we prove the result for invertible Q; the proof then extends to the non invertible case replacing Q by $Q+\pi_{Q_{[0]}}$ in the complex powers. Since $\left(Q_\theta^{-z}\right)_{[0]} = (Q_\theta)_{[0]}^{-z}$ has order $-q_0\, z$ with q_0 the order of $Q_{[0]}$, dropping the subscript θ to simplify notations, we write for any $b \in B$

$$\begin{aligned}
&\mathrm{tr}^Q(A)(b) := \mathrm{fp}_{z=0}\int_{M_b} dx \;{-\!\!\!\!\!\!\int}_{T_x^* M_b} \mathrm{tr}_x\left(\sigma_{A\,Q^{-z}}(b,x,\xi)\right)\\
&= \mathrm{fp}_{z=0}\int_{M_b} dx \;{-\!\!\!\!\!\!\int}_{T_x^* M_b} đ\xi\, \mathrm{tr}_x\left(\sigma_{A\,Q^{-z}}(b,x,\xi) - |\xi|^{-q_0 z}\sigma_A(b,x,\xi)\right) +\\
&+ \mathrm{fp}_{z=0}\int_{M_b} dx \;{-\!\!\!\!\!\!\int}_{T_x^* M_b} đ\xi\, |\xi|^{-q_0\, z}\mathrm{tr}_x\sigma_A(b,x,\xi)\\
&= \mathrm{fp}_{z=0}\int_{M_b} dx \;{-\!\!\!\!\!\!\int}_{T_x^* M_b} đ\xi\, \mathrm{tr}_x\left(\sigma_{A\,Q^{-z}}(b,x,\xi) - |\xi|^{-q_0\, z}\sigma_A(b,x,\xi)\right)\\
&\qquad\text{by Lemma 6.1}\\
&= \mathrm{Res}_{z=0}\int_{M_b} dx \;{-\!\!\!\!\!\!\int}_{T_x^* M_b} đ\xi\, \frac{\mathrm{tr}_x\left(\sigma_{A\,Q^{-z}}(b,x,\xi)\right) - |\xi|^{-q_0\, z}\mathrm{tr}_x\left(\sigma_A(b,x,\xi)\right)}{z}.
\end{aligned}$$

Applying Theorem 5.1 to

$$\sigma_z(b,x,\xi) := \frac{\mathrm{tr}_x\sigma_{A\,Q^{-z}}(b,x,\xi) - |\xi|^{-q_0\, z}\mathrm{tr}_x\sigma_A(b,x,\xi)}{z}$$

then yields for any $d \in \{1,\cdots,\dim B\}$

$$\begin{aligned}
&\mathrm{tr}^Q(A)(b)_{[k]}\\
&= \mathrm{Res}_{z=0}\int_{M_b} dx \;{-\!\!\!\!\!\!\int}_{T_x^* M_b} đ\xi\; \mathrm{tr}_x\sigma_z(b,x,\xi)_{[k]} = -\frac{1}{\left(\alpha_{[k]}(b)\right)'(0)}\\
&\left[\int_{M_b} dx \int_{S_x^* M_b} đ\xi \left[\frac{\mathrm{tr}_x\sigma_{A\,Q^{-z}}(b,x,\xi) - |\xi|^{-q_0\, z}\mathrm{tr}_x\sigma_A(b,x,\xi)}{z}\right]_{|z=0}\right]_{[k]}
\end{aligned}$$

$$= \frac{1}{q_0}\left[\int_{M_b} dx \int_{S_x^* M_b} đ\xi \right.$$

$$\left. \frac{d}{dz}\left[\mathrm{tr}_x \sigma_{A\,Q^{-z}}(b,x,\xi) - |\xi|^{-q_0 z}\mathrm{tr}_x\sigma_{A\,Q^{-z}}(b,x,\xi)\right]_{|z=0}\right]_{[k]}$$

$$\text{since} \quad \alpha_{[k]}(z) = \mathrm{ord}(\sigma_z)_{[k]} = -q_0\, z + \alpha_{[k]}(0)$$

$$\text{and} \quad \text{since} \quad \left[\mathrm{tr}_x\sigma_{A\,Q^{-z}} - |\xi|^{-q_0\,z}\mathrm{tr}_x\sigma_A\right]_{|z=0} = 0$$

$$= -\frac{1}{q_0}\int_{M_b}\int_{S_x^* M_b} đ\xi\left[\mathrm{tr}_x\sigma_{\log Q\, A}(b,x,\xi) - q_0 \log|\xi|\mathrm{tr}_x\sigma_A(b,x,\xi)\right]_{[k]}$$

$$= -\frac{1}{q_0}\left[\mathrm{res}\,(A\,\log Q)\,(b)\right]_{[k]}.$$

□

7. Chern-Weil forms associated with a superconnection

Definition 7.1. A super connection (introduced by Quillen [16], see also [1], [3]) on $\pi_*\mathcal{E}$ adapted to a smooth family of formally self-adjoint elliptic ΨDOs $P \in \mathcal{A}^0(B, C\ell^q(\mathcal{E}))$ with odd parity is a classical ΨDO $\mathbb{A}$ on $\mathcal{A}(B,\pi_*\mathcal{E})$ of odd parity with respect to the $\mathbb{Z}_2$-grading such that:

$$\mathbb{A}(\omega\cdot\sigma) = d\omega\wedge\sigma + (-1)^{|\omega|}\omega\wedge\mathbb{A}(\sigma) \quad \forall\omega\in\mathcal{A}(B), \sigma\in\mathcal{A}(B,\pi_*\mathcal{E})$$

and

$$\mathbb{A}_{[0]} := P$$

where as before, $\mathbb{A} = \sum_{i=0}^{\dim B}\mathbb{A}_{[i]}$ and $\mathbb{A}_{[i]} : \mathcal{A}^*(B,\pi_*\mathcal{E}) \mapsto \mathcal{A}^{*+i}(B,\pi_*\mathcal{E})$.

The curvature of a super connection $\mathbb{A}$ is given by $\mathbb{A}^2 \in \mathcal{A}(B, C\ell(\mathcal{E}))$. Notice that $\mathbb{A}^2_{[0]} = P^2$ so that $\mathbb{A}^2$ is elliptic with spectral cut π .We know from the previous paragraphs that provided $\mathrm{Ker}\,\mathbb{A}^2(b)_{[0]} = \mathrm{Ker}\,P(b)$ is independent of b:

$$\zeta(\mathbb{A}^{2k}, \mathbb{A}^2 + \pi_P, z) := \mathrm{TR}\left(\mathbb{A}^{2k}\left(\mathbb{A}^2+\pi_P\right)^{-z}\right)$$

$-\pi_P$ will denote the orthogonal projection onto the kernel of P–, is a ΨDO valued form in $\mathcal{A}(B, C\ell(\mathcal{E}))$ so that we can define its finite part:

$$\mathrm{tr}^{\mathbb{A}^2}(\mathbb{A}^{2k}) := \zeta(\mathbb{A}^{2k}, \mathbb{A}^2+\pi_P, 0)|^{mer}.$$

Theorem 7.1. *Let $\mathbb{A}$ be a super connection on $\pi_*\mathcal{E}$ adapted to a smooth family of formally self-adjoint elliptic ΨDOs $P \in \mathcal{A}^0(B, C\ell^p(\mathcal{E}))$ of odd parity which satisfies assumption (8). Let us further assume that the kernel* $\mathrm{Ker}\,\mathbb{A}^2(b)_{[0]} = \mathrm{Ker}\,P(b)$ *is independent of b.*
Then for any non negative integer k,

(1) *the associated Chern forms*

$$\mathrm{c}_k(\mathbb{A}) := \mathrm{tr}^{\mathbb{A}^2}(\mathbb{A}^{2k})$$

are closed forms on B *which are cohomologous in de Rham cohomology to* $\mathrm{tr}(\mathbb{A}^{2k}\,\pi_P)$.

(2) *The corresponding Chern-Weil classes are independent of the scaling of* $\mathbb{A}$ *with fixed kernel and we have the following transgression formula*

$$\partial_t \mathrm{c}_k(\mathbb{A}_t) = d\tau_k(\mathbb{A}_t)$$

where

$$\tau_k(\mathbb{A}_t) = k\,\mathrm{tr}^{\mathbb{A}_t^2}\left(\dot{\mathbb{A}}_t\,\mathbb{A}_t^{2(k-1)}\right) - \frac{1}{p}\,\mathrm{res}\left(\dot{\mathbb{A}}_t\,(\mathbb{A}_t^2+\pi_P)^{k-1}\right)$$

for any smooth one parameter family $\mathbb{A}_t$ *of superconnections associated with* P *of order* p. (3) *If* P *has scalar leading symbol and if* $\mathbb{A}(b)$ *is a differential operator at each point* $b \in B$ *then the Chern-Weil classes relate to the Chern character by*

$$\mathrm{fp}_{t=0}\left(\partial_t^k \mathrm{tr}\left(e^{-t\,\mathbb{A}^2}\right)\right) = (-1)^k c_k(\mathbb{A}).$$

(4) *If* $\mathbb{A}(b)$ *is a differential operator at each point* $b \in B$ *then the associated Chern forms have a local description in terms of the Wodzicki residue:*

$$c_k(\mathbb{A}) = -\frac{1}{2p}\mathrm{res}\left(\mathbb{A}^{2k}\,\log(\mathbb{A}^2+\pi_P)\right). \tag{23}$$

Moreover, τ_k *is also local and we have:*

$$\tau_k(\mathbb{A}_t) = -\frac{k}{2p}\,\mathrm{res}\left(\dot{\mathbb{A}}_t\,(\mathbb{A}_t^2+\pi_P)^{k-1}\,\log(\mathbb{A}_t^2+\pi_P)\right) - \frac{1}{p}\,\mathrm{res}\left(\dot{\mathbb{A}}_t\,(\mathbb{A}_t^2+\pi_P)^{k-1}\right).$$

Proof. (Ad 1) Theorem 6.1 applied to $A = \mathbb{A}^{2k}$ and $Q = \mathbb{A}^2$ with $q_0 = 2p$ yields the closedness. Indeed, using the fact that $\nabla = \mathbb{A}$ commutes with integer powers of $\mathbb{A}$ and with $e^{-\mathbb{A}^2}$, we have:

$$\begin{aligned} d\ \mathrm{tr}^{\mathbb{A}^2}(\mathbb{A}^{2k}) &= \mathrm{fp}_{\epsilon\to 0}\left(d\,\mathrm{tr}(\mathbb{A}^{2k}e^{-\epsilon\mathbb{A}^2})\right) \\ &= \mathrm{fp}_{\epsilon\to 0}\mathrm{tr}([\mathbb{A},\mathbb{A}^{2k}]\,e^{-\epsilon\mathbb{A}^2}) + \mathrm{fp}_{\epsilon\to 0}\mathrm{tr}(\mathbb{A}^{2k}\,[\mathbb{A},e^{-\epsilon\mathbb{A}^2}]) = 0, \end{aligned}$$

since $d \circ \mathrm{tr} = \mathrm{tr} \circ \mathbb{A}$.

Furthermore, from [14] we know that $\zeta(\mathbb{A}^2,-k) := \zeta(\mathbb{A}^{2k},\mathbb{A}^2,z)|_{z=0}^{mer}$ is exact, so that $\mathrm{tr}^{\mathbb{A}^2}(\mathbb{A}^{2k}) = \zeta(\mathbb{A}^{2k},\mathbb{A}^2,z)|_{z=0}^{mer} + \mathrm{tr}(\mathbb{A}^{2k}\,\pi_P)$ is cohomologous to $\mathrm{tr}(\mathbb{A}^{2k}\,\pi_P)$.

(Ad 2) Applying Theorem 6.1 to $\nabla = \partial_t$, $A = \mathbb{A}_t^{2k}$, $Q := \mathbb{A}_t^2$ with $\mathbb{A}_t$ a smooth family of superconnections parametrised by $\mathbb{R}$ associated with a family P_t with constant kernel and corresponding projection π_P we get

$$\partial_t c_k(\mathbb{A}_t) = \mathrm{tr}^{\mathbb{A}_t^2}\left(\partial_t \mathbb{A}_t^{2k}\right) - \frac{1}{2p}\mathrm{res}\left(\mathbb{A}_t^{2k}\,\partial_t\,\log(\mathbb{A}_t^2+\pi_P)\right)$$

$$= \sum_{i=1}^{k} \mathrm{tr}^{\mathbb{A}_t^2}\left(\mathbb{A}_t^{2(i-1)}\,[\mathbb{A}_t, \dot{\mathbb{A}}_t]\mathbb{A}_t^{2(k-i)}\right)$$

$$-\frac{1}{p}\int_0^1 \mathrm{res}\left(\mathbb{A}_t^{2k}\,(\mathbb{A}_t^2+\pi_P)^{-1-\lambda}\,[\mathbb{A}_t, \dot{\mathbb{A}}_t]\,(\mathbb{A}_t^2+\pi_P)^{\lambda}\right)d\lambda$$

$$= \sum_{i=1}^{k} \mathrm{tr}^{\mathbb{A}_t^2}\left(\mathbb{A}_t^{2(i-1)}\,[\mathbb{A}_t, \dot{\mathbb{A}}_t]\mathbb{A}_t^{2(k-i)}\right)$$

$$-\frac{1}{p}\int_0^1 \mathrm{res}\left((\mathbb{A}_t^2+\pi_P)^k\,(\mathbb{A}_t^2+\pi_P)^{-1-\lambda}\,[\mathbb{A}_t, \dot{\mathbb{A}}_t]\,(\mathbb{A}_t^2+\pi_P)^{\lambda}\right)d\lambda$$

$$= k\,\mathrm{tr}^{\mathbb{A}_t^2}\left(\left[\mathbb{A}_t, \dot{\mathbb{A}}_t\,\mathbb{A}_t^{2(k-1)}\right]\right)$$

$$-\frac{1}{p}\mathrm{res}\left(\left[\mathbb{A}_t, \dot{\mathbb{A}}_t\,(\mathbb{A}_t^2+\pi_P)^k\,(\mathbb{A}_t^2+\pi_P)^{-1}\right]\right)$$

$$= k\,d\,\mathrm{tr}^{\mathbb{A}_t^2}\left(\dot{\mathbb{A}}_t\,\mathbb{A}_t^{2(k-1)}\right) - \frac{1}{p}\,d\,\mathrm{res}\left(\dot{\mathbb{A}}_t\,(\mathbb{A}_t^2+\pi_P)^{k-1}\right)$$

where we have used the fact that $\partial_t\mathbb{A}_t^2 = [\mathbb{A}_t, \dot{\mathbb{A}}_t]$ as well as the cyclicity of the Wodzicki residue combined with the fact that it vanishes on finite rank operators (which we used to replace$\mathbb{A}_t^2$ by $\mathbb{A}_t^2+\pi_P$ in the third equality).
(Ad 3) First of all, since $\partial_t e^{-t\mathbb{A}^2} = -\int_0^t ds\, e^{-t-s\mathbb{A}^2}\mathbb{A}^2 e^{-s\,\mathbb{A}^2}$ (see e.g. formula (2.6) in [3]) and since the exponential $e^{-t\mathbb{A}^2}$ commutes with any power of $\mathbb{A}$ we have:

$$\partial_t^k \mathrm{ch}(t\mathbb{A}) = \partial_t^k \mathrm{tr}(e^{-t\mathbb{A}^2}) = (-1)^k \mathrm{tr}\left(\mathbb{A}^{2k}\,e^{-t\mathbb{A}^2}\right).$$

Since weighted traces coincide with the ordinary trace on trace-class operators and since the operator valued form $e^{-t\mathbb{A}^2}$ is trace-class for positive t as a consequence of the ellipticity of the self-adjoint operator P and we have

$$\mathrm{fp}_{t=0}\,\partial_t^k\,\mathrm{tr}\left(e^{-t\,\mathbb{A}^2}\right) = (-1)^k\,\mathrm{fp}_{t=0}\mathrm{tr}\left(\mathbb{A}^{2k}\,e^{-t\,\mathbb{A}^2}\right)$$
$$= (-1)^k \mathrm{tr}^{\mathbb{A}^2}\left(\mathbb{A}^{2k}\right) = (-1)^k c_k(\mathbb{A}).$$

Here we have used the fact that the leading symbol of P is scalar to make sense of the heat-kernel regularised trace $\mathrm{fp}_{t=0}\mathrm{tr}\left(\mathbb{A}^{2k}\,e^{-t\,\mathbb{A}^2}\right)$ and formula

(22) to identify the weighted trace with the heat-kernel regularised trace since, by assumption, all the operators involved are differential operators. (Ad 4) Applying Theorem 6.2 to $A = \mathbb{A}^{2k}$, $Q := \mathbb{A}^2$ then yields the local formula for Chern forms announced in the last part of the theorem. In that case, τ_k is also local and we have

$$\begin{aligned}\tau_k(\mathbb{A}_t) &= -\frac{k}{2p}\,\mathrm{res}\left(\dot{\mathbb{A}}_t\,\mathbb{A}_t^{2(k-1)}\,\log(\mathbb{A}_t^2+\pi_P)\right)\\ &\quad -\frac{1}{p}\,\mathrm{res}\left(\dot{\mathbb{A}}_t\,(\mathbb{A}_t^2+\pi_P)^{k-1}\right)\\ &= -\frac{k}{2p}\,\mathrm{res}\left(\dot{\mathbb{A}}_t\,(\mathbb{A}_t^2+\pi_p)^{k-1}\,\log(\mathbb{A}_t^2+\pi_P)\right)\\ &\qquad -\frac{1}{p}\,\mathrm{res}\left(\dot{\mathbb{A}}_t\,(\mathbb{A}_t^2+\pi_P)^{k-1}\right)\end{aligned}$$

using here again, the fact that the Wodzicki residue vanishes on finite rank operators in order to replace $\mathbb{A}_t^2$ by $\mathbb{A}_t^2+\pi_P$. □

References

1. J.-M. Bismut, *The Atiyah-Singer index theorem for families of Dirac operators: two equation proofs*, Invent. Math. **83**, 91-151 (1986).
2. J.-M. Bismut, D. Freed, *The analysis of elliptic families I and II*, Comm. Math. Phys. **106**, 159-176 and Comm. Math. Phys. **107**, 103-163 (1986).
3. N. Berline, E. Getzler, M. Vergne, *Heat Kernels and Dirac Operators.* Grundlehren der Mathematischen Wissenschaften **298**, Springer Verlag, Berlin 1992.
4. A. Cardona, C. Ducourtioux, J.-P. Magnot, S. Paycha, *Weighted traces on algebras of pseudo-differential operators and the geometry of loop groups*, Infin. Dimens. Anal. Quantum Probab. Relat. Top. **5** n.4, 503–540 (2002)
5. G. Grubb, *A resolvent approach to traces and zeta Laurent expansions*, AMS Contemp. Math. Proc. **366**, 67–93 (2005). See also arXiv: math.AP/0311081.
6. N. Higson, *The local index formula in noncommutative geometry*, Trieste Lecture Notes (2004); *The residue index theorem of Connes and Moscovici*, Clay Mathematics Proceedings (2004). (http://www.math.psu.edu/higson/ResearchPapers.html)
7. M. Kontsevich, S. Vishik, *Geometry of determinants of elliptic operators*, Funct. Anal. on the Eve of the XXI century, Vol I, Progress in Mathematics **131**, 173–197 (1994) ; *Determinants of elliptic pseudodifferential operators*, Max Planck Preprint (1994).
8. M. Lesch, *On the non commutative residue for pseudo-differential operators with log-polyhomogeneous symbols*, Ann. Global Anal. Geom. **17**, 151–187 (1998).
9. R. Melrose, V. Nistor, *Homology of pseudo-differential operators I. Manifolds with boundary*, arXiv: funct-an/960605.

10. K. Okikiolu, *The multiplicative anomaly for determinants of elliptic operators*, Duke Math. J. **79**, 723–750 (1995) ; K. Okikiolu, *The Campbell-Hausdorff theorem for elliptic operators and a related trace formula*, Duke Math. J. **79**, 687–722 (1995).
11. S. Paycha *Renormalised traces as a looking glass into infinite-dimensional geometry*, Infin. Dimens. Anal. Quantum Probab. Relat. Top. **4**, n.2 221–266 (2001).
12. S. Paycha, S. Scott, *An Laurent expansion for regularised integrals of holomorphic symbols*, Geom. Funct. Anal., to appear. arXiv: math.AP/0506211.
13. S. Scott, *The residue determinant*, Commun. Part. Diff. Eqns, **30**, 483–507 (2005).
14. S. Scott, *Zeta-Chern forms and the local family index theorem*, Trans. Amer. Math. Soc., to appear. arXiv: math.DG/0406294.
15. R.T. Seeley, *Complex powers of an elliptic operator, Singular integrals*, Proc. Symp. Pure Math., Chicago, Amer. Math. Soc., Providence 288–307 (1966).
16. D. Quillen, *Superconnections and the Chern character*, Topology **24**, 89–95 (1985).
17. M. Wodzicki, *Non commutative residue,* In: Lecture Notes in Math. **1283**, Springer Verlag 1987.

Received by the editors September 16, 2005; Revised December 23, 2005

Part III

Heat Kernel Calculations and Surgery

Analysis, Geometry and Topology
of Elliptic Operators, pp. 107–140

NON-LAPLACE TYPE OPERATORS ON MANIFOLDS WITH BOUNDARY

IVAN G. AVRAMIDI
Department of Mathematics
New Mexico Institute of Mining and Technology
Socorro, NM 87801, USA
iavramid@nmt.edu

Dedicated to Krzysztof P. Wojciechowski on his 50th birthday

We study second-order elliptic partial differential operators acting on sections of vector bundles over a compact manifold with boundary with a non-scalar positive definite leading symbol. Such operators, called non-Laplace type operators, appear, in particular, in gauge field theories, string theory as well as models of non-commutative gravity theories, when instead of a Riemannian metric there is a matrix valued self-adjoint symmetric two-tensor that plays the role of a "non-commutative" metric. It is well known that there is a small-time asymptotic expansion of the trace of the corresponding heat kernel in half-integer powers of time. We initiate the development of a systematic approach for the explicit calculation of these coefficients, construct the corresponding parametrix of the heat equation and compute explicitly the first two heat trace coefficients.

2000 *Mathematics Subject Classification.* Primary 58J50; Secondary 58J32

1. Introduction

Elliptic differential operators on manifolds play a very important role in mathematical physics, geometric analysis, differential geometry and quantum theory. Of special interest are the resolvents and the spectral functions of elliptic operators; the most important spectral functions being the trace of the heat kernel and the zeta function, which determine, in particular, the functional determinants of differential operators (see the books Gilkey [29], Booss-Bavnbek and Wojciechowski [19], Berline, Getzler and Vergne [18], Avramidi [5] and the reviews Avramidi [4, 6], Vassilevich [43]).

In particular, in quantum field theory and statistical physics the resolvent determines the Green functions, the correlation functions and the propagators of quantum fields, and the functional determinant determines the effective action and the partition function (see, for example [5]). In spectral geometry, one is interested, following Kac [36], in the question: "Does the spectrum of the scalar Laplacian determine the geometry of a manifold" or, more generally, "*To what extent does the spectrum of a differential operator on a manifold determine the geometry of the manifold?*" Of course, the answer to Kac's and other questions depends on the differential operator. Most of the studies in spectral geometry and spectral asymptotics are restricted to so-called Laplace type operators. These are second-order partial differential operators acting on sections of a vector bundle with a positive definite *scalar leading symbol*.

Since, in general, it is impossible to find the spectrum of a differential operator exactly, one studies the asymptotic properties of the spectrum, so-called *spectral asymptotics*, which are best described by the asymptotic expansion of the trace of the heat kernel. If $L : C^\infty(V) \to C^\infty(V)$ is a self-adjoint elliptic second-order partial differential operator with a positive definite leading symbol acting on smooth sections of a vector bundle V over a compact n-dimensional manifold M, then according to Greiner [33] and [29] there is a small-time asymptotic expansion as $t \to 0$

$$\mathrm{Tr}_{L^2} \exp(-tL) \sim (4\pi)^{-n/2} \sum_{k=0}^{\infty} t^{(k-n)/2} A_k\,. \tag{1.1}$$

The coefficients A_k are called the global *heat invariants* (in mathematical literature they are usually called the Minakshisundaram-Pleijel coefficients; in physics literature, they are also called HMDS (Hadamard-Minakshisundaram-De Witt-Seeley) coefficients, or Schwinger-De Witt coefficients).

The heat invariants are spectral invariants of the operator L that encode the information about the asymptotic properties of the spectrum. They are of great importance in spectral geometry and find extensive applications in physics, where they describe renormalization and quantum corrections to the effective action in quantum field theory and the thermal corrections to the high-temperature expansion in statistical physics among many other things. They describe real physical effects and, therefore, the knowledge of these coefficients in *explicit closed form* is important in physics. One would like to have formulas for some lower-order coefficients to be able to study those effects.

The proof of the existence of the asymptotic expansion (1.1) has been a great achievement in geometric analysis. Now it is a well known fact, at least in the smooth category for compact manifolds. This is not the subject of our interest. The main objective in the study of spectral asymptotics (in spectral geometry and quantum field theory) is, rather, the *explicit calculation* of the heat invariants A_k in *invariant geometric terms*.

The approach of Greiner [33] and Seeley [41] is a very powerful general analytical procedure for analyzing the structure of the asymptotic expansion based on the theory of pseudo-differential operators and the calculus of symbols of operators (we will call it *symbolic approach* for symplicity). This approach *can be used* for calculation of the heat invariants explicitly in terms of the jets of the symbol of the operator; it provides an iterative procedure for such a calculation. However, as far as we know, because of the technical complexity and, most importantly, lack of the manifest covariance, such analytical tools *have never been used* for the actual calculation of the explicit form of the heat invariants in an invariant geometric form. As a matter of fact, the symbolic method has only been used to prove the existence of the asymptotic expansion and the general structure of the heat invariants (like their dependence on the jets of the symbol of the operator) (see the reviews [6, 43] and Kirsten [38]). To the best of our knowledge there is no exlicit formula even for the low-order coefficients A_1 and A_2 for a general non-Laplace type operator.

The development of the analysis needed to discuss elliptic boundary value problems is beyond the scope of this paper. We shall simply use the well known results about the existence of the heat trace asymptotics of elliptic boundary value problems from the classical papers of Greiner [33] and Seeley [41] (see also the books of Grubb [35] and [19]). Our approach can be best described by Greiner's own words [33], pp. 165–166,: *"the asymptotic expansion can be obtained by more classical methods. Namely, one constructs the Taylor expansion for the classical parametrix [of the heat equation] . . . and iterates it to obtain the Green's operator. This yields, at least formally, the asymptotic expansion for [the trace of the heat kernel]"*. This is the approach exploited by McKean and Singer [39] for a Laplace type operator and it is this approach that we will use in the present paper for non-Laplace type operators. However, contrary to [33] and [41] we do not use any Riemannian metrics but, instead, work directly with densities, so that our final answers are automatically invariant. Greiner [33], pp. 166, also points out that *"Of course, at the moment it is not clear which representation will yield more easily to geometric interpretation."*

In spectral geometry as well as in physics the motivation and the goals of the study of spectral asymptotics are quite different from those in analysis. The analytic works are primarily interested in the existense and the type of the asymptotic expansion, but not necessarily in the explicit form of the coefficients of the expansion. In spectral geometry one is interested in the *explicit form* of the spectral invariants and their relation to geometry. One considers various special cases when some invariant topological and geometrical constraints are imposed, say, on the Riemannian structure (or on the connection of a vector bundle). Some of these conditions are: positive (negative, or zero) scalar curvature, or positive (negative) sectional curvature, Ricci-flat metrics, Einstein spaces, symmetric spaces, Kaehler manifolds etc. Such conditions lead to very specific consequences for the heat invariants which are obvious in the geometric invariant form but which are hidden in a non-invariant symbolic formula obtained in local coordinates. For example, if the scalar curvature is zero, then for the Laplacian on a manifold without boundary $A_2 = 0$. Such a conclusion cannot be reached until one realizes that the integrand of A_2 is precisely the scalar curvature. There are, of course, many more examples like this.

Another property that does not become manifest at all in the symbolic approach is the behavior of the heat invariants under the conformal transformation of the Riemannian structure and the gauge transformations. This is a very important property that is heavily used in the functorial approach of Branson and Gilkey [22] (see also Branson, Gilkey, Kirsten and Vassilevich [24]), but which is not used at all in the symbolic approach. For conformally covariant operators the symbolic calculus is exactly the same as for non-conformally covariant ones with similar results because the conformal covariance only concerns the low-order terms of the symbol but not its leading symbol. However, the conformal invariance leads to profound consequences for the heat invariants, zeta-function and the functional determinant (see Branson [20]).

The calculation of the explicit form of the heat invariants is a separate important and complicated problem that requires special calculational techniques. The systematic explicit calculation of heat kernel coefficients was initiated by Gilkey [28] (see also [29, 43, 38, 6] and references therein). A review of various algorithms for calculation of the heat kernel coefficients is presented in Avramidi and Schimming [17]. The two most effective methods that have been successfully used for the actual calculation of the heat invariants are: 1) the functorial method of Branson and Gilkey [22] (see also [24, 29]), which is based on the invariance theory, behavior of the heat trace under conformal transformations and some special case calculations, and 2) the method of local Taylor expansion in normal coordinates (which is essentially equivalent to the geometric covariant Taylor expansions of Avramidi [2, 1]). The results of both of these methods are directly obtained in an invariant geometric form. The symbolic calculus approach, despite being a powerful analytical tool, fails to provide such invariant results. It gives answers in local coordinates that are not invariant and cannot be made invariant directly. For high-order coefficients the problem of converting such results in a geometric invariant form is hopeless. One cannot even decide whether a particular coefficient is zero or not.

One of the main problems in the study of spectral asymptotics is to develop a procedure that respects all the invariance transformations (diffeomorphisms and gauge transformations in the physics language) of the differential operator. Symbolic calculus gives an answer in terms of jets of the symbol of the operator in some local coordinates. Thus there remains a very important problem of converting these local expressions to global geometric invariant structures, like polynomials in curvatures and their covariant derivatives. For a general coefficient this problem becomes unmanagable; it is simply exponentially bad in the order of the heat kernel coefficient. The number of the jets of the symbol is much greater than the number of invariant structures of given order. This problem is so bad that it is, in fact, much easier to compute the coefficients by some other methods that directly give an invariant answer than to use the results of the symbolic approach. To our knowledge, none of the results for the explicit form of the spectral invariants were obtained by using the symbolic calculus.

Every problem in geometric analysis has two aspects: an analytical aspect and a geometric aspect. In the study of spectral asymptotics of differential operators the analytic aspect has been succesfully solved in the classical works of Greiner [33] and Seeley [41] and others (see [35, 19]).

The geometric aspect of the problem for Laplace type operators is now also well understood due to the work of Gilkey [28] and many others (see [29, 43, 38, 18, 6] and references therein). The leading symbol of a Laplace type operator naturally defines a Riemannian metric on the manifold, which enables one to employ powerful methods of differential geometry. In other words, the Riemannian structure on a manifold is determined by a Laplace type operator. We take this fact seriously: geometry (Riemannian structure) is determined by analysis (differential operator). In some sense, analysis is primary and geometry is secondary. What kind of geometry is generated does, of course, depend on the differential operator. *A Laplace type differential operator generates the Riemannian geometry.*

As a result, much is known about the spectral asymptotics of Laplace type operators, both on manifolds without boundary and on manifolds with boundary, with various boundary conditions, such as Dirichlet, Neumann, Robin, mixed, oblique, Zaremba etc. On manifolds without boundary all odd coefficients vanish, $A_{2k+1} = 0$, and all even coefficients A_{2k} up to A_8 have been computed in our PhD thesis [1], which was published later as a book [5] (see also [29, 2, 15], Avramidi [10], Yajima, Higasida, Kawano, Kubota, Kamo and Tokuo [44], the reviews [4, 6, 43] and references therein). By using our method [2] Yajima et al. [44] computed the coefficient A_{10} recently. Of course, this remarkable progress can only be achieved by employing modern computer algorithms (the authors of [44] used a Mathematica package). The main reason for this progress is that the heat kernel coefficients are polynomial in the jets of the symbol of the operator (which can be expressed in terms of curvatures and their covariant derivatives); it is essentially an algebraic problem.

On manifolds with boundary, the heat invariants depend on the boundary conditions. For the classical boundary conditions, like Dirichlet, Neumann, Robin, and mixed combination thereof on vector bundles, the coefficients A_k have been explicitly computed up to A_5 (see, for example, Kirsten [37], Avramidi [3] and [22, 24]).

A more general scheme, called oblique boundary value problem (see Grubb [34] and Gilkey and Smith [32, 31]), which includes tangential derivatives along the boundary, was studied by Avramidi and Esposito [14, 16, 15] and Dowker and Kirsten [26, 27]. This problem is not automatically elliptic like the classical boundary problems; there is a certain condition on the leading symbol of the boundary operator that ensures the strong ellipticity of the problem. As a result, the heat invariants are no longer polynomial in the jets of the boundary operator, which makes this problem much more difficult to handle. So far, in the general case only the coefficient A_1 is known [14, 16]. In a particular Abelian case the coefficients A_2 and A_3 have been computed in [27].

A discontinuous boundary value problem, the so-called Zaremba problem, which includes Dirichlet boundary conditions on one part of the boundary and Neumann boundary conditions on another part of the boundary, was studied recently by Avramidi [10], Seeley [42] and Dowker, Gilkey and Kirsten [25]. Because this problem is not smooth, the analysis becomes much more subtle (see [10, 42] and references therein). In particular, there is a singular subset of codimension 2 on which the boundary operator is discontinuous, and, one has to put an additional boundary condition that fixes the behavior at that set. Seeley [42] showed that there are no logarithmic terms in the asymptotic expansion of the trace of the heat kernel, which are possible on general grounds, and that the heat invariants do depend on the boundary condition at the singular set; the neglect of that simple fact lead to some controversy on the coefficient A_2 in the past until this question was finally settled in [42, 10].

Contrary to the Laplace type operators, *there are no systematic effective methods for an explicit calculation of the heat invariants for second-order operators which are not of Laplace type*. Such operators appear in so-called *matrix geometry* (see Avramidi [7, 8, 9, 11]), when instead of a single Riemannian metric there is a matrix-valued symmetric 2-tensor, which we call a "non-commutative metric". Matrix geometry is motivated by the relativistic interpretation of gauge theories and is intimately related to Finsler geometry (rather a collection of Finsler geometries) (see [7, 8, 9]). For an introduction to Finsler geometry see Rund [40].

Of course, the existence and the form of the asymptotic expansion of the heat kernel is well established for a very large class of operators, including all self-adjoint elliptic partial differential operators with positive definite leading symbol; it is essentially the same for all second-order operators, whether of Laplace type or not. However, a *non-Laplace type operator does not induce a unique Riemannian metric* on the manifold. Of course, one can pick any Riemannian metric and work with it, but this is not natural; it does not reflect the properties of the differential operator and its leading symbol. Therefore, it is useless to try to use a Riemannian structure to cast the heat invariants in an invariant form. Rather, a non-Laplace type operator defines a collection of Finsler geometries (a matrix geometry in the terminology of [7, 8, 9, 11]). Therefore, it is the *matrix geometry that should be used to study the geometric structure of the spectral invariants of non-Laplace type operators*. This fact complicates the calculation of spectral asymptotics significantly. Of course, the general classical algorithms described in [33, 41] still apply.

Three decades ago Greiner [33], p. 164, indicated that *"the problem of interpreting these coefficients geometrically remains open"*. There has not been much progress in this direction. In this sense, the study of geometric aspects of spectral asymptotics of non-Laplace type operators is just beginning and the corresponding methodology is still underdeveloped in comparison with the Laplace type theory. The only exception to this is the case of exterior p-forms, which is pretty simple and, therefore, is well understood now (see Gilkey, Branson and Fulling [30], Branson, Gilkey and Pierzchalski [23] and Branson [21]). Thus, the *geometric aspect of the spectral asymptotics of non-Laplace type operators remains an open problem.*

A first step in this direction was made by Avramidi and Branson in the papers [12, 13]. We studied a subclass of so-called natural non-Laplace type operators on Riemannian manifolds, which appear, for example, in the study of spin-tensor quantum gauge fields. The natural non-Laplace type operators are a special case of non-Laplace type operators whose leading symbol is built in a universal, polynomial way, using tensor product and contraction from the Riemannian metric, its inverse, together with (if applicable) the volume form and/or the fundamental tensor-spinor. These operators act on sections of spin-tensor bundles. These bundles may be characterized as those appearing as direct summands of iterated tensor products of the tangent, the cotangent and the spinor bundles (see sect 2.1). Alternatively, they may be described abstractly as bundles associated to representations of the spin group. These are extremely interesting and important bundles, as they describe the fields in quantum field theory. The connection on the spin-tensor bundles is built in a canonical way from the Levi-Civita connection. The symbols of natural operators are constructed from the jets of the Riemannian metric, the leading symbols being constructed just from the metric. In this case, even if the leading symbol is not scalar, its determinant is a polynomial in $|\xi|^2 = g^{\mu\nu}(x)\xi_\mu\xi_\nu$, and, therefore, its eigenvalues are functions of $|\xi|$ only. This allows one to use the Riemannian geometry and simplifies the study of such operators significantly.

For non-Laplace operators on manifolds without boundary even the invariant A_4 is not known, in general (for some partial results see [12, 13, 9, 11] and the review [6]). For natural non-Laplace type differential operators on manifolds without boundary the coefficients A_0 and A_2 were computed in [12]. For general non-Laplace type operators they were computed in our papers [9, 11].

The primary goal of the present work is to generalize this study to general non-Laplace type operators on manifolds with boundary. We introduce a "non-commutative" Dirac operator as a first-order elliptic partial differential operator such that its square is a second-order self-adjoint elliptic operator with positive definite leading symbol (not necessarily of Laplace type) and study the spectral asymptotics of these operators with Dirichlet boundary conditions.

This paper is organized as follows. In Section 2 we describe the construction of non-Laplace type operators. In Section 2.1 we define natural non-Laplace type operators in the context of Stein-Weiss operators [21]. In Section 2.2 we describe a class of non-Laplace type operators that appear in matrix geometry following [9, 11]; we develop what can be called the non-commutative exterior calculus and construct first-order and second-order invariant differential operators. In Section 2.3 we describe the general setup of the Dirichlet boundary value problem for such an operator and introduce necessary tools for the analysis of the ellipticity condition. In Section 3 we review the spectral asymptotics of elliptic operators both from the heat kernel and the resolvent point of view. In Section 4 we develop a formal technique for calculation of the heat kernel asymptotic expansion. In Section 4.1 the interior coefficients A_0 and A_2 are computed (which are the same as for the manifolds without boundary), and in the Section 4.2 we compute the boundary coefficient A_1.

2. Non-Laplace type operators

2.1. *Natural non-Laplace type operators*

Natural non-Laplace type operators can be constructed as follows [21]. Let M be a smooth compact orientable n-dimensional spin manifold (with or without boundary). Let $\mathcal{S}$ be the spinor bundle over a spin manifold M and

$$\mathcal{V} = TM \otimes \cdots \otimes TM \otimes T^*M \otimes \cdots \otimes T^*M \otimes \mathcal{S} \tag{2.1}$$

be a spin-tensor vector bundle corresponding to a representation of the spin group $\mathrm{Spin}(n)$ and $\nabla : C^\infty(\mathcal{V}) \to C^\infty(T^*M \otimes \mathcal{V})$ be a connection on $\mathcal{V}$. Then the decomposition

$$T^*M \otimes \mathcal{V} = \mathcal{W}_1 \oplus \cdots \oplus \mathcal{W}_s \tag{2.2}$$

of the bundle $T^*M \otimes \mathcal{V}$ into its irreducible components $\mathcal{W}_1, \ldots, \mathcal{W}_s$ defines the projections $P_j : T^*M \otimes \mathcal{V} \to \mathcal{W}_j$ and the first-order differential operators

$$G_j = P_j \nabla : C^\infty(\mathcal{V}) \to C^\infty(\mathcal{W}_j), \tag{2.3}$$

called Stein-Weiss operators (or simply the gradients). The number s of gradients is a representation-theoretic invariant of the bundle $\mathcal{V}$.

Then every first-order $\mathrm{Spin}(n)$-invariant differential operator

$$D : C^\infty(\mathcal{V}) \to C^\infty(\mathcal{V}) \tag{2.4}$$

is a direct sum of the gradients

$$D = c_1 G_1 + \cdots + c_s G_s = P\nabla\,, \tag{2.5}$$

where c_j are some real constants and

$$P = \sum_{j=1}^{s} c_j P_j\,, \tag{2.6}$$

and the second-order operators $L : C^\infty(\mathcal{V}) \to C^\infty(\mathcal{V})$ defined by

$$L = D^* D = \nabla^* P^2 \nabla = \sum_{j=1}^{s} c_j^2 G_j^* G_j \tag{2.7}$$

are natural non-Laplace type operators. If all $c_j \neq 0$, then L is elliptic and has a positive definite leading symbol.

2.2. *Non-commutative Laplacian and Dirac operator in matrix geometry*

Let M be a smooth compact orientable n-dimensional spin manifold with smooth boundary ∂M. We label the local coordinates x^μ on the manifold M by Greek indices which run over $1, \ldots, n$, and the local coordinates $\hat{x}^i$ on the boundary ∂M by Latin indices which run over $1, \ldots, n-1$. We use the standard coordinate bases for the tangent and the cotangent bundles. The components of tensors over M in the coordinaate basis will be labeled by Greek indices and the components of tensors over ∂M in the coordinate basis will be labeled by Latin indices. We also use the standard Einstein summation convention for repeated indices.

Let $\mathcal{S}$ be now an arbitrary N-dimensional complex vector bundle over M (non necessarily the spinor bundle) with a positive definite Hermitean inner product $\langle\ ,\ \rangle$, $\mathcal{S}^*$ be its dual bundle and $\mathrm{End}(\mathcal{S})$ be the bundle of linear endomorphisms of the vector bundle S. Further, let $\mathrm{Aut}(\mathcal{S})$ be the group of automorphisms of the vector bundle $\mathcal{S}$ and $G(\mathcal{S})$ be the group of unitary endomorphisms of the bundle $\mathcal{S}$. We will call the unitary endomorphisms of the bundle $\mathcal{S}$ simply gauge transformations.

Let TM and T^*M be the tangent and the contangent bundles. We introduce the following notation for the vector bundles of vector-valued and endomorphism-valued p-forms and p-vectors

$$\Lambda_p = (\wedge^p T^*M) \otimes \mathcal{S}\,, \qquad \Lambda^p = (\wedge^p TM) \otimes \mathcal{S}\,, \tag{2.8}$$

$$E_p = (\wedge^p T^*M) \otimes \mathrm{End}(\mathcal{S})\,, \qquad E^p = (\wedge^p TM) \otimes \mathrm{End}(\mathcal{S})\,. \tag{2.9}$$

We will also consider vector bundles of densities of different weights over the manifold M. For each bundle we indicate the weight explicitly in the notation of the vector bundle; for example, $\mathcal{S}[w]$ is a vector bundle of densities of weight w.

Since M is orientable there is the standard volume form vol $= dx = dx^1 \wedge \cdots \wedge dx^n$ given by the standard Lebesgue measure in a local chart. The volume form is, of course, a density of weight 1, and, hence, is a section of the bundle $E_n[1]$. The components of the volume form in a local coordinate basis are given by the completely anti-symmetric Levi-Civita symbol $\varepsilon_{\mu_1\dots\mu_n}$. The n-vector dual to the volume form is a density of weight (-1) and, hence, is a section of the bundle $E^n[-1]$. Its components are given by the contravariant Levi-Civita symbol $\varepsilon^{\mu_1\cdots\mu_n}$. These objects naturally define the maps

$$\varepsilon : \Lambda^p[w] \to \Lambda_{n-p}[w+1]\,, \qquad \tilde{\varepsilon} : \Lambda_p[w] \to \Lambda^{n-p}[w-1]\,. \tag{2.10}$$

It is not difficult to see that $\varepsilon\tilde{\varepsilon} = \tilde{\varepsilon}\varepsilon = (-1)^{p(n-p)}\mathrm{Id}\,.$

Further, we define the diffeomorphism-invariant L^2-inner product on the space $C^\infty\left(\Lambda_p\left[\frac{1}{2}\right]\right)$ of smooth endomorphism-valued p-form densities of weight $\frac{1}{2}$ by

$$(\psi, \varphi) = \int\limits_M dx\, \langle \psi(x), \varphi(x) \rangle\,. \tag{2.11}$$

The completion of $C^\infty\left(\Lambda_p\left[\frac{1}{2}\right]\right)$ in this norm defines the Hilbert space $L^2\left(\Lambda_p\left[\frac{1}{2}\right]\right)$.

Suppose we are given a map $\Gamma : T^*M \to \mathrm{End}(\mathcal{S})$ determined by a self-adjoint endomorphism-valued vector $\Gamma \in C^\infty\left(TM \otimes \mathrm{End}(\mathcal{S})[0]\right)$, which is described locally by a matrix-valued vector Γ^μ. Let us define an endomorphism-valued tensor $a \in C^\infty\left(TM \otimes TM \otimes \mathrm{End}(\mathcal{S})[0]\right)$ by

$$a(\xi_1, \xi_2) = \frac{1}{2}\left[\Gamma(\xi_1)\Gamma(\xi_2) + \Gamma(\xi_2)\Gamma(\xi_1)\right]\,, \tag{2.12}$$

where $\xi_1, \xi_2 \in T^*M$. Then a is self-adjoint and symmetric

$$a(\xi_1, \xi_2) = a(\xi_2, \xi_1)\,, \qquad \overline{a(\xi_1, \xi_2)} = a(\xi_2, \xi_1)\,. \tag{2.13}$$

One of our main assumptions about the tensor a is that it defines an isomorphism

$$a : T^*M \otimes \mathcal{S} \to TM \otimes \mathcal{S}\,. \tag{2.14}$$

Let us consider the endomorphism

$$H(x, \xi) = a(\xi, \xi) = [\Gamma(\xi)]^2\,, \tag{2.15}$$

with $x \in M$, and $\xi \in T^*_x M$ being a cotangent vector. Our second assumption is that this endomorphism is positive definite, i.e. $H(x, \xi) > 0$ for any point x of the manifold M and $\xi \neq 0$. This endomorphism is self-adjoint and, therefore, all its eigenvalues are real and positive for $\xi \neq 0$. We call the endomorphism-valued tensor a the *non-commutative metric* and the components Γ^μ of the endomorphism-valued vector Γ the *non-commutative Dirac matrices*.

This construction determines a collection of Finsler geometries [8, 11]. Assume, for simplicity, that the matrix $H(x, \xi) = a(\xi, \xi)$ has distinct eigenvalues: $h_{(a)}(x, \xi)$, $a = 1, \dots, N$. Each eigenvalue defines a Hamilton-Jacobi equation

$$h_{(a)}(x, \partial S) = m^2_{(a)}\,, \tag{2.16}$$

where $m_{(a)}$ are some constants, a Hamiltonian system

$$\frac{dx^\mu}{dt} = \frac{1}{2}\frac{\partial}{\partial \xi_\mu} h_{(a)}(x, \xi)\,, \tag{2.17}$$

$$\frac{d\xi^\mu}{dt} = -\frac{1}{2}\frac{\partial}{\partial x^\mu} h_{(a)}(x, \xi)\,, \tag{2.18}$$

(the coefficient $1/2$ is introduced here for convenience) and a positive definite Finsler metric

$$g^{\mu\nu}_{(a)}(x, \xi) = \frac{1}{2}\frac{\partial^2 h_{(a)}}{\partial \xi_\mu \partial \xi_\nu}\,. \tag{2.19}$$

Moreover, each eigenvalue is a positive homogeneous function of ξ of degree 2 and, therefore, the Finsler metric is a homogeneous function of ξ of degree 0. This leads to a number of identities, in particular,

$$h_{(a)}(x, \xi) = g^{\mu\nu}_{(a)}(x, \xi)\xi_\mu \xi_\nu \quad \text{and} \quad \dot{x}^\mu = g^{\mu\nu}_{(a)}(x, \xi)\xi_\nu\,. \tag{2.20}$$

Next, one defines the inverse (covariant) Finsler metrics

$$g_{(a)\,\mu\nu}(x, \dot{x}) g^{\nu\alpha}_{(a)}(x, \xi) = \delta^\alpha_\mu\,, \tag{2.21}$$

the interval

$$ds^2_{(a)} = g_{(a)\,\mu\nu}(x,\dot{x})\,dx^\mu dx^\nu\,, \tag{2.22}$$

connections, curvatures etc (for details, see [40]). Thus, a *non-Laplace type operator generates a collection of Finsler geometries.*

The isomorphism a naturally defines a map $A : \Lambda_p \to \Lambda^p$, by

$$(A\varphi)^{\mu_1\cdots\mu_p} = A^{\mu_1\cdots\mu_p\nu_1\cdots\nu_p}\varphi_{\nu_1\cdots\nu_p}\,, \tag{2.23}$$

where

$$A^{\mu_1\cdots\mu_p\nu_1\cdots\nu_p} = \delta^{[\mu_1}_{\alpha_1}\cdots\delta^{\mu_p]}_{\alpha_p}\delta^{[\nu_1}_{\beta_1}\cdots\delta^{\nu_p]}_{\beta_p}a^{\alpha_1\beta_1}\cdots a^{\alpha_p\beta_p}\,, \tag{2.24}$$

and the square brackets denote the complete antisymmetrization over the indices included. We will assume that these maps are isomorphisms as well. Then the inverse operator $A^{-1} : \Lambda^p \to \Lambda_p$, is defined by

$$(A^{-1}\varphi)_{\mu_1\cdots\mu_p} = (A^{-1})_{\mu_1\cdots\mu_p\nu_1\cdots\nu_p}\varphi^{\nu_1\cdots\nu_p}\,, \tag{2.25}$$

where A^{-1} is determined by the equation

$$(A^{-1})_{\mu_1\cdots\mu_p\nu_1\cdots\nu_p}A^{\nu_1\cdots\nu_p\alpha_1\cdots\alpha_p} = \delta^{\alpha_1}_{[\mu_1}\cdots\delta^{\alpha_p}_{\mu_p]}\,. \tag{2.26}$$

This can be used further to define the natural inner product on the space of p-forms Λ_p via

$$\langle\psi,\varphi\rangle = \frac{1}{p!}\bar{\psi}_{\mu_1\cdots\mu_p}A^{\mu_1\cdots\mu_p\nu_1\cdots\nu_p}\varphi_{\nu_1\cdots\nu_p}\,. \tag{2.27}$$

Let d be the exterior derivative on p-form densities of weight 0

$$d : C^\infty(\Lambda_p[0]) \to C^\infty(\Lambda_{p+1}[0]) \tag{2.28}$$

and $\tilde{d}$ be the coderivative on p-vector densities of weight 1

$$\tilde{d} = (-1)^{np+1}\tilde{\varepsilon}d\varepsilon :\ C^\infty(\Lambda^p[1]) \to C^\infty(\Lambda^{p-1}[1])\,. \tag{2.29}$$

These operators are invariant differential operators defined without a Riemannian metric. They take the following form in local coordinates

$$(d\varphi)_{\mu_1\cdots\mu_{p+1}} = (p+1)\partial_{[\mu_1}\varphi_{\mu_2\cdots\mu_p]} \quad\text{and}\quad (\tilde{d}\varphi)^{\mu_1\cdots\mu_{p-1}} = \partial_\mu\varphi^{\mu\mu_1\cdots\mu_{p-1}}\,. \tag{2.30}$$

Now, let $\mathcal{B} \in C^\infty\left(T^*M \otimes \mathrm{End}(\mathcal{S})[0]\right)$ be a smooth anti-self-adjoint endomorphism-valued connection 1-form on the bundle $\mathcal{S}$, defined by the matrix-valued covector $\mathcal{B}_\mu$. Such a section naturally defines the maps:

$$\mathcal{B} : \Lambda_p\left[\tfrac{1}{2}\right] \to \Lambda_{p+1}\left[\tfrac{1}{2}\right] \tag{2.31}$$

and

$$\tilde{\mathcal{B}} = (-1)^{np+1}\tilde{\varepsilon}\mathcal{B}\varepsilon : \Lambda^p\left[\tfrac{1}{2}\right] \to \Lambda^{p-1}\left[\tfrac{1}{2}\right] \tag{2.32}$$

given locally by

$$(\mathcal{B}\varphi)_{\mu_1\cdots\mu_{p+1}} = (p+1)\mathcal{B}_{[\mu_1}\varphi_{\mu_2\cdots\mu_{p+1}]} \quad \text{and} \quad (\tilde{\mathcal{B}}\varphi)^{\mu_1\cdots\mu_{p-1}} = \mathcal{B}_\mu\varphi^{\mu\mu_1\cdots\mu_{p-1}}\,. \tag{2.33}$$

Finally, we introduce a self-adjoint non-degenerate endomorphism-valued density $\rho \in C^\infty\left(\mathrm{End}(\mathcal{S})\left[\tfrac{1}{2}\right]\right)$ of weight $\frac{1}{2}$. Then ρ^2 has weight 1 and plays the role of a *non-commutative measure*.

This enables us to define the covariant exterior derivative of p-form densities of weight $\frac{1}{2}$

$$\mathcal{D} : C^\infty\left(\Lambda_p\left[\tfrac{1}{2}\right]\right) \to C^\infty\left(\Lambda_{p+1}\left[\tfrac{1}{2}\right]\right). \tag{2.34}$$

and the covariant coderivative of p-vector densities of weight $\frac{1}{2}$

$$\tilde{\mathcal{D}} = (-1)^{np+1}\tilde{\varepsilon}\mathcal{D}\varepsilon :\ C^\infty\left(\Lambda^p\left[\tfrac{1}{2}\right]\right) \to C^\infty\left(\Lambda^{p-1}\left[\tfrac{1}{2}\right]\right), \tag{2.35}$$

by

$$\mathcal{D} = \rho(d + \mathcal{B})\rho^{-1}, \qquad \tilde{\mathcal{D}} = \rho^{-1}(\tilde{d} + \tilde{\mathcal{B}})\rho\,. \tag{2.36}$$

These operators transform covariantly under both the diffeomorphisms and the gauge transformations.

The formal adjoint of the operator $\mathcal{D}$

$$\bar{\mathcal{D}} : C^\infty\left(\Lambda_p\left[\tfrac{1}{2}\right]\right) \to C^\infty\left(\Lambda_{p-1}\left[\tfrac{1}{2}\right]\right), \tag{2.37}$$

has the form

$$\bar{\mathcal{D}} = -A^{-1}\rho^{-1}(\tilde{d} + \tilde{\mathcal{B}})\rho A\,, \tag{2.38}$$

By making use of these operators we define a second-order operator (that can be called the *non-commutative Laplacian*)

$$\Delta :\ C^\infty\left(\Lambda_p\left[\tfrac{1}{2}\right]\right) \to C^\infty\left(\Lambda_p\left[\tfrac{1}{2}\right]\right), \tag{2.39}$$

by

$$\Delta = -\bar{\mathcal{D}}\mathcal{D} - \mathcal{D}\bar{\mathcal{D}}\,. \tag{2.40}$$

In the special case $p = 0$ the non-commutative Laplacian Δ reads

$$\Delta = \rho^{-1}(\tilde{d} + \tilde{\mathcal{B}})\rho A \rho(d + \mathcal{B})\rho^{-1}, \tag{2.41}$$

which in local coordinates has the form

$$\Delta = \rho^{-1}(\partial_\mu + \mathcal{B}_\mu)\rho a^{\mu\nu}\rho(\partial_\nu + \mathcal{B}_\nu)\rho^{-1}\,. \tag{2.42}$$

Next, notice that the endomorphism-valued vector Γ introduced above naturally defines the maps

$$\Gamma : C^\infty\left(\Lambda^p\left[\tfrac{1}{2}\right]\right) \to C^\infty\left(\Lambda^{p+1}\left[\tfrac{1}{2}\right]\right) \tag{2.43}$$

and

$$\tilde{\Gamma} = (-1)^{np+1}\varepsilon\Gamma\tilde{\varepsilon} : C^\infty\left(\Lambda_p\left[\tfrac{1}{2}\right]\right) \to C^\infty\left(\Lambda_{p-1}\left[\tfrac{1}{2}\right]\right) \tag{2.44}$$

given locally by

$$(\Gamma\varphi)^{\mu_1\dots\mu_{p+1}} = (p+1)\Gamma^{[\mu_1}\varphi^{\mu_2\dots\mu_{p+1}]}, \qquad (\tilde{\Gamma}\varphi)_{\mu_1\dots\mu_{p-1}} = \Gamma^\mu\varphi_{\mu\mu_1\dots\mu_{p-1}} \,. \tag{2.45}$$

Therefore, we can define a first-order invariant differential operator (that can be called the *non-commutative Dirac operator*)

$$D : C^\infty\left(\Lambda_p\left[\tfrac{1}{2}\right]\right) \to C^\infty\left(\Lambda_p\left[\tfrac{1}{2}\right]\right) \tag{2.46}$$

by

$$D = i\tilde{\Gamma}\mathcal{D} = i\tilde{\Gamma}\rho(d+\mathcal{B})\rho^{-1}\,, \tag{2.47}$$

where, of course, $i = \sqrt{-1}$. The formal adjoint of this operator is

$$\bar{D} = iA^{-1}\tilde{\mathcal{D}}\Gamma A = iA^{-1}\rho^{-1}(\tilde{d}+\tilde{\mathcal{B}})\rho\Gamma A\,. \tag{2.48}$$

These operators can be used to define second order differential operators $D\bar{D}$ and $\bar{D}D$.

In the case $p = 0$ these operators have the following form in local coordinates

$$D = i\Gamma^\mu\rho(\partial_\mu + \mathcal{B}_\mu)\rho^{-1}\,, \qquad \bar{D} = i\rho^{-1}(\partial_\nu + \mathcal{B}_\nu)\rho\Gamma^\nu\,, \tag{2.49}$$

and, therefore, the second-order operators $D\,\bar{D}$ and $\bar{D}\,D$ read

$$D\,\bar{D} = -\Gamma^\mu\rho(\partial_\mu + \mathcal{B}_\mu)\rho^{-2}(\partial_\nu + \mathcal{B}_\nu)\rho\Gamma^\nu\,, \tag{2.50}$$

$$\bar{D}\,D = -\rho^{-1}(\partial_\nu + \mathcal{B}_\nu)\rho\Gamma^\nu\Gamma^\mu\rho(\partial_\mu + \mathcal{B}_\mu)\rho^{-1}\,. \tag{2.51}$$

In the present paper we will primarily study the second-order operators Δ, $\bar{D}\,D$ and $D\,\bar{D}$ in the case $p = 0$, that is,

$$\Delta, \bar{D}\,D, D\,\bar{D} : C^\infty\left(\mathcal{S}\left[\tfrac{1}{2}\right]\right) \to C^\infty\left(\mathcal{S}\left[\tfrac{1}{2}\right]\right)\,.$$

These are all formally self-adjoint operators by construction. This means that they are symmetric on smooth sections of the bundle $\mathcal{S}\left[\frac{1}{2}\right]$ with compact support in the interior of M (that is, sections that vanish together with all their derivatives on the boundary ∂M).

The leading symbols of all these operators are equal to the matrix $H(x,\xi) = a(\xi,\xi)$, i.e.

$$\sigma_L(\Delta; x,\xi) = \sigma_L(\bar{D}\,D; x,\xi) = \sigma_L(D\,\bar{D}; x,\xi) = H(x,\xi) = a(\xi,\xi)\,, \quad (2.52)$$

where $\xi \in T_x^*M$. By our main assumption about the non-commuting metric the leading symbol is self-adjoint and positive definite in the interior of the manifold. Therefore, the leading symbol is invertible (or elliptic) in the interior of M. Notice that the leading symbol is non-scalar, in general. That is why such operators are called *non-Laplace type operators*.

2.3. *Elliptic boundary value problem*

Let us consider a neighborhood of the boundary ∂M in M. Let $x = (x^\mu)$ be the local coordinates in this neighborhood. The boundary is a smooth hypersurface without boundary. Therefore, there must exist a local diffeomorphism

$$r = r(x) \qquad \hat{x}^i = \hat{x}^i(x)\,, \qquad i = 1,\dots,n-1\,, \quad (2.53)$$

and the inverse diffeomorphism

$$x^\mu = x^\mu(r,\hat{x})\,, \qquad \mu = 1,\dots,n\,, \quad (2.54)$$

such that

$$r(x) = 0 \quad \text{for any } x \in \partial M\,, \qquad r(x) > 0 \quad \text{for any } x \notin \partial M\,, \quad (2.55)$$

and the vector $\partial_r = \partial/\partial r$ is *transversal* (nowhere tangent) to the boundary ∂M. Then the coordinates $\hat{x}^i$ are local coordinates on the boundary ∂M.

Let $\delta > 0$. We define a disjoint decomposition of the manifold

$$M = M_{\text{int}} \cup M_{\text{bnd}}\,, \quad (2.56)$$

where $M_{\text{bnd}} = \{x \in M \mid r(x) < \delta\}$ is a δ-neighborhood of the boundary and $M_{\text{int}} = M \setminus M_{\text{bnd}}$ is the part of the interior of the manifold on a finite distance from the boundary.

For $r = 0$, that is, $x \in \partial M$, the vectors $\{\hat{\partial}_i = \partial/\partial\hat{x}^i\}$ are tangent to the boundary and give the local coordinate basis for the tangent space $T_x\partial M$. The set of vectors $\{\partial_r, \hat{\partial}_i\}$ gives the local coordinate basis for the tangent space T_xM in M_{bnd}. Similarly, the 1-forms $d\hat{x}^i$ determine the local coordinate basis for the cotangent space $T_x^*\partial M$, and the 1-forms $\{dr, d\hat{x}^i\}$ give the local coordinate basis for the cotangent space T_x^*M in M_{bnd}.

We fix the orientation of the boundary by requiring the Jacobian of this diffeomorphism to be positive, in other words, for any $x \in M_{\text{bnd}}$

$$J(x) = \text{vol}\,(\partial_r, \hat{\partial}_1, \dots, \hat{\partial}_{n-1}) > 0\,. \quad (2.57)$$

Let $\varphi \in C^\infty(TM[1])$ be a smooth vector density of weight 1. Then Stokes' Theorem has the form

$$\int\limits_M dx\,\tilde{d}\varphi = \int\limits_{\partial M} d\hat{x}\,N(\varphi)\,, \tag{2.58}$$

where N is a 1-form defined by

$$\begin{aligned} N(\varphi) &= \mathrm{vol}\,(\varphi, \hat{\partial}_1, \dots, \hat{\partial}_{n-1}) = \frac{1}{J} dr(\varphi) \\ &= \varepsilon_{\mu\nu_1\dots\nu_{n-1}} \frac{\partial x^{\nu_1}}{\partial \hat{x}^1} \cdots \frac{\partial x^{\nu_{n-1}}}{\partial \hat{x}^{n-1}} \varphi^\mu = \frac{1}{J}\frac{\partial r}{\partial x^\mu}\varphi^\mu\,. \end{aligned} \tag{2.59}$$

Notice that this formula is valid for densities, and there is no need for a Riemannian metric.

We will study in the present paper, for simplicity, the Dirichlet boundary conditions $\varphi\big|_{\partial M} = 0$. By integration by parts it is not difficult to see that all operators Δ, $D\,\bar{D}$ and $\bar{D}\,D$ are symmetric on smooth sections of the bundle $\mathcal{S}\left[\frac{1}{2}\right]$ satisfying the boundary conditions. One can show that these operators are essentially self-adjoint, that is, their closure is self-adjoint and, hence, they have unique self-adjoint extensions to $L^2\left(\mathcal{S}\left[\frac{1}{2}\right]\right)$.

Let L be one of the operators $\bar{D}\,D, D\,\bar{D}\,, \Delta$ with the Dirichlet boundary conditions. Our primary interest in this paper is the study of elliptic boundary value problems. Ellipticity means invertibility up to a compact operator in appropriate functional spaces (see, for example, [19, 31, 29]). This is, roughly speaking, a condition that implies local invertibility. For a boundary value problem it has two components: i) in the interior of the manifold, and ii) at the boundary.

An operator L is elliptic in the interior of the manifold if for any interior point $x \in M$ and for any nonzero cotangent vector $\xi \in T_x^*M$, $\xi \neq 0$, its leading symbol $\sigma_L(L; x, \xi)$ is invertible. Since all operators $D\,\bar{D}$, $\bar{D}\,D$ and Δ all have positive leading symbols, namely $H(x,\xi)$, they are elliptic in the interior of the manifold.

At the boundary ∂M of the manifold we use the coordinates $(r, \hat{x})$ and define a split of the cotangent bundle $T^*M = \mathbb{R} \oplus T^*\partial M$, so that $\xi = (\xi_\mu) = (\omega, \hat{\xi}) \in T_x^*M$, where $\omega \in \mathbb{R}$ and $\hat{\xi} = (\xi_j) \in T_x^*\partial M$.

Let $\lambda \in \mathbb{C}\setminus\mathbb{R}_+$ be a complex number that does not lie on the positive real axis and $H(r, \hat{x}, \omega, \hat{\xi})$ be the leading symbol of the operator L. We substitute $r = 0$ and $\omega \mapsto -i\partial_r$ and consider the following second-order ordinary differential equation on the half-line, i.e. $r \in \mathbb{R}_+$,

$$\Big[H(0, \hat{x}, -i\partial_r, \hat{\xi}) - \lambda\Big]\,\varphi = 0\,, \tag{2.60}$$

with an asymptotic condition

$$\lim_{r\to\infty} \varphi = 0\,. \tag{2.61}$$

Let $\hat{\mathcal{S}} = \mathcal{S}|_{\partial M}$ be the restriction of the vector bundle $\mathcal{S}$ to the boundary. The operator L with Dirichlet boundary conditions is elliptic if for each boundary point $\hat{x} \in \partial M$, each $\hat{\xi} \in T^*_{\hat{x}}\partial M$, each $\lambda \in \mathbb{C} \setminus \mathbb{R}_+$, such that $\hat{\xi}$ and λ are not both zero, and each $f \in C^\infty\left(\hat{\mathcal{S}}\left[\frac{1}{2}\right]\right)$ there is a unique solution $\varphi(\lambda, r, \hat{\xi})$ to the equation (2.60) subject to the asymptotic condition (2.61) at infinity and the boundary condition at $r = 0$

$$\varphi(\lambda, 0, \hat{\xi}) = f\,. \tag{2.62}$$

We have

$$\begin{aligned} H(0, \hat{x}, \omega, \hat{\xi}) &= [A(\hat{x})\omega + C(\hat{x}, \hat{\xi})]^2 \\ &= A^2(\hat{x})\omega^2 + B(\hat{x}, \hat{\xi})\omega + C^2(\hat{x}, \hat{\xi})\,, \end{aligned} \tag{2.63}$$

where A, B, and C are self-adjoint matrices defined by

$$A(\hat{x}) = \Gamma(dr)\,, \qquad C(\hat{x}, \hat{\xi}) = \Gamma(d\hat{x}^j)\hat{\xi}_j\,, \tag{2.64}$$

$$B(\hat{x}, \hat{\xi}) = A(\hat{x})C(\hat{x}, \hat{\xi}) + C(\hat{x}, \hat{\xi})A(\hat{x})\,. \tag{2.65}$$

Then the differential equation (2.60) has the form

$$\left(-A^2\partial_r^2 - iB\partial_r + C^2 - \lambda\mathbb{I}\right)\varphi = 0\,. \tag{2.66}$$

We notice that the matrix $\left[(A\omega + C)^2 - \lambda\mathbb{I}\right]$ is non-degenerate when ω is real and λ and $\hat{\xi}$ are not both zero, i.e. $(\lambda, \hat{\xi}) \neq (0, 0)$. Moreover, when λ is a negative real number, then this matrix is self-adjoint and positive definite for real ω. Therefore, we can define

$$\Phi(\lambda, y, \hat{\xi}) = \int\limits_{-\infty}^{\infty} \frac{d\omega}{2\pi} e^{i\omega y} R_\lambda(\omega, \hat{\xi})\,, \tag{2.67}$$

where

$$R_\lambda(\omega, \hat{\xi}) = \left\{[A(\hat{x})\omega + C(\hat{x}, \hat{\xi})]^2 - \lambda\mathbb{I}\right\}^{-1}\,. \tag{2.68}$$

The matrix $\Phi(\lambda, y, \hat{\xi})$ is well defined for any $y \in \mathbb{R}$. It: i) vanishes at infinity,

$$\lim_{y \to \pm\infty} \Phi(\lambda, y, \hat{\xi}) = 0\,, \tag{2.69}$$

ii) satisfies the symmetry relations

$$\overline{\Phi(\lambda, y, \hat{\xi})} = \Phi(\bar{\lambda}, -y, \hat{\xi})\,, \qquad \Phi(\lambda, y, -\hat{\xi}) = \Phi(\lambda, -y, \hat{\xi})\,, \tag{2.70}$$

iii) is homogeneous, i.e. for any $t > 0$,

$$\Phi\left(\frac{\lambda}{t}, \sqrt{t}\, y, \frac{\hat{\xi}}{\sqrt{t}}\right) = t^{1/2}\Phi(\lambda, y, \hat{\xi}) , \tag{2.71}$$

iv) is continuous at zero with a well defined value at $y = 0$

$$\Phi_0(\lambda, \hat{\xi}) = \Phi(\lambda, 0, \hat{\xi}) = \int\limits_{-\infty}^{\infty} \frac{d\omega}{2\pi} R_\lambda(\omega, \hat{\xi}) , \tag{2.72}$$

v) has a discontinuous derivative $\partial_y \Phi(\lambda, y, \hat{\xi})$ at $y = 0$ with a finite jump.

We also notice that the matrix $\Phi_0(\lambda, \hat{\xi})$ is an even function of $\hat{\xi}$ and is self-adjoint for real λ, i.e.

$$\Phi_0(\lambda, -\hat{\xi}) = \Phi_0(\lambda, \hat{\xi}) , \qquad \overline{\Phi_0(\lambda, \hat{\xi})} = \Phi_0(\bar{\lambda}, \hat{\xi}) . \tag{2.73}$$

Moreover, for real negative λ the matrix Φ_0 is positive and, therefore, non-degenerate. More generally, it is non-degenerate for $\mathrm{Re}\,\lambda < w$, where w is a sufficiently large negative constant.

In an important particular case, when $B = AC + CA = 0$, one can compute explicitly

$$\Phi(\lambda, y, \hat{\xi}) = \frac{1}{2} A^{-1}\mu^{-1}e^{-\mu|y|}A^{-1} , \qquad \Phi_0(\lambda, \hat{\xi}) = \frac{1}{2} A^{-1}\mu^{-1}A^{-1} , \tag{2.74}$$

where $\mu = \sqrt{A^{-1}(C^2 - \lambda\mathbb{I})A^{-1}}$, defined as an analytical continuation in λ of a positive square root of a self-adjoint matrix when $\lambda \in \mathbb{R}_-$.

One can prove now that the eq. (2.66) with initial condition (2.62) and the asymptotic condition at infinity (2.61) has a unique solution given by

$$\varphi(\lambda, r, \hat{\xi}) = \Phi(\lambda, r, \hat{\xi})[\Phi_0(\lambda, \hat{\xi})]^{-1} f . \tag{2.75}$$

Thus, the Dirichlet boundary value problem for our operator is elliptic.

3. Spectral asymptotics

3.1. *Heat kernel*

Let L be a self-adjoint elliptic second-order partial differential operator acting on smooth sections of the bundle $\mathcal{S}\left[\frac{1}{2}\right]$ over a compact manifold M with boundary ∂M with positive definite leading symbol and with some boundary conditions

$$B\varphi|_{\partial M} = 0 , \tag{3.1}$$

with some boundary operator B. It is well known that such an operator has a discrete real spectrum $\{\lambda_k\}_{k=1}^{\infty}$ bounded from below [29], i.e.,

$$\lambda_1 \le \lambda_2 \le \cdots \le \lambda_k \le \lambda_{k+1} \le \cdots . \tag{3.2}$$

Furthermore: i) each eigenspace is finite-dimensional, ii) the eigensections are smooth sections of the bundle $\mathcal{S}\left[\frac{1}{2}\right]$, and iii) the set of eigensections $\{\varphi_k\}_{k=1}^{\infty}$ forms an orthonormal basis in $L^2\left(\mathcal{S}\left[\frac{1}{2}\right]\right)$.

For $t > 0$ the heat semigroup

$$\exp(-tL) : L^2\left(\mathcal{S}\left[\tfrac{1}{2}\right]\right) \to L^2\left(\mathcal{S}\left[\tfrac{1}{2}\right]\right) \tag{3.3}$$

is a bounded operator. The integral kernel of this operator, called the heat kernel, is given by

$$U(t;x,x') = \sum_{k=1}^{\infty} e^{-t\lambda_k} \varphi_k \otimes \bar{\varphi}_k(x') \,, \tag{3.4}$$

where each eigenvalue is counted with its multiplicity. The heat kernel satisfies the heat equation

$$(\partial_t + L)U(t;x,x') = 0 \tag{3.5}$$

with the initial condition

$$U(0^+;x,x') = \delta(x,x') \,, \tag{3.6}$$

where $\delta(x,x')$ is the Dirac distribution, as well as the boundary conditions

$$B_x U(t;x,x')\Big|_{x\in\partial M} = 0 \,, \tag{3.7}$$

and the self-adjointness condition

$$\overline{U(t;x,x')} = U(t;x',x) \,. \tag{3.8}$$

The heat kernel $U(t) = \exp(-tL)$ is intimately related to the resolvent $G(\lambda) = (L-\lambda)^{-1}$. Let λ be a complex number with a sufficiently large negative real part, $\mathrm{Re}\lambda << 0$. Then the resolvent and the heat kernel are related by the Laplace transform

$$G(\lambda) = \int_0^{\infty} dt\; e^{t\lambda}\, U(t) \,, \tag{3.9}$$

$$U(t) = \frac{1}{2\pi i} \int_{w-i\infty}^{w+i\infty} d\lambda\; e^{-t\lambda}\, G(\lambda) \,, \tag{3.10}$$

where w is a sufficiently large negative real number, $w << 0$.

The resolvent satisfies the equation

$$(L - \lambda\mathbb{I})G(\lambda;x,x') = \delta(x,x') \tag{3.11}$$

with the boundary condition

$$B_x G(\lambda; x, x')\big|_{x\in\partial M} = 0\,, \tag{3.12}$$

and the self-adjointness condition

$$\overline{G(\lambda; x, x')} = G(\bar{\lambda}; x', x)\,. \tag{3.13}$$

The integral kernel of the resolvent reads

$$G(\lambda; x, x') = \sum_{k=1}^{\infty} \frac{1}{\lambda_k - \lambda}\, \varphi_k \otimes \bar{\varphi}_k(x')\,, \tag{3.14}$$

where each eigenvalue is counted with its multiplicity.

For $t > 0$ the heat kernel $U(t; x, x')$ is a smooth section of the exterior tensor product bundle $\mathcal{S}\left[\frac{1}{2}\right] \boxtimes \mathcal{S}^*\left[\frac{1}{2}\right]$; that is, it is a two-point density of weight $\frac{1}{2}$ at each point. In particular, it is a smooth function near the diagonal of $M \times M$ and has a well defined diagonal value $U(t; x, x)$. The diagonal is, of course, a smooth section of the bundle $\mathcal{S}\,[1]$, a density of weight 1.

Moreover, the heat semigroup is a trace-class operator with a well defined L^2-trace

$$\mathrm{Tr}_{\,L^2} \exp(-tL) = \int_M dx\; \mathrm{tr}_{\,S} U(t; x, x)\,, \tag{3.15}$$

where $\mathrm{tr}_{\,S}$ is the trace over the fiber vector space S of the vector bundle $\mathcal{S}$. The trace of the heat kernel is a spectral invariant of the operator L since

$$\mathrm{Tr}_{\,L^2} \exp(-tL) = \sum_{k=1}^{\infty} e^{-t\lambda_k}\,. \tag{3.16}$$

Since the diagonal is a density of weight 1 the trace $\mathrm{Tr}_{\,L^2} \exp(-tL)$ is invariant under diffeomorphisms.

This enables one to define other spectral functions by integral transforms of the trace of the heat kernel. In particular, the zeta function, $\zeta(L; s, \lambda)$, is defined as follows. Let λ be a complex parameter with $\mathrm{Re}\,\lambda < \lambda_1$, so that the operator $(L - \lambda)$ is positive. Then for any $s \in \mathbb{C}$ such that $\mathrm{Re}\, s > n/2$ the trace of the operator $(L - \lambda)^{-s}$ is well defined and determines the zeta function,

$$\zeta(L; s, \lambda) = \mathrm{Tr}_{\,L^2}(L - \lambda)^{-s} = \frac{1}{\Gamma(s)} \int_0^{\infty} dt\; t^{s-1}\, e^{t\lambda}\, \mathrm{Tr}_{\,L^2} \exp(-tL)\,. \tag{3.17}$$

The zeta function enables one to define further the regularized determinant of the operator $(L - \lambda)$ by

$$\frac{\partial}{\partial s} \zeta(L; s, \lambda)\Big|_{s=0} = -\log \mathrm{Det}\,(L - \lambda)\,. \tag{3.18}$$

There is an asymptotic expansion as $t \to 0$ of the trace of the heat kernel [29] (see also [2, 4, 5, 6, 43])

$$\mathrm{Tr}_{L^2} \exp(-tL) \sim (4\pi)^{-n/2} \sum_{k=0}^{\infty} t^{(k-n)/2} A_k(L)\,. \tag{3.19}$$

The coefficients $A_k(L)$, called the global heat invariants, are spectral invariants determined by the integrals over the manifold M and the boundary ∂M of some scalar densities $a_k(L;x)$ and $b_k(L;\hat{x})$, called local heat invariants, viz.

$$A_k(L) = \int\limits_M dx\; a_k(L;x) + \int\limits_{\partial M} d\hat{x}\; b_k(L;\hat{x})\,. \tag{3.20}$$

The local heat invariants $a_k(L;x)$ and $b_k(L;\hat{x})$ are constructed polynomially from the jets of the symbol of the operator L; the boundary coefficients b_k depend, of course, on the boundary conditions and the geometry of the boundary as well.

Contrary to the heat kernel, the resolvent is singular at the diagonal and does not have a well defined trace. However, the derivatives of the resolvent do. Let $m \geq n/2$. Then the trace $\mathrm{Tr}_{L^2}(\partial_\lambda)^m G(\lambda)$ is well defined and has the asymptotic expansion as $\lambda \to -\infty$

$$\mathrm{Tr}_{L^2} \frac{\partial^m}{\partial\lambda^m} G(\lambda) \sim (4\pi)^{-n/2} \sum_{k=0}^{\infty} \Gamma\left[(k-n+2m+2)/2\right] (-\lambda)^{(n-k-2m-2)/2} A_k(L)\,. \tag{3.21}$$

Therefore, one can use either the heat kernel or the resolvent to compute the coefficients A_k.

3.2. *Index of noncommutative Dirac operator*

Notice that the operator Δ can have a finite number of negative eigenvalues, whereas the spectrum of the operators $\bar{D}\,D$ and $D\,\bar{D}$ is non-negative. Moreover, one can easily show that all non-zero eigenvalues of the operators $\bar{D}\,D$ and $D\,\bar{D}$ are equal

$$\lambda_k(\bar{D}\,D) = \lambda_k(D\,\bar{D}) \qquad \text{if } \lambda_k(\bar{D}\,D) > 0\,. \tag{3.22}$$

Therefore, there is a well defined index

$$\mathrm{Ind}(D) = \dim \mathrm{Ker}\,(\bar{D}\,D) - \dim \mathrm{Ker}\,(D\,\bar{D})\,, \tag{3.23}$$

which is equal to the difference of the number of zero modes of the operators $\bar{D}\,D$ and $D\,\bar{D}$.

This leads to the fact that the difference of the heat traces for the operators $\bar{D}\,D$ and $D\,\bar{D}$ determines the index

$$\mathrm{Tr}_{\,L^2}\exp(-t\,\bar{D}\,D) - \mathrm{Tr}_{\,L^2}\exp(-t\,D\,\bar{D}\,) = \mathrm{Ind}(D)\,. \tag{3.24}$$

This means that the spectral invariants of the operators $\bar{D}\,D$ and $D\,\bar{D}$ are equal except for the invariant A_n which determines the index

$$A_k(\,\bar{D}\,D) = A_k(D\,\bar{D}\,) \qquad \text{for } k \neq n\,, \tag{3.25}$$

and

$$A_n(\,\bar{D}\,D) - A_n(D\,\bar{D}\,) = (4\pi)^{n/2}\mathrm{Ind}(D)\,. \tag{3.26}$$

Thus, for $n > 2$ the spectral invariants A_0, A_1 and A_2 of the operators $\bar{D}\,D$ and $D\,\bar{D}$ are equal. Therefore, we can pick any of these operators to compute these invariants. Of course, the spectral invariants of the noncommutaative Laplacian Δ are, in general, different. However, since the operators $\bar{D}\,D$ and $D\,\bar{D}$ have the same leading symbol as the operator Δ there must exist a corresponding Lichnerowicz-Weitzenbock formula (for the spinor bundle see, for example, [18]), which means that the spectral invariants of these operators must be related.

4. Heat invariants

4.1. *Interior coefficients*

The heat kernel in the interior part is constructed as follows. We fix a point $x_0 \in M_{\rm int}$ in the interior of the manifold and consider a neighborhood of x_0 disjoint from the boundary layer $M_{\rm bnd}$ covered by a single patch of local coordinates. We introduce a scaling parameter $\varepsilon > 0$ and scale the variables according to

$$x^\mu \mapsto x_0^\mu + \varepsilon(x^\mu - x_0^\mu)\,, \qquad x'^\mu \mapsto x_0^\mu + \varepsilon(x'^\mu - x_0^\mu)\,, \qquad t \mapsto \varepsilon^2 t\,, \tag{4.1}$$

so that

$$\partial_\mu \mapsto \frac{1}{\varepsilon}\partial_\mu\,, \qquad \partial_t \mapsto \frac{1}{\varepsilon^2}\partial_t\,. \tag{4.2}$$

Then the differential operator $L(\hat{x}, \hat{\partial})$ is scaled according to

$$L \mapsto L_\varepsilon \sim \sum_{k=0}^{\infty} \varepsilon^{k-2} L_k^{\rm int}\,, \tag{4.3}$$

where $L_k^{\rm int}$ are second-order differential operators with homogeneous symbols. Next, we expand the scaled heat kernel in $M_{\rm int}$, which we denote by $U_\varepsilon^{\rm int}$ in a power series in ε

$$U_\varepsilon^{\rm int} \sim \sum_{k=0}^{\infty} \varepsilon^{2-n+k} U_k^{\rm int}\,, \tag{4.4}$$

and substitute into the scaled version of the heat equation. By equating the like powers of ε we get an infinite set of recursive differential equations determining all the coefficients $U_k^{\rm int}$.

The leading order operator $L_0^{\rm int}$ is an operator with constant coefficients determined by the leading symbol

$$L_0^{\rm int} = H(x_0, -i\partial)\,. \tag{4.5}$$

The leading-order heat kernel $U_0^{\rm int}$ can be easily obtained by the Fourier transform

$$U_0^{\rm int}(t;x,x') = \int\limits_{\mathbb{R}^n} \frac{d\xi}{(2\pi)^n} e^{i\xi(x-x')-tH(x_0,\xi)}\,, \tag{4.6}$$

where $\xi(x-x') = \xi_\mu(x^\mu - x'^\mu)$.

The higher-order coefficients $U_k^{\rm int}$, $k \geq 1$, are determined from the recursive equations

$$(\partial_t + L_0^{\rm int})U_k^{\rm int} = -\sum_{j=1}^{k} L_j^{\rm int} U_{k-j}^{\rm int}\,, \tag{4.7}$$

with the initial condition

$$U_k^{\rm int}(0;x,x') = 0\,. \tag{4.8}$$

This expansion is nothing but the decomposition of the heat kernel into the homogeneous parts with respect to the variables $(x-x_0)$, $(x'-x_0)$, and $\sqrt{t}$. That is,

$$U_k^{\rm int}(t;x,x') = t^{(k-n)/2} U_k^{\rm int}\left(1; x_0 + \frac{(x-x_0)}{\sqrt{t}}, x_0 + \frac{(x'-x_0)}{\sqrt{t}}\right)\,. \tag{4.9}$$

In particular, the heat kernel diagonal at the point x_0 scales by

$$U_k^{\rm int}(t;x_0,x_0) = t^{(k-n)/2} U_k^{\rm int}\left(1;x_0,x_0\right)\,. \tag{4.10}$$

To compute the contribution of these coefficients to the trace of the heat kernel we need to compute the integral of the diagonal of the heat kernel $U^{\rm int}(t;x,x)$ over the interior part of the manifold $M_{\rm int}$. By using the homogeneity property (4.10) we obtain

$$\int\limits_{M_{\rm int}} dx\, {\rm tr}_S U^{\rm int}(t;x,x) \sim \sum_{k=0}^{\infty} t^{(k-n)/2} \int\limits_{M_{\rm int}} dx\, {\rm tr}_S U_k^{\rm int}\left(1;x,x\right)\,. \tag{4.11}$$

Next, we take the limit as $\delta \to 0$. Then the integrals over the interior part $M_{\rm int}$ become the integrals over the whole manifold M and give all the interior coefficients $a_k(L)$ in the global heat kernel coefficients $A_k(L)$.

Instead of this rigorous procedure, we present below a pragmatic formal approach that enables one to compute all interior coefficients in a much easier and compact form. Of course, both approaches are equivalent and give the same answers.

First, we present the heat kernel diagonal for the operator $L = \bar{D}\, D$ in the form

$$U^{\rm int}(t;x,x) = \int\limits_{\mathbb{R}^n} \frac{d\xi}{(2\pi)^n} e^{-i\xi x} \exp(-t\,\bar{D}\,D) e^{i\xi x}\,, \tag{4.12}$$

where $\xi x = \xi_\mu x^\mu$, which can be transformed to

$$U^{\rm int}(t;x,x) = \int\limits_{\mathbb{R}^n} \frac{d\xi}{(2\pi)^n} \exp\left[-t\left(H + K + \bar{D}\,D\right)\right] \cdot \mathbb{I}\,, \tag{4.13}$$

where $H = [\Gamma(\xi)]^2$ is the leading symbol of the operator $\bar{D}\,D$, and K is a first-order self-adjoint operator defined by

$$K = -\Gamma(\xi) D - \bar{D}\,\Gamma(\xi)\,. \tag{4.14}$$

Here the operators in the exponent act on the unity matrix $\mathbb{I}$ from the left.

By changing the integration variable $\xi \to t^{-1/2}\xi$ we obtain

$$U^{\rm int}(t;x,x) = (4\pi t)^{-n/2} \int\limits_{\mathbb{R}^n} \frac{d\xi}{\pi^{n/2}} \exp\left(-H - \sqrt{t}\,K - t\,\bar{D}\,D\right) \cdot \mathbb{I}\,. \tag{4.15}$$

Now, the coefficients of the asymptotic expansion of this integral in powers of $t^{1/2}$ as $t \to 0$ determine the interior heat kernel coefficients $a_k(L)$ via

$$\mathrm{tr}\,_S U^{\rm int}(t;x,x) \sim (4\pi)^{-n/2} \sum_{k=0}^{\infty} t^{(k-n)/2} a_k(L)\,. \tag{4.16}$$

By using the Volterra series

$$\exp(A+B) = e^A + \sum_{k=1}^{\infty} \int\limits_0^1 d\tau_k \int\limits_0^{\tau_k} d\tau_{k-1} \cdots \int\limits_0^{\tau_2} d\tau_1$$

$$\times\, e^{(1-\tau_k)A} B e^{(\tau_k - \tau_{k-1})A} \cdots e^{(\tau_2-\tau_1)A} B e^{\tau_1 A}\,, \tag{4.17}$$

we get

$$\exp\left(-H-\sqrt{t}\,K-t\,\bar{D}\,D\right)=e^{-H}-t^{1/2}\int_0^1 d\tau_1 e^{-(1-\tau_1)H}Ke^{-\tau_1 H}$$
$$+t\left[\int_0^1 d\tau_2\int_0^{\tau_2} d\tau_1 e^{-(1-\tau_2)H}Ke^{-(\tau_2-\tau_1)H}Ke^{-\tau_1 H}\right.$$
$$\left.-\int_0^1 d\tau_1 e^{-(1-\tau_1)H}\,\bar{D}\,De^{-\tau_1 H}\right]+O(t^2)\,. \qquad (4.18)$$

Now, since K is linear in ξ the term proportional to $t^{1/2}$ vanishes after integration over ξ. Thus, we obtain the first three interior coefficients of the asymptotic expansion of the heat kernel diagonal in the form

$$a_0(L)=\int_{\mathbb{R}^n}\frac{d\xi}{\pi^{n/2}}\,\mathrm{tr}_S\,e^{-H}\,, \qquad (4.19)$$

$$a_1(L)=0\,, \qquad (4.20)$$

$$a_2(L)=\int_{\mathbb{R}^n}\frac{d\xi}{\pi^{n/2}}\,\mathrm{tr}_S\left[\int_0^1 d\tau_2\int_0^{\tau_2} d\tau_1 e^{-(1-\tau_2)H}Ke^{-(\tau_2-\tau_1)H}Ke^{-\tau_1 H}\right.$$
$$\left.-\int_0^1 d\tau_1 e^{-(1-\tau_1)H}\,\bar{D}\,De^{-\tau_1 H}\right]. \qquad (4.21)$$

4.2. *Boundary coefficients*

On manifolds with boundary, as far as we know, the coefficients A_k have not been studied at all, so, even A_1 is not known. In the present paper we are going to compute the coefficient A_1 on manifolds with boundary for the operators $\bar{D}D$ and $D\bar{D}$. The coefficient A_0 is, of course, the same as for the manifolds without boundary. We will follow the general framework for computation of the heat kernel asymptotics outlined in [10, 15].

The procedures for the resolvent and the heat kernel are very similar. One can, of course, use either of them. We will describe below the construction of the heat kernel.

The main idea can be described as follows. Recall that we decomposed the manifold into a neighborhood of the boundary $M_{\rm bnd}$ and the interior part $M_{\rm int}$. We can use now different approximations for the heat kernel in different domains.

Strictly speaking one has to use 'smooth characteristic functions' of those domains (partition of unity) to glue them together in a smooth way. Then, one has to control the order of the remainder terms in the limit $t \to 0^+$ and their dependence on δ (the size of the boundary layer). However, since we are only interested in the trace of the heat kernel, this is not needed here and we will not worry about such subtle details. We can compute the asymptotic expansion as $t \to 0$ of the corresponding integrals in each domain and then take the limit $\delta \to 0$.

The origin of the boundary terms in the heat trace asymptotics can be explained as follows. The heat kernel of an elliptic boundary value problem in $M_{\rm bnd}$ has exponentially small terms like $\exp(-r^2/t)$ as $t \to 0$. These terms do not contribute to the asymptotic expansion of the diagonal of the heat kernel as $t \to 0$. However, they behave like distributions near the boundary (recall that $r > 0$ inside the manifold and $r = 0$ on the boundary). Therefore, the integral over $M_{\rm bnd}$, more precisely, the limit $\lim_{\delta\to 0} \int\limits_{\partial M} d\hat{x} \int\limits_0^\delta dr(\dots)$ does contribute to the asymptotic expansion of the trace of the heat kernel with coefficients in form of integrals over the boundary. It is this phenomenon that leads to the boundary terms in the global heat invariants.

The heat kernel in the boundary layer $M_{\rm bnd}$ is constructed as follows. We fix a point $\hat{x}_0 \in \partial M$ on the boundary and choose coordinates as described above in section 2.2. Let $\varepsilon > 0$ be a positive real parameter. We use it as a scaling parameter; at the very end of the calculation it will be set to 1. Now we scale the coordinates according to

$$\hat{x}^j \mapsto \hat{x}^j_0 + \varepsilon(\hat{x}^j - \hat{x}^j_0)\,, \qquad \hat{x}'^j \mapsto \hat{x}^j_0 + \varepsilon(\hat{x}'^j - \hat{x}^j_0)\,, \tag{4.22}$$

$$r \mapsto \varepsilon r\,, \qquad r' \mapsto \varepsilon r'\,, \qquad t \mapsto \varepsilon^2 t\,. \tag{4.23}$$

The differential operators are scaled correspondingly by

$$\hat{\partial}_j \mapsto \frac{1}{\varepsilon}\hat{\partial}_j\,, \qquad \partial_r \mapsto \frac{1}{\varepsilon}\partial_r\,, \qquad \partial_t \mapsto \frac{1}{\varepsilon^2}\partial_t\,. \tag{4.24}$$

Let $L(r, \hat{x}, \partial_r, \hat{\partial})$ be the operator under consideration. The scaled operator, which we denoted by L_ε, has the following formal power series expansion in ε

$$L \mapsto L_\varepsilon \sim \sum_{k=0}^{\infty} \varepsilon^{k-2} L^{\rm bnd}_k\,, \tag{4.25}$$

where L_k are second-order differential operators with homogeneous symbols. The leading order operator is determined by the leading symbol

$$L^{\rm bnd}_0 = H(0, \hat{x}_0, -i\partial_r, -i\hat{\partial})\,. \tag{4.26}$$

This is a differential operator with constant coefficients.

Next, we expand the scaled heat kernel in $M_{\rm bnd}$, which we denote by $U^{\rm bnd}_\varepsilon$ in a power series in ε

$$U^{\rm bnd}_\varepsilon \sim \sum_{k=0}^\infty \varepsilon^{2-n+k} U^{\rm bnd}_k\,, \tag{4.27}$$

and substitute into the scaled version of the heat equation and the boundary conditions. By equating the like powers of ε we get an infinite set of recursive differential equations determining all the coefficients $U^{\rm bnd}_k$.

The leading-order heat kernel $U^{\rm bnd}_0$ is determined by the equation

$$(\partial_t + L^{\rm bnd}_0) U^{\rm bnd}_0 = 0 \tag{4.28}$$

with the initial condition

$$U^{\rm bnd}_0(0; r, \hat x, r', \hat x') = \delta(r - r')\delta(\hat x, \hat x')\,, \tag{4.29}$$

the boundary condition

$$U^{\rm bnd}_0(t; 0, \hat x, r', \hat x') = U^{\rm bnd}_0(t; r, \hat x, 0, \hat x') = 0\,, \tag{4.30}$$

and the asymptotic condition

$$\lim_{r\to\infty} U^{\rm bnd}_0(t; r, \hat x, r', \hat x') = \lim_{r'\to\infty} U^{\rm bnd}_0(t; r, \hat x, r', \hat x') = 0\,. \tag{4.31}$$

The higher-order coefficients $U^{\rm bnd}_k$, $k \geq 1$, are determined from the recursive equations

$$(\partial_t + L^{\rm bnd}_0) U^{\rm bnd}_k = -\sum_{j=1}^k L^{\rm bnd}_j U^{\rm bnd}_{k-j}\,, \tag{4.32}$$

with the initial condition

$$U^{\rm bnd}_k(0; r, \hat x, r', \hat x') = 0\,, \tag{4.33}$$

the boundary condition

$$U^{\rm bnd}_k(t; 0, \hat x, r', \hat x') = U^{\rm bnd}_k(t; r, \hat x, 0, \hat x') = 0\,, \tag{4.34}$$

and the asymptotic condition

$$\lim_{r\to\infty} U^{\rm bnd}_0(t; r, \hat x, r', \hat x') = \lim_{r'\to\infty} U^{\rm bnd}_0(t; r, \hat x, r', \hat x') = 0\,. \tag{4.35}$$

This expansion is nothing but the decomposition of the heat kernel into the homogeneous parts with respect to the variables $(\hat x - \hat x_0)$, $(\hat x' - \hat x_0)$, r, r' and $\sqrt{t}$. That is,

$$U^{\rm bnd}_k(t; r, \hat x, r', \hat x') = t^{(k-n)/2} U^{\rm bnd}_k \left(1; \frac{r}{\sqrt{t}}, \hat x_0 + \frac{(\hat x - \hat x_0)}{\sqrt{t}}, \frac{r'}{\sqrt{t}}, \hat x_0 + \frac{(\hat x' - \hat x_0)}{\sqrt{t}}\right)\,. \tag{4.36}$$

In particular, the heat kernel diagonal at the point $(r,\hat{x}_0)$ scales by

$$U_k^{\rm bnd}(t;r,\hat{x}_0,r,\hat{x}_0) = t^{(k-n)/2} U_k^{\rm bnd}\left(1;\frac{r}{\sqrt{t}},\hat{x}_0,\frac{r}{\sqrt{t}},\hat{x}_0\right). \tag{4.37}$$

To compute the contribution of these coefficients to the trace of the heat kernel we need to compute the integral of the diagonal of the heat kernel $U^{\rm bnd}(t;r,\hat{x},r,\hat{x})$ over the boundary layer $M_{\rm bnd}$. This heat kernel diagonal can be decomposed as the sum of two terms, the first coming from the standard interior heat kernel on manifolds without boundary (that does not satisfy the boundary conditions) and the second 'compensating' part, which is the crucial boundary part and whose role is to make the heat kernel to satisfy the boundary conditions (for more details see [10]). The integral of the 'boundary' part over the boundary layer in the limit when the size of the boundary layer goes to zero produces the boundary contributions $b_k(L)$ to the global heat kernel coefficients $A_k(L)$.

By using the homogeneity property (4.37) we obtain

$$\begin{aligned}
\int\limits_{M_{\rm bnd}} dx\,\mathrm{tr}\,_S U^{\rm bnd}(t;x,x) &= \int\limits_{\partial M} d\hat{x}\int\limits_0^{\delta} dr\,\mathrm{tr}\,_S U^{\rm bnd}(t;r,\hat{x},r,\hat{x})\\
&\sim \sum_{k=0}^{\infty} t^{(k-n)/2}\int\limits_{\partial M} d\hat{x}\int\limits_0^{\delta} dr\,\mathrm{tr}\,_S U_k^{\rm bnd}\left(1;\frac{r}{\sqrt{t}},\hat{x},\frac{r}{\sqrt{t}},\hat{x}\right)\\
&\sim \sum_{k=0}^{\infty} t^{(k-n+1)/2}\int\limits_{\partial M} d\hat{x}\int\limits_0^{\delta/\sqrt{t}} du\,\mathrm{tr}\,_S U_k^{\rm bnd}\left(1;u,\hat{x},u,\hat{x}\right)
\end{aligned} \tag{4.38}$$

where $u = r/\sqrt{t}$. Notice the appearance of the extra power of $\sqrt{t}$ in the asymptotic expansion. Of course, if one takes the limit $\lim_{\delta\to 0}$ for a finite t, then all these integrals vanish. However, if one takes the limit $\lim_{t\to 0}$ first for a finite δ, and then the limit $\lim_{\delta\to 0}$, then one gets finite answers for the boundary coefficients $b_k(L)$.

4.2.1. *Leading-order heat kernel*

To compute the coefficient A_1 we just need the leading-order heat kernel $U_0^{\rm bnd}$. We will, in fact, be working in the tangent space $\mathbb{R}_+\times T_{\hat{x}_0}\partial M$ at a point $\hat{x}_0$ on the boundary and reduce our problem to a problem on the half-line. The operator $L_0^{\rm bnd}$ acts on square integrable sections of the vector bundle $\mathcal{S}[\frac{1}{2}]$ in a neighborhood of the point $\hat{x}_0$. We extend the operator appropriately to the space $L^2(\mathcal{S}[\frac{1}{2}], \mathbb{R}_+, \mathbb{R}^{n-1}, dr\, d\hat{x})$ so that it coincides with the initial operator in the neighborhood of the point $\hat{x}_0$. When computing the trace below we set $\hat{x}_0 = \hat{x} = \hat{x}'$.

By using the Laplace transform in the variable t and the Fourier transform in the boundary coordinates $\hat{x}$

$$U_0^{\rm bnd}(t; r, \hat{x}, r', \hat{x}') = \frac{1}{2\pi i} \int\limits_{w-i\infty}^{w+i\infty} d\lambda \int\limits_{\mathbb{R}^{n-1}} \frac{d\hat{\xi}}{(2\pi)^{n-1}} e^{-t\lambda + i\hat{\xi}(\hat{x}-\hat{x}')} F(\lambda, , r, r', \hat{\xi}), \tag{4.39}$$

we obtain an ordinary differential equation

$$\left(-A^2\partial_r^2 - iB\partial_r + C^2 - \lambda\mathbb{I}\right) F(\lambda, r, r', \hat{\xi}) = \mathbb{I}\delta(r - r') \tag{4.40}$$

where the matrices A, B and C are defined in (2.64), (2.65), and are frozen at the point $\hat{x}_0$ (they are constant for the purpose of this calculation), with the boundary condition

$$F(\lambda, 0, r', \hat{\xi}) = F(\lambda, r, 0, \hat{\xi}) = 0 \tag{4.41}$$

the asymptotic condition

$$\lim_{r\to\infty} F(\lambda, r, r', \hat{\xi}) = \lim_{r'\to\infty} F(\lambda, r, r', \hat{\xi}) = 0\,, \tag{4.42}$$

and the self-adjointness condition

$$\overline{F(\lambda, r, r', \hat{\xi})} = F(\bar{\lambda}, r', r, \hat{\xi})\,. \tag{4.43}$$

It is easy to see that F is a homogeneous function

$$F\left(\frac{\lambda}{t}, \sqrt{t}r, \sqrt{t}r', \frac{\hat{\xi}}{\sqrt{t}}, \right) = t^{1/2} F\left(\lambda, r, r', \hat{\xi}\right)\,. \tag{4.44}$$

We decompose the Green function in two parts,

$$F = F_\infty + F_B\,, \tag{4.45}$$

where F_∞ is the part that is valid for the whole real line and F_B is the compensating term. The part F_∞ can be easily obtained by the Fourier transform; it has the form

$$F_\infty(\lambda, r, r', \hat{\xi}) = \Phi(\lambda, r - r', \hat{\xi})\,, \tag{4.46}$$

where $\Phi(\lambda, r, \hat{\xi})$ is defined in (2.67). It is not smooth at the diagonal $r = r'$ and is responsible for the appearance of the delta-function $\delta(r - r')$ on the right-hand side of the eq. (4.40).

The corresponding part of the leading heat kernel is then easily computed to be

$$U^{\rm bnd}_{0,\infty}(t;x,x') = \int\limits_{\mathbb{R}^n} \frac{d\xi}{(2\pi)^n} e^{i\xi(x-x')-tH(x_0,\xi)}\,, \tag{4.47}$$

where $x_0 = (0,\hat{x}_0)$. This part does not contribute to the asymptotics of the trace of the heat kernel in the limit $\delta \to 0$. By rescaling $\xi \mapsto \xi/\sqrt{t}$ we obtain

$$\int\limits_{M_{\rm bnd}} dx\, {\rm tr}_S\, U^{\rm bnd}_{0,\infty}(t;x,x) = (4\pi t)^{-n/2} \int\limits_{M_{\rm bnd}} dx \int\limits_{\mathbb{R}^n} \frac{d\xi}{\pi^{n/2}} {\rm tr}_S\, e^{-H(x,\xi)}\,, \tag{4.48}$$

and in the limit $\delta \to 0$ this integral vanishes.

However, F_∞ does not satisfy the boundary conditions. The role of the boundary part, F_B, is exactly to guarantee that F satisfies the boundary conditions. The function F_B is smooth at the diagonal $r = r'$. It can be presented in the following form

$$F_B(\lambda,r,r',\hat{\xi}) = -\Phi(\lambda,r,\hat{\xi})[\Phi_0(\lambda,\hat{\xi})]^{-1}\Phi(\lambda,-r',\hat{\xi})\,. \tag{4.49}$$

4.2.2. *The coefficient* A_1

The coefficient A_1 is a pure boundary coefficient that is computed by integrating the boundary part $U^{\rm bnd}_{0,B}$ of the heat kernel. We have

$$\int\limits_{M_{\rm bnd}} dx\, {\rm tr}_S\, U^{\rm bnd}_{0,B}(t;x,x) \tag{4.50}$$

$$= \int\limits_{\partial M} d\hat{x} \int\limits_0^\delta dr \int\limits_{\mathbb{R}^{n-1}} \frac{d\hat{\xi}}{(2\pi)^{n-1}} \int\limits_{w-i\infty}^{w+i\infty} \frac{d\lambda}{2\pi i} e^{-t\lambda} {\rm tr}_S\, F_B(\lambda,r,r,\hat{\xi})\,.$$

Now, by rescaling the variables

$$\lambda \mapsto \frac{\lambda}{t}, \qquad r \mapsto \sqrt{t}r, \qquad \hat{\xi} \mapsto \frac{\hat{\xi}}{\sqrt{t}} \tag{4.51}$$

and using the homogeneity property (4.44) we obtain

$$\int\limits_{M_{\rm bnd}} dx\, {\rm tr}_S\, U^{\rm bnd}_{0,B}(t;x,x) \tag{4.52}$$

$$= t^{(1-n)/2} \int\limits_{\partial M} d\hat{x} \int\limits_{\mathbb{R}^{n-1}} \frac{d\hat{\xi}}{(2\pi)^{n-1}} \int\limits_0^{\delta/\sqrt{t}} dr \int\limits_{w-i\infty}^{w+i\infty} \frac{d\lambda}{2\pi i} e^{-\lambda} {\rm tr}_S\, F_B\left(\lambda,r,r,\hat{\xi}\right)\,.$$

Therefore, the coefficient A_1 is given by

$$A_1 = 2\sqrt{\pi}\int\limits_{\partial M} d\hat{x}\int\limits_{\mathbb{R}^{n-1}} \frac{d\hat{\xi}}{\pi^{(n-1)/2}}\int\limits_0^\infty dr \int\limits_{w-i\infty}^{w+i\infty} \frac{d\lambda}{2\pi i} e^{-\lambda} \mathrm{tr}_S\, F_B\left(\lambda, r, r, \hat{\xi}\right). \tag{4.53}$$

Thus, finally, by using eq. (4.49), eliminating the odd functions of $\hat{\xi}$ (since the integrals of them vanish), using the property (2.70) of the function Φ and extending the integration over r from $-\infty$ to $+\infty$ (since the integrand is an even function) we obtain

$$A_1 = \int\limits_{\partial M} d\hat{x}\int\limits_{\mathbb{R}^{n-1}} \frac{d\hat{\xi}}{\pi^{(n-1)/2}} \Psi_1(\hat{\xi}) \tag{4.54}$$

where

$$\Psi_1(\hat{\xi}) = -\frac{\sqrt{\pi}}{2}\int\limits_{-\infty}^{\infty} dr \int\limits_{w-i\infty}^{w+i\infty} \frac{d\lambda}{2\pi i} e^{-\lambda}\, \mathrm{tr}_S\, [\Phi_0(\lambda, \hat{\xi})]^{-1}$$
$$\times\left\{\Phi(\lambda, r, \hat{\xi})\Phi(\lambda, -r, \hat{\xi}) + \Phi(\lambda, -r, \hat{\xi})\Phi(\lambda, r, \hat{\xi})\right\}. \tag{4.55}$$

Recall that w is a sufficiently large negative constant.

Now, using eq. (2.72) and integrating over r we obtain finally

$$\Psi_1(\xi) = -\sqrt{\pi}\int\limits_{w-i\infty}^{w+i\infty} \frac{d\lambda}{2\pi i}\, e^{-\lambda}\, \mathrm{tr}_S\, [\Phi_0(\lambda, \hat{\xi})]^{-1} \frac{\partial}{\partial\lambda}\Phi_0(\lambda, \hat{\xi})$$
$$= -\sqrt{\pi}\int\limits_{w-i\infty}^{w+i\infty} \frac{d\lambda}{2\pi i}\, e^{-\lambda}\, \frac{\partial}{\partial\lambda} \log\det[\Phi_0(\lambda, \hat{\xi})]. \tag{4.56}$$

Thus, the problem is now reduced to the computation of the integral over λ. This is not at all trivial because of the presence of two non-commuting matrices, essentially, $A^{-1}(AC + CA)A^{-1}$ and $A^{-1}(C^2 - \lambda\mathbb{I})A^{-1}$, where the matrices $A = \Gamma^r(\hat{x})$ and $C = \Gamma^j(\hat{x})\hat{\xi}_j$ are defined by (2.64). We will report on this problem in a future work. Here let us just mention that in the particular case when $B = AC + CA = 0$ (for example, this is so in the case of the original Dirac operator) we get

$$\mathrm{tr}_S\, [\Phi_0(\lambda, \hat{\xi})]^{-1} \frac{\partial}{\partial\lambda}\Phi_0(\lambda, \hat{\xi}) = \frac{1}{2}\mathrm{tr}_S\, (C^2 - \lambda\mathbb{I})^{-1}, \tag{4.57}$$

and, therefore, one can compute the integral over λ to obtain

$$A_1 = -\frac{\sqrt{\pi}}{2}\int\limits_{\partial M} d\hat{x}\int\limits_{\mathbb{R}^{n-1}} \frac{d\hat{\xi}}{\pi^{(n-1)/2}}\,\mathrm{tr}_{\,S}\, e^{-[C(\hat{x},\hat{\xi})]^2}\,. \tag{4.58}$$

Of course, for Laplace type operators, when $[C(\hat{x},\hat{\xi})]^2 = \mathbb{I}g^{ij}(\hat{x})\hat{\xi}_i\hat{\xi}_j$, the integral can be computed explicitly, which gives the induced Riemannian volume of the boundary, $A_1 = -(\sqrt{\pi}/2)\,N\,\mathrm{vol}\,(\partial M)$, and coincides with the standard result for Dirichlet Laplacian [29].

Acknowledgement. I would like to thank the organizers of the conference "Boundary Value Problems and Friends", Bernhelm Booss-Bavnbek, Slawomir Klimek and Ryszard Nest, for their kind invitation to present this work. It was a pleasure to contribute to this special issue dedicated to Krzysztof Wojciechowski. We all wish him soon recovery and best of luck in the future.

References

1. I. G. Avramidi, *Covariant methods for the calculation of the effective action in quantum field theory and investigation of higher-derivative quantum gravity*, PhD Thesis, Moscow State University (1987), arXiv:hep-th/9510140
2. I. G. Avramidi, *A covariant technique for the calculation of the one-loop effective action*, Nucl. Phys. **B355** (1991) 712–754 [Erratum Nucl. Phys. **B509** (1998) 557–558].
3. I. G. Avramidi, *A method for calculating the heat kernel for manifolds with boundary*, Yadernaya Fizika **56** (1993) 245–252 [Russian]; Phys. Atomic Nuclei **56** (1993) 138–142 [English].
4. I. G. Avramidi, *Covariant techniques for computation of the heat kernel*, Rev. Math. Phys. **11** (1999) 947–980.
5. I. G. Avramidi, *Heat Kernel and Quantum Gravity*, Lecture Notes in Physics. New Series m: Monographs, 64, Springer-Verlag, Berlin, 2000.
6. I. G. Avramidi, *Heat kernel approach in quantum field theory*, Nuclear Phys. B Proc. Suppl. **104** (2002) 3–32.
7. I. G. Avramidi, *A non-commutative deformation of general relativity*, Phys. Lett. **B576** (2003) 195–198.
8. I. G. Avramidi, *Matrix general relativity: a new look at old problems*, Classical Quantum Gravity **21** (2004) 103–120.
9. I. G. Avramidi, *Gauged gravity via spectral asymptotics of non-Laplace type operators*, J. High Energy Phys. **07** (2004) 030.
10. I. G. Avramidi, *Heat kernel asymptotics of Zaremba boundary value problem*, Math. Phys. Anal. Geom. **7** (2004) 9–46.
11. I. G. Avramidi, *Dirac operator in matrix geometry*, Int. J. Geom. Methods Mod. Phys. **2** (2005) 227–264.
12. I. G. Avramidi and T. Branson, *Heat kernel asymptotics of operators with non-Laplace principal part*, Rev. Math. Phys. **13** (2001) 847–890.

13. I. G. Avramidi and T. Branson, *A discrete leading symbol and spectral asymptotics for natural differential operators*, J. Funct. Anal. **190** (2002) 292–337.
14. I. G. Avramidi and G. Esposito, *Lack of strong ellipticity in Euclidean quantum gravity*, Classical Quantum Gravity **15** (1998) 1141–1152.
15. I. G. Avramidi and G. Esposito, *Gauge theories on manifolds with boundary*, Comm. Math. Phys. **200** (1999) 495–543.
16. I. G. Avramidi and G. Esposito, *Heat kernel asymptotics of the Gilkey-Smith boundary value problem*, in: *Trends in Mathematical Physics*, Eds. V. Alexiades and G. Siopsis, AMS/IP Studies in Advanced Mathematics, vol. 13, American Mathematical Society and International Press, 1999, pp. 15–34.
17. I. G. Avramidi and R. Schimming, *Algorithms for the calculation of the heat kernel coefficients*, in: *Quantum Field Theory under the Influence of External Conditions*, Ed. M. Bordag, Teubner-Texte zur Physik, Band 30, Teubner, Stuttgart, 1996, pp. 150–162.
18. N. Berline, E. Getzler and M. Vergne, *Heat Kernels and Dirac Operators*, Springer-Verlag, Berlin, 1992.
19. B. Booss-Bavnbek and K. P. Wojciechowski, *Elliptic Boundary Problems for Dirac Operators*, Birkhäuser, Boston, 1993.
20. T. Branson, *Sharp inequalities, the functional determinant and the complimentary series*, Trans. Amer. Math. Soc. **347** (1995) 3671–3742.
21. T. Branson, *Stein-Weiss operators and ellipticity*, J. Funct. Anal. **151** (1997) 334–383.
22. T. Branson and P. B. Gilkey, *The asymptotics of the Laplacian on a manifold with boundary*, Comm. Partial Differential Equations **15** (1990) 245–272.
23. T. P. Branson, P. B. Gilkey and A. Pierzchalski, *Heat equation asymptotics of elliptic differential operators with non-scalar leading symbol*, Math. Nachr. **166** (1994) 207–215.
24. T. Branson, P. B. Gilkey, K. Kirsten and D. V. Vassilevich, *Heat kernel asymptotics with mixed boundary conditions*, Nuclear Phys. **B563** (1999) 603–626.
25. J. S. Dowker, P. B. Gilkey and K. Kirsten, *On properties of the asymptotic expansion of the heat trace for the N/D problem*, Internat. J. Math. **12** (2001) 505–517.
26. J. S. Dowker and K. Kirsten, *Heat-kernel coefficients for oblique boundary conditions*, Classical Quantum Gravity **14** (1997) L169–L175.
27. J. S. Dowker and K. Kirsten, *The* $a_{3/2}$ *heat-kernel coefficient for oblique boundary conditions*, Classical Quantum Gravity **16** (1999) 1917–1936.
28. P. B. Gilkey, *The spectral geometry of Riemannian manifold*, J. Differential Geom. **10** (1975) 601–618.
29. P. B. Gilkey, *Invariance Theory, the Heat Equation and the Atiyah-Singer Index Theorem*, CRC Press, Boca Raton, 1995.
30. P. B. Gilkey, T. P. Branson and S. A. Fulling, *Heat equation asymptotics of "nonminimal" operators on differential forms*, J. Math. Phys. **32** (1991) 2089–2091.
31. P. B. Gilkey and L. Smith, *The twisted index problem for manifolds with boundary*, J. Differential Geom. **18** (1983) 393–444.
32. P. B. Gilkey and L. Smith, *The eta invariant for a class of elliptic boundary value problems*, Comm. Pure Appl. Math. **36** (1983) 85–132.
33. P. Greiner, *An asymptotic expansion for the heat equation*, Arch. Rational Mech. Anal. **41** (1971) 163–218.

34. G. Grubb, *Properties of normal boundary value problems for elliptic even-order systems*, Ann. Scuola Norm. Sup. Pisa Cl. Sci. (4) **1** (1974), 1–61.
35. G. Grubb, *Functional Calculus of Pseudodifferential Boundary Problems*, Progress of Mathematics, **65**, Birkhäuser, Boston, 1996.
36. M. Kac, *Can one hear the shape of a drum?*, Amer. Math. Monthly **73** (1966) 1–23.
37. K. Kirsten, *The a_5 heat kernel coefficient on a manifold with boundary*, Classical Quantum Gravity **15** (1998) L5–L12.
38. K. Kirsten, *Spectral Functions in Mathematics and Physics*, CRC Press, Boca Raton, 2001.
39. H. P. McKean, Jr, and I. M. Singer, *Curvature and the eigenvalues of the Laplacian*, J. Differential Geom. **1** (1967) 43–69.
40. H. Rund, *The Differential Geometry of Finsler Spaces*, Nauka, Moscow, 1981 [Russian]; Springer-Verlag, Berlin, 1959 [English].
41. R. T. Seeley, *The resolvent of an elliptic boundary value problem*, Amer. J. Math. **91** (1969) 889–920.
42. R. T. Seeley, *Trace Expansions for the Zaremba Problem*, Comm. Partial Differential Equations **27** (2002) 2403–2421.
43. D. V. Vassilevich, *Heat kernel expansion: user's manual*, Phys. Rep. **388** (2003) 279–360.
44. S. Yajima, Y. Higasida, K. Kawano, S.-I. Kubota, Y. Kamo and S. Tokuo, *Higher coefficients in asymptotic expansion of the heat kernel*, Phys. Rep. Kumamoto Univ. **12** (2004), No 1, 39–62.

Received by the editors September 12, 2005; Revised December 23, 2005

Analysis, Geometry and Topology
of Elliptic Operators, pp. 141–172

ETA INVARIANTS FOR MANIFOLD WITH BOUNDARY

XIANZHE DAI
Math. Dept., UCSB,
Santa Barbara, CA 93106, USA
dai@math.ucsb.edu

Dedicated to Krzysztof P. Wojciechowski on his 50th birthday

For a compact manifold with boundary, M, there are well known local boundary conditions that make the de Rham operator $A = d+\delta$ elliptic, namely the absolute and relative boundary conditions. We study the eta invariants of such elliptic boundary value problems under the metric deformation

$$g_\epsilon = \frac{dx^2}{x^2+\epsilon^2} + g,$$

where $x \in C^\infty(M)$ is, near the boundary, the geodesic distance to the boundary, and g is a Riemannian metric on M which is of product type near the boundary. Under certain acyclicity condition we show that when M is odd dimensional

$$\eta(A_a) = \eta(A_r) = \eta_b(A_0),$$

where the subscript a (r) indicates the absolute (relative) boundary condition, and $\eta_b(A_0)$ is the b-eta invariant of the limiting operator A_0. If M is even dimensional then

$$\eta(A_a) = -\eta(A_r) = \frac{1}{2}\eta(A_{\partial M}).$$

Most of the analysis extends to analytic torsion, yielding

$$\log T_\epsilon(M,\rho) = \log {}^bT(\bar{M},\rho) + r_1(\epsilon) + r_2(\epsilon)\log\epsilon$$

when $\dim M$ is odd, and

$$\log T_\epsilon(M,\rho) = \pm\frac{1}{2}\log T(\partial M,\rho) + r_1(\epsilon) + r_2(\epsilon)\log\epsilon$$

when $\dim M$ is even. Here the sign $\pm$ depends on the choice of the boundary condition and r_1, r_2 vanishes at $\epsilon = 0$.

2000 *Mathematics Subject Classification.* Primary 58J28; Secondary 35J25

Supported in part by the NSF

1. Introduction

The eta invariant for a closed manifold is introduced by Atiyah-Patodi-Singer [1] as the boundary correction term in the index formula for manifold with boundary. It has found many significant applications in diverse fields of mathematics and physics. There are now various works generalizing it to manifolds with boundary. Using his cone method, Cheeger [5] introduced an eta invariant in the context of manifolds with conical singularity. In [9] Gilkey and Smith considered eta invariants for local boundary conditions. On the other hand, Douglas and Wojciechowski defined and studied eta invariants for generalized APS boundary conditions [8] (see also Bunke [4], Lesch-Wojciechowski [12], Müller [19]). Also, in the context of manifolds with asymptotically cylindrical end Melrose introduced a regularized eta invariant, the b-eta invariant [18]. Meanwhile Müller [19] introduced an L^2-eta invariant for manifolds with cylindrical end, which turns out to be the same as the b-eta invariant. We also note that in his work on Casson invariant [23], Taubes used the local boundary condition, while in the subsequent work by others it is the APS boundary condition that is used, see, for example, Yoshida [25]. Thus it is a natural and interesting question to clarify the relationships among the various generalizations.

In the very interesting work [19] Müller considered the relationship between the eta invariants for generalized APS boundary conditions and the L^2-eta (or the b-eta) invariants. Using scattering theory he showed that they are essentially the same. Earlier Douglas and Wojciechowski [8] have considered the situation where the boundary operator is invertible.

In this work we consider the relationship between the eta invariants for local boundary conditions and the b-eta (or L^2-eta) invariants for the (twisted) de Rham operator $A = d + \delta$. Under certain acyclicity condition we show that they are the same. Thus, at least for de Rham operators, the three generalizations of eta invariant to manifolds with boundary, using local boundary condition, generalized APS boundary condition, or L^2 condition, all coincide.

Theorem 1.1. *Let M be a compact manifold with boundary and ξ a flat unitary bundle over M such that $H^*(\partial M, \xi) = 0$ and $Im(H^*(M, \partial M; \xi) \to H^*(M; \xi)) = 0$. Then if* $\dim M$ *is odd we have*

$$\eta(A_a) = \eta(A_r) = \eta_b(A_0),$$

where subscript 'a' ('r') denotes the absolute (relative) boundary condition, and A_0 is the de Rham operator on the complete manifold obtained from

M by attaching an infinite half cylinder. On the other hand, if $\dim M$ *is even, then*

$$\eta(A_a) = -\eta(A_r) = \frac{1}{2}\eta(A_{\partial M}).$$

The theorem is proved by considering the behavior of the eta invariant on the manifold with boundary under a metric degeneration in which the boundary is being 'pushed' to infinity. This is motivated by the work [14] of R. Mazzeo and R. Melrose who studied the behavior of eta invariant on a closed manifold under the metric deformation

$$g_\epsilon = \frac{dx^2}{x^2 + \epsilon^2} + h, \tag{1.1}$$

where x is a defining function for an embedded hypersurface. The limiting metric g_0 for (1.1) is an exact b-metric on the compact manifold with boundary obtained by cutting along the hypersurface. (An exact b-metric gives the manifold with boundary asymptotically cylindrical ends.) Under the assumption that the induced Dirac operator on (a double cover of) the hypersurface is invertible, Mazzeo and Melrose showed that

$$\eta(D_\epsilon) = \eta_b(D_0) + r_1(\epsilon) + r_2(\epsilon)\log\epsilon + \tilde{\eta}(\epsilon), \tag{1.2}$$

where D_ϵ is the Dirac operator associated to the metric g_ϵ, and $\eta_b(D_0)$ is the b-eta invariant of the (b-)Dirac operator D_0 associated to the metric g_0. Also, r_1, r_2 are smooth functions vanishing at $\epsilon = 0$. Finally, $\tilde{\eta}(\epsilon)$ is the signature of the small eigenvalues of D_ϵ. This analysis is extended to analytic torsion by Hassell in [10].

We consider the corresponding case for manifold with boundary and let the boundary play the role of the hypersurface in [14]. We study the eta invariants of elliptic boundary value problems under the metric deformation (1.1). In this case a formula similar to (1.2) holds. We also show that the eta invariant does not change under this deformation.

Another source of inspiration comes from a paper of I. M. Singer, [22], and the subsequent work of Klimek-Wojciechowski [11]. Singer considers the difference of two eta invariants of Dirac operators with local boundary conditions and shows that the limit of the difference under stretching is the log determinant. The result is viewed as an analog of the identity that the difference of the indexes of the two elliptic boundary value problems for Dirac operators is given by the index of the Dirac operator on the boundary. This is given full mathematical treatment and generalized in [11].

The consideration in [22] is motivated by E. Witten's 'adiabatic limit'. For this and related topics we refer to Witten [24], Bismut-Freed [3],

Bismut-Cheeger [2], Cheeger [5], Dai [6], Mazzeo-Melrose [13] and Singer [22].

The idea of studying the behavior of eta invariant under singular degeneration probably goes back to [5] where the particular case of conical degeneration is briefly discussed. Conical degeneration has been discussed to greater extent by R. Seeley and Singer, see Seeley [20] and Seeley-Singer [21].

Finally, let us mention that the same analysis applies to analytic torsions as well, see §3 for the statement of the result (Theorem 3.3).

2. Elliptic boundary value problem and eta invariant

Let M be a compact manifold with boundary and V a vector bundle over M. Let

$$P: \; C^\infty(M, V) \to C^\infty(M, V)$$

be a differential operator of order d and B a boundary condition. By P_B we denote the realization of the boundary value problem (P, B); namely, P_B is the operator P acting on the space of smooth sections verifying $B(\phi|_{\partial M}) = 0$. Let

$$\mathcal{C} = \{z : \; |\mathrm{Re} z| \le |\mathrm{Im} z|\}$$

be the closed 45° cone about the imaginary axis in the complex plane. According to [9], when (P, B) is elliptic with respect to $\mathcal{C}$, P_B has discrete spectrum with finite multiplicity, all except finite of which lie inside $\mathcal{C}$. Let $\{\lambda_i\}$ denote the spectrum of P_B where each spectral value is repeated according to its multiplicity. Gilkey-Smith defined

$$\eta(s, P, B) = \sum_{\mathrm{Re}\lambda_i > 0} \lambda_i^{-s} - \sum_{\mathrm{Re}\lambda_i < 0} (-\lambda_i)^{-s}$$

for $\mathrm{Re} s \gg 0$ and showed that η has a meromorphic extension to the whole complex plane with isolated simple poles. Unlike the case when M is boundaryless, $s = 0$ may be a simple pole here. However, the residue being a local homotopy invariant, one defines the eta invariant

$$\eta(P, B) = \text{finite part of } \eta(s, P, B) \text{ at } 0 = (s\eta(s, P, B))'|_{s=0}.$$

When P_B has no eigenvalues lying inside $\mathcal{C}$, $\eta(s, P, B)$ can be expressed in terms of heat kernel as is the case when $\partial M = \emptyset$:

$$\eta(s, P, B) = \frac{1}{\Gamma(\frac{s+1}{2})} \int_0^\infty t^{\frac{s-1}{2}} \mathrm{Tr}(P_B e^{-tP_B^2}) dt. \tag{2.3}$$

(When P_B does have eigenvalues lying inside $\mathcal{C}$, one just have to treat them separately.) Here $P_B e^{-tP_B^2}$ is defined via functional calculus

$$P_B e^{-tP_B^2} = \frac{-1}{2\pi i}\int_\Gamma (P_B - \lambda)^{-1}\lambda e^{-t\lambda^2}\,d\lambda$$

with Γ an appropriate contour.

Thus defined, this invariant behaves much like the usual eta for manifold without boundary. For example, one has the following variation formula [9]:

Theorem 2.1. *Let (P_u, B) be a smooth one-parameter family which is elliptic with respect to $\mathcal{C}$. Then*

$$\frac{d}{du}\{Res_{s=0}\eta(s, P_u, B)\} = 0.$$

Further, if no eigenvalues lie inside $\mathcal{C}$, then the variation of eta itself is given by a local formula

$$\frac{d}{du}\eta(P_u, B) = \int_M a(y, P_u', P_u)dvol(y) + \int_{\partial M} a(x, P_u', P_u, B)dvol(x),$$

where the $a(y, P_u', P_u)$ and $a(x, P_u', P_u, B)$ are the coefficients of $t^{-1/2}$ in the asymptotic expansion for $tr(P_u' e^{-tP_{u,B}^2})$.

We now specialize to the de Rham operator. Let M be an odd dimensional compact manifold with boundary and g be a Riemannian metric on M which is of product type near the boundary

$$g = dx^2 + g_{\partial M},$$

where x is the geodesic distance to the boundary. Let $\xi \to M$ be the flat bundle associated to a representation $\rho : \pi_1(M) \to O(k)$. By de Rham operator we mean

$$A = d + \delta : \; C^\infty(M; \Lambda(M) \otimes \xi) \to C^\infty(M; \Lambda(M) \otimes \xi). \tag{2.4}$$

At the boundary we have the splitting

$$\Lambda(M) \otimes \xi|_{\partial M} = \Lambda(\partial M) \otimes \xi \oplus \Lambda(\partial M) \otimes \xi \tag{2.5}$$

corresponding to the decomposition for a form $\theta \in C^\infty(M; \Lambda(M) \otimes \xi)$:

$$\theta = \theta_1 + dx \wedge \theta_2, \; \theta_1, \theta_2 \in C^\infty(M; \Lambda(\partial M) \otimes \xi)$$

near the boundary. Define a linear map σ:

$$\sigma(\theta) = \theta_1 - dx \wedge \theta_2.$$

Then σ is self adjoint and $\sigma^2 = 1$. Moreover the splitting (2.5) corresponds to the decomposition into the ± 1-eigenspace of σ.

From the splitting we define two projections

$$P_a,\ P_r:\ C^\infty(\partial M; \Lambda(M)\otimes \xi|_{\partial M}) \to C^\infty(\partial M; \Lambda(\partial M)\otimes \xi),$$
$$P_a(\theta) = \theta_2|_{\partial M}; \qquad P_r(\theta) = \theta_1|_{\partial M}.$$

I.e., P_a is the orthogonal projection onto the -1-eigenspace of σ and P_r the orthogonal projection onto the $+1$-eigenspace. Let A_a (resp. A_r) be the de Rham operator equipped with the boundary condition P_a (resp. P_r). Then A_a, A_r are elliptic boundary value problems; in fact they are also self adjoint. Hence $\eta(A_a)$ and $\eta(A_r)$ can be defined and moreover, because of the self-adjointness, the eta functions are actually regular at 0.

3. Deforming eta invariant

Now, for ϵ a positive parameter, consider the family of metrics

$$g_\epsilon = \frac{dx^2}{x^2+\epsilon^2} + g. \tag{3.6}$$

The limiting metric g_0 is an exact b-metric on M, in the terminology of Melrose [18]. Let $A_{\epsilon,a}$ ($A_{\epsilon,r}$) be the associated elliptic boundary value problems. We note in the passing that the metric deformation (3.6) leaves invariant the projections P_a (P_r), hence the boundary conditions. Let us also denote by A_0 the b-de Rham operator associated with g_0 (see [18]).

Theorem 3.1. *Assume that* $H^*(\partial M, \xi) = 0$ *and* $Im(H^*(M, \partial M; \xi) \to H^*(M;\xi)) = 0$. *Then if* $\dim M$ *is odd*

$$\eta(A_{\epsilon,a}) = \eta_b(A_0) + r_1(\epsilon) + r_2(\epsilon)\log\epsilon, \tag{3.7}$$
$$\eta(A_{\epsilon,r}) = \eta_b(A_0) + r_1(\epsilon) + r_2(\epsilon)\log\epsilon. \tag{3.8}$$

And if $\dim M$ *is even,*

$$\eta(A_{\epsilon,a}) = \frac{1}{2}\eta(A_{\partial M}) + r_1(\epsilon) + r_2(\epsilon)\log\epsilon, \tag{3.9}$$
$$\eta(A_{\epsilon,r}) = -\frac{1}{2}\eta(A_{\partial M}) + r_1(\epsilon) + r_2(\epsilon)\log\epsilon. \tag{3.10}$$

As before, r_1, r_2 *are smooth functions of* ϵ *vanishing at* 0.

Remark. 1. Without the assumption that $H^*(\partial M, \xi) = 0$ and $\mathrm{Im}(H^*(M, \partial M; \xi) \to H^*(M;\xi)) = 0$, the analysis of the small eigenvalues is much more complicated. In [19] this is dealt with via the scattering theory. Similar idea should apply here, which will be treated elsewhere.

Remark. 2. Intuitively the formula can be seen as follows. As $\epsilon \to 0$ the boundary is pushed to the infinity and in the heat kernel the interior contribution and boundary contribution separate. So in the end one is left with a manifold with cylindrical end and an infinite half-cylinder. The b-eta invariant comes from the former, and, depending on the parity of dimension, the contribution from the half-cylinder is either zero or the eta invariant of the boundary. In our proof this intuitive picture is realized geometrically by method of boundary-fibration structure of Melrose [17], [16].

When (M, g) is of product type near the boundary the eta invariant can actually be shown to be invariant under this deformation. Thus we have

Theorem 3.2. *Assume additionally that (M, g) is a product near the boundary. Then $\eta(A_{\epsilon,a}) \equiv \eta(A_a)$ is a constant independent of ϵ. The same is true for $\eta(A_{\epsilon,r})$.*

Proof. Since (M, g) is a product near the boundary we can assume that near the boundary

$$g_\epsilon = \frac{dx^2}{x^2 + \epsilon^2} + g_{\partial M},$$

where $g_{\partial M}$ is a metric on the boundary independent of both x and ϵ. Put $y = \int_0^x \frac{dx}{\sqrt{x^2+\epsilon^2}}$. Then

$$g_\epsilon = dy^2 + g_{\partial M},$$

with $y \in [0, R(\epsilon)]$, $R(\epsilon) = \int_0^1 \frac{dx}{\sqrt{x^2+\epsilon^2}} \to \infty$ as $\epsilon \to 0$. Now choose a diffeomorphism $\varphi_\epsilon : [0,1] \to [0, R(\epsilon)]$ such that $\varphi_\epsilon(t) = t$, $t \in [0, 1/4]$ and $\varphi_\epsilon(t) = t + R(\epsilon) - 1$, $t \in [3/4, 1]$, and $\varphi'_\epsilon(t)$ symmetric with respect to $t = 1/2$. Then

$$g_\epsilon = (\varphi'_\epsilon(t))^2 dt^2 + g_{\partial M}.$$

By Theorem 2.1 the variation of $\eta(A_{\epsilon,a})$ is the same as that of $\eta(A_\epsilon)$, where A_ϵ is the corresponding operator on $\partial M \times S^1$ with the metric $(\varphi'_\epsilon(t))^2 dt^2 + g_{\partial M}$. By the symmetry of $\varphi'_\epsilon(t)$, we have $\eta(A_\epsilon) \equiv 0$. Therefore

$$\frac{d}{d\epsilon}\eta(A_{\epsilon,a}) \equiv 0. \qquad \square$$

Theorem 1.1 follows from Theorem 3.1 and Theorem 3.2.

The same analysis (except the invariance under the deformation) also applies to the analytic torsion. Thus let $T^a_\epsilon(M, \rho)$ ($T^r_\epsilon(M, \rho)$ resp.) denote the analytic torsion associated to the representation $\rho: \pi_1(M) \to O(k)$ and the absolute (relative resp.) boundary condition on M with the metric (1.1).

Theorem 3.3. *Assume that* $H^*(\partial M, \xi) = 0$ *and* $Im(H^*(M, \partial M; \xi) \to H^*(M; \xi)) = 0$. *Then if* $\dim M$ *is odd*

$$\log T_\epsilon(M, \rho) = \log {}^bT(\bar{M}, \rho) + r_1(\epsilon) + r_2(\epsilon) \log \epsilon, \tag{3.11}$$

where ${}^bT(\bar{M}, \rho)$ *is the analytic torsion for manifold with cylindrical end (the b-torsion [18]). Here the analytic torsion on* M *is with respect to either of the boundary conditions. If* $\dim M$ *is even*

$$\log T^a_\epsilon(M, \rho) = \frac{1}{2} \log T(\partial M, \rho) + r_1(\epsilon) + r_2(\epsilon) \log \epsilon, \tag{3.12}$$

and

$$\log T^r_\epsilon(M, \rho) = -\frac{1}{2} \log T(\partial M, \rho) + r_1(\epsilon) + r_2(\epsilon) \log \epsilon. \tag{3.13}$$

The proof of Theorem 3.1 will be deferred to the last section, after the study of the uniform structure of the heat kernels involved. The rest of the paper is organized as follows. After the model case of the half infinite cylinder is discussed, we first show that for ϵ sufficiently small, the spectrum of $A_{\epsilon,a}$ falls uniformly outside a small neighborhood of the origin. This gives us sufficient control over the large time behavior of the heat kernel. Then for the finite time behavior the uniform structure of the heat kernel is examined by constructing the heat surgery 0-calculus, which is then exploited via Laplace transform to also extract information about the resolvent. The large time behavior of the heat kernel follows. Finally combining all these analyses we prove Theorem 3.1.

4. Computation on the half-cylinder

Our heat operator is defined via functional calculus:

$$e^{-tA_a^2} = \frac{i}{2\pi} \int_\Gamma (A_a - \lambda)^{-1} e^{-t\lambda^2} d\lambda.$$

Clearly, it satisfies the heat equation

$$(\partial_t + A_a^2) e^{-tA_a^2} = 0$$

with the correct initial condition:

$$e^{-tA_a^2}|_{t=0} = \mathrm{Id}.$$

From its definition, and the fact that

$$A_a e^{-tA_a^2} = \frac{i}{2\pi}\int_\Gamma (A_a-\lambda)^{-1}\lambda e^{-t\lambda^2}\,d\lambda,$$

it also satisfies the following boundary conditions:

$$\begin{cases} P_a e^{-tA_a^2}|_{x=0} &= 0 \\ P_a A e^{-tA_a^2}|_{x=0} = 0. \end{cases} \tag{4.14}$$

For our purpose it is easier to deal with heat kernels satisfying such boundary conditions. As we are going to show later that the heat kernels satisfying such boundary conditions are unique, they are the same as defined via functional calculus.

For later purpose, and also to get a flavor of the boundary condition, we now consider the situation on the infinite half-cylinder:

$$H = \partial M \times [0,\infty). \tag{4.15}$$

In this case we have the global decomposition

$$\Lambda^*(H) = \Lambda^*(\partial M)\oplus\Lambda^*(\partial M). \tag{4.16}$$

With respect to this decomposition $\theta_1 + du\wedge\theta_2$ corresponds to (θ_1,θ_2) (where we now use u to denote the variable in $[0,\infty)$. Therefore

$$d = \begin{pmatrix} d_{\partial M} & 0 \\ \partial_u & -d_{\partial M}\end{pmatrix}.$$

Hence

$$A = \gamma\partial_u + \sigma A_{\partial M}, \tag{4.17}$$

where

$$\gamma = \begin{pmatrix} 0 & -1 \\ 1 & 0\end{pmatrix}, \quad \sigma = \begin{pmatrix} 1 & 0 \\ 0 & -1\end{pmatrix}.$$

We consider only A_a^2, the other being similar. Its heat kernel E satisfies (4.14). Write E in terms of the decomposition (4.16):

$$E = \begin{pmatrix} E_{11} & E_{12} \\ E_{21} & E_{22}\end{pmatrix}.$$

Then the equation (4.14) becomes four parabolic boundary value problems:

$$\begin{cases} (\partial_t - \partial_u^2 + A_{\partial M}^2)E_{11} = 0, \\ E_{11}|_{t=0} = \mathrm{Id}, \\ (\partial_u E_{11} - A_{\partial M}E_{21})|_{u=0} = 0. \end{cases} \tag{4.18}$$

$$\begin{cases} (\partial_t - \partial_u^2 + A_{\partial M}^2)E_{12} = 0, \\ E_{12}|_{t=0} = 0, \\ (\partial_u E_{12} - A_{\partial M} E_{22})|_{u=0} = 0. \end{cases} \tag{4.19}$$

$$\begin{cases} (\partial_t - \partial_u^2 + A_{\partial M}^2)E_{21} = 0, \\ E_{21}|_{t=0} = 0, \\ E_{21}|_{u=0} = 0. \end{cases} \tag{4.20}$$

$$\begin{cases} (\partial_t - \partial_u^2 + A_{\partial M}^2)E_{22} = 0, \\ E_{22}|_{t=0} = \mathrm{Id}, \\ E_{22}|_{u=0} = 0. \end{cases} \tag{4.21}$$

The same discussion applies to the heat kernel on our manifold with boundary, restricted to the cylindrical part, since everything is local. From here we have the uniqueness of the heat kernel.

Proposition 4.1. *Let M be a compact Riemannian manifold with boundary, with product metric near the boundary. Let A_a be the de Rham operator equipped with the absolute boundary condition defined above. The heat kernel E satisfying*

$$\begin{cases} (\partial_t + A_a^2)E = 0 \\ E|_{t=0} = Id \\ P_a E|_{u=0} = 0, \ P_a AE|_{u=0} = 0 \end{cases}$$

is unique.

Proof. If E and E' are two heat kernels satisfying the above equations, then $\widetilde{E} = E - E'$ satisfies the same set of equations except the initial condition, which should be replaced by $\widetilde{E}|_{t=0} = 0$. We first look at $\widetilde{E}$ near the boundary where it decomposes into $\widetilde{E}_{11}, \cdots, \widetilde{E}_{22}$ satisfying, respectively, (4.18) - (4.21), but once again with initial conditions replaced by zero ones. Now (4.20), (4.21) are heat equations with Dirichlet boundary condition, therefore by the energy estimate, we have $\widetilde{E}_{21} \equiv 0$, $\widetilde{E}_{22} \equiv 0$ (on the cylindrical part). From this, we find that (4.18), (4.19) reduce to heat equations with Neumann boundary condition. Hence again by the energy estimate we have $\widetilde{E}_{11} \equiv 0$, $\widetilde{E}_{12} \equiv 0$ on the cylindrical part. Now this implies that $\widetilde{E}$ satisfies a heat equation on the whole manifold with completely decoupled Dirichlet and Neumann boundary conditions. Therefore again we invoke the energy estimate to deduce that $\widetilde{E} \equiv 0$ on M. □

We now return to the half-cylinder. The last equation is a Dirichlet problem and can be solved explicitly in terms of the heat kernel on ∂M:

$$E_{22} = f_D(t,u,v)e^{-tA^2_{\partial M}}, \tag{4.22}$$

where

$$f_D(t,u,v) = \frac{1}{\sqrt{4\pi t}}(e^{-(u-v)^2/4t} - e^{-(u+v)^2/4t}).$$

The third equation has the trivial solution $E_{21} = 0$. Hence the first equation (4.18) becomes a Neumann problem while the second also gives the trivial solution:

$$E_{11} = f_N(t,u,v)e^{-tA^2_{\partial M}}, \tag{4.23}$$

where

$$f_N(t,u,v) = \frac{1}{\sqrt{4\pi t}}(e^{-(u-v)^2/4t} + e^{-(u+v)^2/4t}).$$

It follows that

$$e^{-tA^2_a} = e^{-tA^2_{\partial M}} \begin{pmatrix} f_N & 0 \\ 0 & f_D \end{pmatrix}. \tag{4.24}$$

Similarly

$$e^{-tA^2_r} = e^{-tA^2_{\partial M}} \begin{pmatrix} f_D & 0 \\ 0 & f_N \end{pmatrix}. \tag{4.25}$$

We now compute the pointwise trace $tr(A_a e^{-tA^2_a})$. Using (4.17) and (4.24) we find

$$tr(A_a e^{-tA^2_a}) = \frac{1}{\sqrt{\pi t}} e^{-u^2/t} tr(A_{\partial M} e^{-tA^2_{\partial M}}). \tag{4.26}$$

Integrating (4.26) gives

$$Tr(A_a e^{-tA^2_a}) = \frac{1}{2} Tr(A_{\partial M} e^{-tA^2_{\partial M}}). \tag{4.27}$$

Consequently we deduce

Proposition 4.2. *For the infinite half cylinder,*

$$\eta(A_a) = -\eta(A_r) = \frac{1}{2}\eta(A_{\partial M}). \tag{4.28}$$

5. Surgery 0-calculus

The proof of Theorem 3.1 depends essentially on the analysis of the uniform structure of the heat kernels for the elliptic boundary value problems. As in [14] this will be examined from the point of view of boundary-fibration structure (see [17]). That is, a calculus of pseudo-differential operators will be constructed, quite geometrically in the sense that the Schwartz kernels of these operators are to live on a space obtained from the usual space by blowing up certain submanifolds. The blowup resolves, analytically and geometrically, the singularities of the Schwartz kernels of these pseudo-differential operators. The construction in our case, loosely speaking, incorporates the $\mathcal{V}_0$-calculus (see Mazzeo-Melrose [15], [17] and the references therein) into the calculus of [14].

In this section the elliptic part of the calculus will be discussed, leading to the construction of the uniform resolvent and the uniform structure of the spectrum.

5.1. *Single surgery space*

The single surgery space is a natural compactification of the geometric degeneration, and the structure algebra defined on it captures the degeneration of the geometric operator, the de Rham operator in our case here. The space is defined as (Cf. [14] for the blowup notation):

$$X_{s0} = [M \times [0,\ 1]; \partial M \times \{0\}].$$

Here $[0, 1]$ is the parameter space for ϵ.

This is a manifold with corner, with the "trivial" extension boundary at $\epsilon = 1$. The more interesting boundary hypersurfaces are: B_{ss} resulting from the blow up; B_{bb} from the lift of $\{\epsilon = 0\}$; and B_{0b} from the lift of $\partial M \times [0,\ 1]$.

The boundary face B_{bb} is diffeomorphic to M while the interior of B_{ss} is diffeomorphic to the normal bundle of ∂M in M. The two intersect at the corner ∂M. On the other hand, the boundary face B_{0b} is diffeomorphic to $\partial M \times [0,\ 1]$.

Let

$$\beta_{s0} : \ X_{s0} \to X = M \times [0,\ 1]$$

be the blow-down map. Composed with the projection

$$\pi_\epsilon : X \to [0,\ 1]$$

we get a b-fibration map

$$\tilde{\pi}_\epsilon = \pi_\epsilon \circ \beta_{s0} : X_{s0} \longrightarrow [0,\ 1].$$

Note that for $\epsilon > 0$, the fibers of $\tilde{\pi}_\epsilon$ are diffeomorphic to M while at $\epsilon = 0$, $\tilde{\pi}_\epsilon^{-1}(0) = B_{ss} \cup B_{bb}$. This b-fibration captures the metric degeneration. In this picture, the geometric degeneration appears as the creasing of M into M together with the normal bundle of ∂M.

The structure algebra $\mathcal{V}_{s0}(X_{s0})$ is defined as

$$\mathcal{V}_{s0}(X_{s0}) = \{V \in \mathcal{V}_b(X_{s0}); (\tilde{\pi}_\epsilon)_*(V) = 0, \text{ and } V|_{B_{0b}} = 0\}.$$

This determines the structure bundle ${}^{s0}TX_{s0}$ by the equation

$$\mathcal{V}_{s0}(X_{s0}) = C^\infty(X_{s0};\ {}^{s0}TX_{s0}). \tag{5.29}$$

That ${}^{s0}TX_{s0}$ is a well-defined smooth vector bundle over X_{s0} follows from a general statement in [7] (see also [14]). In fact, over the part of X_{s0} where $\epsilon > 0$, ${}^{s0}TX_{s0}$ is simply the pull-back of ${}^{0}TM$ while restricted to B_{bb} it is canonically isomorphic to the b-tangent bundle of this compact manifold with boundary. When restricted to B_{ss} it is canonically isomorphic to the b-tangent bundle of this manifold near the boundary that meets B_{bb} and the 0-tangent bundle near the boundary that meets B_{0b}.

The structure algebra $\mathcal{V}_{s0}(X_{s0})$ is a Lie algebra of vector fields which degenerates in the same manner as the de Rham operator in this geometric degeneration (except at the boundary where the degeneration is *created* for treating the boundary problem). To analyze the de Rham operator via microlocal analysis we first construct from it the space of $s0$-differential operators $\mathrm{Diff}^*_{s0}(M; E, F)$ (E, F vector bundles on X_{s0}) in the usual way. Indeed, the space $\mathrm{Diff}^k_{s0}(M; E, F)$ consists of those differential operators from $C^\infty(X_{s0}; E)$ to $C^\infty(X_{s0}; F)$ which are given, with respect to local basis of E and F, by sums of up to k-fold products of elements of $\mathcal{V}_{s0}(X_{s0})$.

A $s0$-differential operator can be analyzed by its symbol plus the so-called normal homomorphisms. The symbol sort of measures its "interior strength", and is defined as follows. By (5.29) and the natural isomorphism between a vector space and its double dual, a vector field in $\mathcal{V}_{s0}(X_{s0})$ can be naturally identified with a C^∞ function on ${}^{s0}T^*X_{s0}$ that is linear along the fiber. This gives rise to the symbol map

$${}^{s0}\sigma : \mathrm{Diff}^k_{s0}(M; E, F) \to S^k({}^{s0}T^*X_{s0}; \hom(E, F)).$$

The normal homomorphisms, on the other hand, capture the leading terms in the degeneration. These are defined by restriction. The restriction of the Lie algebra $\mathcal{V}_{s0}(X_{s0})$ to the boundary hypersurface B_{bb} gives the full algebra $\mathcal{V}_b(B_{bb})$, the space of vector fields on B_{bb} tangent to the boundary of B_{bb} and its restriction to B_{ss} gives the algebra $\mathcal{V}_{0b}(B_{ss})$, the space of vector fields on B_{ss} tangent to one boundary component, $B_{ss} \cap B_{bb}$, and vanishing at the other, $B_{ss} \cap B_{0b}$. As a consequence, the space $\mathrm{Diff}^*_{s0}(M; E, F)$ comes equipped with the normal homomorphisms

$$\begin{aligned} N_b : \mathrm{Diff}^k_{s0}(M; E, F) &\to \mathrm{Diff}^k_b(M; E, F), \\ N_s : \mathrm{Diff}^k_{s0}(M; E, F) &\to \mathrm{Diff}^k_{0b}(B_{ss}; E, F). \end{aligned}$$

Here the image space Diff^k_b has a normal homomorphism itself, called the indicial homomorphism:

$$I : \mathrm{Diff}^k_b(M; E, F) \to \mathrm{Diff}^k_{I,b}(\partial M \times [0, \infty); E, F),$$

where the space with the subscript I denotes the subspace of $\mathbf{R}^+$-invariant operators. Similarly the indicial operator of an element of $\mathrm{Diff}^k_{0b}(B_{ss}; E, F)$ at the b-boundary ∂M is also an element of $\mathrm{Diff}^k_{I,b}(\partial M \times [0, \infty); E, F)$. The compatibility condition between the normal operators is just

$$N_{\partial M}(P) \stackrel{\text{def}}{=} I(N_b(P)) = I(N_s(P)), \;\; P \in \mathrm{Diff}^k_{s0}(M; E, F),$$

which is a consequence of (5.29).

If we choose local coordinates (x, y) on M near the boundary, where y is a local coordinate on ∂M and x the geodesic distance to the boundary, one obtains defining functions for the various boundary hypersurfaces:

$$\rho_{ss} = \sqrt{x^2 + \epsilon^2}, \quad \rho_{bb} = \frac{\epsilon}{\sqrt{x^2 + \epsilon^2}}, \quad \rho_{0b} = \frac{x}{\sqrt{x^2 + \epsilon^2}}.$$

From (4.17) we have for the de Rham operator A_ϵ

$$A_\epsilon = \gamma \sqrt{x^2 + \epsilon^2} \partial_x + \sigma A_{\partial M} = \gamma \rho_{ss} \partial_x + \sigma A_{\partial M}. \tag{5.30}$$

This is not yet a $s0$-differential operator. However

$$\rho_{0b} A_\epsilon \in \mathrm{Diff}^1_{s0}(M; F),$$

and

$$N_b(\rho_{0b} A_\epsilon) = \rho_{0b} A_0 \in \mathrm{Diff}^1_b(M; F). \tag{5.31}$$

$$N_s(\rho_{0b} A_\epsilon) = \rho_{0b} A_{B_{ss}} \in \mathrm{Diff}^1_{0b}(B_{ss}; F). \tag{5.32}$$

Moreover the restriction at the corner $B_{ss} \cap B_{0b} = \partial M$ is given by

$$R_{\partial M}(\rho_{0b} A_\epsilon) = \rho_{0b} A_{\partial M} \in \mathrm{Diff}^1(\partial M; F). \tag{5.33}$$

5.2. *Double surgery space*

We analyze the degenerating de Rham operator by looking at the resolvent and the singularity of its Schwartz kernel. This is done by constructing a pseudo-differential calculus in which lies the resolvent of the degenerating de Rham operator. This pseudo-differential calculus comes from microlocalizing $s0$-differential operators.

To microlocalize the Lie algebra of vector fields $\mathcal{V}_{s0}(X_{s0})$ we now define the double surgery 0-space, on which live the kernels of surgery 0-operators (or $s0$-operators):

$$\begin{aligned} X^2_{s0,f} = [&M^2 \times [0,\ 1];\ (\partial M)^2 \times \{0\};\ \partial M \times M \times \{0\}; \\ &M \times \partial M \times \{0\};\ \Delta(\partial M) \times [0,1]], \end{aligned}$$

where the subscript f indicates that this is a full blown-up version of the double surgery 0-space. The blow down map will be denoted by β^2_{s0}.

There are seven boundary hypersurfaces besides the trivial extension face $\{\epsilon = 1\}$, which we will ignore. We have B_{ds} from the first blow up; B_{ls}, B_{rs} from the second and third respectively; and B_{0s} from the last blow up. Finally the original boundary hypersurfaces $\{\epsilon = 0\}$, $\partial M \times M \times [0,1]$, and $M \times \partial M \times [0,1]$ lift to boundary hypersurfaces B_{db}, B_{lb} and B_{rb} respectively. Also the diagonal $\Delta(M) \times [0,1]$ lifts to an embedded submanifold Δ_{s0} meeting only B_{ds}, B_{db}, B_{0s} and does so transversally.

Let π_L, π_R denote the projections of $X^2 \stackrel{\text{def}}{=} M^2 \times [0,1]$ onto X by omitting the right and left factors respectively. These lift to b-fibrations

$$\begin{aligned} \pi_{s0,L} &: X^2_{s0,f} \to X_{s0}, \\ \pi_{s0,R} &: X^2_{s0,f} \to X_{s0}. \end{aligned}$$

Both restrict to Δ_{s0} to a diffeomorphism: $\Delta_{s0} \cong X_{s0}$. Moreover, by analyzing the lifting properties of $\mathcal{V}_{s0}(X_{s0})$, it is not hard to see that there is a natural isomorphism:

$$N(\Delta_{s0}) \cong {}^{s0}TX_{s0}. \tag{5.34}$$

Let ρ_{0s} be a defining function of B_{0s}. Define the kernel density bundle KD so that

$$C^\infty(X^2_{s0,f}, KD) = \rho_{0s}^{-n/2} C^\infty(X^2_{s0,f}, \Omega^{1/2}((X^2_{s0,f})).$$

The small surgery 0-calculus is

$$\Psi^m_{s0}(M; E, F) = \rho_{ls}^\infty \rho_{rs}^\infty \rho_{lb}^\infty \rho_{rb}^\infty I^{m-1/4}(X^2_{s0,f}, \Delta_{s0}; Hom(F,E) \otimes KD).$$

This is a microlocalization for $\mathcal{V}_{s0}(X_{s0})$ since $\mathrm{Diff}^*_{s0}(M) \subset \Psi^*_{s0}(M)$. However this calculus is too small to contain the inverses of its elliptic elements. Thus one has to enlarge the calculus to include boundary terms. Let us denote by $\mathcal{A}(X^2_{s0,f}; Hom(F,E) \otimes KD)$ the space of all sections of $Hom(F,E) \otimes KD$ smooth in the interior and conormal to all boundary faces. For a positive number τ define

$$\begin{aligned} &\mathcal{A}^{\tau}_{-}(X^2_{s0,f}; Hom(F,E) \otimes KD) \\ &= \bigcap_{\delta>0} \rho_{ds}^{\tau-\delta} \rho_{db}^{\tau-\delta} \rho_{ls}^{\tau-\delta} \rho_{rs}^{\tau-\delta} \mathcal{A}(X^2_{s0,f}; Hom(F,E) \otimes KD) \\ &= \bigcap_{\delta>0} \epsilon^{\tau-\delta} \mathcal{A}(X^2_{s0,f}; Hom(F,E) \otimes KD). \end{aligned}$$

We call τ the conormal bound for the conormal sections in $\mathcal{A}^{\tau}_{-}$.

Using this notation the residual calculus is defined as

$$\Psi^{\tau}_{s0,res}(M; E, F) = \mathcal{A}^{\tau}_{-}(X^2_{s0,f}; Hom(F,E) \otimes KD) \tag{5.35}$$

This is the space of 'good' error terms in the sense that they vanish at a positive rate at $\epsilon = 0$.

The space of boundary terms is defined as (using the notation of [14])

$$\Psi^{-\infty,\tau}_{s0}(M; E, F) = \mathcal{B}_{dB}\mathcal{A}^{\tau}_{-}(X^2_{s0,f}; Hom(F,E) \otimes KD), \tag{5.36}$$

where $dB = \{ds, db, 0s\}$ and τ is a positive number. Roughly speaking $\Psi^{-\infty,\tau}_{s0}$ consists of all sections smooth in the interior and conormal to the boundary faces (with conormal bound 0) and vanish at rate τ at the boundary faces B_{ls}, B_{rs} and have some partial smoothness up to B_{ds}, B_{db}, B_{0s}.

Now the 'calculus with (conormal) bounds' is defined as

$$\Psi^{m,\tau}_{s0}(M; E, F) = \Psi^{m}_{s0}(M; E, F) + \Psi^{-\infty,\tau}_{s0}(M; E, F). \tag{5.37}$$

Since

$$\Psi^{m}_{s0}(M; E, F) \cap \Psi^{-\infty,\tau}_{s0}(M; E, F) = \Psi^{-\infty}_{s0}(M; E, F),$$

the first thing to note here is that the symbol map for conormal distributions

$${}^{s0}\sigma_m : \Psi^{m}_{s0}(M; E, F) \to S^m({}^{s0}TX^*_{s0}; E, F) \tag{5.38}$$

extends to the whole calculus.

The symbol map alone is not enough to invert the elliptic elements modulo compact errors. The utility of the calculus constructed above lies largely in the existence of additional, non-commutative 'symbols'. These are obtained by restricting the elements to each of the boundary faces B_{ds},

B_{db}, B_{0s}, B_{ls}, B_{rs}. Since an element of $\Psi_{s0}^{m,\tau}$ is required to vanish at a positive rate at the boundary faces B_{ls}, B_{rs}, the restrictions will be trivial there and will be ignored. The only nontrivial ones are at B_{db}, B_{ds}, B_{0s}, called the b-normal homomorphism, the surgery normal homomorphism, and the 0-normal homomorphism respectively.

Clearly the b-normal homomorphism N_b maps onto the b-calculus with conormal bounds on M:

$$N_b : \Psi_{s0}^{m,\tau}(M; E, F) \to \Psi_b^{m,\tau}(M; E, F). \tag{5.39}$$

The name homomorphism indicates that N_b respects the composition (in the sense of operators acting on distributions, see Proposition 5.1). But only the weaker form $N_b(P \circ A) = N_b(P) \circ N_b(A)$, $P \in \mathrm{Diff}_{s0}^*$ will be used here. This will be discussed below (Proposition 5.2).

Similarly the surgery normal homomorphism is a map

$$N_s : \Psi_{s0}^{m,\tau}(M; E, F) \to \Psi_{0b}^{m,\tau}(\bar{H}; E, F). \tag{5.40}$$

Here $\bar{H} = \partial M \times [0,1]$ is the compactification of the half normal bundle of ∂M, or in other words the half infinite cylinder. And the image lies in the 0b-calculus which will be briefly discussed in the next section.

Finally for the 0-normal homomorphism note that B_{0s} can be identified with a natural compactification of the half tangent bundle of M at ∂M lifted to $\partial M \times [0,1]$. By definition then, one finds that N_0 maps onto the conormal distributions conormal to the section of the lifted normal bundle over $\partial M \times [0,1]$ given by $(1, 0, \cdots, 0)$ and which are smooth up to the boundaries.

From definition it is not hard to see that, for an element in $\Psi_{s0}^{m,\tau}$ its various 'symbols' have to be compatible in the sense that restricted to the common corner or the intersection with the diagonal the resulting 'symbols' have to agree. Moreover these are the only obstructions for the existence of surgery 0-calculus with prescribed 'symbols'.

Although defined as distributions the surgery 0-operators can be made to act on distributions on X_{s0}, thus justifying the name. We state the mapping properties in the following

Proposition 5.1. *An element A of $\Psi_{s0}^{m,\tau}(M; E, F)$ defines a bounded linear map*

$$A : \; C^{-\infty}(X_{s0}; E) \to C^{-\infty}(X_{s0}; F)$$

which restricts to

$$A : \; \mathcal{A}_{-}^{r}(X_{s0}; E) \to \mathcal{A}_{-}^{r}(X_{s0}; F),$$

if $r < \tau$. Moreover, if $m \le 0$, $\tau > 0$, then

$$A : L^2(X_{s0}; E \otimes \Omega_{s0}^{1/2}) \to L^2(X_{s0}; F \otimes \Omega_{s0}^{1/2}). \tag{5.41}$$

is also bounded. Here $\Omega_{s0}^{1/2}(X_{s0}) = \rho_{0s}^{-n/2}\Omega^{1/2}(X_{s0})$.

Proof. Recall that the projections π_L, π_R from $X^2 = M^2 \times [0,\ 1]$ to X, obtained by dropping the right and left M factor in X^2 respectively, lift to b-fibrations

$$\tilde{\pi}_L,\ \tilde{\pi}_R : X_{s0,f}^2 \to X_{s0}.$$

Similarly the projection onto the ϵ variable, $\pi_\epsilon :\ X^2 \to [0,\ 1]$, lifts to b-fibration

$$\tilde{\pi}_\epsilon : X_{s0,f}^2 \to [0,\ 1].$$

Now the equation

$$Au = (\tilde{\pi}_L)_*[A \cdot (\tilde{\pi}_R)^*(u)(\tilde{\pi}_\epsilon)^*(|d\epsilon|^{-1/2})]$$

defines the action of $A \in \Psi_{s0}^{m,\tau}(M; E, F)$; the fact that it is well defined is a consequence of the calculus of wave front sets. This proves the first part. The second follows from the calculus of conormal functions (Cf. [14]).

To show the L^2 boundedness, it suffices to show that for $A \in \Psi_{s0}^{-\infty,\tau}$ (by Hörmander's lemma). We decompose A into four pieces, $A = A_1 + A_2 + A_3 + A_4$, where A_1 is supported near B_{0s}; A_2 supported near B_{lb}, but away from B_{0s}; A_3 supported near B_{rb}, but away from B_{0s}; and the final piece A_4 supported away from $B_{lb} \cup B_{os} \cup B_{rb}$.

By its support property, the L^2-boundedness of A_4 is a consequence of [14]. For A_2, A_3, since the result of its action will always have support away from B_{0s}, the L^2-boundedness also follows similarly. The L^2-boundedness of A_1 is a uniform version of the result in [Ma] and can be shown in the same way. □

We now turn to the composition with $s0$-differential operators.

Proposition 5.2. *If $P \in \mathit{Diff}_{s0}^*(M; E, F)$, $A \in \Psi_{s0}^{*,\tau}(M; E, F)$, then $P \circ A \in \Psi_{s0}^{*,\tau}(M; E, F)$. Further*

$$N_b(P \circ A) = N_b(P) \circ N_b(A) \tag{5.42}$$

and similarly for the other homomorphisms.

Proof. Clearly, if $P \in \mathrm{Diff}^k_{s0}$, $A \in \Psi^m_{s0}$, then $P \circ A \in \Psi^{m+k}_{s0}$. Also if $A \in \Psi^{-\infty,\tau}_{s0}$, then $P \circ A \in \Psi^{-\infty,\tau}_{s0}$. Now if $V \in \mathrm{Diff}^1_{s0}$, we have $N_b(V \circ A) = \pi^*_{s0,L}V(A)|_{B_{db}} = \pi^*_{s0,L}V|_{B_{db}}(A|_{B_{db}}) = N_b(V) \circ N_b(A)$. The general case follows. □

The residual space $\Psi^\tau_{s0,res}$ is the space of good error terms. Crucial to our construction is the 'semi-ideal property' that this space satisfies. Define $\mathcal{L}_C(M)$ to be the algebra of bounded operators on $L^2(X_{s0}; \Omega^{1/2}_{s0})$ which depends parametrically and conormally on ϵ, i.e. an element of $\mathcal{L}_C(M)$ is a $B \in \mathcal{L}(L^2(X_{s0}; \Omega^{1/2}_{s0}))$ such that

$$[\epsilon, B] = 0 \text{ and } (\epsilon\frac{\partial}{\partial\epsilon})^k B \in \mathcal{L}(L^2(X_{s0}; \Omega^{1/2}_{s0})) \text{ for all } k \geq 0.$$

Clearly $\Psi^\tau_{s0,res}(M) \subset \mathcal{L}_C(M)$ is a subalgebra, but more is true.

Proposition 5.3. *If $\tau > 0$, then*

$$\Psi^\tau_{s0,res}(M) \cdot \mathcal{L}_C(M) \cdot \Psi^\tau_{s0,res}(M) \subset \Psi^\tau_{s0,res}(M).$$

Proof. Let A, $B \in \Psi^\tau_{s0,res}(M)$, and $K \in \mathcal{L}_C(M)$. We need to examine the kernel of BKA and show that it has the required regularity. For this purpose, we apply the operator BKA to certain weighted delta half-density. For $z \in M$, let $\delta_z \in C^{-\infty}(M; \Omega^{1/2})$ be a delta half-density at z. This gives a continuous map

$$M \ni z \mapsto \delta_z|d\epsilon|^{1/2} \in H^{-k}(X; \Omega^{1/2}), \quad \text{for } k > \frac{n}{2} + 1.$$

The continuity is a consequence of the Sobolev Embedding Theorem. Since this family of half-densities is ϵ-independent, it follows that the lifts to X_{s0} of the following weighted half-densities give rise to a continuous map

$$M \ni z \mapsto (x^2 + \epsilon^2)^{1/4}\epsilon^{\nu - \frac{1}{2}}\delta_z|d\epsilon|^{1/2} \in H^{-k}_b(X_{s0}; \Omega^{1/2}), \quad \forall \nu > 0.$$

By the assumption on K and the mapping properties of $\Psi^\tau_{s0,res}(M)$, we obtain a continuous map

$$z \mapsto \epsilon^{-2\tau'} BKA((x^2 + \epsilon^2)^{1/4}\epsilon^{\nu - \frac{1}{2}}\delta_z|d\epsilon|^{1/2}) \in H^\infty_b(X_{s0}; \Omega^{1/2}), \quad \forall \tau' < \tau.$$

However the space $H^\infty_b(X_{s0}; \Omega^{1/2})$ consists of half-densities of the form $\epsilon^{-1/2}a\mu$ where a is continuous on X_s and μ is a non-vanishing smooth half-density on X_{s0}. This implies that the Schwartz kernel of BKA is of the form

$$\epsilon^{2\tau'} b\mu \otimes \frac{v}{(x^2 + \epsilon^2)^{1/4}} \otimes |d\epsilon|^{-1/2},$$

where b is continuous on $X_{s0} \times M$ and $\tau' < \tau$ arbitrary. Lifting to X^2_{s0} shows that the kernel is the product of $\epsilon^{2\tau'}$ and a continuous section of the kernel density bundle. Since this regularity is clearly stable under the repeated action of $\epsilon\partial_\epsilon$ and of $\mathcal{V}_b(X_{s0})$ lifted from either the left or the right it follows that

$$\Psi^{\tau}_{s0,res}(M) \cdot \mathcal{L}_C(M) \cdot \Psi^{\tau}_{s0,res}(M) \subset \Psi^{2\tau}_{s0,res}(M). \qquad \square$$

For its role in the heat surgery 0-calculus, the reduced double surgery space is defined to be

$$X^2_{s0} = [M^2 \times [0,\ 1];\ (\partial M)^2 \times \{0\};\ \partial M \times M \times \{0\};\ M \times \partial M \times \{0\}].$$

It can be obtained from $X^2_{s0,f}$ by blowing down the boundary face B_{0s}.

The elements of $\Psi^{-\infty}_{s0}(M; E)$ are smoothing operators on M, hence trace class. By Lidsky's theorem the trace is the integral over the diagonal of the point wise trace of the kernel, which can be interpreted as a density:

$$Hom(E) \otimes \Omega^{1/2}(X^2_{s0})|_{\Delta_{s0}} \cong hom(E) \otimes \Omega(X_{s0}).$$

Thus the trace of $A \in \Psi^{-\infty}_{s0}(M; E)$ is, as a function, the push-forward to $[0,\ 1]$ of the density

$$(tr\ A)|_{\Delta_{s0}} \in C^\infty(X_{s0}; \Omega).$$

The following lemma is from [14].

Lemma 5.4. *As a map*

$$Tr:\ \Psi^{-\infty}_{s0}(M; E) \to C^\infty([0,\ 1]) + \log\epsilon\, C^\infty([0,\ 1]).$$

I.e.

$$Tr(A) = r_A(\epsilon) + \log\epsilon\, \tilde{r}_A(\epsilon),$$

for r_A, $\tilde{r}_A$ smooth functions of ϵ. Moreover for the leading terms

$$\tilde{r}_A(0) = \int_{\partial M} (tr\ A)|_{\partial M}, \tag{5.43}$$

$$r_A(0) = b\text{-}Tr(N_s(A)) + b\text{-}Tr(N_b(A)). \tag{5.44}$$

5.3. *The 0b-calculus*

To construct a good parametrix for an elliptic $s0$-operator we need to invert its various normal operators. The normal operator at B_{0s} lands in the 0b-calculus, which we discuss here in somewhat more detail.

Let $\bar{H} = \partial M \times [0,1]$ be the compactified normal bundle of ∂M. The structure algebra $\mathcal{V}_{0b}$ is defined to be the Lie algebra of all vector fields that vanish at $\partial M \times \{0\}$ and tangent to $\partial M \times \{1\}$. The structure bundle ${}^{0b}T\bar{H}$ is defined, as usual, via

$$C^\infty(\bar{H}, {}^{0b}T\bar{H}) = \mathcal{V}_{0b}.$$

From the structure algebra we construct the $0b$-differential operators in the usual way.

To define $0b$-pseudodifferential operators we construct the double $0b$-space

$$\bar{H}^2_{0b} = [\bar{H}^2; \Delta(\partial M) \times \{0\}; (\partial M)^2 \times \{1\}].$$

Denote by Δ_{0b} the lifted diagonal. There are six boundary hypersurfaces for $\bar{H}^2_{0b}$, namely B_{d0}, B_{db} from the blow-up operations respectively; B_{l0}, B_{r0}, B_{lb}, B_{rb} from the lift of the original boundary faces. The lifted diagonal intersects only B_{d0} and B_{db} and does so transversally.

Now the space of $0b$-pseudodifferential operators is defined to be ($\tau > 0$)

$$\Psi^{m,\tau}_{0b}(\bar{H}, \Omega^{1/2}) = \rho^\infty_{l0}\rho^\infty_{r0}\rho^\infty_{lb}\rho^\infty_{rb} I^m(\bar{H}^2_{0b}; \Delta_{0b}; KD) + \mathcal{A}^\tau_-(\bar{H}^2_{0b}; KD),$$

where kernel density bundle KD is defined so that

$$C^\infty(\bar{H}^2_{0b}; KD) = \rho_{d0}^{-n/2} C^\infty(\bar{H}^2_{0b}; \Omega^{1/2}).$$

We will denote $\Psi^\tau(\bar{H}; \Omega^{1/2}) = \Psi^{-\infty,\tau}_{0b}(\bar{H}, \Omega^{1/2}) = \mathcal{A}^\tau_-(\bar{H}^2_{0b}; KD)$. Since this is just a mixture of the 0-calculus and the b-calculus, it is quite clear that their common properties carry over.

Proposition 5.5.

(1) The $0b$-differential operators are $0b$-pseudo-differential operators.

(2) The symbol map is a homomorphism:

$$\sigma_{0b} : \Psi^{m,\tau}_{0b}(\bar{H}, \Omega^{1/2}) \to S^m({}^{0b}T^*\bar{H}).$$

We also have the 0-normal and b-normal homomorphisms:

$$N_0 : \Psi^{m,\tau}_{0b}(\bar{H}; \Omega^{1/2}) \to \Psi^{m,\tau}_0(\bar{H}; \Omega^{1/2}),$$
$$N_b : \Psi^{m,\tau}_{0b}(\bar{H}; \Omega^{1/2}) \to \Psi^{m,\tau}_{b,I}(\bar{H}; \Omega^{1/2}).$$

(3) Elements of $\Psi^{m,\tau}_{0b}(\bar{H}, \Omega^{1/2})$ define continuous linear operators:

$$C^{-\infty}(\bar{H}; \Omega^{1/2}) \to C^{-\infty}(\bar{H}; \Omega^{1/2}),$$
$$\mathcal{A}^r_-(\bar{H}; \Omega^{1/2}) \to \mathcal{A}^r_-(\bar{H}; \Omega^{1/2}).\ (r < \tau)$$

(4) *We also have L^2-continuity:* $A \in \Psi_{0b}^{m,\tau}(\bar{H};\Omega^{1/2})$ *defines a continuous linear map*

$$A : \rho^z H_{0b}^k(\bar{H};KD) \to \rho^z H_{0b}^{k-m}(\bar{H};KD), \quad KD = \rho_{d0}^{-n/2}\Omega^{1/2}.$$

Crucial to our discussion is the so-called semi-ideal property of the residual calculus. Let $\mathcal{L}_c(L^2(\bar{H};KD)) = \cap_z \mathcal{L}(\rho^z L^2(\bar{H},KD))$.

Proposition 5.6. *If* $\tau > 0$, $\Psi^\tau(M)$ *is a semi-ideal in* $\mathcal{L}_c(L^2(\bar{H};KD))$, *i.e. for any* K *a continuous linear operator on* $\rho^z L^2(\bar{H};KD)$ *for all* z *and any* $A,\ B \in \Psi^\tau(M)$,

$$BKA \in \Psi^{2\tau}(M).$$

Proof. To examine the Schwartz kernel of BKA, we apply it to the delta densities. For $z \in \bar{H}$, let $\delta_z \in C^{-\infty}(\bar{H};\Omega^{1/2})$ be the delta half density at z. As a map,

$$\bar{H} \ni z \mapsto \delta_z \in H_b^{-k}(\bar{H};\Omega^{1/2}),\ k > \frac{n}{2}$$

is continuous. It follows that

$$\bar{H} \ni z \mapsto A\delta_z \in H_b^\infty(\bar{H};\Omega^{1/2})$$

is also continuous. Therefore

$$\bar{H} \ni z \mapsto BKA\delta_z \in H_b^\infty(\bar{H};\Omega^{1/2})$$

is continuous as well. But an element in $H_b^\infty(\bar{H};\Omega^{1/2})$ can be written as a continuous section of the half-density bundle divided by the square root of a defining function to the boundaries. This shows that the kernel of BKA can be lifted to $\bar{H}_{0b}$. □

Recall that $A = d + \delta$ is the (twisted) de Rham operator. We use $A_{\bar{H}}$ to denote the de Rham operator on $\bar{H}$. Now we can show

Proposition 5.7. *The resolvent of* $A_{\bar{H}}^2$ *lies in the* 0b*-calculus, i.e.* $\exists \tau > 0$ *such that*

$$(A_{\bar{H}}^2 - \lambda)^{-1} \in \Psi_{0b}^{-2,\tau}(\bar{H};\Omega^{1/2}).$$

Proof. First of all, by taking the Laplace transform of (4.24), we have

$$(A_{\bar{H}}^2 - \lambda)^{-1} \in \mathcal{L}_c(L^2(\bar{H};\Omega^{1/2})).$$

On the other hand, using the $0b$-calculus, one can easily construct left and right parametrices for $D^2_{\bar{H}} - \lambda$:

$$(A^2_{\bar{H}} - \lambda)G_1 = \mathrm{Id} + R_1,$$
$$G_2(A^2_{\bar{H}} - \lambda) = \mathrm{Id} + R_2,$$

with $G_1,\ G_2 \in \Psi^{-2,\tau}_{0b}(\bar{H};\Omega^{1/2}),\ R_1,\ R_2 \in \Psi^{\tau}(\bar{H};\Omega^{1/2})$. Applying $(A^2_{\bar{H}} - \lambda)^{-1}$ to both equations we obtain

$$\begin{aligned}(A^2_{\bar{H}} - \lambda)^{-1} &= G_1 - (A^2_{\bar{H}} - \lambda)^{-1} R_1 \\ &= -G_2 R_1 + R_2 (A^2_{\bar{H}} - \lambda)^{-1} R_1 \in \Psi^{-2,\tau}_{0b}(\bar{H};\Omega^{1/2})\end{aligned}$$

by Proposition 5.6. □

As before, all the constructions and the discussions apply to operators acting on sections of a vector bundle. From now on, we denote by E the vector bundle

$$E = \Lambda(M) \otimes \xi.$$

5.4. *The uniform structure of the resolvent*

Let A_0 denote the (twisted) de Rham operator associated to the exact b-metric g_0 on M. Also, denote by $A_{\epsilon,a}$ ($A_{\epsilon,r}$ resp.) the (twisted) de Rham operator associated to g_ϵ with the absolute (relative resp.) boundary condition. With all the machinery developed so far we can now prove

Proposition 5.8. *Assume that $A_{\partial M}$ is invertible and also 0 is not in the spectrum of A_0. If $\Omega \subset \mathbf{C}$ is an open bounded set with closure disjoint from the spectrum of A_0^2, then for some $\tau > 0$ and $\epsilon_0 > 0$ the resolvent of $A^2_{\epsilon,a}$ ($A^2_{\epsilon,r}$ resp.) is a holomorphic map*

$$\begin{aligned}\Omega &\to \Psi^{-2,\tau}_{s0}(M;E) \\ \lambda &\mapsto \quad R(\lambda).\end{aligned}$$

In particular the spectrum of $A^2_{\epsilon,a}$ ($A^2_{\epsilon,r}$ resp.) falls outside a neighborhood of the imaginary axis.

Proof. One tries to solve the equation

$$\begin{cases} (A^2_{\epsilon,a} - \lambda) R(\lambda) = \mathrm{Id} \\ R(\lambda) \text{ satisfies the boundary condition} \end{cases} \tag{5.45}$$

by solving the corresponding equations for the symbol map and the normal homomorphisms. The symbol for $R(\lambda)$ can be solved via

$$^{s0}\sigma_2(R(\lambda)) = |\xi|^{-2}\mathrm{Id},$$

as does the b-normal homomorphism,

$$N_b(R(\lambda)) = (A_0^2 - \lambda)^{-1} \in \Psi_b^{-2,\tau}(M; E).$$

For the surgery normal homomorphism we note that

$$N_s(R(\lambda)) = (A_{\bar{H}}^2 - \lambda)^{-1}$$

is the solution for the corresponding equation for the half infinite cylinder. By taking the Laplace transform of (4.24) we find

$$N_s(R(\lambda)) \in \Psi_{0b}^{-2,\tau}(\bar{H}; E).$$

Finally the 0-normal homomorphism of $R(\lambda)$ satisfies a family of Laplace equations on the half Euclidean space with the boundary condition. Hence it can be solved similarly as in the half cylinder case.

These solutions for the normal homomorphisms and symbol clearly satisfy the compatibility condition. Thus there exists a family of surgery 0-operators $E'(\lambda) \in \Psi_{s0}^{-2,\tau}(M; E)$ with the correct symbol and normal homomorphisms. This means that $E'(\lambda)$ is already a parametrix for the resolvent family.

To get a better parametrix, note that the interior singularity can be removed in the small calculus. It follows then that there is a correction term $G_0'(\lambda) \in \Psi_{s0}^{-2}(M; E)$ such that $E = E' - G_0'$ is a parametrix in the strong sense that

$$(A_{\epsilon,a}^2 - \lambda)E(\lambda) = \mathrm{Id} - G(\lambda),\ G(\lambda) \in \Psi_{s0,res}^{\tau}(M; E).$$

Now $G(\lambda)$ vanishes at a positive rate at $\epsilon = 0$. Hence where ϵ is small the Neumann series provides an inverse for $\mathrm{Id}-G(\lambda)$ and

$$R(\lambda) = E(\lambda)(\mathrm{Id} - G(\lambda))^{-1}. \qquad \square$$

6. Heat surgery 0-calculus

After the discussion of the elliptic calculus we now turn to the parabolic calculus and examine the uniform structure of the heat kernel.

6.1. *Heat surgery* 0*-operators*

To construct the heat surgery 0-space we note that Δ_{s0} intersects the boundaries of X^2_{s0} at B_{ds}, B_{db}, and at the lift of $(\partial M)^2 \times [0,\, 1]$ which is a corner. This means that in defining the heat surgery 0-space one needs to first blow up the intersection at the corner, which is $\Delta(\partial M) \times [0,\, 1]$:

$$X^2_{hs0} = [X^2_{s0} \times [0, \infty);\, \Delta(\partial M) \times [0,\, 1] \times \{0\}, S;\, \Delta_{s0} \times \{0\}, S],$$

where S is the parabolic bundle $sp(dt)$ (see [14]). The blow down map is denoted by β_h.

For analyzing the normal homomorphisms we look at the structures of the boundary hypersurfaces of X^2_{hs0}. There are three of them lying above $\{t = 0\}$: B_{ff} from the blow up of $\Delta(\partial M) \times [0,\, 1] \times \{0\}$; B_{tf} from the blow up of $\Delta_{s0} \times \{0\}$; and B_{tb}, the lift of $\{t = 0\}$. The first two are fibered over the submanifolds to be blown up. In fact B_{ff} can be viewed as the natural compacitification of the lift to $\partial M \times [0,\, 1]$ of the half tangent bundle of M at ∂M times $[0, \infty)$ and $B_{tf} \cong {}^{s0}TX_{s0}$.

The rest of the boundary hypersurfaces arise from the lift of those of X^2_{s0}. Precisely we have

$$\begin{aligned}
B_{ds}(X^2_{hs0}) &= [B_{ds}(X^2_{s0}) \times [0, \infty); \Delta(\partial M) \times \{0\}, S;\, \Delta_{ds} \times \{0\}, S],\\
B_{db}(X^2_{hs0}) &= [B_{db}(X^2_{s0}) \times [0, \infty); \Delta_{db} \times \{0\}, S],\\
B_{ls}(X^2_{hs0}) &= B_{ls}(X^2_{s0}) \times [0, \infty),\\
B_{rs}(X^2_{hs0}) &= B_{rs}(X^2_{s0}) \times [0, \infty),\\
B_{lb}(X^2_{hs0}) &= [B_{lb}(X^2_{s0}) \times [0, \infty);\, \Delta(\partial M) \times [0,\, 1] \times \{0\}, S],\\
B_{rb}(X^2_{hs0}) &= [B_{rb}(X^2_{s0}) \times [0, \infty);\, \Delta(\partial M) \times [0,\, 1] \times \{0\}, S],
\end{aligned}$$

Note that B_{tf} only meets B_{ff}, B_{tb}, B_{db} and B_{ds}.

The kernels of the heat surgery 0-operators are normalized with respect to the half-density

$$KD_{hs0} = \rho_{ff}^{-(n+2)/2} \rho_{tf}^{-(n+3)/2} \Omega^{1/2}(X^2_{hs0}).$$

Let I denote the index set $\{k_1, k_2, k_3, k_4\}$. The space of the heat surgery 0-operators is defined to be

$$\Psi^I_{hs0} = \rho_{ff}^{k_1} \rho_{tf}^{k_2} \rho_{ds}^{k_3} \rho_{db}^{k_4} \rho_{ls}^{\infty} \rho_{rs}^{\infty} \rho_{tb}^{\infty} C^\infty(X^2_{hs0}; KD_{hs0}).$$

6.2. *Normal homomorphisms*

The normal homomorphism at B_{ff} is defined by dividing by $\rho_{ff}^{k_1}$ and restricting to B_{tf}:

$$N_{hs0;f,k_1}: \Psi_{hs0}^{I} \to \rho_{tf}^{k_2}\rho_{ds}^{k_3}\rho_{tb}^{\infty}C^{\infty}(B_{ff}; KD_{hs0}|_{B_{ff}}).$$

By the previous discussion on the structure of B_{ff} we see that it can also be thought as the front face of the heat 0-space of the half normal bundle of ∂M. Therefore the range of $N_{hs0;f,k_1}$ is also the range of the normal homomorphism of this heat 0-calculus at the front face.

The normal homomorphism at B_{tf}, or the heat homomorphism, is defined similarly.

$$N_{hs0;h,k_2}: \Psi_{hs0}^{I} \to \rho_{ff}^{k_1}\rho_{ds}^{k_3}\rho_{db}^{k_4}\rho_{tb}^{\infty}C^{\infty}(B_{tf}; KD_{hs0}|_{B_{tf}}).$$

Since $KD_{hs0}|_{B_{tf}}$ is canonically isomorphic to the fiber density bundle of ${}^{s0}TX_{s0}$, the heat homomorphism can be rewritten as

$$N_{hs0;h,k_2}: \Psi_{hs0}^{I} \to \rho_{ff}^{k_1}\rho_{ds}^{k_3}\rho_{db}^{k_4}\mathcal{S}({}^{s0}TX_{s0}; \Omega_{\text{fiber}}).$$

Restricting to B_{db} gives us the surgery homomorphism:

$$N_{hs0;b}: \Psi_{hs0}(M) \to \Psi_{hb}(M),$$

while restricting to B_{ds} gives a normal homomorphism which maps onto the heat $0b$-calculus of the compactified half normal bundle of ∂M:

$$N_{hs0;s}: \Psi_{hs0}(M) \to \Psi_{h0b}(N_+(\partial M)).$$

These normal homomorphisms are nontrivial only for $k_3 = 0$, $k_4 = 0$. Moreover if $N_{hs0;b}(A) = 0$ and $N_{hs0;s}(A) = 0$ for $A \in \Psi_{hs0}(M)$ then $A = \epsilon B$ for $B \in \Psi_{hs0}(M)$. This will be used in the construction of the heat kernels.

Individually, each normal homomorphism is surjective. However the normal operators for an element of $\Psi_{hs0}(M)$ have to agree at the common corners. These are the compatibility conditions. On the other hand, since essentially just smooth functions are involved, the compatibility conditions are the only obstructions to the existence of heat surgery 0-operator with given normal operators.

6.3. *Uniform structure of the heat kernel*

It suffices to consider the heat kernel for $A_{\epsilon,a}^2$, the other boundary condition being similar. The proceeding construction enables us to prove the following

Theorem 6.1. *There is a unique* $H \in \Psi^{I}_{hs0}(M;E)$ *where* $I = \{-2,-2,0,0\}$ *such that*

$$x^2(\partial_t + A_\epsilon^2)H = 0 \text{ in } \Psi^{I'}_{hs0}(M;E) \tag{6.46}$$

for $I' = \{-2,0,-2,0\}$, *and*

$$N_{hs0;h,-2}(H) = Id, \tag{6.47}$$

and H *satisfies the boundary condition*

$$P_a H|_{B_{0b}} = 0, \quad P_a A_\epsilon H|_{B_{0b}} = 0.$$

Proof. Equation (6.46) and (6.47) translate to conditions on the four normal operators of H:

$$({}^{s0}\sigma_2(x^2 A_\epsilon^2 - \frac{1}{2}(R+n))N_{hs0;h,-2}(H) = 0, \int_{\text{fiber}} N_{hs0;h,-2}(H) = \text{Id}, \tag{6.48}$$

$$x^2(\partial_t + N_b(A_\epsilon^2))N_{hs0;b}(H) = 0, \tag{6.49}$$

$$\rho_{0b}^2((\partial_t + N_s(A_\epsilon^2))N_{hs0;s}(H) = 0, \tag{6.50}$$

$$s'^2(\partial_{T'} + \Delta_E)N_{hs0;f}(H) = 0. \tag{6.51}$$

Finally the boundary condition translates to boundary conditions for (6.51) and (6.50).

The first equation is a fiber by fiber differential equation and can be solved uniquely subject to the integral condition. Furthermore, because of the compatibility condition, this fixes the integral conditions for (6.49) and (6.50). Thus the two normal operators $N_{hs0;b}(H)$, $N_{hs0;s}(H)$ are necessarily the heat kernels for the elliptic b-differential operator $N_b(A_\epsilon^2)$ and elliptic $0b$-differential operator $N_s(\rho_{0b}^2 A_\epsilon^2)$. As such they are unique and are elements of the corresponding small heat calculus. These two operators have the same indicial family, so using the existence part of the compatibility it follows that there is an element $H' \in \Psi^{I}_{hs0}(M;E)$ satisfying the symbolic conditions (6.48), (6.49), (6.50), (6.51).

This first approximation therefore satisfies

$$x^2(\partial_t + A_\epsilon^2)H' = -\epsilon R_1, \; R_1 \in \Psi^{-3,-3,0,0}_{hs0}(M;E). \tag{6.52}$$

Now we proceed exactly as in [14]. □

7. Large time behavior of heat kernel

The implication of the previous construction of the uniform heat kernel will be further exploited in this section following the ideas of [14].

7.1. *The resolvent near infinity*

The resolvent and the heat kernel are related by the Laplace transform

$$\begin{aligned}(A^2_{\epsilon,a}-\lambda)^{-1} &= \int_0^\infty e^{\lambda t}e^{-tA^2_{\epsilon,a}}dt,\\ e^{-tA^2_{\epsilon,a}} &= \frac{1}{2\pi i}\int_\Gamma e^{-t\lambda}(A^2_{\epsilon,a}-\lambda)^{-1}d\lambda,\end{aligned} \tag{7.53}$$

where Γ is a contour enclosing the spectrum of $A^2_{\epsilon,a}$. This makes it possible to obtain information about one from the other. In fact, the large spectral parameter behavior of the resolvent corresponds to the small time behavior of the heat kernel and the large time behavior of the heat kernel corresponds to the small spectral parameter of the resolvent.

To estimate the resolvent as the spectral parameter tends to infinity outside a sector containing the spectrum we use the discussion of the heat kernel in the last section. Choose $\phi \in C^\infty_c(\mathbf{R})$ with $\phi(t)=1$ in $|t|<1$ and $\phi(t)=0$ in $|t|>2$. Let

$$R_1(\lambda) = \int_0^\infty e^{\lambda t}\phi(t)e^{-tA^2_{\epsilon,a}}dt. \tag{7.54}$$

Then

$$(A^2_{\epsilon,a}-\lambda)R_1(\lambda) = \mathrm{Id} - E_1(\lambda), \tag{7.55}$$

where the error term

$$E_1(\lambda) = \int_0^\infty e^{\lambda t}\phi'(t)e^{-tA^2_{\epsilon,a}}dt \in \Psi^{-\infty}_{s0}(M), \tag{7.56}$$

is in the small calculus, and since $\phi'(t)$ has compact support in $(0,\infty)$, vanishes rapidly as $|\lambda|\to\infty$ in any closed sector in $Re\,\lambda<0$.

To improve on the parametrix we now solve the equation

$$(A^2_{\epsilon,a}-\lambda)R_2(\lambda) = E_1(\lambda) - \epsilon E_2(\lambda). \tag{7.57}$$

This reduces to solving for the resolvent of the normal operators. It follows that we can solve $R_2 \in \Psi^{-\infty,\tau}_{s0}(M)$ with the error $E_2(\lambda) \in \Psi^{-\infty,\tau}_{s0}(M)$. Therefore

$$(A^2_{\epsilon,a}-\lambda)(R_1(\lambda)+R_2(\lambda)) = \mathrm{Id} - \epsilon E_2(\lambda). \tag{7.58}$$

For small ϵ, $\mathrm{Id} - \epsilon E_2(\lambda)$ can be inverted in L^2 by the Neumann series. Writing the inverse as $\mathrm{Id} - S(\lambda)$, one sees that the norm of $S(\lambda)$ is rapidly decreasing as $|\lambda| \to \infty$. Moreover, $S(\lambda)$ is conormal in ϵ, and therefore, belongs to $\mathcal{L}_c$. Again from the Neumann series,

$$S(\lambda) = \epsilon E_2(\lambda) + \epsilon^2 E_2(\lambda) E_2(\lambda) + \epsilon^2 E_2(\lambda) S(\lambda) E_2(\lambda). \tag{7.59}$$

Thus, by the semi-ideal property, $S(\lambda)$ is also in the surgery calculus and rapidly decreasing in λ.

Hence we have

$$(A_{\epsilon,a}^2 - \lambda)^{-1} = R_1(\lambda) + R_1(\lambda) \tag{7.60}$$

with $R_1(\lambda) = R_2(\lambda)(\mathrm{Id} - S(\lambda)) \in \Psi_{s0}^{-\infty,\tau}(M)$ being holomorphic in λ and rapidly decreasing as $|\lambda| \to \infty$.

7.2. *Large time behavior of heat kernel*

We can now determine the large time behavior of the heat kernel. By (7.53), (7.54), (7.60), one has

$$(1 - \phi(t)) e^{-tA_{\epsilon,a}^2} = \frac{1}{2\pi i} \int_\Gamma e^{-t\lambda} R_1(\lambda) d\lambda. \tag{7.61}$$

By our assumption, the contour Γ can be deformed to a contour lying in the right half plane but still below the spectrum. It follows then that $e^{-tA_{\epsilon,a}^2}$ is exponentially decreasing, with all t-derivatives, as $t \to \infty$ with values in $\Psi_{s0}^{-\infty,\tau}(M)$, where $\tau > 0$ is the largest τ for which the resolvent takes values in $\Psi_{s0}^{-\infty,\tau}(M)$ along the new contour.

7.3. *Proof of Theorem 3.1*

Finally we are in a position to prove Theorem 3.1.

Proof. Let $i: \Delta_{hs0} \to X_{hs0}^2$ be the embedding of the lifted diagonal. We have

$$\Delta_{hs0} \cong [X_{s0} \times [0, \infty); B_{0b} \times \{0\}, S] \tag{7.62}$$

which blows down to $X_{s0} \times [0, \infty)$. Denote the blow down map by β. On the other hand, the projection $\pi_\epsilon : M \times [0, 1] \to [0, 1]$ lifts to a b-fibration

$$\pi_{s0} : X_{s0} \to [0, 1].$$

Let us use the same notation to denote the induced b-fibration

$$\pi_{s0} : X_{s0} \times [0, \infty) \to [0, 1] \times [0, \infty).$$

Finally let $\pi_s = \pi_{s0} \circ \beta$ and π_t be the projection $[0, 1] \times [0, \infty) \to [0, 1]$. Then we can rewrite the eta function as

$$\eta(A_{\epsilon,a}, s) = (\pi_t)_*(\pi_s)_*[i^* trF], \ F = \frac{t^{(s-1)/2}}{\Gamma((s+1)/2)} A_{\epsilon,a} e^{-tA^2_{\epsilon,a}}. \qquad (7.63)$$

The polyhomogeneity of $i^* trF$ follows immediately from Theorem 6.1. Now for each $t > 0$ the computation of the pushforward $(\pi_s)_*$ falls into the realm of Lemma 5.4. To apply this result we must compute the three terms in (5.43). By (5.32) the leading term for the log term is the integral of $\mathrm{tr}(Ae^{-tA_a^2})|_{\partial M}$ where A lives on the half infinite cylinder. By (4.24) and (4.17),

$$\mathrm{tr}(Ae^{-tA_a^2}) = \mathrm{tr}(\gamma\partial_x e^{-tA_a^2}) + \mathrm{tr}(\sigma A_{\partial M} e^{-tA_a^2}) = 0.$$

Here the second term is identically zero by the splitting (4.16). Thus, the leading log term vanishes. The leading coefficient for the other term is given by

$$b\text{-}Tr(N_b(i^*F)) + b\text{-}Tr(N_s(i^*F)).$$

It follows that, when $\dim M$ is odd,

$$\eta(A_{\epsilon,a}) = b\text{-}Tr(N_b(i^*F)) + b\text{-}Tr(N_s(i^*F)) + r_1(\epsilon) + r_2(\epsilon)\log\epsilon.$$

The first term is by definition $\eta_b(A_0)$, while the second one is computed in Proposition 4.2. Note that $\eta(A_{\partial M}) = 0$ in this case.

For the even dimensional case, the term $tr(A_{\partial M} e^{-t\Delta_{\partial M}})$ no longer vanishes and it gives rise to the eta invariant for $A_{\partial M}$, whereas the b-eta term vanishes because of the parity of the dimension. □

We now explain how the analysis extends to the analytic torsion. The analytic torsion is defined in terms of the zeta function

$$\zeta_T(s) = \frac{1}{\Gamma(s)} \int_0^\infty t^{s-1} \mathrm{Tr}_s(Ne^{-t\underline{\Delta}})dt, \ \Re s >> 0,$$

where Tr_s is the supertrace associated to the usual $\mathbb{Z}_2$-grading via even/odd degree, and N is the number operator acting as multiplication by k on k-forms. Also, $\underline{\Delta}$ denotes the Laplacian restricted to the orthocomplement of its null space. In our situation, the acyclicity condition rules out the null space and so it is just the Laplacian. This zeta function extends to a meromorphic function on the entire complex plane with $s = 0$ a regular value. We define the analytic torsion of Ray and Singer by

$$\log T(M, \rho) = \zeta_T'(0).$$

For the half infinite cylinder, using (4.24), (4.25), one derives for $\Delta = A_a^2$

$$\mathrm{tr}_s(Ne^{-t\Delta}) = \frac{1}{\sqrt{\pi t}} e^{-u^2/t} \mathrm{tr}_s^{\partial M}(N_{\partial M} e^{-t\Delta_{\partial M}}) - \frac{1}{\sqrt{4\pi t}}(1 + e^{-u^2/t}) \mathrm{tr}_s^{\partial M}(e^{-t\Delta_{\partial M}}).$$

Hence

$$\mathrm{Tr}_s(Ne^{-t\Delta}) = \frac{1}{2} \mathrm{Tr}_s^{\partial M}(N_{\partial M} e^{-t\Delta_{\partial M}}).$$

Here we have used the fact that $\mathrm{Tr}_s^{\partial M}(e^{-t\Delta_{\partial M}}) = \chi(\partial M, \xi) = 0$ by our assumption. It follows then that for half infinite cylinder,

$$T_a(M, \rho) = \frac{1}{2} T(\partial M, \rho),$$

and similarly

$$T_r(M, \rho) = -\frac{1}{2} T(\partial M, \rho).$$

Even though the analytic torsion is defined in terms of analytic continuation, it has an explicit heat kernel representation involving the coefficients of the asymptotic expansion of $\mathrm{Tr}_s(Ne^{-t\Delta})$, see for example Dai-Melrose [7] where all negative powers except $t^{-1/2}$ are shown to vanish. Using this, one can proceed as before and derive the results for the analytic torsion.

Acknowledgements. We would like to thank Jeff Cheeger, Dan Freed, Richard Melrose, John Lott and Peter Gilkey for illuminating discussions. The author also acknowledge gratefully the influence of the ideas and the work of Krzysztof.

References

1. M. F. Atiyah, V. K. Patodi, and I. M. Singer. Spectral asymmetry and riemannian geometry. I, II, III. *Math. Proc. Cambridge Philos. Soc.* 77(1975),43–69, 78(1975):405–432, 79(1976),71–99.
2. J.-M. Bismut and J. Cheeger. η-invariants and their adiabatic limits. *J. Amer. Math. Soc.*, 2(1989),33–70.
3. J.-M. Bismut and D. S. Freed. The analysis of elliptic families I, II. *Comm. Math. Phys.* 106(1986),159–167, 107(1986),103–163.
4. U. Bunke. On the gluing problem for the η–invariant. *J. Differential Geom.*, 41(1995),397–448.
5. J. Cheeger. Eta invariants, the adiabatic approximation and conical singularities. *J. Differential Geom.*, 26(1987),175–211.
6. X. Dai. Adiabatic limits, nonmultiplicativity of signature, and Leray spectral sequence. *J. Amer. Math. Soc.*, 4(1991),265–321.

7. X. Dai and R. B. Melrose. Adiabatic limit, heat kernel and analytic torsion. preprint.
8. R. G. Douglas and K. P. Wojciechowski. Adiabatic limits of the η–invariants, the odd–dimensional Atiyah–Patodi–Singer problem. *Comm. Math. Phys.*, 142(1991),139–168.
9. P. Gilkey and L. Smith. The eta invariant for a class of elliptic boundary value problems. *Comm. Pure Appl. Math.*, 36(1983),85–131.
10. A. Hassell. Analytic surgery and analytic torsion. *Comm. Anal. Geom.*, 6(1998),255–289.
11. S. Klimek and K. P. Wojciechowski. Adiabatic cobordism theorems for analytic torsion and η-invariant. *J. Funct. Anal.*, 136(1996),269–293.
12. M. Lesch and K. P. Wojciechowski. On the η–invariant of generalized Atiyah–Patodi–Singer boundary value problems. *Illinois J. Math.*, 40(1996),30–46.
13. R. Mazzeo and R. Melrose. The adiabatic limit, hodge cohomology and Leray's spectral sequence for a fibration. *J. Differential Geom.*, 31(1990),185–213.
14. R. Mazzeo and R. Melrose. Analytic surgery and the eta invariant. *Geom. Funct. Anal.*, 5(1995),14–75.
15. R. Mazzeo and R. Melrose. Meromorphic extension of the resolvent on complete spaces with asymptotically constant negative curvature. *J. Funct. Anal.*, 75(1987),260–310.
16. R. Melrose. Differential analysis on manifolds with corners. preprint.
17. R. Melrose. Pseudodifferential operators, corners and singular limits. In: *Proceedings of the International Congress of Mathematicians 1990*, I(1991),217–234, Math. Soc. Japan, Tokyo.
18. R. Melrose. The Atiyah-Patodi-Singer index theorem. Research Notes in Mathematics, 4. A K Peters, Ltd., Wellesley, MA, 1993.
19. W. Müller. Eta invariants and manifolds with boundary. *J. Differential Geom.*, 40(1994),311–377.
20. R. Seeley. Conic degeneration of the Gauss-Bonnet operator. Festschrift on the occasion of the 70th birthday of Shmuel Agmon. *J. Anal. Math.*, 59(1992),205–215.
21. R. Seeley and I. M. Singer. Extending $\overline{\partial}$ to singular Riemann surfaces. *J. Geom. Phys.*, 5(1988),121–136.
22. I. M. Singer. Index and eta invariant. In: S. T. Yau, editor, *Proceedings of the conference on string theory (San Diego, Calif., 1986)*, 239–258, Adv. Ser. Math. Phys., 1, World Sci. Publishing, Singapore, 1987.
23. C. H. Taubes. Casson's invariant and gauge theory. *J. Differential Geom.*, 31(1990),547–599.
24. E. Witten. Global gravitational anomalies. *Comm. Math. Phys.*, 100(1985),197–229.
25. T. Yoshida. Floer homology and splittings of manifolds. *Ann. of Math.*, 134(1991),277–323.

Received by the editors September 12, 2005; Revised December 28, 2005

Analysis, Geometry and Topology
of Elliptic Operators, pp. 173–214

HEAT KERNELS OF THE SUB-LAPLACIAN AND THE LAPLACIAN ON NILPOTENT LIE GROUPS

KENRO FURUTANI

Department of Mathematics
Faculty of Science and Technology
Science University of Tokyo
2641 Noda, Chiba (278-8510) Japan
furutani_kenro@ma.noda.tus.ac.jp

Dedicated to Krzysztof P. Wojciechowski on his 50th birthday

Since the heat kernel of the sub-Laplacian on Heisenberg group was constructed in an explicit integral form by A. Hulanicki, we have several ways to construct the heat kernel for the sub-Laplacian and the Laplacian on 2-step nilpotent Lie groups. In this note we explain a method effectively employed by Beals-Gaveau-Greiner, the so called complex Hamilton-Jacobi theory, and illustrate the construction of the heat kernel for general 2-step cases. We discuss the solution of the generalized Hamilton-Jacobi equation and a quantity similar to the van Vleck determinant and their roles in the integral expression of the heat kernel. We expect this method will work also for 3-step cases to construct the heat kernel together with the theory of elliptic functions. So as an example, we consider the solution of the generalized Hamilton-Jacobi equation for the lowest dimensional 3-step nilpotent Lie group (Engel group). Then we discuss a hierarchy of heat kernels for the three dimensional Heisenberg group and Heisenberg manifolds as a simple example.

2000 *Mathematics Subject Classification.* Primary 35K05; Secondary 22E25

1. Introduction

The existence of the heat kernel is guaranteed for a fairly large class of second order operators in the frame work based on hypoellipticity and the spectral decomposition theorem. Our concern here is to construct the heat kernel in an explicit form for invariant sub-Laplacians and Laplacians on nilpotent Lie groups in terms of a certain class of special functions.

Since A. Hulanicki ([14]) constructed the heat kernel for the invariant sub-Laplacian on the Heisenberg group by a probability theoretic method,

Work partially supported by the Grant-in-Aid for Scientific Research (No. 17540202 C), The Ministry of Education, Culture, Sports, Science and Technology, Japan.

we have several ways to construct the heat kernel for the sub-Laplacian (and the Laplacian) on 2-step nilpotent Lie groups (cf. Gaveau [9], Beals, Gaveau and Greiner [2], [3], [5], Klinger [15], Furutani [7] and references therein), and for any 2-step nilpotent Lie group the heat kernel for the invariant sub-Laplacian is expressed in an integral form on a subspace in the complexified cotangent bundle of the Lie group with the integrand consisting of hyperbolic functions. However, until now, no explicit construction of the heat kernel has been given for 3-step nilpotent Lie groups, even for the lowest dimensional case.

This is a mostly expository article toward the construction of the heat kernel of the sub-Laplacian (and of the Laplacian) on 3-step nilpotent Lie group.

In §2 we start from the fundamental property of nilpotent Lie groups and we introduce the invariant sub-Laplacian. In §3 we explain the complex Hamilton-Jacobi theory for 2-step nilpotent Lie groups and construct the heat kernel for the sub-Laplacian and for the Laplacian in two ways. In §4 we discuss the Hamiltonian system and a solution of generalized Hamilton-Jacobi equation as a first step to construct the heat kernel for a sub-Laplacian on the 4-dimensional 3-step nilpotent Lie group, which is the lowest dimensional 3-step group. In §5 we discuss a "hierarchy" of heat kernels, by which we mean that heat kernels on homogeneous spaces of nilpotent Lie groups are obtained in terms of fiber integration of heat kernels on the whole group. Here we only illustrate concrete calculations for the 3-dimensional Heisenberg group, as a simple example.

2. Sub-Laplacian on nilpotent Lie groups and the heat kernel

2.1. *Nilpotent Lie groups and problems*

Let G be a connected and simply connected nilpotent Lie group. The basic fact about such nilpotent Lie groups is that the exponential map

$$\exp : \mathfrak{g} \longrightarrow G$$

is a diffeomorphism. Owing to this fact we can work on the linear space $\mathfrak{g}$ instead of the group G together with the help of the Campbell-Hausdorff formula :

$$\begin{aligned}&\exp X \cdot \exp Y \\ &\quad = \exp(X + Y + 1/2[X,Y] + 1/12[[X,Y], Y - X] + \cdots).\end{aligned}$$

Under the identification $\exp : \mathfrak{g} \xrightarrow{\sim} G$ the group law written on $\mathfrak{g}$ is,

if $\exp X \cdot \exp Y = \exp Z$, then we write $X * Y = Z$, and Z is given by
$Z = X + Y + 1/2[X, Y] + 1/12[[X, Y], Y - X] + \cdots$ (finite sum).

Let $\{X_i\}_{i=1}^m$ be linearly independent elements of the Lie algebra $\mathfrak{g}$, and $\widetilde{X}_i$ corresponding left invariant vector fields on the group G ($\cong \mathfrak{g}$).

Let

$$-\Delta_{sub} = \frac{1}{2}\sum_{i=1}^{m} \widetilde{X}_i^2, \ m \le \dim G = n$$

be a second order differential operator.

Proposition 2.1. *In general, if the vector fields* $\{\widetilde{X}\}_{i=1}^m$ *and a finite number of their brackets generate the whole tangent space at each point, then* Δ_{sub} *is a hypo-elliptic operator. Of course for left invariant vector fields it is equivalent to assume that* $\{X\}_{i=1}^m$ *and their brackets generate the whole Lie algebra* $\mathfrak{g}$.
In this case we call the operator Δ_{sub} *the sub-Laplacian, and we shall assume that the subbundle spanned by* $\{X_i\}$ *is equipped with the inner product which makes* $\{X_i\}$ *an orthonormal basis of this subbundle.*

Note that the invariant vector field $\widetilde{X}$ is anti-symmetric ($X \in \mathfrak{g}$):

$$\int_G \widetilde{X}(f)(x)g(x)\mathbf{d}x = -\int_G f(x)\widetilde{X}(g)(x)\mathbf{d}x, \ f, \ g \ \in C_0^\infty(G),$$

and so the operator Δ_{sub} is symmetric and positive (**dx** is a fixed Haar measure on G, which coincides with the Lebesgue measure on the Euclidean space $\mathfrak{g}$ under the identification $\exp : \mathfrak{g} \xrightarrow{\sim} G$). Moreover we have

Theorem 2.1. Δ_{sub} *is essentially selfadjoint on* $C_0^\infty(G)$.

This property follows from a theorem by Strichartz [18]:

Theorem 2.2. *If a sub-Riemannian metric on a non-compact manifold can be extended to a complete Riemannian metric, then the sub-Laplacian is essentially selfadjoint on the space of support compact smooth functions.*

Here we call a manifold M sub-Riemannian, if there is a subbundle E with an inner product in the tangent bundle $T(M)$ such that the vector fields taking values in E and their finite brackets generate the whole tangent space at each point.

So we do not distinguish Δ_{sub} on $C_0^\infty(G)$ and its unique selfadjoint realization in $L_2(G,\ \mathbf{d}x)$. Then by the spectral decomposition theorem

$$\Delta_{sub} = \int_0^\infty \lambda dE_\lambda,\ (\{E_\lambda\}\ \text{: the spectral measure})$$

we know that the heat kernel $K(t;\widetilde{g},\ g)$ is the kernel distribution of the operator

$$e^{-t\Delta_{sub}} = \int_0^\infty \lambda dE_\lambda : L_2(G) \to L_2(G)\ (\ t > 0),$$

which is smooth because

(a) Δ_{sub} is hypoelliptic,
(b) for any integer k, $\Delta_{sub}^{\ k} \circ e^{-t\Delta_{sub}} = e^{-t\Delta_{sub}} \circ \Delta_{sub}^{\ k}$ is defined on $L_2(G)$ and $e^{-t\Delta_{sub}}$ maps continuously $L_2(G)$ to $\bigcap_{k=1}^{\infty}\left(\text{ domain of } \Delta_{sub}^{\ k}\right) = C^\infty(G)$,
(c) $e^{-t\Delta_{sub}}$ can be extended from $L_2(G)$ to the whole space of distributions $\mathcal{D}'(G)$ on G, and by $e^{-t\Delta_{sub}} = e^{-t/2\Delta_{sub}} \circ e^{-t/2\Delta_{sub}}$, it is in fact a map from $\mathcal{D}'(G)$ to $C^\infty(G)$.

Now our interests are:

[I] Spectral decomposition of the sub-Laplacian (and of the Laplacian) in a *multiplication form*, that is, to find a measure space $(\mathbf{X},\ \mathbf{dm})$, a (positive) function φ on $\mathbf{X}$ and a unitary transformation

$$\mathbf{U} : L_2(G) \overset{\sim}{\to} L_2(\mathbf{X},\ \mathbf{dm}) \text{ such that } \Delta_{sub} = \mathbf{U}^{-1} \circ \mathbf{M}_\varphi \circ \mathbf{U},$$

where $\mathbf{M}_\varphi : L_2(\mathbf{X},\ \mathbf{dm}) \ni f \mapsto \varphi f \in L_2(\mathbf{X},\ \mathbf{dm})$ is a multiplication operator with the function φ.

[II] *Explicit* construction of the heat kernel $K(t;x,y)$ of the sub-Laplacian Δ_{sub} (and of the Laplacian):

$$\left(\Delta_{sub} + \frac{\partial}{\partial t}\right) K(t;x,y) = 0\,,$$

$$\lim_{t\to 0} \int_G K(t;x,y) f(y)\mathbf{d}y = f(x),\ (\mathbf{d}y \text{ is the Haar measure})\,;$$

here we know that $K(t;x,y)$ is of the form

$$K(t;x,y) = k_t(y^{-1} * x),$$

where $k_t(x)$ is a smooth function on $\mathbb{R}_+ \times G$ because of the (left) invariance of the heat kernel:

$$K(t; g\cdot x, g\cdot y) = K(t; x, y), \text{ for all } g, x, y \in G.$$

[III] *Explicit* expression of the Green function and/of fundamental solutions of Δ_{sub} (and Δ).

There are so many papers concerning the hypoelliptic operators on nilpotent Lie groups (see Hörmander [13], Rothshild and Stein [17] and papers cited therein). Our concern is to construct the heat kernel in terms of special functions, such as Hermite functions, hyperbolic functions (for 2-step cases). Especially for the 3-step cases it will be worked out by elliptic functions, since the bicharacteristic flow for the sub-Laplacian is solved in terms of elliptic functions.

Remark 2.1. If we have a spectral decomposition of Δ (or Δ_{sub}) in the multiplication form, then the heat kernel is expressed as

$$e^{-t\Delta} = \mathbf{U} \circ e^{-t\mathbf{M}_\varphi} \circ \mathbf{U}^{-1},$$

just like the case of Euclidean spaces where the operator $\mathbf{U}$ is the Fourier transformation. So [**I**] gives [**II**], but not in the opposite way.

Then if we have the heat kernel

$$K(t; x, y) \in C^\infty(\mathbb{R}^+ \times G \times G),$$

the Green function $G(x, y)$ can be expressed as

$$\int_G G(x,y)f(y)dy = \int_0^\infty \int_G K(t; x, y)f(y)dydt.$$

Even if we have an *explicit* expression of the heat kernel, this does not give the spectral decomposition of the (sub-)Laplacian in a multiplication form directly. For this purpose we are still required to find a measure space $(\mathbf{X}, \mathbf{dm})$, a unitary transformation between $L_2(G)$ and $L_2(\mathbf{X}, \mathbf{dm})$ and a positive function on $\mathbf{X}$. With these the Laplacian (sub-Laplacian) is expressed as a multiplication operator.

2.2. *Spectral decomposition and heat kernel*

Here we give an *explicit* multiplication form of the Laplacian for a certain class of 2-step nilpotent Lie groups including Heisenberg groups (cf. [7]).

Let $\mathfrak{g}$ be a nilpotent Lie algebra of 2-step such that

$$\begin{aligned} &\mathfrak{g} = \mathfrak{g}_+ \oplus \mathfrak{g}_- \oplus \mathfrak{z}, \ [\mathfrak{g}_+, \mathfrak{g}_-] = \mathfrak{z}, \ [\mathfrak{g}_\pm, \ \mathfrak{g}_\pm] = 0 \\ &\mathfrak{z} \text{ is the center and} \\ &n = \dim \mathfrak{g}_+ = \dim \mathfrak{g}_-, \ \dim \mathfrak{z} = d. \end{aligned} \tag{1}$$

Let $\{X_i\}_{i=1}^n$, $\{Y_i\}_{i=1}^n$ and $\{Z_k\}_{k=1}^d$ be a basis of $\mathfrak{g}_+$, $\mathfrak{g}_-$ and $\mathfrak{z}$ respectively with the structure constants $C_{i,j}^k$:

$$[X_i, \ Y_j] = \sum_{k=1}^{d} C_{i,j}^k \ Z_k \, ,$$

and all other brackets are zero.

Put the matrix $C(\lambda) \, : \, \lambda \in \mathfrak{z}^*$,

$$C(\lambda)_{i,j} = \sum_{k=1}^{d} C_{i,j}^k \, \lambda(Z_k). \tag{2}$$

We assume that the matrix $C(\lambda)\,{}^tC(\lambda)$ is diagonal:

$$C(\lambda)\,{}^tC(\lambda) = \begin{pmatrix} c_1(\lambda) & & & \cdots & 0 \\ 0 & c_2(\lambda) & & \cdots & \\ & & \cdots & & \\ \cdots & & & \cdots & \\ 0 & \cdots & & & c_n(\lambda) \end{pmatrix}$$

and also we assume that all diagonal elements $c_i(\lambda)$ are non-degenerate (positive) bilinear forms on $\mathfrak{z}$.

We identify the Lie algebra

$$\mathfrak{g} = \mathfrak{g}_+ \oplus \mathfrak{g}_- \oplus \mathfrak{z}$$

satisfying above conditions and the Lie Group

$$\exp : \mathfrak{g}_+ \times \mathfrak{g}_- \times \mathfrak{z} \cong G$$

through the exponential map. We denote the element g in G by

$$g = \sum x_i X_i + \sum y_i Y_i + \sum z_k Z_k = (x, y, z).$$

The multiplication law is given by

$$g * \widetilde{g} = g + \widetilde{g} + \frac{1}{2}[g, \ \widetilde{g}].$$

The Laplacian Δ is given by the formula:

$$
\begin{aligned}
-\Delta = &\sum \frac{\partial^2}{\partial x_i^2} + \frac{\partial^2}{\partial y_i^2} - \sum C_{i,j}^k y_j \frac{\partial^2}{\partial x_i \partial z_k} + \sum C_{i,j}^k x_i \frac{\partial^2}{\partial y_j \partial z_k} \\
&+ \frac{1}{2} \sum_i \sum_{k_1<k_2} \sum_{j_1,j_2} C_{i,j_1}^{k_1} C_{i,j_2}^{k_2} y_{j_1} y_{j_2} \frac{\partial^2}{\partial z_{k_1} \partial z_{k_2}} \\
&\qquad + \frac{1}{2} \sum_j \sum_{k_1<k_2} \sum_{i_1,i_2} C_{i_1,j}^{k_1} C_{i_2,j}^{k_2} x_{i_1} x_{i_2} \frac{\partial^2}{\partial z_{k_1} \partial z_{k_2}} \\
&+ \sum_k \left(1 + \frac{1}{4} \left\{ \sum_{i=1}^n \left(\sum_{j=1}^n y_j C_{i,j}^k \right)^2 + \sum_{j=1}^n \left(\sum_{i=1}^n x_i C_{i,j}^k \right)^2 \right\} \right) \frac{\partial^2}{\partial z_k^2} .
\end{aligned}
$$

Now we list two examples of the Lie algebra satisfying the above assumptions.

Example 2.1. *Heisenberg algebra* of any dimension.

Example 2.2. *Heisenberg type algebra* whose dimension of the center is $\equiv 0 (mod\, 4)$, that is, $\mathfrak{g}$ is 2-step and equipped with an inner product $< \bullet, \bullet >$ such that for any $Z \in [\mathfrak{g}, \mathfrak{g}] = \mathfrak{z}$, the map $j(Z) : \mathfrak{z}^\perp \to \mathfrak{z}^\perp$ defined by

$$< j(Z)(X), Y > = < Z, [X, Y] >$$

satisfies

$$j(Z)^2 = - < Z, Z > Id.$$

For the Laplacian of this group we have the explicit spectral decomposition in a multiplication form (cf. [7]):

Theorem 2.3. *There exists a measure space* $(\mathbf{X}, \mathbf{dm})$*, a positive function* φ *on* $\mathbf{X}$ *and a unitary operator* $\mathbf{U} : L_2(G) \cong L_2(\mathbf{X}, \mathbf{dm})$ *such that*

$$\mathbf{U}^{-1} \circ \mathbf{M}_\varphi \circ \mathbf{U} = \Delta.$$

From the explicit form of the function φ and the measure $\mathbf{dm}$ (see below (3) and (6)) we get easily

Corollary 2.1. *The spectrum of the Laplacian* Δ *is* $[0, \infty]$ *and is purely continuous.*

Although we do not give the proof of this theorem, we describe the measure space $(\mathbf{X},\ \mathbf{dm})$, the unitary transformation $\mathbf{U}$ and the positive function φ on $\mathbf{X}$ appearing in the above theorem.

Let $\mathbf{k} = (k_1, \cdots, k_n),\ k_i \in \mathbb{N},\ k_i \geq 0$ be a multi-index and define the measure space

$$(\mathbf{X_k}, \mathbf{dm_k}) = \left(\mathfrak{g}_+ \times \mathfrak{z}^*\backslash\{0\}, \prod_{i=1}^{n} \sqrt{c_i(\lambda)}dvd\lambda\right) \tag{3}$$

and put

$$(\mathbf{X}, \mathbf{dm}) = \coprod_{\mathbf{k}\in\mathbb{N}^n} (\mathbf{X_k}, \mathbf{dm_k}),\ \text{(direct sum)}. \tag{4}$$

Let $\mathcal{F}$ be a partial Fourier transformation

$$\mathcal{F} : C_0^\infty(\mathfrak{g}_+ \times \mathfrak{g}_- \times \mathfrak{z}) \to C^\infty(\mathfrak{g}_+ \times \mathfrak{g}_-^* \times \mathfrak{z}^*),$$
$$(\mathcal{F}f)(x, \xi, \eta) = (2\pi)^{-(n+d)/2} \int e^{\sqrt{-1}(<\xi,y>+<\eta,z>)} f(x, y, z)dydz,$$

and

$$\mathbf{R} : L_2(\mathfrak{g}_+ \times \mathfrak{g}_-^* \times \mathfrak{z}^*) \longrightarrow L_2(\mathfrak{g}_+ \times \mathfrak{g}_-^* \times (\mathfrak{z}^*\backslash\{0\}))$$

the restriction map. Also let

$$\Phi : \mathfrak{g}_+ \times \mathfrak{g}_+ \times (\mathfrak{z}^*\backslash\{0\}) \longrightarrow \mathfrak{g}_+ \times \mathfrak{g}_-^* \times (\mathfrak{z}^*\backslash\{0\})$$

denote the diffeomorphism defined by

$$\Phi(v, w, \lambda) = (x, \xi, \eta), \quad x = v - w, \quad \eta = \lambda,$$
$$< \xi, y > = -\frac{1}{2} < \lambda, [v + w, y] > \ (= \mathbf{T}_\lambda(v + w)(y),\ y \in \mathfrak{g}_-).$$

Here $\mathbf{T}_\lambda : \mathfrak{g}_+ \to \mathfrak{g}_-^*$ is an isomorphism for any $\lambda \neq 0$. Then the composition $\mathbf{K} = \Phi^* \circ \mathbf{R} \circ \mathcal{F}$ is a unitary transformation:

$$\mathbf{K} = \Phi^* \circ \mathbf{R} \circ \mathcal{F} : L_2(\mathfrak{g}_+ \times \mathfrak{g}_- \times \mathfrak{z})\ (=\ L_2(G))$$
$$\longrightarrow L_2\Big(\mathfrak{g}_+ \times \mathfrak{g}_+ \times (\mathfrak{z}^*\backslash\{0\}), \prod_{i=1}^{n} \sqrt{c_i(\lambda)}dvdwd\lambda\Big).$$

Next let the first order differential operators $\mathbf{S}_i$ be

$$\mathbf{S}_i = \frac{\partial}{\partial w_i} - \sqrt{c_i(\lambda)}w_i,\ i = 1, \ldots, n$$

and put the functions

$$h_i(w,\lambda) = e^{-\frac{1}{2}\sqrt{c_i(\lambda)}w_i^2}.$$

For a fixed $\lambda \neq 0$ and each multi-index $\mathbf{k} = (k_1, \ldots, k_n)$ we denote by $h(w, \lambda, \mathbf{k})$ the Hermite function of the variables $w \in \mathfrak{g}_+$:

$$h(w, \lambda, \mathbf{k}) = (\mathbf{S}_1^{k_1} h_1)(w, \lambda) \cdots (\mathbf{S}_n^{k_n} h_n)(w, \lambda).$$

Now for a function $f \in C_0^\infty(\mathfrak{g}_+ \times \mathfrak{g}_+ \times (\mathfrak{z}^* \backslash \{0\}))$, we define

$$\mathbf{E}(f)(v, \lambda, \mathbf{k}) = \frac{1}{N_{\mathbf{k}}(\lambda)} \int_{\mathfrak{g}_+} f(v, w, \lambda) h(w, \lambda, \mathbf{k}) dw \ \in C^\infty(\mathfrak{g}_+ \times (\mathfrak{z}^* \backslash \{0\})),$$

where $N_{\mathbf{k}}(\lambda)$ is the L_2-norm of the function $h(w, \lambda, \mathbf{k})$:

$$N_{\mathbf{k}}(\lambda)^2 = \int_{\mathfrak{g}_+} |h(w, \lambda, \mathbf{k})|^2 \, dw_1 \cdots dw_n = 2^{|\mathbf{k}|} \cdot \mathbf{k}! \cdot \pi^{n/2} \cdot \prod_{i=1}^n c_i(\lambda)^{\frac{k_i - 1}{2}}.$$

Then the operator $\mathbf{E}$ is extended to a unitary transformation from $L_2(\mathfrak{g}_+ \times \mathfrak{g}_+ \times (\mathfrak{z}^* \backslash \{0\}))$ to $L_2(\mathbf{X}, \mathbf{dm})$, and the unitary operator $\mathbf{U}$ is defined by

$$\mathbf{U} = \mathbf{E} \circ \Phi^* \circ \mathbf{R} \circ \mathcal{F}. \tag{5}$$

Finally, let a function φ on $\mathbf{X}$ be

$$\varphi(v, \lambda, \mathbf{k}) = \|\lambda\|^2 + \sum_{i=1}^n (2k_i + 1)\sqrt{c_i(\lambda)}, \ (v, \lambda) \in \mathfrak{g}_+ \times (\mathfrak{z}^* \backslash \{0\}). \tag{6}$$

With these data we have an explicit integral expression of the heat kernel by calculating the kernel distribution of the composition operator

$$\mathbf{U}^{-1} \circ e^{-t\mathbf{M}_\varphi} \circ \mathbf{U} :$$

Theorem 2.4.

$$\begin{aligned} K(t; g, \widetilde{g}) &= K(t; x, y, z, \widetilde{x}, \widetilde{y}, \widetilde{z}) \qquad (7) \\ &= (2\pi)^{-(n+d)/2} \int_{\mathfrak{z}^*} e^{\sqrt{-1} < \eta, \widetilde{z} - z + 1/2[\widetilde{x}, y] - 1/2[x, \widetilde{y}] >} \\ &\times e^{-t\|\eta\|^2} \cdot \prod_{i=1}^n \frac{\sqrt{c_i(\eta)}}{2 \sinh t\sqrt{c_i(\eta)}} \cdot e^{-\frac{\sqrt{c_i(\eta)}}{4} \cdot \frac{\cosh t\sqrt{c_i(\eta)}}{\sinh t\sqrt{c_i(\eta)}} \cdot \{(x_i - \widetilde{x}_i)^2 + (y_i - \widetilde{y}_i)^2\}} \, d\eta. \end{aligned}$$

To calculate the integrals included in the formula $\mathbf{U}^{-1} \circ e^{-t\mathcal{M}_\varphi} \circ \mathbf{U}$, a formula called Mehler's formula (cf. Thangaveln [19]) for the generating function of Hermite polynomials is important. Such a formula is also important in the construction by the probability theoretic method given by A. Hulanicki [14].

Of course this result coincides with the expression given by the method of the complex Hamilton-Jacobi theory by Beals, Gaveau and Greiner in [2] (also see Beals, Gaveau and Greiner [3] and [4]), where they do not need such a formula for the generating function of Hermite polynomials. They obtained the same formula directly. Their method starts by assuming that the heat kernel $k_t(g)$ has an integral expression (see (9) below) which reflects the physical phenomena, that is, the heat flows *mostly* along the *geodesics* starting from the identity element (the δ-function is put as the identity element at the time $= 0$) and the total amount ($= k_t(g)$) should be summed up ($=$ integrated) over a certain class of geodesics arriving at the point g at a time t from *somewhere*. This class of geodesics is determined by solving the Hamiltonian system (bicharacteristic flow) under an initial-boundary condition, i.e., we assume that the coordinates in $\mathfrak{g}/[\mathfrak{g},\mathfrak{g}]$ are zero at $t = 0$ and that the end point g is arbitrarily given in the space G. In the Euclidean case there is only one such geodesic arriving at the point g under this condition, so no integration is taken and we have the well known formula. However in the nilpotent (non-abelian) cases we must consider geodesics whose initial points will not be the identity element. These will be parametrized by the dual space $[\mathfrak{g},\mathfrak{g}]^*$ both for the Laplacian and sub-Laplacian cases and in fact they mostly are in the complexified space. The reason why we need to consider such geodesics is that in our curved space (although topologically it is Euclidean) the wave front set of the δ-function influences the points in the direction $[\mathfrak{g},\mathfrak{g}]$. Of course this argument will not be enough to study the construction of the heat kernel under the assumption that it has a prescribed integral form. However, together with the result in Theorem 2.4 of the heat kernel for a special class of nilpotent Lie groups, here following [3], we take as our point of departure that the heat kernel of general (two step) nilpotent Lie groups will be of an integral form[a] with the action function f and the volume element W such that

$$K(t;(x,z),(\widetilde{x},\widetilde{z})) = k_t((\widetilde{x},\widetilde{z})^{-1} * (x,z)), \tag{8}$$

$$k_t(x,z) = \frac{1}{t^N}\int_\tau e^{-\frac{f(x,z,\tau)}{t}} W(x,z,\tau)d\tau,\ \ (x,z) = g \in \left(\mathfrak{g}/[\mathfrak{g},\mathfrak{g}] \times [\mathfrak{g},\mathfrak{g}]\right), \tag{9}$$

with a specific order $N = \frac{1}{2}\dim \mathfrak{g}/[\mathfrak{g},\mathfrak{g}] + \dim[\mathfrak{g},\mathfrak{g}]$.

[a]A coordinate change is needed in Theorem 2.4 to arrive at the form (9).

Consequently the function $f(g,\tau)$ should include all the informations of the *real* geodesics when $t \downarrow 0$ and the volume element $W(g,\tau)$ will reflect the amount of energy flowing through geodesics arriving at the point g.[b]

In the next section we determine these functions, and in fact we will know that it gives us the heat kernel for general two step nilpotent Lie groups.

3. Complex Hamilton-Jacobi theory

3.1. *Heat kernel of sub-Laplacian*

Let G be an $(n+d)$-dimensional connected and simply connected 2-step nilpotent Lie group with the center $\mathfrak{z} = [\mathfrak{g},\mathfrak{g}]$, $\dim \mathfrak{z} = d$.

We identify T^*G with $\mathfrak{g}\times\mathfrak{g}^*$. Let $\{X_i\}_{i=1}^n$ be a basis of a complement of the derived algebra $[\mathfrak{g},\mathfrak{g}]$ and denote the coordinates on $\mathfrak{g}\times\mathfrak{g}^*$ by $(x,z;\xi,\theta)$ by fixing a suitable basis $\{Z_k\}_{k=1}^d$ in the center $\mathfrak{z} = [\mathfrak{g},\mathfrak{g}]$.

Let $[X_i,X_j] = 2\sum\limits_{k=1}^{d} a_{ij}^k Z_k (a_{ij} = -a_{ji})$, and let $\Omega(\theta)$ [c] be a $d\times d$ matrix with the entries $\Omega(\theta)_{ij} = \sum\limits_{k=1}^{d} a_{ij}^k \theta_k$.

To each X_j, we denote by $\widetilde{X}_j = \dfrac{\partial}{\partial x_j} + \sum\limits_{i=1}^{n}\sum\limits_{k=1}^{d} a_{ij}^k x_i \dfrac{\partial}{\partial z_k}$ the corresponding left invariant vector field on the group G. Then the sum

$$-\Delta_{sub} = \frac{1}{2}\sum_{i=1}^{n} \widetilde{X}_i^2$$

is a sub-Laplacian which satisfies the "Hörmander condition" for the hypoellipticity.

Let H be the Hamiltonian of this sub-Laplacian Δ_{sub}:

$$H(x,z;\xi,\theta) = \frac{1}{2}\sum_{j=1}^{n}\Big(\xi_j + \sum_{i=1}^{n}\sum_{k=1}^{d} a_{ij}^k x_i\theta_k\Big)^2 = \frac{1}{2}\sum_{j}\Big(\xi_j + \sum_{i}\Omega(\theta)_{ij}\cdot x_i\Big)^2.$$

We consider the Hamiltonian system

$$\begin{cases} \dot{x} = H_\xi = \xi - \Omega(\theta)x, & \dot{z} = H_\theta\,, \\ \dot{\xi} = -H_x, \quad \dot{\theta} = -H_z \equiv 0, & \end{cases} \tag{10}$$

[b]In this note we do not discuss these aspects. See [3] for the Heisenberg group case.

[c] $\Omega(\theta) = \begin{pmatrix} 0 & C(\theta) \\ -{}^tC(\theta) & 0 \end{pmatrix}$, where $C(\theta)$ is the matrix defined in (2).

with the *initial-boundary conditions* such that

$$\begin{cases} x(0) = 0 & x(s) = x = (x_1, \cdots, x_n) \in \mathbb{R}^n, \\ & z(s) = z = (z_1, \cdots, z_d) \in \mathbb{R}^d, \\ \theta(0) = \sqrt{-1}\tau & \tau = (\tau_1, \cdots, \tau_d) \in \mathbb{R}^d, \end{cases} \tag{11}$$

where $s \in \mathbb{R}$, x and z are arbitrarily given.

Since $\dot{x}(t) = e^{-2t\Omega(\theta)}\xi(0)$, by integrating the equation $\Omega(\theta)\dot{x}(t) = \Omega(\theta)e^{-2t\Omega(\theta)}\xi(0)$, we have

$$\Omega(\theta)x(t) = -1/2\Big(e^{-2t\Omega(\theta)} - Id\Big)\xi(0).$$

Now by the condition that the value $\theta \equiv \theta(0) = \sqrt{-1}\tau$ is purely imaginary, the matrix $\sqrt{-1}\Omega(\tau)$ is selfadjoint. Hence the matrix

$$\frac{\sqrt{-1}s\Omega(\tau)}{\sinh\sqrt{-1}s\Omega(\tau)} = \frac{1}{2\pi\sqrt{-1}}\int_\Gamma \frac{\lambda}{\sinh\lambda}\Big(\lambda - \sqrt{-1}s\Omega(\tau)\Big)^{-1} d\lambda$$

is well defined and invertible for any $s \in \mathbb{R}$ and $\tau \in \mathbb{R}^d$, so that we have one to one correspondence between $\xi(0)$ and x:

$$\xi(0) = e^{s\sqrt{-1}\Omega(\tau)} \cdot \frac{\sqrt{-1}\Omega(\tau)}{\sinh s\sqrt{-1}\Omega(\tau)} \cdot x, \; s \neq 0. \tag{12}$$

The contour Γ is taken suitably surrounding the spectrum of the matrix $s\sqrt{-1}\Omega(\tau)$.

Now we solve the initial value problem:

$$\begin{cases} \dot{x}_j(t) = H_{\xi_j} = \xi_j + \sqrt{-1}\sum_{i,k} a^k_{ij}x_i\tau_k = \xi_j + \sqrt{-1}\sum_i \Omega(\tau)_{ij}x_i, \\ \dot{\xi}_i(t) \; = -H_{x_i} = -\sqrt{-1}\sum_j\Big(\xi_j + \sqrt{-1}\sum_\ell \Omega(\tau)_{\ell j}x_\ell\Big)\cdot\Omega(\tau)_{ij} \end{cases} \tag{13}$$

with the initial conditions

$$\begin{cases} x(0) = 0 \\ \xi(0) = e^{s\sqrt{-1}\Omega(\tau)} \cdot \frac{\sqrt{-1}\Omega(\tau)}{\sinh s\sqrt{-1}\Omega(\tau)} x. \end{cases} \tag{14}$$

Then we have the solutions:

$$x(t) = x(t; s, x, \tau) = e^{(s-t)\sqrt{-1}\Omega(\tau)} \frac{\sinh t\sqrt{-1}\Omega(\tau)}{\sinh s\sqrt{-1}\Omega(\tau)} \cdot x$$

$$\begin{aligned}
\xi(t) &= \xi(t, s, x, \tau) \\
&= \frac{\sqrt{-1}\Omega(\tau)}{\sinh s\sqrt{-1}\Omega(\tau)} \cdot e^{s\sqrt{-1}\Omega(\tau)} \Big(Id - e^{-t\sqrt{-1}\Omega(\tau)} \sinh t\sqrt{-1}\Omega(\tau)\Big) \cdot x \\
&= \Big(e^{-t\sqrt{-1}\Omega(\tau)} \cosh t\sqrt{-1}\Omega(\tau)\Big) \cdot \Big(e^{s\sqrt{-1}\Omega(\tau)} \frac{\sqrt{-1}\Omega(\tau)}{\sinh s\sqrt{-1}\Omega(\tau)}\Big) x \\
&= \Big(e^{-t\sqrt{-1}\Omega(\tau)} \cosh t\sqrt{-1}\Omega(\tau)\Big) \xi(0).
\end{aligned}$$

These give us solutions for the initial-boundary problem (10) under the condition (11) together with the solutions

$$z_k(t) = z_k + \int_s^t \sum_j \left(\Big(e^{-2u\sqrt{-1}\Omega(\tau)} \xi(0)\Big)_j \cdot \sum_i a_{ij}^k x_i(u) \right) du, \ k = 1, \cdots, d$$

$$\theta(t) \equiv \sqrt{-1}\tau, \ \tau = (\tau_1, \cdots, \tau_d) \in \mathbb{R}^d.$$

We do not give the final form of the solutions $z_k(t) (k = 1, \cdots, d)$, but it will be seen from the expression that the functions $z_k(t)$ are determined uniquely and we do not need the explicit form of the solutions $z_k(t)$ in the following calculations.

Let $g = g(s; x, z, \tau) \in C^\infty(\mathbb{R} \times \mathbb{R}^n \times \mathbb{R}^d \times \mathbb{R}^d)$ be the *complex action integral*

$$\begin{aligned}
g(s; x, z, \tau) &= \sqrt{-1} \sum_{i=1}^d \tau_i, z_i(0; s, x, z, \tau) \\
&\quad + \int_0^s < \xi(t), \dot{x}(t) > + < \theta(t), \dot{z}(t) > -H(x(t), z(t); \xi(t), \theta(t)) dt,
\end{aligned}$$

then g satisfies the usual Hamilton-Jacobi equation (see Appendix A):

Proposition 3.1.

(a)

$$\frac{\partial g}{\partial s} + H(x, z; \nabla g) = 0.$$

(b) Also the function g satisfies a relation

$$g(s; x, z, \ell \cdot \tau) = \frac{1}{\ell} \cdot g(1; x, z, \tau).$$

Determination of the function $g(s;x,z,\tau)$.

Put $\zeta(t)=\dot{x}(t)=\xi(t)-\sqrt{-1}\Omega(\tau)x$, then

$$\dot{\zeta}(t)=-2\sqrt{-1}\Omega(\tau)\zeta(t)$$

and we have

$$\begin{aligned}
g(s;x,z,\tau) &= \sqrt{-1}\sum_{i=1}^{d}\tau_i\cdot z_i(0;s,x,z,\tau)\\
&+\int_0^s <\xi(t),\ \dot{x}(t)> + <\theta(t),\ \dot{z}(t)> -H(x(t),z(t);\xi(t),\theta(t))dt\\
&=\sqrt{-1}\sum_{i=1}^{d}\tau_i\cdot z_i+\int_0^s <\xi(t),\ \dot{x}(t)> -\frac{1}{2}<\zeta(t),\ \zeta(t)> dt\\
&=\sqrt{-1}\sum_{i=1}^{d}\tau_i\cdot z_i+\int_0^s <\zeta(t)+\sqrt{-1}\Omega(\tau)x,\ \zeta(t)> -\frac{1}{2}<\zeta(t),\ \zeta(t)> dt\\
&=\sqrt{-1}\sum_{i=1}^{d}\tau_i\cdot z_i+\int_0^s \frac{1}{2}<\zeta(t),\ \zeta(t)> - <x,\ \sqrt{-1}\Omega(\tau)\zeta(t)> dt\\
&=\sqrt{-1}\sum_{i=1}^{d}\tau_i\cdot z_i+\int_0^s 1/2<\zeta(t),\ \zeta(t)> +\frac{1}{2}<x,\ \dot{\zeta}(t)> dt\\
&=\sqrt{-1}\sum_{i=1}^{d}\tau_i\cdot z_i+\frac{1}{2}<x,\zeta>\Big|_0^s\\
&=\sqrt{-1}\sum_{i=1}^{d}\tau_i\cdot z_i+\frac{1}{2}\Big\langle\sqrt{-1}\Omega(\tau)\coth\big(\sqrt{-1}s\Omega(\tau)\big)\cdot x,\ x\Big\rangle.
\end{aligned}$$

Now let $f=f(x,z,\tau)$ be

$$f(x,z,\tau)=g(1;x,z,\tau)=\sqrt{-1}\sum_{i=1}^{d}\tau_i\cdot z_i+\frac{1}{2}\Big\langle\sqrt{-1}\Omega(\tau)\coth(\sqrt{-1}\Omega(\tau))\cdot x,\ x\Big\rangle,$$

then f satisfies

$$\frac{f(x,z,s\cdot\tau)}{s}=g(s;x,z,\tau)$$

and is a solution of the following equation, called generalized Hamilton-Jacobi equation:

$$H(x,z;\nabla f)+\sum_{i=1}^{d}\tau_i\frac{\partial f}{\partial\tau_i}=f(x,z,\tau_1,\cdots,\tau_d).$$

The heat kernel $K(t;(x,z),(\widetilde{x},\widetilde{z})) = k_t((\widetilde{x},\widetilde{z})^{-1}*(x,z))$ ($*$ is the product of the group G) is given by a function $k_t(x,z)$ of the form

$$k_t(x,z) = \frac{1}{t^{n/2+d}} \int_{\mathbb{R}^d} e^{-\frac{f(x,z,\tau)}{t}} \cdot W(\tau)d\tau, \qquad d = \dim\ [\mathfrak{g},\mathfrak{g}],$$

if we have a function $W(x,z,\tau)$ which is a solution of the equation, called the *transport equation* (Appendix B):

$$\sum \tau_i \frac{\partial W}{\partial \tau_i} + \sum_j \widetilde{X}_j(f)\widetilde{X}_j(W) - \left(\Delta_{sub}(f) + \frac{d}{2}\right) \cdot W = 0. \tag{15}$$

By noting that

$$\begin{aligned} -\Delta_{sub}(f) &= \frac{1}{2}\mathrm{tr}\Big(\sqrt{-1}\Omega(\tau)\coth\big(\sqrt{-1}\Omega(\tau)\big)\Big) \\ &= \frac{1}{2}\cdot \mathrm{tr}\Big(\int_\Gamma \lambda \cdot \frac{\cosh\lambda}{\sinh\lambda} \cdot \big(\lambda - \sqrt{-1}\Omega(\tau)\big)^{-1} d\lambda\Big) \end{aligned} \tag{16}$$

does not depend on the space variables (x,z), we may have a solution of this transport equation (15) in the class of the functions $W(x,z,\tau) = W(\tau)$. In fact the square root of the Jacobian of the correspondence (12) is a solution of the transport equation (15):

Proposition 3.2. *Let*

$$W(\tau) = \left(\det e^{\sqrt{-1}\Omega(\tau)} \cdot \frac{\sqrt{-1}\Omega(\tau)}{\sinh\sqrt{-1}\Omega(\tau)}\right)^{\frac{1}{2}} = \left(\det \frac{\sqrt{-1}\Omega(\tau)}{\sinh\sqrt{-1}\Omega(\tau)}\right)^{\frac{1}{2}},$$

where the branch is chosen such that $W(0) = 1$, then the function $W(\tau)$ is a solution of the transport equation (15).

Proof. Let $\sigma(t) = W(t\tau)^2 = \det\Big(\dfrac{\sqrt{-1}t\Omega(\tau)}{\sinh\sqrt{-1}t\Omega(\tau)}\Big)$, then

$$\begin{aligned} \frac{d}{dt}\sigma(t) &= 2W(t\tau)\sum_{k=1}^{d} \tau_k \left(\frac{\partial W}{\partial \tau_k}\right)(t\tau) \\ &= \mathrm{tr}\left(\frac{d}{dt}\left(\frac{\sqrt{-1}t\Omega(\tau)}{\sinh\sqrt{-1}t\Omega(\tau)}\right)\cdot\left(\frac{\sqrt{-1}t\Omega(\tau)}{\sinh\sqrt{-1}t\Omega(\tau)}\right)^{-1}\right)\cdot\sigma(t). \end{aligned}$$

By making use of the resolvent equation we have

$$
\begin{aligned}
&\sigma(t)^{-1}\cdot\frac{d}{dt}\sigma(t)\\
&=\operatorname{tr}\left(\int_\Gamma \lambda\cdot\frac{d}{d\lambda}\left(\frac{\lambda}{\sinh\lambda}\right)\cdot\frac{\sinh\lambda}{\lambda}\cdot\left(\lambda-\sqrt{-1}t\Omega(\tau)\right)^{-1}d\lambda\right)\\
&=\operatorname{tr}\left(\int_\Gamma \lambda\cdot\frac{\sinh\lambda-\lambda\cosh\lambda}{\sinh^2\lambda}\cdot\frac{\sinh\lambda}{\lambda}\cdot\left(\lambda-\sqrt{-1}t\Omega(\tau)\right)^{-1}d\lambda\right),
\end{aligned}
$$

and so

$$
\begin{aligned}
&\sum_{k=1}^{d}\tau_k\frac{\partial W}{\partial\tau_k}(\tau)\\
&=\frac{1}{2}\operatorname{tr}\left(\int_\Gamma \lambda\cdot\frac{\sinh\lambda-\lambda\cosh\lambda}{\sinh^2\lambda}\cdot\frac{\sinh\lambda}{\lambda}\cdot\left(\lambda-\sqrt{-1}\Omega(\tau)\right)^{-1}d\lambda\right)\cdot W(\tau).
\end{aligned}
$$

Hence we have

$$
\begin{aligned}
&\sum\tau_k\frac{\partial W}{\partial\tau_k}(\tau)-\Delta_{sub}(f)\cdot W(\tau)\\
&=\frac{1}{2}\operatorname{tr}\left(\int_\Gamma \lambda\cdot\frac{\sinh\lambda-\lambda\cosh\lambda}{\sinh^2\lambda}\cdot\frac{\sinh\lambda}{\lambda}\cdot\left(\lambda-\sqrt{-1}\Omega(\tau)\right)^{-1}d\lambda\right)\cdot W(\tau),\\
&\qquad\qquad+\frac{1}{2}\operatorname{tr}\left(\int_\Gamma \lambda\cdot\frac{\cosh\lambda}{\sinh\lambda}\cdot\left(\lambda-\sqrt{-1}\Omega(\tau)\right)^{-1}d\lambda\right)\cdot W(\tau)\\
&=\frac{d}{2}\cdot W(\tau),
\end{aligned}
$$

which shows that $W(\tau)$ is a solution of the transport equation (15). □

Remark 3.1. The function $W(\tau)$ is similar to the van Vleck determinant (See van-Vleck [20], and a recent book by de Gosson [10]).

Hence the function $k_t(x,z)$ is given by the integral:

Theorem 3.1.

$$
k_t(x,z)=\frac{1}{t^{n/2+d}}\int_{\mathbb{R}^d}e^{-\frac{f(x,z,\tau)}{t}}\cdot\left(\det\frac{\sqrt{-1}\Omega(\tau)}{\sinh\sqrt{-1}\Omega(\tau)}\right)^{1/2}d\tau,
$$

By the arguments in Appendix B we know that

$$
\left(\Delta_{sub}+\frac{\partial}{\partial t}\right)k_t(x,z)=0.
$$

Also by the asymptotic behaviours of the integrands

- $W(\tau) = O(\|\tau\|^{-j})$, $j > 0$ *is arbitrary,*
- *the bilinear form* $<\left(\sqrt{-1}\Omega(\tau)\coth\left(\sqrt{-1}\Omega(t\tau)\right)\right)\cdot x,\, x>$

  is (strictly) *positive definite and*

  $<\left(\sqrt{-1}\Omega(\tau)\coth\left(\sqrt{-1}\Omega(t\tau)\right)\right)\cdot x,\, x>= O(\|\tau\|\|x\|^2)$

  (*since non-zero eigenvalues of* $\sqrt{-1}\Omega(\tau)$ *are proportional to* $\|\tau\|$),

the Fourier inversion formula implies that

$$\lim_{t\to 0}\frac{1}{t^{n/2+d}}\int_{\mathfrak{g}}\int_{\mathbb{R}^d} e^{-\frac{f(x,z,\tau)}{t}}\cdot\left(\det\frac{\sqrt{-1}\Omega(\tau)}{\sinh\sqrt{-1}\Omega(\tau)}\right)^{1/2} d\tau\varphi(x,z)dxdz$$
$$= \varphi(0,0)$$

for all $\varphi \in C_0^\infty(\mathfrak{g})$.

Remark 3.2. By continuing analytically the integrand of the heat kernel with respect to the variable τ to the whole complex space, we obtain more precise information of the asymptotic property (when $t \downarrow 0$) of the heat kernel (see [4] for the Heisenberg group case).

3.2. *Heat kernel for the Laplacian*

Let Δ *be a Laplacian on* G:

$$\Delta = \Delta_{sub} - 1/2\sum_{k=1}^{d}\widetilde{Z}_k^2,\ \widetilde{Z}_k = \frac{\partial}{\partial z_k},$$

that is we are assuming that G *is equipped with the left invariant metric such that* $\{X_i\}$ *and* $\{Z_k\}$ *are an orthonormal basis at the identity element, and then the operator above is the Laplacian with respect to this Riemannian metric.*

Since

$$[\Delta_{sub}, \widetilde{Z}_k] = 0,\ (Z_k \in \text{ center of } \mathfrak{g})$$

the heat kernel $K_\Delta(t;(x,z),(\widetilde{x},\widetilde{z}))$ for the Laplacian Δ is the kernel distribution ($\in C^\infty(\mathbb{R}_+\times G\times G)$) of the composed operator:

$$e^{-t\Delta_{sub}}\circ\left(\mathrm{Id}\otimes e^{-t/2\sum\widetilde{Z}_k^2}\right).$$

Here Id denotes the identity operator on the space $L_2(\mathfrak{g}/[\mathfrak{g},\mathfrak{g}])$, and we regard

$$L_2(G) \cong L_2(\mathfrak{g}/[\mathfrak{g},\mathfrak{g}])\widehat{\otimes}L_2([\mathfrak{g},\mathfrak{g}])$$

by identifying

$$[\{X_i\}] \cong \mathfrak{g}/[\mathfrak{g},\mathfrak{g}], \text{ where } [\{X_i\}] \text{ is the subspace spanned by } \{X_i\}.$$

Proposition 3.3.

$$\begin{aligned}
&K_\Delta(t;(x,z),(\widetilde{x},\widetilde{z}))\\
&= \int_{center} K(t;(x,z),(\widetilde{x},y))\cdot \frac{1}{(2\pi t)^{d/2}} e^{-\frac{\|y-\widetilde{z}\|^2}{2t}}dy\\
&= \frac{1}{t^{n/2+d}}\int_{\mathbb{R}^d} e^{-\frac{f^\Delta((\widetilde{x},\widetilde{z})^{-1}*(x,z),\tau)}{t}}W(\tau)d\tau\\
&= \frac{1}{t^{n/2+d}}\int_{\mathbb{R}^d} e^{-\frac{f^\Delta(x-\widetilde{x},z-\widetilde{z}-1/2[\widetilde{x},x],\tau)}{t}}W(\tau)d\tau,
\end{aligned}$$

where

$$\begin{aligned}
&f^\Delta(x,z,\tau)\\
&= \sqrt{-1}<\tau,z> +1/2 <\sqrt{-1}\Omega(\tau)\bigl(\coth\sqrt{-1}\Omega(\tau)\bigr)\cdot x,\ x> +1/2\|\tau\|^2.
\end{aligned}$$

Remark 3.3. We will show later that the function f^Δ is the complex action integral for the Laplacian (see (19)).

Proof. Since $K_\Delta(t;(x,z),(\widetilde{x},\widetilde{z}))$ is of a form $K_\Delta(t;(x,z),(\widetilde{x},\widetilde{z})) = k_t^\Delta((\widetilde{x},\widetilde{z})^{-1}*(x,z))$ with a function $k_t^\Delta(x,z)\in C^\infty(\mathbb{R}_+\times\mathfrak{g})$, it will be enough to express this function $k_t^\Delta(x,z)$:

$$\begin{aligned}
k_t^\Delta(x,z) &= \int_{\text{center}} K(t;(x,z),(0,y))\cdot\frac{1}{(2\pi t)^{d/2}}e^{-\frac{\|y\|^2}{2t}}dy\\
&= \frac{1}{t^{n/2+d}}\int_{\mathbb{R}^d}\int_{\text{center}} e^{-\frac{\sqrt{-1}<\tau,z-y>+1/2<\sqrt{-1}\Omega(\tau)\coth\sqrt{-1}\Omega(\tau)\cdot x,x>}{t}}W(\tau)\\
&\qquad\qquad\times\frac{1}{(2\pi t)^{d/2}}e^{-\frac{\|y\|^2}{2t}}d\tau dy\\
&= \frac{1}{t^{n/2+d}}\cdot\frac{1}{(2\pi t)^{d/2}}\int_{\mathbb{R}^d}\int_{\text{center}} e^{-\frac{\|y\|^2}{2t}}e^{\frac{\sqrt{-1}}{t}<\tau,y>}dy\\
&\qquad\times e^{-\frac{\sqrt{-1}<\tau,z>+1/2<\sqrt{-1}\Omega(\tau)\coth\sqrt{-1}\Omega(\tau)\cdot x,x>}{t}}W(\tau)d\tau\\
&= \frac{1}{t^{n/2+d}}\int_{\mathbb{R}^d} e^{-\frac{\|\tau\|^2}{2t}}\cdot e^{-\frac{\sqrt{-1}<\tau,z>+1/2<\sqrt{-1}\Omega(\tau)\coth\sqrt{-1}\Omega(\tau)\cdot x,x>}{t}}W(\tau)d\tau\\
&= \frac{1}{t^{n/2+d}}\int_{\mathbb{R}^d} e^{-\frac{f^\Delta(x,z,\tau)}{t}}W(\tau)d\tau.
\end{aligned}$$

□

Of course this formula coincides with the earlier particular case (§2 Theorem 2.4) and is the heat kernel we wanted to construct.

We can also construct the heat kernel for the Laplacian by the complex Hamilton-Jacobi method. So we describe the Hamiltonian system and relating quantities corresponding to the case of the sub-Laplacian:

(a) *Hamiltonian* H^Δ:

$$\begin{aligned}
&H^\Delta(x,z;\xi,\theta)\\
&= \frac{1}{2}\left(\sum_{j=1}^{n}(\xi_j + \sum_{i=1}^{n}\sum_{k=1}^{d} a_{ij}^k x_i\theta_k)^2 + \sum_{k=1}^{d}\theta_k^2\right)\\
&= \frac{1}{2}\left(\sum_{j}(\xi_j + \sum_{i}\Omega(\theta)_{ij}\cdot x_i)^2 + \sum_{k=1}^{d}\theta_k^2\right)\\
&= H(x,z;\xi,\theta) + \frac{1}{2}\sum_{k=1}^{d}\theta_k^2.
\end{aligned}$$

(b) *Hamiltonian system*:

$$\begin{cases} \dot{x}_i = H^\Delta_{\xi_i} = H_{\xi_i} = \xi - \Omega(\theta)x, \quad \dot{z}_k = H^\Delta_{\theta_k} = H_{\theta_k} + \theta_k\,, \\ \dot{\xi}_j = -H^\Delta_{x_j} = -H_{x_j}, \quad \dot{\theta}_k = -H^\Delta_z \equiv 0. \end{cases} \tag{17}$$

(c) *Initial-boundary conditions*:

$$\begin{cases} x(0) = 0 \qquad x(s) = x = (x_1,\cdots,x_n) \in \mathbb{R}^n, \\ \qquad\qquad\quad z(s) = z = (z_1,\cdots,z_d) \in \mathbb{R}^d, \\ \qquad\qquad\quad \theta(0) = \sqrt{-1}\tau, \qquad \tau = (\tau_1,\cdots,\tau_d) \in \mathbb{R}^d, \end{cases} \tag{18}$$

where $s \in \mathbb{R}$, x and z are arbitrarily given.

In the above Hamiltonian system (17), all equations other than the second coincide with corresponding equations in (10). So we have the same solutions $x^\Delta(t) = x(t)$ and $\xi^\Delta(t) = \xi(t)$:

$$\begin{aligned}
x^\Delta(t) &= x^\Delta(t;s,x,\tau) = x(t;s,x,\tau)\\
&= e^{(s-t)\sqrt{-1}\Omega(\tau)}\frac{\sinh t\sqrt{-1}\Omega(\tau)}{\sinh s\sqrt{-1}\Omega(\tau)}\cdot x
\end{aligned}$$

and

$$\begin{aligned}
\xi^{\Delta}(t) &= \xi(t) = \xi(t; s, x, \tau) \\
&= \frac{\sqrt{-1}\Omega(\tau)}{\sinh s\sqrt{-1}\Omega(\tau)} \cdot e^{s\sqrt{-1}\Omega(\tau)}\left(Id - e^{-t\sqrt{-1}\Omega(\tau)} \sinh t\sqrt{-1}\Omega(\tau)\right) \cdot x \\
&= \left(e^{-t\sqrt{-1}\Omega(\tau)} \cosh t\sqrt{-1}\Omega(\tau)\right) \cdot \left(e^{s\sqrt{-1}\Omega(\tau)} \frac{\sqrt{-1}\Omega(\tau)}{\sinh s\sqrt{-1}\Omega(\tau)}\right) x \\
&= \left(e^{-t\sqrt{-1}\Omega(\tau)} \cosh t\sqrt{-1}\Omega(\tau)\right) \xi(0)
\end{aligned}$$

as the system (13) under the initial conditions (14). Moreover, we have the solution $z^{\Delta}(t) = z(t) + \sqrt{-1}(t - s)\tau_k$.

Now the complex action integral

$$g^{\Delta} \in C^{\infty}(\mathbb{R} \times \mathbb{R}^n \times \mathbb{R}^d \times \mathbb{R}^d)$$

is given by

$$\begin{aligned}
&g^{\Delta}(s; x, z, \tau) \\
&= \sqrt{-1}\sum_{i=1}^{d} \tau_i \cdot z_i^{\Delta}(0; s, x, z, \tau) \\
&\qquad + \int_0^s < \xi^{\Delta}(t), \dot{x}^{\Delta}(t) > + < \theta(t), \dot{z}^{\Delta}(t) > \\
&\qquad\qquad - H^{\Delta}(x^{\Delta}(t), z^{\Delta}(t); \xi^{\Delta}(t), \theta(t))dt \\
&= \sqrt{-1}\sum_{i=1}^{d} \tau_i \cdot z_i \\
&\qquad + \int_0^s < \xi(t), \dot{x}(t) > - \left(H(x(t), z(t); \xi(t), \theta(t)) + 1/2 \sum_{k=1}^{d} \theta_k(t)^2\right) dt \\
&= g(s; x, z, \tau) + \frac{s}{2}\sum_{k=1}^{d} \tau_k^2. \qquad (19)
\end{aligned}$$

Then g^{Δ} satisfies the usual Hamilton-Jacobi equation:

Proposition 3.4.

(a)

$$\frac{\partial g^{\Delta}}{\partial s} + H^{\Delta}(x, z; \nabla g^{\Delta}) = 0.$$

(b) The function g^Δ satisfies the relation

$$g^\Delta(s; x, z, \ell \cdot \tau) = \frac{1}{\ell} \cdot g^\Delta(1; x, z, \tau).$$

Hence by the same reason as in the case of the sub-Laplacian the function $f^\Delta(x, z, \tau) = g^\Delta(1; x, z, \tau)$ satisfies the generalized Hamilton-Jacobi equation or we can prove directly:

$$\begin{aligned} &H^\Delta(x, z; \nabla f^\Delta) + \sum_{i=1}^{d} \tau_i \frac{\partial f^\Delta}{\partial \tau_i} \\ &= H(x, z; \nabla f) - 1/2 \sum_{k=1}^{d} \tau_k^2 + \sum_{i=1}^{d} \tau_i \frac{\partial f}{\partial \tau_i} + \sum_{k=1}^{2} \tau_k^2 \\ &= f(x, z; \tau) + 1/2 \sum \tau_k^2 = f^\Delta(x, z; \tau). \end{aligned}$$

The Laplacian $\Delta(f^\Delta)$ is

$$\Delta(f^\Delta) = \Delta(f + 1/2 \sum \tau_k^2) = \Delta_{sub}(f).$$

From that we know that the same volume element $W(\tau)$ is the solution of the transport equation:

$$\begin{aligned} &\sum \tau_i \frac{\partial W}{\partial \tau_i} + \sum_j \widetilde{X}_j(f^\Delta) \widetilde{X}_j(W) \\ &\qquad + \sum_k \widetilde{Z}_k(f^\Delta) \widetilde{Z}_k(W) - \left(\Delta(f^\Delta) + \frac{d}{2} \right) \cdot W \\ &= \sum \tau_i \frac{\partial W}{\partial \tau_i} - \left(\Delta_{sub}(f) + \frac{d}{2} \right) \cdot W = 0 \qquad (20) \end{aligned}$$

$$(21)$$

Hence the heat kernel K^Δ for the Laplacian is given

Proposition 3.5.

$$K^\Delta(t; (\widetilde{x}, \widetilde{z}), (x, z)) = \frac{1}{t^{n/2+d}} \int_{\mathbb{R}^d} e^{-f^\Delta(x - \widetilde{x}, z - \widetilde{z} - 1/2[\widetilde{x}, x])} W(\tau) d\tau. \qquad (22)$$

Remark 3.4. This form coincides with the one we obtained earlier (Proposition 3.3). Both in the expressions in Theorem 3.1 (sub-Laplacian case) and Proposition 3.5 (Laplacian case) the integrands of $k_t(g)$ and $k_t^\Delta(g)$ are defined on $G \times \mathbb{R}^d$. We may identify them with the characteristic variety of the sub-Laplacian $Cha = \{H = 0\} = \{(x, z; \xi, \theta) \mid H(x, z; \xi, \theta) = 0\}$ through the map

$$G \times \mathbb{R}^d \ni (x, z, \tau) \mapsto (x, z; \Omega(\tau)x, \tau) \in T^*G.$$

The characteristic variety is a subbundle in T^*G and the integral formula of the heat kernel can be seen as the fiber integration of the d-form

$$\frac{1}{t^{n/2+d}} e^{-\frac{f(x,z,\tau)}{t}} W(\tau) d\tau \quad \text{and} \quad \frac{1}{t^{n/2+d}} e^{-\frac{f^{\Delta}(x,z,\tau)}{t}} W(\tau) d\tau$$

on the characteristic variety Cha.

On the other hand, for our special 2-step cases in §2 (Theorem 2.3) the domain of the integration for expressing the heat kernel was the dual of the center $= [\mathfrak{g}, \mathfrak{g}]$. It parametrizes the irreducible representations of G which appear in the description of the unitary transformation $U : L_2(G) \xrightarrow{\sim} L_2(\mathbf{X})$.

4. A 3-step nilpotent Lie group

So far we have looked at the general 2-step cases for constructing heat kernels of sub-Laplacians and Laplacians. It seems that there are no explicit expressions of the heat kernel for any 3-step cases until now. So we discuss a possibility of the complex Hamilton-Jacobi theory to construct the heat kernel for the lowest 3-step nilpotent Lie group G_4 (called Engel group):

$$G_4 = \left\{ \begin{pmatrix} 1 & x & x^2/2 & z \\ 0 & 1 & x & w \\ 0 & 0 & 1 & y \\ 0 & 0 & 0 & 1 \end{pmatrix} \;\middle|\; x, y, w, z \in \mathbb{R} \right\}$$

and its Lie algebra

$$\mathfrak{g}_4 = \left\{ \begin{pmatrix} 0 & x & 0 & z \\ 0 & 0 & x & w \\ 0 & 0 & 0 & y \\ 0 & 0 & 0 & 0 \end{pmatrix} \;\middle|\; x, y, w, z \in \mathbb{R} \right\}.$$

Let $\{X, Y, W, Z\}$ be a basis of $\mathfrak{g}_4$:

$$X = \begin{pmatrix} 0\,1\,0\,0 \\ 0\,0\,1\,0 \\ 0\,0\,0\,0 \\ 0\,0\,0\,0 \end{pmatrix}, \; Y = \begin{pmatrix} 0\,0\,0\,0 \\ 0\,0\,0\,0 \\ 0\,0\,0\,1 \\ 0\,0\,0\,0 \end{pmatrix},$$

$$W = \begin{pmatrix} 0\,0\,0\,0 \\ 0\,0\,0\,1 \\ 0\,0\,0\,0 \\ 0\,0\,0\,0 \end{pmatrix}, \; Z = \begin{pmatrix} 0\,0\,0\,1 \\ 0\,0\,0\,0 \\ 0\,0\,0\,0 \\ 0\,0\,0\,0 \end{pmatrix},$$

$[X, Y] = W$, $[X, W] = Z$ and all other brackets are zero.

The left invariant vector fields corresponding to each X, Y, W, Z are given by

$$\widetilde{X} = \frac{\partial}{\partial x}, \; \widetilde{Y} = \frac{\partial}{\partial y} + x\frac{\partial}{\partial w} + \frac{x^2}{2}\frac{\partial}{\partial z}$$
$$\widetilde{W} = \frac{\partial}{\partial w}, \; \widetilde{Z} = \frac{\partial}{\partial z},$$

and so the left-invariant sub-Laplacians and the Laplacian are

$$-\Delta_2 = -\Delta_{sub} = \frac{1}{2}\left(\widetilde{X}^2 + \widetilde{Y}^2\right)$$
$$-\Delta_1 = \frac{1}{2}\left(\widetilde{X}^2 + \widetilde{Y}^2 + \widetilde{W}^2\right)$$
$$\Delta_0 = Laplacian = -\frac{1}{2}\left(\widetilde{X}^2 + \widetilde{Y}^2 + \widetilde{W}^2 + \widetilde{Z}^2\right).$$

Now the Hamiltonian and the Hamiltonian system for Δ_{sub} are written as:

$$(x, y, w, z; \xi, \eta, \tau, \theta): \text{ coordinate on } T^*G_4 \cong \mathbb{R}^4 \times \mathbb{R}^4$$

$$H(x, y, w, z; \xi, \eta, \tau, \theta) = 1/2\left(\xi^2 + (\eta + \tau \cdot x + 1/2 \cdot \theta \cdot x^2)^2\right)$$

$$\begin{cases} \dot{x} = \xi, & \dot{y} = \eta + \tau x + 1/2 \cdot \theta x^2 \\ \dot{w} = \dot{y}x, & \dot{z} = 1/2 \cdot \dot{y}x^2 \\ \xi = -\dot{y}(\theta x + \tau), & \dot{\eta} = 0 \\ \dot{\tau} = 0, & \dot{\theta} = 0. \end{cases}$$

We shall solve this system under the initial-boundary conditions:

$$\begin{cases} x(0) = 0, \; x(s) = x, \; y(0) = 0, \; y(s) = y, \\ w(s) = w, \; z(s) = z, \\ \tau(0) = \tau_0(\equiv \tau(t)), \; \theta(0) = \theta_0(\equiv \theta(t)), \end{cases}$$

where τ_0 and θ_0 take purely imaginary values, while $s,\ x,\ y \in \mathbb{R}$ are arbitrarily given.

Since we have solutions to the above system under usual initial conditions, if we temporarily put $\xi(0)=\xi_0$ and $\eta(0)=\eta_0 \equiv \eta(t)$, we have

$$\theta_0 x(t) + \tau_0 = 2\sqrt{\alpha}\cdot \mathbf{sn}\left(t\sqrt{\beta}+C_0, \sqrt{-1}\sqrt{\alpha/\beta}\right),$$

where $\mathbf{sn}(u, \sqrt{-1}\sqrt{\alpha/\beta})$ is the Jacobi's $\mathbf{sn}$ function with the modulus $k = \sqrt{-1}\sqrt{\alpha/\beta}$:

$$u = \int_0^{\mathbf{sn}(u,\ k)} \frac{1}{\sqrt{(1-t^2)(1-k^2t^2)}}dt,$$

$$C_0 = \int_0^{\frac{\tau_0}{2\sqrt{\alpha}}} \frac{1}{\sqrt{(1-t^2)\left(1+\frac{\alpha}{\beta}t^2\right)}}dt,$$

i.e.,

$$\mathbf{sn}(C_0, \sqrt{-1}\sqrt{\alpha/\beta}) = \frac{\tau_0}{2\sqrt{\alpha}},\quad \alpha = \frac{\tau_0^2}{4} + \frac{\theta_0}{2}\left(\sqrt{\xi_0^2+\eta_0^2} - \eta_0\right),$$

$$\beta = -\frac{\tau_0^2}{4} + \frac{\theta_0}{2}\left(\eta_0 + \sqrt{\xi_0^2+\eta_0^2}\right),$$

and

$$y(t) = \frac{2\alpha}{\theta_0}\int_0^t \mathbf{sn}^2\left(\ell\sqrt{\beta}+C_0, \sqrt{-1}\sqrt{\alpha/\beta}\right)d\ell + \frac{t(2\theta_0\eta_0 - \tau_0^2)}{2\theta_0}.$$

Solution of the generalized Hamilton-Jacobi equation:

$$\begin{aligned}
g^E(s;x,y,w,z;\tau_0,\theta_0) &= \tau_0 w(0) + \theta_0 z(0)\\
&\quad + \int_0^s \xi(t)\dot{x}(t) + \eta(t)\dot{y}(t) + \tau(t)\dot{w}(t) + \theta(t)\dot{z}(t)\\
&\quad - H(x(t),y(t),w(t),z(t);\xi(t),\eta(t),\tau(t),\theta(t))dt\\
&= (\tau_0 w + \theta_0 z) + \int_0^s \frac{1}{2}\left(\dot{x}(t)^2 + \dot{y}(t)^2\right) + \eta_0\dot{y}(t) - \dot{y}(t)^2 dt\\
&= (\tau_0 w + \theta_0 z) + \frac{s}{2}\left(\xi_0^2+\eta_0^2\right) + \eta_0 y - \int_0^s \dot{y}(t)^2 dt\\
&= (\tau_0 w + \theta_0 z) + \frac{s}{2}\left(\xi_0^2+\eta_0^2\right) + \eta_0 y\\
&\quad - \frac{1}{4\theta_0^2}\int_0^s (\theta_0 x + \tau_0)^4 + 4\theta_0(2\theta_0\eta_0 - \tau_0^2)\dot{y}(t) - (2\theta_0\eta_0 - \tau_0^2)^2 dt
\end{aligned}$$

$$= (\tau_0 w + \theta_0 z) + \frac{s}{2}\Big(\xi_0^2 + \eta_0^2\Big) + \frac{(\tau_0^2 - \theta_0\eta_0)}{\theta_0} y + \frac{(2\theta_0\eta_0 - \tau_0^2)^2}{4\theta_0^2} s$$

$$- \frac{1}{4\theta_0^2}\int_0^s \Big(2\sqrt{\alpha}\,\mathbf{sn}(t\sqrt{\beta} + C_0)\Big)^4 dt.$$

For a further explicit expression of the complex action integral g^E here we recall an integral formula for the **sn** function:

Put

$$E(u) = u - \int_0^u \mathbf{sn}^2(v)dv,$$

then we have (cf. Lawden [16])

$$\int \mathbf{sn}^4(u,k)du = \frac{1}{3k^4}\Big[(2+k^2)u - 2(1+k^2)E(u) + k^2\mathbf{sn}(u)\frac{d}{du}\mathbf{sn}(u)\Big].$$

By making use of this formula, the last integral in the above expression of the function g^E is evaluated as

$$- \frac{1}{4\theta_0^2}\int_0^s \Big(2\sqrt{\alpha}\,\mathbf{sn}(t\sqrt{\beta} + C_0)\Big)^4 dt$$

$$= -\frac{4\alpha^2}{\theta_0^2\sqrt{\beta}} \cdot \frac{1}{3k^4}\Big[(2+k^2)s\sqrt{\beta} - 2(1+k^2)\Big(E(s\sqrt{\beta}+C_0) - E(C_0)\Big)$$

$$+ k^2 \cdot \mathbf{sn}(s\sqrt{\beta}+C_0)\Big(\frac{d\,\mathbf{sn}}{du}\Big)(s\sqrt{\beta}+C_0) - k^2 \cdot \mathbf{sn}(C_0)\Big(\frac{d\,\mathbf{sn}}{du}\Big)(C_0)\Big]$$

$$= -\frac{4\alpha^2}{\theta_0^2\sqrt{\beta}} \cdot \frac{1}{3k^4}\Big[(2+k^2)s\sqrt{\beta} - 2(1+k^2)$$

$$\Big\{s\sqrt{\beta} - \frac{\sqrt{\beta}}{4\alpha}(2\theta_0 y - (2\theta_0\eta_0 - \tau_0^2)s)\Big\} + k^2 \cdot \frac{\theta_0 x + \tau_0}{2\sqrt{\alpha}}$$

$$\cdot \sqrt{\Big(1 - \frac{(\theta_0 x + \tau_0)^2}{4\alpha}\Big)\Big(1 + \frac{(\theta_0 x + \tau_0)^2}{4\beta}\Big)} - k^2 \cdot \frac{\tau_0}{2\sqrt{\alpha}}\frac{\theta_0\xi_0}{2}\Big]$$

$$= \frac{4\alpha^2}{3\theta_0^2 k^4}\Big(k^2 s + \frac{(2\theta_0\eta_0 - \tau_0^2)^2}{4\alpha\beta}s\Big) - \frac{4\alpha}{3\theta_0 k^4}(1+k^2)y$$

$$- \frac{2}{3\theta_0} \cdot \Big\{(\theta_0 x + \tau_0) \cdot \sqrt{\Big(\alpha - \frac{(\theta_0 x + \tau_0)^2}{4}\Big)\Big(\beta + \frac{(\theta_0 x + \tau_0)^2}{4}\Big)} - \tau_0 \cdot \frac{\theta_0\xi_0}{2}\Big\}$$

$$= \Big(\frac{1}{3\theta_0^2} \cdot \frac{\beta}{\alpha} \cdot (2\eta_0\theta_0 - \tau_0^2)^2 - \frac{4}{3\theta_0^2} \cdot (\alpha\beta)\Big)s + \frac{4}{3\theta_0} \cdot \frac{\beta}{\alpha}(\alpha - \beta)y$$

$$+ \frac{2}{3\theta_0^2} \cdot \Big\{(\theta_0 x + \tau_0) \cdot \sqrt{\Big(\alpha - \frac{(\theta_0 x + \tau_0)^2}{4}\Big)\Big(\beta + \frac{(\theta_0 x + \tau_0)^2}{4}\Big)} - \tau_0 \cdot \frac{\theta_0\xi_0}{2}\Big\}.$$

Now we have

$$g^E(s;x,y,w,z;\tau_0,\theta_0) = (\tau_0 w + \theta_0 z) + \frac{s}{2}(\xi_0^2+\eta_0^2)$$
$$+\left\{\frac{1}{3\theta_0^2}\cdot\frac{\beta}{\alpha}\cdot(2\eta_0\theta_0-\tau_0^2)^2 - \frac{4}{3\theta_0^2}\cdot(\alpha\beta)+\frac{(2\theta_0\eta_0-\tau_0^2)^2}{4\theta_0^2}\right\}s$$
$$+\left\{\frac{4}{3\theta_0}\cdot\frac{\beta}{\alpha}(\alpha-\beta)+\frac{(\tau_0^2-\eta_0\theta_0)}{\theta_0}\right\}y$$
$$+\frac{2}{3\theta_0^2}\cdot\left\{(\theta_0 x+\tau_0)\cdot\sqrt{\left(\alpha-\frac{(\theta_0 x+\tau_0)^2}{4}\right)\left(\beta+\frac{(\theta_0 x+\tau_0)^2}{4}\right)}-\tau_0\cdot\frac{\theta_0\xi_0}{2}\right\}$$
$$=(\tau_0 w+\theta_0 z)+\frac{1}{3}\left(\frac{\tau_0^2-2\theta_0\eta_0}{\theta_0}\right)^2\cdot\left(1-\frac{1}{k^2}\right)\cdot s+\frac{1}{6}\left(\xi_0^2+\eta_0^2\right)\cdot s$$
$$+\left\{\left(1-\frac{2}{3k^2}\right)\left(\frac{\tau_0^2-\eta_0\theta_0}{\theta_0}\right)+\frac{2}{3k^2}\eta_0\right\}\cdot y$$
$$+\frac{2}{3\theta_0^2}\cdot\left\{(\theta_0 x+\tau_0)\cdot\sqrt{\frac{\theta_0^2}{4}(\xi_0^2+\eta_0^2)-\left(\frac{(\theta_0 x+\tau_0)^2+2\theta_0\eta_0-\tau_0^2}{2\theta_0}\right)^2}\right\}.$$

Although this is the final form of the complex action integral g^E in terms of the initial data, this is not the final form for the initial-boundary data. For the moment the correspondence $(x,y)\mapsto(\xi(0),\eta(0))$ is not explicit so that we can not erase the terms including ξ_0 and η_0. Moreover in this case we must solve the *generalized transport equation* of the following form (see (B.5)):

$$-\,dg^E\wedge\Delta_{sub}(\mathbf{V})=\widetilde{X}(f)\cdot\widetilde{X}(d\mathbf{V})+\widetilde{Y}(f)\cdot\widetilde{Y}(d\mathbf{V})$$
$$+\,\tau\frac{\partial}{\partial\tau}(d\mathbf{V})+\theta\frac{\partial}{\partial\theta}(d\mathbf{V})-\left(\Delta_{sub}(g^E)+1\right)d\mathbf{V}, \tag{23}$$

where dg^E and $d\mathbf{V}$ are the exterior derivatives (with respect to the variables (τ_0,θ_0)) of the solution of the generalized Hamilton-Jacobi equation and a one-form $\mathbf{V}$ on

$$Cha=\{H=0\}=\{(x,y,w,z;\xi,\eta,\sqrt{-1}\tau,\sqrt{-1}\theta)\in G_4\times\mathbb{C}^4$$
$$\mid\ \xi=0,\ \eta=-\sqrt{-1}(\tau\cdot x+1/2\cdot\theta x^2),\ \tau,\ \theta\in\mathbb{R}\}$$
$$=\coprod_{(x,y,w,z)}\{(\xi,\eta,\tau,\theta)\mid\xi=0,\ \eta=-\sqrt{-1}(\tau\cdot x+1/2\cdot\theta x^2),\tau,\ \theta\in\mathbb{R}\}$$
$$=\coprod Cha_{(x,y,w,z)}$$

with the group variables (= space variables) being treated as parameters (see Appendix B).

Now we expect that the solution of this equation (23) can be found, if we could describe the correspondence

$$(x, y) \mapsto (\xi(0), \eta(0))$$

explicitly, and then the heat kernel $K(t; \widetilde{x}, \widetilde{y}, \widetilde{w}, \widetilde{z}, x, y, w, z)$ *will be given by a function* k_t*:*

$$k_t(x, y, w, z) = \frac{1}{t^3} \int_{Cha_{(x,y,w,z)}} e^{-\frac{g^E(1;x,y,w,z;\tau,\theta)}{t}} d\mathbf{V},$$

$$K(t; \widetilde{x}, \widetilde{y}, \widetilde{w}, \widetilde{z}, x, y, w, z)$$
$$= k_t\Big(x - \widetilde{x}, y - \widetilde{y}, w - \widetilde{w} + \widetilde{x}(\widetilde{y} - y), z - \widetilde{z} + \widetilde{x}(\widetilde{w} - w) + \frac{\widetilde{x}^2}{2}(y - \widetilde{y})\Big).$$

Here we fix the order of $\frac{1}{t}$ *as*

$$\frac{1}{2} \dim \mathfrak{g}/[\mathfrak{g}, \mathfrak{g}] + \dim[\mathfrak{g}, \mathfrak{g}] = 3,$$

and the integration is taken with respect to the variables $(\tau, \theta) \in \mathbb{R}^2$.

Remark 4.1. We note an aspect of our problem from the representation theoretic point, that is we explain what kind of equations we must solve to obtain informations on the heat kernel or rather spectral data for the compact nilmanifolds.

According to the Kirillov theory we know what kinds of unitary representations appear in the (right) regular representation $\mathcal{R}$ of the nilpotent Lie group G to $L_2(\Gamma\backslash G)$ where Γ is a uniform discrete subgroup of G (see Corwin and Greenleaf [6]):

$$\mathcal{R} = \sum m_\rho[\rho], \quad \rho: \text{ irreducible representation}, \quad m_\rho: \text{ multiplicity of } \rho$$

and the problem reduces to solving the second order equation

$$\sum \rho_*(X_i)^2 + \sum \rho_*(Y_j)^2 + \sum \rho_*(Z_k)^2 = 0. \tag{24}$$

We note here a difficulty to determine the spectrum in explicit forms for the above case G_4 with, say for a uniform discrete subgroup Γ given by

$$\Gamma = \left\{ \begin{pmatrix} 1 & k & k^2/2 & n \\ 0 & 1 & k & m \\ 0 & 0 & 1 & \ell \\ 0 & 0 & 0 & 1 \end{pmatrix} \;\middle|\; k, \ell, m, n \in \mathbb{Z} \right\}.$$

To determine the spectrum of the Laplacian (and of the sub-Laplacian) on $\Gamma\backslash G_4$ explicitly, it is not enough to decompose the space $L_2(\Gamma\backslash G_4)$

into irreducible subspaces. We must solve the eigenvalue problem for the equation (24) in $L_2(\mathbb{R})$, and a typical one of them is the equation with a quartic potential

$$-\frac{d^2}{dx^2} + x^4.$$

This equation was studied by Voros [21] (also see Voros [22]) precisely, however we have no explicit expressions of eigenvalues and eigenfunctions of this equation. For 2-step cases we have the well known and completely solvable equation (harmonic oscillator) as a corresponding equation to this equation:

$$-\frac{d^2}{dx^2} + x^2.$$

5. Hierarchy of heat kernels

Finally we discuss a relation between heat kernels on the group and their homogeneous spaces through a most simple example, that is, the case of the 3-dimensional Heisenberg group H_3:

Let H_3 be the 3-dimensional Heisenberg group realized in the space of real 3×3 matrices:

$$H_3 = \left\{ \begin{pmatrix} 1 & x & z \\ 0 & 1 & y \\ 0 & 0 & 1 \end{pmatrix} \right\}.$$

By identifying H_3 with the Lie algebra $\mathfrak{h}_3$ through the exponential map

$$\exp : \mathfrak{h}_3 \to H_3, \; (x, y, z) \mapsto \begin{pmatrix} 1 & x & z + 1/2xy \\ 0 & 1 & y \\ 0 & 0 & 1 \end{pmatrix},$$

we get the corresponding left invariant vector field to each element X and Y in the Lie algebra, where

$$X = \begin{pmatrix} 0 & 1 & 0 \\ 0 & 0 & 0 \\ 0 & 0 & 0 \end{pmatrix} \text{ and } Y = \begin{pmatrix} 0 & 0 & 0 \\ 0 & 0 & 1 \\ 0 & 0 & 0 \end{pmatrix}.$$

$\widetilde{X}$ is expressed as $\dfrac{\partial}{\partial x} - y\dfrac{\partial}{\partial z}$ and $\widetilde{Y}$ is expressed as $\dfrac{\partial}{\partial y} + x\dfrac{\partial}{\partial z}$.

Let $k_t(g) = k_t(x, y, z)$ be

$$\frac{1}{(2\pi t)^2} \int_{\mathbb{R}} e^{-\frac{1}{t}\left\{\sqrt{-1}\tau z + 1/2\left(\tau \coth \tau \cdot (x^2+y^2)\right)\right\}} \frac{\tau}{\sinh \tau} d\tau,$$

then

Theorem 5.1. *The heat kernel* $K(t;\widetilde{g},g) = K(t;\widetilde{x},\widetilde{y},\widetilde{z};x,y,z)$ *of the sub-Laplacian*

$$-\Delta_{sub} = \frac{1}{2}\left(\widetilde{X}^2 + \widetilde{Y}^2\right)$$

is

$$K(t;\widetilde{g},g) = k_t(\widetilde{g}^{-1}\cdot g) = \frac{1}{(2\pi t)^2}\times$$
$$\times \int_{\mathbb{R}} e^{-\frac{1}{t}\left(\sqrt{-1}\tau(z-\widetilde{z}+\widetilde{y}x-\widetilde{x}y)+1/2\left(\tau\coth\tau\cdot\left((x-\widetilde{x})^2+(y-\widetilde{y})^2\right)\right)\right)}\frac{\tau}{\sinh\tau}d\tau.$$

Now let $N_Z = \{tZ\}_{\mathbb{R}}$ be the center of the group H_3, then by the projection map

$$\rho : H_3 \to N_Z\backslash H_3 \cong \mathbb{R}^2,$$
$$(x,y,z) \mapsto (x,y)$$

we have $\rho_*(\widetilde{X}) = \frac{\partial}{\partial x}$ and $\rho_*(\widetilde{Y}) = \frac{\partial}{\partial y}$. So the sub-Laplacian on H_3 descends to the usual Laplacian on $\mathbb{R}^2$. The fiber integral

$$\rho_*\Big(K(t;\widetilde{g},\bullet)dx\wedge dy\wedge dz\Big)(x,y)$$

is the Euclidean heat kernel:

$$\rho_*\Big(K(t;\widetilde{g},\bullet)dx\wedge dy\wedge dz\Big)(x,y)$$
$$= \frac{1}{(2\pi t)^2}\left(\int_{\mathbb{R}}\int_{\mathbb{R}} e^{-\frac{1}{t}\left\{\sqrt{-1}\tau(s+\widetilde{y}x-\widetilde{x}y)+1/2\left(\tau\coth\tau\cdot\left((x-\widetilde{x})^2+(y-\widetilde{y})^2\right)\right)\right\}}\right.$$
$$\left.\frac{\tau}{\sinh\tau}d\tau ds\right)dx\wedge dy$$
$$= \frac{1}{2\pi t}e^{-\frac{1}{t}\left\{(x-\widetilde{x})^2+(y-\widetilde{y})^2\right\}}dx\wedge dy.$$

Here $\rho_*\Big(K(t;h\cdot\widetilde{g},\bullet)dx\wedge dy\wedge dz\Big)(x,y) = \rho_*\Big(K(t;\widetilde{g},\bullet)dx\wedge dy\wedge dz\Big)(x,y)$ for any $h = tZ$ because of the invariance $K(t;h\cdot\widetilde{g},h\cdot g) = K(t;\widetilde{g},g)$, $h\in H_3$:

Next, let N_X be a subgroup $N_X = \{tX\}_{\mathbb{R}}$, then by the projection map $\mathfrak{q} : H_3 \to N_X\backslash H_3 = \mathbb{R}^2$, $(x,y,z)\mapsto(u,v) = (y, z-xy)$ the vector fields $\widetilde{X}$ and $\widetilde{Y}$ descend to $-2u\dfrac{\partial}{\partial v}$ and $\dfrac{\partial}{\partial u}$ respectively, The resulting sub-elliptic operator

$$-4u^2\frac{\partial^2}{\partial v^2} - \frac{\partial^2}{\partial u^2}$$

is the Grusin operator.

In this case we regard

$$N_X \times \mathbb{R} \cong H_3$$

through the map

$$(s, u, v) \mapsto (x, y, z), \ x = s, \ y = u, \ z = u + sv.$$

Then

Theorem 5.2. *The fiber integration* $\mathfrak{q}_*\Big(K(t;\widetilde{g},\bullet)dx\wedge dy\wedge dz\Big)(u,v)$ *is given by*

$$\begin{aligned}
&\mathfrak{q}_*\Big(K(t;0,\widetilde{u},\widetilde{v};\bullet)dx\wedge dy\wedge dz\Big)(u,v)\\
&=\left(\int_{\mathbb{R}} K(t;0,\widetilde{u},\widetilde{v};s,u,v+su)ds\right)du\wedge dv\\
&=\frac{1}{(2\pi t)^{(3/2)}}\left(\int_{\mathbb{R}} e^{-\frac{\sqrt{-1}\tau(v-\widetilde{v})}{t}}e^{-\frac{\tau}{2t}\left\{(u+\widetilde{u})^2\cdot\tanh\tau+(u-\widetilde{u})^2\cdot\coth\tau\right\}}\right.\\
&\qquad\left.\sqrt{\frac{\tau}{\cosh\tau\cdot\sinh\tau}}d\tau\right)du\wedge dv
\end{aligned}$$

and gives the heat kernel of the Grusin operator.

Acknowledgement. The author would like to express his hearty thanks to professor Peter Greiner (University of Toronto) for his encouragement and various suggestions on the role of the complex Hamilton-Jacobi theory for the construction of the heat kernel during his stay at Toronto in the fall of 2004.

Appendix A. A solution of the Hamilton-Jacobi equation

Let $H(x,y;\xi,\eta)$ be a polynomial of the variables $(x,y;\xi,\eta)\in\mathbb{R}^m\times\mathbb{R}^d\times\mathbb{R}^m\times\mathbb{R}^d$, degree two with respect to the variables ξ and η. We regard the variables ξ and η as the dual variables of x and y and we allow that the variables take complex values. Especially we will assume that the polynomial H is a principal symbol of an invariant (sub-)Laplacian on nilpotent Lie groups.

We consider the Hamiltonian system:

$$\begin{cases} \dot{x} = H_\xi, \ \ \dot{y} = H_\eta \\ \dot{\xi} = -H_x, \ \dot{\eta} = -H_y \end{cases}$$

with the initial-boundary conditions

$$\begin{cases} x(0) = 0, \quad x(s) = x, \quad y(s) = y \\ \eta(0) = \tau \,. \end{cases}$$

We assume that there exists an open domain $\mathcal{B}$ in $\mathbb{C}^d$ such that for any $s \in \mathbb{R}$, $(x, y) \in \mathbb{R}^m \times \mathbb{R}^d$ and $\tau \in \mathcal{B}$ there exists a unique global solutions of the above system

$$X(t) = X(t; s, x, y, \tau), \ Y(t) = Y(t; s, x, y, \tau)$$
$$\xi(t) = \xi(t; s, x, y, \tau), \ \eta(t) = \eta(t; s, x, y, \tau),$$

all of which are smooth with respect to the parameters (s, x, y, τ). Note that the solution curve $(X(t), Y(t), \xi(t), \eta(t))$ will not be included in the real space anymore.

Let g be a function defined by the integral:

$$g(x, y; s, \tau) = \sum_{j=1}^{d} \tau_j Y_j(0; s, x, y, \tau)$$
$$+ \int_0^s \sum \xi_i(t)\dot{X}_i(t) + \sum \eta_j(t)\dot{Y}_j(t) - H(X(t), Y(t); \xi(t), \eta(t))dt.$$

Then g satisfies the Hamilton-Jacobi equation:

Proposition A.1.

$$\frac{\partial g}{\partial s} + H(x, y; \nabla g) = 0 \tag{A.1}$$

Proof. This is proved by explicitly calculating the derivatives:

$$\frac{\partial g}{\partial s}(x, y; s, \tau), \qquad \frac{\partial g}{\partial x}(x, y; s, \tau), \qquad \text{and} \qquad \frac{\partial g}{\partial y}(x, y; s, \tau).$$

First we show

$$\frac{\partial g}{\partial s}(x, y; s, \tau) + H(x, y; \xi(s; s, x, y, \tau), \eta(s; s, x, y, \tau)) = 0. \tag{A.2}$$

$$\frac{\partial g}{\partial s}(x, y; s, \tau) = \sum_{j=1}^{d} \tau_j \frac{\partial Y_j}{\partial s}(0; s, x, y, \tau)$$
$$+ \sum \xi_i(s)\dot{X}_i(s) + \sum \eta(s)_j \dot{Y}_j(s) - H(x, y; \xi(s; s, x, y, \tau), \eta(s; s, x, y, \tau))$$
$$+ \int_0^s \Big(\sum \frac{\partial \xi_i}{\partial s}(t; s, x, y, \tau)\dot{X}_i(t; s, x, y, \tau)$$

$$
\begin{aligned}
&\quad + \sum \xi_i(t;s,x,y,\tau)\frac{\partial \dot{X}_i}{\partial s}(t;s,x,y,\tau) \\
&\quad + \sum \frac{\partial \eta_j}{\partial s}(t;s,x,y,\tau)\dot{Y}_j(t;s,x,y,\tau) + \sum \eta_j(t;s,x,y,\tau)\frac{\partial \dot{Y}_j}{\partial s}(t;s,x,y,\tau) \\
&\quad - \sum \frac{\partial H}{\partial x_i}(X(t),Y(t);\xi(t),\eta(t))\frac{\partial X_i}{\partial s}(t;s,x,y,\tau) \\
&\quad - \sum \frac{\partial H}{\partial y_j}(X(t),Y(t);\xi(t),\eta(t))\frac{\partial Y_j}{\partial s}(t;s,x,y,\tau) \\
&\quad - \sum \frac{\partial H}{\partial \xi_i}(X(t),Y(t);\xi(t),\eta(t))\frac{\partial \xi_i}{\partial s}(t;s,x,y,\tau) \\
&\quad - \sum \frac{\partial H}{\partial \eta_j}(X(t),Y(t);\xi(t),\eta(t))\frac{\partial \eta_j}{\partial s}(t;s,x,y,\tau)\Bigg)\, dt \\
&= \sum_{j=1}^{d} \tau_j \frac{\partial Y_j}{\partial s}(0;s,x,y,\tau) + \sum \xi_i(s)\dot{X}_i(s) \\
&+ \sum \eta(s)_j \dot{Y}_j(s) - H(x,y;\xi(s;s,x,y,\tau),\eta(s;s,x,y,\tau)) \\
&+ \int_0^s \Bigg\{ \sum \Bigg(\frac{\partial \xi_i}{\partial s}(t;s,x,y,\tau)\dot{X}_i(t;s,x,y,\tau) \\
&\qquad + \xi_i(t;s,x,y,\tau)\frac{\partial \dot{X}_i}{\partial s}(t;s,x,y,\tau)\Bigg) \\
&+ \sum \Bigg(\frac{\partial \eta_j}{\partial s}(t;s,x,y,\tau)\dot{Y}_j(t;s,x,y,\tau) + \eta_j(t;s,x,y,\tau)\frac{\partial \dot{Y}_j}{\partial s}(t;s,x,y,\tau)\Bigg) \\
&+ \sum \dot{\xi}_i(t;s,x,y,\tau)\frac{\partial X_i}{\partial s}(t;s,x,y,\tau) + \sum \dot{\eta}_j(t;s,x,y,\tau)\frac{\partial Y_j}{\partial s}(t;s,x,y,\tau) \\
&- \sum \dot{X}_i(t;s,x,y,\tau)\frac{\partial \xi_i}{\partial s}(t;s,x,y,\tau) - \sum \dot{Y}_j(t;s,x,y,\tau)\frac{\partial \eta_j}{\partial s}(t;s,x,y,\tau)\Bigg\}\, dt \\
&= \sum_{j=1}^{d} \tau_j \frac{\partial Y_j}{\partial s}(0;s,x,y,\tau) \\
&+ \sum \xi_i(s)\dot{X}_i(s) + \sum \eta(s)_j \dot{Y}_j(s) - H(x,y;\xi(s;s,x,y,\tau),\eta(s;s,x,y,\tau)) \\
&+ \sum \xi_i(t)\frac{\partial X_i}{\partial s}(t;s,x,y,\tau)\Big|_0^s + \sum \eta_j(t)\frac{\partial Y_j}{\partial s}(t;s,x,y,\tau)\Big|_0^s .
\end{aligned}
$$

Now from the initial-boundary conditions $X(0;s,x,y,\tau) = 0,$

$X(s; s, x, y, \tau) = x$ and $Y(s; s, x, y, \eta) = y$, we have

$$\dot{X}_i(s; s, x, y, \tau) + \frac{\partial X_i}{\partial s}(s; s, x, y, \tau) = 0,$$
$$\dot{Y}_j(s; s, x, y, \tau) + \frac{\partial Y_j}{\partial s}(s; s, x, y, \tau) = 0,$$
$$\frac{\partial X_i}{\partial s}(0; s, x, y, \tau) = 0,$$

and from that we finally have (A.2).

Next we show

$$\frac{\partial g}{\partial x_k}(x, y; s, \tau) = \xi_i(s; s, x, y, \tau) \tag{A.3}$$

and

$$\frac{\partial g}{\partial y_l}(x, y; s, \tau) = \eta_l(s; s, x, y, \tau). \tag{A.4}$$

Then

$$\frac{\partial g}{\partial x_k}(x, y; s, \tau) = \sum_{j=1}^{d} \tau_j \frac{\partial Y_j}{\partial x_k}(0; s, x, y, \tau)$$
$$+ \int_0^s \Bigg(\sum \frac{\partial \xi_i}{\partial x_k}(t; s, x, y, \tau) \dot{X}_i(t; s, x, y, \tau)$$
$$+ \sum \xi_i(t; s, x, y, \tau) \frac{\partial \dot{X}_i}{\partial x_k}(t; s, x, y, \tau)$$
$$+ \sum \frac{\partial \eta_j}{\partial x_k}(t; s, x, y, \tau) \dot{Y}_j(t; s, x, y, \tau) + \sum \eta_j(t; s, x, y, \tau) \frac{\partial \dot{Y}_j}{\partial x_k}(t; s, x, y, \tau)$$
$$- \sum \frac{\partial H}{\partial x_i}(X(t), Y(t); \xi(t), \eta(t)) \frac{\partial X_i}{\partial x_k}(t; s, x, y, \tau)$$
$$- \sum \frac{\partial H}{\partial y_j}(X(t), Y(t); \xi(t), \eta(t)) \frac{\partial Y_j}{\partial x_k}(t; s, x, y, \tau)$$
$$- \sum \frac{\partial H}{\partial \xi_i}(X(t), Y(t); \xi(t), \eta(t)) \frac{\partial \xi_i}{\partial x_k}(t; s, x, y, \tau)$$
$$- \sum \frac{\partial H}{\partial \eta_j}(X(t), Y(t); \xi(t), \eta(t)) \frac{\partial \eta_j}{\partial x_k}(t; s, x, y, \tau) \Bigg) dt$$
$$= \sum_{j=1}^{d} \tau_j \frac{\partial Y_j}{\partial x_k}(0; s, x, y, \tau) + \int_0^s \Bigg(\sum \frac{\partial \xi_i}{\partial x_k}(t; s, x, y, \tau) \dot{X}_i(t; s, x, y, \tau)$$
$$+ \sum \xi_i(t; s, x, y, \tau) \frac{\partial \dot{X}_i}{\partial x_k}(t; s, x, y, \tau)$$

$$+\sum \frac{\partial \eta_j}{\partial x_k}(t;s,x,y,\tau)\dot{Y}_j(t;s,x,y,\tau)+\sum \eta_j(t;s,x,y,\tau)\frac{\partial \dot{Y}_j}{\partial x_k}(t;s,x,y,\tau)$$
$$+\sum \dot{\xi}_i(t;s,x,y,\tau)\frac{\partial X_i}{\partial x_k}(t;s,x,y,\tau)+\sum \dot{\eta}_j(t;s,x,y,\tau)\frac{\partial Y_j}{\partial x_k}(t;s,x,y,\tau)$$
$$-\sum \dot{X}_i(t;s,x,y,\tau)\frac{\partial \xi_i}{\partial x_k}(t;s,x,y,\tau)$$
$$\left.-\sum \dot{Y}_j(t;s,x,y,\tau)\frac{\partial \eta_j}{\partial x_k}(t;s,x,y,\tau)\right)dt$$
$$=\sum_{j=1}^{d}\tau_j\frac{\partial Y_j}{\partial x_k}(0;s,x,y,\tau)\sum \xi_i(t;s,x,y,\tau)\frac{\partial X_i}{\partial x_k}\Big|_0^s+\sum \eta_j(t;s,x,y,\tau)\frac{\partial Y_j}{\partial x_k}\Big|_0^s$$
$$=\xi_k(s;s,x,y,\tau),$$

since again by the initial-boundary conditions

$$\frac{\partial X_i}{\partial x_k}(s;s,x,y,\tau)=\frac{\partial x_i}{\partial x_k}=\delta_{i,k},\quad \frac{\partial X_i}{\partial x_k}(0;s,x,y,\tau)=0,\quad \text{and}$$
$$\frac{\partial Y_j}{\partial x_k}(s;s,x,y,\tau)=\frac{\partial y_j}{\partial x_k}=0.$$

By the similar data

$$\frac{\partial Y_j}{\partial y_l}(s;s,x,y,\tau)=\frac{\partial y_j}{\partial y_l}=\delta_{j,l},\quad \frac{\partial X_i}{\partial y_l}(0;s,x,y,\tau)=0,\quad \text{and}$$
$$\frac{\partial X_i}{\partial y_l}(s;s,x,y,\tau)=\frac{\partial x_i}{\partial y_l}=0,$$

we have

$$\frac{\partial g}{\partial y_l}(x,y;s,\tau)=\sum_{j=1}^{d}\tau_j\frac{\partial Y_j}{\partial y_l}(0;s,x,y,\tau)$$
$$+\int_0^s\left(\sum \frac{\partial \xi_i}{\partial y_l}(t;s,x,y,\tau)\dot{X}_i(t;s,x,y,\tau)+\sum \xi_i(t;s,x,y,\tau)\frac{\partial \dot{X}_i}{\partial y_l}(t;s,x,y,\tau)\right.$$
$$+\sum \frac{\partial \eta_j}{\partial y_l}(t;s,x,y,\tau)\dot{Y}_j(t;s,x,y,\tau)+\sum \eta_j(t;s,x,y,\tau)\frac{\partial \dot{Y}_j}{\partial y_l}(t;s,x,y,\tau)$$
$$-\sum \frac{\partial H}{\partial x_i}(X(t),Y(t);\xi(t),\eta(t))\frac{\partial X_i}{\partial y_l}(t;s,x,y,\tau)$$
$$-\sum \frac{\partial H}{\partial y_j}(X(t),Y(t);\xi(t),\eta(t))\frac{\partial Y_j}{\partial y_l}(t;s,x,y,\tau)$$
$$-\sum \frac{\partial H}{\partial \xi_i}(X(t),Y(t);\xi(t),\eta(t))\frac{\partial \xi_i}{\partial y_l}(t;s,x,y,\tau)$$

$$
\begin{aligned}
&-\sum \frac{\partial H}{\partial \eta_j}(X(t),Y(t);\xi(t),\eta(t))\frac{\partial \eta_j}{\partial y_l}(t;s,x,y,\tau)\Bigg)\,dt \\
&=\sum_{j=1}^{d}\tau_j\frac{\partial Y_j}{\partial y_l}(0;s,x,y,\tau) \\
&+\int_0^s\Bigg(\sum\frac{\partial \xi_i}{\partial y_l}(t;s,x,y,\tau)\dot{X}_i(t;s,x,y,\tau)+\sum \xi_i(t;s,x,y,\tau)\frac{\partial \dot{X}_i}{\partial y_l}(t;s,x,y,\tau) \\
&+\sum\frac{\partial \eta_j}{\partial y_l}(t;s,x,y,\tau)\dot{Y}_j(t;s,x,y,\tau)+\sum \eta_j(t;s,x,y,\tau)\frac{\partial \dot{Y}_j}{\partial y_l}(t;s,x,y,\tau) \\
&+\sum\dot{\xi}_i(t;s,x,y,\tau)\frac{\partial X_i}{\partial y_l}(t;s,x,y,\tau)+\sum \dot{\eta}_j(t;s,x,y,\tau)\frac{\partial Y_j}{\partial y_l}(t;s,x,y,\tau) \\
&-\sum \frac{\partial H}{\partial \xi_i}(X(t),Y(t);\xi(t),\eta(t))\frac{\partial \xi_i}{\partial y_l}(t;s,x,y,\tau) \\
&-\sum \frac{\partial H}{\partial \eta_j}(X(t),Y(t);\xi(t),\eta(t))\frac{\partial \eta_j}{\partial y_l}(t;s,x,y,\tau)\Bigg)\,dt \\
&=\sum_{j=1}^{d}\tau_j\frac{\partial Y_j}{\partial y_l}(0;s,x,y,\tau) \\
&+\sum \xi_i(t;s,x,y,\tau)\frac{\partial X_i}{\partial y_l}(t;s,x,y,\tau)\Big|_0^s+\sum \eta_j(t;s,x,y,\tau)\frac{\partial Y_j}{\partial y_l}(t;s,x,y,\tau)\Big|_0^s \\
&=\eta_l(s;s,x,y,\tau).
\end{aligned}
$$

Hence we have that $g=g(x,y;s,\tau)$ satisfies

$$\frac{\partial g}{\partial s}+H(x,y;\nabla g)=0. \qquad \square$$

Appendix B. Generalized transport equation

Let G be an n-dimensional connected and simply connected nilpotent Lie group, and $\{X_i\}_{i=1}^m$ be a system of linearly independent elements in $\mathfrak{g}\backslash[\mathfrak{g},\mathfrak{g}]$, where $m=\dim\mathfrak{g}/[\mathfrak{g},\mathfrak{g}]$.

Let

$$-\Delta_{sub}=\frac{1}{2}\sum_{i=1}^{m}\widetilde{X}_i^2$$

be the sum of the left invariant vector fields on the group G, then Δ_{sub} is a sub-Laplacian satisfying the Hörmander condition of the hypoellipticity.

Let H be the Hamiltonian of this sub-Laplacian Δ_{sub}:

$$H(x; \nabla f) = \frac{1}{2}\sum_{i=1}^{m} X_i(f)^2 \tag{B.1}$$

and f be a solution of the generalized Hamilton-Jacobi equation

$$H(x; \nabla f) + \sum_{i=1}^{\ell} \tau_i \frac{\partial f}{\partial \tau_i} = f(x; \tau_1, \cdots, \tau_\ell). \tag{B.2}$$

We assume that the heat kernel $K(t; x, \widetilde{x})$ takes the form

$$\begin{aligned} K(t; x, \widetilde{x}) &= k_t(\widetilde{x}^{-1} * x), \\ k_t(x) &= \frac{1}{t^N}\int_{\mathbb{R}^\ell} e^{-\frac{f(x,\tau)}{t}} W(x,\tau) d\tau. \end{aligned} \tag{B.3}$$

Let the characteristic variety of Δ_{sub} be $Cha = \{H = 0\}$. This is a subbundle in T^*G and is trivialized by the subspace $[\mathfrak{g}, \mathfrak{g}]$ (see Remark 3.4). We regard that the dimension ℓ of the variable τ above is dim $Cha -$ dim G $=$ dim $[\mathfrak{g}, \mathfrak{g}]$. However the calculations below are valid for any $N > 0$ and $\ell > 0$. So in the calculation we do not specify the order N. The true order N should be fixed as $N = \frac{1}{2}\dim \mathfrak{g}/[\mathfrak{g}, \mathfrak{g}] + \dim[\mathfrak{g}, \mathfrak{g}] = m/2 + \ell$.

Note that we need to assume (and it will be reasonable to assume) that the integrand $e^{-\frac{f(x,\tau)}{t}} W(x, \tau)$ *will decrease in a suitable order for the validity of the partial integrations, when* $|\tau| \to \infty$ *(see Theorem 3.1). Also later we will assume that the function* W *has a special form.*

Then we have

$$\begin{aligned} \Delta_{sub}(e^{-\frac{f}{t}} W) &= \frac{-1}{t^2} H(x; \nabla f) \cdot W \cdot e^{-\frac{f}{t}} \\ &- \frac{1}{t}\Big(\Delta_{sub}(f) \cdot W - \sum_{i=1}^{m} X_i(f) X_i(W)\Big) e^{-\frac{f}{t}} + \Delta_{sub}(W) e^{-\frac{f}{t}}, \end{aligned}$$

and

$$\begin{aligned} &\frac{\partial}{\partial t}\Big(\frac{1}{t^N}\int_{\mathbb{R}^\ell} e^{-\frac{f}{t}} \cdot W d\tau_1 \cdots d\tau_\ell\Big) \\ &= -\frac{1}{t^N}\int_{\mathbb{R}^\ell} \frac{N}{t} \cdot e^{-\frac{f}{t}} \cdot W d\tau_1 \cdots d\tau_\ell + \frac{1}{t^N}\int_{\mathbb{R}^\ell} \frac{f}{t^2} \cdot e^{-\frac{f}{t}} \cdot W d\tau_1 \cdots d\tau_\ell. \end{aligned}$$

Then

$$-\left(\Delta_{sub} + \frac{\partial}{\partial t}\right)\Big(k_t(x)\Big) = \frac{1}{t^N}\left\{\frac{1}{t^2}\int_{\mathbb{R}^\ell} (H(x; \nabla f) - f) \cdot e^{-\frac{f}{t}} \cdot W d\tau\right.$$

$$- \frac{1}{t} \int_{\mathbb{R}^\ell} \left(\sum_i \widetilde{X}_i(f) \widetilde{X}_i(W) + (-\Delta_{sub}(f) - N) \cdot W \right) e^{-\frac{f}{t}} d\tau$$

$$- \int_{\mathbb{R}^\ell} \Delta_{sub}(W) e^{-\frac{f}{t}} d\tau \Big\}$$

$$= \frac{1}{t^N} \Big\{ - \frac{1}{t^2} \int_{\mathbb{R}^\ell} \sum \tau_i \frac{\partial f}{\partial \tau_i} \cdot e^{-\frac{f}{t}} \cdot W d\tau$$

$$- \frac{1}{t} \int_{\mathbb{R}^\ell} \left(\sum_i \widetilde{X}_i(f) \widetilde{X}_i(W) - (\Delta_{sub}(f) + N) \cdot W \right) e^{-\frac{f}{t}} d\tau$$

$$- \int_{\mathbb{R}^\ell} \Delta_{sub}(W) e^{-\frac{f}{t}} d\tau \Big\}$$

$$= \frac{1}{t^N} \left\{ \frac{1}{t^2} \cdot A_2 - \frac{1}{t} \cdot A_1 - A_0 \right\}.$$

Now we assume

$$W d\tau = df \wedge V^{\mathrm{d}} \tag{B.4}$$

with an $(\ell - 1)$–form

$$V = \sum_{\alpha=1}^{\ell} (-1)^{\alpha-1} V_\alpha d\tau_1 \wedge \cdots \wedge \widehat{d\tau_\alpha} \wedge \cdots \wedge d\tau_\ell.$$

The coefficients V_i are, in general, functions both of space variables $\{x_i\}$ and the variables $\{\tau_j\}$ in the characteristic variety. So the operation $\Delta_{sub}(V)$ is defined as

$$\Delta_{sub}(V) = \sum_{\alpha=1}^{\ell} (-1)^{\alpha-1} \Delta_{sub}(V_\alpha) d\tau_1 \wedge \cdots \wedge \widehat{d\tau_\alpha} \wedge \cdots \wedge d\tau_\ell.$$

Under this assumption of the function W we calculate the terms A_1 and A_2 in the following way:

$$\frac{1}{t^2} \cdot A_2$$

$$= - \frac{1}{t^2} \int_{\mathbb{R}^\ell} \left(\sum \tau_i \frac{\partial f}{\partial \tau_i} \right) \left(e^{-\frac{f}{t}} \right) df \wedge V$$

$$= \frac{1}{t} \int_{\mathbb{R}^\ell} \left(\sum \tau_i \frac{\partial f}{\partial \tau_i} \right) d \left(e^{-\frac{f}{t}} \right) \wedge V = - \frac{1}{t} \int_{\mathbb{R}^\ell} e^{-\frac{f}{t}} d \left(\sum \tau_i \frac{\partial f}{\partial \tau_i} \cdot V \right)$$

[d] P. Greiner worked out this form for a special case of $\ell = 1$. See Remark B.1.

$$= -\frac{1}{t}\int_{\mathbb{R}^\ell} e^{-\frac{f}{t}}\left(df\wedge V + \left(\sum_j\sum_i \tau_i \frac{\partial^2 f}{\partial\tau_i\partial\tau_j}V_j\right)d\tau + \left(\sum_i \tau_i\frac{\partial f}{\partial \tau_i}\right)dV\right)$$

$$= \int_{\mathbb{R}^\ell} d\left(e^{-f/t}\right)\wedge V - \frac{1}{t}\int_{\mathbb{R}^\ell} e^{-f/t}\left(\sum_j\sum_i \tau_i\frac{\partial^2 f}{\partial\tau_i\partial\tau_j}V_j\right)d\tau$$

$$+ \int_{\mathbb{R}^\ell}\left(\sum \tau_i \frac{\partial e^{-f/t}}{\partial \tau_i}\right)\cdot\left(\sum_j \frac{\partial V_j}{\partial\tau_j}\right)d\tau$$

$$= -\frac{1}{t}\int_{\mathbb{R}^\ell} e^{-f/t}\left(\sum_j\sum_i \tau_i\frac{\partial^2 f}{\partial\tau_i\partial\tau_j}V_j\right)d\tau$$

$$- \int_{\mathbb{R}^\ell} e^{-f/t}\left(\sum_i\sum_j \tau_i\frac{\partial^2 V_j}{\partial\tau_i\partial\tau_j}\right)d\tau - (\ell+1)\int_{\mathbb{R}^\ell} e^{-f/t}dV.$$

$$\frac{1}{t}A_1$$

$$= \frac{1}{t}\int_{\mathbb{R}^\ell}\sum_i \widetilde{X}_i(f)\widetilde{X}_i\left(\sum_j \frac{\partial f}{\partial\tau_j}V_j\right)e^{-f/t}d\tau$$

$$- \frac{1}{t}\int_{\mathbb{R}^\ell}(\Delta_{sub}(f)+N)e^{-f/t}df\wedge V$$

$$= \frac{1}{t}\int_{\mathbb{R}^\ell}\sum_j\left(\frac{\partial}{\partial\tau_j}H(x;\nabla f)\right)\cdot V_j\cdot e^{-f/t}d\tau$$

$$- \int_{\mathbb{R}^\ell}\sum_j\sum_i \widetilde{X}_i(f)\widetilde{X}_i(V_j)\frac{\partial e^{-f/t}}{\partial\tau_j}d\tau + \int_{\mathbb{R}^\ell}(\Delta_{sub}(f)+N)d\left(e^{-f/t}\right)\wedge V$$

$$= \frac{1}{t}\int_{\mathbb{R}^\ell} e^{-f/t}\sum_j \frac{\partial H(x;\nabla f)}{\partial\tau_j}V_j d\tau$$

$$+ \int_{\mathbb{R}^\ell} e^{-f/t}\sum_j\sum_i \frac{\partial}{\partial\tau_j}\left(\widetilde{X}_i(f)\widetilde{X}_i(V_j)\right)d\tau - \int_{\mathbb{R}^\ell} e^{-f/t}d\Big((\Delta_{sub}(f)+N)V\Big).$$

Since

$$\frac{\partial}{\partial\tau_j}H(x;\nabla f) + \sum_i \tau_i\frac{\partial^2 f}{\partial\tau_j\partial\tau_i} = 0$$

for any j, now $-A_0+\frac{1}{t^2}A_2-\frac{1}{t}A_1$ equals to

$$
\begin{aligned}
&-A_0+\frac{1}{t^2}A_2-\frac{1}{t}A_1=\int -\Delta_{sub}\Big(\sum\frac{\partial f}{\partial\tau_j}V_j\Big)e^{-f/t}dt-(\ell+1)\int e^{-f/t}dV\\
&-\int e^{-f/t}\Big(\sum\sum\tau_i\frac{\partial^2 V_j}{\partial\tau_i\partial\tau_j}\Big)d\tau-\int e^{-f/t}\sum\sum\frac{\partial}{\partial\tau_j}\Big(\widetilde{X}_i(f)\widetilde{X}_i(V_j)\Big)d\tau\\
&\qquad\qquad+\int e^{-f/t}d\left(\left(\Delta_{sub}(f)+N\right)V\right)\\
&=-\int e^{-f/t}\Delta_{sub}\Big(\sum\frac{\partial f}{\partial\tau_j}V_j\Big)d\tau+(N-\ell-1)\int e^{-f/t}dV\\
&\qquad+\int e^{-f/t}d(\Delta_{sub}(f)V)-\int e^{-f/t}\sum\sum\tau_i\frac{\partial^2 V_j}{\partial\tau_i\partial\tau_j}d\tau\\
&\qquad\quad-\int e^{-f/t}\Big(\sum\sum\frac{\partial}{\partial\tau_j}\Big(\widetilde{X}_i(f)\widetilde{X}_i(V_j)\Big)\Big)d\tau\\
&=-\int e^{-f/t}\sum\Big(\Delta_{sub}\Big(\frac{\partial f}{\partial\tau_j}\Big)V_j+\frac{\partial f}{\partial\tau_j}\Delta_{sub}(V_j)\Big)d\tau\\
&\quad+(N-\ell-1)\int e^{-f/t}dV+\int e^{-f/t}d(\Delta_{sub}(f)V)\\
&\qquad-\int e^{-f/t}\sum\sum\tau_i\frac{\partial^2 V_j}{\partial\tau_i\partial\tau_j}d\tau-\int e^{-f/t}\sum\widetilde{X}_i(f)\widetilde{X}_i\Big(\sum\frac{\partial V_j}{\partial\tau_j}\Big)d\tau\\
&=-\int e^{-f/t}df\wedge\Delta_{sub}(V)-\int e^{-f/t}\left(-\Delta_{sub}(f)-N+\ell+1\right)dV\\
&\quad-\int e^{-f/t}\sum\sum\tau_i\frac{\partial^2 V_j}{\partial\tau_i\partial\tau_j}d\tau-\int e^{-f/t}\sum\widetilde{X}_i(f)\widetilde{X}_i\Big(\sum\frac{\partial V_j}{\partial\tau_j}\Big)d\tau\\
&=-\int e^{-f/t}df\wedge\Delta_{sub}(V)+\int e^{-f/t}\Delta_{sub}(f)dV+(N-\ell-1)\int e^{-f/t}dV\\
&\qquad-\int e^{-f/t}\mathfrak{D}(dV)-\int e^{-f/t}\sum\widetilde{X}_i(f)\widetilde{X}_i(dV),
\end{aligned}
$$

thus we have the *generalized transport equation*

$$
\begin{aligned}
&df\wedge\Delta_{sub}(V)+\sum\widetilde{X}_i(f)\cdot\widetilde{X}_i(dV)\\
&\qquad\qquad+\mathfrak{D}(dV)+\left(\Delta_{sub}(f)+N-\ell-1\right)dV=0,
\end{aligned}
\tag{B.5}
$$

where $\mathfrak{D}(V)$ is defined by

$$
\mathfrak{D}(V)=\sum\tau_i\frac{\partial}{\partial\tau_i}(V)=\sum\sum(-1)^{j-1}\tau_i\frac{\partial V_j}{\partial\tau_i}d\tau_1\wedge\cdots\wedge\widehat{d\tau_j}\wedge\cdots\wedge\tau_\ell\,,
$$

and we have

$$
\mathfrak{D}(dV)=d\mathfrak{D}(V)-dV.
$$

Remark B.1. Fundamental solutions have volume elements which do not depend on the dimension of missing directions (see [5]). The result is similar for the heat kernel volume element W if we assume that the volume element is of the form $W d\tau = df \wedge V$ (see (B.4)). When the number of missing directions is 1 (= dim $[\mathfrak{g}, \mathfrak{g}]$ for our nilpotent Lie group case), this was already worked out by P. Greiner in [12].

Again we consider

$$
\begin{aligned}
&\left(\Delta_{sub} + \frac{\partial}{\partial t}\right)\left(k_t(x)\right) \\
&= \frac{1}{t^N}\left\{\frac{-1}{t^2}\int_{\mathbb{R}^\ell}(H(x;\nabla f) - f)\cdot e^{-\frac{f}{t}}\cdot W d\tau \right.\\
&\qquad + \frac{1}{t}\int_{\mathbb{R}^\ell}\left(\sum_i \widetilde{X}_i(f)\widetilde{X}_i(W) - (\Delta_{sub}(f) + N)\cdot W\right) e^{-\frac{f}{t}} d\tau \\
&\qquad\qquad \left. + \int_{\mathbb{R}^\ell}\Delta_{sub}(W) e^{-\frac{f}{t}} d\tau\right\} \\
&= \frac{1}{t^N}\left\{\frac{1}{t^2}\int_{\mathbb{R}^\ell}\sum \tau_i \frac{\partial f}{\partial \tau_i}\cdot e^{-\frac{f}{t}}\cdot W d\tau \right.\\
&\qquad + \frac{1}{t}\int_{\mathbb{R}^\ell}\left(\sum_i \widetilde{X}_i(f)\widetilde{X}_i(W) - (\Delta_{sub}(f) + N)\cdot W\right) e^{-\frac{f}{t}} d\tau \\
&\qquad\qquad \left. + \int_{\mathbb{R}^\ell}\Delta_{sub}(W) e^{-\frac{f}{t}} d\tau\right\} \\
&= \frac{-1}{t^N}\left\{\frac{1}{t}\int_{\mathbb{R}^\ell}\sum \tau_i \frac{\partial}{\partial \tau_i}\left(e^{-\frac{f}{t}}\right)\cdot W d\tau \right.\\
&\qquad + \frac{1}{t}\int_{\mathbb{R}^\ell}\left(\sum_i \widetilde{X}_i(f)\widetilde{X}_i(W) - (\Delta_{sub}(f) + N)\cdot W\right) e^{-\frac{f}{t}} d\tau \\
&\qquad\qquad \left. + \int_{\mathbb{R}^\ell}\Delta_{sub}(W) e^{-\frac{f}{t}} d\tau\right\} \\
&= \frac{1}{t^N}\left\{\frac{1}{t}\int_{\mathbb{R}^\ell} e^{-\frac{f}{t}}\cdot \sum \frac{\partial}{\partial \tau_i}(\tau_i W) d\tau \right.\\
&\qquad \frac{1}{t}\int_{\mathbb{R}^\ell}\left(\sum_i \widetilde{X}_i(f)\widetilde{X}_i(W) - (\Delta_{sub}(f) + N)\cdot W\right) e^{-\frac{f}{t}} d\tau \\
&\qquad\qquad \left. + \int_{\mathbb{R}^\ell}\Delta_{sub}(W) e^{-\frac{f}{t}} d\tau\right\}
\end{aligned}
$$

$$= \frac{1}{t^N}\left\{\frac{1}{t}\int_{\mathbb{R}^\ell} e^{-\frac{f}{t}} \cdot \sum \tau_i \frac{\partial W}{\partial \tau_i} d\tau \right.$$
$$+ \frac{1}{t}\int_{\mathbb{R}^\ell}\left(\sum_i \widetilde{X}_i(f)\widetilde{X}_i(W) - (\Delta_{sub}(f) + N - \ell)\cdot W\right) e^{-\frac{f}{t}} d\tau$$
$$\left. + \int_{\mathbb{R}^\ell} \Delta_{sub}(W) e^{-\frac{f}{t}} d\tau \right\} .$$

So if the function W does not depend on the space variables, then

$$\sum \tau_i \frac{\partial W}{\partial \tau_i} - (\Delta_{sub}(f) + N - \ell)\, W = 0 \tag{B.6}$$

is the first order transport equation.

References

1. R. Beals and P. C. Greiner: *Calculus on Heisenberg Manifolds*, Ann. Math. Studies, **119**, Princeton Univ. Press, Princeton (1998).
2. R. Beals, B. Gaveau and P. Greiner: *The Green function of model step two hypoelliptic operators and the analysis of certain tangential Cauchy Riemann complexes*, Adv. Math. **121** (1996), 288–345.
3. —, —, —: *Hamilton-Jacobi theory and the heat kernel on Heisenberg groups*, J. Math. Pures Appl. **79** (2000), 633–689.
4. —, —, —: *Complex Hamiltonian mechanics and parametrises for subelliptic Laplacians*, I, II, III, Bull. Sci. Math. **121** (1997), 1–36, 97–149, 195–259.
5. —, —, —: *On a geometric formula for the fundamental solution of subelliptic Laplacians*, Math. Nachr. **181** (1996), 81–163.
6. L. Corwin and F. P. Greenleaf: *Character formula and spectra of compact nilmanifolds*, J. Funct. Anal. **21** (1976), 123–154.
7. K. Furutani: *The heat kernel and the spectrum of a class of nilmanifolds*, Comm. Partial Differential Equations **21**, No. 3&4 (1996), 423–438.
8. K. Furutani and S. de Gosson: *Determinant of Laplacians on Heisenberg manifolds*, J. Geom. Phys. **48** (2003), No. 2-3, 438–479.
9. B. Gaveau: *Principe de moindre action, propagation de la chaleur et estimeée sous-elliptiques sur certains groupes nilpotents*, Acta Math. **139** (1977), 95–153.
10. M. de Gosson: *The Principles of Newtonian and Quantum Mechanics*, Imperial College Press, London, 2001.
11. C. Gordon and E. Wilson: *The spectrum of the Laplacian on Riemannian Heisenberg manifolds*, Michigan Math. J. **33** (1986), 253–271.
12. P. Greiner: *On the heat kernel volume element*, 2000 (unpublished manuscript).
13. L. Hörmander: *Hypoelliptic second order differential equations*, Acta Math. **119** (1967), 147–171.

14. A. Hulanicki: *The distribution of energy in the Brownian motion in the Gaussian field and analytic-hypoellipticity of certain subelliptic operators on the Heisenberg group*, Studia Math. **56** (1976), 165–173.
15. A. Klinger: *New derivation of the Heisenberg kernel*, Comm. Partial Differential Equations **22** (1997), 2051–2060.
16. D. F. Lawden: *Elliptic Functions and Applications*, Applied Mathematical Sciences **80**, Springer-Verlag (1989).
17. L. P. Rothshild and E. M. Stein: *Hypoelliptic differential operators and nilpotent Lie groups*, Acta Math. **137**, No. 3/4 (1976), 247–320.
18. R. S. Strichartz: *Sub-Riemannian geometry*, J. Differential Geom. **24**, No 2 (1986), 221–263.
19. S. Thangavelu: *Lectures on Hermite and Laguerre Expansions*, Mathematical Notes **42**, Princeton University Press (1993).
20. J. H. van Vleck: *The correspondence principle in the statistical interpretation of quantum mechanics*, Proc. Nat. Acad. Sci. U.S.A. **14**, No. 176 (1928), 178–188.
21. A. Voros: *Zeta functions of the quartic (and homogeneous anharmonic) oscillators*, In: Lecture Notes in Math. **925**, Springer (1982), 184–208.
22. —-: *The return of the quartic oscillator: The complex WKB method*, Ann. Inst. H. Poincaré **39**, No. 3 (1983), 211–338.
23. E. T. Whittaker and G. N. Watson: *A Course of Modern Analysis*, Cambridge University Press (1927).

Received by the editors August 18, 2005; Revised October 15, 2005

Analysis, Geometry and Topology
of Elliptic Operators, pp. 215–234

REMARKS ON NONLOCAL TRACE EXPANSION COEFFICIENTS

GERD GRUBB
Mathematics Department University of Copenhagen
Universitetsparken 5
2100 Copenhagen, Denmark
grubb@math.ku.dk

Dedicated to Krzysztof P. Wojciechowski on his 50th birthday

In a recent work, Paycha and Scott establish formulas for all the Laurent coefficients of $\mathrm{Tr}(AP^{-s})$ at the possible poles. In particular, they show a formula for the zero'th coefficient at $s = 0$, in terms of two functions generalizing, respectively, the Kontsevich-Vishik canonical trace density, and the Wodzicki-Guillemin noncommutative residue density of an associated operator. The purpose of this note is to provide a proof of that formula relying entirely on resolvent techniques (for the sake of possible generalizations to situations where powers are not an easy tool). — We also give some corrections to transition formulas used in our earlier works.

2000 *Mathematics Subject Classification.* Primary 58J42; Secondary 35S05, 58J35

1. Introduction

In an interesting new work [PS], Sylvie Paycha and Simon Scott have obtained formulas for all the coefficients in Laurent expansions of zeta functions $\zeta(A, P, s) = \mathrm{Tr}(AP^{-s})$ around the poles, in terms of combinations of finite part integrals and residue type integrals, of associated logarithmic symbols. We consider classical pseudodifferential operators (ψdo's) A and P of order $\sigma \in \mathbb{R}$ resp. $m \in \mathbb{R}_+$ acting in a Hermitian vector bundle E over a closed n-dimensional manifold X, P being elliptic with principal symbol eigenvalues in $\mathbb{C} \setminus \mathbb{R}_-$. The basic formula is the following formula for $C_0(A, P)$, where $C_0(A, P) - \mathrm{Tr}(A\Pi_0(P))$ is the regular value of $\zeta(A, P, s)$ at $s = 0$:

$$C_0(A, P) = \int_X \left(\mathrm{TR}_x(A) - \tfrac{1}{m}\,\mathrm{res}_{x,0}(A \log P)\right) dx. \qquad (1)$$

The integrand is defined in a local coordinate system by:

$$\mathrm{TR}_x(A) = ⨍ \operatorname{tr} a(x,\xi)\, đ\xi, \ \operatorname{res}_{x,0}(A\log P) = \int_{|\xi|=1} \operatorname{tr} r_{-n,0}(x,\xi) đS(\xi), \quad (2)$$

where $⨍ a(x,\xi)\, đ\xi$ is a finite part integral, and r is the symbol of $R = A\log P$, of the form

$$r(x,\xi) \sim \sum_{j\geq 0,\, l=0,1} r_{\sigma-j,l}(x,\xi)(\log[\xi])^l\,;$$

$r_{\sigma-j,l}$ homogeneous in ξ of degree $\sigma - j$ for $|\xi| \geq 1$, $[\xi]$ positive equal to $|\xi|$ for $|\xi| \geq 1$. (Here $r_{-n,0}$ is set equal to 0 when $\sigma - j$ does not hit $-n$.) Moreover, the expression $\big(\mathrm{TR}_x(A) - \operatorname{res}_{x,0}(A\log P)\big)\, dx$ has an invariant meaning as a density on X, although its two terms individually do not so in general. (In these formulas we use the conventions $đ\xi = (2\pi)^{-n} d\xi$, $đS(\xi) = (2\pi)^{-n} dS(\xi)$, where $dS(\xi)$ indicates the usual surface measure on the unit sphere. tr indicates fiber trace.)

Formula (1) generalizes the formula

$$C_0(A,P) = \mathrm{TR}\,A$$

(the canonical trace), which holds in particular cases, cf. Kontesevich and Vishik [KV], Lesch [L], Grubb [G4]. The general formula (1) is shown in [PS] by use of holomorphic families of ψdo's (depending holomorphically on their complex order z); in particular complex powers of P. The purpose of this note is to derive it by methods relying on the knowledge of the resolvent $(P-\lambda)^{-1}$. This is meant to facilitate generalizations to manifolds with boundary, where powers of operators are not an easy tool.

— We take the opportunity here to correct, in the appendix, some inaccuracies in earlier papers, mainly concerning the relations between expansion coefficients in resolvent traces and zeta functions.

2. Preliminaries

Recall the expansion formulas for the resolvent kernel and trace, worked out in local coordinates in Grubb and Seeley [GS1], when $N > \frac{\sigma+n}{m}$:

$$K\big(A(P-\lambda)^{-N}, x, x\big) \sim \sum_{j\in\mathbb{N}} \tilde{c}_j^{(N)}(x)(-\lambda)^{\frac{\sigma+n-j}{m}-N} + \sum_{k\in\mathbb{N}} \big(\tilde{c}_k^{(N)\prime}(x)\log(-\lambda) + \tilde{c}_k^{(N)\prime\prime}(x)\big)(-\lambda)^{-k-N},$$

$$\operatorname{Tr}\bigl(A(P-\lambda)^{-N}\bigr) \sim \sum_{j\in\mathbb{N}} \tilde{c}_j^{(N)}(-\lambda)^{\frac{\sigma+n-j}{m}-N} + \sum_{k\in\mathbb{N}} \bigl(\tilde{c}_k^{(N)\prime}\log(-\lambda) + \tilde{c}_k^{(N)\prime\prime}\bigr)(-\lambda)^{-k-N}, \quad (3)$$

for $\lambda \to \infty$ on rays in a sector around $\mathbb{R}_-$. The second formula is deduced from the first one by integrating the fiber trace in x. We denote $\{0,1,2,\dots\} = \mathbb{N}$. (More precisely, [GS1] covers the cases where m is integer; the noninteger cases are included in Loya [Lo], Grubb and Hansen [GH].)

The $\tilde{c}_k^{(N)\prime}(x)$ and $\tilde{c}_k^{(N)\prime}$ vanish when

$$\sigma + n + mk \notin \mathbb{N};$$

this holds for all k when m is integer and σ is noninteger. When

$$\sigma + n + mk = j \in \mathbb{N},$$

$\tilde{c}_j^{(N)}(x)$ and $\tilde{c}_k^{(N)\prime\prime}(x)$ are both coefficients of the power $(-\lambda)^{-k-N}$; their individual values depend on the localization used (as worked out in detail in [G4]), and it is only the sum $\tilde{c}_j^{(N)} + \tilde{c}_k^{(N)\prime\prime}$ that has an invariant meaning.

The coefficients depend on N; when $N = 1$, we drop the upper index (N). We are particularly interested in the coefficient of $(-\lambda)^{-N}$, for which we shall use the notation

$$\widetilde{C}_0^{(N)}(A,P) = \tilde{c}_{\sigma+n}^{(N)} + \tilde{c}_0^{(N)\prime\prime}; \quad (4)$$

here we have for convenience set

$$\tilde{c}_{\sigma+n}^{(N)} = 0 \text{ when } \sigma + n \notin \mathbb{N}. \quad (5)$$

Recall that there is, equivalently to the expansion in (3), an expansion formula for complex powers:

$$\Gamma(s)\operatorname{Tr}(AP^{-s}) \sim \sum_{j\in\mathbb{N}} \frac{c_j}{s+\frac{j-\sigma-n}{m}} - \frac{\operatorname{Tr}(A\Pi_0(P))}{s} + \sum_{k\in\mathbb{N}} \Bigl(\frac{c_k'}{(s+k)^2} + \frac{c_k''}{s+k}\Bigr). \quad (6)$$

This means that $\Gamma(s)\operatorname{Tr}(AP^{-s})$, defined as a holomorphic function for $\operatorname{Re} s > \frac{\sigma+n}{m}$, extends meromorphically to $\mathbb{C}$ with the pole structure indicated in the right hand side. Here

$$\Pi_0(P) = \tfrac{i}{2\pi}\int_{|\lambda|=\varepsilon} (P-\lambda)^{-1}\,d\lambda$$

is the projection onto the generalized nullspace of P (on which P^{-s} is taken to be zero). Again, the c'_k vanish when $\sigma + n + mk \notin \mathbb{N}$. We denote

$$C_0(A,P) = c_{\sigma+n} + c''_0, \tag{7}$$

the basic coefficient (setting $c_{\sigma+n}$ equal to 0 when $\sigma + n \notin \mathbb{N}$).

The transition between (3) and (6) is accounted for e.g. in Grubb and Seeley [GS2], Prop. 2.9, (3.21). The coefficient sets in (3) and (6) are derivable from one another. The coefficients $\tilde{c}_j^{(N)}$ and c_j, resp. $\tilde{c}_k^{(N)\prime}$ and c'_k, are proportional by universal nonzero constants. This holds also for $\tilde{c}_k^{(N)\prime\prime}$ and c''_k, when the c'_k vanish. In general, there are linear formulas for the transitions between $\{\tilde{c}_k^{(N)\prime}, \tilde{c}_k^{(N)\prime\prime}\}$ and $\{c'_k, c''_k\}$. (For $N = 1$, [GS2], Cor. 2.10, would imply that $\tilde{c}''_k$ and c''_k are proportional in general, but in fact, the formulas for the $\tilde{a}_{j,l}$ given there are only correct for $l = m_j$, whereas for $l < m_j$ there is an effect from derivatives of the gamma function that was overlooked.) One has in particular that $\tilde{c}_0^{(N)\prime} = c'_0$ for all N.

Division of (6) by $\Gamma(s)$ gives the pole structure of $\zeta(A,P,s)$:

$$\zeta(A,P,s) = \operatorname{Tr}(AP^{-s}) \sim \sum_{j\in\mathbb{N}} \frac{c'''_j}{s + \frac{j-\sigma-n}{m}}, \tag{8}$$

where c'''_j is proportional to c_j if $\frac{j-\sigma-n}{m} \notin \mathbb{N}$, and c'''_j is proportional to c'_k if $\frac{j-\sigma-n}{m} = k \in \mathbb{N}$.

One can study the Laurent series expansions of $\zeta(A,P,s)$ at the poles by use of (3). We now restrict the attention to the possible pole at $s = 0$.

Write the Laurent expansion at 0 as follows:

$$\begin{aligned}\zeta(A,P,s) = C_{-1}(A,P)s^{-1} &+ (C_0(A,P) - \operatorname{Tr}(A\Pi_0(P)))s^0 \\ &+ \sum_{l\geq 1} C_l(A,P)s^l.\end{aligned} \tag{9}$$

It is known from Wodzicki [W], Guillemin [Gu], that

$$C_{-1}(A,P) = c'_0 = \tfrac{1}{m}\operatorname{res}(A), \text{ independently of } P; \tag{10}$$

it vanishes if $\sigma + n \notin \mathbb{N}$ or the symbols have certain parity properties. From [KV], [L], [G4] we have that

$$C_0(A,P) = \operatorname{TR} A$$

when $\sigma + n \notin \mathbb{N}$, and in certain parity cases (given in [KV] for n odd, [G4] for n even, more details at the end of Section 4). Also $C_1(A,P)$ is of interest, since the zeta determinant of P satisfies

$$\log\det P = -C_1(I,P) = C_0(\log P, P) \tag{11}$$

(cf. Okikiolu [O], [G4]); here it is useful to know that expansions like (3) but with higher powers of $\log(-\lambda)$ hold if A is log-polyhomogeneous, cf. [L] and [G4].

We have for general $N \geq 1$:

Lemma 2.1. *When $N > \frac{\sigma+n}{m}$ (so that $A(P-\lambda)^{-N}$ is trace-class), then*

$$\begin{aligned} \tilde{c}_0^{(N)\prime} &= c_0' = \tfrac{1}{m}\operatorname{res} A, \\ \widetilde{C}_0^{(N)}(A,P) &= C_0(A,P) - \alpha_N c_0', \text{ where} \\ \alpha_N &= 1 + \tfrac{1}{2} + \dots + \tfrac{1}{N-1}. \end{aligned} \tag{12}$$

Proof. Denote $N-1 = M$, then

$$(P-\lambda)^{-N} = (P-\lambda)^{-M-1} = \tfrac{1}{M!}\partial_\lambda^M (P-\lambda)^{-1}. \tag{13}$$

The transition from (3) to information on $\zeta(A,P,s)$ is obtained by use of the formula

$$AP^{-s} = \tfrac{M!}{(s-M)\dots(s-1)}\tfrac{i}{2\pi}\int_{\mathcal{C}} \lambda^{M-s}\tfrac{1}{M!}\partial_\lambda^M A(P-\lambda)^{-1}(I-\Pi_0(P))\,d\lambda, \tag{14}$$

where $\mathcal{C}$ is a curve in $\mathbb{C}\setminus\overline{\mathbb{R}}_-$ around the nonzero spectrum of P. Here we can take traces on both sides and apply [GS2], Prop. 2.9, to

$$f(\lambda) = \operatorname{Tr}\big(A\tfrac{1}{M!}\partial_\lambda^M(P-\lambda)^{-1}(I-\Pi_0(P))\big), \tag{15}$$

defining

$$\varrho(s) = \tfrac{i}{2\pi}\int_{\mathcal{C}} \lambda^{-s} f(\lambda)\,d\lambda, \tag{16}$$

for $\operatorname{Re} s$ large, and extending meromorphically. Then

$$\zeta(A,P,s) = \tfrac{M!}{(s-M)\dots(s-1)}\varrho(s-M). \tag{17}$$

(Note that $\zeta(A,P,s) = \varrho(s)$ if $N=1$.) The expansion coefficients of $f(\lambda)$ in powers and log-powers are universally proportional to the pole coefficients of

$$\psi(s) = \frac{\pi}{\sin(\pi s)}\varrho(s)$$

at simple and double poles, for each index, as accounted for in [GS2], Prop. 2.9.

When we apply this to $\zeta(A,P,s)$, we must take the factors

$$g_M(s) = \tfrac{M!}{(s-M)\dots(s-1)}$$

and $\frac{1}{\pi}\sin(\pi(s-M))$ into account. We have

$$\zeta(A,P,s) = g_M(s)\tfrac{1}{\pi}\sin(\pi(s-M))\psi(s-M). \tag{18}$$

By [GS2], Prop. 2.9, a pair of terms $a(-\lambda)^{-M-1}\log(-\lambda) + b(-\lambda)^{-M-1}$ in the expansion of $f(\lambda)$ carries over to the pair of terms $\frac{a}{(s+M)^2} + \frac{b}{s+M}$ in the pole structure of $\psi(s)$, whereby

$$\psi(s-M) = \frac{a}{s^2} + \frac{b}{s} + O(1), \text{ for } s \to 0. \tag{19}$$

Now it is easily checked that

$$\begin{aligned} \tfrac{1}{\pi}\sin(\pi(s-M)) &= (-1)^M(s + cs^3 + O(s^5)), \\ g_M(s) &= (-1)^M(1 + (1 + \tfrac{1}{2} + \cdots + \tfrac{1}{M})s + O(s^2)), \end{aligned} \tag{20}$$

for $s \to 0$. Then, with α_N defined in (12),

$$\begin{aligned} &\zeta(A,P,s) \\ &= (s + cs^3 + O(s^5))(1 + \alpha_{M+1}s + O(s^2))\Big(\frac{a}{s^2} + \frac{b}{s} + O(1)\Big) \\ &= \frac{a}{s} + (b + \alpha_{M+1}a) + O(s), \end{aligned} \tag{21}$$

for $s \to 0$. For $f(\lambda)$ in (16) we have this situation with

$$a = \tilde{c}_0^{(N)\prime} \quad \text{and} \quad b = \widetilde{C}_0^{(N)}(A,P) - \operatorname{Tr}(A\Pi_0(P)),$$

so (21) holds with these values. In view of (9), (10), this shows (12). □

If one writes $f(\lambda)$ in the lemma as a sum $f_1(\lambda)+f_2(\lambda)$, where f_1 has the sum over j in (3), respectively f_2 has the sum over k in (3), as asymptotic expansions, the lemma can be applied to f_1 and f_2 separately, relating their coefficients to those of the poles of the corresponding functions of s.

Remark 2.1. Formula (12) gives a correction to our earlier papers [G1–5] and Grubb and Schrohe [GSc1–2], where it was taken for granted that $C_0(A,P)$ would equal $\tilde{c}_{\sigma+n}^{(N)} + \tilde{c}_0^{(N)\prime\prime}$ for any N. Fortunately, the correction has no consequence for the results in those papers, which were either concerned with the value of $C_0(A,P)$ when $c_0' = 0$, or its value *modulo local terms* (c_0' is local), or values of combined expressions where c_0'-contributions cancel out. More on corrections in the appendix.

In [GS2], Cor. 2.10 was not used in the argumentation, which was based directly on Prop. 2.9 and the primary knowledge of zeta expansions.

We shall now analyze $C_0(A,P)$ further, showing (1) by resolvent considerations. Our proof is based on an explicit calculation of one simple special case, together with the use of the trace defect formula

$$C_0(A,P) - C_0(A,P') = -\tfrac{1}{m}\operatorname{res}(A(\log P - \log P')). \tag{22}$$

This formula is well-known from considerations of complex powers of P ([O], [KV], Melrose and Nistor [MN]), but can also be derived directly from resolvent considerations [G6].

3. The trace defect formula for general orders

In [G6], the arguments for (22) are given in detail in cases where $m > \sigma + n$, whereas more general cases are briefly explained by reference to Remark 3.12 there. For completeness, we give the explanation in detail here. This is a minor technical point that may be skipped in a first reading. Denote as in [G6]

$$S_\lambda = A((P-\lambda)^{-1} - (P'-\lambda)^{-1}), \tag{23}$$

where P and P' are of order $m > 0$; then (cf. (13))

$$A((P-\lambda)^{-N} - (P'-\lambda)^{-N}) = \tfrac{\partial_\lambda^{N-1}}{(N-1)!} S_\lambda \equiv S_\lambda^{(N)}. \tag{24}$$

S_λ and $S_\lambda^{(N)}$ have symbols $s(x,\xi,\lambda)$ respectively

$$s^{(N)}(x,\xi,\lambda) = \tfrac{\partial_\lambda^{N-1}}{(N-1)!} s(x,\xi,\lambda)$$

in local trivializations.

The difference of the two expansions (3) with P resp. P' inserted satisfies

$$\begin{aligned} \operatorname{Tr}(S_\lambda^{(N)}) &\sim \sum_{j\in\mathbb{N}} \tilde{s}_j^{(N)} (-\lambda)^{\frac{\sigma+n-j}{m}-N} \\ &+ \tilde{s}_0^{(N)\prime\prime}(-\lambda)^{-N} + \sum_{k\ge 1} \big(\tilde{s}_k^{(N)\prime}\log(-\lambda) + \tilde{s}_k^{(N)\prime\prime}\big)(-\lambda)^{-k-N}; \end{aligned} \tag{25}$$

in view of Lemma 2.1, the contributions from $\operatorname{res} A$ cancel out and the coefficient of $(-\lambda)^{-N}$ equals $C_0(A,P) - C_0(A,P')$.

The symbol $s(x,\xi,\lambda)$ is analyzed in [G6], Prop. 2.1. For $s^{(N)}$, we conclude that the homogeneous terms have at least $N+1$ factors of the form $(p_m-\lambda)^{-1}$ or $(p'_m-\lambda)^{-1}$, hence the strictly homogeneous version of the symbol of order $\sigma - Nm - j$ satisfies

$$|s^{(N)h}_{\sigma-Nm-j}(x,\xi,\lambda)| \le c(|\xi|^m + |\lambda|)^{-N-1}|\xi|^{\sigma+m-j}, \tag{26}$$

being integrable at $\xi = 0$ for $j < n + \sigma + m$, $\lambda \neq 0$. Then the kernel and trace of $S_\lambda^{(N)}$ have expansions

$$\begin{aligned} K(S_\lambda^{(N)}, x, x) &= \sum_{j<\sigma+m+n} \tilde{s}_j^{(N)}(x)(-\lambda)^{\frac{n+\sigma-j}{m}-N} + O(|\lambda|^{-N-1+\varepsilon}), \\ \operatorname{Tr} S_\lambda^{(N)} &= \sum_{j<\sigma+m+n} \tilde{s}_j^{(N)}(-\lambda)^{\frac{n+\sigma-j}{m}-N} + O(|\lambda|^{-N-1+\varepsilon}), \end{aligned} \tag{27}$$

where the terms for $j < \sigma + m + n$ are calculated from the strictly homogeneous symbols (for $\lambda \in \mathbb{R}_-$):

$$\int_{\mathbb{R}^n} s^{(N)h}_{\sigma-Nm-j}(x,\xi,\lambda)\, d\!\!{}^-\xi = (-\lambda)^{\frac{n+\sigma-j}{m}-N} \int_{\mathbb{R}^n} s^{(N)h}_{\sigma-Nm-j}(x,\eta,-1)\, d\!\!{}^-\eta, \tag{28}$$

so that

$$\tilde{s}_j^{(N)}(x) = \int_{\mathbb{R}^n} s^{(N)h}_{\sigma-Nm-j}(x,\xi,-1)\, d\!\!{}^-\xi, \quad \tilde{s}_j^{(N)} = \int \operatorname{tr} \tilde{s}_j^{(N)}(x)\, dx. \tag{29}$$

When $\sigma + n \notin \mathbb{N}$, there is no term with $(-\lambda)^{-N}$ in the expansion of $\operatorname{Tr} S_\lambda^{(N)}$, and (22) holds trivially.

When $\sigma + n \in \mathbb{N}$, the coefficient of $(-\lambda)^{-N}$ in $\operatorname{Tr} S_\lambda^{(N)}$ equals

$$C_0(A,P) - C_0(A,P') = \tilde{s}^{(N)}_{\sigma+n} = \int\int_{\mathbb{R}^n} \operatorname{tr} s^{(N)h}_{-Nm-n}(x,\xi,-1)\, d\!\!{}^-\xi dx. \tag{30}$$

This demonstrates that the term is local, and gives a means to calculate it (as indicated in [G6], Rem. 3.12): Note that

$$s^{(N)h}_{-Nm-n}(x,\xi,\lambda) = \tfrac{\partial_\lambda^{N-1}}{(N-1)!} s^h_{-m-n}(x,\xi,\lambda).$$

Since s^h_{-m-n} satisfies (26) with $N = 1$, $j = \sigma + n$, it is integrable over $\mathbb{R}^n$ when $\lambda \neq 0$. Here

$$(-\lambda)^{-1} \int_{\mathbb{R}^n} s^h_{-m-n}(x,\xi,-1)\, d\!\!{}^-\xi = \int_{\mathbb{R}^n} s^h_{-m-n}(x,\xi,\lambda)\, d\!\!{}^-\xi \tag{31}$$

for $\lambda \in \mathbb{R}_-$. Moreover, the integral from (28) satisfies

$$\begin{aligned} \int_{\mathbb{R}^n} s^{(N)h}_{-Nm-n}(x,\xi,\lambda)\, d\!\!{}^-\xi &= \tfrac{\partial_\lambda^{N-1}}{(N-1)!} \int_{\mathbb{R}^n} s^h_{-m-n}(x,\xi,\lambda)\, d\!\!{}^-\xi \\ &= \tfrac{\partial_\lambda^{N-1}}{(N-1)!} \Big[(-\lambda)^{-1} \int_{\mathbb{R}^n} s^h_{-m-n}(x,\xi,-1)\, d\!\!{}^-\xi\Big] \\ &= (-\lambda)^{-N} \int_{\mathbb{R}^n} s^h_{-m-n}(x,\xi,-1)\, d\!\!{}^-\xi, \end{aligned} \tag{32}$$

which implies

$$\tilde{s}^{(N)}_{\sigma+n}(x) = \int_{\mathbb{R}^n} s^{(N)h}_{-Nm-n}(x,\xi,-1)\, d\!\!{}^-\xi = \int_{\mathbb{R}^n} s^h_{-m-n}(x,\xi,-1)\, d\!\!{}^-\xi. \tag{33}$$

The latter in turned into the residue integrand for $-\frac{1}{m}\operatorname{res}(A(\log P - \log P'))$ by Lemmas 1.2 and 1.3 of [G6], as already done in Section 2 there.

We conclude:

Theorem 3.1. *Let P and P' be classical elliptic ψdo's of order $m \in \mathbb{R}_+$ and such that the principal symbol has no eigenvalues on $\mathbb{R}_-$, let A be a classical ψdo of order σ, and let $S_\lambda = A((P-\lambda)^{-1} - (P'-\lambda)^{-1})$ and $F = A(\log P - \log P')$ with symbols s resp. f.*

Consider the case $\sigma + n \in \mathbb{N}$. Then

$$C_0(A,P) - C_0(A,P') = \int_X \operatorname{tr} \tilde{s}_{\sigma+n}(x)\, dx = -\tfrac{1}{m}\operatorname{res}(A(\log P - \log P')) \quad (34)$$

where, for each x, in local coordinates,

$$\tilde{s}_{\sigma+n}(x) = \int_{\mathbb{R}^n} s^h_{-m-n}(x,\xi,-1)\, d\!\!{}^-\xi = -\tfrac{1}{m}\int_{|\xi|=1} f_{-n}(x,\xi)\, d\!\!{}^-S(\xi). \quad (35)$$

When $\sigma + n \notin \mathbb{N}$, the identities hold trivially (with zero values).

It follows moreover:

Corollary 3.1. *If, in Theorem 3.1, P' is replaced by an operator of a different order $m' > 0$, then one has:*

$$C_0(A,P) - C_0(A,P') = -\operatorname{res}(A(\tfrac{1}{m}\log P - \tfrac{1}{m'}\log P')). \quad (36)$$

Proof. Let P_0 be an elliptic, selfadjoint positive ψdo of order m, and define $P_0^{m'/m}$ by spectral calculus; it is an elliptic, selfadjoint positive ψdo of order m'. Then by the definition of the zeta function, $C_0(A,P_0) = C_0(A,P_0^{m'/m})$. Applications of Theorem 3.1 with P, P_0 and with $P', P_0^{m'/m}$ give:

$$\begin{aligned}
C_0(A,P) - C_0(A,P') &= (C_0(A,P) - C_0(A,P_0)) \\
&\quad + (C_0(A,P_0) - C_0(A,P_0^{m'/m})) + (C_0(A,P_0^{m'/m}) - C_0(A,P')) \\
&= -\operatorname{res}(A(\tfrac{1}{m}\log P - \tfrac{1}{m}\log P_0)) - \operatorname{res}(A(\tfrac{1}{m'}\log P_0^{m'/m} - \tfrac{1}{m'}\log P')) \\
&= -\operatorname{res}(A(\tfrac{1}{m}\log P - \tfrac{1}{m'}\log P')),
\end{aligned}$$

since $\frac{1}{m'}\log P_0^{m'/m} = \frac{1}{m}\log P_0$. □

4. A formula for the zero'th coefficient

Our strategy for calculating $C_0(A,P)$ is to use (36) in combination with an exact calculation for a very special choice P_0 of P, namely

$$P_0 = ((-\Delta)^{m/2} + 1)I_M \text{ with symbol } p_0 = (|\xi|^m + 1)I_M, \quad (37)$$

in suitable local coordinates; here I_M is the $M \times M$ identity matrix (understood in the following), $M = \dim E$, and m is even.

Let A be given, of order $\sigma \in \mathbb{R}$, then we take $m > \sigma + n$. Let $\Phi_j : E|_{U_j} \to V_j \times \mathbb{C}^M$, $j = 1, \dots, J$, be an atlas of trivializations with base maps κ_j from $U_j \subset X$ to $V_j \subset \mathbb{R}^n$, let $\{\psi_j\}_{1\le j\le J}$ be an associated partition of unity (with $\psi_j \in C_0^\infty(U_j)$), and let $\varphi_j \in C_0^\infty(U_j)$ with $\varphi_j = 1$ on $\operatorname{supp}\psi_j$. Then

$$A = \sum_{1\le j\le J} \psi_j A = \sum_{1\le j\le J} \psi_j A \varphi_j + \sum_{1\le j\le J} \psi_j A(1-\varphi_j), \tag{38}$$

where the last sum is a ψdo of order $-\infty$; for this the formula (1) is well-known, since $C_0(B,P) = \operatorname{Tr} B$ when B is of order $< -n$. So it remains to treat each of the terms $\psi_j A\varphi_j$. Consider e.g. $\psi_1 A \varphi_1$. We could have assumed from the start that X was already covered by a family of open subsets $U_{j0} \subset\subset U_j$. Thus it is no restriction to assume that ψ_1 and φ_1 are supported in $U_{10} \subset\subset U_1$, where $U_{10}, U_2, \dots, U_J$ cover X.

Replace U_j by $U_j' = U_j \setminus \overline{U}_{10}$ for $j \ge 2$, and write $U_1 = U_1'$, then $\{U_j'\}_{1\le j\le J}$ also covers X. Let $\{\psi_j'\}_{1\le j\le J}$ be an associated partition of unity, and let $\varphi_j' \in C_0^\infty(U_j')$ with $\varphi_j' = 1$ on $\operatorname{supp}\psi_j'$. By construction, $\psi_1' = 1$ on U_{10}. We use the Φ_j and κ_j on these subsets (setting $\kappa_j(U_j') = V_j'$, $\kappa_j(U_{10}) = V_{10}$), and denote the induced mappings for sections by Φ_j^*.

Now the auxiliary operator P is taken to act as follows:

$$Pu = \sum_{1\le j\le J} \varphi_j' [P_0((\psi_j' u)\circ \Phi_j^{*-1})]\circ \Phi_j^*. \tag{39}$$

It is elliptic with positive definite principal symbol, and for sections supported in U_{10}, it acts like P_0 when carried over to V_{10} (being a differential operator, it is local). The resolvent $(P-\lambda)^{-1}$, defined for large λ on the rays in $\mathbb{C} \setminus \mathbb{R}_+$, is of course not local, but its symbol in the local chart $V_1 \times \mathbb{C}^M$ is, for $x \in V_{10}$, equivalent with the symbol $(|\xi|^m + 1 - \lambda)^{-1} I_M$ of $(P_0 - \lambda)^{-1}$. For resolvents of differential operators, $q(x,\xi,\lambda) \sim q_0(x,\xi,\lambda)$ means that the difference is of order $-\infty$ and $O(\lambda^{-N})$ for any N (the symbols are strongly polyhomogeneous). This difference does not affect the coefficient of $(-\lambda)^{-1}$ that we are after.

Let $a(x,\xi)$ denote the symbol of $\psi_1 A \varphi_1$ carried over to $V_1 \times \mathbb{C}^M$; it vanishes for $x \notin V_{10}$. Then the symbol of $\psi_1 A \varphi_1 (P-\lambda)^{-1}$ on V_1 is equivalent with $a(x,\xi)(|\xi|^m + 1 - \lambda)^{-1}$ (with an error that is $O(\lambda^{-N})$, any N); we use that the symbol composition here gives only one (product) term.

To find the coefficient of $(-\lambda)^{-1}$ in the expansion of $\operatorname{Tr}(\psi_1 A \varphi_1 (P-\lambda)^{-1})$, we now just have to analyze the diagonal kernel calculated in V_1:

$$\begin{aligned} K(\psi_1 A\varphi_1(P-\lambda)^{-1},x,x) &\sim \int_{\mathbb{R}^n} a(x,\xi)(|\xi|^m+1-\lambda)^{-1}\, d\!\!{}^-\!\xi \\ &\sim \sum_{j\in\mathbb{N}} \tilde{c}_j(x)(-\lambda)^{\frac{\sigma+n-j}{m}-1} + \sum_{k\in\mathbb{N}} \big(\tilde{c}'_k(x)\log(-\lambda)+\tilde{c}''_k(x)\big)(-\lambda)^{-k-1}. \end{aligned} \tag{40}$$

Here the value can be found explicitly, as follows.

Set $a_{-n}=0$ if $\sigma+n\notin\mathbb{N}$, and decompose the symbol a in three pieces $a_{>-n}$, a_{-n} and $a_{<-n}$, where

$$\begin{aligned} a_{>-n}(x,\xi) &= \sum_{0\le j<\sigma+n} a_{\sigma-j}(x,\xi), \\ a_{<-n}(x,\xi) &= a(x,\xi)-a_{-n}(x,\xi)-a_{>-n}(x,\xi). \end{aligned} \tag{41}$$

The symbol terms $a_{\sigma-j}(x,\xi)$ are assumed to be C^∞ in (x,ξ) and homogeneous of degree $\sigma-j$ in ξ for $|\xi|\ge 1$. The strictly homogeneous version $a^h_{\sigma-j}$ is homogeneous for $\xi\ne 0$ and coincides with $a_{\sigma-j}$ for $|\xi|\ge 1$. For the terms in $a_{>-n}$, the strictly homogeneous versions are integrable in ξ at $\xi=0$.

We recall that $⨍ a(x,\xi)\, d\!\!{}^-\!\xi$ is defined for each x as a finite part integral (in the sense of Hadamard), namely the constant term in the asymptotic expansion of $\int_{|\xi|\le R} a(x,\xi)\, d\!\!{}^-\!\xi$ in powers R^{-m_j} and $\log R$, for $R\to\infty$. Here

$$\begin{aligned} ⨍ a_{\sigma-j}(x,\xi)\, d\!\!{}^-\!\xi &= \int_{|\xi|\le 1} (a_{\sigma-j}(x,\xi)-a^h_{\sigma-j}(x,\xi))\, d\!\!{}^-\!\xi, \text{ for } \sigma-j>-n, \\ ⨍ a_{-n}(x,\xi)\, d\!\!{}^-\!\xi &= \int_{|\xi|\le 1} a_{-n}(x,\xi)\, d\!\!{}^-\!\xi, \\ ⨍ a_{<-n}(x,\xi)\, d\!\!{}^-\!\xi &= \int_{\mathbb{R}^n} a_{<-n}(x,\xi)\, d\!\!{}^-\!\xi; \end{aligned} \tag{42}$$

as one can check using polar coordinates (the formulas are special cases of [G4], (1.18)).

Lemma 4.1. *For $a_{\sigma-j}(x,\xi)$ with $\sigma-j>-n$,*

$$\begin{aligned} &\int_{\mathbb{R}^n} a_{\sigma-j}(x,\xi)(|\xi|^m-\lambda)^{-1}\, d\!\!{}^-\!\xi \\ &\qquad = (-\lambda)^{\frac{\sigma+n-j}{m}-1}\int_{\mathbb{R}^n} a^h_{\sigma-j}(x,\xi)(|\xi|^m+1)^{-1}\, d\!\!{}^-\!\xi \\ &\qquad\quad + (-\lambda)^{-1} ⨍ a_{\sigma-j}(x,\xi)\, d\!\!{}^-\!\xi + O(\lambda^{-2}), \end{aligned} \tag{43}$$

for $\lambda \to \infty$ *on rays in* $\mathbb{C} \setminus \mathbb{R}_+$. *Also* $\int_{\mathbb{R}^n} a_{\sigma-j}(x,\xi)(|\xi|^m + 1 - \lambda)^{-1}\, đ\xi$ *has an expansion in powers of* $(-\lambda)$ *plus* $o(\lambda^{-1})$*; here the coefficient of* $(-\lambda)^{-1}$ *is likewise* $⨍ a_{\sigma-j}(x,\xi)\, đ\xi$.

For $a_{-n}(x,\xi)$ *one has:*

$$\begin{aligned}\int_{\mathbb{R}^n} & a_{-n}(x,\xi)(|\xi|^m - \lambda)^{-1}\, đ\xi \\ &= \tfrac{1}{m}(-\lambda)^{-1}\log(-\lambda)\int_{|\xi|=1} a_{-n}(x,\xi)\, đS(\xi) \\ &\quad + (-\lambda)^{-1} ⨍ a_{-n}(x,\xi)\, đ\xi + O(\lambda^{-2}),\end{aligned} \tag{44}$$

for $\lambda \to \infty$ *on rays in* $\mathbb{C}\setminus\mathbb{R}_+$. $\int_{\mathbb{R}^n} a_{-n}(x,\xi)(|\xi|^m + 1 - \lambda)^{-1}\, đ\xi$ *has a similar expansion, the coefficient of* $(-\lambda)^{-1}$ *again being* $⨍ a_{-n}(x,\xi)\, đ\xi$.

Proof. By homogeneity, we have for $\lambda \in \mathbb{C} \setminus \overline{\mathbb{R}}_+$, writing $\lambda = -|\lambda|e^{i\theta}$, $|\theta| < \pi$,

$$\begin{aligned}\int_{\mathbb{R}^n} & a^h_{\sigma-j}(x,\xi)(|\xi|^m + |\lambda|e^{i\theta})^{-1}\, đ\xi \\ &= |\lambda|^{\frac{\sigma-j+n}{m}-1}\int_{\mathbb{R}^n} a^h_{\sigma-j}(x,\eta)(|\eta|^m + e^{i\theta})^{-1}\, đ\eta.\end{aligned} \tag{45}$$

This equals the first term in the right hand side of (43) if $\theta = 0$, and the identity extends analytically to general λ. Moreover, since

$$\begin{aligned}(|\xi|^m - \lambda)^{-1} &= (-\lambda)^{-1}(1 - |\xi|^m/\lambda) \\ &= (-\lambda)^{-1}\sum_{k\in\mathbb{N}}(|\xi|^m/\lambda)^k \text{ for } |\lambda| \ge 2,\ |\xi| \le 1,\end{aligned} \tag{46}$$

we find that for $|\lambda| \ge 2$,

$$\begin{aligned}\int_{\mathbb{R}^n} & (a_{\sigma-j}(x,\xi) - a^h_{\sigma-j}(x,\xi))(|\xi|^m - \lambda)^{-1}\, đ\xi \\ &= \int_{|\xi|\le 1}(a_{\sigma-j}(x,\xi) - a^h_{\sigma-j}(x,\xi))(|\xi|^m - \lambda)^{-1}\, đ\xi \\ &= (-\lambda)^{-1}\int_{|\xi|\le 1}(a_{\sigma-j}(x,\xi) - a^h_{\sigma-j}(x,\xi))\, đ\xi + O(\lambda^{-2}).\end{aligned} \tag{47}$$

This shows (43), in view of (42).

For the next observation, we use that the preceding results give an expansion in powers of $1-\lambda$; then since

$$
\begin{aligned}
(1-\lambda)^k &= \sum_{0\le l\le k} b_l \lambda^l \text{ when } k\in\mathbb{N},\\
(1-\lambda)^s &= (-\lambda)^s + \sum_{l\ge 1} b_l \lambda^{s-l} \text{ when } |\lambda|\ge 2,\ s\notin\mathbb{N},
\end{aligned} \tag{48}
$$

only $⨍ a_{\sigma-j}$ contributes to the coefficient of $(-\lambda)^{-1}$.

Now consider $a_{-n}(x,\xi)$; again we can let $\theta=0$. Here we write

$$
\begin{aligned}
\int_{\mathbb{R}^n} & a_{-n}(x,\xi)(|\xi|^m-\lambda)^{-1}\, đ\xi\\
&= \int_{|\xi|\ge 1} a^h_{-n}(x,\xi)(|\xi|^m+|\lambda|)^{-1}\, đ\xi\\
&\quad + \int_{|\xi|\le 1} a_{-n}(x,\xi)(|\xi|^m-\lambda)^{-1}\, đ\xi.
\end{aligned} \tag{49}
$$

The first term gives

$$
\begin{aligned}
\int_{|\xi|\ge 1} & a^h_{-n}(x,\xi)(|\xi|^m+|\lambda|)^{-1}\, đ\xi\\
&= |\lambda|^{-1}\int_{|\eta|\ge|\lambda|^{-1/m}} a^h_{-n}(x,\eta)(|\eta|^m+1)^{-1}\, đ\eta\\
&= |\lambda|^{-1}\int_{r\ge|\lambda|^{-1/m}} r^{-1}(r^m+1)^{-1}\, dr \int_{|\xi|=1} a_{-n}(x,\xi)\, đS(\xi)\\
&= \tfrac{1}{m}(-\lambda)^{-1}\log(-\lambda)\int_{|\xi|=1} a_{-n}(x,\xi)\, đS(\xi) + O(\lambda^{-2}),
\end{aligned} \tag{50}
$$

since, with $s=r^m$,

$$
\int r^{-1}(r^m+1)^{-1}\, dr = \tfrac{1}{m}\int s^{-1}(s+1)^{-1}\, ds = \tfrac{1}{m}(\log s - \log(s+1)),
$$

where $\log(s+1)=O(s)$. The second term gives, as in (47),

$$
\int_{|\xi|\le 1} a_{-n}(x,\xi)(|\xi|^m-\lambda)^{-1}\, đ\xi = (-\lambda)^{-1}\int_{|\xi|\le 1} a_{-n}(x,\xi)\, đ\xi + O(\lambda^{-2}). \tag{51}
$$

This implies (44), in view of (42). The last statement follows using (48). □

For $a_{<-n}$, it is very well known that

$$
\begin{aligned}
\int_{\mathbb{R}^n} a_{<-n}(x,\xi)(|\xi|^m-\lambda)^{-1}\, đ\xi &= (-\lambda)^{-1}\int_{\mathbb{R}^n} a_{<-n}(x,\xi)\, đ\xi + o(\lambda^{-1}),\\
&= (-\lambda)^{-1} ⨍ a_{<-n}(x,\xi)\, đ\xi + o(\lambda^{-1});
\end{aligned} \tag{52}
$$

also here,

$$\int_{\mathbb{R}^n} a_{<-n}(x,\xi)(|\xi|^m+1-\lambda)^{-1}\,đ\xi = (-\lambda)^{-1} ⨍ a_{<-n}(x,\xi)\,đ\xi + o(\lambda^{-1}) \quad (53)$$

follows by use of (48).

Collecting the informations, we have:

Proposition 4.1. *The coefficient of $(-\lambda)^{-1}$ in the expansion (40) for $K(\psi_1 A\varphi_1(P-\lambda)^{-1},x,x)$ equals*

$$\tilde{c}_n(x) + \tilde{c}_0''(x) = ⨍ a(x,\xi)\,đ\xi. \quad (54)$$

In the same localization, when we calculate $\psi_1 A\varphi_1 \log P$ by a Cauchy integral (as in [G6]), the localized piece will give $\psi_1 A\varphi_1 \operatorname{OP}(\log(|\xi|^m+1))$. The symbol of this operator is

$$r(x,\xi) = a(x,\xi)\log(|\xi|^m+1), \quad (55)$$

which has an expansion

$$r(x,\xi) = a(x,\xi)(m\log|\xi| - |\xi|^{-m} - \sum_{j\ge 2} d_j|\xi|^{-jm}), \quad (56)$$

convergent for $|\xi|\ge 2$. Inserting the expansion of a in homogeneous terms, we find since $m>\sigma+n$ that the full term of order $-n$ in $r(x,\xi)$ is $a_{-n}m\log|\xi|$, with no log-free part. So

$$\operatorname{res}_{x,0} r = \int_{|\xi|=1} \operatorname{tr} r_{-n,0}(x,\xi)đS(\xi) = 0. \quad (57)$$

It follows that the coefficient of $(-\lambda)^{-1}$ in the trace expansion of $\psi_1 A\varphi_1(P-\lambda)^{-1}$ is

$$\int_{\mathbb{R}^n} ⨍ \operatorname{tr} a(x,\xi)\,đ\xi dx = \int_{\mathbb{R}^n} (\operatorname{TR}_x(\psi_1 A\varphi_1) - \tfrac{1}{m}\operatorname{res}_{x,0}(\psi_1 A\varphi_1 \log P))\,dx, \quad (58)$$

using that $\operatorname{res}_{x,0}(\psi_1 A\varphi_1\log P)$ is 0. This shows formula (1) in this very particular case:

$$C_0(\psi_1 A\varphi_1, P) = \int_{\mathbb{R}^n} (\operatorname{TR}_x(\psi_1 A\varphi_1) - \tfrac{1}{m}\operatorname{res}_{x,0}(\psi_1 A\varphi_1 \log P))\,dx. \quad (59)$$

Now, to find $C_0(\psi_1 A\varphi_1, P')$ for a general operator P' of order $m'\in\mathbb{R}_+$, we combine (59) with the trace defect formula (36). This gives, in the considered local coordinates:

$$\begin{aligned} C_0(\psi_1 A\varphi_1, P') &= C_0(\psi_1 A\varphi_1, P) + C_0(\psi_1 A\varphi_1, P') - C_0(\psi_1 A\varphi_1, P) \\ &= \int_{\mathbb{R}^n} (\mathrm{TR}_x(\psi_1 A\varphi_1) - \tfrac{1}{m}\operatorname{res}_{x,0}(\psi_1 A\varphi_1 \log P))\, dx \\ &\quad - \operatorname{res}(\psi_1 A\varphi_1(\tfrac{1}{m'}\log P' - \tfrac{1}{m}\log P)) \\ &= \int_{\mathbb{R}^n} (\mathrm{TR}_x(\psi_1 A\varphi_1) - \tfrac{1}{m'}\operatorname{res}_{x,0}(\psi_1 A\varphi_1 \log P'))\, dx. \end{aligned} \tag{60}$$

To this we can add:

$$\begin{aligned} &C_0(\psi_1 A(1-\varphi_1), P') = \mathrm{Tr}(\psi_1 A(1-\varphi_1)) \\ &= \int (\mathrm{TR}_x(\psi_1 A(1-\varphi_1)) - \tfrac{1}{m'}\operatorname{res}_{x,0}(\psi_1 A(1-\varphi_1)\log P'))\, dx, \end{aligned} \tag{61}$$

where both terms have a meaning on X; TR_x defines the ordinary trace integral and $\operatorname{res}_{x,0}$ is zero.

The method applies likewise to all the other terms $\psi_j A\varphi_j$. Collecting the terms, and relabelling P' of order m' as P of order m, we have found:

Theorem 4.1. *Let A be a classical ψdo of order $\sigma \in \mathbb{R}$, and let P be a classical elliptic ψdo's of order $m \in \mathbb{R}_+$ such that the principal symbol has no eigenvalues on $\mathbb{R}_-$. We have in local coordinates as used above:*

$$\begin{aligned} C_0(A, P) &= \sum_{1\le j\le J} C_0(\psi_j A, P), \quad \text{where} \\ C_0(\psi_j A, P) &= \int (\mathrm{TR}_x(\psi_j A) - \tfrac{1}{m}\operatorname{res}_{x,0}(\psi_j A \log P))\, dx \end{aligned} \tag{62}$$

(the contribution from $\psi_j A\varphi_j$ defined in the corresponding local chart and that from $\psi_j A(1-\varphi_j)$ defined as an ordinary trace).

Note that $C_0(A, P)$ is independent of how we localize, so the expression resulting from (62) is independent of the choice of localization.

The logarithm is here defined by cutting the complex plane along $\mathbb{R}_-$. If P is given with another ray free of eigenvalues, the formulas hold with the logarithm defined to be cut along this ray. (We do not here study the issue of how these expressions depend on the ray.)

The invariance of the density $\big(\mathrm{TR}_x(A) - \frac{1}{m}\operatorname{res}_{x,0}(A\log P)\big)\, dx$ in the formula (1) is verified in [PS] also by a direct calculation. We note that (1) gives back the known formula

$$C_0(A, P) = \mathrm{TR}(A) \tag{63}$$

in cases where $\operatorname{res}_{x,0}(A\log P)$ vanishes. This is so when $\sigma + n \notin \mathbb{N}$ ([KV], [L]), and also in cases $\sigma + n \in \mathbb{N}$ with parity properties ([KV], [G4]): When $\sigma \in \mathbb{Z}$, we say that A has even-even alternating parity (in short: is even-even), resp. has even-odd alternating parity (in short: is even-odd), when

$$\begin{aligned} a_{\sigma-j}(x,-\xi) &= (-1)^{\sigma-j} a_{\sigma-j}(x,\xi), \text{ resp.} \\ a_{\sigma-j}(x,-\xi) &= (-1)^{\sigma-j-1} a_{\sigma-j}(x,\xi), \end{aligned} \tag{64}$$

for $|\xi| \geq 1$, all j. When P is even-even of even order m, then the classical part of $\log P$ is even-even. Then if (a) or (b) is satisfied:

(a) A is even-even and n is odd,

(b) A is even-odd and n is even,

$\operatorname{res}_{x,0}(A\log P)$ vanishes, $\operatorname{TR}_x A\,dx$ is a globally defined density, and (63) holds. [KV] treats the case (a), calling the even-even operators odd-class (perhaps because they have a canonical trace in odd dimension). The statements on $\operatorname{TR}_x A\,dx$ are extended to log-polyhomogeneous operators in [G4].

Observe a general consequence:

Corollary 4.1. *When A has order $\sigma \in \mathbb{Z}$ and satisfies (a) or (b), then $\operatorname{res}_{x,0}(A\log P)\,dx$ defines a global density for any P.*

Proof. In these cases, since $\operatorname{TR}_x A\,dx$ defines a global density, the other summand in $\big(\operatorname{TR}_x(A) - \frac{1}{m}\operatorname{res}_{x,0}(A\log P)\big)\,dx$ must do so too. (Note that P is not subject to order or parity restrictions here.) □

Appendix A. Corrections to earlier papers

Correction to [GS2]: In Corollary 2.10 on page 45, the formulas in (2.38) for the expansion coefficients $\tilde{a}_{j,l}$ are true only for $l = m_j$. For $l < m_j$, the $\tilde{a}_{j,l}$ depend on the full set $\{a_{j,l} \mid 0 \leq j \leq m_j\}$. This is so, because the Taylor expansion of $\Gamma(1-s)^{-1}$ must be taken into account.

Hence in the comparison of the expansion of $\operatorname{Tr}(A(P-\lambda)^{-1})$ with $\zeta(A,P,s)$, only the primary coefficient at each pole of $\Gamma(s)\zeta(A,P,s)$ is directly proportional to a coefficient in $\operatorname{Tr}(A(P-\lambda)^{-1})$. Similar statements hold for comparisons with $\operatorname{Tr}(A(P-\lambda)^{-N})$. This has lead to systematic inaccuracies in a number of subsequent works, however without substantial damage to the results in general.

We explain the needed correction in detail for [G4] and then list the related modifications needed in other papers (including a few additional corrections).

Corrections to [G4]:

1) The statements on page 69 linking the coefficients in (1.1) with the coefficients in (1.2) with the same index by universal proportionality factors is incorrect if $\nu+n \in \mathbb{N}$; the direct proportionality holds only for the primary pole coefficients, not for the next Laurent coefficient at each pole. Instead, at the second-order poles $-k, k \in \mathbb{N}$, there are universal linear transition formulas linking the coefficient set for $(-\lambda)^{-k-N}\log(-\lambda)$ and $(-\lambda)^{-k-N}$ with the coefficient set for $(s+k)^{-2}$ and $(s+k)^{-1}$.

This follows from [GS2, Prop. 2.9], (3.21), as explained in Lemma 2.1 of the present paper. The coefficients of $\operatorname{Tr}(A(P-\lambda)^{-N})$ at integer powers are directly proportional to the Laurent coefficients of the meromorphic function $\psi(s)$, where (with $N-1$ denoted M)

$$\begin{aligned} \zeta(A,P,s) &= \tfrac{M!}{(s-M)\dots(s-1)}\tfrac{1}{\pi}\sin(\pi(s-M))\psi(s-M), \\ \Gamma(s)\zeta(A,P,s) &= \tfrac{M!}{\Gamma(M-s)}\psi(s-M). \end{aligned} \tag{A.1}$$

(Cf. (18), use that $\frac{1}{\pi}\sin(\pi(s-M)) = (-1)^M/[\Gamma(s-1)\Gamma(1-s)]$.) In calculations of Laurent series at the poles, the Taylor expansion of the factor in front of $\psi(s)$ effects the higher terms.

Specifically in [G4], the sentence "The coefficients $\tilde{c}_j$ and c_j, $\tilde{c}'_k$ and c'_k, resp. $\tilde{c}''_k$ and c''_k are proportional by universal nonzero constants." should be replaced by: "The coefficients $\tilde{c}_j$ and c_j, resp. $\tilde{c}'_k$ and c'_k, are proportional by universal nonzero constants. When the c'_k vanish (e.g., when $\nu+n \notin \mathbb{N}$), the same holds for $\tilde{c}''_k$ and c''_k. More generally, the pair $\{\tilde{c}'_k, \tilde{c}''_k\}$ is for each k universally related to the pair $\{c'_k, c''_k\}$ in a linear way." The statement "$\tilde{c}''_0 = c''_0$" should be replaced by "$\tilde{c}''_0 = c''_0$ when $c'_0 = 0$", and the description of $C_0(A,P)$ in terms of resolvent trace expansion coefficients should be replaced by the description given in the present paper in Section 2.

However, since this changes the formula for $C_0(A,P)$ only by a multiple of $\operatorname{res} A$, the results of [G4] on $C_0(A,P)$ remain valid, because they are concerned with cases where $\operatorname{res} A = 0$. The statements Th. 1.3 (ii), Cor. 1.5 (ii) on the vanishing of all log-coefficients in parity cases still imply the vanishing of all double poles in (1.2).

In Section 3, the coefficients in (3.32) are linked with those in (3.30) in a more complicated way than stated, where only the leading coefficient at a pole is directly proportional to a coefficient in (3.30). But again, the results for parity cases remain valid since the needed correction terms vanish in these cases.

2) Page 79, remove the factor 2 (twice) in formula (1.44).

3) Page 84, formulas (3.9) and (3.10): The sums over k' should be removed, and so should the additional term $-P^{-s-1}$ in the first line. So $\mathcal{P}_l(P) = (-\log P)^l$ for all l.

4) Page 91, in formula (3.47), $-P^{-1}$ should be removed.

Correction to [G1]: Page 92, lines 7–8 from below, replace "The coefficients in (9.10) are proportional to those in (9.9) by universal factors." by "The unprimed coefficients in (9.10) are proportional to those in (9.9) by universal factors. For each k, the pair $\{\tilde{a}_{i,k}, \tilde{a}'_{i,k}\}$ (resp. $\{\tilde{b}_{i,k}, \tilde{b}'_{i,k}\}$) is universally related to the pair $\{a_{i,k}, a'_{i,k}\}$ (resp. $\{b_{i,k}, b'_{i,k}\}$) in a linear way."

Corrections to [G2]: Page 4, lines 5–7 from below, replace "The coefficients $\tilde{a}_k$, $\tilde{a}'_k$ and $\tilde{a}''_k$ are proportional to the coefficients a_k, a'_k and a''_k in (0.1) (respectively) by universal nonzero proportionality factor (depending on r)." should be replaced by: "The coefficients $\tilde{a}_k$ and $\tilde{a}'_k$ are proportional to the coefficients a_k and a'_k in (0.1) (respectively) by universal nonzero proportionality factor (depending on r). For each $k \geq 0$, the pair $\{\tilde{a}'_k, \tilde{a}''_k\}$ is universally related to the pair $\{a'_k, a''_k\}$ in a linear way."

Corrections to [G3]: Page 262, lines 13–14 should be changed as for [G2] above. In line 15, "$a''_0(F) = \tilde{a}''_0(F)$" should be replaced by: "$a''_0(F) = \tilde{a}''_0(F)$ if $\tilde{a}'_0(F) = 0$". There are some consequential reformulations in Sections 4 and 5, which do not endanger the results since the a''_0 terms are characterized in general modulo local contributions (and a'_0 is such), with precise statements only when $a'_0 = 0$.

Corrections to [G5]: The statement on Page 44, lines 11–12 from below "There are some universal proportionality factors linking the coefficients $\tilde{c}_j$ and c_j, $\tilde{c}'_k$ and c'_k, resp. $\tilde{c}''_k$ and c''_k" should be changed as indicated for [G4].

Correction to [GSc1]: Page 171, line 1, "The coefficients $\tilde{c}_j$, $\tilde{c}'_l$, $\tilde{c}''_l$ are proportional to the coefficients c_j, c'_l , c''_l by universal constants." should be replaced by "The coefficients $\tilde{c}_j$, $\tilde{c}'_l$ are proportional to the coefficients c_j, c'_l by universal constants."

Corrections to [GSc2]: The definition of $C_0(A, P)$ on page 1644 and the statements in lines 6–8 on page 1645 should be modified as for [G4]. This has no consequences for the results, which are mainly concerned with the trace definitions *modulo local contributions*, with exact formulas established only when the residue corrections vanish.

Corrections to [G6]: When the order m of the auxiliary operator P_1 is odd, the classical part of the symbol of $\log P_1$ does not satisfy the transmission condition, so the formulas referring to the residue definition of [FGLS] are only valid when m is even. This goes for the right-hand side in formula (3.41) of Theorem 3.10, which can however be interpreted in a more general sense when m is odd, since the local estimates in the proof remain valid. Similar remarks hold for formula (4.32) in Theorem 4.5 and (5.9) in Theorem 5.2.

References

FGLS. B. V. Fedosov, F. Golse, E. Leichtnam and E. Schrohe, *The noncommutative residue for manifolds with boundary*, J. Funct. Anal. **142**, 1–31 (1996).

G1. G. Grubb, *Trace expansions for pseudodifferential boundary problems for Dirac-type operators and more general systems*, Arkiv f. Mat. **37**, 45–86, (1999).

G2. G. Grubb, *Logarithmic terms in trace expansions of Atiyah-Patodi-Singer problems*, Ann. Global Anal. Geom. **24**, 1–51 (2003).

G3. G. Grubb, *Spectral boundary conditions for generalizations of Laplace and Dirac operators*, Comm. Math. Phys. **240**, 243–280 (2003).

G4. G. Grubb, *A resolvent approach to traces and zeta Laurent expansions*, Contemp. Math. **366**, 67–93 (2005). Corrected in arXiv: math.AP/0311081.

G5. G. Grubb, *Analysis of invariants associated with spectral boundary problems for elliptic operators*, Contemp. Math. **366**, 43–64 (2005).

G6. G. Grubb, *On the logarithm component in trace defect formulas*, Comm. Partial Differential Equations **30**, 1671–1716 (2005).

GH. G. Grubb and L. Hansen *Complex powers of resolvents of pseudodifferential operators*, Comm. Partial Differential Equations **27**, 2333–2361 (2002).

GSc1. G. Grubb and E. Schrohe, *Trace expansions and the noncommutative residue for manifolds with boundary*, J. Reine Angew. Math. **536**, 167–207 (2001).

GSc2. G. Grubb and E. Schrohe, *Traces and quasi-traces on the Boutet de Monvel algebra*, Ann. Inst. Fourier **54**, 1641–1696 (2004).

GS1. G. Grubb and R. Seeley, *Weakly parametric pseudodifferential operators and Atiyah-Patodi-Singer boundary problems*, Invent. Math. **121**, 481–529 (1995).

GS2. G. Grubb and R. Seeley, *Zeta and eta functions for Atiyah-Patodi-Singer operators*, J. Geom. Anal. **6**, 31–77 (1996).

Gu. V. Guillemin, *A new proof of Weyl's formula on the asymptotic distribution of eigenvalues*, Adv. Math. **102**, 184–201 (1985).

KV. M. Kontsevich and S. Vishik, *Geometry of determinants of elliptic operators*, Functional Analysis on the Eve of the 21'st Century, Vol. I (New Brunswick, N.J. 1993), Progr. Math. 131, Birkhäuser, Boston, 173–197 (1995).

L. M. Lesch, *On the noncommutative residue for pseudodifferential operators with log-polyhomogeneous symbols*, Ann. Global Anal. Geom. **17**, 151–187 (1999).

Lo. P. Loya, *The structure of the resolvent of elliptic pseudodifferential operators*, J. Funct. Anal. **184**, 77–134 (2001).

MN. R. Melrose and V. Nistor, *Homology of pseudodifferential operators I. Manifolds with boundary*, arXiv: funct-an/9606005.
O. K. Okikiolu, *The multiplicative anomaly for determinants of elliptic operators*, Duke Math. J. **79**, 723–750 (1995).
PS. S. Paycha and S. Scott, *A Laurent expansion for regularized integrals of holomorphic symbols*, to appear in Geom. Funct. Anal., arXiv: math.AP/0506211.
W. M. Wodzicki, *Local invariants of spectral asymmetry*, Invent. Math. **75**, 143–178 (1984).

Received by the editors August 13, 2005; Revised January 10, 2006

Analysis, Geometry and Topology
of Elliptic Operators, pp. 235–262

AN ANOMALY FORMULA FOR L^2-ANALYTIC TORSIONS ON MANIFOLDS WITH BOUNDARY

XIAONAN MA

Centre de Mathématiques Laurent Schwartz
UMR 7640 du CNRS
Ecole Polytechnique
91128 Palaiseau Cedex, France
ma@math.polytechnique.fr

WEIPING ZHANG

Chern Institute of Mathematics & LPMC
Nankai University
Tianjin 300071, P.R. China
weiping@nankai.edu.cn

Dedicated to Krzysztof P. Wojciechowski on his 50th birthday

We extend the definition, in the extended cohomology framework, of the L^2-analytic torsion for covering spaces due to Braverman-Carey-Farber-Mathai to the case of manifolds with boundary, and prove an associated anomaly formula. Our main result may be viewed as a common generalization of the anomaly formula for Ray-Singer analytic torsion for manifolds with boundary proved by Brüning-Ma, as well as the anomaly formula for L^2-analytic torsions for closed manifolds proved by Zhang. It generalizes also an earlier result of Lück-Schick, without the assumptions on the unitary representations as well as the technical "determinant class condition".

2000 *Mathematics Subject Classification.* Primary 58J52; Secondary 19K56

1. Introduction

Let F be a unitary flat vector bundle on a closed Riemannian manifold X. Ray and Singer [27] defined an analytic torsion associated to (X, F) and proved that it does not depend on the Riemannian metric on X. Moreover, they conjectured that this analytic torsion coincides with the classical Reidemeister torsion defined using a triangulation on X (cf. Milnor [21]). This conjecture was later proved in the celebrated papers of Cheeger [10] and Müller [22]. Müller generalized this result in [23] to the case where F

is a unimodular flat vector bundle on X. Inspired by the considerations of Quillen [25], Bismut and Zhang [2] reformulated the above Cheeger-Müller theorem as an equality between the Reidemeister and Ray-Singer metrics defined on the determinant of cohomology, and proved an extension of it to the case of general flat vector bundles over X. The method used in [2] is different from those of Cheeger and Müller in that it makes use of a deformation by Morse functions introduced by Witten [30] on the de Rham complex. In particular, as an intermediate step, an important anomaly formula for Ray-Singer metrics has been established in [2], Theorem 0.1.

Recall that Ray and Singer [27] also defined the analytic torsion, in the unitary flat vector bundle case, for manifolds with boundary. Moreover, Cheeger [10] raised the question of computing the corresponding metric anomaly. This question was studied by Dai and Fang [11] for the case of unitary flat vector bundle, while a complete answer, valid for the general case of arbitrary flat vector bundles, is recently obtained by Brüning and Ma [4].

The purpose of this paper is to generalize the main results in [4] to the case of L^2-analytic torsions on infinite Galois covering spaces of manifolds with boundary. We recall that the L^2-torsions were first introduced, for closed manifolds, by Carey, Mathai and Lott in [9], [15] and [20], under the assumptions that the L^2-Betti numbers vanish and that certain technical "determinant class condition" (the more precise definition of "determinant class condition" indeed appears later in [7]) is satisfied. The later condition is satisfied if the Novikov-Shubin [24] invariants are positive. In [6] and [19], extensions to manifolds with boundary, in the case of unitary flat bundle case, have been studied. In [6], only the case of product metric near boundary has been considered, while in [19], Lück and Schick also considered the case of non-product metric near boundary.

Carey, Farber and Mathai [8] showed that the condition on the vanishing of the L^2-Betti numbers can be relaxed. This is achieved by constructing the determinant line of the reduced L^2-cohomology and defining the L^2-torsions as elements of the determinant line.

Recently, Braverman, Carey, Farber and Mathai [3] showed that if one considers the *extended* L^2-cohomology in the sense of Farber (cf. [13]) instead of the usually used *reduced* L^2-cohomology, then one can naturally define the L^2-analytic torsion as an L^2-element on the associated determinant lines, without requiring the "determinant class condition".

In this paper, we first generalize the construction in [3] to the case of manifolds with boundary, to define L^2-analytic torsions, in the case of manifolds with boundary, for arbitrary flat vector bundles and arbitrary Riemannian metric on the base manifold, without using the "determinant class condition". We then prove an anomaly formula of these L^2-analytic torsions. The main result can be thought of as a common generalization of the anomaly formula for Ray-Singer analytic torsion for manifolds with boundary proved by Brüning-Ma [4, 5], as well as the anomaly formula for L^2-analytic torsions for closed manifolds proved by Zhang [32]. It generalizes also [19], Theorem 7.6, without the assumptions on the flatness of the metrics on F, and on the technical "determinant class condition". In particular, it provides a positive answer to a question mentioned in [18], Page 190.

This paper is organized as follows. In Section 2, we recall from [3] the definition of the determinant line of extended cohomology of a finite length Hilbert cochain $\mathcal{A}$-complex with $\mathcal{A}$ a finite von Neumann algebra, as well as the definition of the L^2-torsion element lying in this determinant line. In Section 3, we construct the L^2-analytic torsion element, in the case of manifolds with boundary, by extending the construction in [3], and establish an anomaly formula for it.

2. L^2-torsion on the determinant of extended cohomology

In this section, we recall from [3] the definition of the L^2-torsion element which lies in the determinant of the extended cohomology associated to a finite length Hilbert cochain complex.

This section is organized as follows. In Section 2.1, we recall the definition of the extended cohomology of a finite length Hilbert cochain complex over a finite von Neumann algebra carrying a finite, normal and faithful trace. In Section 2.2, we recall the definition of the determinant of a finitely generated Hilbert module over a finite von Neumann algebra. In Section 2.3, we recall the definition of the L^2-torsion element of a finite length Hilbert cochain complex.

2.1. *Extended cohomology of a finite length Hilbert cochain complex*

Let $\mathcal{A}$ be a finite von Neumann algebra carrying a fixed finite, normal and faithful trace

$$\tau : \mathcal{A} \to \mathbf{C},$$

cf. [12], §I.6. Let * denote the canonical involution on $\mathcal{A}$ defined by taking adjoint. Let $l^2(\mathcal{A})$ denote the Hilbert space completion of $\mathcal{A}$ with respect to the inner product given by the trace

$$\langle a, b\rangle = \tau(b^* a). \tag{2.1}$$

A finitely generated Hilbert module over $\mathcal{A}$ is a Hilbert space M admitting a continuous left $\mathcal{A}$-structure (with respect to the norm topology on $\mathcal{A}$) such that there exists an isometric $\mathcal{A}$-linear embedding of M into $l^2(\mathcal{A}) \otimes H$, for some finite dimensional Hilbert space H.

Let (C^*, ∂) be a finite length Hilbert cochain complex over $\mathcal{A}$,

$$(C^*, \partial) : 0 \to C^0 \overset{\partial_0}{\to} C^1 \overset{\partial_1}{\to} \cdots \overset{\partial_{n-1}}{\to} C^n \to 0, \tag{2.2}$$

where each C^i, $0 \le i \le n$, is a finitely generated Hilbert module over $\mathcal{A}$ and the coboundary maps are bounded $\mathcal{A}$-linear operators. Since the image spaces of these coboundary maps need not be closed, the tautological cohomology of (C^*, ∂) need not be a Hilbert space. This is why in general one studies the *reduced* cohomology of (C^*, ∂), which is defined by

$$H^*(C^*, \partial) = \bigoplus_{i=0}^{n} H^i(C^*, \partial), \tag{2.3}$$

with

$$H^i(C^*, \partial) = \ker(\partial_i)/\overline{\mathrm{im}(\partial_{i-1})}, \quad 0 \le i \le n, \tag{2.4}$$

where one takes obviously that $\partial_{-1} = 0$ and $\partial_n = 0$.

On the other hand, there are still ways to extract more information from (C^*, ∂), rather than just from $H^*(C^*, \partial)$. One such is to consider the *extended* cohomology in the sense of Farber (cf. [13] and [3]), which is defined by

$$\mathcal{H}^*(C^*, \partial) = \bigoplus_{i=0}^{n} \mathcal{H}^i(C^*, \partial), \tag{2.5}$$

with

$$\mathcal{H}^i(C^*, \partial) = (\partial_{i-1} : C^{i-1} \to \ker(\partial_i)), \ 0 \le i \le n, \tag{2.6}$$

where $(\partial_{i-1} : C^{i-1} \to \ker(\partial_i))$, $0 \le i \le n$, lie in an abelian extended category. It consists of two parts: the projective part which is exactly the reduced cohomology defined in (2.3), as well as a torsion part

$$\mathcal{T}(\mathcal{H}^*(C^*, \partial)) = \oplus_{i=0}^{n} \mathcal{T}(\mathcal{H}^i(C^*, \partial))$$

defined as an element in the above abelian extended category, with

$$\mathcal{T}(\mathcal{H}^i(C^*, \partial)) = (\partial_{i-1} : C^{i-1} \to \overline{\mathrm{im}(\partial_{i-1})}), \ 0 \le i \le n. \tag{2.7}$$

More precisely, one has

$$\mathcal{H}^*(C^*, \partial) = H^*(C^*, \partial) \oplus \mathcal{T}(\mathcal{H}^*(C^*, \partial)), \tag{2.8}$$

with

$$\mathcal{H}^i(C^*, \partial) = H^i(C^*, \partial) \oplus \mathcal{T}(\mathcal{H}^i(C^*, \partial)), \quad 0 \le i \le n. \tag{2.9}$$

We refer to [13] and [3] for more details about the definition and basic properties of the above mentioned abelian extended category as well as the extended cohomology.

2.2. *The determinant of a finitely generated Hilbert module*

Let M be a finitely generated Hilbert module over $\mathcal{A}$. Let $GL(M)$ denote the set of all bounded $\mathcal{A}$-linear automorphisms of M. Let $\mathcal{C}_M$ denote the set of all inner products on M such that if $\langle\ ,\ \rangle \in \mathcal{C}_M$, then there exists $A \in GL(M)$ such that

$$\langle u, v\rangle = \langle Au, v\rangle_M, \quad \text{for any } u,\ v \in M, \tag{2.10}$$

with $\langle\ ,\ \rangle_M$ being the original inner product on M.

Following [8] and [3], we define the determinant line $\det M$ of M to be the real one dimensional vector space generated by symbols $\langle\ ,\ \rangle$, one for each element in $\mathcal{C}_M$ such that if $\langle\ ,\ \rangle_1$ and $\langle\ ,\ \rangle_2$ are two elements of $\mathcal{C}_M$ with

$$\langle u, v\rangle_2 = \langle Au, v\rangle_1, \quad \text{for any } u,\ v \in M, \tag{2.11}$$

for some $A \in GL(M)$, then as elements in $\det M$, one has

$$\langle\ ,\ \rangle_2 = \mathrm{Det}_\tau(A)^{-1/2} \cdot \langle\ ,\ \rangle_1, \tag{2.12}$$

where $\mathrm{Det}_\tau(A)$ is the Fuglede-Kadison determinant [14] of A.

For the sake of completeness, we recall the definition of $\mathrm{Det}_\tau(A)$ for any $A \in GL(M)$ and its basic properties from [8] and [3].

Let A_t, $0 \le t \le 1$, be a continuous piecewise smooth path $A_t \in GL(M)$ such that $A_0 = I$ and $A_1 = A$. The existence of such a path is clear as $GL(M)$ is known to be pathwise connected. Then define as in [8], (13) and [3], (2.7) that

$$\log \mathrm{Det}_\tau(A) = \int_0^1 \mathrm{Re}\left(\mathrm{Tr}_\tau[A_t^{-1}A_t']\right) dt, \tag{2.13}$$

where A_t' is the derivative of A_t with respect to t, while Tr_τ is the canonically induced trace on the commutant of M (cf. [8], Proposition 1.8).

It has been proved in [8] that the right hand side of (2.13) does not depend on the choice of the path A_t, $0 \leq t \leq 1$. Moreover, we recall the following basic properties taken from [8], Theorem 1.10 and [3], Theorem 2.11.

Proposition 2.1. *The function,*

$$\mathrm{Det}_\tau : GL(M) \to \mathbf{R}^{>0}, \tag{2.14}$$

called the Fuglede-Kadison determinant of A*, satisfies,*
(a) Det_τ *is a group homomorphism, that is,*

$$\mathrm{Det}_\tau(AB) = \mathrm{Det}_\tau(A) \cdot \mathrm{Det}_\tau(B), \quad \text{for} \quad A,\ B \in GL(M); \tag{2.15}$$

(b) *If* I *is the identity element in* $GL(M)$*, then*

$$\mathrm{Det}_\tau(\lambda I) = |\lambda|^{\tau(I)} \quad \text{for} \quad \lambda \in \mathbf{C},\ \lambda \neq 0; \tag{2.16}$$

(c) *One has*

$$\mathrm{Det}_{\lambda\tau}(A) = \mathrm{Det}_\tau(A)^\lambda \quad \text{for} \quad \lambda \in \mathbf{R}^{>0}; \tag{2.17}$$

(d) Det_τ *is continuous as a map* $GL(M) \to \mathbf{R}^{>0}$*, where* $GL(M)$ *is supplied with the norm topology;*
(e) *If* A_t*,* $t \in [0, 1]$*, is a continuous piecewise smooth path in* $GL(M)$*, then*

$$\log \left[\frac{\mathrm{Det}_\tau(A_1)}{\mathrm{Det}_\tau(A_0)} \right] = \int_0^1 \mathrm{Re}\left(\mathrm{Tr}_\tau[A_t^{-1} A_t'] \right) dt; \tag{2.18}$$

(f) *Let* M*,* N *be two finitely generated Hilbert modules over* $\mathcal{A}$*. Let* $A \in GL(M)$*,* $B \in GL(N)$ *and let*

$$\gamma : N \to M$$

be a bounded $\mathcal{A}$*-linear homomorphism. We extend* A*,* B*,* γ *to obvious endomorphisms on* $M \oplus N$ *by taking* $A|_N = 0$*,* $B|_M = 0$ *and* $\gamma|_M = 0$*. Then* $A + B + \gamma \in GL(M \oplus N)$ *and*

$$\mathrm{Det}_\tau(A + B + \gamma) = \mathrm{Det}_\tau(A) \cdot \mathrm{Det}_\tau(B). \tag{2.19}$$

Now we come back to the determinant line $\det M$. Clearly, $\det M$ has a canonical orientation as the transition coefficient $\mathrm{Det}_\tau(A)^{-1/2}$ is always positive.

Following [8], (2.3), for any bounded $\mathcal{A}$-linear isomorphism $f : M \to N$ between two finitely generated Hilbert modules over $\mathcal{A}$, there induces canonically an isomorphism of determinant lines $f_* : \det M \to \det N$, which preserves the orientations. Moreover, one has the following property which is recalled from [8], Proposition 2.5.

Proposition 2.2. *If $f \in GL(M)$, then the induced isomorphism f_* : $\det M \to \det M$ coincides with the multiplication by $\mathrm{Det}_\tau(f) \in \mathbf{R}^{>0}$.*

Remark 2.1. Following [8] and [3], one thinks of elements of $\det M$ as "densities" on M. In the $\mathcal{A} = \mathbf{C}$ case, this is dual to the considerations in [2] where one uses metrics on determinant lines instead of "volume forms".

2.3. *Extended cohomology and the torsion element of a finite length cochain complex of Hilbert modules*

Let (C^*, ∂) be a finite length Hilbert cochain complex over $\mathcal{A}$

$$(C^*, \partial) : 0 \to C^0 \overset{\partial_0}{\to} C^1 \overset{\partial_1}{\to} \cdots \overset{\partial_{n-1}}{\to} C^n \to 0 \tag{2.20}$$

as in (2.2). Let

$$\mathcal{H}^*(C^*, \partial) = \sum_{i=0}^{n} \mathcal{H}^i(C^*, \partial)$$

denote the corresponding extended cohomology defined in (2.5), which admits the splitting to projective and torsion parts as in (2.7)-(2.9).

Following [3], we define for each $0 \leq i \leq n$ that

$$\det \mathcal{H}^i(C^*, \partial) := \det H^i(C^*, \partial) \otimes \det \mathcal{T}(\mathcal{H}^i(C^*, \partial)) \tag{2.21}$$

with

$$\det \mathcal{T}(\mathcal{H}^i(C^*, \partial)) := \det \overline{\mathrm{im}(\partial_{i-1})} \otimes (\det C^{i-1})^* \otimes \det \ker(\partial_{i-1}). \tag{2.22}$$

Definition 2.1. (i) We define the determinant line of (C^*, ∂) to be

$$\det(C^*, \partial) = \bigotimes_{i=0}^{n} \left(\det C^i\right)^{(-1)^i}. \tag{2.23}$$

(ii) We define the determinant line of $\mathcal{H}^*(C^*, \partial)$ to be

$$\det \mathcal{H}^*(C^*, \partial) = \bigotimes_{i=0}^{n} \left(\det \mathcal{H}^i(C^*, \partial)\right)^{(-1)^i}. \tag{2.24}$$

The following result is recalled from [3], Proposition 7.2.

Proposition 2.3. *The cochain complex* (2.20) *defines a canonical isomorphism*

$$\nu_{(C^*, \partial)} : \det(C^*, \partial) \to \det \mathcal{H}^*(C^*, \partial). \tag{2.25}$$

For each $0 \le i \le n$, the (fixed) inner product on C^i determines an element $\sigma_i \in \det C^i$. They together determine an element

$$\sigma = \prod_{i=0}^{n} \sigma_i^{(-1)^i} \in \det(C^*, \partial). \tag{2.26}$$

Definition 2.2. (cf. [3], Definition 7.5) The positive element

$$\rho_{(C^*,\partial)} = \nu_{(C^*,\partial)}(\sigma) \in \det \mathcal{H}^*(C^*, \partial) \tag{2.27}$$

is called the torsion element of the cochain complex (C^*, ∂).

For any other **Z**-graded inner product $\langle\ ,\ \rangle' \in \mathcal{C}_C$, that is, there exists $A_i \in GL(C^i)$ for any $0 \le i \le n$ such that

$$\langle u, v\rangle_i' = \langle A_i u, v\rangle \quad \text{for any} \quad u,\ v \in C^i, \tag{2.28}$$

let $\rho'_{(C^*,\partial)}$ denote the corresponding torsion element in $\det \mathcal{H}^*(C^*, \partial)$. Then one has the following anomaly formula for the torsion elements in $\det \mathcal{H}^*(C^*, \partial)$.

Proposition 2.4. *The following identity holds in* $\det \mathcal{H}^*(C^*, \partial)$,

$$\rho'_{(C^*,\partial)} = \rho_{(C^*,\partial)} \prod_{i=0}^{n} \mathrm{Det}_\tau(A_i)^{\frac{(-1)^{i+1}}{2}}. \tag{2.29}$$

Proof. Let σ_i' be the corresponding element in $\det C^i$. From (2.28), one has by definition (cf. (2.12))

$$\sigma_i' = \mathrm{Det}_\tau(A_i)^{-1/2}\sigma_i. \tag{2.30}$$

From Proposition 2.3 and from (2.26), (2.27) and (2.30), one gets (2.29).□

For any $0 \le i \le n$, let

$$\partial_i^* : C^{i+1} \to C^i$$

denote the adjoint of ∂^i with respect to the inner products on C^i and C^{i+1}.

Let

$$\partial = \sum_{i=1}^{n} \partial_i : C^* \to C^*, \qquad \partial^* = \sum_{i=1}^{n} \partial_i^* : C^* \to C^*$$

denote the induced homomorphisms on C^*. Then

$$\Box = (\partial + \partial^*)^2 \tag{2.31}$$

preserves each C^i. Let $\Box_i$ denote the restriction of $\Box$ on C^i.

Now consider the special case where the cochain complex (C^*, ∂) is acyclic, i.e., for any $0 \leq i \leq n$, $\mathrm{im}(\partial_i) = \ker(\partial_{i+1})$ (In particular, this implies that $\mathrm{im}(\partial_i)$ is closed in C^{i+1}). Then the torsion element $\rho_{(C^*,\partial)} = \nu_{(C^*,\partial)}(\sigma) \in \det \mathcal{H}^*(C^*, \partial) \simeq \mathbf{R}$ can be thought of as a positive real number.

The following result has been proved in [3], Proposition 7.8.

Proposition 2.5. *If the cochain complex (C^*, ∂) is acyclic, then the following identity holds,*

$$\log \rho_{(C^*,\partial)} = \frac{1}{2} \sum_{i=0}^{n} (-1)^{i+1} i \log \mathrm{Det}_\tau(\Box_i). \tag{2.32}$$

We refer to [3] for more complete discussions about the torsion elements in determinant lines.

3. Infinite covering spaces and the L^2-Ray-Singer torsion on the determinant of extended de Rham cohomology

In this section, we define the L^2-analytic torsion element for the infinite covering space of manifolds with boundary, and prove an anomaly formula for it.

This section is organized as follows. In Section 3.1, we define, in the case of manifolds with boundary, the extended de Rham cohomology associated to a lifted flat vector bundle on an infinite covering space. In Section 3.2, we define the L^2-analytic torsion element, in the manifolds with boundary case, as an element in the determinant of the extended de Rham cohomology. In Section 3.3, we state an anomaly formula, in the case of manifolds with boundary, about the L^2-analytic torsion element. In Section 3.4, we study the variational formula for the heat kernel. The anomaly formula is then proved in Section 3.5.

3.1. *Infinite covering spaces and the extended de Rham cohomology*

Let $\Gamma \to \widetilde{M} \xrightarrow{\pi} M$ be a Galois covering of a compact manifold M with boundary ∂M, with $\dim M = n$. Then $\widetilde{M}$ is a manifold with boundary $\partial\widetilde{M}$, which is a Γ-covering of ∂M. We make the assumption that Γ is an infinite group, as the case of finite group has been dealt with for example in [16] and [17].

Let (F, ∇^F) be a complex flat vector bundle over M carrying the flat connection ∇^F. Let g^F be a Hermitian metric on F. Let $(\widetilde{F}, \nabla^{\widetilde{F}})$ denote the naturally lifted flat vector bundle over $\widetilde{M}$ obtained as the pullback of (F, ∇^F) through the covering map π. Let $g^{\widetilde{F}}$ be the naturally lifted Hermitian metric on $\widetilde{F}$.

Let $\mathcal{N}(\Gamma)$ be the von Neumann algebra associated to Γ generated by the left regular representations on $l^2(\Gamma) \equiv l^2(\mathcal{N}(\Gamma))$. The canonical finite faithful trace on $\mathcal{N}(\Gamma)$ is given by the following formulas,

$$\begin{aligned} \tau_{\mathcal{N}(\Gamma)}(L_\alpha) =& 0, \text{ if } \alpha \neq 1, \\ & 1, \text{ if } \alpha = 1, \end{aligned} \tag{3.1}$$

where L_α denote the left action of $\alpha \in \Gamma$ on $l^2(\Gamma)$. It induces canonically a trace on the commutant of any finitely generated Hilbert $\mathcal{N}(\Gamma)$-module (cf. [8], Proposition 1.8), which will be denoted by $\mathrm{Tr}_{\mathcal{N}}$.

For any $0 \leq i \leq n$, denote

$$\Omega^i(\widetilde{M}, \widetilde{F}) = \Gamma(\Lambda^i(T^*\widetilde{M}) \otimes \widetilde{F}), \qquad \Omega^*(\widetilde{M}, \widetilde{F}) = \bigoplus_{i=0}^{n} \Omega^i(\widetilde{M}, \widetilde{F}). \tag{3.2}$$

Let $d^{\widetilde{F}}$ denote the natural exterior differential on $\Omega^*(\widetilde{M}, \widetilde{F})$ induced from $\nabla^{\widetilde{F}}$ which maps each $\Omega^i(\widetilde{M}, \widetilde{F})$, $0 \leq i \leq n$, into $\Omega^{i+1}(\widetilde{M}, \widetilde{F})$.

Let g^{TM} be a Riemannian metric on TM. Let $g^{T\partial M}$ be its restricted metric on $T\partial M$. Let $g^{T\widetilde{M}}$ be the lifted Riemannian metric on $T\widetilde{M}$ and denote by $\langle \cdot, \cdot \rangle_{\Lambda(T^*\widetilde{M}) \otimes \widetilde{F}}$ the induced Hermitian metric on $\Lambda(T^*\widetilde{M}) \otimes \widetilde{F}$. Let $o(T\widetilde{M})$ be the orientation bundle of $T\widetilde{M}$, and let $dv_{\widetilde{M}}$ be the Riemannian volume element on $(T\widetilde{M}, g^{T\widetilde{M}})$, then we can view $dv_{\widetilde{M}}$ as a section of $\Lambda^n(T^*\widetilde{M}) \otimes o(T\widetilde{M})$. The metrics $g^{T\widetilde{M}}$, $g^{\widetilde{F}}$ determine a canonical inner product on each $\Omega^i(\widetilde{M}, \widetilde{F})$, $0 \leq i \leq n$ as follow,

$$\langle \sigma, \sigma' \rangle := \int_X \langle \sigma, \sigma' \rangle_{\Lambda(T^*\widetilde{M}) \otimes \widetilde{F}} dv_{\widetilde{M}} \quad \text{for } \sigma, \sigma' \in \Omega(\widetilde{M}, \widetilde{F}). \tag{3.3}$$

Let $L^2(\Omega^i(\widetilde{M}, \widetilde{F}))$, $0 \leq i \leq n$, denote the Hilbert spaces obtained from the corresponding L^2-completion.

Let $g^{T\partial\widetilde{M}}$ be the metric on $T\partial\widetilde{M}$ lifted from $g^{T\partial M}$. We identify the normal bundle $N_{\partial\widetilde{M}}$ to $\partial\widetilde{M}$ in $\widetilde{M}$ with the orthogonal complement of $T\partial\widetilde{M}$ in $T\widetilde{M}|_{\partial\widetilde{M}}$.

Denote by $e_n = \widetilde{e}_{\mathrm{n}}$ the inward pointing unit normal vector field along $\partial\widetilde{M}$. We also put, with $i(\cdot)$ the notation of interior multiplication,

$$\Omega^i_{\mathrm{a}}(\widetilde{M}, \widetilde{F}) = \{\sigma \in \Omega^i(\widetilde{M}, \widetilde{F}); \quad i(\widetilde{e}_{\mathrm{n}})\sigma = i(\widetilde{e}_{\mathrm{n}})(d^{\widetilde{F}}\sigma) = 0 \text{ on } \partial\widetilde{M}\}. \tag{3.4}$$

Let $d_a^{\widetilde{F}}$ be the closure of $d^{\widetilde{F}}$ with respect to the (absolute) boundary condition (3.4). Then

$$d_a^{\widetilde{F}} : L^2(\Omega^*(\widetilde{M}, \widetilde{F})) \to L^2(\Omega^*(\widetilde{M}, \widetilde{F}))$$

is an unbounded operator. Let

$$d_a^{\widetilde{F}*} : L^2(\Omega^*(\widetilde{M}, \widetilde{F})) \to L^2(\Omega^*(\widetilde{M}, \widetilde{F}))$$

be the adjoint of it. Set

$$\widetilde{D}_a = d_a^{\widetilde{F}} + d_a^{\widetilde{F}*}. \tag{3.5}$$

For any $\mathcal{I} \subseteq \mathbf{R}$ and $0 \leq i \leq n$, denote by

$$L^2_{a,\mathcal{I}}(\Omega^i(\widetilde{M}, \widetilde{F})) \subseteq L^2(\Omega^i(\widetilde{M}, \widetilde{F})) \tag{3.6}$$

the image of the spectral projection $P_{\mathcal{I},i} : L^2(\Omega^i(\widetilde{M}, \widetilde{F})) \to L^2(\Omega^i(\widetilde{M}, \widetilde{F}))$ of $\widetilde{D}_a^2|_{L^2(\Omega^i(\widetilde{M}, \widetilde{F}))}$ corresponding to $\mathcal{I}$.

The following result generalizes a theorem of Shubin [28], Theorem 5.1 which has been recalled in [32], Theorem 3.1.

Theorem 3.1. *Fix $\varepsilon > 0$. Then for any $0 \leq i \leq n$,*

(i) $L^2_{a,[0,\varepsilon]}(\Omega^i(\widetilde{M}, \widetilde{F})) \subset \Omega^i_a(\widetilde{M}, \widetilde{F})$, *i.e.,* $L^2_{a,[0,\varepsilon]}(\Omega^i(\widetilde{M}, \widetilde{F}))$ *consists of smooth forms verifying the boundary condition* (3.4)*;*

(ii) *When carrying the induced metric from that of* $L^2(\Omega^i(\widetilde{M}, \widetilde{F}))$, $L^2_{a,[0,\varepsilon]}(\Omega^i(\widetilde{M}, \widetilde{F}))$ *is a finitely generated Hilbert module over* $\mathcal{N}(\Gamma)$.

Proof. (i) As in [28], we make use of elliptic estimates. Fix any $\lambda > 0$, from the standard elliptic estimate, one knows that

$$(\widetilde{D}_a^2 + \lambda)^{-1} : L^2_{a,[0,\varepsilon]}(\Omega^i(\widetilde{M}, \widetilde{F})) \to L^2_{a,[0,\varepsilon]}(\Omega^i(\widetilde{M}, \widetilde{F})) \tag{3.7}$$

is a well-defined, onto map which increases the degree of differentiability by two. By applying this to powers of $(\widetilde{D}_a^2 + \lambda)^{-1}$, we see then any element in $L^2_{a,[0,\varepsilon]}(\Omega^i(\widetilde{M}, \widetilde{F}))$ is smooth and verifies the boundary condition (3.4).

(ii) By simple smooth deformations and the homotopy invariance of the finite rank property (cf. [28]), we need only to deal with case where g^{TM} and g^F are of product structure near ∂M. Now in the case where all structures are a product near the boundary, one can proceed as in [17] to reduce the problem to the double of $\widetilde{M}$, on which one can apply the result of Shubin [28], Theorem 5.1. □

Now consider the finite length cochain complex of $\mathcal{N}(\Gamma)$-Hilbert modules

$$(L^2_{a,[0,\varepsilon]}(\Omega^*(\widetilde{M},\widetilde{F})),d_a^{\widetilde{F}}):0\rightarrow L^2_{a,[0,\varepsilon]}(\Omega^0(\widetilde{M},\widetilde{F}))\overset{d_a^{\widetilde{F}}}{\rightarrow} L^2_{a,[0,\varepsilon]}(\Omega^1(\widetilde{M},\widetilde{F}))\rightarrow\cdots\overset{d_a^{\widetilde{F}}}{\rightarrow} L^2_{a,[0,\varepsilon]}(\Omega^n(\widetilde{M},\widetilde{F}))\rightarrow 0. \quad (3.8)$$

It is easy to verify that the extended cohomology of $(L^2_{a,[0,\varepsilon]}(\Omega^*(\widetilde{M},\widetilde{F})),d_a^{\widetilde{F}})$ is independent of $\varepsilon>0$. For if $\varepsilon'>\varepsilon>0$, the sub-complex $(L^2_{a,(\varepsilon,\varepsilon']}(\Omega^*(\widetilde{M},\widetilde{F})),d_a^{\widetilde{F}})$ of $(L^2_{a,[0,\varepsilon']}(\Omega^*(\widetilde{M},\widetilde{F})),d_a^{\widetilde{F}})$ is acyclic. Moreover, it is easy to verify that this extended cohomology, up to bounded $\mathcal{N}(\Gamma)$-linear isomorphisms, does not depend on the choice of the metrics g^{TM} and g^F on TM and F respectively. We denote it by $\mathcal{H}^{(2)}_{a,\mathrm{dR}}(\Omega^*(\widetilde{M},\widetilde{F}),d_a^{\widetilde{F}})$.

Definition 3.1. The extended cohomology $\mathcal{H}^{(2)}_{a,\mathrm{dR}}(\Omega^*(\widetilde{M},\widetilde{F}),d_a^{\widetilde{F}})$ defined above is called the L^2-extended (absolute) de Rham cohomology associated to $\widetilde{M}$ and F.

3.2. *L^2-Ray-Singer torsion on the determinant of the extended de Rham cohomology*

We continue the discussion of the above subsection.

In view of Definition 2.2, for any $\varepsilon>0$, the finite length cochain complex of $\mathcal{N}(\Gamma)$-Hilbert modules $(L^2_{a,[0,\varepsilon]}(\Omega^*(\widetilde{M},\widetilde{F})),d_a^{\widetilde{F}})$ in (3.8) determines a torsion element in $\det\mathcal{H}^{(2)}_{a,\mathrm{dR}}(\Omega^*(\widetilde{M},\widetilde{F}),d_a^{\widetilde{F}})$. We denote this torsion element by $T_{a,[0,\varepsilon]}(\widetilde{M},F,g^{TM},g^F)$.

By proceeding as in [3], Section 12.2, for any $s\in\mathbf{C}$ with $\mathrm{Re}(s)>\frac{n}{2}$ and for $0\le i\le n$, set

$$\zeta^i_{a,(\varepsilon,+\infty)}(s)=\frac{1}{\Gamma(s)}\int_0^{+\infty}t^{s-1}\mathrm{Tr}_{\mathcal{N}}\left[\exp\left(-t\widetilde{D}_a^2|_{L^2_{a,(\varepsilon,+\infty)}(\Omega^i(\widetilde{M},\widetilde{F}))}\right)\right]dt. \quad (3.9)$$

Then $\zeta^i_{a,(\varepsilon,+\infty)}(s)$ is analytic in s for $\mathrm{Re}(s)>\frac{n}{2}$. Moreover, by using [19], Lemma 1.3, one finds that $\zeta^i_{(\varepsilon,+\infty)}(s)$ can be extended to a meromorphic function on $\mathbf{C}$ which is holomorphic at $s=0$.

Let

$$T_{a,(\varepsilon,+\infty)}(\widetilde{M},F,g^{TM},g^F)\in\mathbf{R}^+$$

be defined by

$$\log T_{a,(\varepsilon,+\infty)}(\widetilde{M},F,g^{TM},g^F)=\frac{1}{2}\sum_{i=0}^n(-1)^i i\left.\frac{\partial\zeta^i_{a,(\varepsilon,+\infty)}(s)}{\partial s}\right|_{s=0}. \quad (3.10)$$

By proceeding as in [3], Lemma 12.4, one knows that the product

$$T_{a,[0,\varepsilon]}(\widetilde{M},F,g^{TM},g^F)\cdot T_{a,(\varepsilon,+\infty)}(\widetilde{M},F,g^{TM},g^F)$$

in $\det \mathcal{H}^{(2)}_{a,\mathrm{dR}}(\Omega^*(\widetilde{M},\widetilde{F}),d_a^{\widetilde{F}})$ does not depend on $\varepsilon>0$.

Definition 3.2. The L^2-Ray-Singer (or L^2-analytic) torsion element associated to $(\widetilde{M},F,g^{TM},g^F)$ is the positive element in the determinant of the extended de Rham cohomology $\mathcal{H}^{(2)}_{a,\mathrm{dR}}(\Omega^*(\widetilde{M},\widetilde{F}),d_a^{\widetilde{F}})$ defined by

$$\begin{aligned} T^{(2)}_{a,RS}&(\widetilde{M},F,g^{TM},g^F)\\ &=T_{a,[0,\varepsilon]}(\widetilde{M},F,g^{TM},g^F)\cdot T_{a,(\varepsilon,+\infty)}(\widetilde{M},F,g^{TM},g^F). \end{aligned}\tag{3.11}$$

3.3. *An anomaly formula for the L^2-Ray-Singer torsion elements*

We continue the discussion of the above subsection.

For $\Gamma=\{1\}$, the above construction gives us the usual torsion element $T_{a,RS}(\widetilde{M},F,g^{TM},g^F)$ which is dual to the Ray-Singer metric discussed in [2], [4], Def. 1.2 and [5], Def. 4.3.

For convenience of notation we use $\mathrm{l.i.m.}_{t\to 0}F_t$ to denote the constant term in an asymptotic expansion F_t with respect to the parameter t.

We can now state the main result of this paper.

Let g_u^{TM} (resp. g_u^F), $0\le u\le 1$, be a smooth path of metrics on TM (resp. F). Let $*_u$ be the usual Hodge star operator associated to g^{TM} for the $F=\mathbf{C}$ case (cf. [31], Chapter 4).

Theorem 3.2. *The following identity holds,*

$$\begin{aligned} \tfrac{\partial}{\partial u}\left(\log T^{(2)}_{a,RS}(\widetilde{M},F,g_u^{TM},g_u^F)\right)&=\tfrac{\partial}{\partial u}\left(\log T_{a,RS}(M,F,g_u^{TM},g_u^F)\right)\\ &=-\frac{1}{2}\mathrm{l.i.m.}_{t\to 0}\mathrm{Tr}_s\left[\left(*_u^{-1}\frac{\partial *_u}{\partial u}+(h_u^F)^{-1}\frac{\partial h_u^F}{\partial u}\right)e^{-tD^2_{u,a}}\right]. \end{aligned}\tag{3.12}$$

Remark 3.1. If M is a compact manifold without boundary, then Theorem 3.2 is [32], (3.80). If we assume moreover $\Gamma=\{1\}$, then it is [2], Theorem 4.14

Remark 3.2. If $g_u^F=g^F$ is a fixed flat metric on F (i.e. (F,∇^F,g^F) is an unitary flat bundle), the first equation of (3.12) was obtained in [19], Theorem 7.6 under certain technical "determinant class condition". Thus Theorem 3.2 generalizes [19], Theorem 7.6, without the assumptions on the flatness of g^F, and on the technical "determinant class condition".

The second equation of (3.12) is [10], Theorem 3.27 and [27], Theorem 7.3 when $g_u^F = g^F$ is a fixed flat metric on F. For a general family of metrics (g_u^{TM}, g_u^F), the second equation of (3.12) was proved in [5], Theorem 4.5. From [4], Theorem 0.2 and [5], Theorem 0.1, §5.5, we get immediately the anomaly formula for $\log T^{(2)}_{a,RS}(\widetilde{M}, F, g^{TM}, g^F)$ which differs by a factor $-\frac{1}{2}$, as the torsion element is dual to the Ray-Singer metric. We left the details to the readers.

3.4. *Variational formula for the heat kernel*

The results in this subsection were essentially obtained in [10], Theorems 3.10, 3.27 and [27], Prop. 6.1, Theorem 7.3 when $g_u^F = g^F$ is a fixed flat metric on F. In [5], §4.2, it is observed that their proof works also for any Hermitian metric on F. Our main point here is a reformulation of these results in the spirit of the proof of [1], Theorem 1.18 in the covering case.

Let $*^{\widetilde{F}}$ be the Hodge operator

$$*^{\widetilde{F}} : \Lambda(T^*\widetilde{M}) \otimes \widetilde{F} \to \Lambda(T^*\widetilde{M}) \otimes \widetilde{F}^* \otimes o(T\widetilde{M})$$

defined by

$$(\sigma \wedge *^{\widetilde{F}}\sigma')_F = \langle \sigma, \sigma' \rangle_{\Lambda(T^*\widetilde{M})\otimes\widetilde{F}}\, dv_{\widetilde{M}}.$$

A direct verification shows that, when acting on $\Omega^i(\widetilde{M}, \widetilde{F})$, one has

$$d_u^{\widetilde{F}*} = (-1)^i (*_u^{\widetilde{F}})^{-1} d^{\widetilde{F}\otimes o(T\widetilde{M})} *_u^{\widetilde{F}} . \tag{3.13}$$

We only consider orthonormal frames $\{e_i\}_{i=1}^n$ of $T\widetilde{M}$ with the property that near the boundary V, $e_n =: \widetilde{e}_n$ is the inward pointing unit normal at any boundary point and $\{e_i\}_{i=1}^{n-1}$ is an orthonormal basis of $T\partial\widetilde{M}$. Let $\{e^i\}$ be the corresponding dual frame of $T^*\widetilde{M}$.

Let $e^{-t\widetilde{D}_a^2}(x, z)$, $(x, z \in \widetilde{M})$, be the smooth kernel of the operator $e^{-t\widetilde{D}_a^2}$ with respect to $dv_{\widetilde{M}}(z)$. Then

$$e^{-t\widetilde{D}_a^2}(x, z) \in \oplus_{k=0}^n (\Lambda^k(T^*\widetilde{M}) \otimes \widetilde{F})_x \otimes (\Lambda^k(T^*\widetilde{M}) \otimes \widetilde{F})_z^* .$$

We denote by $e^{-t\widetilde{D}_a^2}(x, z)_k$ the component of $e^{-t\widetilde{D}_a^2}(x, z)$ on $(\Lambda^k(T^*\widetilde{M}) \otimes \widetilde{F})_x \otimes (\Lambda^k(T^*\widetilde{M}) \otimes \widetilde{F})_z^*$. By using the metric $\langle\cdot,\cdot\rangle_{\Lambda(T^*\widetilde{M})\otimes\widetilde{F}}$, we will identify $(\Lambda(T^*\widetilde{M}) \otimes \widetilde{F})^*$ to $\Lambda(T^*\widetilde{M}) \otimes \widetilde{F}$, thus the operations $d_z^{\widetilde{F}}$, $d_z^{\widetilde{F}*}$ act naturally on $e^{-t\widetilde{D}_a^2}(x, z)$.

Lemma 3.1. *For $\sigma \in \Omega(\widetilde{M}, \widetilde{F}) \cap L^2(\Omega(\widetilde{M}, \widetilde{F}))$, we have*

$$\begin{aligned} e^{-t\widetilde{D}^2_a} d^{\widetilde{F}} \sigma &= d^{\widetilde{F}}_a e^{-t\widetilde{D}^2_a} \sigma, \\ e^{-t\widetilde{D}^2_a} d^{\widetilde{F}*} \sigma &= d^{\widetilde{F}*}_a e^{-t\widetilde{D}^2_a} \sigma + \int_{\partial\widetilde{M}} e^{-t\widetilde{D}^2_a}(\cdot, y) i(e_n)\sigma(y) dv_{\partial\widetilde{M}}(y). \end{aligned} \tag{3.14}$$

Especially,

$$d^{\widetilde{F}}_x e^{-t\widetilde{D}^2_a}(x,z)_k = d^{\widetilde{F}*}_z e^{-t\widetilde{D}^2_a}(x,z)_{k+1}. \tag{3.15}$$

Proof. At first, by the identification of the orientation bundle $o(T\widetilde{M})$ and $o(T\partial\widetilde{M})$ in [5], §1.3, for $\sigma, \sigma' \in \Omega(\widetilde{M}, \widetilde{F}) \cap L^2(\Omega(\widetilde{M}, \widetilde{F}))$,

$$\begin{aligned} \langle d^{\widetilde{F}}\sigma, \sigma' \rangle &= \int_{\widetilde{M}} ((d^{\widetilde{F}}\sigma) \wedge *^{\widetilde{F}}\sigma')_{\widetilde{F}} = \langle \sigma, d^{\widetilde{F}*}\sigma' \rangle + \int_{\partial\widetilde{M}} (\sigma \wedge *^{\widetilde{F}}\sigma')_{\widetilde{F}} \\ &= \langle \sigma, d^{\widetilde{F}*}\sigma' \rangle - \int_{\partial\widetilde{M}} \langle e^n \wedge \sigma, \sigma' \rangle (y) dv_{\partial\widetilde{M}}(y). \end{aligned} \tag{3.16}$$

As $d^{\widetilde{F}*}_a$, $d^{\widetilde{F}}_a$ commute with $\widetilde{D}^2_a$, they also commute with $e^{-t\widetilde{D}^2_a}$. Thus for

$$\sigma \in \Omega(\widetilde{M}, \widetilde{F}) \cap L^2(\Omega(\widetilde{M}, \widetilde{F})), \qquad \sigma' \in \Omega_a(\widetilde{M}, \widetilde{F}) \cap L^2(\Omega(\widetilde{M}, \widetilde{F})),$$

by (3.16)

$$\begin{aligned} &\langle d^{\widetilde{F}}_a e^{-t\widetilde{D}^2_a}\sigma, \sigma' \rangle = \langle e^{-t\widetilde{D}^2_a}\sigma, d^{\widetilde{F}*}_a \sigma' \rangle = \langle \sigma, e^{-t\widetilde{D}^2_a} d^{\widetilde{F}*}_a \sigma' \rangle \\ &= \langle \sigma, d^{\widetilde{F}*}_a e^{-t\widetilde{D}^2_a} \sigma' \rangle = \langle d^{\widetilde{F}}\sigma, e^{-t\widetilde{D}^2_a}\sigma' \rangle = \langle e^{-t\widetilde{D}^2_a} d^{\widetilde{F}}\sigma, \sigma' \rangle. \end{aligned} \tag{3.17}$$

and

$$\begin{aligned} \langle d^{\widetilde{F}*}_a e^{-t\widetilde{D}^2_a}\sigma, \sigma' \rangle &= \langle \sigma, e^{-t\widetilde{D}^2_a} d^{\widetilde{F}}_a \sigma' \rangle = \langle \sigma, d^{\widetilde{F}}_a e^{-t\widetilde{D}^2_a}\sigma' \rangle \\ &= \langle d^{\widetilde{F}*}\sigma, e^{-t\widetilde{D}^2_a}\sigma' \rangle - \int_{\partial\widetilde{M}} \langle \sigma, e^n \wedge (e^{-t\widetilde{D}^2_a}\sigma') \rangle (y) dv_{\partial\widetilde{M}}(y) \\ &= \langle e^{-t\widetilde{D}^2_a} d^{\widetilde{F}*}\sigma, \sigma' \rangle - \left\langle \int_{\partial\widetilde{M}} e^{-t\widetilde{D}^2_a}(\cdot, y) i(e_n)\sigma(y) dv_{\partial\widetilde{M}}(y), \sigma' \right\rangle. \end{aligned} \tag{3.18}$$

From (3.17), (3.18), we get (3.14).

Now for $\sigma \in \Omega_a(\widetilde{M}, \widetilde{F}) \cap L^2(\Omega(\widetilde{M}, \widetilde{F}))$, by (3.16) and (3.17),

$$\begin{aligned} \int_{\widetilde{M}} (d^{\widetilde{F}*}_z e^{-t\widetilde{D}^2_a}(x,z))\sigma(z) dv_{\widetilde{M}}(z) &= \int_{\widetilde{M}} e^{-t\widetilde{D}^2_a}(x,z)(d^{\widetilde{F}}\sigma)(z) dv_{\widetilde{M}}(z) \\ &= (d^{\widetilde{F}} e^{-t\widetilde{D}^2_a}\sigma)(x) = \int_{\widetilde{M}} d^{\widetilde{F}}_x e^{-t\widetilde{D}^2_a}(x,z)\sigma(z) dv_{\widetilde{M}}(z). \end{aligned} \tag{3.19}$$

From (3.19), we get (3.15). The proof of Lemma 3.1 is complete. □

Let g_u^{TM} (resp. g_u^F), $0 \le u \le 1$, be a smooth path of metrics on TM (resp. F). When dealing with objects associated with (g_u^{TM}, g_u^F), we will use a subscript "u" to indicate. While at $u = 0$ we usually omit this subscript indication. In particular, we will use $\langle\, , \rangle_{\widetilde{M},u}$, $\langle\, , \rangle_{\partial\widetilde{M},u}$ to denote the product on $\widetilde{M}$, $\partial\widetilde{M}$ with respect to $dv_{\widetilde{M},u}$, $dv_{\partial\widetilde{M},u}$. Then one has

$$Q_u := (*_u^{\widetilde{F}})^{-1}\frac{\partial *_u^{\widetilde{F}}}{\partial u} = (*_u)^{-1}\frac{\partial *_u}{\partial u} + (g_u^{\widetilde{F}})^{-1}\frac{\partial g_u^{\widetilde{F}}}{\partial u}. \tag{3.20}$$

In what follow, all operations are applied to the variable z when we do not specify them.

Lemma 3.2.

$$\begin{aligned}
&(*_{u,w}^F)^{-1}\frac{\partial}{\partial u}\left(*_{u,w}^F e^{-t\widetilde{D}_{u,a}^2}(x,w)\right)\\
&\quad= \int_0^t \Big\{-\Big\langle [d^F, [d_u^{\widetilde{F}*}, Q_u]] e^{-(t-s)\widetilde{D}_{u,a}^2}(x,z), e^{-s\widetilde{D}_{u,a}^2}(z,w)\Big\rangle_{\widetilde{M},u}\\
&\qquad + \Big\langle i(e_n)Q_u d^F e^{-(t-s)\widetilde{D}_{u,a}^2}(x,z), e^{-s\widetilde{D}_{u,a}^2}(z,w)\Big\rangle_{\partial\widetilde{M},u}\\
&\qquad + \Big\langle i(e_n)Q_u e^{-(t-s)\widetilde{D}_{u,a}^2}(x,z), d_u^{\widetilde{F}*} e^{-s\widetilde{D}_{u,a}^2}(z,w)\Big\rangle_{\partial\widetilde{M},u}\Big\}.
\end{aligned} \tag{3.21}$$

Proof. We only need to prove (3.21) for $u = 0$. At first, by [5], (4.10), we have

$$i(e_{n,u})d_u^{\widetilde{F}*}\sigma|_{\partial\widetilde{M}} = 0 \quad \text{if } i(e_{n,u})\sigma|_{\partial\widetilde{M}} = 0. \tag{3.22}$$

We know also that for $\sigma \in \Omega(\widetilde{M}, \widetilde{F}) \cap L^2(\Omega(\widetilde{M}, \widetilde{F}))$

$$\begin{aligned}
&\lim_{s\to 0}\int_{\widetilde{M}} e^{-s\widetilde{D}_{u,a}^2}(x,z)\sigma(z)dv_{\widetilde{M},0}(z)\\
&\quad= \lim_{s\to 0}\int_{\widetilde{M}}\left(e^{-s\widetilde{D}_{u,a}^2}(x,z)\wedge *_u^{\widetilde{F}}((*_u^{\widetilde{F}})^{-1} *_0^{\widetilde{F}}\sigma)(z)\right)_F\\
&\quad= \lim_{s\to 0}\left(e^{-s\widetilde{D}_{u,a}^2}(*_u^{\widetilde{F}})^{-1} *_0^{\widetilde{F}}\sigma\right)(x) = ((*_u^{\widetilde{F}})^{-1} *_0^{\widetilde{F}}\sigma)(x).
\end{aligned} \tag{3.23}$$

By (3.16), (3.22) and (3.23), we get

$$
\begin{aligned}
e^{-t\widetilde{D}^2_{u,a}}(x,w) &- *^{-1}_{u,w} *_{0,w}\, e^{-t\widetilde{D}^2_{0,a}}(x,w) \\
&= -\int_0^t \frac{\partial}{\partial s}\left\langle e^{-(t-s)\widetilde{D}^2_{u,a}}(x,z), e^{-s\widetilde{D}^2_{0,a}}(z,w)\right\rangle_{\widetilde{M},0} \\
&= \int_0^t \Bigg[\left\langle \frac{\partial}{\partial t} e^{-(t-s)\widetilde{D}^2_{u,a}}(x,z), e^{-s\widetilde{D}^2_{0,a}}(z,w)\right\rangle_{\widetilde{M},0} \\
&\quad + \left\langle e^{-(t-s)\widetilde{D}^2_{u,a}}(x,z), \widetilde{D}^2_{0,a} e^{-s\widetilde{D}^2_{0,a}}(z,w)\right\rangle_{\widetilde{M},0}\Bigg] \\
= \int_0^t \Bigg[&\left\langle (\frac{\partial}{\partial t} + \widetilde{D}^2_{0,a}) e^{-(t-s)\widetilde{D}^2_{u,a}}(x,z), e^{-s\widetilde{D}^2_{0,a}}(z,w)\right\rangle_{\widetilde{M},0} \\
&- \left\langle d^{\widetilde{F}} e^{-(t-s)\widetilde{D}^2_{u,a}}(x,z), e_0^n \wedge e^{-s\widetilde{D}^2_{0,a}}(z,w)\right\rangle_{\partial\widetilde{M},0} \\
&- \left\langle e^{-(t-s)\widetilde{D}^2_{u,a}}(x,z), e_0^n \wedge d_0^{\widetilde{F}*} e^{-s\widetilde{D}^2_{0,a}}(z,w)\right\rangle_{\partial\widetilde{M},0}\Bigg].
\end{aligned}
\tag{3.24}
$$

From our definition of $e^{-t\widetilde{D}^2_{u,a}}$, we have

$$
\begin{aligned}
&\Big((\frac{\partial}{\partial t} + \widetilde{D}^2_{u,a}) e^{-t\widetilde{D}^2_{u,a}}\Big)(x,z) = 0, \qquad\qquad (3.25)\\
&\Big(i(e_{n,u}) e^{-t\widetilde{D}^2_{u,a}}\Big)(x,z) = \Big(i(e_{n,u}) d^{\widetilde{F}} e^{-t\widetilde{D}^2_{u,a}}\Big)(x,z) = 0, \text{ for } x \in \partial\widetilde{M}.
\end{aligned}
$$

From the explicit construction of the operator $e^{-t\widetilde{D}^2_{u,a}}$, as observed in [19], p. 560, it is differentiable with respect to u, as $\widetilde{M}$ has bounded geometry. We get

$$
\begin{aligned}
&\Big((\frac{\partial}{\partial t} + \widetilde{D}^2_{u,a}) \frac{\partial}{\partial u} e^{-t\widetilde{D}^2_{u,a}} + (\frac{\partial}{\partial u} \widetilde{D}^2_{u,a}) e^{-t\widetilde{D}^2_{u,a}}\Big)(x,z) = 0, \\
&\Big(i\,(\tfrac{\partial}{\partial u} e_{n,u})\, e^{-t\widetilde{D}^2_{u,a}} + i(e_{n,u}) \tfrac{\partial}{\partial u} e^{-t\widetilde{D}^2_{u,a}}\Big)(x,z) = 0 \text{ for } x \in \partial\widetilde{M}, \\
&\Big(i\,(\tfrac{\partial}{\partial u} e_{n,u})\, d^{\widetilde{F}} e^{-t\widetilde{D}^2_{u,a}} + i(e_{n,u}) d^{\widetilde{F}}\, \tfrac{\partial}{\partial u} e^{-t\widetilde{D}^2_{u,a}}\Big)(x,z) = 0 \text{ for } x \in \partial\widetilde{M}.
\end{aligned}
\tag{3.26}
$$

By (3.26), differentiating (3.24) with respect to u and setting $u = 0$

gives

$$(*^F_{0,w})^{-1}\tfrac{\partial}{\partial u}(*^F_{u,w}e^{-t\widetilde{D}^2_{u,a}})(x,w)$$
$$= \int_0^t \Big[-\Big\langle (\frac{\partial}{\partial u}\widetilde{D}^2_{u,a})e^{-(t-s)\widetilde{D}^2_{u,a}}(x,z), e^{-s\widetilde{D}^2_{0,a}}(z,w)\Big\rangle_{\widetilde{M},0}$$
$$+ \Big\langle i\left(\tfrac{\partial}{\partial u}e_{n,u}\right) d^{\widetilde{F}} e^{-(t-s)\widetilde{D}^2_{0,a}}(x,z), e^{-s\widetilde{D}^2_{0,a}}(z,w)\Big\rangle_{\partial\widetilde{M},0}$$
$$+ \Big\langle i\left(\tfrac{\partial}{\partial u}e_{n,u}\right) e^{-(t-s)\widetilde{D}^2_{0,a}}(x,z), d^{\widetilde{F}*} e^{-s\widetilde{D}^2_{0,a}}(z,w)\Big\rangle_{\partial\widetilde{M},0}\Big]. \quad (3.27)$$

From (3.13) and (3.20), one gets

$$\frac{\partial}{\partial u}d^{\widetilde{F}*}_u = [d^{\widetilde{F}*}_u, Q_u], \quad \frac{\partial}{\partial u}\widetilde{D}^2_{u,a} = [d^F, [d^{\widetilde{F}*}_u, Q_u]]. \quad (3.28)$$

Set

$$\dot{g}^{TX}_u := (g^{TX}_u)^{-1}(\frac{\partial}{\partial u}g^{TX}_u). \quad (3.29)$$

Observe that $\frac{\partial}{\partial u}e_{j,u}$, $e_{j,u} \in T\partial\widetilde{M}$ for $j < n$, thus we compute that

$$\tfrac{\partial}{\partial u}e_{n,u} = -\sum_{j=1}^{n}\langle \dot{g}^{TX}_u e_{j,u}, e_{n,u}\rangle_{g^{TX}_u} e_{j,u} + \frac{1}{2}\langle \dot{g}^{TX}_u e_{n,u}, e_{n,u}\rangle e_{n,u}. \quad (3.30)$$

By [2], Prop. 4.15, we have,

$$*^{-1}_u\frac{\partial *_u}{\partial u} = -\frac{1}{2}\sum_{1\le j,k\le n}\left\langle e_j, \dot{g}^{TX}_u e_k\right\rangle_{g^{TX}_u}(e^j\wedge i(e_k) - i(e_j)\wedge e^k). \quad (3.31)$$

From (3.30) and (3.31), we get at $u = 0$,

$$i(\tfrac{\partial}{\partial u}e_{n,u}) = \left[i(e_n), *^{-1}_u\frac{\partial *_u}{\partial u}\right] + \frac{1}{2}\langle \dot{g}^{TX}_0 e_n, e_n\rangle i(e_n). \quad (3.32)$$

From (3.25), (3.27), (3.28) and (3.32), we get (3.21). □

Let N denote the number operator on $\Omega^*(\widetilde{M},\widetilde{F})$ acting by multiplication by i on $\Omega^i(\widetilde{M},\widetilde{F})$. It extends to obvious actions on L^2-completions.

Let $\mathrm{Tr}_{\mathcal{N},s}[\cdot] = \mathrm{Tr}_{\mathcal{N}}[(-1)^N\cdot]$ be the supertrace in the sense of Quillen[26], taking on bounded $\mathcal{N}(\Gamma)$-linear operators acting on $\Omega^*(\widetilde{M},\widetilde{F})$ as well as their L^2-completions. In what follows we will also adopt the notation in [26] of supercommutators.

Theorem 3.3. *We have the following identity,*

$$\tfrac{\partial}{\partial u}\mathrm{Tr}_{\mathcal{N},s}\left[Ne^{-t\widetilde{D}^2_{u,a}}\right] = t\tfrac{\partial}{\partial t}\mathrm{Tr}_{\mathcal{N},s}\left[Q_u e^{-t\widetilde{D}^2_{u,a}}\right]. \quad (3.33)$$

Proof. Let U be a fundamental domain of the covering $\pi : \widetilde{M} \to M$, and let $U_1 = \overline{U} \cap \pi^{-1}(\partial M)$. Observe first that for any Γ-equivariant smooth operator P acting on $\Omega(\widetilde{M}, \widetilde{F})$, if we denote by $P(x, z)$ the smooth kernel of P with respect to $dv_{\widetilde{M},u}(z)$, then

$$\mathrm{Tr}_{\mathcal{N},s}[P] = \int_U (*_u^{\widetilde{F}} P(x,x))_F. \tag{3.34}$$

Thus when we apply (3.34) to (3.21), and reverse the order of integration on the right hand side of (3.34), then use (3.15), (3.31) and the fact that N preserves the boundary condition, we get

$$\begin{aligned}\frac{\partial}{\partial u}\mathrm{Tr}_{\mathcal{N},s}\left[N\exp\left(-t\widetilde{D}_u^2\right)\right] &= -t\mathrm{Tr}_{\mathcal{N},s}\left[[d^{\widetilde{F}}, [d^{\widetilde{F}*}, Q_u]]Ne^{-t\widetilde{D}^2_{u,a}}\right] \\ &+ t\int_{U_1} \mathrm{Tr}_s[((i(e_n)Q_u d^{\widetilde{F}})_{x'} Ne^{-t\widetilde{D}^2_{u,a}})(w,x')|_{w=x'}]dv_{\partial\widetilde{M},u}(x') \\ &+ t\int_{U_1} \mathrm{Tr}_s[N(Q_u e^n d^{\widetilde{F}*})_{x'} e^{-t\widetilde{D}^2_{u,a}}(x',w)|_{w=x'}]dv_{\partial\widetilde{M},u}(x').\end{aligned} \tag{3.35}$$

By (3.15) and (3.31),

$$\begin{aligned}&\mathrm{Tr}_s\left[(Q_u e^n d^{\widetilde{F}*})_{x'} Ne^{-t\widetilde{D}^2_{u,a}}(x',w)|_{w=x'}\right] \\ &\qquad = -\mathrm{Tr}_s\left[(i(e_n)Q_u N d^{\widetilde{F}} e^{-t\widetilde{D}^2_{u,a}})(x',x')\right].\end{aligned} \tag{3.36}$$

From (3.14) and the fact that $d^{\widetilde{F}}e^{-s\widetilde{D}^2_{u,a}}$, $d^{\widetilde{F}*}e^{-s\widetilde{D}^2_{u,a}}$ are smooth Γ-equivariant operators, we see that for any Γ-equivariant differential operator P which changes the $\mathbf{Z}_2$-grading on $\Omega(\widetilde{M}, \widetilde{F})$, we have

$$\begin{aligned}\mathrm{Tr}_{\mathcal{N},s}\left[d^{\widetilde{F}}Pe^{-t\widetilde{D}^2_{u,a}}\right] &= \mathrm{Tr}_{\mathcal{N},s}\left[e^{-s\widetilde{D}^2_{u,a}}d^{\widetilde{F}}Pe^{-(t-s)\widetilde{D}^2_{u,a}}\right] \\ &= \mathrm{Tr}_{\mathcal{N},s}\left[d_a^{\widetilde{F}}e^{-s\widetilde{D}^2_{u,a}}Pe^{-(t-s)\widetilde{D}^2_{u,a}}\right] \\ = -\mathrm{Tr}_{\mathcal{N},s}\left[Pe^{-(t-s)\widetilde{D}^2_{u,a}}d_a^{\widetilde{F}}e^{-s\widetilde{D}^2_{u,a}}\right] &= -\mathrm{Tr}_{\mathcal{N},s}\left[Pd_a^{\widetilde{F}}e^{-t\widetilde{D}^2_{u,a}}\right],\end{aligned} \tag{3.37}$$

and in the same way

$$\begin{aligned}
\mathrm{Tr}_{\mathcal{N},s}\left[d^{\widetilde{F}*}Pe^{-t\widetilde{D}^2_{u,a}}\right] &= \mathrm{Tr}_{\mathcal{N},s}\left[e^{-s\widetilde{D}^2_{u,a}}d^{\widetilde{F}*}Pe^{-(t-s)\widetilde{D}^2_{u,a}}\right] \\
&= \mathrm{Tr}_{\mathcal{N},s}\Big[d_a^{\widetilde{F}*}e^{-s\widetilde{D}^2_{u,a}}Pe^{-(t-s)\widetilde{D}^2_{u,a}} \\
&\quad + \int_{\partial\widetilde{M}} e^{-sD_a^2}(\cdot,z)i(e_n)Pe^{-(t-s)\widetilde{D}^2_{u,a}}(z,\cdot)dv_{\partial\widetilde{M}}(z)\Big] \\
&= -\mathrm{Tr}_{\mathcal{N},s}\left[Pd_a^{\widetilde{F}*}e^{-t\widetilde{D}^2_{u,a}}\right] \\
&\quad + \int_{U_1}\mathrm{Tr}_s\left[i(e_n)Pe^{-t\widetilde{D}^2_{u,a}}(x',x')\right]dv_{\partial\widetilde{M},u}(x').
\end{aligned} \tag{3.38}$$

We also have

$$[d^{\widetilde{F}},N] = -d^{\widetilde{F}}, \quad [d^{\widetilde{F}*},N] = d^{\widetilde{F}*}. \tag{3.39}$$

From (3.35)-(3.39), we get

$$\begin{aligned}
&\tfrac{\partial}{\partial u}\mathrm{Tr}_{\mathcal{N},s}\left[Ne^{-t\widetilde{D}^2_{u,a}}\right] \\
&= -\mathrm{Tr}_{\mathcal{N},s}\left[Q_u(Nd^{\widetilde{F}}d^{\widetilde{F}*} + d^{\widetilde{F}*}Nd^{\widetilde{F}} - d^{\widetilde{F}}Nd^{\widetilde{F}*} - d^{\widetilde{F}*}d^{\widetilde{F}}N)e^{-t\widetilde{D}^2_{u,a}}\right] \\
&\qquad = -\mathrm{Tr}_{\mathcal{N},s}\left[Q_u\widetilde{D}^2_{u,a}e^{-t\widetilde{D}^2_{u,a}}\right] = \tfrac{\partial}{\partial t}\mathrm{Tr}_{\mathcal{N},s}\left[Q_ue^{-t\widetilde{D}^2_{u,a}}\right].
\end{aligned} \tag{3.40}$$

The proof of Theorem is complete. □

3.5. *A proof of Theorem 3.2*

First, by proceeding as in the beginning of [32], Section 3.4, one gives a slightly more flexible formula of the L^2-Ray-Singer torsion element $T^{(2)}_{a,RS}(\widetilde{M},F,g^{TM},g^F)$ defined in (3.11).

For any $c > 0$, let

$$(C^*, d_a^{\widetilde{F}}) \subset (\Omega_a^*(\widetilde{M},\widetilde{F}), d^{\widetilde{F}})$$

be a finite length $\mathcal{N}(\Gamma)$-Hilbert cochain subcomplex of $(L_a^2(\Omega^*(\widetilde{M},\widetilde{F})), d_a^{\widetilde{F}})$ such that $(L^2_{a,[0,c]}(\Omega^*(\widetilde{M},\widetilde{F})), d_a^{\widetilde{F}})$ is a subcomplex of $(C^*, d_a^{\widetilde{F}})$. That is, as $\mathcal{N}(\Gamma)$-Hilbert cochain complexes, one has

$$(L^2_{a,[0,c]}(\Omega^*(\widetilde{M},\widetilde{F})), d_a^{\widetilde{F}}) \subseteq (C^*, d_a^{\widetilde{F}}). \tag{3.41}$$

Let $d_{C^*}^{\widetilde{F}*} : C^* \to C^*$ be the formal adjoint of $d_a^{\widetilde{F}} : C^* \to C^*$ with respect to the induced Hilbert metric on C^* from that of $L^2(\Omega^*(\widetilde{M},\widetilde{F}))$. Set

$$\begin{aligned}
D_{C^*} &= d_a^{\widetilde{F}} + d_{C^*}^{\widetilde{F}*}, \\
D^2_{C^*} &= (d_a^{\widetilde{F}} + d_{C^*}^{\widetilde{F}*})^2 = d_{C^*}^{\widetilde{F}*}d_a^{\widetilde{F}} + d_a^{\widetilde{F}}d_{C^*}^{\widetilde{F}*} : C^* \to C^*.
\end{aligned} \tag{3.42}$$

Then $D^2_{C^*}$ preserves the **Z**-grading of C^*. Moreover, one has

$$D^2_{C^*} = \widetilde{D}^2_a : L^2_{a,[0,c]}(\Omega^*(\widetilde{M},\widetilde{F})) \to L^2_{a,[0,c]}(\Omega^*(\widetilde{M},\widetilde{F})). \tag{3.43}$$

For any $0 \le i \le n$, let $D^2_{C^i}$ denote the restriction of $D^2_{C^*}$ on C^i.

By (3.41) it is clear that the extended cohomology of $(C^*, d_a^{\widetilde{F}})$ is identical to that of $(L^2_{a,[0,c]}(\Omega^*(\widetilde{M},\widetilde{F})), d_a^{\widetilde{F}})$. That is, one has

$$\mathcal{H}^*(C^*, d_a^{\widetilde{F}}) \equiv \mathcal{H}^{(2)}_{a,\mathrm{dR}}(\Omega^*(\widetilde{M},\widetilde{F}), d_a^{\widetilde{F}}). \tag{3.44}$$

From (3.44), one sees that $(C^*, d_a^{\widetilde{F}})$ induces canonically an L^2-torsion element in $\det \mathcal{H}^{(2)}_{a,\mathrm{dR}}(\Omega^*(\widetilde{M},\widetilde{F}), d_a^{\widetilde{F}})$. We denote it by

$$T_{(C^*, d_a^{\widetilde{F}})} \in \det \mathcal{H}^{(2)}_{a,\mathrm{dR}}(\Omega^*(\widetilde{M},\widetilde{F}), d_a^{\widetilde{F}}). \tag{3.45}$$

For any $s \in \mathbf{C}$ with $\mathrm{Re}(s) > \frac{n}{2}$ and for $0 \le i \le n$, set

$$\zeta^i_{C^*,\perp}(s) = \frac{1}{\Gamma(s)} \int_0^{+\infty} t^{s-1} \Big(\mathrm{Tr}_{\mathcal{N}} \Big[\exp\Big(-t\widetilde{D}^2_a|_{L^2(\Omega^i(\widetilde{M},\widetilde{F}))} \Big) \Big] - \mathrm{Tr}_{\mathcal{N}} \big[\exp\big(-tD^2_{C^i} \big) \big] \Big)\, dt. \tag{3.46}$$

Then one sees easily that each $\zeta^i_{C^*,\perp}(s)$, $0 \le i \le n$, is a holomorphic function for $\mathrm{Re}(s) > \frac{n}{2}$ and can be extended to a meromorphic function on $\mathbf{C}$ which is holomorphic at $s = 0$. Let $T_{(C^*, d_a^{\widetilde{F}}),\perp} \in \mathbf{R}^+$ be defined by

$$\log T_{(C^*, d_a^{\widetilde{F}}),\perp} = \frac{1}{2} \sum_{i=0}^{n} (-1)^i i \left. \frac{\partial \zeta^i_{C^*,\perp}(s)}{\partial s} \right|_{s=0}. \tag{3.47}$$

The following analogue of [32], Proposition 3.6 can be proved in the same way as there.

Proposition 3.1. *There holds in* $\det \mathcal{H}^{(2)}_{a,\mathrm{dR}}(\Omega^*(\widetilde{M},\widetilde{F}), d^{\widetilde{F}})$ *the following identity,*

$$T^{(2)}_{a,RS}(\widetilde{M}, F, g^{TM}, g^F) = T_{(C^*, d_a^{\widetilde{F}})} \cdot T_{(C^*, d_a^{\widetilde{F}}),\perp}. \tag{3.48}$$

We now come to the proof of Theorem 3.2.

Let g^{TM}_u (resp. g^F_u), $0 \le u \le 1$, be a smooth path of metrics on TM (resp. F) such that $g^{TM}_0 = g^{TM}$, $g^{TM}_1 = g'^{TM}$ (resp. $g^F_0 = g^F$, $g^F_1 = g'^F$).

We now state the following analogue of [32], Proposition 3.7.

Proposition 3.2. *For any* $u_0 \in [0,1]$, *there exists* $k_0 > 0$ *such that for any* $k > k_0$, *one can construct a family of finite length* $\mathcal{N}(\Gamma)$*-Hilbert cochain subcomplex* $(C^*(u), d^{\widetilde{F}}_{u,a})$ *of* $(\Omega^*_{u,a}(\widetilde{M},\widetilde{F}), d^{\widetilde{F}})$ *such that*

(i) *One has the inclusion relation of cochain complexes*

$$(L^2_{u,a,[0,1]}(\Omega^*(\widetilde{M},\widetilde{F})), d^{\widetilde{F}}_{u,a}) \subseteq (C^*(u), d^{\widetilde{F}}_{u,a}); \tag{3.49}$$

(ii) *The cochain complex* $(C^*(u), d^{\widetilde{F}}_{u,a})$ *depends smoothly on* $u \in [0,1]$, *and* $(C^*(u_0), d^{\widetilde{F}}_{u_0,a}) = (L^2_{u_0,a,[0,k]}(\Omega^*(\widetilde{M},\widetilde{F})), d^{\widetilde{F}})$.

Proof. Proposition 3.2 can be proved in the same way as in [32], Proposition 3.7 where we take $u_0 = 0$, with easy modifications with respect to the appearance of the boundary ∂M. The only places need to take more care are listed as follows:

1. One notes here that the analogue of [32], (3.32) still holds here, as by Theorem 3.1, $\mathrm{Im}(P_{[0,k],u})$ consists of smooth forms. Thus $d^{\widetilde{F}}_{u,a}$ acts on them just as usual $d^{\widetilde{F}}$, not depending on u. By setting $d^{\widetilde{F}}$ to be $d^{\widetilde{F}}_{u,a}$ in an analogue of [32], (3.35), one can complete the proof of (i) easily.

2. For the proof of (ii), one needs to modify the proof of [32], Lemma 3.8. Here, one needs to take care about the analogue of [32], (3.39). For such an analogue holds, we need to assume that $x \in \Omega^*_{0,a}(\widetilde{M},\widetilde{F})$. Indeed, if we fix a Γ-invariant first Sobolev norm denoted by $\|\cdot\|_1$, then it is easy to see that there exist A_1, $B_1 > 0$ such that for any smooth form $x \in \Omega^*(\widetilde{M},\widetilde{F})$ and any $u \in [0,1]$, one has

$$\|\widetilde{D}_u x\|_{0,u} \leq A_1\|x\|_1 + B_1\|x\|_0, \tag{3.50}$$

while there exist A_2, $B_2 > 0$ such that for any $x \in \Omega^*_{0,a}(\widetilde{M},\widetilde{F})$, one has

$$A_2\|x\|_1 - B_2\|x\|_0 \leq \|\widetilde{D}x\|_0. \tag{3.51}$$

From (3.50) and (3.51), one sees that there exist A, $B > 0$ such that $x \in \Omega^*_{0,a}(\widetilde{M},\widetilde{F})$, one has

$$\|\widetilde{D}_u x\|_{0,u} \leq A\|\widetilde{D}x\|_0 + B\|x\|_0, \tag{3.52}$$

which is exactly the analogue of [32], (3.39) we need.

One can then proceed as in [32], Proof of Lemma 3.8 to complete the proof of (ii). □

We now come back to the proof of Theorem 3.2 for $u = 0$.

By (3.48), one gets that for any $0 \leq u \leq 1$,

$$T^{(2)}_{RS}(\widetilde{M}, F, g^{TM}_u, g^F_u) = T_{(C^*(u),d^{\widetilde{F}})} \cdot T_{(C^*(u),d^{\widetilde{F}}),\perp}. \tag{3.53}$$

For any $s \in \mathbf{C}$ with $\mathrm{Re}(s) > \frac{n}{2}$ and $0 \leq u \leq 1$, set

$$\theta_u(s) = \sum_{i=0}^{n} (-1)^i i \zeta^i_{C^*(u),\perp}(s). \tag{3.54}$$

From (3.46) and (3.54), one can rewrite $\theta_u(s)$ as

$$\theta_u(s) = \frac{1}{\Gamma(s)} \int_0^{+\infty} t^{s-1} \Big(\mathrm{Tr}_{\mathcal{N},s} \Big[N \exp\Big(-t\widetilde{D}^2_{u,a} \Big) \Big] - \mathrm{Tr}_{\mathcal{N},s} \Big[N \exp\Big(-t D^2_{C^*(u)} \Big) \Big] \Big)\, dt. \tag{3.55}$$

For any $0 \leq u \leq 1$, let $P_{C^*(u)}$ denote the orthogonal projection from $L^2_u(\Omega^*(\widetilde{M}, \widetilde{F}))$ onto $C^*(u)$. Then by Proposition 3.2, $P_{C^*(u)}$ depends smoothly on $u \in [0,1]$. Moreover, one has

$$d^{\widetilde{F}} P_{C^*(u)} = P_{C^*(u)} d^{\widetilde{F}} P_{C^*(u)}. \tag{3.56}$$

Let $d^{\widetilde{F}*}_{C^*(u)} : C^*(u) \to C^*(u)$ be the formal adjoint of

$$d^{\widetilde{F}}_{C^*(u)} = P_{C^*(u)} d^{\widetilde{F}} P_{C^*(u)} : C^*(u) \to C^*(u). \tag{3.57}$$

Then in view of (3.56), one has

$$d^{\widetilde{F}*}_{C^*(u)} = P_{C^*(u)} d^{\widetilde{F}*}_{u,a} P_{C^*(u)} = P_{C^*(u)} d^{\widetilde{F}*}_{u,a}. \tag{3.58}$$

Set

$$\widetilde{D}_{C^*(u)} = d^{\widetilde{F}}_{C^*(u)} + d^{\widetilde{F}*}_{C^*(u)}. \tag{3.59}$$

One has, similar as in (3.39), that

$$[\widetilde{D}_{C^*(u)}, N] = -d^{\widetilde{F}}_{C^*(u)} + d^{\widetilde{F}*}_{C^*(u)}. \tag{3.60}$$

In order to have a formula for $\frac{\partial}{\partial u} d^{\widetilde{F}*}_{C^*(u)}$ similar to (3.28), by using (3.28) and (3.58), we compute

$$\begin{aligned}
\tfrac{\partial}{\partial u} d^{\widetilde{F}*}_{C^*(u)} &= \tfrac{\partial}{\partial u} \Big(P_{C^*(u)} d^{\widetilde{F}*}_u \Big) = \big(\tfrac{\partial}{\partial u} P_{C^*(u)} \big) d^{\widetilde{F}*}_u + P_{C^*(u)} \tfrac{\partial}{\partial u} d^{\widetilde{F}*}_u \\
&= \big(\tfrac{\partial}{\partial u} P_{C^*(u)} \big) d^{\widetilde{F}*}_u + P_{C^*(u)} [d^{\widetilde{F}*}_u, Q_u] \\
&= \big(\tfrac{\partial}{\partial u} P_{C^*(u)} \big) d^{\widetilde{F}*}_u + P_{C^*(u)} d^{\widetilde{F}*}_u Q_u - P_{C^*(u)} Q_u d^{\widetilde{F}*}_u \\
&= [d^{\widetilde{F}*}_{C^*(u)}, Q_u] + \big(\tfrac{\partial}{\partial u} P_{C^*(u)} \big) d^{\widetilde{F}*}_u + Q_u P_{C^*(u)} d^{\widetilde{F}*}_u - P_{C^*(u)} Q_u d^{\widetilde{F}*}_u.
\end{aligned} \tag{3.61}$$

Since $C^*(u)$, $0 \leq u \leq 1$, are finitely generated Hilbert modules, by using (3.60), (3.61), as in [32], (3.63), one deduces

$$\begin{aligned}
&\frac{\partial}{\partial u}\mathrm{Tr}_{\mathcal{N},s}\left[N\exp\left(-t\widetilde{D}^2_{C^*(u)}\right)\right] = -t\mathrm{Tr}_{\mathcal{N},s}\left[N\frac{\partial\widetilde{D}^2_{C^*(u)}}{\partial u}\exp\left(-t\widetilde{D}^2_{C^*(u)}\right)\right]\\
&\qquad = -t\mathrm{Tr}_{\mathcal{N},s}\left[\left[N,\widetilde{D}_{C^*(u)}\right]\frac{\partial d^{\widetilde{F}*}_{C^*(u)}}{\partial u}\exp\left(-t\widetilde{D}^2_{C^*(u)}\right)\right]\\
&= t\frac{\partial}{\partial t}\mathrm{Tr}_{\mathcal{N},s}\left[Q_u\exp\left(-t\widetilde{D}^2_{C^*(u)}\right)\right] - t\mathrm{Tr}_{\mathcal{N},s}\left[\left(d^{\widetilde{F}}_{C^*(u)} - d^{\widetilde{F}*}_{C^*(u)}\right)\right.\\
&\left.\left(\frac{\partial P_{C^*(u)}}{\partial u}d^{\widetilde{F}*}_u + Q_uP_{C^*(u)}d^{\widetilde{F}*}_u - P_{C^*(u)}Q_ud^{\widetilde{F}*}_u\right)\exp\left(-t\widetilde{D}^2_{C^*(u)}\right)\right]\\
&= t\frac{\partial}{\partial t}\mathrm{Tr}_{\mathcal{N},s}\left[Q_u\exp\left(-t\widetilde{D}^2_{C^*(u)}\right)\right] - t\mathrm{Tr}_{\mathcal{N},s}\left[\left(d^{\widetilde{F}}_{C^*(u)} - d^{\widetilde{F}*}_{C^*(u)}\right)\right.\\
&\left.\left(P_{C^*(u)}\frac{\partial P_{C^*(u)}}{\partial u}d^{\widetilde{F}*}_uP_{C^*(u)} + Q_u[P_{C^*(u)}, d^{\widetilde{F}*}_u]\right)\exp\left(-t\widetilde{D}^2_{C^*(u)}\right)\right]. \qquad (3.62)
\end{aligned}$$

Denote for $0 \leq u \leq 1$ that

$$\begin{aligned}
f(u) = &\left(d^{\widetilde{F}}_{C^*(u)} - d^{\widetilde{F}*}_{C^*(u)}\right)\cdot\ldots\\
&\ldots\cdot\left(P_{C^*(u)}\frac{\partial P_{C^*(u)}}{\partial u}d^{\widetilde{F}*}_uP_{C^*(u)} + Q_u[P_{C^*(u)}, d^{\widetilde{F}*}_u]\right). \qquad (3.63)
\end{aligned}$$

Since $C^*(u)$ contains $L^2_{u,[0,1]}(\Omega^*(\widetilde{M},\widetilde{F}))$ for $0 \leq u \leq 1$ (cf. (3.49)), one sees that when $t \to +\infty$,

$$\mathrm{Tr}_{\mathcal{N},s}\left[Q_u\exp\left(-t\widetilde{D}^2_{u,a}\right)\right] - \mathrm{Tr}_{\mathcal{N},s}\left[Q_u\exp\left(-t\widetilde{D}^2_{C^*(u)}\right)\right]$$

is of exponential decay.

On the other hand, since, when restricted to the subcomplex $(L^2_{u,a,[0,1]}(\Omega^*(\widetilde{M},\widetilde{F})), d^{\widetilde{F}}_{u,a})$ of $(C^*(u), d^{\widetilde{F}})$, $d^{\widetilde{F}*}_u$ commutes with $P_{C^*(u)}$, while

$$P_{C^*(u)}\frac{\partial P_{C^*(u)}}{\partial u}P_{C^*(u)} = 0, \qquad (3.64)$$

from (3.63), (3.64) one gets

$$f(u)|_{L^2_{u,[0,1]}(\Omega^*(\widetilde{M},\widetilde{F}))} = 0. \qquad (3.65)$$

From (3.49) and (3.65), one sees that as $t \to +\infty$,

$$\mathrm{Tr}_{\mathcal{N},s}\left[f(u)\exp\left(-t\widetilde{D}^2_{C^*(u)}\right)\right]$$

is of exponential decay.

By (3.33), (3.55), (3.62), (3.63) and (3.65), we have for Re(s) large enough that

$$\begin{aligned}\frac{\partial\theta_u(s)}{\partial u} &= \frac{1}{\Gamma(s)}\int_0^{+\infty} t^s\frac{\partial}{\partial t}\left(\mathrm{Tr}_{\mathcal{N},s}\left[Q_u\exp\left(-t\widetilde{D}_{u,a}^2\right)\right]\right.\\ &\qquad \left.-\mathrm{Tr}_{\mathcal{N},s}\left[Q_u\exp\left(-t\widetilde{D}_{C^*(u)}^2\right)\right]\right)dt\\ &\quad -\frac{1}{\Gamma(s)}\int_0^{+\infty} t^s\mathrm{Tr}_{\mathcal{N},s}\left[f(u)\exp\left(-t\widetilde{D}_{C^*(u)}^2\right)\right]dt\\ &= \frac{-s}{\Gamma(s)}\int_0^{+\infty} t^{s-1}\left(\mathrm{Tr}_{\mathcal{N},s}\left[Q_u\exp\left(-t\widetilde{D}_{u,a}^2\right)\right]\right.\\ &\qquad \left.-\mathrm{Tr}_{\mathcal{N},s}\left[Q_u\exp\left(-t\widetilde{D}_{C^*(u)}^2\right)\right]\right)dt\\ &\quad -\frac{1}{\Gamma(s)}\int_0^{+\infty} t^s\mathrm{Tr}_{\mathcal{N},s}\left[f(u)\exp\left(-t\widetilde{D}_{C^*(u)}^2\right)\right]dt. \end{aligned} \tag{3.66}$$

Now by using the finite propagation speed of solutions of hyperbolic equations (cf. [29], §2.8, §6.1), we know from [19], Theorem 2.26 that as $t\to 0^+$, for any positive integer l one has an asymptotic expansion

$$\begin{aligned}\mathrm{Tr}_{\mathcal{N},s}\left[Q_u\exp\left(-t\widetilde{D}_{u,a}^2\right)\right] &= \mathrm{Tr}_s\left[Q_u\exp\left(-tD_{u,a}^2\right)\right]+o(u^{l/2}) \\ &= \sum_{j=-n}^{l} M_{j,u}t^{j/2}+o(u^{l/2}).\end{aligned} \tag{3.67}$$

From (3.66) and (3.67), one finds that for any $0\le u\le 1$, one has

$$\begin{aligned}\frac{\partial}{\partial u}\left(\left.\frac{\partial\theta_u(s)}{\partial s}\right|_{s=0}\right) &= -M_{0,u}+\mathrm{Tr}_{\mathcal{N},s}\left[Q_uP_{C^*(u)}\right]\\ &\quad -\int_0^{+\infty}\mathrm{Tr}_{\mathcal{N},s}\left[f(u)\exp\left(-t\widetilde{D}_{C^*(u)}^2\right)\right]dt.\end{aligned} \tag{3.68}$$

Now observe that we are applying Proposition 3.2 for $u_0=0$, thus one has, as in [32], (3.70),

$$(C^*(0),d^{\widetilde{F}}) = (L^2_{0,a,[0,k]}(\Omega^*(\widetilde{M},\widetilde{F})),d_a^{\widetilde{F}}). \tag{3.69}$$

Thus one again has the fact that $d_u^{\widetilde{F}*}$ commutes with $P_{C^*(u)}$, which, together with (3.64), implies that

$$f(0)=0. \tag{3.70}$$

From (3.47), (3.54), (3.68) and (3.70), one finds

$$\frac{\partial \log T_{(C^*(u),d^{\tilde{F}}),\perp}}{\partial u}\bigg|_{u=0} = -\frac{M_{0,0}}{2} + \frac{1}{2}\mathrm{Tr}_{\mathcal{N},s}\left[Q_0 P_{C^*(0)}\right]. \tag{3.71}$$

Now let us consider the variation of $T_{(C^*(u),d^{\tilde{F}})}$ near $u = 0$. Observe that for any

$$\omega,\ \omega' \in C^*(0) = L^2_{a,[0,k]}(\Omega^*(\widetilde{M}, \widetilde{F})),$$

the induced inner product of them in $C^*(u)$ is given by

$$\begin{aligned}\langle P_{C^*(u)}\omega, P_{C^*(u)}\omega'\rangle_u = \langle \omega, P_{C^*(u)}\omega'\rangle_u &= \int_{\widetilde{M}} \left(\omega \wedge *^{\tilde{F}}_u P_{C^*(u)}\omega'\right)_{\tilde{F}} \\ &= \left\langle \omega, \left(*^{\tilde{F}}\right)^{-1} *^{\tilde{F}}_u P_{C^*(u)}\omega' \right\rangle. \end{aligned} \tag{3.72}$$

Set for $0 \le u \le 1$ that

$$A_u = P_{C^*(0)} \left(*^{\tilde{F}}\right)^{-1} *^{\tilde{F}}_u P_{C^*(u)} P_{C^*(0)} : C^*(0) \to C^*(0). \tag{3.73}$$

From (2.27)-(2.29), (3.45), (3.72) and (3.73), one finds,

$$\log \frac{T_{(C^*(u),d^{\tilde{F}})}}{T_{(C^*(0),d^{\tilde{F}})}} = -\frac{1}{2}\sum_{i=0}^{n}(-1)^i \log \mathrm{Det}_{\tau_{\mathcal{N}(\Gamma)}} \left(A_u|_{C^i(0)}\right). \tag{3.74}$$

From (2.18) and (3.74), one deduces

$$\frac{\partial}{\partial u} \log \frac{T_{(C^*(u),d^{\tilde{F}})}}{T_{(C^*(0),d^{\tilde{F}})}} = -\frac{1}{2}\mathrm{Tr}_{\mathcal{N},s}\left[A_u^{-1}\frac{\partial A_u}{\partial u}\right]. \tag{3.75}$$

By (3.73), one sees directly that

$$A_u|_{u=0} = \mathrm{Id}|_{C^*(0)}. \tag{3.76}$$

From (3.20), (3.64), (3.73), (3.75) and (3.76), one finds

$$\begin{aligned}\frac{\partial}{\partial u}\bigg|_{u=0} \log \frac{T_{(C^*(u),d^{\tilde{F}})}}{T_{(C^*(0),d^{\tilde{F}})}} &= -\frac{1}{2}\mathrm{Tr}_{\mathcal{N},s}\left[P_{C^*(0)} \left(*^{\tilde{F}}\right)^{-1} \frac{\partial *^{\tilde{F}}_u}{\partial u}\bigg|_{u=0} P_{C^*(0)}\right] \\ &= -\frac{1}{2}\mathrm{Tr}_{\mathcal{N},s}\left[Q_0 P_{C^*(0)}\right]. \end{aligned} \tag{3.77}$$

From (3.53), (3.71) and (3.77), one gets

$$\frac{\partial}{\partial u}\bigg|_{u=0} \log \frac{T^{(2)}_{RS}(\widetilde{M}, F, g^{TM}_u, g^F_u)}{T^{(2)}_{RS}(\widetilde{M}, F, g^{TM}, g^F)} = -\frac{M_{0,0}}{2}. \tag{3.78}$$

Since (3.78) holds for arbitrary (g^{TM}, g^F), one gets indeed that for any $0 \le u \le 1$,

$$\frac{\partial}{\partial u} \log \frac{T^{(2)}_{RS}(\widetilde{M}, F, g^{TM}_u, g^F_u)}{T^{(2)}_{RS}(\widetilde{M}, F, g^{TM}, g^F)} = -\frac{M_{0,u}}{2}. \tag{3.79}$$

Now by using (3.67), one sees that for any $0 \le u \le 1$, $M_{0,u}$ is exactly the same quantity appears in [4], (7), (8) and [5], Theorem 4.5, where a similar result is proved for the usual Ray-Singer metrics.

The proof of Theorem 3.2 is complete. □

Remark 3.3. If for any $u \in [0,1]$, $\mathrm{Spec}(\widetilde{D}^2_{u,a})$ contains a non-empty gap, then the proof of Theorem 3.2 can be simplified a lot. Here we did not make this assumption as usually $\mathrm{Spec}(\widetilde{D}^2_{u,a})$, $u \in [0,1]$, may not be discrete when Γ is an infinite group.

Acknowledgement. This work was partially supported by MOEC and the 973 project of MOSTC. Part of this work was done while the first author was visiting the Nankai Institute (now the Chern Institute) of Mathematics during July-August of 2005. He would like to thank the Nankai Institute of Mathematics for the hospitality.

References

1. J.-M. Bismut, H. Gillet and C. Soulé, Analytic torsion and holomorphic determinant bundles III. *Comm. Math. Phys.* 115 (1998), 301-351.
2. J.-M. Bismut and W. Zhang, An extension of a theorem by Cheeger and Müller. *Astérisque* Tom. 205, Paris, (1992).
3. M. Braverman, A. Carey, M. Farber and V. Mathai, L^2 torsion without the determinant class condition and extended L^2 cohomology. *Comm. Contemp. Math.* 7 (2005), 421-462.
4. J. Brüning and X. Ma, An anomaly formula for Ray-Singer metrics on manifolds with boundary. *C. R. Acad. Sci. Paris, Ser. I*, 335 (2002), 603-608.
5. J. Brüning and X. Ma, An anomaly formula for Ray-Singer metrics on manifolds with boundary. To appear in *Geom. Funct. Anal.*
6. D. Burghelea, L. Friedlander and T. Kappeler, Torsions for manifolds with boundary and glueing formulas. *Math. Nachr.* 208 (1999), 31-91.
7. D. Burghelea, L. Friedlander, T. Kappeler and P. McDonald, Analytic and Reidemeister torsion for representations in finite type Hilbert modules. *Geom. Funct. Anal.* 6 (1996), 751-859.
8. A. Carey, M. Farber and V. Mathai, Determinant lines, von Neumann algebras and L^2 torsion. *J. Reine Angew. Math.* 484 (1997), 153-181.
9. A. Carey and V. Mathai, L^2-torsion invariants. *J. Funct. Anal.* 110 (1992), 337-409.

10. J. Cheeger, Analytic torsion and the heat equation. *Ann. of Math.* 109 (1979), 259-332.
11. X. Dai and H. Fang, Analytic torsion and R-torsion for manifolds with boundary. *Asian J. Math.* 4 (2000), 695-714.
12. J. Dixmier, von Neumann Algebras. *North-Holland Publishing Co.*, Amsterdam-New York, 1981. xxxviii+437 pp.
13. M. Farber, Von Neumann categories and extended L^2-cohomology. *K-Theory* 15 (1998), 347-405.
14. B. Fuglede and R. V. Kadison, Determinant theory in finite factors. *Ann. of Math.* 55 (1952), 520-530.
15. J. Lott, Heat kernels on covering spaces and topological invariants. *J. Diff. Geom.* 35 (1992), 471-510.
16. J. Lott and M. Rothenberg, Analytic torsion for group actions. *J. Diff. Geom.* 34 (1991), 431-481.
17. W. Lück, Analytic and topological torsion for manifolds with boundary and symmetry. *J. Diff. Geom.* 37 (1993), 263-322.
18. W. Lück, *L^2-Invariants: Theory and Applications to Geometry and K-Theory.* Springer-Verlag Berlin Heidelberg 2002.
19. W. Lück and T. Schick, L^2-torsion of hyperbolic manifolds of finite volume. *Geom. Funct. Anal.* 9 (1999), 518-567.
20. V. Mathai, L^2-analytic torsion. *J. Funct. Anal.* 107 (1992), 369-386.
21. J. Milnor, Whitehead torsion. *Bull. Amer. Math. Soc.* 72 (1966), 358-426.
22. W. Müller, Analytic torsion and the R-torsion of Riemannian manifolds. *Adv. in Math.* 28 (1978), 233-305.
23. W. Müller, Analytic torsion and the R-torsion for unimodular representations. *J. Amer. Math. Soc.* 6 (1993), 721-753.
24. S. P. Novikov and M. A. Shubin, Morse theory and von Neumann invariants on non-simply connected manifolds. *Uspehi. Mat. Nauk* 41 (1986), 222-223.
25. D. Quillen, Determinants of Cauchy-Riemann operators over a Riemann surface. *Funct. Anal. Appl.* 14 (1985), 31-34.
26. D. Quillen, Superconnections and the Chern character. *Topology* 24 (1985), 89-95.
27. D. B. Ray and I. M. Singer, *R*-torsion and the Laplacian on Riemannian manifolds. *Adv. in Math.* 7 (1971), 145-210.
28. M. Shubin, De Rham theorem for extended L^2-cohomology. *Voronezh Winter Mathematical School,* Amer. Math. Soc., 1998, pp. 217-231.
29. Taylor M., *Partial Differential Equations I. Basic Theory.* Applied Mathematical sciences 115, Springer 1996.
30. E. Witten, Supersymmetry and Morse theory. *J. Diff. Geom.* 17 (1982), 661-692.
31. W. Zhang, *Lectures on Chern-Weil Theory and Witten Deformations,* Nankai Tracks in Mathematics, Vol. 4. World Scientific, Singapore, 2001.
32. W. Zhang, An extended Cheeger-Müller theorem for covering spaces. *Topology* 44 (2005), 1093-1131.

Received by the editors September 6, 2005; Revised December 13, 2005

Analysis, Geometry and Topology
of Elliptic Operators, pp. 263–294

CONFORMAL ANOMALIES VIA CANONICAL TRACES

SYLVIE PAYCHA
Laboratoire de Mathématiques
Université Blaise Pascal (Clermont II)
63177 Aubière Cedex, France
spaycha@math.univ-bpclermont.fr

STEVEN ROSENBERG
Department of Mathematics and Statistics
Boston University
Boston, MA 02215, USA
sr@math.bu.edu

Dedicated to our friend Krzysztof Wojciechowski

Using Laurent expansions of canonical traces of holomorphic families of classical pseudodifferential operators, we define functionals on the space of Riemannian metrics and investigate their conformal properties, thereby giving a unified description of several conformal invariants and anomalies.

2000 *Mathematics Subject Classification.* Primary 58J40; Secondary 53B20, 58J28, 58J52

1. Introduction

In this paper, we use the Kontsevich-Vishik canonical trace to produce a series of conformal spectral invariants (or covariants or anomalies) associated to conformally covariant pseudodifferential operators. Although only one covariant is new, the use of canonical traces provides a systematic treatment of these covariants.

The search for conformal anomalies is motivated by both string theory and conformal geometry. Historically, the variation of functionals $\mathcal{F}$ on the space of Riemannian metrics $\mathrm{Met}(M)$ on a closed manifold M under conformal transformations:

$$g \mapsto e^{2f}g, \quad f \in C^\infty(M, \mathbb{R})$$

has been a topic of interest to both mathematicians and physicists going back at least to Hermann Weyl (see Duff [Du] for a historical review of the physics literature, and Chang [C] for a survey of recent work in mathematics). In physics, the study of conformal invariants underwent a revival in the early 1980s with Polyakov's work [Pol] on the conformal anomaly of bosonic strings, one of the motivating factors behind the development of determinant line bundles in mathematics.

The conformal anomaly of a Fréchet differentiable map $\mathcal{F} : \mathrm{Met}(M) \to \mathbb{C}$ at a given (background) metric g is the differential at 0 of $\mathcal{F}_g : C^\infty(M, \mathbb{R}) \to \mathbb{C}$, $\mathcal{F}_g(f) := \mathcal{F}(e^{2f}\, g)$. Thus the conformal anomaly in the direction f is

$$\delta_f \mathcal{F}_g := d\mathcal{F}_g(0).f = \frac{d}{dt}\Big|_{t=0} \mathcal{F}(e^{2tf}\, g).$$

A functional $\mathcal{F}$ is conformally invariant if $\delta_f \mathcal{F}_g = 0$ for any Riemannian metric g and any smooth function f. If $\delta_f \mathcal{F}_g = \int_M f(x) \overline{\delta_x \mathcal{F}_g}(x) \mathrm{dvol}_g(x)$, then $\overline{\delta_x \mathcal{F}_g}(x)$ is called the the local (conformal) anomaly of $\mathcal{F}_g$ (or equivalently of $\mathcal{F}$ in the background metric g). A functional $\mathcal{F}(g, x)$ on $\mathrm{Met}(M) \times M$ is conformally covariant if, roughly speaking, $\delta_f \mathcal{F}$ does not depend on derivatives of f and g.

Conformal anomalies arise naturally in quantum field theory. A conformally invariant classical action $\mathcal{A}(g)$ in a background metric g, for example the string theory or nonlinear sigma model action, does not usually lead to a conformally invariant effective action $\mathcal{W}(g)$, since the quantization procedure breaks the conformal invariance and hence gives rise to a conformal anomaly. In particular, in string theory the conformal invariance persists after quantization only in specific critical dimensions.

From a path integral point of view, the conformal anomaly of the quantized action is often said to arise from a lack of conformal invariance of the formal measure on the configuration space of the QFT. Whatever this means, we can detect the source of the conformal anomaly in the quantization procedure. In order to formally reduce the path integral to a Gaussian integral, one writes the classical action as a quadratic expression $\mathcal{A}(g)(\phi) = \langle A_g \phi, \phi \rangle_g$ where ϕ is a field, typically a tensor on M, A_g a differential operator on tensors and $\langle \cdot, \cdot \rangle_g$ the inner product induced by g. Because this inner product is not conformally invariant, the conformal invariance of $\mathcal{A}(g)$ usually translates to a conformal covariance of the operator A_g. An operator A_g is *conformally covariant* of bidegree $(a, b) \in \mathbb{R}^2$ if

$$A_{\bar{g}} = e^{-b\,f} A_g e^{a\,f}, \tag{1}$$

for $\bar{g} = e^{2f}g$. Thus this first step, which turns a conformally invariant quantity (the classical action) to a conformally covariant operator, already breaks the conformal invariance.

The second step in the computation of the path integral uses an Ansatz to give a meaning to the formal determinants that arise from the Gaussian integration. Mimicing finite dimensional computations, the effective action derived from a formal integration over the configuration space $\mathcal{C}$ is

$$e^{-\frac{1}{2}\mathcal{W}(g)} := \int_{\mathcal{C}} e^{-\frac{1}{2}\mathcal{A}(g)(\phi)}\,\mathcal{D}\phi = \text{"det"}(A_g)^{-\frac{1}{2}}.$$

If there were a well defined determinant "det" on differential operators with the usual properties, (1) would yield

$$\begin{aligned}\text{"det"}(A_{e^{2f}g}) &= \text{"det"}(e^{-b\,f}A_g e^{a\,f})\\ &= \text{"det"}(e^{-b\,f})\,\text{"det"}(A_g)\,\text{"det"}(e^{a\,f})\\ &= \text{"det"}(e^{(a-b)\,f})\,\text{"det"}(A_g),\end{aligned}$$

where $e^{c\,f}$ is treated as a multiplication operator for $c \in \mathbb{R}$. Hence, even if a "good" determinant exists, the effective action $\mathcal{W}(g)$ would still suffer a conformal anomaly, since A_g is only conformally covariant:

$$\delta_f \mathcal{W}(g) = \delta_f \log \text{"det"}(A_g) = \delta_f \log \text{"det"}(e^{(a-b)\,f}) = (a-b)\,\text{"tr"}(f),$$

where "tr" is a hypothetical trace associated to "det".

The ζ-determinant $\det_\zeta$ on operators is used by both physicists and mathematicans as an Ersatz for the usual determinant on matrices. Since the work of Wodzicki and Kontsevich–Vishik, we know the ζ-determinant has a multiplicative anomaly, which fortunately does not affect our rather specific situation. Indeed, the above heuristic derivation holds (Branson-Orsted [BO], Parker-Rosenberg [PR], Rosenberg [R]):

$$\delta_f \,\log\det_\zeta(A_g) = (a-b)\,\mathrm{tr}^{A_g}(f),$$

if one replaces "tr"(f) with $\mathrm{tr}^{A_g}(f)$, the finite part in the heat-operator expansion $\mathrm{tr}(f\,e^{-\epsilon A_g})$ when $\epsilon \to 0$. (Here and whenever the heat operator $e^{-\epsilon A_g}$ appears, we assume that A_g is elliptic with non-negative leading symbol.) In summary, the regularization procedures involved in the ζ-determinant and the finite part of the heat-operator expansion are *not* responsible for the conformal anomaly of the effective action $\mathcal{W}(g)$; the conformal anomaly appears as soon as one uses the *conformally covariant* operator A_g associated to the originally *conformally invariant* action $\mathcal{A}(g)$.

These QFT arguments lead to the search for conformally covariant operators and associated spectral conformal covariants. There are four types of conformal covariants in the literature, in order of computational difficulty: (i) local covariants, those that depend only on the metric at a fixed point; (ii) global invariants which are the integrals of (noncovariant) local quantities; (iii) global invariants which are not integrals of local expressions, but whose variation in any metric direction is local; (iv) global invariants which are not integrals of local expression, and whose variation in conformal directions is nonlocal. All four types have examples associated to spectral ζ- and η-functions, as we now explain.

For (i), the residue at $z = 1$ of the local zeta function $\zeta_{A_g}(z, x)$, which turns out to be proportional to the local Wodzicki residue $\text{res}_x(A_g^{-1})$, is a pointwise conformal covariant for a conformally covariant operator A_g, under certain ellipticity and positivity conditions on the operator [PR]. (A classical example of a pointwise invariant is the length of the Weyl tensor [We].) For (ii), the value at $z = 0$ of the global ζ-function $\zeta_{A_g}(z)$ of a conformally covariant operator A_g is conformally invariant, again for certain operators, which may be pseudodifferential [PR, R]:

$$\delta_f \zeta_{A_g}(0) = 0.$$

In hindsight, this can be predicted by thinking of $\zeta_{A_g}(0)$ as an Ersatz for "tr"(Id) in the heuristic notation above. It is well known that $\zeta_{A_g}(0)$ is the integral of the finite part of the pointwise heat kernel of A_g (up to the nonlocal conformally invariant term dim $\text{Ker}(A_g)$). When A_g is a differential operator, $\zeta_{A_g}(0) = -\frac{1}{\text{ord}(A_g)}\text{res}(\log A_g)$, so the conformal invariance of $\zeta_{A_g}(0)$ is equivalent to the conformal invariance of the exotic determinant introduced by Wodzicki for zero order classical pseudodiffferential operators and extended by Scott [Sc] to the residue determinant $\det_{\text{res}}(A_g) = e^{\text{res}(\log A_g)}$ on operators of any order. This gives another description of $\zeta_{A_g}(0)$ as the integral of a local quantity, namely the local Wodzicki residue of the logarithm of A_g.

Jumping to (iv), conformal anomalies arising from ζ-determinants of conformally covariant operators vanish in certain cases, for one has [PR, R]

$$\delta_f\, \zeta'_{A_g}(0) = -\delta_f \log \det_\zeta(A_g) = (a-b)\int_M f(x) a_n(A_g, x) \text{dvol}_g(x),$$

where as $\epsilon \to 0$

$$\text{tr}\left(e^{-\epsilon A_g}\right) \sim \sum_{j=0}^{\infty} \left(\int_M a_j(A_g, x)\text{dvol}_g(x)\right) \epsilon^{\frac{j-n}{2\alpha}},$$

for $\alpha = \text{ord}(A_g)$, $n = \dim(M)$; here we assume A_g has all but finitely many eigenvalues nonnegative. The nonlocal nature of the functional determinant and its variation is well known; however, the above formula shows it gives rise to a local conformal anomaly $(a-b)a_n(A_g,x)$. In particular, $\zeta'_{A_g}(0)$ yields a conformal invariant in odd dimensions, as $a_n(A_g)$ then vanishes. The conformal anomaly $\delta_f \log \det_\zeta(\Delta_g)$, where Δ_g is the Laplace-Beltrami operator on a closed Riemannian surface, is responsible for the conformal anomaly in bosonic string theory; since the coefficients a, b depend on the dimension of the manifold and the rank of auxiliary tensor bundles, combinations of such conformal anomalies cancel in certain critical dimensions, viz. the cancellation of conformal anomalies in 26 dimensions for bosonic string theory [Pol]. Further work on the conformal anomaly of functional determinants is in work of Branson and Orsted [B1, B2, BO].

For (iii), if A_g is a self-adjoint invertible elliptic operator, the phase of its ζ-determinant can be expressed in terms of the η-invariant $\eta_{A_g}(0)$ by

$$\det_\zeta(A_g) := \det_\zeta(|A_g|)\cdot e^{i\frac{\pi}{2}\left(\zeta_{|A_g|}(0)-\eta_{A_g}(0)\right)}.$$

Again, only in certain dimensions is the phase conformally invariant; namely if $\dim(M)$ and $\mathrm{ord}(A_g)$ have opposite parity [R].

We will study these four types of conformal anomalies and covariants in the common framework of variations of Kontsevich-Vishik functionals of conformally covariant operators. Whereas previous work on conformal anomalies uses heat kernel expansions, we use ζ-function techniques instead. Our starting point is canonical traces, which are cut-off integrals of symbols of non-integer order pseudodifferential operators, which extend to Laurent expansions of cut-off integrals of holomorphic families of symbols. These coefficients are universal expressions in the symbol expansion of the family (Paycha-Scott [PS]), so their regularity properties and their variation in terms of external parameters (here the metric) are easily controlled. We thereby avoid some technical difficulties in the variation of heat kernel asymptotic expansions. The main result of the paper is that the coefficients of the Laurent expansions give explicit conformal anomalies.

In more detail, the three functionals $\zeta_{A_g}(0)$, $\zeta'_{A_g}(0)$ and $\eta_{A_g}(0)$ are all A_g-weighted traces in the notation of the first author [P2], namely $\mathrm{tr}^{A_g}(I)$, $\mathrm{tr}^{A_g}(\log A_g)$ and $\mathrm{tr}^{A_g}(A_g\,|A_g|^{-1})$ respectively. Here, for a weight Q (i.e. an admissible positive order elliptic operator), the Q-weighted trace $\mathrm{tr}^Q(A)$ of a classical pseudodifferential operator A is the finite part at $z=0$ of the meromorphic map $z\mapsto \mathrm{TR}(A\,Q^{-z})$ (up to a factor depending on the kernel of Q), where TR is the Kontsevich-Vishik canonical trace on noninteger order operators extending the usual trace on smoothing operators [KV]. (This definition of weighted trace is equivalent to previous ones [P2] by the discussion after Def. 3.) Thus all our spectral invariants are examples of canonical traces.

If the conformally covariant operator A_g is a weight, we may define functionals given by meromorphic functions $z \mapsto \mathcal{F}_h(g)(z) := \mathrm{TR}(h(A_g)\, A_g^{-z})$ where h is a real or complex valued function defined on a subset $W \subset \mathbb{C}$. In particular, the functionals $\zeta_{A_g}(z)$ and $\eta_{A_g}(z)$ correspond to choosing $h(\lambda) = 1$ (with $W = \mathbb{C}$) and $h(\lambda) = \frac{\lambda}{|\lambda|}$ (with $W = \mathbb{R}/\{0\}$). Using results on the coefficients in the Laurent expansion [PS] for $z \mapsto \mathcal{F}_h(g)(z)$ at $z = 0$, we derive the conformal anomaly of these meromorphic functionals (Theorem 3.1):

$$\begin{aligned}&\delta_f \mathrm{TR}(h(A_g)\, A_g^{-z})\\ &\qquad = (a-b)\mathrm{TR}\left(f\, h'(A_g)\, A_g^{-z+1}\right) - z\,(a-b)\mathrm{TR}\left(f\, h(A_g)\, A_g^{-z}\right).\end{aligned}$$

This formula strongly depends on the tracial nature of the canonical trace TR on noninteger order operators (7). Identifying the coefficients on either side, we get a hierarchy of functionals and their conformal anomalies, the first one involving the Wodzicki residue res:

$$\begin{aligned}
\delta_f \mathrm{res}(h(A_g)) &= (a-b)\,\mathrm{res}\,(f\, h'(A_g)\, A_g)\,;\\
\delta_f \mathrm{tr}^{A_g}(h(A_g)) &= (a-b)\,\mathrm{tr}^{A_g}\,(f\, h'(A_g)\, A_g) + \frac{a-b}{\alpha}\,\mathrm{res}\,(f\, h(A_g))\,;\\
\delta_f \mathrm{tr}^{A_g}(h(A_g)\,\log A_g)) &= (a-b)\,\mathrm{tr}^{A_g}\,(f\, h'(A_g)\, A_g\, \log A_g)\\
&\quad + \frac{b-a}{\alpha}\,\mathrm{tr}^{A_g}\,(f\, h(A_g))\,;\\
&\ \ \vdots\\
\delta_f \mathrm{tr}^{A_g}(h(A_g)\,\log^j A_g)) &= (a-b)\,\mathrm{tr}^{A_g}\left(f\, h'(A_g)\, A_g\, \log^j A_g\right)\\
&\quad + j\,\frac{a-b}{\alpha}\,\mathrm{tr}^{A_g}\left(f\, h(A_g)\, \log^{j-1} A_g\right).
\end{aligned}$$

Different choices for h lead to conformal covariants/anomalies of the four types mentioned above (Theorem 3.2). Applying this to explicit geometric conformally covariant operators such as the Dirac, Paneitz and Peterson operators (see §2.2) yields conformal anomalies and covariants, including a new example associated to the heat kernel asymptotics of conformally covariant pseudodifferential operators. The Laurent approach provides a natural hierarchy among these invariants/covariants: the most divergent term in the Laurent expansion is a conformal invariant; if this global invariant vanishes in a particular case, then the new "most divergent" term, if it is of the form $\int_M \mathcal{I}(g,x)\mathrm{dvol}_g(x)$ tends to give rise to a local conformal anomaly proportional to $\mathcal{I}(g,x)$.

2. Regularized traces

In this section, we recall known results on regularized traces and the Wodzicki residue, and give some extensions to families of operators.

2.1. *Preliminaries*

Let $E \to M$ be a hermitian vector bundle over a closed Riemannian n-manifold M, and let $Cl(M,E)$ denote the algebra of classical pseudodifferential operators (ΨDOs) acting on smooth sections of E. $S^*M \subset T^*M$ denotes the unit cosphere bundle, and tr_x denotes the trace on the fiber E_x of E over $x \in M$.

Definition 2.1. A positive order elliptic operator $Q \in Cl(M,E)$ is *admissible* if there is an angle with vertex 0 which contains the spectrum of the leading symbol $\sigma_L(Q)$ of Q. A choice of a half line $L_\theta = \{re^{i\theta}, r>0\}$ which does not intersect the spectrum of Q (which is discrete since M is compact) is a *spectral cut* for Q, and θ is an *Agmon angle.* An admissible operator is also called a *weight.* $Ell^{adm}_{>0}(M,E)$ (resp. $Ell^{*,adm}_{>0}(M,E)$) is the class of admissible (resp. invertible admissible) elliptic operators of positive order in $C\ell(M,E)$.

Examples of admissible elliptic operators are classical ΨDOs with positive leading symbol such as generalized Laplacians and formally self-adjoint elliptic classical ΨDOs such as Dirac operators in odd dimensions.

An admissible invertible elliptic operator of positive order and with spectral cut L_θ has well-defined complex powers (Seeley [Se]) defined for $\mathrm{Re}(z)$ sufficiently negative by the contour integral

$$Q_\theta^z := \frac{i}{2\pi}\int_{C_\theta} \lambda^z\,(Q-\lambda I)^{-1}\,d\lambda$$

where C_θ is a contour encircling L_θ. One then extends the complex power Q_θ^z to any half plane $\mathrm{Re}\ z<k, k\in\mathbb{N}$ via the formula $Q_\theta^k Q_\theta^{z-k} = Q_\theta^z$. These complex powers clearly depend on the choice of spectral cut. Setting $z=0$, we get

$$Q_\theta^0 = I - \Pi_Q = \frac{i}{2\pi}\int_{C_\theta} (Q-\lambda I)^{-1}\,d\lambda,$$

where Π_Q is the projection onto the generalized kernel of Q. The logarithm of Q, which also depends on the spectral cut, is defined by

$$\log_\theta Q := \frac{d}{dz}\Big|_{z=0} Q_\theta^z.$$

This dependence will be omitted from the notation from now on.

2.2. *The Wodzicki residue*

Let $A \in Cl(M,E)$ have order α and symbol $\sigma(A)(x,\xi) \sim \sum_{j=0}^\infty \psi(\xi)\,\sigma_{\alpha-j}(A)(x,\xi)$, where $\sigma_{\alpha-j}$ is the positively homogeneous component of order $a-j$ and ψ is a smooth cut-off function which is one outside a ball around 0 and vanishes on a smaller such ball. Let $dx = dx^1\wedge\ldots\wedge dx^n$ be the locally defined coordinate form on M, and let $d\xi$ be the volume form on T^*M (or the restriction of $d\xi$ to the unit cosphere bundle $S^*M \subset T^*M$ or to the unit cosphere S_x^*M at a fixed $x \in M$). Then

$$\mathrm{res}_x(A)dx := \left(\int_{S_x^*M} \mathrm{tr}_x\sigma_{-n}(A)(x,\xi)\,d\xi\right) dx,$$

is (nontrivially) a global top degree form on M whose integral

$$\mathrm{res}(A) := \frac{1}{(2\pi)^n}\int_M \mathrm{res}_x(A)\,dx$$

is the *Wodzicki residue* [Wo] *of* A (see Kassel [K], Lesch [L] for a review and further development).

The Wodzicki residue has several striking properties. From its definition, the Wodzicki residue vanishes on differential operators and operators of nonintegral order, but it is nonzero in general. The Wodzicki residue is local, in that it is integral over M of a density which is computed pointwise from a homogeneous component of the symbol. Most importantly, the Wodzicki residue is cyclic on $CL(M,E)$ in the following sense:

$$\mathrm{res}([A\,,B]) = 0, \qquad \text{for all } A, B \in Cl(M,E).$$

The Wodzicki residue extends to logarithms of admissible elliptic operators Q by

$$\begin{aligned}\mathrm{res}(\log Q) &:= \frac{1}{(2\pi)^n}\int_M \mathrm{res}_x(\log Q)dx\\ &:= \frac{1}{(2\pi)^n}\int_{S^*M} \mathrm{tr}_x\sigma_{-n}(\log Q)(x,\xi)d\xi dx\end{aligned}$$

(Okikiolu [O]). More generally, given $A \in C\ell(M,E)$, if

$$\mathrm{res}_x(A \log Q)\, dx := \left(\int_{S_x^*M} \mathrm{tr}_x \sigma_{-n}(A \log Q)(x,\xi) d\xi \right) dx$$

defines a global form on M, we can integrate it over M to define

$$\mathrm{res}(A \log Q) := \frac{1}{(2\pi)^n} \int_{S^*M} \mathrm{tr}_x \sigma_{-n}(A \log Q)(x,\xi) d\xi dx.$$

This holds in particular if A is a differential operator [PS], Thm. 2.5.

The cyclicity of the Wodzicki residue partially extends to logarithmic operators. The Wodzicki residue vanishes on brackets of the type $[A, B \log_\theta Q]$ where $A, B \in C\ell(M,E)$, $Q \in Ell_0^{*,adm}(M,E)$, and $[A,B]$ is a differential operator [O], [PS], Thm. 4.9.

2.3. *The canonical trace*

By a procedure well known to physicists and mathematicians (see Paycha [P1] for a review), a classical symbol σ on $\mathbb{R}^n$, has a cut-off integral in momentum space $\{\xi\}$. To set the notation, let ψ be the cutoff function of §2.2, and set

$$\sigma_{(N)}(x,\xi) := \sigma(x,\xi) - \sum_{j=0}^{N} \psi(\xi)\, \sigma_{\alpha(z)-j}(x,\xi).$$

Proposition 2.1. *Let σ be a classical symbol on an open subset $U \subset \mathbb{R}^n$ of order α. For $x \in U$, let $B_x^*(R) \subset T_x^*U$ be the ball of radius R centered at 0. As $R \to \infty$,*

$$\int_{B_x^*(R)} \mathrm{tr}_x \sigma(x,\xi)\, d\xi \sim \sum_{j=0, \alpha-j+n\neq 0}^{\infty} a_{\sigma,j}(x)\, R^{\alpha-j+n} + b_\sigma(x) \log R + c_\sigma(x), \tag{2}$$

with

$$a_{\sigma,j}(x) = \frac{\int_{S_x^*U} \mathrm{tr}_x(\sigma_{\alpha-j}(x,\xi))\, d\xi}{\alpha - j + n}; \quad b_\sigma(x) = \int_{S_x^*U} \mathrm{tr}_x \sigma_{-n}(x,\xi)\, d\xi$$

and with finite part/cut-off integral

$$\begin{aligned} c_\sigma(x) &:= -\!\!\!\!\!\!\int_{T_x^*U} \mathrm{tr}_x \sigma(x,\xi)\, d\xi \\ &:= \mathrm{fp}_{R\to\infty} \int_{B_x^*(R)} \mathrm{tr}_x \sigma(x,\xi)\, d\xi \\ &= \int_{T_x^*U} \mathrm{tr}_x \sigma_{(N)}(x,\xi)\, d\xi + \sum_{j=0}^{N} \int_{B_x^*(1)} \psi(\xi)\, \mathrm{tr}_x \sigma_{\alpha-j}(x,\xi)\, d\xi \\ &\qquad - \sum_{j=0, \alpha-j+n\neq 0}^{\infty} a_{\sigma,j}(x). \end{aligned} \tag{3}$$

The finite part is independent of reparametrization of R provided $b_\sigma(x)$ vanishes.

Whenever α is nonintegral, via a partition of unity on M one can patch the local cut-off integrals $-\!\!\!\!\!\!\int_{T_x^*U} \mathrm{tr}_x \sigma_A(x,\xi)\, d\xi$ into a cut-off integral

$$\omega_{KV}(A)(x) = -\!\!\!\!\!\!\int_{T_x^*M} \mathrm{tr}_x \sigma_A(x,\xi)\, d\xi \tag{4}$$

on T_x^*M and then integrate over M to get the Kontsevich-Vishik *canonical trace* [KV]

$$\begin{aligned} \mathrm{TR}(A) &:= \frac{1}{(2\pi)^n} \int_M \omega_{KV}(A)(x)\, dx \\ &= \frac{1}{(2\pi)^n} \int_M dx -\!\!\!\!\!\!\int_{T_x^*M} \mathrm{tr}_x \sigma(A)(x,\xi)\, d\xi. \end{aligned} \tag{5}$$

We consider holomorphic families of classical symbols [KV].

Definition 2.2. A family of complex valued classical symbols $z \mapsto \sigma(z)$ on an open subset U of $\mathbb{R}^n$ is *holomorphic* on a subset $W \subset \mathbb{C}$ if:

1. The order $\alpha(z)$ of $\sigma(z)$ is holomorphic[a] in $z \in W$;
2. For any nonnegative integer j, the map $(z, x, \xi) \mapsto \sigma(z)_{\alpha(z)-j}(x,\xi)$ is holomorphic in z and the map $z \mapsto (\sigma(z))_{\alpha(z)-j}$ is a continuous map from W to $C^\infty(T^*U)$ in the standard topology on $C^\infty(T^*U)$.
3. For $N \gg 0$, the truncated kernel

$$K(z)^{(N)}(x,y) := \int_{T_x^*U} e^{i\xi\cdot(x-y)} \sigma(z)_{(N)}(x,\xi) d\xi$$

[a] i.e. differentiable in z

defines a holomorphic map $W \to C^{k(N)}(U \times U), z \mapsto K(z)^{(N)}$ for some $k(N)$ with $\lim_{N\to\infty} k(N) = \infty$.

A family $A(z) \in Cl(M,E)$ of classical ΨDOs is holomorphic for $z \in W \subset \mathbb{C}$ if it is defined in any local trivialization by a holomorphic family of classical symbols $\sigma_{A(z)}$.

The cut-off integral $⨍_{T_x^*U} \mathrm{tr}_x\sigma(z)(x,\xi)d\xi$ is defined whenever the order $\alpha(z)$ is nonintegral. The following extends results of Kontsevich-Vishik on the explicit Laurent expansions of holomorphic families [PS, Thm. 2.4].

Proposition 2.2. *Let $\sigma(z)$ be a holomorphic family of classical symbols on an open set $U \subset \mathbb{R}^n$ of linear order $\alpha(z) = \alpha'(0)\,z + \alpha(0)$ with $\alpha'(0) \neq 0$. Then the map $z \mapsto ⨍_{T_x^*U} \mathrm{tr}_x\sigma(z)(x,\xi)d\xi$ is meromorphic with Laurent expansion at $z = 0$ given by*

$$\begin{aligned}⨍_{T_x^*U} \mathrm{tr}_x\sigma(z)(x,\xi)d\xi = {} & \left(-\frac{1}{\alpha'(0)}\int_{S_x^*U} \mathrm{tr}_x\sigma(0)_{-n}(x,\xi)d\xi\right)\cdot\frac{1}{z} \\ & + \sum_{k=0}^{K}\frac{z^k}{k!}\left(⨍_{T_x^*U} \mathrm{tr}_x\sigma^{(k)}(0)(x,\xi)d\xi \right. \\ & \left. - \frac{1}{\alpha'(0)(k+1)}\int_{S_x^*U} \mathrm{tr}_x\sigma^{(k+1)}(0)_{-n}(x,\xi)d\xi\right) \\ & + \mathrm{O}(z^K),\end{aligned}$$

for $K \geq 0$.

Applying this to the symbols $\sigma_{A(z)}$ of a holomorphic family $A(z)$ of classical ΨDOs, taking the fibrewise trace and replacing U by M via a partition of the unity provides an analogous formula for the first $k+1$ terms of the Laurent expansion around 0 of $\omega_{KV}(A(z))(x)$ defined by (4) with A replaced by $A(z)$ and hence, after integration over M, of the canonical trace $\mathrm{TR}(A(z))$.

Remark 2.1. (i) Even though $\sigma(z)$ is a classical symbol, $\sigma^{(k)}(0)$ need not be [PS].

(ii) Since $\omega_{KV}(A(z))(x)dx = \left(⨍_{T_x^*M} \mathrm{tr}_x\sigma_{A(z)}(x,\xi)\,d\xi\right)dx$ defines a global form, the coefficient of z^k in the Laurent expansion of $\omega_{KV}(A(z))(x)dx$ also gives rise to a global form.

(iii) For a classical ΨDO A of order α, the operator $A(z) = AQ_\theta^{-z}$ defines a holomorphic family of classical ΨDOs of order $\alpha(z) = -qz + \alpha$. From Proposition 2.2, we recover the well known result relating the Wodzicki residue to a complex residue:

$$\mathrm{Res}_{z=0}\omega_{KV}(AQ^{-z})(x) = -\frac{1}{q}\mathrm{res}_x(A), \tag{6}$$

which after integration over M yields

$$\mathrm{Res}_{z=0}\mathrm{TR}(AQ^{-z}) = -\frac{1}{q}\mathrm{res}(A).$$

If A is a differential operator, $\int\!\!\!\!-_{T_x^*M}\mathrm{tr}_x\sigma(0)(x,\xi)d\xi = \int\!\!\!\!-_{T_x^*M}\mathrm{tr}_x\sigma_{A(0)}(x,\xi)d\xi$ vanishes. Therefore $\left(\int_{S_x^*M}\mathrm{tr}_x\left(\sigma'(0)\right)_{-n}(x,\xi)\right)d\xi dx$ defines a global form, whose integral is $-\mathrm{res}\left(A\log_\theta Q\right)$ [PS].

2.4. *Weighted traces*

For a weight Q with spectral cut and a nonnegative integer k, set

$$\mathcal{A}^k(M,E) := \{\sum_{j=0}^{k} A_j \log^j Q,\ A_j \in Cl(M,E), \quad 0 \le j \le k\}.$$

Operators in $\mathcal{A}^k(M,E)$ coincide with Lesch's log-polyhomogeneous operators [L]. $\mathcal{A}^k(M,E)$ is in fact independent of the choice of Q (Ducourtioux [D, PS]) and coincides with the class $Cl^{*,k}(M,E)$ of Lesch [L]. Note that $\mathcal{A}^0(M,E) = Cl(M,E)$. The order of $A_j \log^j Q$ is defined to be the order of A_j.

Cut-off integrals extend [L] to symbols of operators in $\mathcal{A}^k(M,E)$, once (2) is extended to include the terms $d_{\sigma,j}\log^j R, j = 1,\dots,k+1$. As for classical operators, for a noninteger order $A \in \mathcal{A}^k$, $\omega_{KV}(A)(x)\,dx := \int\!\!\!\!-_{T_x^*M}\mathrm{tr}_x(\sigma(A)(x,\xi)\,d\xi$ defines a global form, and one can define the canonical trace $\mathrm{TR}(A)$ by (5). The linear functional TR is cyclic:

$$\mathrm{TR}([A,B]) = 0, \quad \begin{array}{l}\text{for all } A \in \mathcal{A}^k(M,E), B \in \mathcal{A}^j(M,E),\\ \mathrm{ord}(A) + \mathrm{ord}(B) \notin \mathbb{Z}.\end{array} \tag{7}$$

Weighted traces are defined by the finite part in the Laurent expansion of the canonical trace of a holomorphic family; this is in contrast to the Wodzicki residue, which occurs as the residue in the Laurent expansion.

Definition 2.3. For $A \in \mathcal{A}^k(M, E)$, the *Q-weighted trace* of A is

$$\begin{aligned}\operatorname{tr}^Q(A) &:= \operatorname{fp}_{z=0}\operatorname{TR}(AQ^{-z}) + \operatorname{tr}(A\,\Pi_Q)\\ &:= \lim_{z\to 0}\left(\operatorname{TR}(AQ^{-z}) - \sum_{j=0}^{k}\frac{a_{j+1}}{z^{j+1}}\right) + \operatorname{tr}(A\,\Pi_Q),\end{aligned}$$

where a_{j+1} is the residue of $\operatorname{TR}(AQ^{-z})$ of order $j+1$.

The existence of the Laurent expansion is known [L]. As usual, this definition depends on a choice of spectral cut for Q. For $A \in C\ell(M, E)$, the weighted trace can also be defined by the finite part of $\operatorname{tr}(AQ^{-z})$, where tr is the ordinary operator trace. Indeed, for $\operatorname{Re}(z) \gg 0$, AQ^{-z} is trace-class, in which case $\operatorname{TR}(AQ^{-z}) = \operatorname{tr}(AQ^{-z})$. The known meromorphic continuation of the right hand side (Grubb and Seeley [GS]) gives the equivalence of the two definitions. We prefer our current definition of the weighted trace, since $\operatorname{TR}(AQ^{-z})$ is well defined outside a countable set of poles, and hence does not require a meromorphic continuation.

Weighted traces do not have the local properties of the Wodzicki residue in general. For example, a formally self-adjoint, positive order, invertible elliptic operator $A \in C\ell(M, E)$ is admissible with Agmon angle $\theta = \frac{\pi}{2}$, as is its modulus $|A| := \sqrt{A^*A}$, which has positive leading symbol. Then $A\,|A|^{-1} \in C\ell(M, E)$, and we can set

$$\eta_A(z) := \operatorname{TR}\left(A\,|A|^{-z-1}\right).$$

The η-invariant of A is given by its finite part:

$$\eta_A(0) := \operatorname{tr}^{|A|}(A|A|^{-1}), \tag{8}$$

which is not local in general.

Remark 2.2. The map $z \mapsto \eta_A(z)$ is holomorphic at $z = 0$ since the Wodzicki residue of a ΨDO projection such as $\operatorname{res}(A\,|A|^{-1})$ vanishes. It follows that $\eta_A(0) = \operatorname{tr}^A(A|A|^{-1})$, *i.e.* the η-invariant can be defined using the easier A as a weight (Cardona, Ducourtioux and Paycha [CDP], Prop. 1).

For differential operators A, $\operatorname{res}(A\,\log Q)$ is well defined [PS], Thm. 3.7, and

$$\operatorname{tr}^Q(A) = -\frac{1}{q}\operatorname{res}(A\,\log Q) = -\frac{1}{q}\int_M \operatorname{res}_x(A\,\log Q)dx. \tag{9}$$

In this case, $\mathrm{tr}^Q(A)$ has a partial locality as an integral of $\sigma_{-n}(A\log Q)$. In particular, for $A = I$ we have

$$\mathrm{tr}^Q(I) = -\frac{1}{q}\mathrm{res}(\log Q), \tag{10}$$

an expression related to the exotic determinant $\det_{res}(Q) = e^{\mathrm{res}(\log Q)}$ [Sc] (and references therein). In turn,

$$\mathrm{tr}^Q(I) - \mathrm{tr}(\Pi_Q) = \zeta_Q(0), \tag{11}$$

where the zeta function is given by the usual meromorphic continuation of

$$\zeta_Q(z) = \mathrm{TR}(Q^{-z}) = \mathrm{tr}(Q^{-z}),$$

which is well defined for $\mathrm{Re}(z) > \frac{n}{q}$ $(n = \dim(M),\ q = \mathrm{ord}(Q) > 0)$.

Since the Wodzicki residue vanishes for differential operators Q, $\zeta_Q(z)$ is holomorphic at $z = 0$, and an easy computation yields

$$\zeta_Q'(0) = -\mathrm{tr}^Q(\log Q) \tag{12}$$

for an invertible weight Q.

In summary, the key spectral invariants $\eta_A(0), \zeta_Q(0), \zeta_Q'(0)$ all occur as weighted traces.

The following proposition will be used in §2.

Proposition 2.3. *Let $A \in C\ell(M,E)$ and let Q be an invertible weight. We have the Laurent expansion*

$$\mathrm{TR}(A\,Q^{-z}) = \frac{\mathrm{res}(A)}{q\,z} + \sum_{j=0}^{J} \frac{(-1)^j}{j!}\,\mathrm{tr}^Q(A\,\log^j Q)\,z^j + \mathrm{o}(z^J).$$

Proof. By Remark 2.1, the map $z \mapsto \mathrm{TR}(A\,Q^{-z})$ is meromorphic with a simple pole at $z = 0$ with residue $\frac{\mathrm{res}(A)}{q}$, so

$$\mathrm{TR}(A\,Q^{-z}) = \frac{\mathrm{res}(A)}{q\,z} + \sum_{j=0}^{J} a_j(A,Q)\,z^j + \mathrm{o}(z^J).$$

Since Laurent expansions can be differentiated term by term away from their poles, we obtain

$$\begin{aligned}\mathrm{tr}^Q(A\,\log^j Q) &= \mathrm{fp}_{z=0}\mathrm{TR}(A\,\log^j Q\,Q^{-z}) = (-1)^j\,\mathrm{fp}_{z=0}\left(\partial_z^j \mathrm{TR}(A\,Q^{-z})\right)\\ &= (-1)^j\,j!\,a_j(A,Q).\end{aligned}$$

□

Remark 2.3. $z \mapsto \mathrm{TR}(A \log^j Q\, Q^{-z})$ has a Laurent expansion [L] with poles of order at most $j+1$:

$$\mathrm{TR}(A \log^j Q\, Q^{-z}) = \sum_{l=1}^{j+1} \frac{b_{j,l}(A,Q)}{z^l} + \sum_{i=0}^{k} a_{j,i}(A,Q)\, z^i + \mathrm{o}(z^k).$$

The a and b coefficients are related. For example, the identity $\partial_z \mathrm{TR}(A \log^j Q Q^{-z}) = -\mathrm{TR}(A \log Q^{j+1} Q^{-z})$ implies

$$a_{j+1,i} = -(i+1) a_{j,i+1}(A,Q), \quad b_{j+1,l+1}(A,Q) = l\, b_{j,l}(A,Q).$$

2.5. *Differentiable families of canonical traces*

The definition of a C^k differentiable family of classical symbols is completely analogous to the the holomorphic definition. Namely, the one-parameter family of symbols $t \mapsto \sigma_t, t \in \mathbb{R}$, with σ_t defined on an open set $U \subset \mathbb{R}^n$, is C^k for a fixed $k \in \mathbb{Z}^+$ if (i) the order α_t of σ_t is C^k in t, (ii) each homogeneous component $\sigma_{t,\alpha_t - j}(x,\xi)$ is C^k in t, (iii) for $N \gg 0$ and $K_t^{(N)}(x,y)$ the truncated kernel, the map $U \to C^{K(N)}$, $t \mapsto K_t^{(N)}(x,y)$ is C^k for some $K(N)$ with $\lim_{N\to\infty} K(N) = \infty$. A family $t \mapsto A_t$ of classical ΨDOs is C^k if it is defined in any local trivialization by C^k family of symbols.

Remark 2.4. By (iii), for a C^k family $t \mapsto \sigma_t$, the map $t \mapsto (\sigma_t)_{(N)}$ is also C^k in $C^\infty(T^*U)$, and then by (ii), the family $t \mapsto \sigma_t$ is C^k in the usual sense. As a consequence, for a C^k family $t \mapsto \sigma_t$ and for any compact set $K \subset T^*U$, $t \mapsto \|\partial_t^k \sigma_t\|_K := \sup_{(x,\xi)\in K} |\partial_t^k \sigma_t(x,\xi)|$ is continuous and hence uniformly bounded on any interval $[t_0 - \eta, t_0 + \eta]$, $\eta > 0$, as are the homogeneous components $(\sigma_t)_{\alpha - j}$, with $\alpha = \alpha_t$. Moreover, the minus one order symbol $(|\xi| + 1)^{N-\alpha} \left(\partial_t^k \sigma_t\right)_{(N)} (x,\xi)$ is bounded on T_x^*U and gives rise to a continuous map

$$\begin{aligned} t \mapsto (|\xi|+1)^{N-\alpha} \| \left(\partial_t^k \sigma_t\right)_{(N)} \|_{T_x^*U} \\ := \sup_{\xi \in T_x^* U} \left((|\xi|+1)^{N-\alpha} | \left(\partial_t^k \sigma_t\right)_{(N)} (x,\xi)| \right). \end{aligned}$$

Hence $(|\xi| + 1)^{N-\alpha} | \left(\partial_t^k \sigma_t\right)_{(N)} |$ is uniformly bounded above on $[t_0 - \eta, t_0 + \eta]$ by a constant $C_{t_0,\eta,(N)}$. Therefore, for fixed x the map $\xi \mapsto | \left(\partial_t^k \sigma_t\right)_{(N)} (x,\xi)|$ is bounded above by a map $\xi \mapsto C_{t_0,\eta,(N)} (|\xi| + 1)^{\alpha - N}$, which lies in $L^1(T_x^*U)$ for $N \gg 0$.

This remark implies that the cut-off integral and the canonical trace commute with differentiation as long as the symbols and operators have constant noninteger order.

Theorem 2.1.

1. *Let* $t \mapsto \sigma_t$ *be a* C^1 *family of symbols on* U *with constant noninteger order* α. *Then*

$$\frac{d}{dt} \,⨍_{T_x^*U} \mathrm{tr}_x \sigma_t(x,\xi)\, d\xi = ⨍_{T_x^*U} \mathrm{tr}_x \dot{\sigma}_t(x,\xi),$$

where $\dot{\sigma}_t = \frac{d}{dt}\sigma_t$.

2. *Let* $t \mapsto A_t \in C\ell(M,E)$ *be a* C^1 *family of constant noninteger order operators. Then*

$$\frac{d}{dt}\mathrm{TR}(A_t) = \mathrm{TR}(\dot{A}_t). \tag{13}$$

3. *Assume that for fixed* t, $z \mapsto \sigma_t(z)$ *is a holomorphic family of classical symbols on* U *parametrized by* $z \in W \subset \mathbb{C}$ *with holomorphic order* $\alpha(z)$ *independent of* t *and that* $t \mapsto \sigma_t(z)$ *is a* C^1 *family for fixed* $z \in W$. *Then* $z \mapsto ⨍_{T_x^*U} \mathrm{tr}_x \sigma_t(z)(x,\xi)\, d\xi$ *and* $z \mapsto ⨍_{T_x^*U} \mathrm{tr}_x \dot{\sigma}_t(z)(x,\xi)\, d\xi$ *are meromorphic in* z, *and the Laurent expansion of* $⨍_{T_x^*U} \mathrm{tr}_x \dot{\sigma}_t(z)(x,\xi)\, d\xi$ *around* $z = 0$ *is obtained by term by term* t*-differentiation of the Laurent expansion of* $⨍_{T_x^*U} \mathrm{tr}_x \sigma_t(z)(x,\xi)\, d\xi$.
4. *Assume that for fixed* t, $z \mapsto A_t(z) \in Cl(M,E)$ *defines a holomorphic family on* $W \subset \mathbb{C}$ *with holomorphic order* $\alpha(z)$ *independent of* t, *and assume that* $t \mapsto \partial_z^k|_{z=0} A_t(z)$ *is a* C^1 *family for* $k \in \mathbb{Z}^{\geq 0}$. *Then* $z \mapsto \mathrm{TR}(A_t(z))$ *and* $z \mapsto \mathrm{TR}(\dot{A}_t(z))$ *are meromorphic in* z, *and the Laurent expansion of* $\mathrm{TR}(\dot{A}_t(z))$ *around* $z = 0$ *is obtained by term by term* t*-differentiation of the Laurent expansion of* $\mathrm{TR}(A_t(z))$.

Proof. 1. Once we justify pushing the derivative past the integral, by (3) (and noting that we may choose N independent of t by our assumption on

$\alpha(z)$), we have

$$\begin{aligned}
&\frac{d}{dt}⨍_{T_x^*U} \mathrm{tr}_x\sigma_t(x,\xi)\,d\xi \\
&= \frac{d}{dt}\int_{T_x^*U} \mathrm{tr}_x\,(\sigma_t)_{(N)}\,(x,\xi)\,d\xi + \frac{d}{dt}\sum_{j=0}^{N}\int_{B_x^*U}\psi(\xi)\mathrm{tr}_x\,(\sigma_t)_{\alpha-j}\,(x,\xi)\,d\xi \\
&\quad +\frac{d}{dt}\sum_{j=0,\alpha-j+n\neq 0}^{\infty}\frac{\int_{S_x^*U}\mathrm{tr}_x\,(\sigma_t)_{\alpha-j}\,(x,\xi)\,d_S\xi}{\alpha+n-j} \\
&= \int_{T_x^*U}\frac{d}{dt}\mathrm{tr}_x\,(\sigma_t)_{(N)}\,(x,\xi)\,d\xi + \sum_{j=0}^{N}\int_{B_x^*U}\psi(\xi)\frac{d}{dt}\mathrm{tr}_x\,(\sigma_t)_{\alpha-j}\,(x,\xi)\,d\xi \\
&\quad + \sum_{j=0,\alpha-j+n\neq 0}^{\infty}\frac{\int_{S_x^*U}\frac{d}{dt}\mathrm{tr}_x\,(\sigma_t)_{\alpha-j}\,(x,\xi)\,d_S\xi}{\alpha+n-j} \\
&= ⨍_{T_x^*U}\mathrm{tr}_x\dot{\sigma}_t(x,\xi)\,d\xi,
\end{aligned}$$

Recall that for $\xi \in A \subset \mathbb{R}^n$, $t_0 \in \mathbb{R}$ and $\epsilon > 0$, if $|t - t_0| \leq \epsilon$ implies $|\frac{d}{dt}f(t,\xi)| \leq g(\xi)$ with $g \in L^1(A)$, then $\frac{d}{dt}|_{t=t_0}\int_A f(t,\xi)\,d\xi = \int_A \frac{d}{dt}|_{t=t_0} f(t,\xi)\,d\xi$. This applies to the compact subsets $A = B_x^*U$ and $A = S_x^*U$ and $f(t,\xi) = \mathrm{tr}_x\,(\sigma_t)_{\alpha-j}\,(x,\xi)$ and to $A = T_x^*U$ and $f(t,\xi) = \mathrm{tr}_x\,(\sigma_t)_{(N)}\,(x,\xi)$ (where the the required uniform estimates follow from Remark 2.4 with $k=1$).

2. By 1,

$$\begin{aligned}
\frac{d}{dt}⨍_{T_x^*M}\mathrm{tr}_x\sigma(A_t)(x,\xi)\,d\xi &= ⨍_{T_x^*M}\mathrm{tr}_x\dot{\sigma}(A_t)(x,\xi)\,d\xi \\
&= ⨍_{T_x^*M}\mathrm{tr}_x\sigma(\dot{A}_t)(x,\xi)\,d\xi. \qquad (14)
\end{aligned}$$

Since A_t has constant order α, $\dot{A}_t$ has order α modulo integers. Therefore $\dot{A}_t$ has noninteger order, so $\left(⨍_{T_x^*M}\mathrm{tr}_x\sigma(\dot{A}_t)(x,\xi)\,d\xi\right)\,dx$ is a global form on M and $\mathrm{TR}(\dot{A}) = \frac{1}{(2\pi)^n}\int_M dx\,⨍_{T_x^*M}\mathrm{tr}_x\sigma(\dot{A})(x,\xi)\,d\xi$ is well defined. Since $\mathrm{TR}(A) = \frac{1}{(2\pi)^n}\int_M dx\,⨍_{T_x^*M}\mathrm{tr}_x\sigma(A)(x,\xi)\,d\xi$, integrating (14) over M yields (13).

4. We now prove 4, leaving the similar proof of 3 to the reader. If $A_t(z)$ has noninteger order, by 2

$$\frac{d}{dt}\mathrm{TR}(A_t(z)) = \mathrm{TR}(\dot{A}_t(z))$$

except at the poles, so this is an equality of meromorphic functions. By Proposition 2.2, the coefficients of the Laurent expansion on either side can be expressed in terms of the cut-off integral of $\mathrm{tr}_x\sigma\left(\partial_z^k A_t(z)|_{z=0}\right)$ (resp. $\mathrm{tr}_x\sigma(\partial_z^k \dot{A}_t(z)|_{z=0})$) on T_x^*U and ordinary integrals over compact sets of the $-n$ component of $\mathrm{tr}_x\partial_z^j\sigma(A_t(z))|_{z=0}$ (resp. $\mathrm{tr}_x\partial_z^j\sigma(\dot{A}_t(z))|_{z=0}$), $j \in \mathbb{Z}^{\geq 0}$. As above $t \mapsto \left(\partial_z^k A_t(z)\right)|_{z=0}$ is C^1, so we can push the derivative past the integral as desired. □

Let $h : W \subset \mathbb{C} \to \mathbb{R}$ be a C^1 map such that

$$h(A) := \frac{i}{2\pi}\int_{C_\theta} h(\lambda)\,(A-\lambda)^{-1}\,d\lambda; \quad h'(A) := \frac{i}{2\pi}\int_{C_\theta} h'(\lambda)\,(A-\lambda)^{-1}\,d\lambda$$

takes any weight A to $h(A), h'(A) \in C\ell(M,E)$. Here θ is an Agmon angle for A and C_θ is the associated contour (where we assume $C_\theta \subset W$). Examples of such maps are

$$h(z) = \frac{z}{|z|}, \quad W = \mathbb{R}^*; \quad h(z) = z_\theta^c, \quad W = \mathbb{C}; \quad h(z) = \frac{z}{|z|}, \quad W = \mathbb{R}^*.$$

for fixed $c \in \mathbb{R}$.

Proposition 2.4. *Let $t \mapsto A_t$ be a differentiable family of weights of constant noninteger order and with common Agmon angle. Then*

$$\frac{d}{dt}\mathrm{TR}\left(h(A_t)\,A_t^{-z}\right) = \mathrm{TR}\left(h'(A_t)\,\dot{A}_t\,A_t^{-z}\right) - z\,\mathrm{TR}\left(h(A_t)\,\dot{A}_t\,A_t^{-z-1}\right). \tag{15}$$

This is equivalent to the following set of equations:

$$\frac{d}{dt}\mathrm{res}\left(h(A_t)\right) = \mathrm{res}\left(h'(A_t)\,\dot{A}_t\right), \tag{16}$$

$$\frac{d}{dt}\mathrm{tr}^{A_t}\left(h(A_t)\right) = \mathrm{tr}^{A_t}\left(h'(A_t)\,\dot{A}_t\right) - \frac{1}{q}\mathrm{res}\left(h(A_t)\,\dot{A}_t\,A_t^{-1}\right), \tag{17}$$

$$\frac{d}{dt}\mathrm{tr}^{A_t}\left(h(A_t)\log^j A_t\right) = \mathrm{tr}^{A_t}\left(h'(A_t)\,\dot{A}_t\,\log^j A_t\right) + j\,\mathrm{tr}^{A_t}\left(h(A_t)\,\dot{A}_t\,A_t^{-1}\log^{j-1} A_t\right) \tag{18}$$

for $j \in \mathbb{Z}^+$.

Proof. Applying Theorem 2.1 gives the following equalities of meromorphic functions:

$$
\begin{aligned}
\frac{d}{dt}\mathrm{TR}\left(h(A_t)\,A_t^{-z}\right) &= \mathrm{TR}\left(\frac{d}{dt}\left(h(A_t)\,A_t^{-z}\right)\right) \qquad (19)\\
&= \mathrm{TR}\left(\frac{d}{dt}\left(h(A_t)\right)A_t^{-z}\right) + \mathrm{TR}\left(h(A_t)\,\frac{d}{dt}\left(A_t^{-z}\right)\right)\\
&= -\frac{i}{2\pi}\mathrm{TR}\left(\int_{C_\theta} h(\lambda)\,(A_t-\lambda)^{-1}\,\dot{A}_t\,(A_t-\lambda)^{-1}\,d\lambda\,A_t^{-z}\right)\\
&\qquad -\frac{i}{2\pi}\mathrm{TR}\left(h(A_t)\int_{C_\theta}\lambda^{-z}\,(A_t-\lambda)^{-1}\,\dot{A}_t\,(A_t-\lambda)^{-1}\,d\lambda\right)\\
&= -\frac{i}{2\pi}\mathrm{TR}\left(\left(\int_{C_\theta} h(\lambda)\,(A_t-\lambda)^{-2}\,d\lambda\right)\dot{A}_t\,A_t^{-z}\right)\\
&\qquad -\frac{i}{2\pi}\mathrm{TR}\left(h(A_t)\left(\int_{C_\theta}\lambda^{-z}\,(A_t-\lambda)^{-2}\,d\lambda\right)\dot{A}_t\right) \qquad (20)\\
&= \frac{i}{2\pi}\mathrm{TR}\left(\left(\int_{C_\theta} h'(\lambda)\,(A_t-\lambda)^{-1}\,d\lambda\right)\dot{A}_t\,A_t^{-z}\right)\\
&\qquad -\frac{i\,z}{2\pi}\mathrm{TR}\left(h(A_t)\left(\int_{C_\theta}\lambda^{-z-1}\,(A_t-\lambda)^{-1}\,d\lambda\right)\dot{A}_t\right) \qquad (21)\\
&= \mathrm{TR}\left(h'(A_t)\,\dot{A}_t\,A_t^{-z}\right) - z\mathrm{TR}\left(h(A_t)\,A_t^{-z-1}\,\dot{A}_t\right)\\
&= \mathrm{TR}\left(h'(A_t)\,\dot{A}_t\,A_t^{-z}\right) - z\mathrm{TR}\left(h(A_t)\,\dot{A}_t\,A_t^{-z-1}\right). \qquad (22)
\end{aligned}
$$

In (20), (22), we use the cyclicity of TR on noninteger order operators, and in (21) we use integration by parts. This proves (15).

By Theorem 2.1.3, the Laurent expansion of $\frac{d}{dt}\mathrm{TR}\left(h(A_t)\,A_t^{-z}\right)$ is the term by term derivative of the Laurent expansion of $\mathrm{TR}\left(h(A_t)\,A_t^{-z}\right)$. The rest of the Proposition then follows from identifying the coefficients in the Laurent expansions in (15) and using Proposition 2.3. □

3. Conformal invariants and anomalies

In this part of the paper, we use canonical traces to build functionals of conformally covariant operators and study their conformal properties.

3.1. *The conformal anomaly and associated two-tensor*

Let M be a closed Riemannian manifold and $\mathrm{Met}(M)$ denote the space of Riemannian metrics on M. $\mathrm{Met}(M)$ is trivially a Fréchet manifold as the open cone of positive definite symmetric (covariant) two-tensors inside the Fréchet space

$$C^\infty(T^*M \otimes_s T^*M) := \{h \in C^\infty(T^*M \otimes T^*M) : h_{ab} = h_{ba}\}$$

of all smooth symmetric two-tensors. The Weyl group $W(M) := \{e^f : f \in C^\infty(M)\}$ acts smoothly on $\mathrm{Met}(M)$ by Weyl transformations

$$W(g, f) = \bar{g} := e^{2f} g,$$

and given a reference metric $g \in \mathrm{Met}(M)$, a functional $\mathcal{F} : \mathrm{Met}(M) \to \mathbb{C}$ induces a map

$$\mathcal{F}_g = \mathcal{F} \circ W(g, \cdot) : C^\infty(M) \to \mathbb{C}, \quad f \mapsto \mathcal{F}(e^{2f} g).$$

Definition 3.1. A functional $\mathcal{F}$ on $\mathrm{Met}(M)$ is *conformally invariant for a reference metric* g if $\mathcal{F}_g$ is constant on a conformal class, i.e.

$$\mathcal{F}(e^{2f} g) = \mathcal{F}(g) \quad \text{for all } f \in C^\infty(M).$$

A functional $\mathcal{F}$ on $\mathrm{Met}(M)$ is *conformally invariant* if it is conformally invariant for all reference metrics. A functional $\mathcal{F} : \mathrm{Met}(M) \times M \to \mathbb{C}$ is called a *pointwise conformal covariant of weight* w if

$$\mathcal{F}(e^{2f} g, x) = w \cdot f(x) \mathcal{F}(g, x) \quad \text{for all } f \in C^\infty(M), \quad \text{for all } x \in M.$$

For conformal covariants, we always assume that $\mathcal{F}(g, x)$ is given by a universal formula in the components of g and their derivatives at x.

For a fixed Riemannian metric $g = (g_{ab})$, $C^\infty(M)$ has the L^2 metric

$$(f, \tilde{f})_g = \int_M f(x) \tilde{f}(x) \mathrm{dvol}_g(x).$$

This extends to the L^2 metric on $\mathrm{Met}(M)$ given by

$$\langle h, k \rangle_g := \int_M g^{ac}(x) g^{bd}(x)\, h_{ab}(x)\, k_{cd}(x)\, \mathrm{dvol}_g(x), \tag{23}$$

with $(g^{ab}) = (g_{ab})^{-1}$. The L^2 metric induces a weak L^2-topology on $\mathrm{Met}(M)$, and $L^2(T^*M \otimes_s T^*M)$, the L^2-closure of $C^\infty(T^*M \otimes_s T^*M)$ with respect to $\langle\ ,\ \rangle_g$, is independent of the choice of g up to Hilbert space isomorphism. The choice of a reference metric yields the inner product (23) on the tangent space $T_g\mathrm{Met}(M) = C^\infty(T^*M \otimes_s T^*M)$, giving the weak L^2 Riemannian metric on $\mathrm{Met}(M)$, and forming the completion of each tangent space.

The metric g allows us to contract a two-tensor via

$$\mathrm{tr}_g(h) := h_b^b = g^{ab} h_{ab}.$$

The various inner products are related as follows:

Lemma 3.1. *For* $g \in \mathrm{Met}(M), h \in C^\infty(T^*M \otimes_s T^*M)$ *and* $f \in C^\infty(M)$, *we have*

$$\langle h, f\, g\rangle_g = \left(\mathrm{tr}_g(h), f\right)_g .$$

Proof. We have

$$\begin{aligned}\langle h, f\, g\rangle_g &= \int_M f(x)\, g^{ac}(x) g^{bd}(x)\, h_{ab}(x)\, g_{cd}(x)\, \mathrm{dvol}_g(x)\\ &= \int_M f(x)\, g^{ab}(x)\, h_{ab}(x)\, \mathrm{dvol}_g(x)\\ &= \left(\mathrm{tr}_g(h), f\right)_g .\end{aligned}$$

□

A functional $\mathcal{F} : \mathrm{Met}(M) \to \mathbb{C}$ which is Fréchet differentiable has a differential

$$d\mathcal{F}(g) : T_g\mathrm{Met}(M) = C^\infty(T^*M \otimes_s T^*M) \to \mathbb{C},$$

$$d\mathcal{F}(g).h := \frac{d}{dt}\Big|_{t=0} \frac{\mathcal{F}(g+th) - \mathcal{F}(g)}{t}.$$

For such an $\mathcal{F}$, the differentiability of the Weyl map implies that the composition $\mathcal{F}_g : C^\infty(M) \to \mathbb{C}$ is differentiable at 0 with differential $d\mathcal{F}_g(0) : T_0 C^\infty(M) = C^\infty(M) \to \mathbb{C}$.

Definition 3.2. The *conformal anomaly* for the reference metric g of a differentiable functional $\mathcal{F}$ on $\mathrm{Met}(M)$ is $d\mathcal{F}_g(0)$. In physics notation, the conformal anomaly in the direction $f \in C^\infty(M)$ is

$$\begin{aligned}\delta_f \mathcal{F}_g &:= d\mathcal{F}_g(0).f = d\mathcal{F}(g).2f\, g\\ &= \lim_{t=0} \frac{\mathcal{F}(g+2tfg) - \mathcal{F}(g)}{t} = \frac{d}{dt}\Big|_{t=0} \mathcal{F}(e^{2tf} g).\end{aligned}$$

Remark 3.1. $\mathcal{F}$ is conformally invariant if and only if $d\mathcal{F}_g(0).f = 0$ for all $g \in \mathrm{Met}(M), f \in C^\infty(M)$.

If the differential $d\mathcal{F}(g) : C^\infty(T^*M \otimes_s T^*M) \to \mathbb{C}$ extends to a continuous functional $\overline{d\mathcal{F}(g)} : L^2(T^*M \otimes_s T^*M) \to \mathbb{C}$, then by Riesz's lemma there is a unique two-tensor $T_g(\mathcal{F})$ with

$$\overline{d\mathcal{F}(g)}.h = \langle h, T_g(\mathcal{F})\rangle_g, \qquad \text{for all } h \in L^2(T^*M \otimes_s T^*M).$$

$T_g(\mathcal{F})$ is precisely the L^2 gradient of $\mathcal{F}$ at g.

Proposition 3.1. *Let $\mathcal{F}$ be a functional on* $\mathrm{Met}(M)$ *which is differentiable at the metric g and whose differential $d\mathcal{F}(g)$ extends to a continuous functional $\overline{d\mathcal{F}(g)} : L^2(T^*M \otimes_s T^*M) \to \mathbb{C}$. Then the differential $d\mathcal{F}_g(0)$ also extends to a continuous functional $\overline{d\mathcal{F}_g(0)} : L^2(M) \to \mathbb{C}$. Identifying the conformal anomaly at g with a function in $L^2(M)$, we have*

$$\overline{d\mathcal{F}_g(0)} = 2\,\mathrm{tr}_g\left(T_g(\mathcal{F})\right).$$

In particular, the functional $\mathcal{F}$ is conformally invariant iff $\mathrm{tr}_g\left(T_g(\mathcal{F})\right) = 0$ *for all metrics g.*

Proof. The differential $d(\mathcal{F}_g)_0$ extends to a continuous functional because

$$d\mathcal{F}_g(0).f = d\mathcal{F}(g)(2f\ g) \Rightarrow \overline{d\mathcal{F}_g(0)}.f = \overline{d\mathcal{F}(g)}.2f\ g$$

By Lemma 3.1,

$$\overline{d\mathcal{F}_g(0)}.f = \overline{d\mathcal{F}(g)}.2f\ g = \langle T_g(\mathcal{F}), 2f\,g\rangle_g = 2\left(\mathrm{tr}_g(T_g(\mathcal{F})), f\right)_g,$$

as desired. □

Definition 3.3. Under the assumptions of the Proposition, the function

$$x \mapsto \delta_x \mathcal{F}_g := 2\mathrm{tr}_g\left(T_g(\mathcal{F})\right)(x)$$

is called the *local anomaly* of the functional $\mathcal{F}$ at the reference metric g.

Example 3.1. In field theory, for a classical action $\mathcal{A}$ on a configuration space $\mathrm{Conf}(M)$ with respect to a background metric g, $\mathcal{A} : \phi \mapsto \mathcal{A}(g)(\phi)$, where $\mathrm{Conf}(M)$ is usually a space of tensors on M, the associated two-tensor $T_g(\mathcal{A})$ is called the *stress-energy momentum tensor.* In the path integral approach to quantum field theory, the *effective action* $\mathcal{W}(g)$ is the average over the configuration space of the exponentiated classical action

$$\mathcal{W}(g) := \langle \mathcal{A}(g)\rangle := -\log\langle e^{-\mathcal{A}(g)}\rangle,$$

where

$$\langle \mathcal{F}\rangle = \int_{\mathrm{Conf}(M)} \mathcal{F}(\phi)\,\mathcal{D}\phi$$

is the average of $\mathcal{F}$ over the fields ϕ with respect to some heuristic volume measure $\mathcal{D}\phi$ on $\mathrm{Conf}(M)$. The associated two-tensor $T_g(\mathcal{W})$ is interpreted as the *quantized stress-energy* momentum tensor and denoted by

$$T_g(\mathcal{W}) = T_g\left(\langle \mathcal{A}\rangle\right).$$

The local conformal anomaly associated to $d\mathcal{A}_g(f) = 2\left(\operatorname{tr}_g\left(T_g\left(\mathcal{A}\right)\right), f\right)_g$ is defined to be [Du]

$$\operatorname{tr}_g T_g\left(\langle\mathcal{A}\rangle\right) - \langle \operatorname{tr}_g T_g(\mathcal{A})\rangle.$$

If the classical action is conformally invariant, as in string theory, $\operatorname{tr}_g T_g(\mathcal{A}) = 0$ and the local conformal anomaly reduces to $\operatorname{tr}_g T_g\left(\langle\mathcal{A}\rangle\right)$.

In general, the classical action is quadratic: $\mathcal{A}(g)(\phi) = \langle A_g\phi, \phi\rangle_g$, where $\langle\cdot,\cdot\rangle_g$ is the inner product on the tensor fields ϕ induced by the metric g. A_g is a geometric differential operator, i.e. an operator depending smoothly on the metric g (via the curvature, for example). For bosonic strings, the fields are $\mathbb{R}^d$-valued smooth functions on a Riemann surface M, and A_g is the Laplace-Beltrami operator. As pointed out in the introduction, even if $\mathcal{A}(g)$ is conformally invariant, A_g is usually only conformally covariant.

3.2. *Conformally covariant operators*

Given a vector bundle E over a closed manifold M, we consider maps

$$\operatorname{Met}(M) \to Cl(M, E), \quad g \mapsto A_g.$$

Definition 3.4. The operator $A_g \in Cl(M, E)$ associated to a Riemannian metric g is *conformally covariant* of bidegree (a, b) if the pointwise scaling of the metric $\bar{g} = e^{2f} g$, for $f \in C^\infty(M, \mathbb{R})$ yields

$$A_{\bar{g}} = e^{-bf} A_g e^{af} = e^{(a-b)f}\, A'_g, \quad \text{for } A'_g := e^{-af}\, A_g\, e^{af}, \tag{24}$$

for constants $a, b \in \mathbb{R}$.

We survey known conformally covariant differential and pseudodifferential operators; more details are in Chang [C].

Operators of order 1. (Hitchin [H]) For M^n spin, the Dirac operator $\mathbb{D}_g := \gamma^i \cdot \nabla_i^g$ is a conformally covariant operator of bidegree $\left(\frac{n-1}{2}, \frac{n+1}{2}\right)$.

Operators of order 2. If $\dim(M) = 2$, the Laplace-Beltrami operator Δ_g is conformally covariant of bidegree $(0, 2)$. It is well known that in dimension two

$$R_{\bar{g}} = e^{-2f}\left(R_g + 2\Delta_g f\right), \tag{25}$$

where R_g is the scalar curvature, and by the Gauss-Bonnet theorem

$$\int_M R_g \,\mathrm{dvol}_g = 2\pi\chi(M), \tag{26}$$

with the Euler characteristic $\chi(M)$ (much more than) a conformal invariant.

On a Riemannian manifold of dimension n, the Yamabe operator, also called the conformal Laplacian,

$$L_g := \Delta_g + c_n R_g,$$

is a conformally covariant operator of bidegree $\left(\frac{n-2}{2}, \frac{n+2}{2}\right)$, where $c_n := \frac{n-2}{4(n-1)}$.

Operators of order 4. (Paneitz [Pa, BO]) In dimension n, the Paneitz operators

$$P_g^n := \tilde{P}_g^n + (n-4)Q_g^n$$

are conformally covariant scalar operators of bidegree $\left(\frac{n-4}{2}, \frac{n+4}{2}\right)$. Here $\tilde{P}_g^n := \Delta_g^2 + d^*\left((n-2)\, J_g\, g - 4A_g\cdot\right) d$ with

$$J_g := \frac{R_g}{2(n-1)}, A_g = \frac{Ric_g - \frac{R_g}{n} g}{n-2} + \frac{J_g}{n} g,$$

$A_g\cdot$ the homomorphism on T^*M given by $\phi = (\phi_i) \mapsto (A_g)_i^j\, \phi_j$, and $Q_g^n := \frac{n\, J_g^2 - 4|A_g|^2 + 2\Delta_g\, J_g}{4}$ is Branson's Q-curvature [B1], a local scalar invariant that is a polynomial in the coefficients of the metric tensor and its inverse, the scalar curvature and the Christoffel symbols. Note that $A_g = \frac{1}{n} J_g\, g$ precisely when g is Einstein.

The Q-curvature generalizes the scalar curvature R_g in the following sense. On a 4-manifold, we have

$$Q_{\bar{g}}^4 = e^{-4f}\left(Q_g^4 + \frac{1}{2} P_g^4 f\right)$$

(cf. (25)), and $\int_M Q_g^4 \text{dvol}_g$ is a conformal invariant (cf. (26)), as is $\int_M Q_g^n \text{dvol}_g$ in even dimensions [B2].

Operators of order $2k$. (Graham, Jenne, Mason and Sparling [GJMS]) Fix $k \in \mathbb{Z}^+$ and assume either n is odd or $k \leq n$. There are conformally covariant (self-adjoint) scalar differential operators $P_{g,k}^n$ of bidegree $\left(\frac{n-2k}{2}, \frac{n+2k}{2}\right)$ such that the leading part of $P_{g,k}^n$ is Δ_g^k and such that $P_{g,k}^n = \Delta_g^k$ on $\mathbb{R}^n$ with the Euclidean metric.

$P_{g,k}^n$ generalizes P_g^n, since $P_g^n = P_{g,2}^n$, and satisfies

$$P_{g,k}^n = \tilde{P}_g^n + \frac{n-2k}{2} Q_g^n$$

where $\tilde{P}_g^n = d^* S_g^n d$ for a natural differential operator S_g^n on 1-forms.

Note that $P_{g,k}^n$ has bidegree (a, b) with $b - a = 2k$ independent of the dimension and in particular has bidegree $(0, 2k)$ in dimension $2k$.

Pseudodifferential Operators. (Branson and Gover [BG], Petersen [Pe]) Peterson has constructed ΨDOs, $P_{g,k}^n, k \in \mathbb{C}$, of order $2\mathrm{Re}(k)$ and bidgree $((n-2k)/2, (n+2k)/2)$ on manifolds of dimension $n \geq 3$ with the property that $P_{\bar{g},k}^n - e^{-bf} P_{g,k}^n e^{af}$ is a smoothing operator. Thus any conformal covariant built from the total symbol of $P_{g,k}^n$ is a conformal covariant of $P_{g,k}^n$ itself. The family $P_{g,k}^n$ contains the previously discovered conformally covariant ΨDOs associated to conformal boundary value problems [BG].

3.3. *A hierarchy of functionals and their conformal anomalies*

Since the known conformal invariants for conformally covariant operators A_g

$$\zeta_{A_g}(0) = \mathrm{tr}^{A_g}(I), \quad \log \det_\zeta(A_g) = \mathrm{tr}^{A_g}(\log A_g), \quad \eta_{A_g} = \mathrm{tr}^{A_g}\left(A_g\, |A_g|^{-1}\right)$$

arise as weighted/canonical traces by (8), (11), (12), it is natural to look for a general prescription to derive conformal invariants from canonical traces.

Let $A_g \in Cl(M, E)$ be an operator associated to a Riemannian metric g on M. For $f \in C^\infty(M, \mathbb{R})$, set $g_t := e^{2ft} g, t \in \mathbb{R}$, and set $A_t = A_{g_t}$. We always assume that the map $g \mapsto A_g$ is smooth in the appropriate topologies, so that A_t is automatically a smooth curve in $C\ell(M, E)$.

Lemma 3.2. *$A_g \in C\ell(M, E)$ is conformally covariant of bidegree (a, b) if and only if for all $f \in C^\infty(M, \mathbb{R})$,*

$$\dot{A}_t = (a - b)\, f\, A_t - a\,[f, A_t]. \tag{27}$$

Proof. This follows from differentiating (24) applied to the family g_t. □

Theorem 3.1. *Let A_g be a conformally covariant weight of bidegree (a, b) and whose order α and spectral cut θ are independent of the metric g. The meromorphic map*

$$\mathcal{F}_h(g) : z \mapsto \mathrm{TR}\left(h(A_g)\, A_g^{-z}\right)$$

has conformal anomaly

$$\begin{aligned}\delta_f \mathrm{TR}\left(h(A_g)\, A_g^{-z}\right) &= (a - b)\,\mathrm{TR}\left(f\, h'(A_g)\, A_g^{-z+1}\right) \\ &\quad - z(a - b)\mathrm{TR}\left(f\, h(A_g)\, A_g^{-z}\right)\end{aligned} \tag{28}$$

as an identity of meromorphic functions.

This is equivalent to the following system of equations.

1. *The conformal anomaly of* $\mathrm{res}(h(A_g))$*:*

$$\delta_f \mathrm{res}\,(h(A_g)) = (a-b)\,\mathrm{res}\,(f\,h'(A_g)\,A_g)\,. \tag{29}$$

2. *The conformal anomaly of* $\mathrm{tr}^{A_g}(h(A_g))$*:*

$$\delta_f \mathrm{tr}^{A_g}\,(h(A_g)) = (a-b)\,\mathrm{tr}^{A_g}\,(f\,h'(A_g)\,A_g) + \frac{b-a}{\alpha}\,\mathrm{res}(f\,h(A_g)), \tag{30}$$

3. *The conformal anomaly of* $\mathrm{tr}^{A_g}(h(A_g)\,\log^j A_g)$ *for* $j \in \mathbb{Z}^+$*:*

$$\begin{aligned} &\delta_f \mathrm{tr}^{A_g}\left(h(A_g)\,\log^j A_g\right) \\ &= (a-b)\,\mathrm{tr}^{A_g}\left(f\,h'(A_g)\,A_g\,\log^j A_g\right) \\ &\qquad + j\,(a-b)\,\mathrm{tr}^{A_g}\left(f\,h(A_g)\,\log^{j-1} A_g\right). \end{aligned} \tag{31}$$

Proof. Equations (28), (29),(30), (31) follow from (15), (16), (17), (18), respectively. In the computation, we use the cyclicity of TR on noninteger order operators, which eliminates the second term on the right hand side of (27). □

We collect these formulas for special choices of h. We assume A_g and hence $A_{\bar{g}}$ is invertible. This allows us to ignore terms depending on the kernel of A_g, which can be easily treated as in the proof of part 1 below. All invariants and covariants are understood to be conformal.

Corollary 3.1. *We have the following conformal anomalies for conformally covariant weights* A_g *of order* α*:*

1. *Anomalies associated to* $h \equiv 1$*:*

$$\begin{aligned} \delta_f \zeta_{A_g}(0) &= \delta_f \mathrm{tr}^{A_g}(I) = 0, \\ \delta_f \zeta'_{A_g}(0) &= -\delta_f \mathrm{tr}^{A_g}\,(\log\ A_g) = -(a-b)\mathrm{tr}^{A_g}(f). \end{aligned} \tag{32}$$

Hence $\zeta_{A_g}(0) = -\frac{1}{\alpha}\mathrm{res}(\log A_g)$ *is an invariant.* $\zeta'_{A_g}(0)$ *has local anomaly* $\frac{a-b}{\alpha}\mathrm{res}_x(\log A_g)$ *and is an invariant whenever* $\mathrm{res}_x(\log A_g)$ *vanishes for all* $x \in M$.

2. *Anomalies associated to* $h(\lambda) = \lambda$*:*

$$\delta_f \text{res}(A_g) = (a-b)\text{res}(fA_g),$$

$$\delta_f \text{tr}^{A_g}(A_g) = (a-b)\text{tr}^{A_g}(fA_g) + \frac{b-a}{\alpha}\text{res}(fA_g), \quad (33)$$

$$\delta_f \text{tr}^{A_g}(A_g \log A_g) = (a-b)\text{tr}^{A_g}(fA_g \log A_g) + (a-b)\text{tr}^{A_g}(fA_g). \quad (34)$$

$\text{res}_x(A_g)$ *is a pointwise covariant of weight* $a-b$*. If* A_g *is a differential operator, then* $\text{res}_x(A_g)$ *vanishes and* $\text{tr}^{A_g}(A_g) = -\frac{1}{\alpha}\text{res}(A_g \log A_g)$ *has local anomaly given by* $\frac{b-a}{\alpha}\text{res}_x(A_g \log A_g)$*.*

3. *Anomalies associated to* $h(\lambda) = \lambda^c, c \in \mathbb{R}$*: assuming* A_g *is admissible for fixed c, we have*

$$\delta_f \text{res}(A_g^c) = c(a-b)\text{res}(fA_g^c),$$

$$\delta_f \text{tr}^{A_g}(A_g^c) = c(a-b)\text{tr}^{A_g}(fA_g^c) + \frac{b-a}{\alpha}\text{res}(fA_g^c). \quad (35)$$

If A_g^c *is a differential operator (in particular, if* A_g *is differential and* $c \in \mathbb{Z}^+$*), then* $\text{res}_x(A_g^c)$ *vanishes and* $\text{tr}^{A_g}(A_g^c) = -\frac{1}{\alpha}\text{res}(A_g^c \log A_g)$ *has local anomaly given by* $\frac{c(b-a)}{\alpha}\text{res}_x(A_g^c \log A_g)$*.*

4. *Anomalies associated to* $h(\lambda) = \lambda/|\lambda|$ *and invertible* A_g*:*

$$\delta_f \text{tr}^{A_g}(A_g/|A_g|) = \frac{b-a}{\alpha}\text{res}\left(f\frac{A_g}{|A_g|}\right). \quad (36)$$

Proof. Much of the Corollary follows immediately from the Theorem. In 1, the invariance of $\text{res}_x(\log A_g)$ follows from (10). The last statement follows from $\text{tr}^{A_g}(f) = -\frac{1}{\alpha}\text{res}(f \log A_g)$ (9), since multiplication by f is a differential operator, and (12). If $\text{Ker}(A)$ is nontrivial, $\zeta_{A_g}(0)$ is still a conformal invariant: by (11), (32), the new terms cause no trouble, as $\text{tr}(\Pi_A) = \dim(\text{Ker } A)$ is a conformal invariant and $\text{res}(I) = 0$.

For the statement about $\text{res}_x(A_g)$ in 2, if ϕ is a smooth function on M, then $\phi \cdot A_g$ is conformally covariant if A_g is. By (33),

$$\begin{aligned}
&\delta_f \int_M \phi(x)\left(\int_{S_x^*M} \text{tr}_x \sigma_{-n}(A_g)(x,\xi)d\xi\right)dx \\
&= \delta_f \text{res}(\phi \cdot A_g) = (a-b)\text{res}(f \cdot \phi \cdot A_g) \\
&= (a-b)\int_M f(x)\phi(x)\left(\int_{S_x^*M} \text{tr}_x \sigma_{-n}(A_g)(x,\xi)d\xi\right)dx.
\end{aligned}$$

Letting ϕ approach a delta function at x and using the compactness of M to push this limit past δ_f gives

$$\delta_f \mathrm{res}_x(A_g) = (a-b)f(x)\mathrm{res}_x(A_g).$$

The last statement in 2 follows as in the proof of 1 from $\mathrm{tr}^{A_g}(A_g) = -\frac{1}{\alpha}\mathrm{res}(A_g \log A_g)$. In 3, the last statement follows from $\mathrm{tr}^{A_g}(fA_g^c) = -\frac{1}{\alpha}\mathrm{res}(fA_g^c \log A_g)$. □

Remark 3.2. The conformal anomaly in string theory boils down to a finite linear combination of local conformal anomalies $\frac{b-a}{\alpha}\mathrm{res}_x(\log A_g)$ where the A_g are Laplacians on forms. Their bidegree involves the dimension of spacetime, so this local conformal anomaly vanishes for a certain well chosen dimension.

As stated in the introduction, the Corollary and the Laurent expansion of Theorem 3.1 provide a natural hierarchy among these invariants/covariants. The most divergent term in the Laurent expansion is a conformal invariant; if this global invariant vanishes in a particular case, then the new "most divergent" term, if it is of the form $\int_M \mathcal{I}(g,x)\mathrm{dvol}_g(x)$, tends to give rise to a local conformal anomaly proportional to $\mathcal{I}(g,x)$. This is confirmed by the more refined analysis for weights with nonnegative leading order symbol, i.e. weights with smoothing heat kernels.

Lemma 3.3. *Let A_g be a weight of order α with nonnegative leading symbol and let the heat kernel for $A = A_g$ have the asymptotic expansion* [GS]

$$\mathrm{tr}_x e_A(\epsilon, x, x) \sim \sum_{j=0}^{\infty} a_j(A,x)\epsilon^{\frac{j-n}{\alpha}} + \sum_{k=0}^{\infty} b_k(A,x)\epsilon^k \log\epsilon + \sum_{\ell=1}^{\infty} c_\ell(A,x)\epsilon^\ell. \quad (37)$$

Then

$$\mathrm{res}_x(A^k) = \begin{cases} (-1)^{k+1}k!\alpha b_k(A,x),\ k \in \mathbb{Z}^{\geq 0}, \\ \frac{\alpha}{(-k-1)!}a_{n+\alpha k}(A,x),\ k \in \mathbb{Z}^-, \end{cases} \quad (38)$$

with the understanding that $a_{n+\alpha k} = 0$ if $\alpha k \notin \mathbb{Z}$. In particular, $\mathrm{res}_x(A) = \alpha b_1(A,x)$.

The last sum in (37) appears only if $(j-n)/\alpha$ is never integral. In particular, this sum does not appear if A_g is a differential operator.

Proof. Setting $\zeta_A(z,x) := \omega_{KV}(A^{-z})(x)$ we have

$$\mathrm{Res}_{z=-k}\zeta_A(z,x) = \mathrm{Res}_{z=0}\omega_{KV}(A^{-z+k})(x) = -\frac{1}{\alpha'(0)}\mathrm{res}_x(A^k) = \frac{\mathrm{res}_x(A^k)}{\alpha},$$

since $\alpha(z) = \alpha(-z+k)$ (see (4), (6)), where we use the same symbol for an operator and its kernel and assume A is invertible for simplicity. Let us compute this complex residue. We have

$$\begin{aligned}\zeta_A(z,x) &= \frac{1}{\Gamma(z)}\int_0^\infty t^{z-1}\mathrm{tr}_x e_A(t,x,x)dt \\ &\sim \frac{1}{\Gamma(z)}\int_0^1 t^{z-1}\left(\sum_{j=0}^\infty a_j(A,x)t^{\frac{j-n}{\alpha}} + \sum_{k=0}^\infty b_k(A,x)t^k\log t\right.\\ &\qquad \left. + \sum_{\ell=0}^\infty c_\ell(A,x)t^\ell\right)dt + \frac{1}{\Gamma(z)}\int_1^\infty t^{z-1}\mathrm{tr}_x e_A(t,x,x)dt.\end{aligned}$$

Since the last term is analytic in z, an easy integration on the first term yields the result. In particular, $\mathrm{res}_x(A) = \alpha b_1$. □

To state the final theorem, let the kernel $\tilde{e}(\epsilon,x,y)$ of $Ae^{-\epsilon|A|}$ have the asymptotic expansion

$$\mathrm{tr}_x\tilde{e}(\epsilon,x,x) \sim \sum_{j=0}^\infty \tilde{a}_j(A,x)\epsilon^{\frac{j-\alpha-n}{n}} + \sum_{k=0}^\infty \tilde{b}_k(A,x)\epsilon^k\log\epsilon + \sum_{\ell=1}^\infty \tilde{c}_\ell(A,x)\epsilon^\ell.$$

Set $a_j(A) = \int_M a_j(A,x)\mathrm{dvol}_g(x)$, etc.

Theorem 3.2. *Let $A = A_g$ be a conformally covariant weight of bidegree (a,b), with nonnegative leading order symbol, and whose order α is independent of the metric g.*

1. *$a_n(A) + c_0(A)$ is a conformal invariant.*
2. *We have*

$$\begin{aligned}\delta_f \log\det{}_\zeta A &= -\delta_f(x)\zeta_A'(0)\\ &= (a-b)\int_M f(x)\left(a_n(A,x)+c_0(A,x)\right)\mathrm{dvol}_g(x),\end{aligned}$$

so $\log\det_\zeta(A)$ has local conformal anomaly given by $(b-a)(a_n(A,x)+c_0(A,x))$. In particular, $\det_\zeta A$ is a conformal invariant if A is a differential operator and $\dim(M)$ is odd.

3. *$b_1(A,x)$ is a a pointwise conformal covariant of weight $a-b$.*
4. *$a_{n-\alpha}(A,x)$ is a pointwise conformal covariant of weight $b-a$.*

5. *We have*

$$\begin{aligned}&\delta_f \mathrm{tr}^A(A)\\&= -\delta_f(c_1(A) + a_{n+\alpha}(A))\\&= (b-a)\int_M f(x)\left[c_1(A,x) + a_{n+\alpha}(A,x) - b_1(A,x)\right] \mathrm{dvol}_g(x),\end{aligned}$$

so that $\mathrm{tr}^A(A)$ *has a local conformal anomaly given by* $(b-a)\left(c_1(A,x) + a_{n+\alpha}(A,x) - b_1(A,x)\right)$ *(with the understanding that* $a_{n+\alpha}(A,x) = 0$ *if* $\alpha \notin \mathbb{Z}$*, and* $c_1(A,x) = 0$ *if* A *is differential). In particular, if* A *is a differential operator, then* $\mathrm{tr}^A(A)$ *has a local conformal anomaly given by* $(b-a)a_{n+\alpha}(A,x)$*, and if* A *is a noninteger order* ΨDO*, it has a local conformal anomaly given by* $(b-a)c_1(A,x)$*.*

6. *The results of 5 generalize to* $\mathrm{tr}^A(A^k)$ *for* $k \in \mathbb{Z}^+$*, replacing* c_1 *by* c_k*,* $a_{n+\alpha}$ *by* $a_{n+\alpha k}$*,* b_1 *by* b_k*, and* $(a-b)$ *by* $k(a-b)$*.*
7. $\delta_f \eta_A(0) = \frac{b-a}{\alpha}\mathrm{res}\left(f\frac{A}{|A|}\right) = -(b-a)\int_M f(x)\tilde{a}_n(A,x)$*. In particular,* $\eta_A(0)$ *is a conformal invariant if* n *and* α *have opposite parity.*

Proof. 1. This follows from the first point in Corollary 3.1 and the fact that $\zeta_A(0) = a_n(A) + c_0(A)$.

2. This follows from (12), (32), and the fact that $\mathrm{tr}^A(f) = \int_M f(x)(a_n(A,x) + c_0(A,x))$. It is well known that only the $\epsilon^{k-\frac{\dim(M)}{2}}$ terms in the heat kernel asymptotics are nonzero for differential operators, so $a_n(A,x) = 0$ in odd dimensions.

3. This was shown in the second point of Corollary 3.1.

4. If A is conformally covariant, so is A^{-1}. The result now follows from 3 applied to $c = -1$, which yields $\delta_f \mathrm{res}(A^{-1}) = (b-a)\mathrm{res}(fA^{-1})$, and the Lemma.

5. $\mathrm{tr}^A(fA) = \mathrm{f.p.}_{\epsilon=0}\mathrm{tr}(fAe^{-\epsilon A})$ is the finite part of $\mathrm{tr}(fAe^{-\epsilon A}) = -\partial_\epsilon \mathrm{tr}(fe^{-\epsilon A})$ as $\epsilon \to 0$, so $\mathrm{tr}^A(fA) = -\int_M f(x)(a_{n+\alpha}(A,x) + c_1(A,x))$. The first statement now follows from the second equation in (33) and the fact that $\frac{1}{\alpha}\mathrm{res}_x(A) = -b_1(A,x)$ (38). If A is a differential operator, then c_1 is replaced by $a_{n+\alpha}$ and $\mathrm{res}_x(fA) = 0$. If A is a non-integral order ΨDO, then again $\mathrm{res}_x(fA) = 0$.

6. By the first equation in (35), $\delta_f \mathrm{res}(A^k) = k(a-b)\mathrm{res}(fA^k), k \in \mathbb{Z}^+$. The results for b_k follow as in 5, using (38). We can use the second equation in (35) and $\mathrm{tr}^A(fA^k) = (-1)^k \mathrm{f.p.}_{\epsilon=0}\partial_\epsilon^k \mathrm{tr}(fe^{-\epsilon A})$ to prove the result for the a and c coefficients.

7. The first equality follows from (36) and Remark 2.2. For the second

equality, we have

$$\operatorname{res}(A/|A|) = \operatorname{res}_{s=0}\operatorname{tr}(A|A|^{-1}|A|^{-s}) = \operatorname{res}_{s=1}\operatorname{tr}(A|A|^{-s}).$$

Using the pointwise version of the Mellin transform $A|A|^{-s} = \Gamma(s)^{-1}\int_0^\infty t^{s-1}Ae^{-t|A|}dt$ as in the Lemma, we get $\operatorname{res}\left(f\frac{A}{|A|}\right) = -(b-a)\int_M f(x)\tilde{a}_n(A,x)$. The last statement follows from a careful computation [R] (Prop. 3) of the the residue of $A/|A|$. Note that since $|A|$ has nonnegative symbol, this restriction on the symbol of A can be dropped here. □

Remark 3.3. (i) 1, 2, and 4 are known for the conformal Laplacian [BO, PR]. 3 is new to our knowledge. Related results for contact geometry are in Ponge [Pon].

(ii) The results above involving Wodzicki residues can be proved directly, where the cyclicity is valid for all order operators.

References

B1. T. Branson, "Sharp inequalities, the functional determinant and the complementary series," *Trans. Amer. Math. Soc.* **347** (1995), 3671–3742; "Differential oprators canonically associated to a conformal structure," *Math. Scand.* **57** (1985), 293–345.

B2. T. Branson, "The Functional Determinant," *Global Analysis Research Center Lecture Note Series*, **4**, Seoul National University (1993).

BG. T. Branson, A. Gover, "Conformally invariant non-local operators," *Pacific J. Math.* **201** (2001), 19-60.

BO. T. Branson, B. Orsted, "Explicit functional determinants in four dimensions," *Proc. Amer. Math. Soc.* **113** (1991), 669-682.

CDP. A. Cardona, C. Ducourtioux, S. Paycha, "From tracial anomalies to anomalies in quantum field theory," *Comm. Math. Phys.* **242** (2003), 31-65.

C. S.-Y. A. Chang, "Conformal invariants and partial differential equations," *Bull. Amer. Math. Soc.* **42** (2005), 365-394.

D. C. Ducourtioux, "Weighted traces on pseudodifferential operators and associated determinants," PhD thesis, Université Blaise Pascal (Clermont-Ferrand) (2001).

Du. M. J. Duff, "Twenty years of the Weyl anomaly," *Classical Quantum Gravity* (1994), 1387–1403.

GJMS. C. R. Graham, R. Jenne, L. Mason, G. A. J. Sparling, "Conformally invariant powers of the Laplacian I: Existence", *J. London Math. Soc.* **46** (1992), 557–565.

GS. G. Grubb, R. Seeley, "Weakly parametric pseudodifferential operators and Atiyah-Patodi-Singer boundary problems," *Invent. Math.* **121** (1995), 481–529.

H. N. Hitchin, "Harmonic spinors," *Adv. in Math.* **14** (1974), 1-55.

K. C. Kassel, "Le résidu non commutatif [d'apres M. Wodzicki]", *Sém. Bourbaki, Astérisque* **177-178** (1989), 199–229.

KV. M. Kontsevich, S. Vishik, "Geometry of determinants of elliptic operators." In *Functional Analysis on the Eve of the XXI century, Vol. I,* Progress in Mathematics **131** (1993), Birkhäuser Boston, Boston, MA, 1995, p. 173-197; "Determinants of elliptic pseudodifferential operators," Preprint, Max–Planck Institute, Bonn (1994).

O. K. Okikiolu, "The multiplicative anomaly for determinants of elliptic operators," *Duke Math J.* **79** (1995), 723–749.

L. M. Lesch, "On the non commutative residue for pseudodifferential operators with log-polyhomogeneous symbols," *Ann. Global Anal. and Geom.* **17** (1999), 151–187.

Pa. S. Paneitz, "A quartic conformally covariant differential operator for arbitrary pseudo-Riemannian manifolds," Preprint 1983.

PR. T. Parker, S. Rosenberg, "Invariants of conformal Laplacians," *J. Differential Geom.* **25** (1987), 199-222.

P1. S. Paycha, "From heat-operators to anomalies; a walk through various regularization techniques in mathematics and physics," Emmy Noether Lectures, Göttingen, 2003, `http://www.math.uni-goettingen.de`.

P2. S. Paycha, "Renormalized traces as a geometric tool." In: *Geometric Methods for Quantum Field Theory (Villa de Leyva, 1999)*, World Scientific Publishing, River Edge, NJ, 2001, p. 293–360.

PS. S. Paycha, S. Scott, "A Laurent expansion for regularised integrals of holomorphic symbols," Preprint 2005, to appear in *Geom. Funct. Anal.*

Pe. L. Peterson, "Conformally covariant pseudo-differential operators," *Differential Geom. Appl.* **13** (2000), 197-211.

Pol. S. Polyakov, "Quantum geometry of bosonic strings," *Phys. Lett. B* **103** (1981), 207–210.

Pon. R. Ponge, "Calcul hypoelliptique sur les variétés de Heisenberg, résidu non commutatif et géométrie pseudo-hermitienne," PhD Thesis, Univ. Paris-Sud (Orsay), 2000.

R. S. Rosenberg, "The determinant of a conformally invariant operator," *J. London Math. Soc.* **36** (1987), 553–568.

Sc. S. Scott "The residue determinant," to appear in *Comm. Partial Differential Equations*, `math.AP/0406268`.

Se. R. Seeley, "Complex powers of an elliptic operator." In: *Singular Integrals (Proc. Sympos. Pure Math., Chicago, Ill., 1966)*, Amer. Math. Soc., Providence, R.I., p. 288–307.

We. H. Weyl, "Reine Infinitesimalgeometrie," *Math. Zeit.* **2** (1918), 384-411.

Wo. M. Wodzicki, "Noncommutative residue. I. Fundamentals." In: *K-theory, Arithmetic and Geometry (Moscow, 1984–1986)*, Lecture Notes in Math., 1289, Springer, Berlin, 1987, p. 320-399.

Received by the editors July 28, 2005; Revised December 26, 2005

Part IV

Noncommutative Geometry

Analysis, Geometry and Topology
of Elliptic Operators, pp. 297–352

AN ANALYTIC APPROACH TO SPECTRAL FLOW IN VON NEUMANN ALGEBRAS

MOULAY-TAHAR BENAMEUR

UMR 7122 du CNRS, Université de Metz
Ile du Saulcy, Metz, France
benameur@math.univ-metz.fr

ALAN L. CAREY

Mathematical Sciences Institute, Australian National University
Canberra ACT, 0200 Australia
acarey@maths.anu.edu.au

JOHN PHILLIPS

Department of Mathematics and Statistics, University of Victoria
Victoria, B.C. V8W 3P4, Canada
phillips@math.uvic.ca

ADAM RENNIE

Institute for Mathematical Sciences, Department of Mathematics
Universitetsparken 5, DK-2100 Copenhagen, Denmark
rennie@math.ku.dk

FYODOR A. SUKOCHEV

School of Informatics and Engineering, Flinders University
Bedford Park S.A 5042 Australia
sukochev@infoeng.flinders.edu.au

KRZYSZTOF P. WOJCIECHOWSKI

Department of Mathematics, IUPUI (Indiana/Purdue)
Indianapolis, IN, 46202-3216, USA
kwojciechowski@math.iupui.edu

This research is supported by the ARC (Australia), NSERC (Canada) and an early career grant from the University of Newcastle.

Dedicated to Krzysztof P. Wojciechowski on his 50th birthday from his co–autho

The analytic approach to spectral flow is about ten years old. In that time it has evolved to cover an ever wider range of examples. The most critical extension was to replace Fredholm operators in the classical sense by Breuer-Fredholm operators in a semifinite von Neumann algebra. The latter have continuous spectrum so that the notion of spectral flow turns out to be rather more difficult to deal with. However quite remarkably there is a uniform approach in which the proofs do not depend on discreteness of the spectrum of the operators in question. The first part of this paper gives a brief account of this theory extending and refining earlier results. It is then applied in the latter parts of the paper to a series of examples. One of the most powerful tools is an integral formula for spectral flow first analysed in the classical setting by Getzler and extended to Breuer-Fredholm operators by some of the current authors. This integral formula was known for Dirac operators in a variety of forms ever since the fundamental papers of Atiyah, Patodi and Singer. One of the purposes of this exposition is to make contact with this early work so that one can understand the recent developments in a proper historical context. In addition we show how to derive these spectral flow formulae in the setting of Dirac operators on (non-compact) covering spaces of a compact spin manifold using the adiabatic method. This answers a question of Mathai connecting Atiyah's L^2-index theorem to our analytic spectral flow. Finally we relate our work to that of Coburn, Douglas, Schaeffer and Singer on Toeplitz operators with almost periodic symbol. We generalise their work to cover the case of matrix valued almost periodic symbols on $\mathbf{R}^N$ using some ideas of Shubin. This provides us with an opportunity to describe the deepest part of the theory namely the semifinite local index theorem in noncommutative geometry. This theorem, which gives a formula for spectral flow was recently proved by some of the present authors. It provides a far-reaching generalisation of the original 1995 result of Connes and Moscovici.

2000 *Mathematics Subject Classification.* Primary 19K56; Secondary 58J20, 46L80, 58J30

1. Introduction

Spectral flow is normally associated with paths of operators with discrete spectrum such as Dirac operators on compact manifolds. Even then it is only in the last decade that analytic definitions have been introduced (previously the definitions were topological). Recently it has been discovered that if one takes an analytic approach to spectral flow then one can handle examples where the operators may have zero in the continuous spectrum.

The aim of this article is to give a discussion of spectral flow in as general an analytic setting as is currently feasible. In fact we consider unbounded operators affiliated to a semifinite von Neumann algebra and give examples where the phenomenon of spectral flow for paths of such operators occurs quite naturally. There has been a lot published recently on this subject, which is rather technical although the ideas can be explained reasonably simply. This article is thus partly a review of this theory aimed at exposing these recent results to a wider audience. As the early papers dealt with von Neumann algebras with trivial centre (factors) and the more general situation of non-trivial centre was only recently completely understood we also felt that it was timely to collect the basic definitions and results in one place. Moreover we have rounded out the account with some additional new results and some carefully chosen illustrative examples.

The methods we use are motivated by noncommutative geometry however our results may be stated without using that language. The novel feature of spectral flow for operators affiliated to a general semifinite von Neumann algebra is that the operators in question may have zero in their continuous spectrum. It is thus rather surprising that spectral flow can even be defined in this situation.

We focus on spectral flow for a continuous path of self adjoint unbounded Breuer-Fredholm operators $\{D(s) = D_0 + A(s)\}$ for $s \in [0,1]$ in the sense that $A(s)$ is a norm continuous family of bounded self adjoint operators in a fixed semifinite von Neumann algebra $\mathcal{N}$ and $D(s)$ is affiliated to $\mathcal{N}$ for all $s \in [0,1]$ (we will elaborate on all of this terminology in subsequent Sections). We restrict to the paths of bounded perturbations because the analytic theory is complete and many interesting examples exist. The wider question of paths where the domain and the Hilbert space $H(s)$, on which $D(s)$ is densely defined, varies with s is still under investigation (see the article by Furutani [39] for motivation). This situation may arise on manifolds with boundary where one varies the metric and is a difficult problem unless one makes very specific assumptions. An approach to this question has been introduced by Leichtnam and Piazza [45] building on ideas of Dai and Zhang [33] which in turn is based on unpublished work of F. Wu. It works for Dirac type operators in both the case of closed manifolds and the case of (possibly noncompact) covering spaces. This new notion is that of spectral section.

Spectral sections enable one to define spectral flow as an invariant in the K-theory of a certain C^*-subalgebra of the von Neumann algebras that we consider in this article. We have chosen not to discuss it here because, although it can handle the case where the space $H(s)$ varies with s we feel that the theory is not yet in final form. Moreover it reduces in the von Neumann setting to Phillips approach. Another omission is a discussion of the topological meaning of spectral flow in the general analytic setting. We refer the reader to the work of Getzler [40], Booß-Bavnbek et al [9], and Lesch [47].

While our aim is to put in one place all of the basic ideas we do not include complete proofs instead referring where necessary to the literature. Thus we start with a summary of Fredholm theory in a general semifinite von Neumann algebra $\mathcal{N}$ with a fixed faithful semifinite trace τ. We refer to such operators as 'τ-Breuer-Fredholm' because we can trace the origins of the theory to Breuer [11, 12] but we need to refine his theory to take account of non-uniqueness of the trace τ on a von Neumann algebra with non-trivial centre. In this setting we discuss Phillips' analytic approach to spectral flow for paths of bounded self adjoint Breuer-Fredholm operators in $\mathcal{N}$. Then we include some simple analytic examples that show the theory is non-trivial.

The theory for paths $\{D(s)\}_{s\in[0,1]}$ of self adjoint unbounded operators proceeds via the map $s \mapsto D(s) \mapsto D(s)(1+D(s)^2)^{-1/2}$. When $\{D(s)\}$ is a norm-continuous path of perturbations (of the kind considered above) of $D(0)$, an unbounded self adjoint τ-Breuer-Fredholm operator, then its image under this map is a continuous path in the space of bounded self adjoint τ-Breuer-Fredholm operators [15]. Although (in the case $\mathcal{N} = \mathcal{B}(\mathcal{H})$) spectral flow can be defined directly for such paths of unbounded operators [9], we can also define spectral flow in terms of the corresponding path of bounded self adjoint operators.

The second half of the paper is about analytic formulae for spectral flow that have appeared in the literature. After reviewing these formulae we relate them to classical theory via a study of spectral flow of generalised Dirac operators on compact manifolds without boundary and their covering spaces. A question first raised by Mathai [49] is settled by relating spectral flow to the L^2 index theorem. The deepest result in the theory is the semifinite local index theorem which we illustrate by application to an example of spectral flow for differential operators with almost periodic coefficients. This is inspired by work of Shubin [62] who initiated this line of enquiry. The generalisation to semifinite von Neumann algebras of the local index theorem of Connes and Moscovici [32] was achieved in papers of some of the present authors [18, 19] and has other interesting applications (not included here) for example see Pask et al [54].

2. Preliminaries

2.1. *Notation*

Our basic reference for von Neumann algebras is Dixmier [34] where many of the concepts we discuss here are described in detail. For the theory of ideals of compact operators in a semifinite von Neumann algebra we refer to Fack et al [38] and Dodds et al [35]. Throughout this paper we will consider $\mathcal{N}$, a semifinite von Neumann algebra (of type I_∞ or II_∞ or mixed type) acting on a separable Hilbert space $\mathcal{H}$. We will denote by τ a fixed faithful, normal semifinite trace on $\mathcal{N}$ (with the usual normalization if $\mathcal{N}$ is a type I_∞ factor). The norm-closed 2-sided ideal in $\mathcal{N}$ generated by the projections of finite trace (usually called τ-finite projections) will be denoted by $\mathcal{K}_{\tau\mathcal{N}}$ or just $\mathcal{K}_\mathcal{N}$ to lighten the notation. The quotient algebra $\mathcal{N}/\mathcal{K}_\mathcal{N}$ will be denoted by $\mathcal{Q}_\mathcal{N}$ and will be called the (generalized) Calkin algebra. We will let π denote the quotient mapping $\mathcal{N} \to \mathcal{Q}_\mathcal{N}$.

We will let $\mathcal{F}$ denote the space of all τ-Breuer-Fredholm operators in $\mathcal{N}$, *i.e.* ,

$$\mathcal{F} = \{T \in \mathcal{N} \mid \pi(T) \text{ is invertible in } \mathcal{Q}_\mathcal{N}\}.$$

We denote by $\mathcal{F}^{sa}$ the space of self adjoint operators in $\mathcal{F}$. The more interesting part of the space of self adjoint τ-Breuer-Fredholm operators in $\mathcal{N}$ will be denoted by $\mathcal{F}^{sa}_*$, *i.e.*,

$$\mathcal{F}^{sa}_* = \{T \in \mathcal{F} \mid T = T^* \text{ and } \pi(T) \text{ is neither positive nor negative}\}.$$

2.2. *Some history*

For $\mathcal{N}$ being the algebra of bounded operators on $\mathcal{H}$, i.e. the type I_∞ factor case, Atiyah and Lusztig [2, 3] have defined the *spectral flow* of a continuous path in $\mathcal{F}^{sa}_*$ to be the number of eigenvalues (counted with multiplicities) which pass through 0 in the positive direction minus the number which pass through 0 in the negative direction as one moves from the initial point of the path to the final point. This definition is appealing geometrically as an "intersection number" and has been made precise [40, 10, 57] although it cannot easily be generalised beyond the type I_∞ factor. Other motivating remarks may be found in Booß-Bavnbek et al [8, 9]. More importantly, there is no obvious generalization of this definition if the algebra $\mathcal{N}$ is of type II_∞, where the spectrum of a self adjoint Breuer-Fredholm operator is not discrete in a neighbourhood of zero. J. Kaminker has described this as the problem of counting "moving globs of spectrum".

In his 1993 Ph.D. thesis, V.S. Perera [55, 56] gave a definition of the spectral flow of a *loop* in $\mathcal{F}^{sa}_*$ for a II_∞ factor, $\mathcal{N}$. He showed that the space, $\Omega(\mathcal{F}^{sa}_*)$, of loops based at a unitary $(2P-1)$ in $\mathcal{F}^{sa}_*$, is homotopy equivalent to the space, $\mathcal{F}$, of all Breuer-Fredholm operators in the II_∞ factor, $P\mathcal{N}P$. Since Breuer [11, 12] showed that the index map $\mathcal{F} \longrightarrow \mathbf{R}$

classifies the connected components of $\mathcal{F}$, Perera defines spectral flow as the composition $sf : \Omega(\mathcal{F}_*^{sa}) \longrightarrow \mathcal{F} \longrightarrow \mathbf{R}$ and so obtains the isomorphism $\pi_1(\mathcal{F}_*^{sa}) \cong \mathbf{R}$. He also showed that this gives the "heuristically correct" answer for a simple family of loops.

While this is an important and elegant result, it has a couple of weaknesses. Firstly, since the map sf is not defined directly and constructively on individual loops it is not clear why spectral flow is counting "moving globs of spectrum". Secondly, in the nonfactor setting where the von Neumann algebra may have summands of finite type the map may not extend to paths which are not loops in any sensible way: in a finite algebra (see 5.1) there can be paths with nonzero spectral flow, but every loop has zero spectral flow.

Phillips' approach [57, 58] is the following. Let χ denote the characteristic function of the interval $[0, \infty)$. If $\{B_t\}$ is any continuous path in $\mathcal{F}_*^{sa}$, then $\{\chi(B_t)\}$ is a <u>dis</u>continuous path of projections whose discontinuities arise precisely because of spectral flow. For example, if $t_1 < t_2$ are neighbouring path parameters and **if** the projections $P_i = \chi(B_{t_i})$ commute, then the spectral flow from t_1 to t_2 should be trace$(P_2 - P_1P_2)$ minus trace$(P_1 - P_1P_2)$ (= amount of nonnegative spectrum gained minus amount of nonnegative spectrum lost). However, this is clearly the index of the operator $P_1P_2 : P_2(H) \to P_1(H)$. If these projections do <u>not</u> commute then one can still make sense of this index provided $\pi(P_1) = \pi(P_2)$ in the Calkin algebra. This notion was called *essential codimension* by Brown, Douglas and Fillmore [13] in the type I_∞ case and denoted by $ec(P_1, P_2)$. Perera [55, 56] defined the obvious extension of this concept to II_∞ factors and used it to explain why his definition of spectral flow gives the "right" answer in a representative family of simple loops. Phillips' [58, 57] new ingredient is the fact that the operator $P_1P_2 : P_2(H) \to P_1(H)$ is always a τ-Breuer-Fredholm operator provided $||\pi(P_1) - \pi(P_2)|| < 1$. While Phillips only proved this in the case of a factor, we observed in Carey et al [21] that it works for a general semifinite von Neumann algebra. We will explain the proof in the next Section, and show that the condition $||\pi(P_1) - \pi(P_2)|| < 1$ is necessary and sufficient for P_1P_2 to be τ-Breuer-Fredholm.

Since we can (easily) show that the mapping $t \mapsto \pi(\chi(B_t))$ is continuous, we can partition the parameter interval $a = t_0 < t_1 < \cdots < t_k = b$ so that on each small subinterval the projections $\pi(\chi(B_t))$ are all close.

Letting $P_i = \chi(B_{t_i})$ for $i = 0, 1, \cdots, k$ we then define:

$$sf(\{B_t\}) = \sum_{i=1}^{k} \text{Ind}(P_{i-1}P_i).$$

With a little effort this works equally well in both the type I_∞ and II_∞ settings and agrees with all previous definitions of spectral flow where they exist. A simple lemma is the key to showing that sf is well-defined and (path-) homotopy invariant. Defining $\text{Hom}(\mathcal{F}_*^{sa})$ to be the homotopy groupoid of $\mathcal{F}_*^{sa}$, Phillips proved the following theorem in the case of a factor. It extends to the general semifinite case [19].

Theorem 2.1. *If $\mathcal{N}$ is a general semifinite von Neumann algebra then sf as defined above is a homomorphism from $Hom(\mathcal{F}_*^{sa})$ to $\mathbf{R}$ which restricts to an isomorphism of $\pi_1(\mathcal{F}_*^{sa})$ with $\mathbf{Z}$ (respectively $\mathbf{R}$) when $\mathcal{N}$ is a factor of type I_∞ (respectively, type II_∞).*

We note that to show that sf is one-to-one on $\pi_1(\mathcal{F}_*^{sa})$ one must rely on Perera's result that $\Omega(\mathcal{F}_*^{sa}) \simeq \mathcal{F}$. We also remark that in paragraphs 7, 8 and 9 of the introduction to the Atiyah-Patodi-Singer paper [3] the authors appear to be hinting at the existence of a notion of spectral flow (for paths of self adjoint Breuer-Fredholm operators in a II_∞ factor) to be used as a possible tool in an alternate proof of their index theorem for flat bundles. In some sense this hope is realised by the generalisation [18, 19] of the Connes-Moscovici local index formula to the semifinite von Neumann algebra setting.

3. Breuer-Fredholm theory

The standard references for Breuer-Fredholm operators in a general semifinite von Neumann algebra are in Breuer [11, 12]. In earlier work of some of the current authors [19] this theory was extended to handle Breuer-Fredholm operators in a skew-corner $P\mathcal{N}Q$ in the general semifinite situation with a fixed (scalar) trace τ in both the bounded and unbounded cases. All of the expected results hold but their proofs are a little more subtle. The most difficult case, index theory for unbounded Breuer-Fredholm operators will not be covered here. However, in order to handle more cases (including the case of τ-finite von Neumann algebras), we allow our operators to vary within all of $\mathcal{F}^{sa}$ and not just in $\mathcal{F}_*^{sa}$.

If H_1 is a subspace of H, we denote the projection onto the closure of H_1 by $[H_1]$.

Definition 3.1. Let P and Q be projections (not necessarily infinite and not necessarily equivalent) in $\mathcal{N}$ and let $T \in P\mathcal{N}Q$. We let $\ker_Q(T) = \ker\left(T_{|Q(H)}\right) = \ker(T) \cap Q(H)$. The operator $T \in P\mathcal{N}Q$ is called $(P \cdot Q)$ *τ-Fredholm* if and only if
(1) $[\ker_Q(T])$ and $[\ker_P(T^*)]$ are τ-finite in $\mathcal{N}$, and
(2) there exists a projection $P_1 \leq P$ in $\mathcal{N}$ with $P - P_1$ τ-finite in $\mathcal{N}$ and $P_1(H) \subseteq T(H)$.
In this case, we define the *$(P \cdot Q)$-index of T* to be the number:

$$\mathrm{Ind}_{(P.Q)}(T) = \tau\left[\ker_Q(T)\right] - \tau\left[\ker_P(T^*)\right].$$

We will henceforth abbreviate this terminology to τ-Fredholm or sometimes Breuer-Fredholm and drop the $(P \cdot Q)$ when there is no danger of confusion. We observe that if $P = Q$ then this is just the definition of τ-Fredholm used in Phillips et al [59] in the semifinite von Neumann algebra, $Q\mathcal{N}Q$, with the trace being the restriction of τ to $Q\mathcal{N}Q$.

We summarize the general situation of τ-Fredholm operators with different domain and range [19]. We re-iterate that the **order** of proving the usual results is crucial in developing the skew-corner case, as the various projections are neither equivalent nor infinite in general.

Lemma 3.1. *Let $T \in P\mathcal{N}Q$. Then,* (1) *If T is $(P \cdot Q)$-Fredholm, then T^* is $(Q \cdot P)$-Fredholm and $Ind(T^*) = -Ind(T)$. If $T = V|T|$ is the polar decomposition, then V is $(P \cdot Q)$-Fredholm with $Ind(V) = Ind(T)$ and $|T|$ is $(Q \cdot Q)$-Fredholm of index* 0.
(2) *The set of all $(P \cdot Q)$-Fredholm operators in $P\mathcal{N}Q$ is open in the norm topology.*

Definition 3.2. If $T \in P\mathcal{N}Q$, then a **parametrix** for T is an operator $S \in Q\mathcal{N}P$ satisfying $ST = Q + k_1$ and $TS = P + k_2$ where $k_1 \in \mathcal{K}_{Q\mathcal{N}Q}$ and $k_2 \in \mathcal{K}_{P\mathcal{N}P}$.

Lemma 3.2. *If the usual assumptions on $\mathcal{N}$ are satisfied, then $T \in P\mathcal{N}Q$ is $(P \cdot Q)$-Fredholm if and only if T has a parametrix $S \in Q\mathcal{N}P$. Moreover, any such parametrix is $(Q \cdot P)$-Fredholm.*

Prop 3.1. Let G, P, Q be projections in $\mathcal{N}$ and let $T \in P\mathcal{N}Q$ be $(P \cdot Q)$-Fredholm and $S \in G\mathcal{N}P$ be $(G \cdot P)$-Fredholm, respectively. Then, ST is $(G \cdot Q)$-Fredholm and $Ind(ST) = Ind(S) + Ind(T)$.

This proof carefully adapts the original ideas of Breuer [12] in a crucial way. Finally one is easily able to deduce the following expected results.

Corollary 3.1. *(Invariance properties of the $(P\cdot Q)$-Index) Let $T \in P\mathcal{N}Q$.*
(1) If T is $(P\cdot Q)$-Fredholm then there exists $\delta > 0$ so that if $S \in P\mathcal{N}Q$ and $\| T - S \| < \delta$ then S is $(P\cdot Q)$-Fredholm and $Ind(S) = Ind(T)$.
(2) If T is $(P\cdot Q)$-Fredholm and $k \in P\mathcal{K}_{\mathcal{N}}Q$ then $T + k$ is $(P\cdot Q)$-Fredholm and $Ind(T + k) = Ind(T)$.

4. The analytic definition of spectral flow

4.1. *Essential codimension*

If P, Q are projections (not necessarily infinite) in the semifinite von Neumann algebra $\mathcal{N}$ we wish to define the *essential codimension* of P in Q whenever $||\pi(P) - \pi(Q)|| < 1$, where $\pi : \mathcal{N} \to \mathcal{Q}_{\mathcal{N}}$ is the Calkin map. Once we show that the operator $PQ \in P\mathcal{N}Q$ is a τ-Fredholm operator in the sense of Section 3 then we will define the essential codimension of P in Q to be $\mathrm{Ind}(PQ)$. In case $\mathcal{N} = \mathcal{B}(\mathcal{H})$ a related result to the following lemma appears in Proposition 3.1 of Avron et al [5] where one of their conditions is in terms of essential spectrum. Our one condition is in terms of the norm, and the proof is very different.

Lemma 4.1. *If P, Q are projections in the semifinite von Neumann algebra $\mathcal{N}$ and $\pi : \mathcal{N} \to \mathcal{Q}_{\mathcal{N}}$ is the Calkin map, then $PQ \in P\mathcal{N}Q$ is $(P\cdot Q) - \tau$-Fredholm if and only if $||\pi(P) - \pi(Q)|| < 1$.*

Proof. Suppose $||\pi(Q) - \pi(P)|| < 1$. Then since

$$||\pi(PQP) - \pi(P)|| \leq ||\pi(Q) - \pi(P)|| < 1$$

and $\pi(P)\,(\mathcal{N}/\mathcal{K}_{\mathcal{N}})\,\pi(P) = (P\mathcal{N}P)/\mathcal{K}_{P\mathcal{N}P}$ we see that PQP is a τ-Fredholm operator in $P\mathcal{N}P$. Thus, $\ker_P(QP) \subseteq \ker_P(PQP)$ and so $[\ker_P(QP)] \leq [\ker_P(PQP)]$ where the latter is a finite projection in $P\mathcal{N}P$. Similarly, $[\ker_Q(PQ)]$ is a finite projection in $Q\mathcal{N}Q$. Since the range of PQ contains the range of PQP, and since this latter operator is τ-Fredholm in $P\mathcal{N}P$, there is a projection $P_1 \leq P$ so that $\tau(P - P_1) < \infty$ and the range of P_1 is contained in the range of PQ. That is, PQ is $(P\cdot Q)$-Fredholm.

On the other hand, if PQ is τ-Fredholm then PQP is a positive τ-Fredholm operator in $P\mathcal{N}P$. Letting $p = \pi(P)$ and $q = \pi(Q)$, we see that pqp is an invertible positive operator in $p\mathcal{Q}_{\mathcal{N}}p$ which is $\leq p$, so $||p - pqp|| < 1$. Similarly, $||q - qpq|| < 1$. Now,

$$(p - q)^3 = [p - pqp] - [q - qpq]$$

is the difference of two positive operators, so that:

$$-[q - qpq] \leq (p-q)^3 \leq [p - pqp].$$

Hence,

$$||(p-q)^3|| \leq Max\{||p - pqp||, ||q - qpq||\} < 1.$$

That is,

$$||\pi(P) - \pi(Q)|| = ||(p-q)^3||^{1/3} < 1. \qquad \square$$

Definition 4.1. If P and Q are projections in $\mathcal{N}$ and if $||\pi(P) - \pi(Q)|| < 1$ then the *essential codimension of* P *in* Q, denoted $ec(P, Q)$, is the number $\mathrm{Ind}(PQ) = \mathrm{Ind}_{(P \cdot Q)}(PQ)$. If $P \leq Q$ it is exactly the codimension of P in Q.

Lemma 4.2. *If* P_1, P_2, P_3 *are projections in* $\mathcal{N}$ *and if* $||\pi(P_1) - \pi(P_2)|| < \frac{1}{2}$ *and* $||\pi(P_2) - \pi(P_3)|| < \frac{1}{2}$ *then* $ec(P_1, P_3) = ec(P_1, P_2) + ec(P_2, P_3)$.

Proof. Since we also have $||\pi(P_1) - \pi(P_3)|| < 1$, the terms in the equation are all defined by Lemma 4.1. Translating the equation into the language of index and using Lemma 3.1 and Proposition 3.1 we see that it suffices to prove that $\mathrm{Ind}\left((P_1P_3)^*(P_1P_2P_3)\right) = 0$. But,

$$||\pi\left((P_1P_3)^*(P_1P_2P_3)\right) - \pi(P_3)|| = ||\pi(P_3P_1P_2P_3) - \pi(P_3)||$$

$$\leq ||\pi(P_1P_2) - \pi(P_3)|| \leq ||\pi(P_1P_2) - \pi(P_2)|| + ||\pi(P_2) - \pi(P_3)||$$

$$\leq ||\pi(P_1) - \pi(P_2)|| + ||\pi(P_2) - \pi(P_3)|| < 1.$$

Thus, there is a compact k in $P_3\mathcal{N}P_3$ with $||P_3P_1P_2P_3 + k - P_3|| < 1$. Hence, $\mathrm{Ind}(P_3P_1P_2P_3) = \mathrm{Ind}(P_3P_1P_2P_3 + k) = 0$ as this latter operator is invertible in $P_3\mathcal{N}P_3$. $\square$

Remarks 4.1. If P and Q are projections in $\mathcal{N}$ with $||P - Q|| < 1$, then $ec(P, Q) = 0$. To see this, note that $||PQP - P|| \leq ||Q - P|| < 1$ so that PQP is invertible in $P\mathcal{N}P$ and hence range $P \supseteq$ range $PQ \supseteq$ range $PQP =$ range P. Thus, range $PQ =$ range P and similarly range $QP =$ range Q so the $(P \cdot Q)$ index of PQ is 0.

4.2. *The general definition*

Recall that $\chi = \chi_{[0,\infty)}$ is the characteristic function of the interval $[0,\infty)$ so that if T is any self adjoint operator in a von Neumann algebra $\mathcal{N}$ then $\chi(T)$ is a projection in $\mathcal{N}$.

Lemma 4.3. *If $\mathcal{N}$ is a von Neumann algebra, $\mathcal{J}$ is a norm closed 2-sided ideal in $\mathcal{N}$, T is a self adjoint operator in $\mathcal{N}$ and $\pi(T)$ is invertible in $\mathcal{N}/\mathcal{J}$ (where $\pi : \mathcal{N} \to \mathcal{N}/\mathcal{J}$ is the quotient mapping), then $\chi(\pi(T)) = \pi(\chi(T))$.*

Proof. Since 0 is not in the spectrum of $\pi(T)$, the left hand side is a well-defined element of the C^*-algebra $\mathcal{N}/\mathcal{J}$. Choose $\epsilon > 0$ so that $[-\epsilon, \epsilon]$ is disjoint from $sp(\pi(T))$. Let $f_1 \geq f_2$ be the following piecewise linear continuous functions on $\mathbf{R}$:

$$f_1(t) = \begin{cases} 1 & \text{if } t \geq 0 \\ \text{linear on } [-\epsilon, 0] \\ 0 & \text{if } t \leq -\epsilon \end{cases}, \qquad f_2(t) = \begin{cases} 1 & \text{if } t \geq \epsilon \\ \text{linear on } [0, \epsilon] \\ 0 & \text{if } t \leq 0 \end{cases}.$$

Now, $f_1 \geq \chi \geq f_2$ on $\mathbf{R}$, but all three functions are equal on $sp(\pi(T))$. Thus,

$$\chi(\pi(T)) = f_1(\pi(T)) = \pi(f_1(T)) \geq \pi(\chi(T)) \geq \pi(f_2(T))$$
$$= f_2(\pi(T)) = \chi(\pi(T)).$$

Hence, $\chi(\pi(T)) = \pi(\chi(T))$. □

Definition 4.2. Let $\mathcal{N}$ be a semifinite von Neumann algebra with fixed semifinite, faithful, normal trace, τ. Let $\mathcal{F}^{sa}$ denote the space of all self adjoint τ-Fredholm operators in $\mathcal{N}$. Let $\{B_t\}$ be any continuous path in $\mathcal{F}^{sa}$ (indexed by some interval $[a, b]$). Then $\{\chi(B_t)\}$ is a (generally discontinuous) path of projections in $\mathcal{N}$. By Lemma 4.3 $\pi(\chi(B_t)) = \chi(\pi(B_t))$ and since the spectra of $\pi(B_t)$ are bounded away from 0, this latter path is continuous. By compactness we can choose a partition $a = t_0 < t_1 < \cdots < t_k = b$ so that for each $i = 1, 2, \cdots, k$

$$\|\pi(\chi(B_t)) - \pi(\chi(B_s))\| < \frac{1}{2} \quad \text{for all } t, s \text{ in } [t_{i-1}, t_i].$$

Letting $P_i = \chi(B_{t_i})$ for $i = 0, 1, \cdots, k$ we define the *spectral flow of the path* $\{B_t\}$ to be the number:

$$sf(\{B_t\}) = \sum_{i=1}^{k} ec(P_{i-1}, P_i).$$

To see that this definition is independent of the partition, it suffices to see that it is invariant under adding a single point to the partition. However, this is exactly the content of Lemma 4.2.

Remarks 4.2. (i) If $\{B_t\}$ is a path in $\mathcal{F}^{sa}$ and if $t \mapsto \chi(B_t)$ is continuous, then $sf(\{B_t\}) = 0$. That is, as expected heuristically, spectral flow can be nontrivial only when the path $t \mapsto \chi(B_t)$ has discontinuities.
(ii) For $T \in \mathcal{F}^{sa}$, let

$$N(T) = \{S \in \mathcal{F}^{sa} \mid ||\pi(\chi(S)) - \pi(\chi(T))|| < \tfrac{1}{4}\}.$$

Then $N(T)$ is open in $\mathcal{F}^{sa}$ since $S \mapsto \pi(\chi(S)) = \chi(\pi(S))$ is continuous on $\mathcal{F}^{sa}$. Moreover, if $S_1, S_2 \in N(T)$, then by the definition of spectral flow, all paths from S_1 to S_2 lying entirely in $N(T)$ have the same spectral flow, namely, $ec(\chi(S_1), \chi(S_2))$.

Prop 4.1. Spectral flow is homotopy invariant, that is, if $\{B_t\}$ and $\{B'_t\}$ are two continuous paths in $\mathcal{F}^{sa}$ with $B_0 = B'_0$ and $B_1 = B'_1$ which are homotopic in $\mathcal{F}^{sa}$ via a homotopy leaving the endpoints fixed, then $sf(\{B_t\}) = sf(\{B'_t\})$.

Proof. Let $H : I \times I \to \mathcal{F}^{sa}$ be a homotopy from $\{B_t\}$ to $\{B'_t\}$. That is, H is continuous, $H(t,0) = B_t$ for all t, $H(t,1) = B'_t$ for all t, $H(0,s) = B_0 = B'_0$ for all s, and $H(1,s) = B_1 = B'_1$ for all s. By compactness we can cover the image of H by a finite number of open sets $\{N_1, \cdots, N_k\}$ as in Remark 4.2. The inverse images of these open sets, $\{H^{-1}(N_1), \cdots, H^{-1}(N_k)\}$ is a finite cover of $I \times I$. Thus, there exists $\epsilon_0 > 0$ (the Lebesgue number of the cover) so that any subset of $I \times I$ of diameter $\leq \epsilon_0$ is contained in some element of this finite cover of $I \times I$. Thus, if we partition $I \times I$ into a grid of squares of diameter $\leq \epsilon_0$, then the image of each square will lie entirely within some N_i. Effectively, this breaks H up into a finite sequence of "short" homotopies by restricting H to $I \times J_i$ where J_i are subintervals of I (of length $\leq \epsilon_0/\sqrt{2}$). These short homotopies have the added property that for fixed J_i we can choose a single partition of I so that for each subinterval J_ℓ of the partition, $H(J_\ell \times J_i)$ is contained in one of $\{N_1, \cdots, N_k\}$. By concentrating on the ith " short homotopy" and relabelling $N_1, \cdots, N_k$ if necessary we can assume H is such a "short homotopy." By definition, the sum of the spectral flows of the lower paths (i.e. along $I \times \{0\}$) is $sf(\{B_t\})$. Since the spectral flows of the vertical paths (i.e. along $\{t_k\} \times J_i$) cancel in pairs, the sum of the spectral flows of the upper paths (i.e., along $I \times \{t_1\}$) equals $sf(\{B'_t\})$ and hence $sf(\{B_t\}) = sf(\{B'_t\})$. □

Examples 4.1. If $\mathcal{N}$ is a II_∞ von Neumann factor with trace τ then it is well-known (and not difficult to prove) that $\mathcal{N}$ contains an abelian von Neumann subalgebra isomorphic to $L^\infty(\mathbf{R})$ with the property that the restriction of the trace τ to $L^\infty(\mathbf{R})$ coincides with the usual trace on $L^\infty(\mathbf{R})$ given by Lebesgue integration. We construct our first examples inside this subalgebra. Let B_0 in $L^\infty(\mathbf{R})$ be the continuous function:

$$B_0(t) = \begin{cases} 1 & \text{if } t \geq 1, \\ t & \text{if } t \in [-1,1], \\ -1 & \text{if } t \leq -1. \end{cases}$$

Let s be any fixed real number. Then for $t \in [0,1]$ let B_t be defined by $B_t(r) = B_0(r + ts)$ for all $r \in \mathbf{R}$. Clearly $\{B_t\}$ is a continuous path in $\mathcal{F}_*^{sa}$. Moreover, $\chi(B_t) = \chi_{[-ts,\infty)}$ which differs from $\chi_{[0,\infty)}$ by the finite projection $\chi_{[-ts,0)}$ if $s > 0$ (or, $\chi_{[0,-ts)}$ if $s < 0$). Thus, $\pi(\chi(B_t))$ is *constant* in $\mathcal{Q}_N$. Hence,

$$P_0 = \chi(B_0) = \chi_{[0,\infty)},\ P_1 = \chi(B_1) = \chi_{[-s,\infty)}$$

and

$$sf(\{B_t\}) = ec(P_0, P_1) = \mathrm{Ind}(P_0 P_1)$$

$$= \tau(P_1 - P_0P_1) - \tau(P_0 - P_0P_1) = s.$$

We note that for these examples the spectral pictures are constant! That is, $sp(B_t) = [-1,1]$ for all t and $sp(\pi(B_t)) = \{-1,1\}$ for all t. Thus, one cannot tell from the spectrum alone (even knowing the multiplicities) what the spectral flow will be.

These examples may seem paradoxical as there exists a (strong-operator topology) continuous path of unitaries $\{U_t\}$ so that $B_t = U_t B_0 U_t^*$. However, there cannot exist a *norm*-continuous path of such unitaries as this would imply that the path $t \mapsto \chi(B_t)$ is a norm-continuous path of projections which it is not since $||\chi(B_t) - \chi(B_s)|| = 1$ if $s \neq t$.

On the other hand, it is not hard to prove that there is a unitary U_1 in $\mathcal{N}$ so that $B_1 = U_1 B_0 U_1^*$. Since the unitary group of $\mathcal{N}$ is connected in the norm topology we can find in $\mathcal{N}$ a norm continuous path $\{U_t\}$ of unitaries for $t \in [1,2]$ so that U_1 is as above and $U_2 = I$. Then we can extend $\{B_t\}$ to a continuous loop based at B_0 by defining $B_t = U_t B_0 U_t^*$ for $t \in [1,2]$.

Since the second half of the loop satisfies $t \mapsto \chi(B_t)$ is norm continuous, its spectral flow is 0 and so $sf\left(\{B_t\}_{[1,2]}\right) = r$.

When $\mathcal{N}$ is a type I_∞ factor we can use a similar construction with $\ell^\infty(\mathbf{Z})$ in place of $L^\infty(\mathbf{R})$ to obtain paths with any given integer as their spectral flow. Of course, these examples will not have a constant spectral picture.

Remarks 4.3. It is clear from the above definition that spectral flow does not change under reparametrization of intervals and is additive when we compose paths by concatenation. Hence, spectral flow defines a groupoid homomorphism from the homotopy groupoid, $\mathrm{Hom}(\mathcal{F}_*^{sa})$ to $\mathbf{Z}$ in the type I_∞ factor case (respectively, to $\mathbf{R}$ in the type II_∞ factor case). By the examples just constructed these homomorphisms are surjective in the case of factors, even when restricted to paths based at a point B_0 in F_*^{sa}, *i.e.*, $sf : \pi_1(\mathcal{F}_*^{sa}) \to \mathbf{Z}$ (respectively, $\mathbf{R}$) is surjective. To see that this group homomorphism is one-to-one on a factor requires the homotopy equivalence $\mathcal{F}_*^{sa} \simeq U(\infty)$ in the type I_∞ factor case or the homotopy equivalence [55, 56] $\Omega(\mathcal{F}_*^{sa}) \simeq \mathcal{F}$ in the type II_∞ factor case: in fact, both results only need the somewhat weaker result, $\Omega(\mathcal{F}_*^{sa}) \simeq \mathcal{F}$.

5. Spectral flow between self adjoint involutions

We now revisit the special case which is naturally suggested by the definition of spectral flow. Choose projections $P, Q \in \mathcal{N}$ such that $||\pi(P) - \pi(Q)|| < 1$ so that QP is τ-Fredholm. Let $B_0 = 2Q - 1, B_1 = 2P - 1$ and introduce the path $B(t) = (1-t)B_0 + tB_1, 0 \le t \le 1$. One can easily show in this case that the path B_t consists of Breuer-Fredholm operators. We are interested in the spectral flow along this path. By Definition 4.2 it is equal to the Breuer-Fredholm index of QP in $P\mathcal{N}Q$. By a careful analysis we will explain why this is the right definition.

First notice that $\ker_P(QP) = \ker(Q) \cap \mathrm{ran}(P)$ and $\ker_Q(PQ) = \ker(P) \cap \mathrm{ran}(Q)$. A simple calculation also yields $\ker_P(QP) \oplus \ker_Q(PQ) \subset \ker(B_0 + B_1)$. Conversely any element v of $\ker(B_0 + B_1)$ satisfies $v = Pv + Qv$ and hence $PQ(Qv) = 0$ and $QP(Pv)v = 0$ implying that $ker(B_0 + B_1) \subset \ker_P(QP) \oplus \ker_Q(PQ)$ (note that it is elementary to check that this is an orthogonal decomposition and in particular that $\ker_P(QP) \cap \ker_Q(PQ) = \{0\}$).

Consequently to see what happens as we flow along $B(t), 0 \le t \le 1$ we initially track what happens in

$$ker(B_0 + B_1) = \ker_P(QP) \oplus \ker_Q(PQ).$$

Now for $v \in \ker_P(QP)$, $B_0 v = -v, B_1 v = v$ so that $B(t)v = (2t-1)v$ and spectrum flows from -1 to 1. Conversely for $v \in \ker_Q(PQ)$ $B_0 v = v, B_1 v = -v$ and $B(t)v = (1-2t)v$. Thus we get flow from 1 to -1.

Hence the spectral flow along the path $\{B_t\}$, denoted $sf\{B_t\}$ is the index of $QP : PH \to QH$ as long as we can show that there cannot be spectral flow coming in some more complex way from 'outside' $\ker(B_0+B_1)$. We analyse this possibility below.

Remarks 5.1. Spectral flow for the path $\{B(t)\}$ actually occurs at one point, namely $t = 1/2$. To see this we note that $B(t)$ has no kernel for $t \neq 1/2$ and $\ker(B(1/2)) = \ker(B_0 + B_1)$. The proof of the former assertion is elementary because if $B(t)v = 0$ then $B_0 B_1 v = -\frac{1-t}{t} v$ so that, taking norms on both sides we deduce that $1 - t \leq t$ or $t \geq 1/2$. Similarly $B_1 B_0 v = -\frac{t}{1-t} v$ so that again taking norms we obtain $t \leq 1/2$. Thus there is only a kernel when $t = 1/2$.

The analysis of this example is helped by the structure of the algebra generated by P and Q. We have:

Lemma 5.1. *Let U be the partial isometry in the polar decomposition of $B_0 + B_1$. Then*
(i) $(B_0 + B_1)B(t) = B(1-t)(B_0 + B_1)$
(ii) $UB(t) = B(1-t)U$ *so that* $UB_0 = B_1 U$

Proof. (i) This is a straightforward calculation.
(ii) From (i) we get $(B_0 + B_1)^2 B(t) = B(t)(B_0 + B_1)^2$ so that

$$UB(t)|B_0 + B_1| = B(1-t)U|B_0 + B_1|$$

and hence on $\ker(B_0 + B_1)^\perp$ equation (ii) of the lemma holds. Because $B(t)$ leaves the kernel of $B_0 + B_1$ invariant both sides of (ii) are zero on this kernel proving the result. □

In the type I factor case one can show there is always a gap in the spectrum of $B_0 + B_1$ about zero. This is because on $\ker(B_0 + B_1)^\perp$, the operator $B_0 + B_1$ is boundedly invertible in the type I factor case so there can be no spectral flow on $\ker(B_0 + B_1)^\perp$. We now show that even in a general semifinite von Neumann algebra that there can be **no** spectral flow when $\ker(B_0 + B_1) = \{0\}$.

Prop 5.1. With the above notation, if $\ker(B_0 + B_1) = \{0\}$, then $sf\{B(t)\} = 0$.

Proof. By assumption

$$\ker(P) \cap \operatorname{ran}(Q) = \{0\} = \ker(Q) \cap \operatorname{ran}(P).$$

Now $\mathcal{B} = \{1, (Q-P)^2, (Q+P)\}''$ is a commutative von Neumann algebra, so that all the spectral projections of $(Q-P)^2$ lie in $\mathcal{B}$. Now $\|(Q-P)^2\| = \|Q-P\|^2 \leq 1$ and by our assumption, 1 is **not** an eigenvalue of $(Q-P)^2$ because

$$(Q-P)^2 x = x \Rightarrow P(Q-P)^2 x = Px \Rightarrow PQPx = 0 \Rightarrow (QP)^2 x = 0$$

$$\Rightarrow (QP)x = 0 \Rightarrow Px \in \ker(Q) \cap \operatorname{ran}(P) = \{0\}.$$

Hence $Px = 0$ and similarly $Qx = 0$. Thus $x = (Q-P)^2 x = 0$, and so $\chi_{\{1\}}((Q-P)^2) = 0$. Now, since $||\pi(P) - \pi(Q)|| < 1$ the spectral projections

$$p_n = \chi_{[1-1/n,1]}((Q-P)^2)$$

are τ-finite for large n, and in the commutative algebra $\mathcal{B}$. Now, the p_n are decreasing to $\chi_{\{1\}}((Q-P)^2) = 0$ and so $\tau(p_n) \to 0$. Let $\epsilon > 0$ and choose n so that $\tau(p_n) < \epsilon$, and note that

$$p_n = \chi_{[-1,-\sqrt{1-1/n}]\cup[\sqrt{1-1/n},1]}((Q-P))$$

so that p_n commute with $Q-P$. Since $p_n \in \mathcal{B}$, it commutes with $Q+P$, and so commutes with both Q and P!

We now decompose our space with respect to $1 = (1-p_n) + p_n$ and note that both $p_n(\mathcal{H})$ and $(1-p_n)(\mathcal{H})$ are left invariant by all the B_t. Hence the spectral flow will be the sum of the spectral flows on these two subspaces. Now, since $\tau(p_n) < \epsilon$, the maximum absolute spectral flow on $p_n(\mathcal{H})$ is $|\tau(p_n)| < \epsilon$.

On the other hand, on $(1-p_n)(\mathcal{H})$ we let $Q_n := (1-p_n)Q(1-p_n) = Q(1-p_n)$ and $P_n := (1-p_n)P(1-p_n) = P(1-p_n)$ so that

$$B_t^n := (1-p_n)B_t(1-p_n) = (1-t)Q_n + tP_n \quad \text{on} \quad (1-p_n)(\mathcal{H}).$$

Now, $\|Q_n - P_n\| = \|(Q_n - P_n)^2\|^{1/2} \leq (1-1/n)^{1/2}$. So $B_0^n - B_t^n = 2t(Q_n - P_n)$ and so for $t \leq 1/2$

$$\|B_0^n - B_t^n\| \leq \|Q_n - P_n\| \leq (1-1/n)^{1/2} =: 1 - \delta_n.$$

So when $t \leq 1/2$ we have

$$\sigma(B_t^n) \subseteq [-1,1] \cap \overline{Ball_{1-\delta_n}(\sigma(B_0))} = [-1,-\delta_n] \cup [\delta_n, 1].$$

For $t > 1/2$ we have $(1-t) \leq 1/2$ and $\|B_1^n - B_t^n\| \leq 1-\delta_n$ so again $\sigma(B_t^n) \subseteq [-1,-\delta_n] \cup [\delta, 1]$. Hence there can be no spectral flow on $(1-p_n)(\mathcal{H})$. Finally, since $|sf(B_t)| < \epsilon$ and $\epsilon > 0$ was arbitrary, $sf(B_t) = 0$. □

5.1. *The case of finite von Neumann algebras*

Let us consider the case where the trace τ on $\mathcal{N}$ is finite so that for any two projections P, Q in $\mathcal{N}$, $P - Q$ is trace class. The size of the positive part of the spectrum of B_0 is $\tau(Q)$ and the size of the positive part of the spectrum of B_1 is $\tau(P)$. Thus it is clear that $\tau(P) - \tau(Q) = \tau(P - Q)$ counts the net amount of spectrum that has moved across zero as one moves along the path $B(t), 0 \le t \le 1$. So the spectral flow is

$$\tau(P - Q) = \frac{1}{2}\tau(B_1 - B_0) = \frac{1}{2}\tau\left(\int_0^1 \frac{d}{dt}B(t)dt\right) = \frac{1}{2}\int_0^1 \tau(\frac{d}{dt}B(t))dt.$$

This simple observation should be compared with later formulae for spectral flow.

Now by Lemma 5.1 there is a partial isometry U with $UB_0 = B_1U$ on $\ker(B_0 + B_1)^\perp$ so that if R is the projection onto this subspace $\tau[R(B_1 - B_0)] = 0$. Thus as before, to calculate $\tau(B_1 - B_0)$, it suffices to work in $\ker(B_0 + B_1) = \ker_Q(PQ) \oplus \ker_P(QP)$ and then it is clear that on this space $\tau(P - Q) = \frac{1}{2}\tau(B_1 - B_0)$ is the τ dimension of $\ker_P(QP)$ minus the τ dimension of $\ker_Q(PQ)$.

5.2. *Example: APS boundary conditions*

Another way of thinking about spectral flow along $\{B(t)\}$ which is familiar from Atiyah et al [3] is to relate it to the index of the differential operator $\frac{\partial}{\partial t} + B(t)$. We will briefly sketch this connection for our example of involutions.

Let us suppose that there is a path $w(t), 0 \le t \le 1$ of vectors in $\mathcal{H}$ such that $w(0) \in \ker(Q)$ and $w(1) \in ran(P)$. That is, $B_0(w(0)) = -w(0)$ and $B_1(w(1)) = w(1)$ so that this path represents some flow of spectrum across zero along the path $B(t), 0 \le t \le 1$. Assume the path is smooth and consider the equation

$$w'(t) + B(t)w(t) = 0. \tag{5.1}$$

By restricting our vectors w to lie in $\ker(B_0 + B_1)$ we can easily solve this equation. First note that from

$$(B_0 + B_1)B(t) = B(1 - t)(B_0 + B_1)$$

we see that $B(t)$, for each t leaves $\ker(B_0 + B_1)$ invariant. We know that for $w = w(0)$ in $\ker_P(QP) = ker(Q) \cap ran(P)$, $B(t)w = (2t - 1)w$, so (5.1) becomes $w'(t) + (2t - 1)w(t) = 0$ which has the solution:

$$w(t) = e^{-(t^2 - t)}w(0)$$

noting that $w(0) = w = w(1)$ satisfies the boundary conditions. Similarly if $w \in \ker_Q(PQ)$ one easily constructs a solution to the adjoint equation

$$-w'(t) + B(t)w(t) = 0.$$

Of course these are APS boundary conditions and we are verifying here that for the differential operator $\mathcal{B} = \frac{\partial}{\partial t} + B(t)$ with APS boundary conditions the index of $\mathcal{B}$ is the spectral flow along the path $B(t), 0 \leq t \leq 1$. More precisely $\mathcal{B}$ is densely defined on $L^2([0,1], \mathcal{H})$ with domain the Sobolev space of $\mathcal{H}$-valued functions on $[0,1]$ with L^2 derivative. Because $B(t)$ leaves $\ker(B_0 + B_1)$ invariant we can solve $\mathcal{B}w = 0$ separately on this space and its orthogonal complement.

Recall Lemma 5.1 where U, the partial isometry in the polar decomposition of $B_0 + B_1$, gives an isometry from $\ker(B_0 + B_1)^\perp$ to itself and satisfies $UB(t) = B(1-t)U$. Suppose then that we have $\mathcal{B}w = 0$ where w takes its values in $\ker(B_0 + B_1)^\perp$. Then $v(t) = Uw(1-t)$ satisfies the adjoint equation $\mathcal{B}^*v = 0$ with the adjoint boundary conditions $v(0) \in Q\mathcal{H}$, $v(1) \in P\mathcal{H}$. In other words, each solution of $\mathcal{B}w = 0$ has a counterpart solution Uw of the adjoint equation and vice versa. Thus, as expected, the net spectral flow on $\ker(B_0 + B_1)^\perp$ must be zero.

6. Spectral flow for unbounded operators

The framework is that of noncommutative geometry in the sense of Alain Connes [26, 27, 28, 29, 31]. However we need to extend this to cover odd unbounded *θ-summable* or finitely-summable *Breuer-Fredholm modules* for a unital Banach $*$-algebra, $\mathcal{A}$. These are pairs $(\mathcal{N}, D)$ where $\mathcal{A}$ is represented in the semifinite von Neumann algebra $\mathcal{N}$ with fixed faithful, normal semifinite trace τ acting on a Hilbert space, H, and D is an unbounded self adjoint operator on H affiliated with $\mathcal{N}$ satisfying: $(1+D^2)^{-1}$ is compact with the additional side condition that either e^{-tD^2} is trace class for all $t > 0$ (θ-summable) or $(1+D^2)^{-1/2} \in \mathcal{L}^n$ for all $n > p$ (with p chosen to be the least real number for which this holds) and $[D, a]$ is bounded for all a in a dense $*$-subalgebra of $\mathcal{A}$. The condition $(1+D^2)^{-1/2} \in \mathcal{L}^n$ is known as n-summability. An alternative terminology is to refer to $(\mathcal{A}, \mathcal{N}, D)$ as a semifinite spectral triple. The theory of spectral triples in a von Neuman algebra was first exposed in Carey et al [15] and further developed by Benameur et al [6] and some of the present authors [20, 17, 18, 19, 66].

If u is a unitary in this dense $*-$subalgebra then

$$uDu^* = D + u[D, u^*] = D + B$$

where B is a bounded self adjoint operator in $\mathcal{N}$. We say D and uDu^* are **gauge equivalent**. The path

$$D_t^u := (1-t)\,D + tuDu^* = D + tB$$

is a "continuous" path of unbounded self adjoint Breuer-Fredholm operators. More precisely,

$$F_t^u := D_t^u \left(1 + (D_t^u)^2\right)^{-1/2}$$

is a norm-continuous path of (bounded) self adjoint τ-Breuer-Fredholm operators. The *spectral flow* along this path $\{F_t^u\}$ (or $\{D_t^u\}$) is defined using the first Section via $sf(\{D_t^u\}) := sf(\{F_t^u\})$. It recovers the pairing of the K-homology class $[D]$ with the K-theory class $[u]$.

We can relate this spectral flow for the path $\{D_t^u\}$ of unbounded Breuer Fredholm operators to the relative index of two projections as follows. Let $\tilde{F}_0^u$ and $\tilde{F}_1^u$ be the partial isometries in the polar decomposition of F_0^u and F_1^u respectively. By convention these extend to unitaries by making them the identity on $\ker(F_0^u)$ and $\ker(F_1^u)$ respectively. We introduce the path $\{\tilde{F}_t^u\}$ where $\tilde{F}_t^u = (1-t)\tilde{F}_0^u + t\tilde{F}_1^u$. We show below that the spectral flow along $\{F_t^u\}$ is in fact equal to the spectral flow along $\{\tilde{F}_t^u\}$.

Prop 6.1. Let $(\mathcal{A}, \mathcal{N}, D)$ be a semifinite spectral triple, and let $u \in \mathcal{A}$ be a unitary. Then the spectral flow from D to uDu^* is $sf(D, uDu^*) = \mathrm{Ind}(PuP)$, where $P := \chi(D)$.

Proof. By the definition $sf(D, uDu^*) := sf(F_D, F_{uDu^*})$, where $F_D := D(1+D^2)^{-1/2}$. With $\tilde{F}_D$ as defined in the paragraph preceding the proposition introduce the non-negative spectral projection P of F_D by $\tilde{F}_D = 2P - 1$, $\tilde{F}_{uDu^*} = 2Q - 1 = 2(uPu^*) - 1$. **If** $\|\pi(P) - \pi(Q)\| < 1$, then by the definition

$$sf(F_D, F_{uDu^*}) := \mathrm{Ind}(PQ) = ec(P, Q).$$

To see that $\|\pi(P) - \pi(Q)\| < 1$, we have [15], $F_D - F_{uDu^*} \in \mathcal{K}_{\tau\mathcal{N}}$ and

$$\tilde{F}_D - F_D = \tilde{F}_D(1 - |F_D|) = \tilde{F}_D(1 - |F_D|^2)(1 + |F_D|)^{-1}$$

$$= \tilde{F}_D(1 + D^2)^{-1}(1 + |F_D|)^{-1} \in \mathcal{K}_{\tau\mathcal{N}}.$$

Hence,

$$2(P - Q) = \tilde{F}_D - \tilde{F}_{uDu^*} = (\tilde{F}_D - F_D) + (F_D - F_{uDu^*}) + (F_{uDu^*} - \tilde{F}_{uDu^*})$$

is also τ-compact, and therefore $\|\pi(P) - \pi(Q)\| = 0 < 1$. By Lemma 4.1 this shows that the operator PQ is $(P \cdot Q)$-Fredholm. Hence, using the formula $\mathrm{Ind}(ST) = \mathrm{Ind}(S) + \mathrm{Ind}(T)$ from Section 3 above, we obtain

$$sf(D, uDu^*) = \mathrm{Ind}(PQ) = \mathrm{Ind}(PuPu^*) = \mathrm{Ind}(PuP). \qquad \square$$

We conclude this Section with a discussion of a theorem of Lesch [46]. Let $\mathcal{A}$ be a unital C^*-algebra with a faithful tracial state τ, $(\pi_\tau, \mathcal{H}_\tau)$ be the GNS representation of $\mathcal{A}$. Let $(\mathcal{A}, \mathbf{R}, \alpha)$ be a τ-invariant C^*-dynamical system. We will identify $\mathcal{A}$ with its image $\pi_\tau(\mathcal{A})$. Let $\mathcal{A} \times_\alpha \mathbf{R}$ be the crossed product, so it acts on $\mathcal{H} = L^2(\mathbf{R}, \mathcal{H}_\tau) = L^2(\mathbf{R}) \otimes \mathcal{H}_\tau$. So we have representations π and λ of $\mathcal{A}$ and $\mathbf{R}$ given as follows: $\pi(a)$ acts on $\xi \in \mathcal{H}$ by $\pi(a)\xi(s) = \alpha_s^{-1}(a)\xi(s)$ and $\lambda_t\xi(s) = \xi(s-t)$. Let $\mathcal{N}$ be the von Neumann algebra generated by $\mathcal{A} \times_\alpha \mathbf{R}$. Clearly, $\lambda = \{\lambda_t\}_{t\in\mathbf{R}}$ is a one-parameter group of unitaries in $\mathcal{N}$. Let D be its infinitesimal generator, that is $\lambda_t = e^{-itD}$, $t \in \mathbf{R}$. We have $\pi(\alpha_t(a)) = \lambda_t\pi(a)\lambda_{-t}$, which is equivalent to $\pi(\delta(a)) = 2\pi i[D, \pi(a)]$, where δ is the infinitesimal generator of α_t and a is in the domain of δ which is a dense *-subalgebra $\mathcal{A}_0$ of $\mathcal{A}$.

In this situation a combination of Proposition 6.1 and earlier work [21] gives the following index theorem of Lesch [46] and Phillips and Raeburn [59].

Theorem 6.1. *The triple $(\mathcal{A}_0, \mathcal{H}, D)$ is a $(1,\infty)$-summable semifinite spectral triple and for any unitary $u \in \mathcal{A}_0$, the operator PuP is Breuer-Fredholm in $P\mathcal{N}P$, and*

$$sf(D, uDu^*) = \mathrm{Ind}(PuP) = \frac{1}{2\pi i}\tau(u\delta(u^*)),$$

where $P = \chi_{[0,\infty)}(D)$.

In the case when $\mathcal{A} = C(\mathbf{T})$, α_t is the rotation by an angle t and τ is the arclength integral on $\mathbf{T}$, then modulo some fiddling with $\mathbf{R}$ vs. $\mathbf{T}$, we infer the classical Gohberg-Krein theorem [41].

Corollary 6.1. *If u is a unitary in $C(\mathbf{T})$ which is continuously differentiable then*

$$sf(D, uDu^*) = \mathrm{Ind}(PuP) = \frac{-1}{2\pi i}\int_{\mathbf{T}} \frac{u'(x)}{u(x)}dx.$$

In the case when $\mathcal{A} = CAP(\mathbf{R}) = C(\mathbf{R}_B)$ (i.e. $\mathcal{A}$ is the C^*-algebra of all uniformly almost periodic functions on $\mathbf{R}$, which we identify with the C^*-algebra of all continuous functions on the Bohr compactification $\mathbf{R}_B$),

α_t is a shift by t and τ is the Haar integral on $\mathbf{R}_B$, we immediately infer the Coburn-Douglas-Schaeffer-Singer theorem.

Corollary 6.2. *If u is a unitary almost periodic continuously differentiable function then*

$$sf(D, uDu^*) = \mathrm{Ind}(PuP) = \lim_{T\to\infty} \frac{-1}{4\pi i T}\int_{-T}^{T} \frac{u'(x)}{u(x)}dx.$$

In the case when $\mathcal{A} = C(\mathbf{T}^2)$, α_t is the Kronecker flow given by the vector field $\partial_x + \theta\partial_y$ with an irrational angle θ and τ is the Haar integral on $\mathbf{T}^2$ we get the following example [15].

Corollary 6.3. *For the unitary element $u(z_1, z_2) = z_2$ from $C(\mathbf{T}^2)$, we have $sf(D, uDu^*) = \mathrm{Ind}(PuP) = \theta$.*

7. Fredholm modules and formulae for spectral flow

In the case of a finite von Neumann algebra with finite trace τ for any projections P and Q, $P - Q$ is trivially trace class. For a general semifinite von Neumann algebra $\mathcal{N}$ arbitrary projections do not satisfy this property. However summability conditions on D guarantee that there is a function f such that $f(P - Q)$ is trace class. In this setting Carey and Phillips [16] extended results of Avron et al [5]. Specifically, provided $f(1)$ is nonvanishing and f is odd with $f(P - Q)$ trace class, then

$$sf\{B(t)\} = \mathrm{Ind}QP = \frac{1}{f(1)}\tau(f(P - Q)).$$

Starting from these results a somewhat lengthy argument produces general formulae for spectral flow in the case of p-summable and Θ-summable unbounded Fredholm modules [15, 16] which we will now describe.

7.1. *The Carey-Phillips formulae for spectral flow*

It was Singer [65] who suggested, in the 1974 Vancouver ICM, that spectral flow and eta invariants were given by integrating a one form. The first paper to systematically exploit this observation was that of Getzler [40]. Getzler's paper provided the inspiration for the following extensions. For paths of Dirac type operators formulae analogous to the ones we describe here go back to the original papers [3], see for example Chapter 8 of Melrose [52].

Let $(\mathcal{A}, \mathcal{H}, D)$ be a θ-summable spectral triple. We focus here on a couple of the main results of Carey et al [16] in the particular case where we compute the spectral flow from D_0 to $D_1 = uD_0u^*$. First we have the formula

$$sf(D, uDu^*) = \frac{1}{\sqrt{\pi}} \int_0^1 \tau(u[D, u^*]e^{-(D+tu[D,u^*])^2})dt.$$

If $(\mathcal{A}, \mathcal{H}, D)$ is n-summable for some $n > 1$ then

$$sf(D, uDu^*) = \frac{1}{C_{n/2}} \int_0^1 \tau(u[D, u^*](1 + (D + tu[D, u^*])^2)^{-n/2})dt,$$

with $C_{n/2} = \int_{-\infty}^{\infty}(1 + x^2)^{-n/2}dx$.

In the type I case the theta summable formula appeared in Geztler [40]. The proof of this formula in general uses a result on spectral flow for **bounded** self adjoint Breuer-Fredholm operators which we will briefly explain.

7.2. *Paths of unbounded Breuer-Fredholm operators*

Our approach to spectral flow for a path of unbounded self adjoint operators affiliated to $\mathcal{N}$ is to introduce the map $D \mapsto F_D = D(1+D^2)^{-1/2}$. By Carey et al [19], Section 3 if $\{D(t)\} = \{D(0) + A(t)\}$ is a path of unbounded self adjoint τ-Breuer-Fredholm operators affiliated to $\mathcal{N}$ where $\{A(s)\}$ is a norm continuous path of bounded self adjoint operators in $\mathcal{N}$, then $\{F_{D(t)}\}$ is a continuous path of self adjoint τ-Breuer-Fredholm operators in $\mathcal{N}$. We then define the spectral flow of the path $\{D(t)\}$ to be the spectral flow of the path $\{F_{D(t)}\}$, and note that in the case $\mathcal{N} = \mathcal{B}(\mathcal{H})$, by Booß-Bavnbek et al [9] one can define $sf(\{D(s)\})$ directly.

The principal difficulty introduced by this point of view is that in practice the map $D(t) \mapsto F_{D(t)}$ is hard to deal with when it comes to proving continuity and differentiability. One of the main features of earlier work [16] was to surmount this hurdle.

It is easier to deal with the map $s \mapsto (\lambda - D(t))^{-1}$ where λ is in the resolvent set of $D(t)$ and to require continuity of this map into the bounded operators in $\mathcal{N}$. This is equivalent to graph norm continuity. It is shown that [9, 47] for $\mathcal{N} = \mathcal{B}(\mathcal{H})$ this resolvent map suffices for a definition of spectral flow for paths of unbounded self adjoint Fredholm operators. Unfortunately the case of general semifinite $\mathcal{N}$ seems beyond the scope of these methods.

We let $\mathcal{M}_0 = \{D = D_0 + A \mid A \in \mathcal{N}_{sa}\}$ be an affine space modelled on $\mathcal{N}_{sa}$. Let $\gamma = \{D_t = D_0 + A(t), a \leq t \leq b\}$ be a piecewise C^1 path in $\mathcal{M}_0$ with D_a and D_b invertible. The spectral flow formula of Getzler [40] when $\mathcal{N} = B(H)$ is

$$\begin{aligned}\mathrm{sf}(D_a, D_b) &= -\int_\gamma \alpha_\epsilon + \frac{1}{2}\eta_\epsilon(D_b) - \frac{1}{2}\eta_\epsilon(D_a) \\ &= \sqrt{\frac{\epsilon}{\pi}}\int_a^b \tau\left(\frac{d}{dt}(D_t)e^{-\epsilon D_t^2}\right)dt + \frac{1}{2}\eta_\epsilon(D_b) - \frac{1}{2}\eta_\epsilon(D_a).\end{aligned}$$

where $\eta_\epsilon(D)$ are approximate eta invariant correction terms:

$$\eta_\epsilon(D) = \frac{1}{\sqrt{\pi}}\int_\epsilon^\infty \tau\left(De^{-tD^2}\right)t^{-1/2}dt$$

and α_ϵ is a one form defined on $\mathcal{N}_{sa}$, the tangent space to $\mathcal{M}_0$, via $\alpha_\epsilon(X) = \sqrt{\frac{\epsilon}{\pi}}\tau(Xe^{-\epsilon D^2})$.

The most general formula in the bounded case deals with a pair of self adjoint τ-Breuer-Fredholm operators $\{F_j,\ j = 1,2\}$, joined by a piecewise C^1 path $\{F_t\}$, $t \in [1,2]$ in a certain affine subspace of the space of all self adjoint τ-Breuer-Fredholm operators. The spectral flow along such a path is given by

$$sf(F_1, F_2) = \frac{1}{C}\int_1^2 \tau\left(\frac{d}{dt}(F_t)|1 - F_t^2|^{-r}e^{-|1-F_t^2|^{-\sigma}}\right)dt + \gamma(F_2) - \gamma(F_1)$$

where the $\gamma(F_j)$ are eta invariant type correction terms, C is a normalization constant depending on the parameters $r \geq 0$ and $\sigma \geq 1$. The affine space in which $\{F_t\}$ live is defined in terms of perturbations of one fixed F_0 and by the condition that $|1 - F_t^2|^{-r}e^{-|1-F_t^2|^{-\sigma}}$ is trace class [16]. To see one important place where such complicated formulae arise, one takes the Getzler expression with $\epsilon = 1$ and where the endpoints are unitarily equivalent so that the end-point correction terms cancel:

$$\frac{1}{\sqrt{\pi}}\int_a^b \tau\left(\frac{d}{dt}(D_t)e^{-D_t^2}\right)dt$$

and does the change of variable $F_t = D_t(1 + D_t^2)^{-1/2}$. Then, $(1 + D_t^2)^{-1} = (1 - F_t^2) = |1 - F_t^2|$, and if one is careless and just differentiates formally (not worrying about the order of the factors), one obtains the expression:

$$\frac{e}{\sqrt{\pi}}\int_a^b \tau\left(\frac{d}{dt}(F_t)|1 - F_t^2|^{-3/2}e^{|1-F_t^2|^{-1}}\right)dt.$$

While the actual details are much more complicated, this is the heuristic essence of the reduction of the unbounded case to the bounded case.

The key observation in this approach is a geometric viewpoint due to Getzler. He noted [40] that in the unbounded case the integrand in the spectral flow formula is a one form on the affine space $\mathcal{M}_0$. This goes back to the observation by Singer that the eta invariant itself is actually a one form. One may also explain this fact from our point of view [16]. In proving the bounded spectral flow formula one uses in a crucial way the fact that $D \to \alpha_D$ where $\alpha_D(X) = \tau(Xe^{|1-F_D^2|^{-1}})$ for X in the tangent space to $\mathcal{M}_0$ at D is an exact one form.

Example. We revisit Corollary 6.3. The straight line path from D to uDu^* is $D_t^u = D + t\theta 1$ for $t \in [0,1]$. As t increases from 0 to 1, the spectral subspaces of the operators D_t^u remain the same, but the spectral values each increase by θ. The spectral subspace of D corresponding to the interval $[-\theta, 0)$, $E = E_{[-\theta,0)}$, is exactly the subspace where the spectral values change from negative to non-negative. By a calculation very similar to the example from Section 4.2, the spectral flow of the path $\{D_t^u\}$ is exactly $\tau(E)$ and since $E = \lambda(\hat{g}) \otimes 1$ where $g = \chi_{[-\theta,0)}$ we have

$$sf(\{D_t^u\}) = \tau(E) = \int_{-\infty}^{\infty} \chi_{[-\theta,0)} dr = \theta.$$

It is also easy to verify directly that θ is the Breuer-Fredholm index of the operator $T_u := PuP$ in the II_∞ factor $P\mathcal{N}P$. Finally, using the formula of Section 7.1 with $n/2 = 1$ we calculate:

$$\frac{1}{\pi}\int_0^1 \tau\left(\frac{d}{dt}(D_t^u)\left(1+(D_t^u)^2\right)^{-1}\right) dt = \frac{1}{\pi}\int_0^1 \tau\left(\theta\left(1+(D+t\theta)^2\right)^{-1}\right) dt$$

$$= \frac{\theta}{\pi}\int_0^1 \left(\int_{-\infty}^{\infty} \frac{1}{1+(r+t\theta)^2} dr\right) dt = \frac{\theta}{\pi}\int_0^1 \left(\int_{-\infty}^{\infty} \frac{1}{1+u^2} du\right) dt = \theta$$

which gives the expected result $\mathrm{Ind}(PuP) = sf(\{D_t^u\}) = \theta$.

8. Spectral flow, adiabatic limits and covering spaces

8.1. *Introductory remarks*

We start with some observations of Mathai [49, 48] on the motivating example which arises from the fundamental paper of Atiyah [1]. Assume that D_0 is a self adjoint Dirac type operator on smooth sections of a bundle over an odd dimensional manifold M. We assume that M is not compact but admits a continuous free action of a discrete group G such that the quotient of M by G is a compact manifold. We assume M is equipped with a G invariant metric in terms of which D_0 is defined and from which we obtain an Hermitian inner product on the sections of the bundle such that D_0 is unbounded and self adjoint on an appropriate domain. In this setting the G action on M lifts to an action by unitary operators on L^2 sections of the bundle. The von Neumann algebra $\mathcal{N}$ we consider is the commutant of the G action on L^2 sections [49]. Consider a path of the form $D_t = D_0 + A(t)$ where $A(t)$ is a bounded self adjoint pseudodifferential operator depending continuously on t in the norm topology and commuting with the G action. Then D_t is self adjoint Breuer-Fredholm operator affiliated to $\mathcal{N}$, and $D_t(1+D_t^2)^{-1/2}$ is a bounded self adjoint Breuer-Fredholm operator in $\mathcal{N}$ for each $t \in \mathbf{R}$.

While $\mathcal{N}$ is a semifinite von Neumann algebra it is not in general a factor. There is a natural trace on $\mathcal{N}$ (considered by Atiyah [1] in his account of the L^2 index theorem) which we now define. It will be with respect to this trace that we calculate spectral flow along $\{D_0 + A(t)\}$ (recall that the type II spectral flow depends on the choice of trace non-trivially when the algebra $\mathcal{N}$ has non-trival centre). On operators with smooth Schwartz kernel $k(x,y)$; $x, y \in M$ the trace τ_G is given by taking the fibrewise trace of the kernel on the diagonal $tr(k(x,x))$ and integrating over a fundamental domain for G. This is the natural trace as may be seen by recognising that the representation of G we obtain here is quasi-equivalent to the regular representation. The regular representation is determined by the standard trace τ_0 which is given on an element $\sum \lambda_g g$ of the group algebra by $\tau_0(\sum \lambda_g g) = \lambda_e$ (with the identity being $e \in G$ and the $\lambda_g \in \mathbf{C}$).

The analysis of spectral flow traditionally proceeds by replacing M by $M \times S^1$ or $M \times [0,1]$ and considering the Dirac operator on this even dimensional manifold as in Mathai [49]. On covering spaces it is believed by the experts that one should be able to recover an analytic spectral flow formula however one cannot easily extract a proof from the literature. The argument we present in this Section shows how a special case of the spectral flow formulae discussed in the previous Section arises naturally from adiabatic limit ideas due originally to Cheeger [22].

8.2. *Easy adiabatic formula in even dimensions (EAF)*

We use an adiabatic process, which leads to the formula for the leading term in the expansion of the (difference of) heat kernels. This is part of an IUPUI preprint [67] which was never submitted for publication. We discuss the simplest possible case (of a compatible Dirac operator with coefficients in an auxiliary bundle) in order to avoid use of elliptic estimates as was done in Cheeger [22]. Therefore our main tool is Duhamel's principle (the expansion of the heat kernels of the Dirac Laplacians with respect to the perturbation terms [10, 51]) and we call our result the *EAF* = *Easy Adiabatic Formula.* (Let us point out that it is not difficult to imitate Cheeger's proof and obtain the *EAF* in complete generality i.e. for the family of Dirac operators with varying first order part.) We present a proof in the case of a closed manifold M and later on outline why our argument holds in the case of a continuous free action of a discrete group.

Let $B : C^\infty(M;S) \to C^\infty(M;S)$ denote a compatible Dirac operator acting on sections of a bundle of Clifford modules S over a closed, odd-dimensional manifold M [10]. Introduce an auxiliary hermitian vector bundle E with hermitian connection ∇, and the operator $B_0 = B \otimes_\nabla Id_E$ (see Palais [53], Chapter IV). Let $g : E \to E$ denote a unitary bundle automorphism, then we can introduce the operator

$$B_1 = gB_0g^{-1} = (Id \otimes g)(B \otimes_\nabla Id_E)(Id \otimes g^{-1}) \ .$$

The difference $T = B_1 - B_0 = [g, B]g^{-1}$ is a bundle endomorphism and we want to present a formula for the spectral flow of the family

$$\{B_u = B_0 + uT\}_{0 \le u \le 1} \ . \tag{8.1}$$

Spectral flow is a homotopy invariant, so we can restrict ourselves to the study of the spectral flow of a smooth family of self adjoint operators over S^1. We introduce a smooth cut-off function $\alpha : \mathbf{R} \to \mathbf{R}$, such that

$$\alpha(u) = \begin{cases} 0 & \text{if } \ u \le 1/4 \ , \\ 1 & \text{if } \ 3/4 \le u \ . \end{cases}$$

We may also assume that there exists a positive constant c , such that

$$\left| \frac{d^k \alpha}{du^k} \right| \le c \cdot u \tag{8.2}$$

for $0 \le u \le 1$, $k = 0, 1, 2$. Now, we consider the family

$$\{B_u = B_0 + \alpha(u)T\} \ , \tag{8.3}$$

which in an obvious way provides us with a family of operators on S^1. We may also consider the corresponding operator $\mathcal{D} = \partial_u + B_u$ on the closed manifold $N = S^1 \times M$ where B_u is given by the formula (8.3). The operator

$\mathcal{D}$ acts on sections of the bundle $[0,1] \times S \otimes E/_{\cong}$, where the identification is given by

$$(1, y; g(y)w) \cong (0, y; w) \quad \text{where} \quad w \in S_y \otimes E_y \quad .$$

Theorem 8.1. *The following formula holds for any* $t > 0$

$$index\ \mathcal{D} = sf\{B_u\} = \sqrt{\frac{t}{\pi}} \cdot \int_0^1 Tr_M \dot{B}_u e^{-tB_u^2} du \ , \tag{8.4}$$

where as usual $\dot{B}_u = \frac{dB}{du}$.

The first equality in (8.4) goes back to the original Atiyah-Patodi-Singer paper [3]. They also proved a formula

$$sf\{B_u\} = \int_0^1 \dot{\eta}_u du \ ,$$

where η_u denotes the η-invariant of the operator B_u . The equality

$$sf\{B_u\} = \sqrt{\frac{t}{\pi}} \cdot \int_0^1 Tr_M \dot{B}_u e^{-tB_u^2} du$$

and more (see below) was proved by the last named author around 1991 and published in one of his IUPUI preprints [67]. The formal paper, which was supposed to contain a new discussion of *Witten's Holonomy Theorem* never appeared and the result eventually resurfaced in the paper by Getzler [40]. We have to mention that it is not difficult to manufacture a more straightforward argument to prove the second equality in (8.4). Here we prove a stronger result that provides the "adiabatic" equality on the level of heat kernels, from which (8.4) and other results not covered in the current paper follow. The original proof was a "toy" model for a simplified version of Cheeger's proof of *Witten's Holonomy Theorem* [22] and proves the corresponding adiabatic equality on the level of the kernels of the corresponding heat operators. Therefore we call this equality the *EAF* = *Easy Adiabatic Formula.*

The *EAF* is obtained by applying the adiabatic process on N in the normal direction to a fibre M . We replace the product Riemannian metric $g = du^2 + g_M$ by a new metric

$$g_\epsilon = \frac{du^2}{\epsilon^2} + g_M = dv^2 + g_M \ , \tag{8.5}$$

and let the positive parameter ϵ run to 0 . The corresponding operator $\mathcal{D}_\epsilon$ has the following representation

$$\mathcal{D}_\epsilon(v, y) = \partial_v + B_{\epsilon v}(y) \ . \tag{8.6}$$

In the equality (8.6) we use a new normal coordinate $v = \frac{u}{\epsilon}$ $(y \in M)$. The operator $\mathcal{D}_\epsilon$ lives on the manifold $N_\epsilon = S^1_\epsilon \times M$, where S^1_ϵ denotes the circle of length $\frac{1}{\epsilon}$. Both *index* and sf do not change their values under the deformation, hence we have the equality

$$index\ \mathcal{D} = index\ \mathcal{D}_\epsilon = sf\{B_{\epsilon v}\} = sf\{B_u\} \ .$$

The famous McKean-Singer equality expresses the index in terms of the kernels of the heat operators

$$index\ \mathcal{D} = index\ \mathcal{D}_\epsilon = Tr\ e^{-t\mathcal{D}^*_\epsilon \mathcal{D}_\epsilon} - Tr\ e^{-t\mathcal{D}_\epsilon \mathcal{D}^*_\epsilon} \ ,$$

for fixed $t > 0$ as $\epsilon \to 0$. Let $k_\epsilon(t; (v_1, y_1), (v_2, y_2))$ (v_i is the coordinate on S^1_ϵ and $y_i \in M$), denote the kernel of the operator $e^{-t\mathcal{D}^*\mathcal{D}_\epsilon} - e^{-t\mathcal{D}_\epsilon\mathcal{D}^*_\epsilon}$. This is the difference of the heat kernels, which are pointwise bounded for fixed time t (see Proposition 8.1), which expands into expansion with respect to the parameter ϵ. The contributions to the leading term from the kernel of the operator $\mathcal{D}^*\mathcal{D}$ and the kernel of the operator $\mathcal{D}\mathcal{D}^*$ cancel each other, hence this term is equal to 0. Theorem 8.2 provides the formula for the second term in the expansion. It follows that the kernel $k_\epsilon(t; (v_1, y_1), (v_2, y_2))$ at the given point is of size of ϵ and the volume of the manifold N_ϵ is equal to $\frac{1}{\epsilon} \cdot vol(M)$ hence at the end we obtain a finite limit

$$\lim_{\epsilon \to 0} \int_{N_\epsilon} tr\ k_\epsilon(t; (v, y), (v, y)) dv dy$$

equal to (8.4).

Now, we present the kernel on N_ϵ, which gives the leading term in the expansion of $k_\epsilon(t; (v_1, y_1), (v_2, y_2))$. Let us fix v_0 the value of the normal coordinate and let $e_{v_0}(t; (v_1, y_1), (v_2, y_2))$ denote the kernel of the heat operator $e^{-tB^2_{\epsilon v_0}}$. We also introduce $\epsilon\alpha'(\epsilon v_0) T e_{v_0}$, kernel of the operator $\epsilon\alpha'(\epsilon v_0) T e^{-tB^2_{\epsilon v_0}}$. To get the final product we have to take the convolution of kernels. If k_1, k_2 denote two time-dependent operators with smooth kernels on M, then $k_1 * k_2(t) = \int_0^t k_1(s) k_2(t-s) ds$ and on the level of the kernels we have the equality

$$k_1 * k_2(t; y_1, y_2) = \int_0^t ds \int_M dz\ k_1(s; y_1, z) k_2(t-s; z, y_2) \ .$$

We introduce the kernel

$$\mathcal{E}_{v_0}(t; (v_1, y_1), (v_2, y_2)) = 2\epsilon\alpha'(\epsilon v_0) e_{\partial_v}(t; v_1, v_2)(e_{v_0} * T e_{v_0})(t; y_1, y_2), \quad (8.7)$$

where $e_{\partial_v}(t; v_1, v_2)$ denotes the kernel of the 1-dimensional heat operator defined on $\mathbf{R}$ by the operator $-\partial_v^2$. Let us also point out that $\epsilon\alpha'(\epsilon v_0) T$ is simply equal to $\epsilon \dot{B}$ (at v_0), where *dot* denotes the derivative with respect to u-variable, so it is the operator $\frac{d}{dv} B_{\epsilon v}\Big|_{v=v_0}$.

At last, we are ready to formulate the *EAF*

Theorem 8.2. *For any $t > 0$ there exists ϵ_0 and a constant $c > 0$ such that for any $0 < \epsilon < \epsilon_0$*

$$\frac{1}{\epsilon} \cdot \| k_\epsilon(t; (v_0, y_1), (v_0, y_2)) - \mathcal{E}_{v_0}(t; (v_0, y_1), (v_0, y_2)) \| \leq c \cdot \sqrt{\frac{\epsilon}{t}} \ . \tag{8.8}$$

Remarks 8.1. (1) It has been already pointed out that we only present the proof of *EAF* for the family (8.3). The method we use works for any family $\{B_u\}_{0 \leq u \leq 1}$, such that for any $0 \leq u \leq 1$ the difference $B_u - B_0$ is an operator of order 0 and

$$B_1 = g B_0 g^{-1} \quad \textit{and} \quad \|B_u - B_0\| \leq c \cdot u \ .$$

If the differeeen between the operators B_u is a $1st$ order operator then we have to follow a more complicated version of the argument as presented in Cheeger's work [22].

(2) The proof we use allows us to replace $\sqrt{\epsilon}$, which appears on the right side of (8.8), by ϵ^r , for any $0 < r < 1$.

Proof. *(2nd Part).* The most technical part of the proof is presented in the next subsection. There we use the Duhamel's Principle to obtain the equality

$$e^{-t\Delta_{1,\epsilon}} - e^{-t\Delta_{2,\epsilon}} = \int_0^t e^{-s\Delta_{1,\epsilon}} (\Delta_{2,\epsilon} - \Delta_{1,\epsilon}) e^{-(t-s)\Delta_{2,\epsilon}} ds$$

$$= \int_0^t [e^{-s\Delta_0}(\Delta_{2,\epsilon} - \Delta_{1,\epsilon}) e^{-(t-s)\Delta_0} + (e^{-s\Delta_{1,\epsilon}} - e^{-s\Delta_0})(\Delta_{2,\epsilon} - \Delta_{1,\epsilon}) e^{-(t-s)\Delta_{2,\epsilon}}] ds$$

$$+ \int_0^t e^{-s\Delta_0} (\Delta_{2,\epsilon} - \Delta_{1,\epsilon})(e^{-(t-s)\Delta_{2,\epsilon}} - e^{-(t-s)\Delta_0}) ds \ ,$$

and we show that the kernels of the second and third terms on the right side are point-wise at most of the size $\epsilon^{\frac{3}{2}}$. Hence, we only have to study the first term and show that it gives the kernel (8.7) as $\epsilon \to 0$.

The operator

$$\int_0^t e^{-s\Delta_0} (\Delta_{2,\epsilon} - \Delta_{1,\epsilon}) e^{-(t-s)\Delta_0} ds \tag{8.9}$$

can be represented as

$$\int_0^t e^{-s(-\partial_v^2)} e^{-sB_{\epsilon v_0}^2} (2\epsilon\alpha'(\epsilon v)T) e^{-(t-s)(-\partial_v^2)} e^{-(t-s)B_{\epsilon v_0}^2} ds =$$

$$\int_0^t e^{-s(-\partial_v^2)} (2\epsilon\alpha'(\epsilon v)) e^{-(t-s)(-\partial_v^2)} e^{-sB_{\epsilon v_0}^2} T e^{-(t-s)B_{\epsilon v_0}^2} ds \quad .$$

First, we study $l_\epsilon(s,t;v_1,v_2)$ the kernel of the part which acts in the normal direction

$$e^{-s(-\partial_v^2)} (2\epsilon\alpha'(\epsilon v)) e^{-(t-s)(-\partial_v^2)} \quad .$$

We have

$$l_\epsilon(s,t;v_0,v_0) = \int_{-\infty}^{+\infty} \frac{e^{-\frac{(v_0-z)^2}{4s}}}{\sqrt{4\pi s}} (2\epsilon\alpha'(\epsilon z)) \frac{e^{-\frac{(v_0-z)^2}{4(t-s)}}}{\sqrt{4\pi(t-s)}} dz \quad .$$

To simplify, we assume $v_0 = 0$ (this will be justified in the next subsection). We obtain

$$l_\epsilon(s,t;v_0,v_0) = \frac{1}{2\pi} \int_{-\infty}^{+\infty} \frac{e^{-\frac{tz^2}{4s(t-s)}}}{\sqrt{s(t-s)}} (\epsilon\alpha'(\epsilon z)) dz \quad .$$

Now, we only have to show that

$$\frac{1}{2\pi} \int_{-\infty}^{+\infty} \frac{e^{-\frac{tz^2}{4s(t-s)}}}{\sqrt{s(t-s)}} (\epsilon\alpha'(\epsilon z) - \epsilon\alpha'(0)) dz| \le c\frac{\epsilon^{\frac{3}{2}}}{\sqrt{t}} \quad .$$

This is the place where we use (8.2). We apply here the special case of a trick used in the next subsection.

First, an elementary computation shows that

$$\int_{|z|>\frac{1}{\sqrt{\epsilon}}} \frac{e^{-\frac{tz^2}{4s(t-s)}}}{\sqrt{s(t-s)}} (\epsilon\alpha'(\epsilon z) - \epsilon\alpha'(0)) dz|$$

is exponentially small with respect to ϵ . We have

$$|\int_{|z|>\frac{1}{\sqrt{\epsilon}}} \frac{e^{-\frac{tz^2}{4s(t-s)}}}{\sqrt{s(t-s)}} (\epsilon\alpha'(\epsilon z) - \epsilon\alpha'(0)) dz| \le c\frac{\epsilon^{\frac{3}{2}}}{\sqrt{t}} | \le \epsilon c_1 \cdot \int_{|z|>\frac{1}{\sqrt{\epsilon}}} \frac{e^{-\frac{tz^2}{4s(t-s)}}}{\sqrt{s(t-s)}} dz$$

$$= \frac{2}{\sqrt{t}} \cdot \int_{r>\sqrt{\frac{t}{4s(t-s)\epsilon}}} e^{-r^2} \le \frac{2}{\sqrt{t}} \cdot e^{-\frac{1}{4t\epsilon}} \le c_1 e^{-\frac{c_2}{t\epsilon}} \quad .$$

It follows now from (8.2) that

$$|\alpha'(\epsilon z) - \alpha'(0)| \le c_3\sqrt{\epsilon}$$

for $z \leq \frac{1}{\sqrt{\epsilon}}$. This gives

$$|\int_{|z|<\frac{1}{\sqrt{\epsilon}}} \frac{e^{-\frac{tz^2}{4s(t-s)}}}{\sqrt{s(t-s)}}(\epsilon\alpha'(\epsilon z) - \epsilon\alpha'(0))dz| \leq c_3\epsilon^{\frac{3}{2}} \cdot \int_{-\infty}^{+\infty} \frac{e^{-\frac{tz^2}{4s(t-s)}}}{\sqrt{s(t-s)}} \leq c_4 \frac{\epsilon^{\frac{3}{2}}}{\sqrt{t}} .$$

We see that up to a term of order $\frac{\epsilon^{\frac{3}{2}}}{\sqrt{t}}$ the kernel $l_\epsilon(s,t;v_1,v_2)$ (for $v_1 = v_2 = v_0$) is equal to

$$\epsilon\alpha'(v_0)\frac{1}{2\pi}\int_{-\infty}^{+\infty} \frac{e^{-\frac{tz^2}{4s(t-s)}}}{\sqrt{s(t-s)}} = \frac{\epsilon\alpha'(v_0)}{\sqrt{t}} ,$$

and $l_\epsilon(s,t;v_1,v_2)$ can be replaced by the kernel of the operator

$$2\epsilon\alpha'(v_0)e^{-t(-\partial_v^2)} .$$

As a result of these estimates the operator (8.9) can be replaced by

$$2\epsilon\alpha'(\epsilon v_0)e^{-t(-\partial_v^2)} \int_0^t e^{-sB^2_{\epsilon v_0}} T e^{-(t-s)B^2_{\epsilon v_0}} ds ,$$

which has kernel equal to

$$2\epsilon\alpha'(\epsilon v_0)e_{\partial_v}(t;v_1,v_2) \int_0^t ds \int_M e_{v_0}(s;y_1,w)T(w)e_{v_0}(t-s;w,y_2)dw ,$$

which is exactly the kernel $\mathcal{E}_{v_0}$. □

Corollary 8.1. *The spectral flow formula (8.4) follows from the EAF.*

Proof. The *EAF* shows that $k_\epsilon(t;(v_1,y_1),(v_2,y_2))$ is equal to

$$2\epsilon\alpha'(\epsilon v_1)e_{\partial_v}(t;v_1,v_2)(e_{v_1} * Te_{v_1})(t;y_1,y_2) + O(\frac{\epsilon^{\frac{3}{2}}}{\sqrt{t}}) .$$

It follows that, for ϵ small enough, we have the equality

$$index\ \mathcal{D} = 2\int_{N_\epsilon} tr\ \epsilon\alpha'(\epsilon v_0)e_{\partial_v}(t;v_0,v_0)(e_{v_0} * Te_{v_0})(t;y,y)dydv_0$$

$$= \int_0^{\frac{1}{\epsilon}} \frac{\epsilon\alpha'(\epsilon v_0)}{\sqrt{\pi t}}dv_0 \int_0^t Tr_M e^{-sB^2_{\epsilon v_0}} T e^{-(t-s)B^2_{\epsilon v_0}} ds$$

$$= \int_0^{\frac{1}{\epsilon}} \frac{\epsilon\alpha'(\epsilon v_0)}{\sqrt{\pi t}}dv_0 \int_0^t Tr_M T e^{-tB^2_{\epsilon v_0}} ds$$

$$= \sqrt{\frac{t}{\pi}} \int_0^{\frac{1}{\epsilon}} \alpha'(\epsilon v_0) Tr_M T e^{-tB^2_{\epsilon v_0}}(\epsilon dv_0) = \sqrt{\frac{t}{\pi}} \int_0^1 Tr_M \dot{B}_u e^{-tB^2_u} du .$$ □

Let us observe that (for compatible Dirac operators)

$$\lim_{t\to 0}\sqrt{\frac{t}{\pi}}\cdot\int_0^1 Tr_M\dot{B}_u e^{-tB_u^2}du = \int_0^1 du \lim_{t\to 0}\sqrt{\frac{t}{\pi}}\cdot Tr_M\dot{B}_u e^{-tB_u^2} = \int_0^1 \dot{\eta}_u du \quad ,$$

where $\eta_u = \frac{1}{2}(dim\ ker(B_u) + \eta_{B_u}(0))$ is the η-invariant of the operator B_u [3].

8.3. *Technicalities*

We write the operator $\mathcal{D}_\epsilon$ as

$$\mathcal{D}_\epsilon = \partial_v + (B_0 + \alpha(\epsilon v_0)T) + (\alpha(\epsilon v) - \alpha(\epsilon v_0))T = \partial_v + B_{\epsilon v_0} + \beta(\epsilon v)T \ . \quad (8.10)$$

In the following we consider the operator $\mathcal{D}_\epsilon$ as an operator living on $\mathbf{R}\times M$. This does not change anything in the proof, but simplifies the computations. Of course there is a problem with the definition of the index of $\mathcal{D}_\epsilon$ in this set-up. Even though $B_{\epsilon v}$ is a constant operator for $v < 0$ and $\frac{1}{\epsilon} \le v$ the index may not be well-defined unless the operator $B_{\epsilon v}$ is invertible for those values of the normal coordinate. Hence, one can think that we perturbed the tangential operator by a small number and the invertibility condition is satisfied. In any case it follows that the integral from the kernel $k_\epsilon(t;(v,y),(v,y))$ over $[0,\frac{1}{\epsilon}]\times M$ gives an integer equal to the index of $\mathcal{D}_\epsilon$ on N_ϵ and the integral over the leftover of $\mathbf{R}\times M$ gives a finite error term, which goes to 0 as $\epsilon \to 0$. Hence it is not difficult to show that

$$index\ \mathcal{D}_\epsilon = \int_{\mathbf{R}\times M} tr\ k_\epsilon(t;(v,y),(v,y))dydv \ ,$$

where on the left side we have the operator on N_ϵ.

We work on $\mathbf{R}\times M$ and we have to show that EAF holds at any given point (v_0, y). After reparametrization, we can assume that $v_0 = 0$ and our operator has the form

$$\mathcal{D}_\epsilon = \partial_v + B_0 + \beta(\epsilon v)T \quad ,$$

where the cut-off function $\beta(v)$ satisfies $\beta(0) = 0$. The corresponding Laplacians are

$$\Delta_{1,\epsilon} = \mathcal{D}_\epsilon^*\mathcal{D}_\epsilon = -\partial_v^2 + B_0^2 + \beta(\epsilon v)(B_0T + TB_0) + \beta^2(\epsilon v)T^2 - \epsilon\beta'(\epsilon v)T$$

$$= -\partial_v^2 + B_0^2 + \beta(\epsilon v)T_1 - \epsilon\beta'(\epsilon v)T \quad ,$$

and

$$\Delta_{2,\epsilon} = \mathcal{D}_\epsilon\mathcal{D}_\epsilon^* = -\partial_v^2 + B_0^2 + \beta(\epsilon v)T_1 + \epsilon\beta'(\epsilon v)T \quad ,$$

where T_1 denotes the 1st order tangential operator $B_0T + TB_0 + \beta(\epsilon v)T^2$.

To evaluate $k_\epsilon(t;(0,y),(0,y))$ we apply Duhamel's Principle [51, 10]:

$$e^{-t\Delta_{1,\epsilon}} - e^{-t\Delta_{2,\epsilon}} = \int_0^t \frac{d}{ds}(e^{-s\Delta_{1,\epsilon}}e^{-(t-s)\Delta_{2,\epsilon}})ds$$

$$= \int_0^t e^{-s\Delta_{1,\epsilon}}(\Delta_{2,\epsilon} - \Delta_{1,\epsilon})e^{-(t-s)\Delta_{2,\epsilon}}ds \quad .$$

The difference $\Delta_{2,\epsilon} - \Delta_{1,\epsilon}$ is equal to the bundle endomorphism $2\epsilon\beta'(\epsilon v)T$ and this is the term which brings the 1st, and the most important, factor of ϵ into the formula. Once again, we simplify the presentation and introduce Laplacian $\Delta_0 = -\partial_v^2 + B_0^2$ and study each summand in the equality

$$e^{-t\Delta_{1,\epsilon}} - e^{-t\Delta_{2,\epsilon}} = (e^{-t\Delta_{1,\epsilon}} - e^{-t\Delta_0}) + (e^{-t\Delta_0} - e^{-t\Delta_{2,\epsilon}}) \quad ,$$

The application of Duhamel's Principle to the first summand leads to the series

$$e^{-t\Delta_{1,\epsilon}} - e^{-t\Delta_0} = \int_0^t e^{-s\Delta_{1,\epsilon}}(\Delta_0 - \Delta_{1,\epsilon})e^{-(t-s)\Delta_0}ds$$

$$\int_0^t e^{-s_1\Delta_0}(\Delta_0 - \Delta_{1,\epsilon})e^{-(t-s_1)\Delta_0}ds_1$$

$$+ \int_0^t ds_1 \int_0^{s_1} ds_2\ e^{-s_2\Delta_{1,\epsilon}}(\Delta_0 - \Delta_{1,\epsilon})e^{-(s_1-s_2)\Delta_0}(\Delta_0 - \Delta_{1,\epsilon})e^{-(t-s_1)\Delta_0}$$

$$= \sum_{k=1}^{\infty} \int_0^t ds_1 ... \int_0^{s_{k-1}} ds_k\ e^{-s_k\Delta_0}(\Delta_0 - \Delta_{1,\epsilon})...(\Delta_0 - \Delta_{1,\epsilon})e^{-(t-s_1)\Delta_0} \ .$$

The result we state now is a standard application of Duhamel's Principle

Prop 8.1. Let us consider the operator $\Delta_R = \Delta_0 + \beta(\epsilon v)R$, where $R : C^\infty(M;S) \to C^\infty(M;S)$ is a tangential differential operator of order 1 . Then, there exists positive constants c_1 and c_2 (independent of ϵ) such that for any sufficiently small ϵ the following estimate holds

$$\|e_R(t;(v_1,y_1),(v_2,y_2))\| \le c_1 t^{-\frac{m+1}{2}} e^{-c_2\frac{d^2(((v_1,y_1),(v_2,y_2))}{2}} \quad , \tag{8.11}$$

where $e_R(t;(v_1,y_1),(v_2,y_2))$ denotes the kernel of $e^{-t\Delta_R}$.

Proof. The estimate (8.11) holds for Δ_0 and we have

$$e^{-t\Delta_R} - e^{-t\Delta_0} = \sum_{k=1}^{\infty} \int_0^t ds_1 ... \int_0^{s_{k-1}} ds_k \, e^{-s_k \Delta_0} (\beta(\epsilon v) R) ... (\beta(\epsilon v) R) e^{-(t-s_1)\Delta_0}$$

$$= \sum_{k=1}^{\infty} \int_0^t ds_1 ... \int_0^{s_{k-1}} ds_k \left(e^{-s_k(-\partial_v^2)} (\beta(\epsilon v)) ..... (\beta(\epsilon v)) e^{-(t-s_1)(-\partial_v^2)} \right) \times$$

$$\left(e^{-s_k B_0^2} R .... R e^{-(t-s_1) B_0^2} \right) .$$

The kernel of the operator in the first bracket is estimated as follows

$$\| \left(e^{-s_k(-\partial_v^2)} (\beta(\epsilon v)) ..... (\beta(\epsilon v)) e^{-(t-s_1)(-\partial_v^2)} \right) (t; v_1, v_2) \| \leq$$

$$\| \left(e^{-s_k(-\partial_v^2)} ... e^{-(t-s_1)(-\partial_v^2)} \right) (t; v_1, v_2) \| ,$$

as $0 \leq \beta(\epsilon v)$. This leads to the estimate

$$\| e_R(t; (v_1, y_1), (v_2, y_2)) - e_0(t; (v_1, y_1), (v_2, y_2)) \| \leq$$

$$\frac{e^{-\frac{(v_1 - v_2)^2}{4t}}}{\sqrt{4\pi t}} \sum_{k=1}^{\infty} \| \left(e^{-s_k B_0^2} R .... R e^{-(t-s_1) B_0^2} \right) (t; y_1, y_2) \| .$$

The series here is estimated in the standard way. The kernel of the operator $Re^{-tB_0^2}$ is bounded by $c_1 t^{-\frac{m+1}{2}} e^{-c_2 \frac{(v_1 - v_2)^2}{t}}$. Therefore we follow [51] and obtain

$$\sum_{k=1}^{\infty} \| \left(e^{-s_k B_0^2} R .... R e^{-(t-s_1) B_0^2} \right) (t; y_1, y_2) \| \leq c_3 t^{-\frac{m-1}{2}} e^{-c_4 \frac{d^2(y_1, y_2)}{t}} .$$

The positive constants above do not depend on ϵ. □

It is no problem to see that the estimate (8.11) from Proposition 8.1 is also satisfied by kernels of the heat operators defined by $\Delta_{1,\epsilon}$ and $\Delta_{2,\epsilon}$, which leads to the following useful property.

Corollary 8.2. *The contribution to the kernel $k_\epsilon(t; (0, y_1), (0, y_2))$ provided by the points distant more than $\frac{1}{\sqrt{\epsilon}}$ from $\{0\} \times M$ may be disregarded.*

Proof. We have

$$\| \left(e^{-t\Delta_{1,\epsilon}} - e^{-t\Delta_{2,\epsilon}}\right)(t;(0,y_1),(0,y_2))\| =$$

$$\| \int_0^t e^{-s\Delta_{1,\epsilon}}(2\epsilon\beta(\epsilon v)T)e^{-(t-s)\Delta_{2,\epsilon}}(t;(0,y_1),(0,y_2))ds\| \leq$$

$$\int_0^t ds \int_{\mathbf{R}\times M} dudz\|e_{1,\epsilon}(s;(0,y_1),(u,z))2\epsilon\beta(\epsilon u)T(u)e_{2,\epsilon}(t-s;(u,z),(0,y))\| \ .$$

We want to show that the integral over $|u| > \frac{1}{\sqrt{\epsilon}}$ is exponentially small with respect to ϵ.

$$\int_0^t ds \int_{|u|>\frac{1}{\sqrt{\epsilon}}} du \int_M dz\|e_{1,\epsilon}(s;(0,y_1),(u,z))\|2\epsilon\beta(\epsilon u)$$

$$\times\|T(u)\|\|e_{2,\epsilon}(t-s;(u,z),(0,y))\|$$

$$\leq c_3\epsilon \int_0^t ds \int_{|u|>\frac{1}{\sqrt{\epsilon}}} du \int_M dz\ c_1 s^{-\frac{m+1}{2}} e^{-c_2\frac{d^2((0,y_1),(u,z))}{s}}(t-s)^{-\frac{m+1}{2}}$$

$$\times e^{-c_2\frac{d^2((u,z),(0,y_2))}{t-s}}$$

$$\leq c_4\epsilon\cdot vol(M)\int_0^t ds \int_{|u|>\frac{1}{\sqrt{\epsilon}}} (s(t-s))^{-\frac{m+1}{2}} e^{-c_2\frac{u^2}{s}} e^{-c_2\frac{u^2}{t-s}} du$$

$$\leq c_4\epsilon\cdot vol(M)\int_0^t ds \int_{|u|>\frac{1}{\sqrt{\epsilon}}} e^{-c_5\frac{u^2}{s}} e^{-c_5\frac{u^2}{t-s}} du$$

$$\leq c_4\epsilon\cdot vol(M)\int_0^t ds \int_{|u|>\frac{1}{\sqrt{\epsilon}}} e^{-c_5\frac{u^2}{t}} du \leq c_6\epsilon\cdot e^{-\frac{c_7}{\epsilon t}} \ .$$

□

Now the idea of the proof of the EAF can be easily understood. The kernel $k_\epsilon(t;(0,y_1),(0,y_2)$ is the kernel of the operator

$$\int_0^t e^{-s\Delta_{1,\epsilon}}(\Delta_{2,\epsilon} - \Delta_{1,\epsilon})e^{-(t-s)\Delta_{2,\epsilon}}ds \ . \tag{8.12}$$

Kernels of both heat operators $e^{-t\Delta_{i,\epsilon}}$, expands into series where the leading term is the kernel of the operator $e^{-t\Delta_0}$. It follows that the leading term in the expansion of the kernel of the operator (8.12) is

$$\int_0^t ds \int_{\mathbf{R}\times M} e_0(s;(0,y_1),(z,w))(2\epsilon\beta'(\epsilon z)T(w))e_0(t-s;(z,w),(0,y_2))dzdw \ . \tag{8.13}$$

This gives us the contribution which appears in the *EAF*.

We have to show that further perturbation brings the consecutive powers of $\sqrt{\epsilon}$ into the picture in order to finish the proof. So let us replace $e_0(s;(0,y_1),(z,w))$ by $e_{1,\epsilon}(s;(0,y_1),(z,w))$ in the formula (8.13), hence

$$\int_0^t e^{-s\Delta_0}(2\epsilon\beta'(\epsilon z)T(w))e^{-(t-s)\Delta_0}ds$$

is replaced by

$$\int_0^t e^{-s\Delta_{1,\epsilon}}(2\epsilon\beta'(\epsilon z)T(w))e^{-(t-s)\Delta_0}ds$$

and the kernel of $e^{-s\Delta_{1,\epsilon}} - e^{-s\Delta_0}$ brings the extra factor $\sqrt{\epsilon}$ as the elementary estimates presented below show. The operator $\Delta_{1,\epsilon}$ is obtained from Δ_0 by adding the correction term of the form $\gamma(\epsilon v)S_1$, where $\gamma(\epsilon v)$ is the cut-off function with the properties specified earlier.

The main estimate

We have

$$\|e_{1,\epsilon}(t;(0,y),(0,w)) - e_0(t;(0,y),(0,q))\|$$

$$= \| \int_0^t ds \int_{|v|\leq\sqrt{\epsilon}} dv \int_M dz\ e_{1,\epsilon}(s;(0,y),(v,z))\gamma(\epsilon v)S_1(z)e_0(t;(v,z),(0,q))\|$$

$$\leq \int_0^t ds \int_{-\frac{1}{\sqrt{\epsilon}}}^{-\frac{1}{\sqrt{\epsilon}}} dv \int_M dz\ \|e_{1,\epsilon}(s;(0,y),(v,z))\|\cdot|\gamma(\epsilon v)|\cdot\|S_1(z)e_0(t;(v,z),(0,q))\|$$

$$\leq c\sqrt{\epsilon}\int_0^t ds \int_{-\frac{1}{\sqrt{\epsilon}}}^{-\frac{1}{\sqrt{\epsilon}}} dv \int_M dz\ c_1 s^{-\frac{m+1}{2}} e^{-c_2\frac{v^2+d^2(y,z)}{s}} c_1(t-s)^{-\frac{m+2}{2}} e^{-c_2\frac{v^2+d^2(z,q)}{t-s}}$$

$$\leq c\sqrt{\epsilon}c_1^2 \int_0^t \frac{1}{\sqrt{t-s}}ds \int_{-\frac{1}{\sqrt{\epsilon}}}^{-\frac{1}{\sqrt{\epsilon}}} dv \frac{e^{-c_2\frac{tv^2}{s(t-s)}}}{\sqrt{s(t-s)}} \int_M dz \frac{e^{-c_2\frac{d^2(y,z)}{s}} e^{-c_2\frac{d^2(z,q)}{t-s}}}{(s(t-s))^{-\frac{m}{2}}}$$

$$\leq c_3 \frac{\sqrt{\epsilon}}{\sqrt{t}} \cdot \int_0^t \frac{1}{\sqrt{t-s}} ds \int_M dz\ (s(t-s))^{-\frac{m}{2}} e^{-c_2 \frac{d^2(y,z)}{s}} e^{-c_2 \frac{d^2(z,q)}{t-s}} \quad .$$

The following elementary inequality is used to estimate the factor $e^{-c_2 \frac{d^2(y,z)}{s}} e^{-c_2 \frac{d^2(z,q)}{t-s}}$,

$$\frac{d^2(y,q)}{t} \leq \frac{d^2(y,z)}{s} + \frac{d^2(z,q)}{t-s} \quad .$$

We have

$$e^{-c_2 \frac{d^2(y,z)}{s}} e^{-c_2 \frac{d^2(z,q)}{t-s}} = e^{-2c_2 \frac{d^2(y,z)}{2s}} e^{-2c_2 \frac{d^2(z,q)}{2(t-s)}} \leq$$

$$e^{-c_2 \frac{d^2(y,q)}{2t}} e^{-c_2 \frac{d^2(y,z)}{2s}} e^{-c_2 \frac{d^2(z,q)}{2(t-s)}} \quad .$$

All this amounts to

$$\|e_R(t;(0,y),(0,w)) - e_0(t;(0,y),(0,q))\|$$

$$\leq c_3 \sqrt{\frac{\epsilon}{t}} \cdot e^{-c_2 \frac{d^2(y,q)}{2t}} \int_0^t \frac{ds}{\sqrt{t-s}} \int_M dz\ (s(t-s))^{-\frac{m}{2}} e^{-c_2 \frac{d^2(y,z)}{2s}} e^{-c_2 \frac{d^2(z,q)}{2(t-s)}}$$

$$\leq c_4 \sqrt{\epsilon} \cdot e^{-c_2 \frac{d^2(y,q)}{2t}} \int_M dz\ (s(t-s))^{-\frac{m}{2}} e^{-c_2 \frac{d^2(y,z)}{2s}} e^{-c_2 \frac{d^2(z,q)}{2(t-s)}}$$

$$\leq c_5 \sqrt{\epsilon} \cdot t^{-\frac{m}{2}} \cdot e^{-c_2 \frac{d^2(y,q)}{2t}} \quad .$$

Now, we can show the fact used in the proof of the EAF in the previous subsection. We have

$$e^{-t\Delta_{1,\epsilon}} - e^{-t\Delta_{2,\epsilon}} = \int_0^t e^{-s\Delta_{1,\epsilon}} (\Delta_{2,\epsilon} - \Delta_{1,\epsilon}) e^{-(t-s)\Delta_{2,\epsilon}} ds$$

$$= \int_0^t [e^{-s\Delta_0}(\Delta_{2,\epsilon} - \Delta_{1,\epsilon}) e^{-(t-s)\Delta_{2,\epsilon}} + (e^{-s\Delta_{1,\epsilon}} - e^{-s\Delta_0})(\Delta_{2,\epsilon} - \Delta_{1,\epsilon}) e^{-(t-s)\Delta_{2,\epsilon}}] ds$$

$$= \int_0^t e^{-s\Delta_0} (\Delta_{2,\epsilon} - \Delta_{1,\epsilon}) e^{-(t-s)\Delta_0} ds$$

$$+ \int_0^t (e^{-s\Delta_{1,\epsilon}} - e^{-s\Delta_0})(\Delta_{2,\epsilon} - \Delta_{1,\epsilon}) e^{-(t-s)\Delta_{2,\epsilon}} ds$$

$$+ \int_0^t e^{-s\Delta_0} (\Delta_{2,\epsilon} - \Delta_{1,\epsilon})(e^{-(t-s)\Delta_{2,\epsilon}} - e^{-(t-s)\Delta_0}) ds \quad .$$

The operator $\Delta_{2,\epsilon} - \Delta_{1,\epsilon} = 2\epsilon\beta'(\epsilon v)T$ brings a factor ϵ and the first term on the right side is of the form $\int_0^t e^{-s\Delta_0}(\Delta_{2,\epsilon} - \Delta_{1,\epsilon})e^{-(t-s)\Delta_0}ds$ and of the ϵ size. The next term on the right side

$$\int_0^t (e^{-s\Delta_{1,\epsilon}} - e^{-s\Delta_0})(\Delta_{2,\epsilon} - \Delta_{1,\epsilon})e^{-(t-s)\Delta_{2,\epsilon}}ds$$

contains additionally the difference $e^{-s\Delta_{1,\epsilon}} - e^{-s\Delta_0}$, hence it is of the size $\epsilon^{\frac{3}{2}}$. This is also the case of the last term $\int_0^t e^{-s\Delta_0}(\Delta_{2,\epsilon} - \Delta_{1,\epsilon})(e^{-(t-s)\Delta_{2,\epsilon}} - e^{-(t-s)\Delta_0})ds$. It follows that, as we take limit as $\epsilon \to 0$, only the integral (over $\mathbf{R} \times M$) from the first term is going to survive. The kernel of this operator at the point $(t, (0, y_1), (0, y_2))$ has the form

$$(e_0 \# (2\epsilon\beta'(\epsilon v)Te_0)(t, (0, y_1), (0, y_2)) \quad ,$$

where $e_0(t; (v_1, y_1), (v_2, y_2))$ denotes kernel of the operator $e^{-tB_0^2}$. This is exactly what we need in order to complete the proof.

8.4. *The EAF for operators on covering spaces*

We let $\tilde{M}$ be the universal covering space for the closed manifold M , with the corresponding fundamental group $G(= \pi_1(M))$. We assume that we are given a G-invariant, compatible Dirac operator on $\tilde{M}$

$$\tilde{B} : C^\infty(\tilde{M}; \tilde{S} \otimes \tilde{E}) \to C^\infty(\tilde{M}; \tilde{S} \otimes \tilde{E}) \quad ,$$

where G acts on $\tilde{E}$ via a representation ρ . The appropriate von Neumann algebra $\mathcal{N}$ is the commutant of the G action and there is a corresponding trace $\tau = \tau_G$ as described by Atiyah [1] and introduced at the beginning of this Section.

We introduce now a G-invariant unitary bundle automorphism $\tilde{h}$ of the auxiliary bundle $\tilde{E}$, and we consider family $\{\tilde{B}_u\}$ as defined in (8.1). This family has a well-defined spectral flow and we want to prove that

$$sf\{\tilde{B}_s\} = index\tilde{A} \tag{8.14}$$

(Here the tilde just denotes the covering space analogues of the operators we introduced before). This is the result which corresponds to Theorem 8.1

Theorem 8.3.

$$index\tilde{A} = sf\{\tilde{B}_s\} = \int_0^1 \dot{\eta}_s^G ds. \tag{8.15}$$

We remind the reader that the η-invariant in this context was studied by Cheeger and Gromov [23], Hurder [44] and later on by Mathai [49] and others. To prove the theorem 8.3 have only to show

$$index_G \tilde{A} = \int_0^1 \sqrt{\frac{t}{\pi}} \cdot \tau_G \dot{B}_s e^{-tB_s^2} ds, \tag{8.16}$$

This however follows easily from the extension of the *EAF* to the present context. We follow the previous argument. The $index_G \tilde{A}$ is equal to

$$\tau_G(e^{-t\tilde{A}^*\tilde{A}} - e^{-t\tilde{A}\tilde{A}^*})$$

(The McKean-Singer formula for the index holds when the von Neumann algebra is not necessarily a factor [19]). We blow up the metric and all the arguments from the compact case come through with the slight modifications.

The only problem we face is that although our trace is defined by the integral over the (compact) fundamental domain we have to integrate over the whole non-compact manifold $\tilde{M}$, when applying *Duhamel's Principle.* First let us notice that the standard point-wise estimate on the heat-kernel (8.11) holds on $\tilde{M}$ [36]. The difficulty follows from the well-known fact that the volume of the ball with a fixed center on $\tilde{M}$ may grow exponentially with the radius of the ball. Therefore we have to be careful with the arguments which lead to the proof of the results which correspond to Proposition 8.1 and Corollary 8.2. Actually, everything works only because the volume growth is at most exponential in the diameter. We omit the estimates which lead to the proof of Proposition 8.1 for the case of the covering space $\tilde{M}$. However to present the flavour of the computations involved we present a modification of the argument used to get Corollary 8.2. We work on the space $\tilde{M}_\epsilon$ (or on $\mathbf{R} \times \tilde{M}$) now, so the proper formulation of the result is as follows.

Corollary 8.3. *Let $y_1, y_2 \in \bar{M}$ then we may disregard the contribution to the kernel $k_\epsilon(t;(0,y_1),(0,y_2))$ provided by the points $(u,z) \in \mathbf{R} \times \tilde{M}$ such that $|u| > \frac{1}{\sqrt{\epsilon}}$ and $d(z,\bar{M}) > 1$.*

Proof. We start as in Section 8.2

$$\|e^{-t\Delta_{1,\epsilon}} - e^{-t\Delta_{2,\epsilon}}\|(t;(0,y_1),(0,y_2)) =$$

$$\| \int_0^t e^{-s\Delta_{1,\epsilon}}(2\epsilon\beta(\epsilon v)T)e^{-(t-s)\Delta_{2,\epsilon}}\|(t;(0,y_1),(0,y_2))ds \leq$$

$$\int_0^t ds \int_{\mathbf{R}\times\tilde{M}} dudz \|e_{1,\epsilon}(s;(0,y_1),(u,z))2\epsilon\beta(\epsilon u)T(u)e_{2,\epsilon}(t-s;(u,z),(0,y))\|.$$

We have to remember that nothing is changed (i.e. we obtain a negligible contribution), while we work on compact pieces of the space $\mathbf{R}\times\tilde{M}$. Here we want to show that the integral over $(u,z)\in\mathbf{R}\times\tilde{M}$ such that $|u|>\frac{1}{\sqrt{\epsilon}}$ and $d(z,\bar{M})>1$ is exponentially small with respect to ϵ .

$$\int_0^t ds \int_{|u|>\frac{1}{\sqrt{\epsilon}}} du \int_{\{z;d(z,\bar{M})>1\}} dz$$

$$\|e_{1,\epsilon}(s;(0,y_1),(u,z))\|2\epsilon\beta(\epsilon u)\|T(u)\|\|e_{2,\epsilon}(t-s;(u,z),(0,y))\|\leq$$

$$c_3\epsilon\cdot\int_0^t ds \int_{|u|>\frac{1}{\sqrt{\epsilon}}} du \int_{\{z;d(z,\bar{M})>1\}} dz e^{-c_2\frac{d^2((0,y_1),(u,z))}{s}} e^{-c_2\frac{d^2((u,z),(0,y_2))}{t-s}} \leq$$

$$c_3\epsilon\cdot\int_0^t ds \int_{|u|>\frac{1}{\sqrt{\epsilon}}} e^{-c_2\frac{u^2}{s}} e^{-c_2\frac{u^2}{t-s}} du \int_{\{z;d(z,\bar{M})>1\}} dz\ e^{-c_2\frac{d^2(z;\bar{M}}{s}} e^{-c_2\frac{d^2(z;\bar{M}}{t-s}}.$$

The first integral on the right side is estimated as in Section 2

$$\int_{|u|>\frac{1}{\sqrt{\epsilon}}} e^{-c_2\frac{u^2}{s}} e^{-c_2\frac{u^2}{t-s}} du \leq$$

$$\int_{|u|>\frac{1}{\sqrt{\epsilon}}} e^{-c_2\frac{u^2}{t}} du \leq \sqrt{\frac{t}{c_2}}\cdot \int_{|v|>\sqrt{\frac{c_2}{t}}} e^{-v^2} dv \leq \sqrt{\frac{t}{c_2}}\cdot e^{-\frac{c_2}{\epsilon t}}.$$

The second integral involves volume of the manifold $\tilde{M}$. Modulo negligible error (up to a contribution from a compact set) we can look at it as the integral over the outside of the ball centered at the fixed point $\bar{y}\in\bar{M}$ with radius $R=1+diam\ \bar{M}$. We do have

$$\int_{\{z;d(z,\bar{y})>R\}} e^{-c_2\frac{d^2(z;\bar{y}}{s}} e^{-c_2\frac{d^2(z;\bar{y}}{t-s}} dz \leq \int_{\{z;d(z,\bar{y})>R\}} e^{-c_2\frac{d^2(z;\bar{y}}{t}} dz \leq$$

$$c_3\int_R^\infty e^{-c_2\frac{r^2}{t}} e^{c_4 r} dr \leq c_5\int_R^\infty e^{-c_6\frac{r^2}{t}} dr \leq$$

$$c_5\sqrt{\frac{t}{c_6}}\cdot\int_{v>\sqrt{\frac{c_6}{t}}R} e^{-v^2} dv \leq c_5\sqrt{\frac{t}{c_6}}\cdot e^{-c_6\frac{R^2}{t}}.$$

□

9. Spectral flow for almost periodic gauge transformations

9.1. *Shubin's framework*

We follow Shubin [62, 63] which in turn extends the original paper of Coburn et al [24, 25]. In this paragraph, we review the definition of the von Neumann algebra which is appropriate for the study of almost periodic operators. Recall that a trigonometric function is a finite linear combination of exponential functions $e_\xi : x \mapsto e^{i<x,\xi>}$. The space $\mathrm{Trig}(\mathbf{R}^n)$ of trigonometric functions is clearly a $*-$subalgebra of the C^*-algebra $C_b(\mathbf{R}^n)$ of continuous bounded functions. The uniform closure of $\mathrm{Trig}(\mathbf{R}^n)$ is thus a C^*-algebra called the algebra of almost periodic functions and denoted $\mathcal{AP}(\mathbf{R}^n)$. Since this C^*-algebra is unital and commutative, it is the C^*-algebra of continuous functions on a compact space $\mathbf{R}^n_B$ which is a compactification of $\mathbf{R}^n$ with respect to the appropriate topology. The compact space $\mathbf{R}^n_B$ is called the Bohr compactification of $\mathbf{R}^n$ or simply the Bohr space. Addition extends to $\mathbf{R}^n_B$ which is a compact abelian group containing $\mathbf{R}^n$ as a dense subgroup. There is a unique normalized Haar measure α_B on $\mathbf{R}^n_B$ such that the family $(e_\xi)_{\xi\in\mathbf{R}^n}$ is orthonormal. Namely, the measure α_B is given for any almost periodic function f on $\mathbf{R}^n$ by:

$$\alpha_B(f) := \lim_{T\to+\infty} \frac{1}{(2T)^n} \int_{(-T,T)^n} f(x)dx.$$

By using the measure α_B one defines the Hilbert space completion $L^2(\mathbf{R}^n_B)$ of $\mathrm{Trig}(\mathbf{R}^n)$. This Hilbert space is called the Besicovich space and it has an orthonormal basis given by $(e_\xi)_{\xi\in\mathbf{R}^n}$. In other words, the Pontryagin dual of $\mathbf{R}^n_B$ is the discrete abelian group $\mathbf{R}^n_d$ and the Fourier transform $\mathcal{F}_B : \ell^2(\mathbf{R}^n_d) \longrightarrow L^2(\mathbf{R}^n_B)$ is given by:

$$\mathcal{F}_B(\delta_\xi) = e_\xi, \qquad \text{with } \delta_\xi(\eta) = \delta_{\xi,\eta},$$

where $\delta_{\xi,\eta}$ is the Kronecker symbol. We shall denote by $\mathcal{F}$ the usual Fourier transform on the abelian group $\mathbf{R}^n$ with its usual Lebesgue measure.

For any $f \in C_b(\mathbf{R}^n)$ we shall denote, for any vector $\lambda \in \mathbf{R}^n$, by $T_\lambda f$ the translated function defined by $(T_\lambda f)(x) = f(x-\lambda)$. Let $f \in C_b(\mathbf{R}^n)$ and let $\epsilon > 0$ be given. A vector $\lambda \in \mathbf{R}^n$ is called an $\epsilon-$ period for f if the uniform norm of $T_\lambda f - f$ is bounded by ϵ, i.e.

$$\|T_\lambda f - f\|_\infty := \sup_{t\in\mathbf{R}^n} |f(t-\lambda) - f(t)| \leq \epsilon.$$

A subset E of $\mathbf{R}^n$ is relatively dense if there exists $T > 0$ such that

$$\forall x \in \mathbf{R}^n, \exists u \in E : u - x \in \left[-\frac{T}{2}, +\frac{T}{2}\right]^n.$$

It is worth pointing out that, for any function $f : \mathbf{R}^n \to \mathbf{C}$, the following properties are equivalent [24]:

- f is an almost periodic function.
- f is a continuous bounded function whose ϵ periods are relatively dense for every $\epsilon > 0$.

It is clear from the second characterization of an almost periodic function that any periodic function is almost periodic. An interesting class of examples arises from the study of quasi-periodic functions. Assume for simplicity that $n = 1$ and let $\alpha = (\alpha_1, \cdots, \alpha_p) \in \mathbf{R}^p$ be a fixed list of real numbers. Then for any summable sequence $c = (c_m)_{m \in \mathbf{Z}^p}$, we get an almost periodic function on $\mathbf{R}$ by setting:

$$\varrho_{c,\alpha}(x) := \sum_{m \in \mathbf{Z}^p} c_m e^{2i\pi <m,\alpha> x}.$$

Then more complicated examples of almost periodic functions arise as limits of periodic or quasi-periodic functions. For instance, the function $\sum_{n \geq 0} a_n \cos(\frac{x}{2^n})$ where $\sum_n |a_n| < +\infty$, is an almost periodic function.

The action of $\mathbf{R}^n$ on $\mathbf{R}^n_B$ by translations yields a topological dynamical system whose naturally associated von Neumann algebra is the crossed product von Neumann algebra $L^\infty(\mathbf{R}^n_B) \rtimes \mathbf{R}^n$. It is more convenient for applications to consider the commutant of this von Neumann algebra denoting it by $\mathcal{N}$. It is also a crossed product. This time it is the von Neumann algebra $L^\infty(\mathbf{R}^n) \rtimes \mathbf{R}^n_d$. The von Neumann algebra $\mathcal{N}$ is a type II_∞ factor with a faithful normal semi-finite trace τ. It can be described as the set of Borel essentially bounded families $(A_\mu)_{\mu \in \mathbf{R}^n_B}$ of bounded operators in $L^2(\mathbf{R}^n)$ which are $\mathbf{R}^n$-equivariant, i.e. such that

$$A_\mu = \sigma_\mu(A_0) = T_{-\mu} A_0 T_\mu, \quad \forall \mu \in \mathbf{R}^n.$$

Here and in the sequel we denote by σ_μ conjugation of any operator with the translation T_μ so that $\sigma_\mu(B) = T_{-\mu} B T_\mu$. If we denote by M_φ the operator of multiplication by a bounded function φ, then examples of such families are given for any λ by the families

$$(\sigma_\mu(M_{e_\lambda}))_{\mu \in \mathbf{R}^n_B}.$$

We choose the Fourier transform

$$\mathcal{F}f(\zeta) = \frac{1}{(2\pi)^{n/2}} \int_{\mathbf{R}^n} e^{ix\zeta} f(x) dx.$$

Now the von Neumann algebra $\mathcal{N}$ can be defined [24, 25] as the double commutant of the set of operators $\{M_{e_\lambda} \otimes M_{e_\lambda}, T_\lambda \otimes 1 | \lambda \in \mathbf{R}^n\}$ on the Hilbert space $\mathcal{H} = L^2(\mathbf{R}^n) \otimes L^2(\mathbf{R}^n_B)$

There is a natural way to imbed the C^*-algebra $\mathcal{A}P(\mathbf{R}^n)$ in $\mathcal{N}$ by setting

$$\pi(f) := (\sigma_\mu(M_f))_{\mu \in \mathbf{R}^n_B}$$

This family then belongs to $\mathcal{N}$ and π is clearly faithful. Viewed as an operator on $\mathcal{H}$, $\pi(f)$ is given by $\pi(f)(g)(x,\mu) = f(x+\mu)g(x,\mu)$. If $B = (B_\mu)_\mu$ is a positive element of $\mathcal{N}$, then we define the expectation $E(B)$ as the Haar integral:

$$E(B) := \int_{\mathbf{R}^n_B} B_\mu d\alpha_B(\mu).$$

Since the family B is translation equivariant and since α_B is translation invariant, the operator $E(B)$ clearly commutes with the translation in $L^2(\mathbf{R}^n)$ and is therefore given by a Fourier multiplier $\widetilde{M}(\varphi_B)$ with φ_B a positive element of $L^\infty(\mathbf{R}^n)$. Recall that the Fourier multiplier $\widetilde{M}(\varphi_B)$ is conjugation of the multiplication operator M_φ by the Fourier transform, i.e. $\widetilde{M}(\varphi_B) = \mathcal{F}^{-1} M_\varphi \mathcal{F}$. When the function φ is for instance in the Schwartz space, the operator $\widetilde{M}(\varphi_B)$ is convolution by the Schwartz function $\frac{1}{(2\pi)^{n/2}} \mathcal{F}^{-1}\varphi$. Hence the expectation E takes values in the von Neumann algebra $\widetilde{M}(L^\infty(\mathbf{R}^n))$, i.e.

$$E : \mathcal{N} \longrightarrow \widetilde{M}(L^\infty(\mathbf{R}^n)).$$

Now, using the usual Lebesgue integral on $\mathbf{R}^n$, we use the normalisation of Coburn et al [24] and introduce the following definition of the trace τ:

$$\tau(B) = \int_{\mathbf{R}^n} \varphi_B(\zeta) d\zeta.$$

Lemma 9.1. [24, 25] *The map τ is, up to constant, the unique positive normal faithful semi-finite trace on $\mathcal{N}$.*

The space $L^1(\mathcal{N}, \tau)$ of trace-class $\tau-$measurable operators with respect to $\mathcal{N}$ is the space of $\tau-$measurable operators T as explained by Fack et al [38] such that $\int_{(0,+\infty)} \mu_s^\tau(T) ds < +\infty$. Here $\mu_s^\tau(T)$ is the $s-$th characteristic value of T [38], for the precise definitions. More generally and for any $p \geq 1$, we shall denote by $L^p(\mathcal{N}, \tau)$ the space of $\tau-$measurable operators T such that $(T^*T)^{p/2} \in L^1(\mathcal{N}, \tau)$. It is well known that the space $L^p(\mathcal{N},\tau)\cap\mathcal{N}$ is a two-sided $*-$ideal in $\mathcal{N}$ that we shall call the $p-$th Schatten ideal of $\mathcal{N}$.

We also consider the Dixmier space $L^{1,\infty}(\mathcal{N},\tau)$ of those operators $T\in\mathcal{N}$ such that

$$\int_0^s \mu_t^\tau(T)dt \sim O(\log(s)).$$

Again, $L^{1,\infty}(\mathcal{N},\tau)$ is a two-sided $*-$ideal in $\mathcal{N}$. There are well defined Dixmier traces τ_ω on $L^{1,\infty}(\mathcal{N},\tau)$ parametrized by limiting processes ω [6, 21].

Consider the trace on the von Neumann algebra $\mathcal{N}$ evaluated on an operator of the form M_aK where a is almost periodic and K is a convolution operator on $L^2(\mathbf{R}^n)$ arising from multiplication by an L^1 function k on the Fourier transform. We have,

$$\tau(M_aK) = \lim_{T\to+\infty}\frac{1}{(2T)^n}\int_{(-T,T)^n} a(x)dx\int_{\mathbf{R}^n} k(\zeta)d\zeta$$

More generally, any pseudodifferential operator A on $L^2(R^n,\mathbf{C}^N)$ with almost periodic coefficients of nonpositive order m acting on $\mathbf{C}^N-$valued functions, can be viewed as a family over $\mathbf{R}^n_B$ of pseudodifferential operators on $\mathbf{R}^n$. To do this first take the symbol a of A, then the operator $\sigma_\mu(A)$ is the pseudodifferential operator with almost periodic coefficients whose symbol is

$$(x,\xi)\longmapsto a(x+\mu,\xi).$$

When $m\le 0$, we get in this way an element of the von Neumann algebra $\mathcal{N}$. We denote by Ψ^0_{AP} the algebra of pseudodifferential operators with almost periodic coefficients and with non positive order. When the order m of A is >0 then the operator $A^\sharp$ given by the family $(\sigma_\mu(A))_{\mu\in\mathbf{R}^n_B}$ is affiliated with the von Neumann algebra $\mathcal{N}$. If the order m of A is $<-n$, then the bounded operator $A^\sharp$ is trace class with respect to the trace τ on the von Neumann algebra $\mathcal{N}\otimes M_N(\mathbf{C})$ [64][Proposition 3.3] and we have:

$$\tau(A^\sharp) = \lim_{T\to+\infty}\frac{1}{(2T)^n}\int_{(-T,+T)^n\times\mathbf{R}^n}\operatorname{tr}(a(x,\zeta))dxd\zeta$$

Indeed, the expectation $E(A^\sharp)$ is a pseudodifferential operator on $\mathbf{R}^n$ with symbol denoted by $E(a)$ and is independent of the $x-$variable, it is given by:

$$E(a)(\zeta) = \lim_{T\to+\infty}\frac{1}{(2T)^n}\int_{(-T,+T)^n} a(x,\zeta)dx.$$

Hence the operator $E(A^\sharp)$ is precisely the Fourier multiplier $\widetilde{M}(E(a))$ and so:

$$\tau(A^\sharp) = \int_{\mathbf{R}^n} \mathrm{tr}(E(a)(\zeta))d\zeta.$$

Let Ψ^∞_{AP} be the space of one step polyhomogeneous classical pseudodifferential operators on $\mathbf{R}^n$ with almost periodic coefficients.

Theorem 9.1. *Let A be a (scalar) pseudodifferential operator with almost periodic coefficients on $\mathbf{R}^n$. We assume that the order m of A is $\leq -n$ and we denote by a_{-n} the $-n$ homogeneous part of the symbol a. Then the operator $A^\sharp$ belongs to the Dixmier ideal $L^{1,\infty}(\mathcal{N},\tau)$. Moreover, the Dixmier trace $\tau_\omega(A^\sharp)$ of $A^\sharp$ associated with a limiting process ω does not depend on ω and is given by the formula:*

$$\tau_\omega(A^\sharp) = \frac{1}{n}\int_{\mathbf{R}^n_B\times \mathbf{S}^{n-1}} a_{-n}(x,\zeta)d\alpha_B(x)d\zeta.$$

Proof. We denote as usual by Δ the Laplace operator on $\mathbf{R}^n$. The operator $A(1+\Delta)^{n/2}$ is then a pseudodifferential operator with almost periodic coefficients and nonpositive order. Hence, the operator $[A(1+\Delta)^{n/2}]^\sharp = A^\sharp(1+\Delta^\sharp)^{n/2}$ belongs to the von Neumann algebra $\mathcal{N}$. Now the operator $(1+\Delta^\sharp)^{-n/2}$ is a Fourier multiplier defined by the function $\zeta \mapsto (1+\zeta^2)^{n/2}$. Hence if, for $\lambda > 0$, E_λ is the spectral projection of the operator $(1+\Delta)^{-n/2}$ corresponding to the interval $(0,\lambda)$ then the operator $1-E_\lambda$ is the Fourier multiplier defined by the function $\zeta \mapsto 1_{(\lambda,+\infty)}(\zeta^2+1)^{n/2})$. It follows that the trace τ of the operator $1-E_\lambda$ is given by

$$\int_{\mathbf{R}^n} 1_{(\lambda,+\infty)}(\frac{1}{(\zeta^2+1)^{n/2}})d\zeta.$$

It is easy to compute this integral and to show that it is proportional to $\frac{1}{\lambda}$. So the infimum of those λ for which $\tau(1-E_\lambda) \leq t$ is precisely proportional to $\frac{1}{t}$. Hence the operator $(1+\Delta^\sharp)^{-n/2}$, and hence A, belongs to the Dixmier ideal $L^{1,\infty}(\mathcal{N},\tau)$.

In order to compute the Dixmier trace of the operator A, we apply Shubin [62][Theorem 10.1] to deduce that the spectral $\tau-$density $N_A(\lambda)$ of A has the asymptotic expansion

$$N_A(\lambda) = \frac{\chi_0(A)}{\lambda}(1+o(1)), \quad \lambda \to +\infty,$$

where $\chi_0(A)$ is given by:

$$\chi_0(A) = \frac{1}{n}\int_{\mathbf{R}^n_B\times \mathbf{S}^{n-1}} a_{-n}(x,\zeta)d\alpha_B(x)d\zeta.$$

Now, if A is positive then by Benameur et al [6][Proposition 1]:

$$\tau_\omega(A) = \lim_{\lambda \to +\infty} \lambda N_A(\lambda) = \chi_0(A).$$

This proves the theorem for positive A. Since the principal symbol map is a homomorphism, we deduce the result for general A. □

The reader familiar with the Wodzicki residue will observe that the normalisation we have chosen for the trace in the von Neumann setting of this Section eliminates a factor of $\frac{1}{(2\pi)^n}$ which occurs at the corresponding point in the type I theory.

10. The odd semifinite local index theorem

The original type I version of this result is due to Connes-Moscovici [32]. There are two new proofs, one due to Higson [43] and one due to Carey et al [18]. The latter argument handles the case of semifinite spectral triples. Quite remarkably this very general odd semifinite local index theorem is proved by starting from the integral formulae for spectral flow that we have described in earlier sections. We do not have the space here to explain how it is done.

We restrict our discussion to a statement of the theorem. First, we require multi-indices $(k_1, ..., k_m)$, $k_i \in \{0, 1, 2, ...\}$, whose length m will always be clear from the context. We write $|k| = k_1 + \cdots + k_m$, and define $\alpha(k)$ by

$$\alpha(k) = 1/k_1!k_2! \cdots k_m!(k_1 + 1)(k_1 + k_2 + 2) \cdots (|k| + m).$$

The numbers $\sigma_{n,j}$ are defined by the equality

$$\prod_{j=0}^{n-1}(z + j + 1/2) = \sum_{j=0}^{n} z^j \sigma_{n,j}$$

with $\sigma_{0,0} = 1$. These are just the elementary symmetric functions of $1/2, 3/2, ..., n - 1/2$.

If $(\mathcal{A}, \mathcal{H}, \mathcal{D})$ is a smooth semifinite spectral triple (ie $\mathcal{A}$ is in the domain of δ^n for all n where $\delta(a) = [(1 + D^2)^{1/2}, a]$) and $T \in \mathcal{N}$, we write $T^{(n)}$ to denote the iterated commutator

$$[\mathcal{D}^2, [\mathcal{D}^2, [\cdots, [\mathcal{D}^2, T] \cdots]]]$$

where we have n commutators with $\mathcal{D}^2$. It follows [18] that operators of the form

$$T_1^{(n_1)} \cdots T_k^{(n_k)} (1 + \mathcal{D}^2)^{-(n_1 + \cdots + n_k)/2}$$

are in $\mathcal{N}$ when $T_i = [\mathcal{D}, a_i]$, or $= a_i$ for $a_i \in \mathcal{A}$.

Definition 10.1. If $(\mathcal{A}, \mathcal{H}, \mathcal{D})$ is a smooth semifinite spectral triple, we call

$$p = \inf\{k \in \mathbf{R} : \tau((1+\mathcal{D}^2)^{-k/2}) < \infty\}$$

the **spectral dimension** of $(\mathcal{A}, \mathcal{H}, \mathcal{D})$. We say that $(\mathcal{A}, \mathcal{H}, \mathcal{D})$ has **isolated spectral dimension** if for b of the form

$$b = a_0[\mathcal{D}, a_1]^{(k_1)} \cdots [\mathcal{D}, a_m]^{(k_m)}(1+\mathcal{D}^2)^{-m/2-|k|}$$

the zeta functions

$$\zeta_b(z-(1-p)/2) = \tau(b(1+\mathcal{D}^2)^{-z+(1-p)/2})$$

have analytic continuations to a deleted neighbourhood of $z = (1-p)/2$.

Now we define, for $(\mathcal{A}, \mathcal{H}, \mathcal{D})$ having isolated spectral dimension and

$$b = a_0[\mathcal{D}, a_1]^{(k_1)} \cdots [\mathcal{D}, a_m]^{(k_m)}(1+\mathcal{D}^2)^{-m/2-|k|}$$

$$\tau_j(b) = res_{z=(1-p)/2}(z-(1-p)/2)^j \zeta_b(z-(1-p)/2).$$

The hypothesis of isolated spectral dimension is clearly necessary here in order to define the residues. The semifinite local index theorem is as follows.

Theorem 10.1. *Let $(\mathcal{A}, \mathcal{H}, \mathcal{D})$ be an odd finitely summable smooth spectral triple with spectral dimension $p \geq 1$. Let $N = [p/2]+1$ where $[\cdot]$ denotes the integer part (so $2N-1$ is the largest odd integer $\leq p+1$), and let $u \in \mathcal{A}$ be unitary. Then if $(\mathcal{A}, \mathcal{H}, \mathcal{D})$ also has isolated spectral dimension then*

$$sf(\mathcal{D}, u^*\mathcal{D}u) = \frac{1}{\sqrt{2\pi i}} \sum_m (-1)^{(m-1)/2} (\frac{(m-1)}{2})! \phi_m(u, u^*, \ldots, u, u^*)$$

where $\phi_m(u, u^*, \ldots, u, u^*)$ is

$$\sum_{|k|=0}^{2N-1-m} \sum_{j=0}^{|k|+(m-1)/2} (-1)^{|k|}\alpha(k)\sigma_{(|k|+(m-1)/2),j}$$

$$\times \tau_j\left(u[\mathcal{D}, u^*]^{(k_1)} \cdots [\mathcal{D}, u]^{(k_m)}(1+\mathcal{D}^2)^{-|k|-m/2}\right),$$

When $[p] = 2n$ is even, the term with $m = 2N-1$ is zero, and for $m = 1, 3, \ldots, 2N-3$, all the top terms with $|k| = 2N-1-m$ are zero.

We aim to compute the terms in this formula for semifinite spectral flow in the case where $\mathcal{D}$ is the Euclidean Dirac operator on the spin bundle $\mathcal{S}$ over $\mathbf{R}^n$ tensored with the trivial bundle rank N bundle and u is a smooth almost periodic function from $\mathbf{R}^n$ to $U(N)$.

11. Almost periodic spectral triple

We now apply the local index theorem to compute spectral flow. We thus assume that n is odd. The von Neumann algebra constructed previously is non-separable and so to avoid a discussion of the non-separable situation we need to slightly modify our approach in this Section. In fact it is sufficient to study the dense countable abelian subgroups of $\mathbf{R}^n$. Let us fix one such, call it D and explain how the theory works for this case. Consider the subalgebra $\mathcal{A}$ of $\mathcal{AP}(\mathbf{R}^n)$ consisting of almost periodic functions generated by e_ξ with $\xi \in D$. We denote by $\mathcal{A}^\infty$ the $*$-subalgebra of $\mathcal{AP}(\mathbf{R}^n)$ consisting of functions in $\mathcal{A}$ which have bounded derivatives of all orders. The von Neumann algebra we now consider is the crossed product algebra of D with $L^\infty(\mathbf{R}^n)$)and is denoted by $\mathcal{M}$. We take the Hilbert space on which this algebra acts to be $B^2_D(\mathbf{R}^n) \otimes L^2(\mathbf{R}^n)$ where $B^2_D(\mathbf{R}^n)$ is the Hilbert space completion of $\mathcal{A}$ where the norm and inner product are given by the restriction of the Haar trace on $\mathcal{AP}^\infty(\mathbf{R}^n)$ to $\mathcal{A}$ (note that $B^2_D(\mathbf{R}^n) \cong \ell^2(D)$). This type II_∞ von Neumann algebra is endowed with a faithful normal semi-finite trace that we denote by τ. (We note that the explicit formula for τ is as given in Section 9.)

The usual Dirac operator on $\mathbf{R}^n$ is denoted by $\eth$. So, if $\mathcal{S}$ carries the spin representation of the Clifford algebra of $\mathbf{R}^n$ then $\eth$ acts on smooth $\mathcal{S}-$valued functions on $\mathbf{R}^n$. The operator $\eth$ is $\mathbf{Z}^n-$periodic and it is affiliated with the von Neumann algebra $\mathcal{M}_\mathcal{S} = \mathcal{M} \otimes \mathrm{End}(\mathcal{S})$. This latter is also a type II_∞ von Neumann algebra with the trace $\tau \otimes \mathrm{tr}$. More generally, for any $N \geq 1$, we shall denote by $\mathcal{M}_{\mathcal{S},N}$ the von Neumann algebra $\mathcal{M} \otimes \mathrm{End}(\mathcal{S} \otimes \mathbf{C}^N)$ with the trace $\tau \otimes \mathrm{tr}$.

The algebra $\mathcal{A}$ and its closure are faithfully represented as $*-$subalgebras of the von Neumann algebra $\mathcal{M}_\mathcal{S}$. In the same way the algebra $\mathcal{A} \otimes M_N(\mathbf{C})$ can be viewed as a $*-$subalgebra of $\mathcal{M}_{\mathcal{S},N}$. More precisely, if $a \in \mathcal{A}$ then the operator $a^\sharp$ defined by:

$$(a^\sharp f)(x,y) := a(x+y)f(x,y), \quad \forall f \in B^2_D(\mathbf{R}^n) \otimes L^2(\mathbf{R}^n),$$

belongs to $\mathcal{M}$. The operator $a^\sharp$ is just the one associated with the zero-th order differential operator corresponding to multiplication by a. The same formula allows to represent $\mathcal{A} \otimes M_N(\mathbf{C})$ in $\mathcal{M}_{\mathcal{S},N}$. For notional simplicity we put $N = 1$ in the next result although we will use a general $N \geq 1$ in the subsequent subsection.

Prop 11.1. The triple $(\mathcal{A}, \mathcal{M}_\mathcal{S}, \eth^\sharp)$ is a semifinite spectral triple of finite dimension equal to n.

Proof. Note that the algebra $\mathcal{A}$ is unital. The differential operator $\eth$ is known to be densely defined, elliptic, periodic and self adjoint on $L^2(\mathbf{R}^n, \mathcal{S})$. Therefore, the operator $\eth^\sharp$ is affiliated with the von Neumann algebra $\mathcal{M}_\mathcal{S}$ and it becomes self adjoint as a densely defined unbounded operator on the Hilbert space $B_D^2(\mathbf{R}^n) \otimes L^2(\mathbf{R}^n, \mathcal{S})$ with $\eth^2 = \Delta I$ where I is the identity operator and Δ is the usual Laplacian. For any smooth bounded almost periodic function f on $\mathbf{R}^n$, with bounded derivatives of all orders, the commutator $[\eth, f]$ is a 0−th order almost periodic differential operator and so $[\eth^\sharp, f]$ belongs to the von Neumann algebra $\mathcal{M}_\mathcal{S}$.

On the other hand, the pseudodifferential operator $T = (\eth^2 + I)^{-1/2}$ is essentially the Fourier multiplier associated with the function $k \mapsto \frac{1}{(\|k\|^2+1)^{1/2}}$. Therefore, its singular numbers $\mu_t(T)$ can be computed explicitly as in the proof of Theorem 9.1 and shown to be proportional to $t^{-1/n}$. □

11.1. *Analysis of terms in the above example*

First we note that the spectral dimension is the dimension n of the underlying Euclidean space and this is assumed to be odd. It follows that the summation over $|k|$ in each term in the preceding theorem is over the range $0 \le |k| \le n - m$. Second we note that m is always odd.

Let us write $e_1, e_2, \ldots, e_n$ for an orthonormal basis of $\mathbf{R}^n$, $c(e_1), c(e_2), \ldots, c(e_n)$ for the corresponding Clifford generators. So we have $c(e_i)c(e_j) + c(e_j)c(e_i) = 2\delta_{ij}1$ and we can write $\eth = \sum ic(e_j) \otimes 1\partial_j$ where 1 just denotes the identity matrix. We let $u \in \mathcal{A}^\infty \otimes \mathrm{End}(\mathbf{C}^N)$ be unitary. Thus $u[\eth, u^*] = \sum ic(e_j) \otimes \partial_j u^*$ The trace is now the product of the trace on the spinor part times the von Neumann trace composed with the matrix trace on the matrices acting on V. This very simple structure enables us to eliminate all but one of the terms in the local index formula by first taking the trace of the product of Clifford generators. Note that the trace on the Clifford algebra in the spin representation is given by

$$Tr_{Spin}(i^n c(e_1)c(e_2)\ldots c(e_n)) = i^{-[(n+1)/2]} 2^{(n-1)/2}$$

and the trace on any product of $0 < k < n$ generators is zero.

A typical term in the local index formula is proportional to

$$\tau_j(u[\eth, u^*]^{(k_1)}[\eth, u]^{(k_2)} \ldots [\eth, u]^{(k_{m-1})}[\eth, u^*]^{(k_m)}(1+\Delta)^{-|k|-m/2}) \qquad (*)$$

This is, up to a sign, a product of factors of the form $(u\eth u^* - \eth)^{(k_l)}$. The Laplacian commutes with $\eth$ so that a typical factor is of the form $\sum_i c(e_i) \otimes g_i$ and the g_i are $N \otimes N$ matrix valued pseudodifferential operators. Since there is always a product of an odd number of factors of this form $\sum_j c(e_j)\otimes g_j$ in a term $(*)$ the trace on the Clifford elements will produce zero unless $m = n$. In that case $|k|$ is forced to be zero.

Thus only one term survives in the local index theorem and that term is (see appendix)

$$\frac{(-1)^{(n-1)/2}}{n2^{(n-1)}}\tau_0(u[\eth, u^*][\eth, u][\eth, u^*][\eth, u]\ldots[\eth, u][\eth, u^*](1+\Delta)^{-n/2})$$

To compute this we first take care of the Clifford algebra. Using the fact that $[\eth, u] = -u[\eth, u^*]u$ we write the formula for the spectral flow as

$$\frac{(-1)^n}{n2^{(n-1)}}\tau_0(([\eth, u]u^*)^n(1+\Delta)^{-n/2})$$

We let Tr_N be the matrix trace on the auxiliary vector space. Now

$$([\eth, u]u^*) = \sum i\partial_j(u)u^* c(dx_j).$$

Writing $f_j = \partial_j(u)u^*$ we then have

$$([\mathcal{D}, u]u^*)^n = i^n \sum_{J=(j_1,\ldots,j_n)} f_{j_1}\cdots f_{j_n} c(e_{j_1})\cdots c(e_{j_n}),$$

where the sum is extended over all multi-indices J. Every term in the sum is a multiple of the volume form, and so has non-zero (spinor) trace. In terms of permutations we have

$$([\mathcal{D}, u]u^*)^n = i^n\left(\sum_{\sigma\in\Sigma^n}(-1)^\sigma f_{\sigma(1)}\cdots f_{\sigma(n)}\right)c(e_1)\cdots c(e_n)$$

$$=: \Omega i^n c(e_1)\cdots c(e_n).$$

In taking the trace we may first take the matrix trace over the Clifford endomorphisms of the spin bundle (with [...] denoting 'the integer part of') and so, with $\tau_S = \tau \times \mathrm{Tr}_N \times \mathrm{Tr}_{Spin}$

$$\tau_0\left(([\mathcal{D}, u]u^*)^n(1+\mathcal{D}^2)^{-n/2}\right) = res_{s=0}\tau_S\left(([\mathcal{D}, u]u^*)^n(1+\mathcal{D}^2)^{-n/2-s}\right)$$

$$= res_{s=0} 2^{(n-1)/2} i^{-[(n+1)/2]} (\tau \times \mathrm{Tr}_N) \left(\Omega (1+\mathcal{D}^2)^{-n/2-s} \right)$$

$$= res_{s=0} \frac{2^{(n-1)/2}}{i^{[(n+1)/2]}} \lim \frac{1}{(2T)^n} \int_{[-T,T]^n} \mathrm{Tr}_N(\Omega) \int_{\mathbf{R}^n} (1+|\xi|^2)^{-n/2-s} d\xi$$

$$= res \frac{2^{(n-1)/2} Vol(S^{n-1})}{i^{[(n+1)/2]}} \lim \frac{1}{(2T)^n} \int_{[-T,T]^n} \mathrm{Tr}_N(\Omega) \int_0^\infty \frac{r^{n-1}}{(1+r^2)^{n/2+s}} dr$$

$$= res_{s=0} \frac{2^{(n-1)/2} Vol(S^{n-1})}{i^{[(n+1)/2]}} \lim \frac{1}{(2T)^n} \int_{[-T,T]^n} \mathrm{Tr}_N(\Omega) \frac{\Gamma(n/2)\Gamma(s)}{2\Gamma(n/2+s)}$$

$$= \frac{2^{(n-1)/2}}{i^{(n+1)/2}} Vol(S^{n-1}) \frac{1}{2} \lim \frac{1}{(2T)^n} \int_{[-T,T]^n} \mathrm{Tr}_N(\Omega).$$

Now

$$Vol(S^{n-1}) = \frac{(4\pi)^{n/2}}{2^{n-1}\Gamma(n/2)}.$$

Putting the previous calculations together gives our final result.

Theorem 11.1. *With the notation as above the spectral flow along any path joining the Dirac operator $\eth$ to its gauge equivalent transform $u\eth u^*$ by an almost periodic $U(N)$ valued function on $\mathbf{R}^n$ is given by the following formula:*

$$sf(\eth, u\eth u^*) = \frac{-i^{-[(n+1)/2]}\pi^{n/2}}{\Gamma(1+n/2)2^{(n+1)/2}} \lim_{T\to\infty} \frac{1}{(2T)^n} \int_{(-T,T)^n} tr_N(\Omega)$$

12. Appendix

12.1. *Coefficients from the Local Index Theorem*

The formula provided by the local index theorem for the special case considered in Section 11 is

$$sf(\mathcal{D}, u^*\mathcal{D}u) = \frac{1}{\sqrt{2\pi i}} \sum_{m=1}^{n} (-1)^{(m-1)/2} (\frac{(m-1)}{2})! \phi_m(u, u^*, \ldots, u, u^*)$$

where $\phi_m(u, u^*, \ldots, u, u^*)$ is given in Theorem 10.1. We already know that we need only compute the top term (degree n) of the local index theorem, because the Clifford trace will kill all the other terms. Since we have a simple spectral triple, the only multi-index $k = (k_1, ..., k_n)$ which arises is zero. In particular, we require $\alpha(0) = \frac{1}{n!}$.

The numbers $\sigma_{m,j}$ are defined by the equality

$$\prod_{l=0}^{m-1} (z + l + 1/2) = \sum_{j=0}^{m} z^j \sigma_{m,j}.$$

These are just the elementary symmetric functions of $1/2, 3/2, ..., m-1/2$. When $m = 0$, this is the empty product, so $\sigma_{0,0} = 1$. For $|k| = 0$ we have $h := |k| + (n-1)/2 = (n-1)/2$ and because we have simple dimension spectrum, we only want $j = 0$. Then $\sigma_{(n-1)/2,0}$ is the coefficient of 1 in the product $\prod_{l=0}^{(n-3)/2}(z + l + 1/2)$. This is the product of all the non-z terms, which is

$$(1/2)(3/2) \times \cdots \times ((n-3)/2 + 1/2) = \frac{1.3.\cdots.(2(n-1)/2 - 1)}{2^{(n-1)/2}}.$$

The reason for writing this so elaborately, is that in this form it is obvious that it is equal to

$$\frac{1}{\sqrt{\pi}}\Gamma((n-1)/2 + 1/2) = \frac{1}{\sqrt{\pi}}\Gamma(n/2).$$

Combining all these calculations gives us

$$\phi_n(a_0, a_1, \ldots, a_n) = \sqrt{2\pi i}\frac{\Gamma(n/2)}{\sqrt{\pi}n!}\tau_0(a_0[\mathcal{D}, a_1]\cdots[\mathcal{D}, a_n](1 + \mathcal{D}^2)^{-n/2}).$$

12.2. *Constants from $Ch_n(u^*)$ and pairing*

When we pair ϕ_n with the Chern character of a unitary, we divide out by $\sqrt{2\pi i}$, which is only in the Chern character of $(\mathcal{A}, \mathcal{H}, \mathcal{D})$ to make it compatible with the Kasparov product. The Chern character of u^* has degree n component

$$(-1)^{(n-1)/2}((n-1)/2))!u \otimes u^* \otimes u \otimes \cdots \otimes u^* \in \mathcal{A}^{\otimes n+1}.$$

So (since $sf(\mathcal{D}, u\mathcal{D}u^*) = \frac{1}{\sqrt{2\pi i}}\phi_n(Ch_n(u^*))$)

$$sf(\mathcal{D}, u\mathcal{D}u^*) = \frac{(-1)^{(n-1)/2}\Gamma(n/2)\Gamma((n+1)/2)}{\sqrt{\pi}n!}$$

$$\times\tau_0(u[\mathcal{D}, u^*]\cdots[\mathcal{D}, u^*](1 + \mathcal{D}^2)^{-n/2})$$

Using the duplication formula for the Gamma function, we can simplify the constant in the last displayed formula. The duplication formula yields

$$\Gamma(n/2)\Gamma(n/2 + 1/2) = \sqrt{\pi}\Gamma(n)2^{-n+1} = \sqrt{\pi}(n-1)!2^{-n+1},$$

and inserting this gives

$$sf(\mathcal{D}, u\mathcal{D}u^*) = \frac{(-1)^{(n-1)/2}}{n2^{(n-1)}}\tau_0(u[\mathcal{D}, u^*]\cdots[\mathcal{D}, u^*](1 + \mathcal{D}^2)^{-n/2}).$$

References

1. M. F. Atiyah, *Elliptic operators, discrete groups and von Neumann algebras*, Astérisque **32-33** (1976), 43-72.
2. M. F. Atiyah, V. Patodi, and I. M. Singer, *Spectral asymmetry and Riemannian geometry, I*, Math. Proc. Cambridge Philos. Soc. **77** (1975), 43–69.
3. M. F. Atiyah, V. Patodi, and I. M. Singer, *Spectral asymmetry and Riemannian geometry. III*, Math. Proc. Cambridge Philos. Soc. **79** (1976), 71–99.
4. M. F. Atiyah and I. M. Singer, *Index theory for skew-adjoint Fredholm operators*, Publ. Math. Inst. Hautes Études Sci. **37** (1969), 5–26.
5. J. Avron, R. Seiler and B. Simon, *The index of a pair of projections*, J. Funct. Anal. **120** (1994), 220–237.
6. M-T. Benameur and T. Fack *Type II non-commutative geometry. I. Dixmier trace in von Neumann algebras*, Adv. Math. **199** (2006), 29–87.
7. J-M. Bismut and D. S. Freed *The analysis of elliptic families II. Dirac operators, eta invariants and the holonomy theorem*, Comm. Math. Phys. **107** (1986), 103-163.
8. B. Booß-Bavnbek and K. Furutani, *The Maslov index: a functional analytic definition and the spectral flow formula*, Tokyo J. Math. **21**, No. 1 (1998), 1-34.
9. B. Booß-Bavnbek, M. Lesch and J. Phillips *Unbounded Fredholm operators and spectral flow*, Canad. J. Math. **57** (2005), 225-250.
10. B. Booß-Bavnbek and K. P. Wojciechowski, *Elliptic boundary problems for Dirac operators*, Birkhäuser, Boston, 1993.
11. M. Breuer, *Fredholm theories in von Neumann algebras. I*, Math. Ann. **178** (1968), 243-254.
12. M. Breuer, *Fredholm theories in von Neumann algebras. II*, Math. Ann. **180** (1969), 313-325.
13. L. G. Brown, R. G. Douglas and P. A. Fillmore, *Extensions of C^*-algebras and K-homology*, Ann. of Math. **105** (1977), 349-363.
14. A. L. Carey and J. Phillips, *Algebras almost commuting with Clifford algebras in a II_∞ factor*, K Theory **4** (1991), 445-478.
15. A. L. Carey and J. Phillips, *Unbounded Fredholm modules and spectral flow*, Canad. J. Math. **50** (1998), 673-718.
16. A. L. Carey and J. Phillips, *Spectral flow in Θ-summable Fredholm modules, eta invariants and the JLO cocycle*, K Theory **31** (2004), 135-194.
17. A. L. Carey, J. Phillips, A. Rennie and F. A. Sukochev, *The Hochschild class of the Chern character of semifinite spectral triples*, J. Funct. Anal. **213** (2004), 111-153.
18. A. L. Carey, J. Phillips, A. Rennie and F. A. Sukochev, *The local index theorem in semifinite von Neumann algebras I: Spectral flow*, Adv. Math. **202** (2006), to appear.
19. A. L. Carey, J. Phillips, A. Rennie and F. A. Sukochev, *The local index theorem in semifinite von Neumann algebras II: the even case*, Adv. Math.

202 (2006), to appear.

20. A. L. Carey, J. Phillips and F. A. Sukochev, *On unbounded p-summable Fredholm modules*, Adv. Math. **151** (2000), 140-163.
21. A. L. Carey, J. Phillips and F. A. Sukochev, *Spectral flow and Dixmier traces*, Adv. Math. **173** (2003), 68-113.
22. J. Cheeger, *η-invariants, the adiabatic approximation and conical singularities*, J. Differential Geom. **26** (1987), 175–221.
23. J. Cheeger and M. Gromov, *Bounds on the von Neumann dimension of L^2-cohomology and the Gauss-Bonnet theorem for open manifolds*, J. Differential Geom. **21** (1985), 1-34.
24. L. A. Coburn, R. G. Douglas, D. G. Schaeffer and I. M. Singer, *C^*-algebras of operators on a half space II. Index theory*, Publ. Math. Inst. Hautes Études Sci. **40** (1971), 69-80.
25. L. A. Coburn, R. D. Moyer and I. M. Singer, *C^*-algebras of almost periodic pseudo-differential operators*, Acta Math. **130** (1973), 279-307.
26. A. Connes, *Noncommutative differential geometry*, Publ. Math. Inst. Hautes Études Sci. **62** (1985), 41-44.
27. A. Connes, *Cyclic cohomology of Banach algebras and characters of θ-summable Fredholm modules*, K Theory **1** (1988), 519-548.
28. A. Connes, *Compact metric spaces, Fredholm modules and hyperfiniteness*, Ergodic Theory Dynamical Systems **9** (1989), 207-220.
29. A. Connes, *Noncommutative geometry*, Acad. Press, San Diego, 1994.
30. A. Connes and H. Moscovici, *Cyclic cohomology, the Novikov conjecture and hyperbolic groups*, Topology **29** (1990), 345-388.
31. A. Connes, *Geometry from the spectral point of view*, Lett. Math. Phys. **34** (1995), 203-238.
32. A. Connes and H. Moscovici, *The local index formula in noncommutative geometry*, Geom. Funct. Anal. **5** (1995), 174-243.
33. X. Dai and W. Zhang, *Higher spectral flow*, J. Funct. Anal. **157** (1998), 432-469.
34. J. Dixmier, *Les algèbres d'opérateurs dans l'espace Hilbertien (Algèbres de von Neumann)*, Gauthier-Villars, Paris, 1969.
35. P. G. Dodds, B. de Pagter, E. M. Semenov and F. A. Sukochev *Symmetric functionals and singular traces*, Positivity **2** (1998), 47-75.
36. H. Donnelly, *Asymptotic expansion for the compact quotients of properly discontinuous group actions*, Illinois J. Math. **23** (1979), 175-221.
37. R. G. Douglas, S. Hurder and J. Kaminker, *Cyclic cocycles, renormalisation and eta-invariants*, Invent. Math. **103** (1991), 101-179.
38. T. Fack and H. Kosaki *Generalised s-numbers of τ-measurable operators*, Pacific J. Math. **123** (1986), 269–300.
39. K. Furutani, *Fredholm–Lagrangian–Grassmannian and the Maslov index*, J. Geom. Phys. **51** (2004), 269–331, math.DG/0311495.
40. E. Getzler, *The odd Chern character in cyclic homology and spectral flow*, Topology **32** (1993), 489-507.
41. I. C. Gohberg, M. G. Krein, *Fundamental aspects of defect numbers, root numbers and indices of linear operators*, Uspekhi Mat. Nauk **12** No. 2

(1957), 43-118.
42. I. C. Gohberg, M. G. Krein, *Introduction to the theory of non-selfadjoint operators*, Translations of Mathematical Monographs **18**, Amer. Math. Soc, 1969.
43. N. Higson, *The local index formula in noncommutative geometry*, Contemporary Developments in Algebraic K-Theory, ICTP Lecture Notes **15** (2003), 444-536.
44. S. Hurder, *Eta invariants and the odd index theorem for coverings*, Contemp. Math. **105** (1990), 47-82.
45. E. Leichtmann and P. Piazza, *Spectral sections and higher Atiyah-Patodi-Singer index theory on Galois coverings*, Geom. Funct. Anal. **8** (1998), 17-58.
46. M. Lesch, *On the index of the infinitesimal generator of a flow*, J. Operator Theory **26** (1991), 73-92.
47. M. Lesch, *The uniqueness of the spectral flow on spaces of unbounded self-adjoint Fredholm operators*, In: Spectral geometry of manifolds with boundary and decomposition of manifolds, Contemp. Math. **366**, Amer. Math. Soc., Providence, RI, 2005, pp. 193–224, math.FA/0401411.
48. V. Mathai, *L^2 invariants of covering spaces*, Geometric Analysis and Lie Theory in Mathematics and Physics, Cambridge University Press, Cambridge, 1998.
49. V. Mathai, *Spectral flow, eta invariants and von Neumann algebras*, J. Funct. Anal. **109** (1992), 442–456.
50. V. Mathai, preprint and private communication.
51. H. McKean and I. M. Singer, *Curvature and eigenvalues of the Laplacian*, J. Differential Geom. **1** (1967), 43–69.
52. R. B. Melrose, *The Atiyah-Patodi-Singer index theorem*, A K Peters Ltd, Wellesley, USA, 1993.
53. R. Palais, *Seminar on the Atiyah-Singer index theorem*, Princeton University Press, Princeton, 1965.
54. D. Pask and A. Rennie, *The noncommutative geometry of graph C^*-algebras*, J. Funct. Anal., to appear.
55. V. S. Perera, *Real valued spectral flow in a type II_∞ factor*, Ph.D. Thesis, IUPUI, 1993 and *Real valued spectral flow*, Contemp. Math. **185** (1995), 307-318.
56. V. S. Perera, *Real valued spectral flow in a type II_∞ factor*, Houston J. Math. **25** (1999), 55-66.
57. J. Phillips, *Self-adjoint Fredholm operators and spectral flow*, Canad. Math. Bull. **39** (1996), 460–467.
58. J. Phillips, *Spectral flow in type I and type II factors-a new approach*, Fields Inst. Commun. **17** (1997), 137–153.
59. J. Phillips and I. F. Raeburn *An index theorem for Toeplitz operators with noncommutative symbol space*, J. Funct. Anal. **120** (1993), 239-263.
60. R. Prinzis, *Traces residuelles et asymptotique du spectre d'opérateurs pseudo-differentiels*, Thèse, Université de Lyon, unpublished.
61. J. Roe *Elliptic operators, topology and asymptotic methods*, Pitman Re-

search Notes in Mathematics **395**, Longman, Harlow, 1998.
62. M. A. Shubin, *Pseudodifferential almost periodic operators and von Neumann algebras*, Trans. Moscow Math. Soc. **1** (1979), 103-166.
63. M. A. Shubin, *Almost periodic functions and partial differential operators*, Russian Math. Surveys **33** No. 2 (1978), 1-52.
64. M. A. Shubin, *Spectral theory and index of elliptic operators with almost periodic coefficients*, Russian Math. Surveys **34** No. 2 (1979), 109-157.
65. I. M. Singer, *Eigenvalues of the Laplacian and invariants of manifolds*, Proceedings of the International Congress, Vancouver 1974, **vol. I**, 187-200.
66. F. A. Sukochev, *Unbounded Fredholm modules and submajorization*, Canad. J. Math. **52** (2000), 849-896.
67. K. P. Wojciechowski, Preprint, IUPUI, Indianapolis, 1991.

Received by the editors August 12, 2005; Revised November 22, 2005

Analysis, Geometry and Topology
of Elliptic Operators, pp. 353–368

ELLIPTIC OPERATORS ON INFINITE GRAPHS

JÓZEF DODZIUK

Ph.D. Program in Mathematics
Graduate Center (CUNY)
New York, NY 10016
jozek@derham.math.qc.edu

Dedicated to Krzysztof P. Wojciechowski on his 50th birthday

We present some applications of ideas from partial differential equations and differential geometry to the study of difference equations on infinite graphs. All operators that we consider are examples of "elliptic operators" as defined by Colin de Verdiere [4]. For such operators, we discuss analogs of inequalities of Cheeger and Harnack and of the maximum principle (in both elliptic and parabolic versions), and apply them to study spectral theory, the ground state and the heat semigroup associated to these operators.
kpwmsc39A1258J05

1. Preliminaries

We consider graphs (without loops or multiple connections) $G = (V, E)$ where V is a set whose elements are called vertices and E, the set of edges, is a subset of the set of two-element subsets of V. For an edge $e = \{x, y\} \in E$, we will denote by $[x, y]$ the *oriented* edge from x to y and write $\overline{E}$ for the set of all oriented edges. We also write $x \sim y$ if $\{x, y\}$ is an edge. All graphs considered will be connected.

By a function on a graph we will mean a mapping $f : V \longrightarrow \mathbb{C}$. By an operator on a graph, we shall always mean an operator acting on functions and follow Colin de Verdière [4] in defining the notion of "self-adjoint, positive, elliptic operator." Observe first that every operator L is given by a matrix $(b_{x,y})$. We require our operators to be local, i.e.

$$b_{x,y} = 0 \qquad \text{if } \{x, y\} \text{ is not an edge and } x \neq y \ .$$

This work is supported in part by a PSC-CUNY Research Grant.

Thus

$$Lf(x) = b_{x,x}f(x) + \sum_{x\sim y} b_{x,y}f(y).$$

The constant functions are annihilated by L if and only if $\sum_{y\sim x} b_{x,y} = -b_{x,x}$ for every $x \in V$. Every local operator L can be rewritten in the form

$$Lf(x) = W(x)f(x) + \sum_{y\sim x} a_{x,y}(f(x) - f(y)) \tag{1}$$

where $W(x) = b_{x,x} + \sum_{y\sim x} b_{x,y}$ and $a_{x,y} = -b_{x,y}$. We will often write $L = A + W$, where A, given by the sum in the formula above, annihilates constant functions and W denotes the operator of multiplication by the function $W(x)$.

Let $\ell^2(V)$ be the space of complex-valued functions f satisfying $\sum_{x\in V} |f(x)|^2 < \infty$ equipped with the standard hermitian inner product

$$(f, g) = \sum_{x\in V} f(x)\overline{g(x)}.$$

We denote by $C_0(V)$ the space of all functions on V with finite support. In order that the operator L be symmetric on $C_0(V)$, i.e. $(Lf, g) = (f, Lg)$ it is necessary and sufficient that $a_{x,y} = \overline{a_{y,x}}$ and $W(x) + \sum_{y\sim x} a_{x,y} \in \mathbb{R}$. We want to think of the operator in (1) as a "Laplacian" plus a potential. Thus, we impose an additional condition on A that will make it positive on $C_0(V)$. Namely, we require that $a_{x,y}$ be real and positive for every edge $\{x, y\}$. We will refer to such operators as *elliptic*, *positive* and *symmetric*. A very important example is the combinatorial Laplacian $A = \Delta$ given by choosing $a_{x,y} = 1$ for every edge,

$$\Delta f(x) = \sum_{x\sim y}(f(x) - f(y)) = m(x)f(x) - \sum_{x\sim y} f(y),$$

where $m(x)$ is the valence of the vertex $x \in V$ i.e. the number of edges emanating from x.

The following lemma sheds some light on the structure of a positive, symmetric operator. First, we need a definition. Let $C(\overline{E})$ denote the space of functions ϕ on *oriented* edges satisfying $\phi([x, y]) = -\phi([y, x])$ for every edge $\{x, y\}$ and let

$$\ell^2(\overline{E}) = \{\phi \in C(\overline{E}) \mid \sum_{\{x,y\}\in E} |\phi([x, y])|^2 < \infty\}.$$

We equip $\ell^2(\overline{E})$ with the natural inner product

$$< \phi, \psi >= \sum_{\{x,y\} \in E} \phi([x,y])\overline{\psi([x,y])}.$$

In addition, given a positive, symmetric operator A as above, define the (possibly unbounded) operator d_A from $\ell^2(V)$ to $\ell^2(\overline{E})$ by

$$d_A f([x,y]) = \sqrt{a_{x,y}}(f(x) - f(y)).$$

Lemma 1.1. *Suppose f and g are two functions on the graph and one of them has finite support. Then*

$$(Af, g) =< d_A f, d_A g > .$$

In particular, if f has finite support, $(Af, f) \geq 0$ with equality if and only if $f \equiv 0$.

Proof. The proof is a simple calculation.

$$\begin{aligned}(Af, g) &= \sum_x \left(\sum_{y \sim x} a_{x,y}(f(x) - f(y)) \right) \overline{g(x)} \\ &= \sum_{\{x,y\} \in E} a_{x,y}(f(x) - f(y))\overline{(g(x) - g(y))} =< d_A f, d_A g >\end{aligned}$$

To justify it note that an edge $\{z, w\}$ contributes to the first sum twice. The contribution is

$$\begin{aligned}a_{z,w}(f(z) - f(w))\overline{g(z)} + a_{w,z}(f(w) - f(z))\overline{g(w)} = \\ a_{z,w}(f(z) - f(w))\overline{(g(z) - g(w))}\end{aligned}$$

since $a_{z,w}$ is symmetric. This proves that the two sums are equal. The statement about strict positivity of (Af, f) follows trivially. □

We wish to consider $L = A + W$ as an unbounded operator on $\ell^2(V)$ and to study its spectrum. In order to obtain a reasonable setup we will require that the potential W be bounded from below by a constant, $W(x) \geq c$ for all $x \in V$. By the lemma above, L is semi-bounded, i.e. $(Lf, f) \geq c(f, f)$ for every $f \in C_0(V)$. By Theorem X.23 of Simon and Reed [12], L then has a distinguished self-adjoint extension, the Friedrichs extension, $\hat{L}$ such that $\lambda_0(\hat{L})$, the bottom of the spectrum of $\hat{L}$, has a variational characterization

$$\lambda_0(\hat{L}) = \inf_{f \in C_0(V) \setminus \{0\}} \frac{(Lf, f)}{(f, f)}. \tag{2}$$

We will abuse the notation and write $\lambda_0(L)$ for $\lambda_0(\hat{L})$.

In general, without any further restrictions, the operator L with domain $C_0(V)$ may have many self-adjoint extensions. The theorem below gives conditions under which L is essentially self-adjoint, i.e. has a unique self-adjoint extension, cf. [12], Theorem X.28.

Theorem 1.1. *Suppose that A is a positive, symmetric and bounded as an operator on $\ell^2(V)$. Let W be bounded from below by a constant. Then $L = A + W$ is essentially self-adjoint on $C_0(V)$.*

Proof. Choose a positive constant κ so that $W + \kappa \geq 1$. By Theorem X.26 of [12], it suffices to show that

$$(A + W + \kappa)^* f = 0 \tag{3}$$

implies that $f = 0$. Taking the inner product of the equation above with the function δ_x ($\delta_x(y) = 1$ if $x = y$ and 0 otherwise), using the definition of the adjoint and Lemma 1.1, we see that (3) is equivalent to

$$(A + W + \kappa)f = 0, \qquad f \in \ell^2(V)$$

where $(A + W + \kappa)f$ is computed pointwise as in (1) with W replaced by $W + \kappa$. Since A is bounded and $C_0(V)$ is dense in $\ell^2(V)$, $(Af, f) \geq 0$ by Lemma 1.1. Therefore, $0 = (Af, f) + ((W + \kappa)f, f) \geq (f, f)$. It follows that $f = 0$ which proves the theorem. □

Remark 1.1. Observe that the condition that A be bounded holds if $a = \sup a_{x,y} < \infty$ and $M = \sup m(x) < \infty$. In fact, in this case $\| A \| \leq 2aM$.

Remark 1.2. We view the Theorem 1.1 as an analog of Theorem X.28 of [12] which applies to a differential operator $-\Delta + V$ on $\mathbb{R}^n$. Clearly, Δ is unbounded but the unboundedness is an infinitesimal effect that does not occur for difference operators on graphs. We view the boundedness of A or the condition $a < \infty$ as a partial replacement of uniform ellipticity, (see Corollary 2.1 below for a proper analog of uniform ellipticity). Similarly, $M < \infty$ is a bounded geometry condition.

We now state two local results. Their continuous analogs - the maximum principle and Harnack's inequality - are discussed at great length in Protter and Weinberger [11]. Let $V_1 \subset V$ be a set of vertices and let G_1 be the full subgraph of G generated by V_1 (i.e. the set of edges of G_1 consists of all edges $\{x, y\}$ of G such that $x, y \in V_1$). Let $\overset{o}{V_1} = \{x \in V_1 \mid y \sim x \quad \text{implies} \quad y \in$

$V_1\}$ and $\partial V_1 = V_1 \backslash \overset{o}{V_1}$. We say that $\overset{o}{V_1}$ is connected if every two of its vertices x, y can be connected by a path of edges $[x_0, x_1], [x_1, x_2], \ldots, [x_{n-1}, x_n]$, $x_0 = x, x_n = y$ with $x_i \in \overset{o}{V_1}$ for $i = 0, 1, \ldots, n$.

Lemma 1.2. *Let $L = A + W$ where A is positive, symmetric and W is nonnegative. Suppose $V_1 \subset V$ is a subset with $\overset{o}{V_1}$ connected. Let f be a function on V_1 such that*

$$Lf(x) = Af(x) + W(x)f(x) \geq 0 \quad \textit{for} \quad x \in \overset{o}{V_1}.$$

If f has a minimum at $x_0 \in \overset{o}{V_1}$ and $f(x_0) \leq 0$ then f is constant on V_1.

Proof. Suppose $x_0 \in \overset{o}{V_1}$ is a minimum and $f(x_0) \leq 0$. Then

$$0 \leq \sum_{y \sim x_0} a_{x_0,y}(f(x_0) - f(y)) + W(x_0)f(x_0) \leq 0$$

since A is positive, x_0 is a minimum, and $W(x_0)f(x_0) \leq 0$. It follows that all terms in the sum above are equal to zero, i.e. $f(y) = f(x_0)$ for every $y \sim x_0$. By connectedness, f is constant. □

Lemma 1.3. *Suppose A and W satisfy the assumptions of Lemma 1.2. Let $V_1 \subset V$, $x \sim y$, $x, y \in \overset{o}{V_1}$. If*

$$Lf = Af + Wf \geq 0 \quad \textit{and} \quad f > 0 \quad \textit{on} \quad V_1$$

then

$$\frac{a_{x,y}}{\left(W(x) + \sum_{z \sim x} a_{x,z}\right)} \leq \frac{f(x)}{f(y)} \leq \frac{\left(W(y) + \sum_{z \sim y} a_{y,z}\right)}{a_{x,y}}.$$

Proof. By symmetry, it suffices to prove one of the two inequalities above. We have

$$(A + W)f(x) = \sum_{z \sim x} a_{x,z}(f(x) - f(z)) + W(x)f(x) \geq 0.$$

Therefore,

$$\left(\sum_{z \sim x} a_{x,z}\right) f(x) + W(x)f(x) \geq \sum_{z \sim x} a_{x,z} f(z) \geq a_{x,y} f(y).$$

This, of course, is equivalent to the lower bound on $f(x)/f(y)$ in the statement of the lemma. □

We refer to Lemma 1.2 as the maximum principle and to Lemma 1.3 as the Harnack inequality. The significance of the Harnack inequality is that it gives a bound of the ratio $f(x)/f(y)$ in terms of the coefficients of the operator but *independent* of the function f.

2. Existence of ground state

In this section we prove, for an operator $L = A+W$ with positive, symmetric A and the potential W bounded from below by a constant, the existence of a ground state, i.e. a positive solution of the equation

$$L\phi = \lambda_0(L)\phi,$$

cf. Pinsky [10] for an extensive discussion in the continuous setting. We assume that the underlying graph G is connected and fix a vertex x_0 as an "origin". Consider the exhaustion $\{G_n\}_{n=1}^{\infty}$ of G where, for every n, G_n is the full subgraph with the vertex set $V_n = \{x \in V \mid d(x_0, x) \leq n\}$. Here, $d(x, y)$ denotes the combinatorial distance between $x, y \in V$, i.e. the length of the shortest path of edges connecting x with y. Clearly, $\overset{o}{V}_n$ is connected for every $n \geq 1$. We will construct a ground state ϕ by solving certain "boundary value problems" on G_n and taking a limit of the solutions. In order to get started we need to review these boundary value problems. Thus, let U be a finite subset of V such that the full subgraph generated by U has connected interior. Let $C_0(U)$ be the space of functions on U that vanish on ∂U. Extending functions in $C_0(U)$ by zero embeds $C_0(U)$ isometrically in $C_0(V)$. We define, for $f \in C_0(U)$, $L_U f \in C_0(U)$ by

$$L_U f(x) = \begin{cases} W(x)f(x) + \sum_{x\sim y} a_{x,y}(f(x) - f(y)) & \text{if } x \in \overset{o}{U}, \\ 0 & \text{if } x \in \partial U. \end{cases}$$

We can define $A_U f \in C_0(U)$ for $f \in C_0(U)$ analogously. The calculation in the proof of Lemma 1.1 shows that A_U and L_U are symmetric operators on $C_0(V)$ and that A_U is strictly positive. It follows that $\lambda_0(L_U)$, the smallest eigenvalue of L_U on $C_0(U)$, has variational characterization

$$\lambda_0(L_U) = \inf_{f\in C_0(U)\setminus\{0\}} \frac{(L_U f, f)}{(f, f)} = \inf_{f\in C_0(U)\setminus\{0\}} \frac{(Lf, f)}{(f, f)} \tag{4}$$

where in the last expression above we identify f with its extension by zero outside U.

Proposition 2.1. *The eigenspace of $\lambda_0(L_U)$ is one-dimensional and every eigenfunction ψ belonging to $\lambda_0(L_U)$ has constant sign in the interior of U.*

Proof. It is enough to consider real-valued functions. Replacing W by $W + c$ with a suitably large c, we can assume that W is nonnegative. Since

$$(L_U f, f) = \sum_{x \sim y,\, x \in U,\, y \in \overset{o}{U}} a_{x,y}(f(x) - f(y))^2 + \sum_{x \in \overset{o}{U}_0} W(x) f(x)^2$$

replacing f by $|f|$ decreases the Rayleigh-Ritz quotient in (4). Therefore, it follows that if ψ is an eigenfunction belonging to $\lambda_0(L_U)$ then $|\psi|$ is one as well. Thus we can assume that there exists a nonnegative eigenfunction ψ. Since the Raylegh-Ritz quotient is nonnegative, $\lambda_0(L_U) \geq 0$. The maximum principle in Lemma 1.2 implies that ψ is strictly positive in $\overset{o}{U}$. Finally, if the eigenspace of $\lambda_0(L_U)$ had two or more dimensions, there would exist another eigenfunction ϕ orthogonal to ψ. Therefore ϕ would have to change sign and be negative at an interior point, but this is impossible by the maximum principle. □

We are now ready to prove

Theorem 2.1. *Consider an operator $L = A + W$ on a connected graph G with positive, symmetric A and the potential W bounded below by a constant. There exists a ground state ϕ for L i.e. a function $\phi > 0$ on V such that*

$$L\phi = \lambda_0 \phi$$

where $\lambda_0 = \lambda_0(L)$ is the bottom of the spectrum of (the Friedrichs extension of) L on G.

Proof. The proof for the case of the combinatorial Laplacian was given in Dodziuk and Mathai [8]. We follow the same line of argument here but remark that exhaustion argument of this kind is applied very often in studying partial differential equations on noncompact domains or domains with non-smooth boundaries as, for example, in [10], Chapter 4. Note first that by adding a suitable constant to the potential W we can assume without any loss of generality that $W > 0$. We use the exhaustion of G by finite subgraphs G_n described above. Let $\lambda_n = \lambda_0(L_{G_n})$ and let ϕ_n be the corresponding positive eigenfunction of L on $C_0(V_n)$ normalized so that $\phi_n(x_0) = 1$. By the variational characterization of eigenvalues and of the bottom of the spectrum (2), (4) we have $\lambda_n \searrow \lambda_0$. Fix a point $y \in V$. Then, there exists $k = k(y)$ such that $y \in \overset{o}{V}_n$ for all $n > k$. Choose a path of length $d(x_0, y)$ that connects x_0 and y. Using the normalization $\phi_n(x_0) = 1$

and applying the local Harnack inequality in Lemma 1.3 to successive edges of the path, we see that the sequence $\phi_n(y)$ is bounded above and below by positive constants that are independent of n. Using the diagonal process, we choose a subsequence $(n_k)_{k=1}^{\infty}$ such that the sequence $(\phi_{n_k}(y))_{k=1}^{\infty}$ converges to the limit $\phi(y)$ of every vertex $y \in V$ and $\phi(y) > 0$. Since $L\phi$ is given by the formula (1) and $\lambda_n \searrow \lambda_0$ we see that ϕ is a positive solution of $L\phi = \lambda_0 \phi$ as required. $\square$

We now need the following lemma to control the behavior at infinity of a ground state under certain additional assumptions.

Corollary 2.1. *Assume that A is symmetric and positive, that the graph G has bounded valence $\sup_{x \in V} m(x) = M < \infty$ and that the operator A is uniformly elliptic in the sense that there exist constants $\gamma, \Gamma > 0$ so that $\gamma \leq a_{x,y} \leq \Gamma$ for every edge $\{x, y\}$. Suppose a function f on V satisfies $Af \geq 0$, $f > 0$. Then, for every $x, y \in V$,*

$$\left(\frac{M\Gamma}{\gamma}\right)^{-d(x,y)} \leq \frac{f(x)}{f(y)} \leq \left(\frac{M\Gamma}{\gamma}\right)^{d(x,y)}.$$

Proof. By Lemma 1.3, $\gamma/M\Gamma \leq f(z)/f(w) \leq M\Gamma/\gamma$ if $z \sim w$. We connect x with y by a path of edges of length $d(x, y)$ and apply these inequalities for every edge along the path. The corollary follows. $\square$

Observe that this is entirely analogous to Theorem 21 in [11].

3. Cheeger's inequality

In this section, we assume that $L = A$ and give a lower bound for the bottom of the spectrum of A on G. This bound originated in Riemannian geometry, cf. Cheeger [3], and has been studied a great deal for the combinatorial Laplacian on graphs, cf. Lubotzky [9], Dodziuk [6], Dodziuk-Kendall [7].

As before, let A be a positive, symmetric elliptic operator on an infinite graph G and let $U \subset V$ be a finite subset. We define

$$h_A(U) = \frac{\sum_{x \in \mathring{U},\, y \in \partial U,\, x \sim y} \sqrt{a_{x,y}}}{\#(U)}, \tag{5}$$

and

$$\beta(G, A) = \inf_U h_A(U) \tag{6}$$

where $\#U$ denotes the number of vertices of U.

Theorem 3.1. *Suppose* $\sup_{x\in V} m(x) = M < \infty$ *The lower bound of the spectrum of* A *on* G *satisfies*

$$\lambda_0(A) \geq \frac{\beta(G,A)^2}{2M}.$$

Proof. We follow the proof of Theorem 2.3 of [6]. Let $(G_n)_{n=1}^{\infty}$ be the exhaustion of G used in the proof of Theorem 2.1. Since $\lambda_n \searrow \lambda_0$ it will suffice to show that $\lambda_n \geq \beta(G,A)^2/2M$ independently of n. We will fix n, set $U = V_n$ and let ϕ be positive eigenfunction of A_U. Observe that by Lemma 1.1 and (4)

$$\lambda_n = \lambda_0(A_U) = \frac{< d_A\phi, d_A\phi >}{(\phi,\phi)} \tag{7}$$

if we extend ϕ by zero outside U. Consider the expression

$$\mathcal{A} = \sum_{\{x,y\}\in E} \sqrt{a_{x,y}}|\phi^2(x) - \phi^2(y)|.$$

By Cauchy-Schwartz inequality we have

$$\begin{aligned}
\mathcal{A} &= \sum_{\{x,y\}} \sqrt{a_{x,y}}|\phi(x) - \phi(y)|\,|\phi(x) + \phi(y)| \\
&\leq \left(\sum_{\{x,y\}} |\phi(x) + \phi(y)|^2\right)^{1/2} \left(\sum_{\{x,y\}} a_{x,y}|\phi(x) - \phi(y)|^2\right)^{1/2} \\
&\leq \sqrt{2}\left(\sum_{\{x,y\}} (\phi^2(x) + \phi^2(y))\right)^{1/2} (d_A\phi, d_A\phi)^{1/2}.
\end{aligned}$$

In $\sum_{\{x,y\}}(\phi^2(x) + \phi^2(y))$, every vertex contributes as many times as the number of edges emanating from it. Hence we get the following upper bound on $\mathcal{A}$.

$$\mathcal{A} \leq \sqrt{2M}\,(\phi,\phi)^{1/2}\,(d_A\phi, d_A\phi)^{1/2}. \tag{8}$$

On the other hand we can estimate $\mathcal{A}$ from below in terms of (ϕ,ϕ) as follows. Let $0 = \nu_0 < \nu_1 < \nu_2 < \ldots < \nu_N$ be the sequence of all values of ϕ^2. Note that, since $A\phi(x) = \lambda_0(U)\phi(x)$ at every interior vertex x and since $\lambda_0(U) > 0$ by (7), every interior vertex x will have a neighbor y such that $\phi(x) > \phi(y)$. Define a set of vertices U_i, $i = 1, 2, \ldots, N$ as follows. A vertex $x \in U$ belongs to U_i if and only if $\phi^2(x) \geq \nu_i$ and let F_i be the full graph generated by the set U_i. Now

$$\mathcal{A} = \sum_{i=1}^{N} \sum_{\phi^2(x)=\nu_i} \sum_{y\sim x\, \phi^2(y)<\nu_i} \sqrt{a_{x,y}}(\phi^2(x) - \phi^2(y)).$$

If $\phi^2 = \nu_i$ and $\phi^2(y) = \nu_{i-k}$ for some $k \in \{1, 2, \ldots, i\}$, then on the one hand,$\phi^2(x) - \phi^2(y) = (\nu_i - \nu_{i-1}) + (\nu_{i-1} - \nu_{i-2} + \ldots (\nu_{i-k+1} - \nu_{i-k})$ and, on the other hand, $x \in \partial U_i \cap \partial U_{i-1} \cap \ldots \cap \partial U_{i-k+1}$. It follows that

$$\mathcal{A} \geq \sum_{i=1}^{N} (\nu_i - \nu_{i-1}) \sum_{y\sim x,\, y\in\partial U_i} \sqrt{a_{x,y}}.$$

Applying (6) we obtain

$$\mathcal{A} \geq h_A(U) \sum_{i=1}^{N} \#U_i(\nu_i - \nu_{i-1}) \geq \beta \sum_{i=1}^{N} \#U_i(\nu_i - \nu_{i-1})$$

with $\beta = \beta(G, A)$. "Summation by parts" now yields

$$\mathcal{A} \geq \beta \left(\nu_N \#U_n + \sum_{i=1}^{N-1} \nu_i(\#U_i - \#U_{i+1}) \right).$$

Observe that $\#U_n$ is the cardinality of the set where $\phi^2 = \nu_N$ while $\#U_i - \#U_{i+1}$ is the number of points where $\phi^2 = \nu_i$. It follows that

$$\mathcal{A} \geq \beta(\phi, \phi).$$

This inequality combined with (7) and (8) gives the desired lower bound□

We remark that one can also bound $\lambda_0(A)$ from above by a related isoperimetric constant. Namely, let χ_U be the characteristic function of a finite set of vertices $U \subset V$. Then

$$\lambda_0(A) \leq \frac{< d_A\chi_U, d_A\chi_U >}{(\chi_U, \chi_U)} = \frac{\sum_{x\sim y, x\in U, y\notin U} a_{x,y}}{\#U}$$

It follows that

$$\lambda_0(A) \leq \beta_1(G, A) = \inf \frac{\sum_{x\sim y, x\in U, y\notin U} a_{x,y}}{\#U}$$

where the infimum is taken over all finite subsets U of V.

Note that for the combinatorial Laplacian Δ, $a_{x,y} \equiv 1$. Thus $\beta(G, A) = \beta_1(G, A)$. In particular, for graphs of bounded valence $\lambda_0(\Delta) = 0$ if and only if $\beta(G, \Delta) = 0$ which is analogous to a result of Buser [2] in the Riemannian setting and is very useful in connection with various characterizations of amenability of discrete, finitely generated groups, cf. Brooks [1].

4. The heat equation

In this section we make several standing assumptions. Namely, we assume that the graph G has bounded valence $\sup_{x \in V} m(x) = M < \infty$; that the potential $W \equiv 0$ i.e. $L = A$; and that $a = \sup_{\{x,y\} \in E} a_{x,y} < \infty$. We shall study the parabolic initial value problem

$$\begin{aligned} Au + \frac{\partial u}{\partial t} &= 0 \\ u(x,0) &= u_0(x) \end{aligned} \tag{9}$$

and the associated heat semigroup using the method of Dodziuk [5] applied previously to the combinatorial Laplacian in [8]. Here $u(x,t)$ is a function of $x \in V$ and $t > 0$, while u_0 is a given function on G. The first equation above will be referred to as the heat equation.

We are going to use the following version of the maximum principle, see [11], Chapter 4 for an analog in the continuous setting.

Lemma 4.1. *Suppose $u(x,t)$ satisfies the inequality $Au + \frac{\partial u}{\partial t} < 0$ on $\overset{o}{U} \times [0,T]$ for a finite subset U of V. Then the maximum of u on $U \times [0,T]$ is attained on the set $U \times \{0\} \cup \partial U \times [0,T]$.*

Proof. Suppose $(x_0, t_0) \in \overset{o}{U} \times (0,T]$ is a maximum. It follows that $\frac{\partial u}{\partial t}(x_0,t_0) \geq 0$ so that $Au(x_0,t_0) < 0$. On the other hand, (1) and positivity of A imply that $Au(x_0,t_0) \geq 0$. The contradiction proves the lemma. □

We use the lemma above to prove the uniqueness of bounded solutions of (9).

Theorem 4.1. *Let $u(x,t)$ be a bounded solution of (9) with the initial condition $|u_0(x)| \leq N_0$. Then u is determined uniquely by u_0 and*

$$|u(x,t)| \leq N_0$$

for all (x,t). Moreover, if a bounded initial condition u_0 is given, then a bounded solution $u(x,t)$ of (9) exists.

Proof. Suppose that $u(x,t)$ is a bounded solution. Let $N_1 = \sup |u(x,t)|$. Fix $x_0 \in V$ and define $r(x) = d(x,x_0)$. By our assumption on the valence and (1)

$$|Ar| \leq aM. \tag{10}$$

Consider an auxiliary function

$$v(x,t) = u(x,t) - N_0 - \frac{N_1}{R}\left(r(x) + a(M+1)t\right),$$

where R is a large parameter. Let $U = B(x_0, R)$ be the set of vertices of V at distance at most R from x_0. The function $v(x,t)$ is nonpositive on the set $U \times \{0\} \cup \partial U \times [0,T]$ and satisfies $(A + \frac{\partial}{\partial t})v < 0$ on $\overset{o}{U} \times [0,T]$ because of (10). Lemma 4.1 implies therefore that $v(x,t) \leq 0$ so that

$$u(x,t) \leq N_0 + \frac{N_1}{R}\left(r(x) + a(M+1)t\right)$$

on $B(x_0, R) \times [0,T]$. Keeping (x,t) fixed and letting R increase without bounds, we see that $u(x,t) \leq N_0$. Applying the same argument to $-u$ yields $|u(x,t)| \leq N_0$. Since $T > 0$ and x were arbitrary, this last inequality holds for all $x \in V$ and $t \geq 0$. Uniqueness follows by considering the difference of two solutions. We postpone the proof of existence of the solution. □

Recall that under our assumption A is a bounded operator on $\ell^2(V)$. Therefore, we can define for $t \geq 0$

$$P_t = e^{-tA} = \sum_{k=0}^{\infty}(-1)^k \frac{t^k A^k}{k!}. \tag{11}$$

Obviously, $u(x,t) = (P_t u_0)(x)$ is a solution of (9) whenever u_0 is in $\ell^2(V)$. Since $\| P_t \| \leq 1$ we see that for every $x \in V$ and $t \geq 0$

$$|u(x,t)| \leq \| u(\cdot,t) \| \leq \| u_0 \|$$

so that $u(x,t)$ is a bounded solution and we get uniqueness. We would like to extend the semigroup P_t to a larger class of functions.

We define $p_t(x,y)$ to be matrix coefficients of the operator P_t, i.e.

$$p_t(x,y) = (P_t \delta_x, \delta_y)$$

where δ_x is the characteristic function of the set $\{x\}$. Similarly, let $A(x,y) = (A\delta_x, \delta_y)$. Since A is self-adjoint both of these matrices are symmetric. Writing $u_0 = \sum_y u_0(y)\delta_y$ and using the symmetry, we see that

$$P_t u_0(x) = (P_t u_0, \delta_x) = \sum_y p_t(x,y) u_0(y) \tag{12}$$

for $u_0 \in \ell^2(V)$. Substituting $u_0 = \delta_y$ we see that $p_t(x,y)$ satisfies the heat equation in variables x,t. We try to extend P_t to functions that are not necessarily in $\ell^2(V)$ by using this formula and verifying the convergence of the series. To do this we shall need an estimate in the lemma below of $p_t(x,y)$ for $t \in [0,T]$ and $d(x,y)$ large.

Lemma 4.2. *For every $T > 0$ there exist a constant $C = C(a, M, T) > 0$ such that*

$$p_t(x,y) \leq \frac{C}{d(x,y)!}$$

for all $t \in [0,T]$.

Proof. Write $A^n(x,y)$ for the matrix coefficient of the n-th power of A. Then $A(x,y) = 0$ if $d(x,y) > 1$ by the locality of A. It follows, that $A^n(x,y) = 0$ if $d(x,y) > n$. Now suppose that $d(x,y) = k$. It follows from (11) that

$$p_t(x,y) = \sum_{n=k}^{\infty} \frac{(-t)^n A^n(x,y)}{n!}. \tag{13}$$

Since the operator A is bounded with $\| A \| \leq 2aM$,

$$|A^n(x,y)| = |(A^n \delta_x, \delta_y)| \leq 2^n a^n M^n.$$

Therefore the series obtained by factoring out $1/k!$ from (13) is easily seen to be uniformly bounded for $t \leq T$. This proves the lemma. □

The lemma says that for t bounded, the heat kernel $p_t(x,y)$ decays very rapidly as the distance $d(x,y)$ goes to infinity. This is a familiar behavior of the heat kernel of a Riemannian manifold of bounded geometry. Thus we can substitute for u_0 in (12) functions of moderate growth so that the series defining $u(x,t)$ converges and produces a solution of (9). In particular, this yields existence of bounded solutions of (9) asserted in Theorem 4.1. More precisely, for bounded initial data $|u_0| \leq c$, we define the solution of (9) by (12) and group the terms as follows

$$\sum_y p_t(x,y)u_0(y) = \sum_{k=0}^{\infty} \left(\sum_{d(x,y)=k} p_t(x,y)u_0(y) \right).$$

By our assumption on the valence, the number of terms in the inner sum is at most M^k. Thus, for a bounded t, the absolute value of the k-th term together with its t derivative is dominated by $(C/k!)M^k c$ because of Lemma 4.2. This shows that the series converges very rapidly and can be differentiated term by term proving existence in Theorem 4.1. For future reference we make the following

Remark 4.1. In the argument above we could have allowed u_0 to grow at a certain rate. For example, the argument goes through if $|u_0(y)| \leq c_1 e^{c_2 d(x,y)}$.

Our next result gives a relation between a ground state and the heat semigroup. It illustrates a technique used frequently in the study of diffusions as, for example, in Sullivan [13], Pinsky [10] and Dodziuk-Mathai [8]. Let $\mathcal{H} = \{u : V \longrightarrow \mathbb{C} \mid u \cdot \phi \in \ell^2(V)\}$. It is a Hilbert space with the inner product $< u, v >= \sum_{x \in V} u(x)\overline{v}(x)\phi^2(x)$. We use the ground state ϕ to transplant the semigroup P_t to $\mathcal{H}$. Namely, define $\tilde{P}_t$ as a bounded self-adjoint operator on $\mathcal{H}$ by

$$\tilde{P}_t = e^{\lambda_0 t}[\phi^{-1}]P_t[\phi] = e^{\lambda_0 t}[\phi^{-1}]e^{-tA}[\phi], \tag{14}$$

where $\lambda_0 = \lambda_0(A)$ and $[f]$ denotes the operator of multiplication by a function f. Observe that for $u_0 \in \mathcal{H}$

$$\tilde{P}_t u_0(x) = e^{\lambda_0 t} \sum_y \frac{1}{\phi(x)} p_t(x,y)\phi(y)u_0(y) \tag{15}$$

by (12). Clearly, $\tilde{P}_t$, $t \geq 0$ is a semigroup with infinitesimal generator

$$-\tilde{A} = -[\phi^{-1}](A - \lambda_0)[\phi].$$

The following calculation gives a local formula for $\tilde{A}$.

$$\begin{aligned}
\tilde{A}u(x) &= \phi^{-1}(x)A(\phi u)(x) - \lambda_0 u(x) \\
&= \phi^{-1}(x)\sum_{y\sim x} a_{x,y}\left(\phi(x)u(x) - \phi(y)u(x)\right) \\
&\quad +\phi^{-1}(x)\sum_{y\sim x} a_{x,y}\left(\phi(y)u(x) - \phi(y)u(y)\right) - \lambda_0 u(x) \\
&= \lambda_0 u(x) + \sum_{y\sim x} a_{x,y}\frac{\phi(y)}{\phi(x)}\left(u(x) - u(y)\right) - \lambda_0 u(x) \\
&= \sum_{y\sim x} a_{x,y}\frac{\phi(y)}{\phi(x)}\left(u(x) - u(y)\right).
\end{aligned} \tag{16}$$

Note that $\tilde{A}$ is different than the local operators considered until now as its coefficients are not symmetric in x, y. We will consider however the initial value problem analogous to (9) for the operator $\tilde{A}$.

Theorem 4.2. *Under the assumptions stated in the beginning of this section, the initial value problem*

$$\tilde{A}u + \frac{\partial u}{\partial t} = 0$$
$$u(x, 0) = u_0(x)$$

has a unique bounded solution $u(x,t)$ for every bounded function u_0.

Proof. The proof is completely analogous to the proof of Theorem 4.1. The uniqueness used only the maximum principle in Lemma 4.1 which in turn depended only on positivity and *not on symmetry* of the coefficients of the operator A. The proof thus applies equally well to the operator $\tilde{A}$ whose coefficients are positive by (16) since the ground state ϕ is positive. Similarly, one proves existence for bounded initial data using the formula (15) and applying Remark 4.1 together with the estimate of Corollary 2.1. □

The following corollary is of independent interest. Its special case was used to derive certain estimates of the heat kernel for the combinatorial Laplacian in [8].

Corollary 4.1. *Under the assumption of this section, the ground state ϕ of A is complete i.e. satisfies*

$$P_t \phi = e^{-\lambda_0 t} \phi.$$

Proof. By the theorem above, $\tilde{P}_t$ applied to the function $u_0 \equiv 1$ is a solution of the equation $\tilde{A}u + \frac{\partial u}{\partial t} = 0$ with the initial data u_0. The function identically equal to one is also a solution. By uniqueness, the two solutions are equal i.e.

$$e^{\lambda_0 t} \sum_y \frac{1}{\phi(x)} p_t(x,y) \phi(y) = 1$$

for all $t > 0, x \in V$. This proves the corollary. □

Acknowledgement. I am very grateful to Radek Wojciechowski for a careful reading of the paper, correcting errors and making suggestions that lead to improvement of exposition.

References

1. R. BROOKS. The fundamental group and the spectrum of the Laplacian. *Comment. Math. Helv.* **56** #4 (1981), 581–598.
2. P. BUSER. A note on the isoperimetric constant. *Ann. Sci. École Norm. Sup.* **15** (1982), 213–230.
3. J. CHEEGER. A lower bound for the smallest eigenvalue of the Laplacian. In: *Problems in Analysis (Papers dedicated to Salomon Bochner, 1969)*, R. C. GUNNING, ed., 195–199. Princeton: Princeton University Press, 1970.
4. Y. COLIN DE VERDIÈRE. *Spectres de graphes*, Cours Spécialisés [Specialized Courses] **4**. Paris: Société Mathématique de France, 1998.
5. J. DODZIUK. Maximum principle for parabolic inequalities and the heat flow on open manifolds. *Indiana Univ. Math. J.* **32** #5 (1983), 703–716.
6. J. DODZIUK. Difference equations, isoperimetric inequality and transience of certain random walks. *Trans. Amer. Math. Soc.* **284** (1984), 787–794.
7. J. DODZIUK & W. KENDALL. Combinatorial Laplacian and isoperimetric inequality. In: *From local times to global geometry, control and physics*, K. D. ELLWORTHY, ed., Pitman Res. Notes Math. Ser. **150**, 68–74. Harlow: Longman Scientific & Technical, 1986.
8. J. DODZIUK & V. MATHAI. Kato's inequality and asymptotic spectral properties for discrete magnetic laplacians. In: *Geometry of the Laplace Operator*, J. JORGENSON & L. WALLING, eds., Contemporary Mathematics. Ubiquitous Heat Kernels, to appear.
9. A. LUBOTZKY. *Discrete groups, expanding graphs and invariant measures*, Progress in Mathematics **125**. Basel: Birkhäuser Verlag, 1994. With an appendix by Jonathan D. Rogawski.
10. R. G. PINSKY. *Positive harmonic functions and diffusion*, Cambridge Studies in Advanced Mathematics **45**. Cambridge: Cambridge University Press, 1995.
11. M. H. PROTTER & H. F. WEINBERGER. *Maximum principles in differential equations*. New York: Springer-Verlag, 1984. Corrected reprint of the 1967 original.
12. M. REED & B. SIMON. *Methods of modern mathematical physics. II. Fourier analysis, self-adjointness*. New York: Academic Press [Harcourt Brace Jovanovich Publishers], 1975.
13. D. SULLIVAN. Related aspects of positivity: λ-potential theory on manifolds, lowest eigenstates, Hausdorff geometry, renormalized Markoff processes.... In: *Aspects of mathematics and its applications*, J. A. BARROSO, ed., North-Holland Math. Library, **34**, 747–779. Amsterdam: North-Holland, 1986.

Received by the editors August 8, 2005; Revised December 22, 2005

Analysis, Geometry and Topology
of Elliptic Operators, pp. 369–382

A NEW KIND OF INDEX THEOREM

RONALD G. DOUGLAS
Department of Mathematics
Texas A&M University
College Station, TX 77843, USA
rdouglas@math.tamu.edu

Dedicated to Krzysztof P. Wojciechowski on his 50th birthday

Index theory has had profound impact on many branches of mathematics. In this note we discuss the context for a new kind of index theorem. We begin, however, with some operator-theoretic results. Berger and Shaw [11] established that finitely cyclic hyponormal operators have trace-class self-commutators. Berger [9] and Voiculescu [31] extended this result to operators whose self-commutators can be expressed as the sum of a positive and a trace-class operator. In this note we show this result can't be extended to operators whose self-commutators can be expressed as the sum of a positive and a $\mathcal{S}_p$-class operator. Then we discuss a conjecture of Arveson [4] on homogeneous submodules of the m-shift Hilbert space H^2_m and propose some refinements of it.
Further, we show how a positive solution would enable one to define K-homology elements for subvarieties in a strongly pseudo-convex domain with smooth boundary using submodules of the corresponding Bergman module. Finally, we discuss how the Chern character of these classes in cyclic cohomology could be defined and indicate what we believe the index to be.

2000 *Mathematics Subject Classification.* Primary 47B99; Secondary 47L80, 47L15, 58J20

1. Introduction

The complex Hilbert space $\mathcal{H}$ is said to be a Hilbert module over the algebra A if $\mathcal{H}$ is a unital module over A. This is equivalent to a representation of A on $\mathcal{H}$. In the last two decades, there has been considerable interest in the study of Hilbert modules for various classes of algebras, in part as an approach to multivariate operator theory. For Douglas and Paulsen [20],

Part of this research was done while the author attended the joint India-AMS Meeting held in Bangalore in December, 2004. His visit was supported by a grant from the DST-NSF Science and Technology Cooperation Programme.

A was assumed to be a function algebra and module multiplication to be bounded. Other authors (e.g. Muhly–Solel [29], Eschmeier–Putinan [23]) have considered other kinds of algebras. More recently, there has been an interest in modules for which A is the algebra of polynomials $\mathbb{C}[\boldsymbol{z}]$ with various assumptions such as (1) coordinate functions act contractively or (2) they act as a spherical contraction. (The bold $\boldsymbol{z}$ will be used to denote points in $\mathbb{C}^m$.) Arveson [2] considered the latter case and identified the m-shift space H^2_m as having particularly nice properties. In the course of his studies, he [4] raised a question about the almost reductivity of the submodules of $H^2_m \otimes \mathbb{C}^k$, for $1 \leq k < \infty$, generated by homogeneous polynomials; that is, modules for which the coordinate multipliers and their adjoints have compact or p-summable cross-commutators. Later Arveson [5] established this result for submodules generated by monomials. (Also, see Arveson [6] and Guo–Wang [27] for some subsequent work on this topic.) Douglas [16] extended this result to a class of commuting weighted shifts which includes the m-shift and Bergman and Hardy modules for the ball.

In this note we discuss Arveson's conjecture in full generality and more. We suggest, in particular, that submodules of Bergman modules over strongly pseudo-convex domains of $\mathbb{C}^m$ with smooth boundary determined by subvarieties are p-reductive for $p > m$. Moreover, in such a case they determine odd K-homology classes (cf. Brown–Douglas–Fillmore [13]) for the space equal to the intersection of the subvariety with the boundary of the domain. Further, one could define a Chern character using the cyclic cohomology of Connes [14]. We conjecture that this class is the one determined by the almost complex structure on the intersection of the subvariety and the boundary. Such a result would be a new kind of index theorem.

We begin by considering some results of Berger [9], which extended his earlier theorem with Shaw [11] in operator theory. The latter result established that self-commutators of hyponormal operators are trace-class in the presence of finite cyclicity.

My interest in the question of almost reductivity was spurred by Arveson and resulted from an ongoing dialogue with him on his work on this topic. This rather unusual note was the subject of conference talks given in 2005 at IUPUI, Penn State and Roskilde University and is presented here to bring to the attention of other researchers, what we believe to be a most promising and interesting topic.

2. Results in operator theory

Recall that the bounded linear operator T on the Hilbert space $\mathcal{H}$ is said to be *hyponormal* if the self-commutator $[T^*, T] = T^*T - TT^*$ is a positive operator. Berger and Shaw [11] demonstrated the surprising result that a finitely cyclic hyponormal operator has a trace-class self-commutator. There is also an estimate of the trace involving the degree of cyclicity and the area of the spectrum of T but that inequality will not concern us at this time.

Recall that an operator X on $\mathcal{H}$ is said to belong to $\mathcal{S}_p$, the Schatten–von Neumann class, for $1 \le p < \infty$, if X is compact and the eigenvalues of $(X^*X)^{1/2}$ belong to ℓ^p (cf. Gohberg–Krein [25]). Now $\mathcal{S}_1$ consists of the trace-class operators. One knows that $\mathcal{S}_p$ is a Banach space with dual space $\mathcal{S}_q$ with $\frac{1}{p} + \frac{1}{q} = 1$, for $1 \le p < \infty$, if we identify $\mathcal{S}_\infty$ with $\mathcal{L}(\mathcal{H})$, the space of all bounded operators on $\mathcal{H}$. Further, $\mathcal{S}_p$ is a two-sided ideal in $\mathcal{S}_\infty = \mathcal{L}(\mathcal{H})$.

In subsequent years, Berger [9] extended the Berger–Shaw Theorem to cover a larger class of operators which is the class we shall consider. (There was also related work by Voiculescu [31] and Carey–Pincus on this class.) For $1 \le p < \infty$, we'll say that an operator T on $\mathcal{H}$ belongs to $\mathcal{A}_p$ if $[T^*, T] = P + C$, where $P \ge 0$ and C is in $\mathcal{S}_p$. Observe that all hyponormal operators are in $\mathcal{A}_p$ as are all operators T for which $[T^*, T]$ is in $\mathcal{S}_p$. Observe also for $p = 1$, that there is a well-defined trace on the self-commutators of operators in $\mathcal{A}_1$ taking values in $(-\infty, \infty]$ and that for T in $\mathcal{A}_1$ we have $[T^*, T]$ trace-class iff this trace is finite.

Finally, we will let $\mathcal{A}_0$ denote the operators T for which $[T^*, T] = P + C$ with P positive and C compact.

The class of hyponormal operators is closed under restriction to invariant subspaces. That is, if T is hyponormal and $\mathcal{V}$ is an invariant subspace for T, then $T|_\mathcal{V}$ is hyponormal. The following lemma shows the same is true for class $\mathcal{A}_p$.

Lemma 2.1. (Berger [9] and Voiculescu [31]) *If T belongs to $\mathcal{A}_p$ for $1 \le p \le \infty$ or $p = 0$, and $\mathcal{V}$ is an invariant subspace for T, then $T|_\mathcal{V}$ is in $\mathcal{A}_p$.*

Proof. If one writes $T = \left(\begin{smallmatrix} A & B \\ 0 & C \end{smallmatrix}\right)$ relative to the decomposition $\mathcal{H} = \mathcal{V} \oplus \mathcal{V}^\perp$, and Q is the orthogonal projection on to $\mathcal{V}$, then

$$[(T|_\mathcal{V})^*, (T|_\mathcal{V})] = Q[T^*, T]Q + QTQ^\perp T^*Q,$$

where $Q^\perp = I - Q$. Since $[T^*, T] = P + C$, with $P \ge 0$ and C in $\mathcal{S}_p$, we have

$$[(T|_{\mathcal{V}})^*, (T|_{\mathcal{V}})] = (QPQ + QTQ^{\perp}T^*Q) + QCQ$$

and the first sum on the right-hand side is positive while QCQ is in $\mathcal{S}_p$. □

The following result is a special case of a result due to Berger [9]. We reproduce the proof since it is short and we believe deserves to be better known.

Proposition 2.1. *If T is in $\mathcal{A}_1$ and $\mathcal{V}$ and $\{\mathcal{V}_n\}$ are invariant subspaces for T such that each $\mathcal{V}_n$ is finite dimensional, $\mathcal{V}_n \subset \mathcal{V}_{n+1}$ for all n and $\bigcup_{n=1}^{\infty} \mathcal{V}_n$ is dense in $\mathcal{V}$, then $[(T|_{\mathcal{V}})^*, (T|_{\mathcal{V}})]$ is trace-class.*

Proof. Let Q and $\{Q_n\}$ be the orthogonal projections onto $\mathcal{V}$ and $\{\mathcal{V}_n\}$, respectively. Then $\{Q_n\}$ converges in the strong operator topology to Q. Adopting the same notation as in the preceding proof for the representation of the self-commutators, we have $[(T|_{\mathcal{V}_n})^*, (T|_{\mathcal{V}_n})] = P_n + C_n$ for each n and $[(T|_{\mathcal{V}})^*, (T|_{\mathcal{V}})] = P + C$. Moreover, the sequence $\{P_n\}$ converges strongly to P while the sequence $\{C_n\}$ converges to C in the norm on $\mathcal{S}_1$.

Since $T|_{\mathcal{V}_n}$ is finite rank, we have $Tr[(T|_{\mathcal{V}_n})^*, (T|_{\mathcal{V}_n})] = 0$ and hence $0 \leq TrP_n \leq \|C_n\|_1$ for all n. Further, we have $\|C_n\|_1 \to \|C\|_1$ which implies that $\overline{\lim}\ TrP_n \leq M < \infty$ and hence $TrP \leq M$ using a variant of Fatou's Lemma. Therefore, P is trace-class from which the result follows. □

Actually Berger proved a stronger result. Suppose we have another invariant subspace $\mathcal{V}_0$ contained in all the $\mathcal{V}_n$ so that the dimension of $\mathcal{V}_n/\mathcal{V}_0$ is finite for all n and $T|_{\mathcal{V}_0}$ is in $\mathcal{A}_1$. Then the preceding argument yields the same conclusion, namely, that the self-commutator of $T|_{\mathcal{V}}$ is in $\mathcal{S}_1$.

We now reframe Proposition 2.1 in a setting which makes the hypotheses more transparent.

Theorem 2.1. *Let T be an operator in $\mathcal{A}_1$ and $\mathcal{V}$ be an invariant subspace for T spanned by generalized eigenvectors for $T|_{\mathcal{V}}$. Then $[(T|_{\mathcal{V}})^*, (T|_{\mathcal{V}})]$ is in $\mathcal{S}_1$.*

Proof. Let $\{f_k\}$ be a sequence of generalized eigenvectors for $T|_{\mathcal{V}}$ which spans $\mathcal{V}$. Further, let $\mathcal{V}_n$ be the invariant subspace for T generated by $\{f_k\}_{k=1}^n$. Then the $\{\mathcal{V}_n\}$ are nested, finite dimensional and their union is dense in $\mathcal{V}$. The result now follows from Proposition 2.1. □

We would like to obtain the analogous result for operators in A_p. In an earlier version of this note we thought we had proved it. Unfortunately, the following example shows it to be false.

Example 2.1. Consider the weighted unilateral shift S_n defined on ℓ^2 with the standard basis $\{e_k\}_{k=1}^{\infty}$ so that

$$S_n e_k = \begin{cases} \sqrt{\frac{k}{n}}\, e_{k+1}, & 1 \le k \le n. \\ e_{k+1}, & n < k. \end{cases}$$

An easy calculation shows that $\|[S_n^*, S_n]\|_p^p = n^{1-p}$ for $1 \le p < \infty$. If $\mathcal{V}_n$ is the subspace of ℓ^2 spanned by $\{e_k\}_{k=n}^{\infty}$, then $S_n\mathcal{V}_n \subset \mathcal{V}_n$ and $\|[(S_n|_{\mathcal{V}_n})^*, (S_n|_{\mathcal{V}_n})]\|_p = 1$ for all n. Moreover, if we set $S = \bigoplus_{n=1}^{\infty} S_n$ acting on $\bigoplus_{n=1}^{\infty} \ell^2$ and $\mathcal{V} = \bigoplus_{n=1}^{\infty} \mathcal{V}_n$, then $\|[S^*, S]\|_p < \infty$ but $\|[(S|_{\mathcal{V}})^*, (S|_{\mathcal{V}})]\|_p = \infty$ for all p, $1 < p < \infty$.

We conclude that S^* is in A_p, $\mathcal{V}_n^{\perp}$ is spanned by generalized eigenvectors for S^* (hence one can construct the desired sequence of finite dimensional approximates for it) but $S^*|_{\mathcal{V}^{\perp}}$ is not p almost reductive. Observe that $\mathcal{V}^{\perp}$ and $\mathcal{V}$ are not finitely cyclic for S^* and S, respectively.

As we indicated above, our original goal was to extend Proposition 2.1 to $\mathcal{A}_p$ and thereby extend Theorem 2.1 to this class. Unfortunately, Example 2.1 shows this is impossible.

Since a finite dimensional invariant subspace for T is spanned by the generalized eigenvectors for it, the hypotheses of the foregoing theorem is the only way to fulfill the condition in Proposition 2.1. Berger introduced the notion of an invariant subspace $\mathcal{V}_0$ being effectually full in $\mathcal{V}$ by requiring the denseness in $\mathcal{V}$ of the set of vectors in $\mathcal{V}$ that some nonzero polynomial in T takes into $\mathcal{V}_0$. This hypothesis enabled him to satisfy the weaker hypotheses we have mentioned earlier.

A question which presents itself at this point is whether Theorem 2.1 might hold for all invariant subspaces of T without any restriction. The following example shows that this is not the case.

Example 2.2. Consider the Bergman space $B^2(\mathbb{D})$ for the unit disk $\mathbb{D}$ which can be defined as the closure of the analytic polynomials in $L^2(\mathbb{D})$

We want to thank W.B. Arveson for pointing out the mistake in the earlier version of this note.

relative to planar Lebesgue measure. Further, consider the operator $T = M_z^*$, where M_z is multiplication by z on $B^2(\mathbb{D})$. Note that T lies in $\mathcal{A}_1$.

Suppose we have invariant subspaces $\mathcal{M}$ and $\mathcal{N}$ for T such that $0 \subset \mathcal{M} \subset \mathcal{N} \subset B^2(\mathbb{D})$. If both $T|_{\mathcal{M}}$ and $T|_{\mathcal{N}}$ are in A_1, then the same is true for the compression of T to the semi-invariant subspace $\mathcal{N}/\mathcal{M}$. (This is essentially Theorem 1 in Douglas [16].) However, one knows (cf. Apostol–Bercovici–Foias–Pearcy [1]) that there is a semi-invariant subspace for the Bergman shift so that the restriction of the Bergman shift to it realizes each proper contraction operator on a separable Hilbert space. Hence we can obtain a semi-invariant subspace for which the self-commutator of the restriction is not even compact.

Here would be a good place to record a result which is a refinement of Theorem 1 of Douglas [16].

Theorem 2.2. *If $\mathcal{M}_1$ and $\mathcal{M}_2$ are essentially reductive modules for the algebra A and $X\colon\ \mathcal{M}_1 \to \mathcal{M}_2$ is a module map having closed range, then both ker X and ran X are essentially reductive.*

Proof. If we write $\mathcal{M}_1 = (\ker X)^\perp \oplus \ker X$ and $\mathcal{M}_2 = \operatorname{ran} X \oplus (\operatorname{ran} X)^\perp$, then X has the form $\left(\begin{smallmatrix} X_0 & 0 \\ 0 & 0 \end{smallmatrix}\right)$, where X_0 is an invertible operator from $(\ker X)^\perp$ onto ran X.

For T any operator between Hilbert spaces, let $\widehat{T}$ denote its image in the corresponding Calkin algebra, that is, modulo the compact operators. Since both $\mathcal{M}_1$ and $\mathcal{M}_2$ are essentially reductive, for φ in A the elements $\hat{A}_\varphi$ and $\widehat{B}_\varphi$ are normal, where A_φ and B_φ denote the operators defined by module multiplication by φ.

Let $A_\varphi = \begin{pmatrix} A_\varphi^{11} & 0 \\ A_\varphi^{21} & A_\varphi^{22} \end{pmatrix}$ and $B_\varphi = \begin{pmatrix} B_\varphi^{11} & B_\varphi^{12} \\ 0 & B_\varphi^{22} \end{pmatrix}$ be the representations of A_φ and B_φ relative to the decompositions of $\mathcal{M}_1$ and $\mathcal{M}_2$, respectively. If we consider the images of all these operators in their respective Calkin algebras, we can apply the Fuglede–Putnam Theorem to conclude that the relationship $\widehat{X}\hat{A}_\varphi = \widehat{B}_\varphi\widehat{X}$ implies that $\hat{A}_\varphi\widehat{X}^* = \widehat{X}^*\widehat{B}_\varphi$ and therefore, we have $\widehat{X}^*\widehat{X}\hat{A}_\varphi = \hat{A}_\varphi\widehat{X}^*\widehat{X}$. This equation in turn implies that $\hat{A}_\varphi^{21}\widehat{X}_0^*\widehat{X}_0 = 0$ and thus $\hat{A}_\varphi^{21} = 0$ since $\widehat{X}_0^*\widehat{X}_0$ is invertible. However, $\hat{A}_\varphi^{21} = 0$ for all φ in A means that $\ker X$ is essentially reductive since $\mathcal{M}_1$ is essentially reductive. Working with $\mathcal{M}_2$ we conclude that $\widehat{X}\widehat{X}^*\widehat{B}_\varphi = \widehat{B}_\varphi\widehat{X}\widehat{X}^*$ and hence $\widehat{X}_0\widehat{X}_0^*\widehat{B}_\varphi^{12} = 0$. Again this implies that ran X is essentially reductive since $\mathcal{M}_2$ is essentially reductive which completes the proof. □

Unfortunately, since no appropriate analogue of the Fuglede–Putnam Theorem is known for the p-summable case, such a proof won't allow us to conclude p-reductivity of the kernel and range if $\mathcal{M}_1$ and $\mathcal{M}_2$ are. However, Arveson [6] gives such a result for a specific class of Hilbert modules and module maps.

3. Almost reductive Hilbert modules

A Hilbert module $\mathcal{M}$ over the algebra $A(\Omega)$ for Ω a bounded domain in $\mathbb{C}^m$ is said to be essentially reductive [20] if all cross-commutators $[M_\varphi^*, M_\psi]$ are compact for φ and ψ in $A(\Omega)$. If these operators actually lie in $\mathcal{L}_p$ for φ and ψ coordinate functions, then $\mathcal{M}$ is said to be p-reductive.

We now show that Theorem 2.1 enables one to settle a question about cross-commutators for some submodules in the multi-variable setting so long as they have trace-class cross-commutators. Since the conjecture of Arveson [4] motivated this study, let us begin by considering it in some detail.

Recall that H_m^2, the m-shift Hilbert space for $1 \le m < \infty$, is defined using the symmetric Fock space and is a module over $\mathbb{C}[\boldsymbol{z}]$. Moreover, Arveson showed that multiplication by each coordinate function Z_i acts contractively on H_m^2 and all cross-commutators $[Z_i^*, Z_j]$ lie in $\mathcal{S}_p$ for $p > m$ and $1 \le i, j \le m$. Arveson conjectured that the restriction operators $Y_i = Z_i|_\mathcal{S}$ and their adjoints also have $\mathcal{S}_p$ cross-commutators for any submodule $\mathcal{S}$ of $H_m^2 \otimes \mathbb{C}^k$ for $1 \le k < \infty$ generated by homogeneous polynomials. Moreover, he established the result for $\mathcal{S}$ generated by monomials. He [6] has also developed a theory of "standard Hilbert modules" in an effort to establish his conjecture. Another proof of the result for monomial submodules was given in Douglas [16] and it also covered certain commuting weighted shifts. Also, Arveson showed that the general case for homogeneous submodules of H_m^2 for $m = 2$ followed from a result of Guo [26]. Finally, a generalization to the case of certain pairs of commuting weighted shifts was recently obtained in Guo and Wang [27].

The simple matrix calculation used in Douglas [16] and the proof of Theorem 2.2 shows that if T_1 and T_2 are two operators on a Hilbert space $\mathcal{H}$ with $\mathcal{G}$ an invariant subspace for both such that $[T_1, T_2^*]$ lies in $\mathcal{S}_p$ for $1 \le p < \infty$ or $p = 0$, then the compressions $S_i = T_i|_\mathcal{G}$ have $[S_1^*, S_2]$ in $\mathcal{S}_p$ iff $[R_1^*, R_2]$ is in $\mathcal{S}_p$ for $R_i = T_i^*|_{\mathcal{G}^\perp}$. Thus we can focus on either $\mathcal{G}$ or $\mathcal{G}^\perp$.

We consider the case of commuting weighted shifts using the notation of Douglas [16]. We have a weight set Λ for the index set A_m, $1 \le m <$

∞, with the Hilbert space $\mathcal{M}_\Lambda$ and the weighted shifts defined by the coordinate functions Z_i, $1 \leq i \leq m$. The weight set Λ satisfies $(*)$ if the shifts are contractive, $(**)$ if all cross-commutators of the coordinate multipliers and their adjoints are compact, and $(**)_p$ if the latter operators lie in $\mathcal{S}_p$. Actually, in the following result one can replace $(*)$ by assuming only $(*)'$ that the Z_i are only bounded.

Theorem 3.1. *If Λ is a weight set satisfying $(*)'$ and $(**)_1$, $\mathcal{S}$ is a submodule of $\mathcal{M}_\Lambda \otimes \mathbb{C}_k$, $1 \leq k < \infty$, so that $\mathcal{S}^\perp$ is generated by polynomials, and $Y_i = Z_i|_\mathcal{S}$, then the cross-commutators $[Y_i^*, Y_j]$ are in $\mathcal{S}_1$ for $1 \leq i, j \leq m$.*

Proof. If we set $T = Z_i^*$ for some fixed i, then $\mathcal{S}^\perp$ is invariant for T. Moreover, $\mathcal{S}^\perp$ is spanned by polynomials. Therefore, T and $\mathcal{S}^\perp$ satisfy the hypotheses of Theorem 2.1 which implies that $[Y_i, Y_i^*]$ lies in $\mathcal{S}_1$ for all $1 \leq i \leq m$. Here we are using the fact that the self-commutator of the restriction of Z_i to $\mathcal{S}$ lies in $\mathcal{S}_1$ iff the same is true for the restriction of Z_i^* to $\mathcal{S}^\perp$.

Now if we take $T = Z_j^* + Z_k^*$ for $1 \leq j \neq k \leq m$, then T and $\mathcal{S}^\perp$ again satisfy the hypotheses of Theorem 2.1. Therefore, we have $[Y_j + Y_k, Y_j^* + Y_k^*]$ lies in $\mathcal{S}_1$. Since $[Y_j, Y_j^*]$ and $[Y_k, Y_k^*]$ lie in $\mathcal{S}_1$, we conclude that the real part of $[Y_j, Y_k^*]$ is in $\mathcal{S}_1$. Repeating the argument for $T = Z_j^* + iZ_k^*$, we see that the imaginary part of $[Y_j, Y_k^*]$ is in $\mathcal{S}_1$ which completes the proof. □

Let $\mathcal{P}_n$ denote the subspace of $\mathbb{C}[z]$ consisting of homogeneous polynomials of degree n. If $\mathcal{S}$ is generated by homogeneous polynomials, then $S = \oplus(\mathcal{S} \cap \mathcal{P}_n)$. This in turn implies that $\mathcal{S}^\perp = \oplus(\mathcal{S}^\perp \cap \mathcal{P}_n)$ and hence $\mathcal{S}^\perp$ is generated by polynomials. Thus Theorem 2.1 applies to homogeneous submodules. Instead of assuming that $\mathcal{S}^\perp$ is generated by polynomials, which are joint generalized eigenvectors for the adjoint of coordinate multipliers, we could assume more generally that $\mathcal{S}^\perp$ is spanned by such vectors.

Observe that we can't consider $\mathcal{M}_\Lambda \otimes \ell^2$, where ℓ^2 is the infinite-dimensional Hilbert space since the cross-commutators on it would no longer be in $\mathcal{S}_1$. However, if we consider a finite direct sum of block weighted shifts satisfying the analogues of $(*)'$, $(**)_1$, then the result does carry over and the blocks could be infinite, so long as the cross-commutators are still in $\mathcal{S}_1$.

While most natural examples of multi-variate Hilbert modules are not 1-reductive, one can obtain a family of nontrivial examples in the context of commuting weighted shifts.

Example 3.1. For $m > 1$, if the weight set is taken to be: $\lambda_{\boldsymbol{\alpha}} = \{(1 + \alpha_1 + \alpha_2 + \cdots + \alpha_m)!\}^{-\delta}$, then $\mathcal{M}_\Lambda$ is 1-reductive if $\delta > \frac{m-1}{2}$ and the Z_i are in $\mathcal{S}_2$ if $\delta > \frac{m}{2}$. Thus $\mathcal{M}_\Lambda$ is a nontrivial example of a 1-reductive Hilbert module for δ satisfying $\frac{m}{2} \geq \delta > \frac{m-1}{2}$ and Theorem 3.1 applies.

As we have indicated, originally we had hoped that Theorem 2.1 would extend to $\mathcal{S}_p$, $p > 1$, but as Example 2.1 indicates, this is not the case. Another approach would be to represent either the submodule or the corresponding quotient module as the kernel or cokernel of a closed module map to which Theorem 2.2 applies. The difficulty here is that the module map must have closed range and we know few conditions that guarantee that.

Since most natural examples of multivariate Hilbert modules are p-reductive only for $p > 1$, this approach reveals little about the validity of Arveson's conjecture in general either for H^2_m or other natural examples. Even though that is the case, let us describe what we believe is a natural setting for the conjecture.

Let Ω be a bounded, strongly pseudo-convex domain in $\mathbb{C}^m$ with smooth boundary and $B^2(\Omega)$ be the Bergman space, that is, the subset of functions f in $L^2(\Omega)$ relative to volume measure for which $\bar{\partial} f = 0$ taken in the sense of distributions. One knows (cf. Taylor [30]) that the module action on $B^2(\Omega)$ by functions holomorphic on a neighborhood of the closure of Ω is p-reductive for $p > m$. That is, cross-commutators of these multiplication operators and their adjoints lie in $\mathcal{S}_p$. For $\mathcal{Z}$ a variety of Ω, let $B^2_{\mathcal{Z}}(\Omega)$ be the functions in $B^2(\Omega)$ that vanish on $\mathcal{Z}$ and let $\mathcal{Q}_{\mathcal{Z}}$ be the quotient module $B^2(\Omega)/B^2_{\mathcal{Z}}(\Omega)$ (cf. [18]). One can show that $\mathcal{Q}_{\mathcal{Z}}$ is a contractive Hilbert module over $A(\Omega)$ with support in the closure of $\mathcal{Z}$. Moreover, since $B^2(\Omega)$ is a kernel Hilbert space and evaluation at $\boldsymbol{z}$ in Ω is continuous, there is a vector $k_{\boldsymbol{z}}$ in $B^2(\Omega)$ for which $f(\boldsymbol{z}) = \langle f, k_{\boldsymbol{z}} \rangle_{B^2(\Omega)}$ for f in $B^2(\Omega)$. The vectors $\{k_{\boldsymbol{z}}\}$ are joint eigenvectors for the adjoint of the operators defined by the module action. Moreover, one has that $\mathcal{Q}_{\mathcal{Z}}$ is the closed span of $\{k_{\boldsymbol{z}} \mid \boldsymbol{z} \in \mathcal{Z}\}$. Therefore, this example satisfies the same kind of hypotheses as in Theorem 3.1.

More generally, one can see that one could consider any submodule of $B^2(\Omega)$ defined as the orthogonal complement of a collection of eigenvectors $\{k_{\boldsymbol{z}}\}$ and their partial derivatives, which are also generalized eigenvectors for the adjoint of module action. These submodules include the closures of ideals in the algebra of functions holomorphic on some neighborhood of the closure of Ω. In particular, one can consider not just the functions that vanish on a subvariety but those that vanish to higher order. Moreover, using the result in Douglas [16] we see that if these submodules are p-reductive for $p > m$, then the quotient module obtained from them are also p-reductive for $p > m$.

Although the evidence for such a result is perhaps scant we are optimistic enough to formulate:

Conjecture 3.1. *If $\mathcal{S}$ is a submodule of $B^2(\Omega)$ such that $\mathcal{S}^\perp$ is spanned by joint generalized eigenvectors for the adjoint of the operators defined by the module action, then both $\mathcal{S}$ and $\mathcal{S}^\perp$ are p-reductive for $p > m$.*

This result, even in the multiplicity one case, would be of considerable interest. For a submodule obtained as the closure of a principal ideal I in $\mathbb{C}[\boldsymbol{z}]$, the result is equivalent to the weighted Bergman space defined for the measure $|q|^2 d$ Vol on Ω being p-reductive for $p > m$, where $q(\boldsymbol{z})$ is a generator for I. However, one might expect, if Conjecture 3.1 holds, for the generalization to finite multiplicity to also be valid.

Conjecture 3.2. *The same conclusion as in Conjecture 3.1 for submodules of $B^2(\Omega) \otimes \mathbb{C}^k$.*

There is an even stronger result possible which would be very useful in our considerations of the following section. (See Douglas [17] and Douglas–Misra [19] for the necessary definitions.)

Conjecture 3.3. *If $\mathcal{M}$ is a finite rank, quasi-free, p-reductive Hilbert module over $A(\Omega)$ and $\mathcal{S}$ is a submodule for which $\mathcal{S}^\perp$ is spanned by generalized eigenvectors for the adjoint of the operators defined by the module action, then $\mathcal{S}$ and $\mathcal{S}^\perp$ are p-reductive.*

It is quite likely that some additional "regularity" hypotheses on $\mathcal{M}$ are necessary for the last conjecture to hold.

There is another way to frame the final conjecture using a notion introduced in Douglas and Misra [17]. Recall that a Hilbert module $\mathcal{M}$ is said to belong to class (PS) if it is spanned by the generalized eigenvectors for the adjoint of the operators defined by the module action.

Conjecture 3.3′. Let $\mathcal{H}$ be a finite rank quasi-free, p-reductive Hilbert module over the algebra $A(\Omega)$. If $\mathcal{M}$ is a submodule of $\mathcal{H}$ such that $\mathcal{H}/\mathcal{M}$ belongs to the class (PS), then $\mathcal{M}$ is p-reductive.

4. *K*-homology classes

Let $\mathcal{H}$ be a p-reductive Hilbert module over the algebra A and $\mathcal{J}(\mathcal{H})$ be the C^*-algebra generated by the operators defined by module multiplication on $\mathcal{H}$ and let $\mathcal{C}(\mathcal{H})$ be the commutator ideal in $\mathcal{J}(\mathcal{H})$. Then $\mathcal{C}(\mathcal{H})$ consists of compact operators and hence $(\mathcal{J}(\mathcal{H}) + \mathcal{K}(\mathcal{H}))/\mathcal{K}(\mathcal{H})$ is a commutative C^*-algebra. Therefore this quotient C^*-algebra is isometrically isomorphic to $C(X_{\mathcal{H}})$ for some compact Hausdorff space $X_{\mathcal{H}}$. In Davidson and Douglas [15], it is shown for A a commutative Banach algebra that $X_{\mathcal{H}}$ can be identified with a closed subset of the maximal ideal space M_A. Similarly, if $A = \mathbb{C}[\boldsymbol{z}]$ and the module action of the coordinate functions are all contractive operators, then one can identify $X_{\mathcal{H}}$ as a closed subset of the unit polydisk ${}^{cl}\mathbb{D}^m$.

In any case, since we have the short exact sequence $0 \to \mathcal{K}(\mathcal{H}) \to \mathcal{J}(\mathcal{H}) + \mathcal{K}(\mathcal{H}) \to C(X_{\mathcal{H}}) \to 0$, one always obtains an odd K-homology element, denoted $[\mathcal{H}]$, in $K_1(X_{\mathcal{H}})$. While we hope to investigate these classes more thoroughly after additional cases of the conjecture have been established, we want to draw attention here to a few natural questions and raise a few more conjectures. Our aim is to show why these are interesting questions. We focus on the case of Bergman spaces over strongly pseudo-convex domains with smooth boundary.

Theorem 4.1. *Let Ω be a bounded strongly pseudo-convex domain in $\mathbb{C}^m$ with smooth boundary, $B^2(\Omega)$ be the Bergman module, $\mathcal{Z}$ be a subvariety of Ω, $B^2_{\mathcal{Z}}(\Omega)$ be the submodule of functions in $B^2(\Omega)$ that vanish on $\mathcal{Z}$ and $\mathcal{Q}_{\mathcal{Z}}$ be the quotient module $B^2(\Omega)/B^2_{\mathcal{Z}}(\Omega)$. If $\mathcal{Q}_{\mathcal{Z}}$ is a p-reductive module for the algebra of functions holomorphic on some neighborhood of ${}^{cl}\Omega$, then $[\mathcal{Q}_{\mathcal{Z}}]$ is in $K_1(\mathcal{Z} \cap \partial\Omega)$.*

Proof. The only thing requiring proof is the fact that $X_{Q_{\mathcal{Z}}} \subseteq \mathcal{Z} \cap \partial\Omega$. This follows from the fact that $X_{B^2(\Omega)} = \partial\Omega$ and that $B^2_{\mathcal{Z}}(\Omega)$ is a Hilbert module over $A(\Omega)/A_{\mathcal{Z}}(\Omega)$. □

The question arises as to which element of $K_1(\mathcal{Z}\cap\partial\Omega)$ is obtained. One can show in some cases such as $\Omega = \mathbb{B}^m$ that it is the fundamental class, taking multiplicity into account, determined by the complex structure on Ω or the spin^c-structure on $\partial\Omega$ (or the negative of these classes) and I conjecture that this is true in general. One problem which arises is that $\partial\Omega \cap \mathcal{Z}$ need not be a manifold.

One can show by various means that the K_1-classes determined by $B^2(\mathbb{B}^m)$ and H^2_m are equal. In fact, the same seems to be true for any kernel Hilbert module over $\mathbb{B}^m$ that is essentially reductive. (An argument showing this fact would follow from Conjecture 3.3.) I suspect the same thing is true for the K_1-classes obtained for a subvariety $\mathcal{Z}$, that is, the K_1-class doesn't depend on the kernel Hilbert module over Ω with which one begins.

Finally, there is one other issue I would like to raise before concluding. We will again frame it in the context of submodules of Bergman modules. Although one can show that $B^2_{\mathcal{Z}}(\Omega)$ is p-reductive for $p > m$, it is not p-reductive for any smaller p. That is, it has the same degree of "smoothness" (cf. [9]) as does $B^2(\Omega)$. However, I don't believe that is the case for $\mathcal{Q}_{\mathcal{Z}}$. In particular, in Douglas [16], I showed that its smoothness depends on the dimension of $\mathcal{Z}$ or the degree of the Hilbert polynomial (cf. Douglas–Yan [22]) for $\mathcal{Q}_{\mathcal{Z}}$. I will formulate one final conjecture, that an analogous result holds in general. We state it only for the case of the unit ball.

Conjecture 4.1. *Let I be an ideal in $\mathbb{C}[\boldsymbol{z}]$ and $\mathcal{S}$ be the submodule obtained from its closure in $B^2(\mathbb{B}^m)$. Then the quotient module $B^2(\mathbb{B}^m)/\mathcal{S}$ is q-reductive for $q >$ dim$(\mathcal{Z} \cap \mathbb{B}^m)$, where $\mathcal{Z}$ is the zero variety of I.*

There is another line of investigation possible here if this conjecture holds. If $\mathcal{Q}_{\mathcal{Z}}$ is p-reductive for $q > \dim(\mathcal{Z} \cap \mathbb{B}^m)$, then it should be possible to define a cyclic cohomology class following Connes [14] which will be the Chern character of $[\mathcal{Q}_{\mathcal{Z}}]$. One interesting question is how this class varies when the subvariety $\mathcal{Z}$ changes. For example, suppose one considers $\mathcal{Z}_c = \{\boldsymbol{z} \in \Omega | p(\boldsymbol{z}) = c\}$ for c in $\mathbb{C}$. As one knows, for some c, $\mathcal{Z}_c \cap \partial\Omega$ will be a manifold while for others, it is not. Moreover, there is also the issue of $\partial\Omega \cap \mathcal{Z}_c$ being a manifold while $\mathcal{Z}_c$ has singularities in Ω.

I thank Paul Baum for discussions on how to define such a K-homology class which is related to our earlier work Baum and Douglas [7].
In Guo and Wang [27], the K-homology class obtained for homogeneous modules in $B^2(\mathbb{B}^2)$ is consistent with this conjecture.
Added in proof: R. Levy has pointed out to the author that his earlier work in [J. Operator Theory 21 (1989), 219–253] and [Acta Math. 158 (1987), 149–188] is relevant to this index question.
Guo and Wang [27] verify this conjecture in case $\dim(Z \cap \mathbb{B}^m) \leq 1$.

One fascinating example to consider would be the presentation of the exotic spheres found by Brieskorn [12]. Recall that he exhibited analytic polynomials for which an exotic sphere is obtained from the intersection of the zero variety of the polynomial in $\mathbb{C}^n$ with spheres of small diameter. Although the precise polynomials he used are not homogeneous, this example indicates that one is likely to obtain interesting varieties in our context.

I believe different techniques will be needed to establish such a conjecture. The result of Douglas and Voiculescu [21] provides a lower bound on p if $[\mathcal{Q}_{\mathcal{Z}}]$ is indeed a fundamental class for $\partial\Omega \cap \mathcal{Z}$.

References

1. C. Apostol, H. Bercovici, C. Foias, and C. M. Pearcy, Invariant subspaces, dilation theory, and the structure of the predual of a dual algebra. I., *J. Funct. Anal.* **63**, 369–404 (1985).
2. W. B. Arveson, Subalgebras of C^*-algebras III: Multivariate operator theory, *Acta Math.* **181**, 159–228 (1998).
3. W. B. Arveson, The curvature invariant of a Hilbert module over $\mathbb{C}[z_1, \ldots, z_d]$, *J. Reine Angew. Math.* **522**, 173–236 (2000).
4. W. B. Arveson, The Dirac operator of a commuting d-tuple, *J. Funct. Anal.* **189**, 53–79 (2002).
5. W. B. Arveson, p-Summable commutators in dimension d, *J. Operator Theory* (to appear).
6. W. B. Arveson, Quotients of standard Hilbert modules, preprint, 2005.
7. P. Baum and R. G. Douglas, K-homology and index theory, In: *Operator Algebras and Applications*, Part 1 (Kingston, Ont., 1980) pp. 117–173, *Proc. Sympos. Pure Math.*, 38, *Amer. Math. Soc.*, Providence, RI, 1982.
8. H. Bercovici, The algebra of multiplication operators on Bergman spaces, *Arch. Math. (Basel)* **48**, 165–174 (1987).
9. C. A. Berger, Intertwined operators and the Pincus principal function, *Integral Equations Operator Theory* **4**, 1–9 (1981).
10. C. A. Berger and M. G. Ben-Jacob, Trace class self-commutators, *Trans. Amer. Math. Soc.* **277**, 75–91 (1983).
11. C. A. Berger and B. I. Shaw, Self-commutators of multi-cyclic hyponormal operators are always trace class, *Bull. Amer. Math. Soc.* **79**, 1193–1199 (1973).
12. E. Brieskorn, Die Auflösung der rationalen Singularitäten holomorpher Abbildungen, *Math. Ann.* **178**, 255–270 1968.
13. L. G. Brown, R. G. Douglas and P. A. Fillmore, Extensions of C^*-algebras and K-homology, *Ann. of Math.* **(2)105**, 265–324 (1977).
14. A. Connes, *Noncommutative Geometry*, Academic Press, San Diego, 1994.
15. K. Davidson and R. G. Douglas, Generalized Berezin transform and commutator ideals, *Pacific J. Math.* (to appear).
16. R. G. Douglas, Essentially reductive Hilbert modules, *J. Operator Theory* (to appear).

17. R. G. Douglas and G. Misra, Quasi-free resolutions of Hilbert modules, *Integral Equations Operator Theory* **47**, 435–456 (2003).
18. R. G. Douglas and G. Misra, On quotient modules, in Operator Theory Advances and Applications, Vol. 127, 203–209, Birkhäuser, Basel, 2001.
19. R. G. Douglas and G. Misra, On quasi-free Hilbert modules, *New York J. Math* **11** (2005), 1–15.
20. R. G. Douglas and V. I. Paulsen, *Hilbert Modules over Function Algebras*, Longman Research Notes, 217, 1989.
21. R. G. Douglas and D. Voiculescu, On the smoothness of sphere extensions, *J. Operator Theory* **6**, 103–111 (1981).
22. R. G. Douglas and K. Yan, Hilbert–Samuel polynomials for Hilbert modules, *Indiana Univ. Math. J.* **42**, 811–820 (1993).
23. J. Eschmeier and M. Putinan, *Spectral Decompositions and Analytic Sheaves*, London Math. Soc. Monographs, New Series, 10, Oxford Science Publ., The Clarendon Press, Oxford University Press, New York, 1996.
24. J. Gleason, S. Richter and C. Sundberg, On the index of invariant subspaces in spaces of analytic functions of several complex variables, preprint.
25. I. C. Gohberg and M. G. Krein, *Introduction to the Theory of Linear Nonselfadjoint Operators*, Vol. 18, Translations of Mathematical Monographs, American Math. Society, Providence, 1969.
26. K. Guo, Defect operators for submodules of H_d^2, *J. Reine Angew. Math.* **573**, 181–209 (2004).
27. K. Guo and K. Wang, Essentially normal Hilbert modules and K-homology, preprint, 2005.
28. D. W. Hadwin and E. A. Nordgen, Extensions of the Berger–Shaw Theorem, *Proc. Amer. Math. Soc.* **102**, 517–525 (1988).
29. P. S. Muhly and B. Solel, Hilbert modules over operator algebras, *Mem. Amer. Math. Soc.* **117** (1995).
30. M. Taylor, *Partial Differential Equations, Basic Theory*, Texts in Applied Mathematics, 23, Springer-Verlag, New York, 1996.
31. D. V. Voiculescu, A note on quasi-triangularity and trace-class commutators, *Acta Sci. Math. (Szeged)* **42**, 195–199 (1980).

Received by the editors July 27, 2005; Revised October 19, 2005

Analysis, Geometry and Topology
of Elliptic Operators, pp. 383–400

A NOTE ON NONCOMMUTATIVE HOLOMORPHIC AND HARMONIC FUNCTIONS ON THE UNIT DISK

SŁAWOMIR KLIMEK
Department of Mathematics
Indiana University Purdue University Indianapolis
402 N. Blackford St.
Indianapolis, IN 46202 USA
sklimek@math.iupui.edu

Dedicated to Krzysztof P. Wojciechowski on his 50th birthday

We study noncommutative versions of holomorphic and harmonic functions on the unit disk.

2000 *Mathematics Subject Classification.* Primary 46L65; Secondary 46L87, 47L80

1. Introduction

The objective of this paper is to determine a complex structure on the noncommutative disk $C(D_q)$, the q-deformation of the unit disk D. This noncommutative disk is a C^*-algebra that is a subalgebra of the quantum group $SU_q(2)$ and can be conveniently described using generators and a (quadratic) relation. It has been studied fairly extensively in the literature - see Klimek, Lesniewski [4], Nagy, Nica [5], Shklyarov, Sinel'shchikov, Vaksman [8], [9], [10], [11], [12], and references therein. In particular, as q varies, the family of quantum disks $C(D_q)$ forms a deformation of the commutative disk, corresponding to $q = 1$. To determine a complex structure on $C(D_q)$ we define and study partial derivatives on $C(D_q)$ and in particular the concept of holomorphic noncommutative functions.

The series of papers by Shklyarov and collaborators is very much related in spirit to our paper but is technically quite different, much more algebraic. A similar study of a complex structure on the noncommutative plane was done by Rochberg and Weaver [7].

An important point of view of this paper is that we work in a concrete representation of $C(D_q)$ in a Hilbert space $H^2(D, d\mu)$ of holomorphic functions on the unit disk D, square integrable with respect to a certain measure. The algebra $C(D_q)$ is in this representation realized as the algebra of Toeplitz operators with continuous symbols. Also, we use this representation to realize different operations (scaling, derivatives, integral...) on $C(D_q)$ as coming from operators in $H^2(D, d\mu)$. This is well suited for operations that are only densely defined as it allows for good control over domains.

It turns out that there are two natural notions of holomorphic structure on the quantum disk which we call weak and strong. Weakly holomorphic noncommutative functions directly correspond to ordinary holomorphic functions while the strongly holomorphic ones come from the scaled disk $\frac{1}{q}D$. We also study noncommutative harmonic functions. Just as ordinary two dimensional harmonic functions, their quantum counterparts on the unit disk can be written as a sum of holomorphic and antiholomorphic part. They exhibit many of the familiar properties like a maximum principle.

The paper is organized as follows. In Section 2 we recall the definition of the quantum disk and in particular we study in depth its representation using Toeplitz operators. Section 3 contains the definition and our study of the properties of the derivatives and the integral on the quantum unit disk $C(D_q)$. Finally in Section 4 we introduce and study quantum holomorphic, antiholomorphic and harmonic functions on the unit disk.

2. Quantum unit disk

In this section we review C^*-algebraic aspects of the quantum unit disk $C(D_q)$. It is defined as the universal unital C^*-algebra generated by a generator z, and its conjugate denoted by $\bar{z}$, and satisfying the following relation: $\bar{z}z = qz\bar{z} + (1-q)$. Symbolically:

$$C(D_q) := < z, \bar{z} \mid \bar{z}z = qz\bar{z} + (1-q) > \tag{1}$$

We will restrict ourself to $0 \leq q < 1$. Let us briefly recall the construction of the universal C^*-algebra. If a is a polynomial in $z, \bar{z}$ we define its norm as the supremum of $||\rho(a)||$ over all Hilbert space representations ρ satisfying the relation. One verifies that this defines a sub-C^*-norm and the corresponding completion mod the null space gives the universal C^*-algebra.

Notice that if $q = 0$ the relation $\bar{z}z = 1$ is the defining relation of the standard Toeplitz algebra $\mathfrak{T}$ - see Fillmore [3]. If $q = 1$ the relation becomes the commutativity statement $\bar{z}z = z\bar{z}$. Additionally, since $||\bar{z}z|| = ||z\bar{z}|| =$

$||z||^2$ we get $||z||^2 = q||z||^2 + (1-q)$, which implies that $||z|| = 1$. It is then natural to define $C(D_1)$ to be the algebra of continuous functions $C(D)$ on the unit disk $D := \{\zeta \in \mathbb{C} : |\zeta| \leq 1\}$.

Theorem 2.1. (see [4]) *Let $\{e_n\}$ be the canonical basis in l_2, $n = 0, 1, 2 \ldots$, and let $U : l_2 \to l_2$ be the following weighted unilateral shift: $Ue_n = \sqrt{1 - q^{n+1}} e_{n+1}$. Then $C(D_q) \cong C^*(U)$, where $C^*(U)$ is the C^*-algebra generated by U.*

Proof. As noted above the case of $q = 0$ is the standard Toeplitz algebra case. If $q > 0$ we calculate explicitly that $U^* e_{n+1} = \sqrt{1 - q^{n+1}}\, e_n$, and consequently $U^*U e_n = (1-q^{n+1})\, e_n$ and $UU^* e_n = (1-q^n)\, e_n$, which verifies that U gives a representation for $C(D_q)$. To verify that this indeed is the defining representation one needs to classify all irreducible representations. This was done in [4]. □

Corollary 2.1. (see [4]) *We have the exact sequence:*

$$0 \longrightarrow \mathcal{K} \longrightarrow C(D_q) \xrightarrow{\sigma} C(\partial D) \longrightarrow 0, \tag{2}$$

where $\mathcal{K}$ is the ideal of compact operators in l_2 and ∂D, the boundary of D, is the unit circle.

Proof. The corollary follows from the general theory of weighted shifts, see Conway [2]. Briefly, the commutator $[U, U^*]$ is compact since its eigenvalues $(1-q)q^n \to 0$ as $n \to \infty$, and, since the C^*-algebra is irreducible, it contains all compact operators. The quotient $C(D_q)/\mathcal{K}$ is generated by the unitary operator $[U]$ the spectrum of which is the full unit circle. For details see [4]. □

In the exact sequence (2) the map $\sigma : C(D_q) \to C(\partial D)$ is called the symbol map.

Proposition 2.1. *The C^*-algebras $C(D_q)$ are isomorphic to each other and for every q, $0 \leq q < 1$, we have $C(D_q) \cong \mathfrak{T}$, where $\mathfrak{T}$ is the Toeplitz algebra.*

Proof. Let $V : l_2 \to l_2$ be the unilateral shift $Ve_n = e_{n+1}$, so that $C^*(V) = \mathfrak{T}$. Notice that $U - V$ is a weighted shift $(U - V)e_n = \lambda_n e_{n+1}$, where weights $\lambda_n \to 0$ as $n \to \infty$. Consequently $U - V$ is a compact operator. The proposition now follows from the exact sequence (2). □

The C^*-algebras $C(D_q)$ are also continuous in q in the following sense:

Theorem 2.2. (see [5]) *The C^*-algebras $C(D_q)$ for $0 \le q < 1$ and $C(D)$ for $q = 1$ form a continuous field of C^*-algebras with the space of cross-sections obtained by completing the space of polynomials in z and $\bar{z}$ with coefficients which are continuous functions of q.*

Proof. This was done by Nagy and Nica in [5] for an even bigger range $-1 \le q \le 1$. □

As the final part of this section we will discuss another useful representation of $C(D_q)$. It is using Toeplitz operators and is implicitly contained in [4] but we work out the details here.

Consider the following measure on the unit disk D:

$$d\mu(\zeta) = \prod_{i\ge 0}(1 - |\zeta|^2 q^{i+1}) \sum_{m\ge 0} q^m \delta_{|\zeta|^2 = q^m}(\zeta), \tag{3}$$

where $\delta_{|\zeta|^2=r^2}$ is the normalized Lebesgue measure on the circle $|\zeta|^2 = r^2$. Let $H^2(D, d\mu) \subset L^2(D, d\mu)$ be the closed subspace consisting of holomorphic functions and let P be the corresponding orthogonal projection. If $f \in C(D)$ we define the Toeplitz operator $T(f) : H^2(D, d\mu) \to H^2(D, d\mu)$ with symbol f, by: $T(f) = PM(f)P$, where $M(f)$ is the multiplication by f.

Theorem 2.3. (see [4]) *With the above notation, the C^*-algebra generated by $\{T(f)\}$, $f \in C(D)$, is naturally isomorphic with $C(D_q)$. The isomorphism is determined by identifications: $z = T(\zeta)$, $\bar{z} = T(\bar{\zeta})$*

Proof. It follows from the definition that $||T(f)|| \le \sup_{\zeta\in D}|f(\zeta)|$. Since polynomials in ζ and $\bar{\zeta}$ are dense in $C(D)$ and $T(\bar{\zeta}^m \zeta^n) = (T(\zeta)^*)^m (T(\zeta))^n$, we see that the algebra generated by Toeplitz operators is in fact generated by the single operator $T(\zeta)$. Next, because $d\mu$ is rotationally invariant, the functions ζ^n are mutually orthogonal and form an unnormalized basis in $H^2(D, d\mu)$. To find an orthonormal basis we compute

$$\begin{aligned}\int |\zeta|^{2n}\, d\mu(\zeta) &= \sum_{m\ge 0} q^m q^{nm} \prod_{i\ge 0}(1 - q^m q^{i+1}) \\ &= \prod_{i\ge 0}(1 - q^{i+1}) \left(1 + \sum_{m\ge 1} \frac{q^{m(n+1)}}{\prod_{1\le k\le m}(1 - q^k)}\right).\end{aligned}$$

Using the Euler's identity:

$$\frac{1}{\prod_{i\geq 0}(1-xq^i)} = 1 + \sum_{m\geq 1} \frac{x^m}{\prod_{1\leq k\leq m}(1-q^k)}, \tag{4}$$

gives $\int |\zeta|^{2n}\, d\mu(\zeta) = 1$ if $n = 0$, and, for $n \geq 1$:

$$\int |\zeta|^{2n}\, d\mu(\zeta) = \frac{\prod_{i\geq 0}(1-q^{i+1})}{\prod_{i\geq 0}(1-q^{n+i+1})} = \prod_{i=0}^{n-1}(1-q^{i+1}).$$

It follows that the measure $d\mu$ is probabilistic and the following is an orthonormal basis in $H^2(D, d\mu)$:

$$e_n = \begin{cases} 1 & \text{if } n = 0 \\ \frac{\zeta^n}{\sqrt{\prod_{i=0}^{n-1}(1-q^{i+1})}} & \text{if } n \geq 1 \end{cases} \tag{5}$$

We now find the matrix elements of $T(\zeta)$ with respect to the basis e_n:

$$\begin{aligned} T(\zeta)e_n = \zeta e_n &= \frac{\zeta^n}{\sqrt{\prod_{i=0}^{n-1}(1-q^{i+1})}} = \sqrt{1-q^{n+1}}\frac{\zeta^n}{\sqrt{\prod_{i=0}^{n}(1-q^{i+1})}} \\ &= \sqrt{1-q^{n+1}}\, e_{n+1}. \end{aligned}$$

So the matrix elements of $T(\zeta)$ are equal to that of U of the structure theorem 2.1, which concludes the proof. □

From now on we will identify $C(D_q)$ with the concrete algebra generated by Toeplitz operators in $H^2(D, d\mu) \subset L^2(D, d\mu)$. For future reference we recall here the definition of the Bergman kernel $K(\zeta, \bar{\eta})$ for $H^2(D, d\mu)$. It is the integral kernel of the projection P so it has the reproducing property:

$$\int K(\zeta, \bar{\eta})\phi(\eta)\, d\mu(\eta) = \phi(\zeta), \tag{6}$$

where $\phi(\zeta) \in H^2(D, d\mu)$. It can be explicitly computed using a basis in $H^2(D, d\mu)$, for example the one given by (5). We obtain:

$$\begin{aligned} K(\zeta, \bar{\eta}) &= \sum_{n=0}^{\infty} e_n(\zeta)\overline{e_n(\eta)} = 1 + \sum_{n\geq 1} \frac{(\zeta\bar{\eta})^n}{\prod_{1\leq k\leq n}(1-q^k)} \\ &= \frac{1}{\prod_{i\geq 0}(1-\zeta\bar{\eta}q^i)}. \end{aligned} \tag{7}$$

In the above we again used the Euler identity (4).

By construction, the space of polynomials in Toeplitz operators is dense in $C(D_q)$. More is actually true as spelled out in the next statement.

Proposition 2.2. *The subspace of Toeplitz operators $T(f)$, $f \in C(D)$ is dense in $C(D_q)$.*

Proof. It follows from the defining relation of $C(D_q)$ that the linear span of $\bar{z}^m z^n$, $m, n \geq 0$, forms a dense subalgebra of $C(D_q)$. Indeed, since $z\bar{z}$ expresses linearly in terms of $\bar{z}z$ we can rearrange any polynomial in $z, \bar{z}$ so that powers of $\bar{z}$ come first. But $\bar{z}^m z^n = T(\bar{\zeta}^m \zeta^n)$ and the claim follows.□

3. Calculus on $C(D_q)$

In this section we introduce calculus on the quantum unit disk. In the following we assume that $q > 0$. Formal aspects of the calculus on $C(D_q)$ can be found in Chu, Ho, Zumino [1] as well as in [8, 9, 10, 11, 12]. We concentrate here on issues of domains for various unbounded operators and we will always identify $C(D_q)$ with the concrete algebra of Toeplitz operators of Theorem 2.3.

Let $\mathcal{D} := \{\phi \in H^2(D, d\mu) : \ \phi(\zeta/q) \in H^2(D, d\mu)\}$. Clearly $\mathcal{D}$ is a dense subspace in $H^2(D, d\mu)$ containing all polynomials, or more generally entire functions. We define a scaling operator $j : H^2(D, d\mu) \to H^2(D, d\mu)$ by the formula:

$$j\phi(\zeta) := \phi(q\zeta).$$

The operator j is bounded, one-to-one and $\mathrm{Ran} j = \mathcal{D}$. Using the defining formula (5) we have

$$je_n = q^n e_n, \tag{8}$$

so that j is a self-adjoint compact operator. Since $z\bar{z}e_n = (1 - q^n)e_n$ and $\bar{z}ze_n = (1 - q^{n+1})e_n$ we have

$$z\bar{z} = 1 - j, \ \ \bar{z}z = 1 - qj. \tag{9}$$

An element $a \in C(D_q)$ is called *scalable* if $J(a) := j^{-1}aj$ is a bounded operator. We have a simple proposition:

Proposition 3.1. *The operator $j^{-1}aj$ is bounded iff a maps $\mathcal{D}$ to $\mathcal{D}$.*

Proof. If a preserves $\mathcal{D}$ then $j^{-1}aj$ is defined everywhere. To show that it is bounded, we use the closed graph theorem which implies that we need to verify that if $x_n \to x$ and $y_n := J(a)x_n \to y$ then $J(a)x = y$. Since j is continuous we have $jy_n = ajx_n \to jy$. But aj is continuous so $ajx_n \to x$ and consequently $ajx = jy$, which is what we wanted. The converse statement is straightforward. □

We write $C_s(D_q)$ for the set of scalable elements of $C(D_q)$. The proposition below shows that $C_s(D_q)$ is a subalgebra of $C(D_q)$ containing $\mathrm{Pol}(D_q)$, the algebra of polynomials in $z, \bar{z}$. However, examples below show that $C_s(D_q)$ is not closed with respect to taking adjoints and inverses.

Proposition 3.2. *With the above notation we have:*

$$J(1) = 1,\ J(\bar{z}) = q\bar{z},\ J(z) = q^{-1}z. \tag{10}$$

If $a, b \in C_s(D_q)$ then $ab \in C_s(D_q)$ and $J(ab) = J(a)J(b)$.

Proof. The proof consist of straightforward computations verifying each of the properties. For this we need explicit formulas for $z, \bar{z}$. Theorem 2.3 implies

$$z\phi(\zeta) = \zeta\phi(\zeta), \tag{11}$$

while the structure Theorem 2.1 says that

$$ze_n = \sqrt{1 - q^{n+1}}\, e_{n+1}, \tag{12}$$

where e_n were defined in (5). Taking the adjoint gives

$$\bar{z}e_n = \sqrt{1 - q^n}\, e_{n-1}, \tag{13}$$

(the right-hand side is defined to be 0 when $n = 0$). This implies that $\bar{z}\zeta^n = (1 - q^n)\zeta^{n-1}$, which in turn gives:

$$\bar{z}\phi(\zeta) = \frac{\phi(\zeta) - \phi(q\zeta)}{\zeta}. \tag{14}$$

A sample calculation verifying one of the statements of the proposition follows:

$$\begin{aligned} J(z)\phi(\zeta) &= j^{-1}zj\phi(\zeta) = zj\phi(\zeta/q) = q^{-1}\zeta j\phi(\zeta/q) \\ &= q^{-1}\zeta\phi(\zeta) = q^{-1}z\phi(\zeta) \end{aligned}$$

□

We are now going to look at examples to illustrate some subtleties of the notion of scalability. First notice that $1 - q\bar{z}$ is invertible since $||q\bar{z}|| = q < 1$. The inverse $a := (1 - q\bar{z})^{-1}$ is clearly in $C(D_q)$ and is scalable because $J(a) = (1 - q^2\bar{z})^{-1}$. However, $a^* = (1 - qz)^{-1}$ is not scalable as $J(a^*) = (1 - z)^{-1}$ is unbounded. Next consider $b := 1 - qz$. Clearly $b \in C(D_q)$, b is scalable, b is invertible, and the inverse of b is in $C(D_q)$. But since $b^{-1} = a^*$, b^{-1} is not scalable.

Next we introduce two operators $\delta, \bar{\delta}$ in $H^2(D, d\mu)$ that will be used to define Dolbeault - type operators $\partial, \bar{\partial}$ on $C(D_q)$. The precise form of $\delta, \bar{\delta}$

is dictated by the desired properties of $\partial, \bar{\partial}$ as described in the Proposition 3.4.

The operators $\delta, \bar{\delta}$ are defined to be unbounded operators in $H^2(D, d\mu)$ with domains both equal to $\mathcal{D}$ and given by the following formulas using $z, \bar{z}, j$:

$$\bar{\delta} = (q-1)^{-1} j^{-1} z = (q-1)^{-1} q^{-1} z j^{-1}, \tag{15}$$

$$\delta = (1-q)^{-1} j^{-1} \bar{z} = (1-q)^{-1} q \bar{z} j^{-1}. \tag{16}$$

Proposition 3.3. *With the above notation, the operators $\delta, \bar{\delta}$ are closed (on $\mathcal{D}$).*

Proof. To show that $\bar{\delta}$ is closed the following needs to be demonstrated: if $\phi_n \to \phi$, $\phi_n \in \mathcal{D}$ and $\psi_n := j^{-1} z \phi_n \to \psi$, then $\phi \in \mathcal{D}$ and $j^{-1} z \phi = \psi$. Applying j to ψ_n and using the continuity of j gives $z\phi_n \to j\psi$. On the other hand, since z is continuous, we have $z\phi_n \to z\phi$. Consequently $z\phi = j\psi$, which means that $z\phi \in \mathcal{D}$ and $j^{-1} z\phi = \psi$. What's left is to show that $\phi \in \mathcal{D}$. Applying $\bar{z}$ to both sides of $z\phi = j\psi$ and using (9) and (10) we obtain

$$\phi = qj(1-qj)^{-1} \bar{z} \psi,$$

which concludes the proof that $\bar{\delta}$ is closed. The proof for δ is analogous with the exception of the fact that $z\bar{z}$ has a kernel. The analog of the above formula works on the orthogonal complement of that kernel. The proof is then concluded by observing that the kernel of $z\bar{z}$ is one dimensional and is contained in $\mathcal{D}$. □

Using the equations (8), (11) and (14), we obtain the following explicit descriptions of the operators $\delta, \bar{\delta}$:

$$\bar{\delta}\phi(\zeta) = (q-1)^{-1} \zeta/q \, \phi(\zeta/q) \tag{17}$$

$$\delta\phi(\zeta) = \frac{\phi(\zeta) - \phi(\zeta/q)}{(1-1/q)\zeta}. \tag{18}$$

Optionally, when working with the operators $\delta, \bar{\delta}$ one can use their matrix elements, obtained using (8), (12) and (13):

$$\bar{\delta} e_n = (q-1)^{-1} q^{-(n+1)} \sqrt{1-q^{n+1}} \, e_{n+1},$$

$$\delta e_n = \frac{1-1/q^n}{1-1/q} \sqrt{1-q^n} \, e_{n-1}.$$

Now we use those operators to define complex structure on $C(D_q)$ - for this we need the analogs of the usual complex derivatives $\partial, \bar{\partial}$. They are defined using scaled commutators with $\delta, \bar{\delta}$ as follows. If a is scalable we define $\bar{\partial}(a)$, $\partial(a)$ to be (in general unbounded) linear operators defined on $\mathcal{D}$ by:

$$\bar{\partial}(a) = \bar{\delta}a - J(a)\bar{\delta},$$

$$\partial(a) = \delta a - J(a)\delta.$$

Proposition 3.1 assures that $\bar{\partial}(a)$, $\partial(a)$ are well defined operators on $\mathcal{D}$. For general a i.e. not necessarily scalable, $\bar{\partial}(a)$, $\partial(a)$ make sense only as quadratic forms - see below. We will use those quadratic forms in the discussion of quantum holomorphic and harmonic functions in the next section.

The following proposition summarizes the main properties of the operators $\partial, \bar{\partial}$.

Proposition 3.4. *With the above notation we have, assuming $a, b \in C_s(D_q)$:*

$$\bar{\partial}(1) = 0,\ \bar{\partial}(\bar{z}) = 1,\ \bar{\partial}(z) = 0,\ \bar{\partial}(ab) = (\bar{\partial}a)b + J(a)(\bar{\partial}b), \tag{19}$$

$$\partial(1) = 0,\ \partial(\bar{z}) = 0,\ \partial(z) = 1,\ \partial(ab) = (\partial a)b + J(a)(\partial b). \tag{20}$$

In particular if $a \in Pol(D_q)$ then $\partial a, \bar{\partial} a \in Pol(D_q)$.

Proof. The proof again consist of straightforward verifications using definitions. Below we show calculations of the action of the operators $\partial, \bar{\partial}$ on $z, \bar{z}$ that utilize commutation relations among $j, z, \bar{z}$. All the manipulations with unbounded operators make sense pointwise on $\mathcal{D}$.

$$\begin{aligned}\partial(z) &= \delta z - J(z)\delta = (1-q)^{-1}(j^{-1}\bar{z}z - q^{-1}zj^{-1}\bar{z}) \\ &= (1-q)^{-1}j^{-1}(\bar{z}z - z\bar{z}) = (1-q)^{-1}j^{-1}(1 - qj - 1 + j) = 1\end{aligned}$$

$$\begin{aligned}\partial(\bar{z}) &= \delta\bar{z} - J(\bar{z})\delta = (1-q)^{-1}(j^{-1}\bar{z}^2 - q\bar{z}j^{-1}\bar{z}) \\ &= (1-q)^{-1}j^{-1}(\bar{z}^2 - \bar{z}^2) = 0\end{aligned}$$

The other two calculations are very similar. □

For future reference we note the formulas for action of $\partial, \bar{\partial}$ on monomials:

$$\partial(\bar{z}^n z^m) = q^{n-m+1}[m]_q \bar{z}^n z^{m-1} \tag{21}$$

and similar:

$$\bar{\partial}(\bar{z}^n z^m) = [n]_q \bar{z}^{n-1} z^m. \tag{22}$$

Here, and later in the paper, we use the notation

$$[n]_q := \frac{1-q^n}{1-q}.$$

Another set of useful formulas follows directly from the definitions:

$$\partial a = (1-q)^{-1} j^{-1}[\bar{z}, a] \tag{23}$$

$$\bar{\partial} a = (q-1)^{-1} j^{-1}[z, a] \tag{24}$$

For example, it follows from those formulas that $\partial a, \bar{\partial} a$ are closable since the domains of the adjoints clearly contain $\mathcal{D}$ and so are dense. Another application of (23) and (24) is in the following definition of the derivatives $\partial a, \bar{\partial} a$ as quadratic forms for a general, not necessarily scalable $a \in C(D_q)$. They are defined on $\mathcal{D}$ as

$$Q_{\partial a}(\phi) := (1-q)^{-1}(j^{-1}\phi, [\bar{z}, a]\phi), \tag{25}$$

$$Q_{\bar{\partial} a}(\phi) := (q-1)^{-1}(j^{-1}\phi, [z, a]\phi). \tag{26}$$

We now turn to the definition and properties of the laplacian on $C(D_q)$. There are two natural choices that we will look at using the formulas above:

$$\bar{\partial}\partial(\bar{z}^n z^m) = q^{n-m+1}[m]_q [n]_q \bar{z}^{n-1} z^{m-1}.$$

Similarly we obtain

$$\partial\bar{\partial}(\bar{z}^n z^m) = q^{n-m}[m]_q [n]_q \bar{z}^{n-1} z^{m-1}.$$

It follows that, at least on $\mathrm{Pol}(D_q)$,

$$\bar{\partial}\partial = q\partial\bar{\partial}. \tag{27}$$

To define $\bar{\partial}\partial$ and $\partial\bar{\partial}$ for a larger class of elements of $C(D_q)$ we proceed similarly to the way we defined $\partial, \bar{\partial}$. Let $\mathcal{D}_2 := \{\phi \in H^2(D, d\mu) : \phi(\zeta/q^2) \in H^2(D, d\mu)\}$. Clearly $\mathcal{D}_2 = \mathrm{Ran} j^2$, $\mathcal{D}_2$ is dense and $\mathcal{D}_2 \subset \mathcal{D}$. Also, just as in Proposition 3.1, if a is scalable and $J(a)$ is scalable then a maps $\mathcal{D}_2$ into $\mathcal{D}_2$. In particular, $z, \bar{z}$ preserve $\mathcal{D}_2$. Consequently, it follows from (15) and

(16) that $\delta, \bar{\delta} : \mathcal{D}_2 \to \mathcal{D}$. Thus if both a and $J(a)$ are scalable then $\bar{\partial}\partial(a)$ and $\partial\bar{\partial}(a)$ make sense as operators on $\mathcal{D}_2$. It can be easily verified that (27) holds in this more general context i.e. if $a, J(a)$ are scalable and $\phi \in \mathcal{D}_2$ then

$$\bar{\partial}\partial(a)\phi = q\partial\bar{\partial}(a)\phi.$$

In particular, the two laplacians have the same kernels and we will use whatever is more convenient when defining harmonic functions as extended kernels in the next section.

The last item in this section is integration on the quantum unit disk. We define the integral $\int_{D_q} : C(D_q) \to \mathbb{R}$ by

$$\int_{D_q} a = \frac{\mathrm{Tr}(aj)}{\mathrm{Tr}(j)} \tag{28}$$

Using (8) we compute:

$$\mathrm{Tr}(j) = \sum_{n=0}^{\infty} q^n = \frac{1}{1-q}$$

Proposition 3.5. *$\int_{D_q}$ is a faithful state on $C(D_q)$ and*

- $\int_{D_q} ab = \int_{D_q} J(b)a$
- $\int_{D_q} J(a) = \int_{D_q} a$

Proof. Everything follows easily from the definitions. □

The integral is easy to work with as demonstrated in the following computation of its value on monomials.

Lemma 3.1. *With the above notation we have*

$$\int_{D_q} \bar{z}^n z^m = \delta_{n,m} \frac{1}{[n+1]_q}$$

Proof. Using the canonical basis in $H^2(D, d\mu)$ we have

$$\int_{D_q} a = (1-q) \sum_{k=0}^{\infty} q^k (e_k, ae_k).$$

It follows that $\int_{D_q} \bar{z}^n z^m = 0$ if $n \neq m$. Using (12) we compute

$$\int_{D_q} \bar{z}^n z^n = (1-q)\sum_{k=0}^{\infty} q^k (e_k, \bar{z}^n z^n e_k) = (1-q)\sum_{k=0}^{\infty} q^k ||z^n e_k||$$

$$= (1-q)\sum_{k=0}^{\infty} q^k (1-q^{k+1})(1-q^{k+2})\ldots(1-q^{k+n}) = \int_0^1 f(y)\, d_q y,$$

where $f(y) = (1-qy)(1-q^2y)\ldots(1-q^ny)$. Here we used the Jackson's integral for a continuous function f which is defined by:

$$\int_0^1 f(y)\, d_q y := (1-q)\sum_{k=0}^{\infty} q^k f(q^k)$$

It has the property

$$\int_0^1 \delta_q g(y)\, d_q y = g(1) - g(0), \tag{29}$$

where

$$\delta_q g(y) = \frac{g(y) - g(qy)}{y - qy}. \tag{30}$$

We use this property in our calculation. For $g(y) = (1-y)(1-qy)\ldots(1-q^ny)$ we compute:

$$\delta_q g(y) = \tfrac{(1-y)(1-qy)\ldots(1-q^ny)-(1-qy)(1-q^2y)\ldots(1-q^{n+1}y)}{y(1-q)}$$
$$= (1-qy)(1-q^2y)\ldots(1-q^ny)\tfrac{1-y-1+q^{n+1}y}{y(1-q)} = -[n+1]_q f(y).$$

It follows that

$$\int_0^1 f(y)\, d_q y = \int_0^1 \delta_q\left(\frac{-g(y)}{[n+1]_q}\right) d_q y = \frac{g(0)-g(1)}{[n+1]_q},$$

which finishes the proof. □

The integral $\int_{D_q}$ and the derivatives $\bar{\partial}, \partial$ are tightly connected, just as in the commutative case. This is illustrated by the following theorem - compare also [8].

Theorem 3.1. (Green's theorem) *If $a \in Pol(D_q)$ then*

$$\int_{D_q} (\bar{\partial} a) = \frac{1}{2\pi i}\int_{\partial D} \sigma(a)(\zeta) d\zeta.$$

Here $\sigma : C(D_q) \to C(S^1)$ is the symbol map.

Proof. It is enough to consider monomials of the following form:

$$a = \bar{z}^{n+1} z^n.$$

Then, $\sigma(a) = \bar{\zeta}^{n+1}\zeta^n = \bar{\zeta} = e^{-i\theta}$, and

$$\frac{1}{2\pi i}\int_{\partial D} \sigma(a)(\zeta)\, d\zeta = \frac{1}{2\pi i}\int_0^{2\pi} e^{-i\theta}\, d(e^{i\theta}) = 1.$$

On the other hand $\partial(\bar{\zeta}^{n+1}\zeta^n) = [n+1]_q \bar{\zeta}^n\zeta^n$ by (22) and

$$\int_{D_q} (\bar{\partial} a) = [n+1]_q \int_{D_q} \bar{\zeta}^n \zeta^n = 1$$

by Lemma 3.1. □

4. Quantum holomorphic and harmonic functions

In this section we define quantum holomorphic and harmonic functions on the quantum unit disk $C(D_q)$. We start with the following definition. An element $a \in C(D_q)$ is called strongly holomorphic if a is scalable and $\bar{\partial} a = 0$. Similarly $a \in C(D_q)$ is called weakly holomorphic if $Q_{\bar{\partial} a}(\phi) = 0$ for all $\phi \in \mathcal{D}$, where the quadratic form $Q_{\bar{\partial} a}$ was defined in (26). In the later definition we do not need to assume scalability of a. We denote by $\mathrm{Hol}(D_q)$ the space of weakly holomorphic elements of $C(D_q)$. We have the following simple proposition:

Proposition 4.1.

- *$a \in C(D_q)$ is weakly holomorphic iff $[z, a] = 0$.*
- *$a \in C(D_q)$ is strongly holomorphic iff a is scalable and weakly holomorphic.*

Proof. The formula (26) and polarization imply the first part of the proposition. The second part is just a rephrasing of the definition. □

There are analogous definitions of antiholomorphic functions. An element $a \in C(D_q)$ is called strongly antiholomorphic if a is scalable and $\partial a = 0$. Similarly $a \in C(D_q)$ is called weakly antiholomorphic if $Q_{\partial a}(\phi) = 0$ for all $\phi \in \mathcal{D}$, where the quadratic form $Q_{\partial a}$ was defined in (25). Because z and $\bar{z}$ scale differently, the following analog of Proposition 4.1 looks a little different.

Proposition 4.2.

- *$a \in C(D_q)$ is weakly antiholomorphic iff $[\bar{z}, a] = 0$.*
- *$a \in C(D_q)$ is strongly antiholomorphic iff a is weakly antiholomorphic.*

Proof. The formula (25) and polarization imply the first part of the proposition. The second part is proved in the theorem below. □

The following is the main result describing holomorphic and antiholomorphic functions on the quantum unit disk. Notice that there is slight asymmetry between the notions of strongly holomorphic and antiholomorphic functions which disappears when $q = 1$.

Theorem 4.1.

(1) *If $f \in C(D)$ is holomorphic inside D then the corresponding Toeplitz operator $T(f) \in C(D_q)$ is weakly holomorphic.*
(2) *If $a \in C(D_q)$ is weakly holomorphic then there exist $f \in C(D)$ which is holomorphic inside D such that $a = T(f)$.*
(3) *(maximum principle) If $a \in Hol(D_q)$ then $||a||_{D_q} = ||\sigma(a)||_{\partial D}$*
(4) *The space $Hol(D_q) \subset C(D_q)$ is a Banach subalgebra isomorphic to the algebra $Hol(D) \subset C(D)$ of continuous functions on D and holomorphic inside D.*
(5) *The above statements are also true when the word holomorphic is replaced by antiholomorphic throughout.*
(6) *If $a \in C(D_q)$ then a is strongly antiholomorphic iff a is weakly antiholomorphic.*

Proof. We prove all items in order stated in the theorem:

1. This follows from Proposition 4.1 since if $f \in C(D)$ is holomorphic then:

$$zT(f)\phi(\zeta) = \zeta f(\zeta)\phi(\zeta) = T(f)z\phi(\zeta).$$

2. For a weakly holomorphic $a \in C(D_q)$ we set $f(\zeta) := a \cdot 1(\zeta) \in H^2(D, d\mu)$. In particular f is holomorphic inside the disk D. Because $[z, a] = 0$, we have inductively $a\zeta^n = f(\zeta)\zeta^n$, so a is equal to the Toeplitz operator $T(f)$ on the dense domain and consequently everywhere. To obtain more information about f we prove the following estimate:

$$\sup_{\zeta \in D} |f(\zeta)| \leq ||T(f)||. \tag{31}$$

To do it we consider the family of functions:

$$\phi_\eta(\zeta) := \frac{K(\bar\eta, \zeta)}{(K(\bar\eta, \eta))^{1/2}},$$

where $K(\zeta, \bar\eta)$ is the reproducing kernel (7). It is easily seen that the functions $\phi_\eta(\zeta)$ belong to $H^2(D, d\mu)$ and have norm 1. Using the reproducing property (6) we compute:

$$\begin{aligned}(\phi_\eta, T(f)\phi_\eta) &= (K(\bar\eta, \eta))^{-1} \int K(\eta, \bar\zeta) K(\zeta, \bar\eta) f(\zeta)\, d\mu(\zeta) \\ &= (K(\bar\eta, \eta))^{-1} K(\bar\eta, \eta) f(\eta) = f(\eta).\end{aligned}$$

It follows that

$$\sup_{\zeta \in D} |f(\zeta)| = \sup_{\zeta \in D} |(\phi_\zeta, T(f)\phi_\zeta)| \le ||T(f)|| < \infty,$$

so f is bounded on D and holomorphic inside it. But $a = T(f)$ belongs to $C(D_q)$ so it is a limit of polynomials which implies that $f \in C(D)$ as claimed.

3. It follows from the definition that $||T(f)|| \le \sup_{\zeta\in D}|f(\zeta)|$. On the other hand if $a = T(f) \in \mathrm{Hol}(D_q)$ the estimate (31) is valid and so $||a|| = ||T(f)|| = \sup_{\zeta\in D} |f(\zeta)|$. But $\sigma(T(f)) = f|_{\partial D}$ and the supremum of $|f(\zeta)|$ is achieved on the boundary ∂D and the statement follows.

4. This is just a rephrasing of the previous items.

5. Notice that Propositions 4.1 and 4.2 imply that if a is weakly antiholomorpic then a^* is weakly holomorphic, so all the statements follow by conjugation.

6. The previous item implies that if a is weakly antiholomorphic then $a = T(f)$ where f is antiholomorphic. For such f we have $J(T(f(\zeta))) = T(f(q\zeta))$, so $T(f)$ is automatically scalable. Consequently, a is strongly antiholomorphic. □

Notice that if f is holomorphic then $J(T(f(\zeta))) = T(f(\frac{1}{q}\zeta))$ so that if $T(f)$ is strongly holomorphic then f extends to a holomorphic function inside the disk $\frac{1}{q}D = \{\zeta \in \mathbb{C} : |\zeta| \le \frac{1}{q}\}$.

The last topic of this paper is the notion of quantum harmonic functions. Just as in our treatment of quantum holomorphic functions there are two natural concepts of quantum harmonic functions called weak and strong. To interpret the equation $\partial(\bar\partial a) = 0$ with as few assumption as possible, we use Proposition 4.2 to make the following definitions. An element

$a \in C(D_q)$ is called strongly harmonic if a is scalable and the derivative $\bar{\partial}a$ is antiholomorphic. Similarly $a \in C(D_q)$ is called weakly harmonic if the bilinear form $Q_{\bar{\partial}a}$ is equal (on $\mathcal{D}$) to a quadratic form coming from an antiholomorphic element of $C(D_q)$. We denote by $\mathrm{Har}(D_q)$ the space of weakly harmonic elements of $C(D_q)$. It turns out that such noncommutative harmonic functions share all the essential properties with their commutative counterparts, see Ransford [6]. This is summarized in the following theorem.

Theorem 4.2.

(1) *An element $a \in C(D_q)$ is strongly harmonic iff a is scalable and weakly harmonic.*

(2) *If $f \in C(D)$ is harmonic inside D then the corresponding Toeplitz operator $T(f) \in C(D_q)$ is weakly harmonic.*

(3) *If $a \in C(D_q)$ is weakly harmonic then there exist $f \in C(D)$ which is harmonic inside D such that $a = T(f)$.*

(4) *An element $a \in C(D_q)$ is weakly harmonic iff it can be written as $a = a_1 + a_2$ where a_1 is weakly holomorphic and a_2 is antiholomorphic.*

(5) *(mean value) If $a = T(f) \in Har(D_q)$ then*

$$\int_{D_q} a = \int_D f(\zeta)\, d^2\zeta = f(0).$$

(6) *(maximum principle) If $a \in Har(D_q)$ then $||a||_{D_q} = ||\sigma(a)||_{\partial D}$.*

(7) *The set $Har(D_q) \subset C(D_q)$ is a closed subspace isomorphic to the Banach space $Har(D) \subset C(D)$ of continuous functions on D and harmonic inside D.*

(8) *(Dirichlet problem) For every $f \in C(\partial D)$ there is a unique weakly harmonic element $a \in C(D_q)$ such that $\sigma(a) = f$. In fact $a = T(Pf)$, where*

$$Pf(\zeta) = \int_0^{2\pi} \frac{1 - |\zeta|^2}{|e^{i\theta} - \zeta|^2} f(e^{i\theta}) \frac{d\theta}{2\pi} \tag{32}$$

(Poisson integral)

(9) *An element $a \in Har(D_q)$ is positive, $a \geq 0$, iff $a = T(f)$, where $f \in C(D)$ is harmonic inside D and $f \geq 0$.*

(10) *(Harnack's theorem) If a_n is an uniformly bounded and increasing sequence of weakly harmonic elements of $C(D_q)$, then $\{a_n\}$ is convergent in norm.*

Proof.

1. This follows from the fact that if a is scalable then the quadratic form $Q_{\bar{\partial}a}$ comes from the operator $\bar{\partial}a$.

2. From the classical harmonic analysis, if $f \in C(D)$ is harmonic inside D then it is a sum $f = g + h$ where $g, h \in C(D)$ and g is holomorphic and h antiholomorphic inside D. The Toeplitz operator $T(g)$ is weakly holomorphic so $Q_{\bar{\partial}(T(f))} = Q_{\bar{\partial}(T(h))}$. But the quadratic form $Q_{\bar{\partial}(T(h))}$ comes from the operator $\bar{\partial}(T(h))$ and, using (22) and (30), we compute:

$$\bar{\partial}(T(h)) = T(\delta_q(h)). \tag{33}$$

But $\delta_q(h)$ is again in $C(D)$ and is antiholomorphic inside D. All of this implies that the quadratic form $Q_{\bar{\partial}(T(f))}$ is coming from an antiholomorphic element of $C(D_q)$ as claimed.

3. From the definition, if $a \in C(D_q)$ is weakly harmonic then the quadratic form $Q_{\bar{\partial}a}$ comes from an antiholomorphic element of $C(D_q)$ which, as we know from Theorem 4.1, is of the form $T(k)$, where $k \in C(D)$ is antiholomorphic inside D. Consider:

$$g(\bar{\zeta}) := (1-q)\bar{\zeta}\sum_{n=0}^{\infty} k(q^n\bar{\zeta}).$$

It is easy to see that, just like k, $g \in C(D)$ and is antiholomorphic inside D. Moreover it is straightforward to verify that $\delta_q(g) = k$. This means, using (33) that $\bar{\partial}(T(g)) = T(k)$. Consequently $Q_{\bar{\partial}(a-T(g))} = 0$, which means that $a - T(g)$ is weakly holomorphic and as we know from Theorem 4.1, it is of the form $T(h)$, where $h \in C(D)$ is holomorphic inside D. We see now that, with the above notation, $a = T(f)$, where $f := g + h$, and $f \in C(D)$ is harmonic inside D as claimed.

4. This is a direct consequence of the proof of the previous two statements.

5. If $f \in C(D)$ is harmonic, it has the following power series expansion inside D:

$$f(\zeta) = f(0) + \sum_{n\geq 1} a_n\zeta^n + \sum_{n\geq 1} b_n\bar{\zeta}^n.$$

It follows that $T(f) = f(0)I + \sum_{n\geq 1} a_n z^n + \sum_{n\geq 1} b_n \bar{z}^n$. Lemma 3.1 implies that $\int_{D_q} T(f) = f(0)$, which is clearly also the value of the integral $\int_D f(\zeta)\, d^2\zeta$.

6. From the item 4, if $a \in C(D_q)$ is weakly harmonic then it can be written as $a = a_1 + a_2$ where a_1 is weakly holomorphic and a_2 is antiholo-

morphic. Now it is enough to apply the maximum principle to both a_1 and a_2.

7. The map $\mathrm{Har}(D) \ni f \to T(f) \in \mathrm{Har}(D_q)$ is linear, one-to-one, and, by item 6, an isometry.

8. For a Toeplitz operator $T(f)$ we have $\sigma(T(f)) = f|_{\partial D}$. This, the items 2, 3, and classical results on harmonic functions [6] on the unit disk imply the thesis.

9. This just says that Toeplitz operators preserve positivity which follows immediately from the definition.

10. This again is a direct consequence of the previous items and the classical Harnack's theorem. □

References

1. Chong-Sun Chu, Pei-Ming Ho, B. Zumino. Some complex quantum manifolds and their geometry, preprint hep-th/9608188, *Quantum fields and quantum space time (Cargse, 1996)*, 281 – 322, NATO Adv. Sci. Inst. Ser. B Phys., 364, Plenum, New York, 1997.
2. J. Conway. *Subnormal Operators*, Pitman, Boston London Melbourne, 1981
3. P.A. Fillmore. *A User's Guide to Operator Algebras*, John Wiley and Sons, New York, 1996
4. S. Klimek, A. Lesniewski. A two-parameter quantum deformation of the unit disc, *J. Funct. Anal.* **115**, (1993), 1 – 23.
5. G. Nagy, A. Nica. On the "quantum disc" and a "non-commutative circle", in: Algebraic Methods on Operator Theory, R. E. Curto, P. E. T. Jorgensen (eds.), Birkhauser, Boston, 1994, 276 – 290.
6. T. Ransford, *Potential Theory in the Complex Plane*, London Math Society Student Texts No. 28, Cambridge University Press, 1995
7. R. Rochberg, N. Weaver. Noncommutative complex analysis and Bargmann-Segal multipliers, *Proc. Amer. Math. Soc.* **129**, (2001), 2679 –2687.
8. D. Shklyarov, S. Sinel'shchikov, L. Vaksman. On function theory in quantum disc: integral representations, E-print: math.QA/9808015.
9. D. Shklyarov, S. Sinel'shchikov, L. Vaksman. On function theory in quantum disc: covariance, E-print: math.QA/9808037.
10. D. Shklyarov, S. Sinel'shchikov, L. Vaksman. On function theory in quantum disc: invariant kernels, E-print: math.QA/9808047.
11. D. Shklyarov, S. Sinel'shchikov, L. Vaksman. On function theory in quantum disc: q-differential equations and Fourier transform, E-print: math.QA/9809002.
12. D. Shklyarov, S. Sinel'shchikov, L. Vaksman. On function theory in quantum disc: q-analogue of Berezin transformation, E-print: math.QA/9809018.

Received by the editors September 21, 2005; revised January 10, 2006.

Analysis, Geometry and Topology
of Elliptic Operators, pp. 401–410

STAR PRODUCTS AND CENTRAL EXTENSIONS

JOUKO MICKELSSON
Department of Mathematics
University of Helsinki
and Mathematical Physics
Royal Institute of Technology, Stockholm
jouko.mickelsson@theophys.kth.se

Dedicated to Krzysztof P. Wojciechowski on his 50th birthday

The purpose of the present note is two-fold. First, to show that deformations of algebras of smooth functions can be used to construct topologically nontrivial standard central extensions of loop groups. Second, to use noncommutative geometry as a regularization of current algebras in higher dimensions with the aim of constructing representations of current algebras.

2000 *Mathematics Subject Classification.* Primary 53D55; Secondary 22E67, 81R10

1. Introduction

The standard central extension of the algebra $L\mathbf{g}$ of smooth loops in a Lie algebra $\mathbf{g}$ of a compact Lie group G defines a central extension by the circle of the smooth loop group LG. An explicit geometric construction for the central extension $\widehat{LG}$ was given by Mickelsson [8]; for an alternative construction see Murray [11]. The method in [8] was to first define a topologically trivial central extension of the group DG of smoooth G valued functions in the unit disk D and then take a quotient by a normal subgroup isomorphic with the group $\mathcal{G}$ of functions which take the value $1 \in G$ on the boundary circle. The central extension of DG is defined by a S^1 valued 2-cocycle. In Section 2 we shall see that we can dispens the 2-cocycle if we use a Moyal product for the functions in the disk. The structure of the loop group on the boundary circle remains undeformed but we need a derminant in $\mathcal{G}$ associated to a trace functional on the Moyal algebra.

The second application of the use of Moyal product for function algebras is related to the problem of constructing nontrivial representations of current algebras arising from hamiltonian anomalies, [9]. The main difficulty comes from the missing Hilbert-Schmidt property of off-diagonal elements of the currents with respect to the energy polarization. This problem does not arise in the case of current algebras on the circle (the lowest energy representations are the highest weight representations of affine Lie algebras). However, in any dimension bigger than one the Hilbert-Schmidt condition fails; this is related to ultraviolet divergencies in perturbative Yang-Mills theory. In one space dimension the divergencies can be removed by normal ordering but in higher dimensions one needs additional subtractions. The (background field dependent) subtractions form an obstruction for constructing true Hilbert space representations; the best what one can achieve is a geometric action on sections of a Hilbert bundle over the space of background fields.

A deformation of the commutative algebra of smooth functions on a manifold can improve the short distance behaviour in quantum field theory. One of the examples is the fuzzy sphere which has been studied in great detail by Grosse and Madore, [3, 7]. In this case the algebra becomes finite-dimensional, avoiding any kind of ultraviolet divergencies. Consequencies for the current algebra representations are illustrated in terms of three examples in Section 3.

The algebra of functions on the disk can be deformed in a variety of ways. A different construction can be found in an article by Lizzi, Vitale, and Zampini [5] which is more close in spirit to the fuzzy sphere algebra in [3, 7].

2. The disk algebra and central extensions of loop groups

Let ω be the standard symplectic form $\omega = dx \wedge dy$ in $\mathbb{R}^2$. Its restriction to the unit disk D in $\mathbb{R}^2$ can be used to define a star product deformation of the algebra $\mathcal{B}$ of complex $n \times n$ matrix valued smooth functions in D, with vanishing normal derivatives to all orders at the boundary S^1,

$$(f * g)(x, y) = e^{\frac{i\nu}{2}(-\partial_x \partial_{y'} + \partial_y \partial_{x'})} f(x, y) g(x', y')|_{x=x', y=y'}, \tag{1}$$

defined as a formal power series in ν. Note that at the boundary the star product is just the pointwise product of functions. Thus the restriction to the boundary gives the trivial formal deformation of the loop algebra. For general background on Moyal product and deformation quantization see Bayen, Flato, Fronsdal, Lichnerowicz, and Sternheimer [1].

Integration over the disk defines a linear functional in $\mathcal{B}$,

$$\mathrm{TR}_\nu(f) = \frac{1}{2\pi\nu}\int_D \operatorname{tr} f(x,y)dxdy, \tag{2}$$

where $'\mathrm{tr}'$ is the matrix trace.

If the functions f, g are constant on the boundary then by integration by parts one observes that

$$\mathrm{TR}_\nu(f * g - g * f) = 0. \tag{3}$$

Otherwise, one has

$$\mathrm{TR}_\nu(f * g - g * f) = \frac{1}{4\pi i}\int_D \operatorname{tr}(df dg - dg df) + \cdots = \frac{1}{2\pi i}\int_{S^1} \operatorname{tr} f dg, \tag{4}$$

where the dots denote terms containing higher derivatives in the radial direction which integrate to zero through integration by parts due to the boundary conditions. Thus TR_ν is a true trace only in the subalgebra $\mathcal{B}_0$ of functions constant on the boundary. We shall also use the complex trace 'TR' defined as the zeroth order term in the formal Laurent series TR_ν. This is likewise a true trace on the algebra of functions vanishing at the boundary.

Any multiple of (2) by a Laurent series in ν is also a trace on the star subalgebra of constant functions on the boundary. However, the choice of the normalization will become apparent later. Actually, any trace is proportional, up to a factor in $\mathbb{C}[\nu^{-1}, \nu]]$, to the trace above, Fedosov [2]. (For a short proof in the manifold case see Gutt and Rawnsley [4].)

Let G be a compact matrix group and $DG \subset \mathcal{B}$ be the group of $n \times n$ matrix valued functions on D, formal power series in ν, which are invertible with respect to the star product and matrix multiplication and such that the boundary values belong to the matrix group G. Note that an inverse exists if and only if the zeroth order term in ν is invertible as an ordinary matrix valued function.

The group DG factorizes to a product of two spaces. The first factor is the set D_0G of zero order functions in DG and the second factor is the group K of functions of the form

$$f = 1 + \nu f_1 + \nu^2 f_2 + \ldots.$$

Note that any f of this type has an inverse as a formal power series in ν. The group K is contractible and it has a uniquely defined logarithmic function taking values in the formal power series without constant term.

We denote by $\mathcal{G}$ the subgroup of DG consisting of functions which are constants equal to the neutral element of G on the boundary circle.

Writing a general element $f \in \mathcal{G}$ as $f = gk$ with $g \in D_0G$ and $k \in K$ we can define the determinant as

$$\det(f) = \det(g) \cdot e^{\mathrm{TR}\,\log(k)} = \det(g) e^{\frac{1}{2\pi}\int_D \mathrm{tr}\, k_1}, \tag{5}$$

The determinant $\det(g)$ is defined as

$$\log\det(g) = \int_0^1 \mathrm{TR}(g(t)^{-1} * \partial_t g(t))\, dt, \tag{6}$$

where $g(t)$ (with $0 \leq t \leq 1$) is a homotopy in D_0G joining the neutral element $g(0)$ to $g = g(1)$. One should remember that the inverse $g(t)^{-1}$ is defined with respect to the star and matrix product, so it contains terms of higher order in ν. This determinant for the star product algebra was introduced by Melrose and Rochon in [10] in connection with a construction of determinant line bundles over pseudodifferential operators.

The expression $\mathrm{TR}(g^{-1} * dg)$ is a closed form on $\mathcal{G}$ by the tracial property of TR and for this reason $\log\det(g)$ depends only on the homotopy class of the path $g(t)$. In order that the determinant is well-defined independent of the path one only needs to check that the integral for generators of $\pi_1(\mathcal{G})$ is equal to a multiple of $2\pi i$:

Theorem 2.1. *Let G be connected and simply connected compact simple matrix Lie group and $f : S^1 \to \mathcal{G}$ be a closed smooth loop. Then the winding number of the determinant $\det(f(t,\cdot))$ around the loop is equal to the integer*

$$\frac{-1}{24\pi^2} \int_{S^1 \times D} tr(f^{-1} df)^3.$$

Here we have identified the parameter space D as a unit sphere S^2 since on the boundary of D all the functions $f \in \mathcal{G}$ take the constant value 1.

Proof. The proof is by a direct computation. We need to select a generator for $\pi_1(\mathcal{G}) \doteq \mathbb{Z}$. Since the topology of the group is determined by the constant part of formal power series in ν, we can assume that $f(t,\cdot)$ is zero order in ν. By the definition of 'TR', we need to compute the term first order in ν in the integral (the zeroth order term vanishes identically since $f^{-1}df$ is traceless)

$$\int_0^1 \mathrm{TR}_\nu(f(t,\cdot)^{-1} * f(t,\cdot))dt.$$

The inverse f^{-1}, as defined with respect to the star product, can be written as

$$g_0 + \nu g_1 + \nu^2 g_2 + \ldots,$$

where g_0 is the pointwise matrix inverse of the function $f(t,\cdot)$ and

$$g_1 = \frac{i}{2} df^{-1} df f^{-1}.$$

Thus

$$\begin{aligned}\int \mathrm{TR}(f^{-1} * \partial_t f)\, dt &= \frac{1}{2\pi} \int \int_D \frac{i}{2} \mathrm{tr}\left(df^{-1} df f^{-1} \partial_t f + df^{-1} d(\partial_t f)\right) dt \\ &= \frac{-1}{12\pi i} \int_{[0,1]\times D} \mathrm{tr}(f^{-1} df)^3\end{aligned}$$

which proves the Theorem. □

We define

$$\widehat{LG} = (DG \times S^1)/N, \tag{7}$$

where N is the normal subgroup consisting of pairs (g, λ) such that $g \in \mathcal{G}$ and $\lambda = \det(g)$.

This is a central extension by the circle S^1 of the loop group LG.

Theorem 2.2. *The Lie algebra of $\widehat{LG}$ is isomorphic as a vector space to the direct sum $L\mathbf{g} \oplus i\mathbb{R}$ with the commutator $[(f,\alpha),(g,\beta)] = ([f,g], c(f,g))$ where $[f,g]$ is the point-wise commutator of Lie algebra valued functions and c is the 2-cocycle*

$$c(f,g) = \frac{1}{2\pi i} \int_{S^1} tr\, f dg. \tag{8}$$

Proof. Let ψ be the local section of the circle bundle $\widehat{LG} \to LG$, defined in a neighborhood of the unit element in the loop group, given by

$$\psi(e^X) = e^{\tilde{X}},$$

where $X : S^1 \to \mathbf{g}$ and $\tilde{X} \in \mathcal{B}$ is equal to X on the boundary. For example, we can fix a smooth function $f(r)$ of the radius r such that $f(0) = 0, f(1) = 1$ and all the derivatives of f vanish at $r = 1$ and put $\tilde{X} = f(r)X$. The exponential is defined by the star product,

$$e^Z = \sum_n \frac{1}{n!} Z * Z * \cdots * Z, \ n \text{ factors}.$$

The section ψ is well-defined in an open set of G valued of loops where the logarithm is defined.

Locally, near the unit element, the central extension $\widehat{LG}$ is a product of an open set of LG with S^1. The local S^1 valued group cocycle is evaluated from

$$\det(\psi(e^{\tilde{X}}) \star \psi(e^{\tilde{Y}}) \star \psi(e^{-\tilde{X}}) \star \psi(e^{-\tilde{Y}})). \tag{9}$$

The Lie algebra cocycle $c(X,Y)$ is then the bilinear term in the expansion of (9) in powers of X,Y. Using the definition (5) of the determinant and the Baker-Campbell-Hausdorff formula

$$e^X e^Y = e^{X+Y+\frac{1}{2}[X,Y]+\dots}$$

we obtain

$$c(X,Y) = \mathrm{TR}[\tilde{X},\tilde{Y}]_\star = \frac{1}{2\pi i}\int_{S^1} \mathrm{tr} X dY. \qquad \square$$

The canonical connection on the loop group LG is given through the S^1 invariant 1-form θ on $\widehat{LG}$,

$$\theta = pr_c(g^{-1}dg), \tag{10}$$

where pr_c is the projection onto the center of the Lie algebra $\widehat{L\mathbf{g}}$. The curvature form Ω of this connection is the left invariant 2-form on LG which at the identity element is given by the cocycle $c : L\mathbf{g} \times L\mathbf{g} \to \mathbb{C}$. The winding number in Theorem 2.1 is then $1/2\pi$ times the integral of Ω over the set of loops $t \mapsto f(t,x)$ parametrized by $x \in D$.

3. Generalization to higher dimensions

The discussion above cannot directly be generalized to higher dimensions. The obstruction is the noninvariance of the boundary conditions. If we have a symplectic manifold with boundary of dimension $2d$ then the space of smooth functions with vanishing normal derivatives at the boundary is not closed in general. This happens already in the case of a disk in $\mathbb{R}^{2d}$ with the standard constant symplectic form in $\mathbb{R}^{2d}$. For this reason we focus only on a special case. Let $M = D \times S$ where S is a closed manifold of dimension $2d-2$ and D is the unit disk in $\mathbb{R}^2$. We assume that the algebra of functions $\mathcal{S}$ on S is equipped with a star product and D comes with a star product as in Section 2. The star product on $\mathcal{S}$ does not need to come from a bidifferential operator related to a symplectic form as in the case of the Moyal product. In fact, we can consider as well a product coming from quantum groups or quantum homogeneous spaces. However, what we need is an 'algebra of functions' possessing a trace functional $\mathrm{tr}_{\mathcal{S}}$. In this case the star product algebra of matrix valued functions on M is replaced by the tensor product of the star algebra of matrix valued functions on the disk and a star algebra $\mathcal{S}$. We can now impose vanishing normal derivatives at the boundary of D.

Example 3.1. The product of the symplectic disk D and a fuzzy sphere S^2_N. The fuzzy sphere is defined as the quotient by an ideal I of the noncommutative associative polynomial algebra in three variables x, y, z with relations $x * y - y * x = z, y * z - z * y = x, z * x - x * z = y$. The two-sided ideal I is generated by the single element $x^2 + y^2 + z^2 + N(N+1)$ where N is a nonnegative integer. Since x, y, z define the Lie algebra of $SU(2)$ the trace is defined as the matrix trace in an irreducible representation of dimension $N(N+1)$. The algebra is simply the algebra of square matrices in dimension $N(N+1)$.

Example 3.2. We can take as $\mathcal{S}$ the algebra of smooth $n \times n$ matrix valued functions in $\mathbb{R}^{2d-2}$ which decay faster than any inverse power of $|x|$ at infinity. The star product is defined as the Moyal product and the trace is the integral of a function over R^{2d-2}. In this case the product can actually be defined analytically, not only as a formal power series in ν. This is because the functions can be interpreted as symbols of infinitely smoothing pseudodifferential operators in $\mathbb{R}^{d-1}$. This is achieved by selecting a Lagrangian polarization $\mathbb{R}^{d-1} \oplus \mathbb{R}^{d-1}$ and interpreting the first $d-1$ variables as momenta and the last $d-1$ variables as coordinates. The algebra $\Psi^{-\infty}$ is a subalgebra of the algebra $\mathbf{g}_1$ of trace-class operators in the Hilbert space $H = L^2(\mathbb{R}^{d-1}, \mathbb{C}^N)$.

The linear functional

$$\mathrm{TR}(f) = \frac{1}{2\pi} \int_D dx\, dy\, \mathrm{tr}_{\mathcal{S}}\, f \tag{11}$$

is a trace in the subalgebra of functions which vanish on the boundary of D. Here $\mathrm{tr}_{\mathcal{S}}$ denotes the combined matrix trace and a trace in in the algebra $\mathcal{S}$.

The determinants are defined by straight-forward generalization of (5). The Lie algebra cocycle for $Map(S^1, \mathcal{S} \otimes \mathbf{g})$ becomes

$$c(f,g) = \frac{1}{2\pi} \int_{S^1} \mathrm{tr}_{\mathcal{S}}\, f dg. \tag{12}$$

In the case of Example 3.1 we get the standard central extension of the loop algebra of smooth maps from S^1 to matrices of size $nN(N+1) \times nN(N+1)$ whereas in the example 3.2 we have a central extension of the loop algebra $L\Psi^{-\infty}$ in the algebra $\Psi^{-\infty}$ of infinitely smoothing $n\times n$ matrix pseudodifferential operators over $\mathbb{R}^{d-1}$.

The Lie algebra cocycle (12) extends to the loop algebra $L\mathbf{g}_1$. A representation for $\widehat{L\mathbf{g}_1}$ is obtained essentially in the same way as for central extensions of the loop algebra $L\mathbf{g}$ based on a finite-dimensional Lie algebra $\mathbf{g}$. A highest weight representation of the Lie algebra $\mathbf{g}_1$ is given by an infinite increasing sequence of integers λ_i, $i \in \mathbb{Z}$, with $\lambda_i = \lambda_\infty$ for $i >> 0$ and $\lambda_i = \lambda_{-\infty} \leq \lambda_\infty$ for $i << 0$. The irreducible integrable highest weight representation corresponding to λ is then characterized by the existence of a cyclic vector v_λ such that

$$e_{ii}v_\lambda = \lambda_i v_\lambda \text{ and } e_{ij}v_\lambda = 0 \text{ for } i > j,$$

where the e_{ij}'s are the Weyl basis vectors in $\mathbf{g}_1$,

$$[e_{ij}, e_{kl}] = \delta_{jk}e_{il} - \delta_{il}e_{kj}.$$

Given the irreducible highest weight representation (λ) of $\mathbf{g}_1$ one obtains an irreducible highest weight representation of the central extension of the loop algebra $L\mathbf{g}_1$ by induction. The representation has a highest weight vector $v_{\lambda,k}$ characterized by

$$e_{ij}v_{\lambda,k} = 0 \text{ for } i > j \text{ and } x^{(n)}v_{\lambda,k} = 0 \text{ for } n < 0,$$

where $x^{(n)} = e^{in\phi}x \in L\mathbf{g}_1$ and k is the value of central element in the representation,

$$c(f,g) = \frac{k}{2\pi}\int_{S^1} \mathrm{tr}(f dg).$$

The representation integrates to an unitary representation of the group $\widehat{LG}$ if k is an integer with $\lambda_\infty - \lambda_{-\infty} \leq k$, see the monograph by Kac [6].

The construction of the central group extension $\widehat{LG}$ for the case of a compact matrix group G can now be extended without any changes to the case when G is the infinite-dimensional Lie group of unitary pseudodifferential operators A such that $A-1$ is trace class.

Example 3.3. We deform the gauge current algebra in 3 space dimensions. First, let $\mathbf{n}$ be the ideal of pseudodifferential operators, on a compact spin manifold M of dimension 3, of degree less or equal to -2. All pseudodifferential operators are taken with matrix coefficients. The matrices act in the tensor product of the spinor bundle and a trivial vector bundle V over M. The finite-dimensional Lie algebra $\mathbf{g}$ of a gauge group G acts in the fibers of V through a matrix representation. For each smooth map $X : M \to \mathbf{g}$ we define a deformed operator

$$\tilde{X} = X + \frac{1}{4(D^2+1)}[D,[D,X]],$$

where D is the Dirac operator on M defined by a fixed metric and spin structure. The difference $\tilde{X} - X$ is a pseudodifferential operator of order -1. One easily checks that

$$[\tilde{X},\tilde{Y}] = \widetilde{[X,Y]} \bmod \mathbf{n}.$$

Denote by $\mathbf{p}$ the Lie algebra of pseudodifferential operators such that the leading symbols of order 0 and -1 are given by the leading symbols of symbols of the deformed operators $\{\tilde{X}|X \in Map(M,\mathbf{g})\}$. Let $\epsilon = D/|D|$. Then $[\epsilon, T]$ is Hilbert-Schmidt for all $T \in \mathbf{p}$.

We have the exact sequence

$$0 \to \mathbf{n} \to \mathbf{p} \to Map(M,\mathbf{g}),$$

where the second map is embedding of Lie algebras and the third map extracts the zero order part X of an element $T = \tilde{X} + z \in \mathbf{p}$, where $z \in \mathbf{n}$.

The Lie algebra $\mathbf{p}$ is a subalgebra of $\mathbf{gl}_{res}$ where the latter consists of bounded operators T in the Hilbert space H such that $[\epsilon, T]$ is Hilbert-Schmidt. The algebra $\mathbf{gl}_{res}$ has a canonical central extension $\widehat{\mathbf{gl}}_{res}$ defined by the cocycle

$$c(X,Y) = \frac{1}{4}\mathrm{tr}\, \epsilon[\epsilon, X][\epsilon, Y].$$

The restriction to $\mathbf{p}$ gives a central extension $\widehat{\mathbf{p}}$ of $\mathbf{p}$. Likewise, we have a central extension $\widehat{\mathbf{n}}$ of $\mathbf{n} \subset \mathbf{gl}_{res}$. Putting these together we have the extension

$$0 \to \widehat{\mathbf{n}} \to \widehat{\mathbf{p}} \to Map(M,\mathbf{g}).$$

The algebra $\widehat{\mathbf{p}}$ has unitary highest weight representations. For example, the Fermionic Fock space $\mathcal{F}$ based on the polarization $H = H_+ \oplus H_-$ carries through canonical quantization a resepresentation of $\widehat{\mathbf{gl}}_{res}$ and thus of $\widehat{\mathbf{p}}$. However, this representation does not preserve the domain of the quantization $\widehat{D}$ of the Dirac operator D.

References

1. F. Bayen, M. Flato, C. Fronsdal, A. Lichnerowicz, and D. Sternheimer: Deformation theory and quantization, I–II. *Ann. Physics* **111**, 61-151 (1978).

2. B.V. Fedosov: *Deformation Quantization and Index Theory*, Section 5.6. Akademie Verlag, Berlin, 1996.
3. H. Grosse and J. Madore: A noncommutative version of the Schwinger model. *Phys. Lett.* **B283**, 218-222 (1992).
4. S. Gutt and J. Rawnsley: Traces for star products on symplectic manifolds. math.QA/0105089. *J. Geom. Phys.* **42**, No. 1-2, 12-18 (2002).
5. F. Lizzi, P. Vitale, A. Zampini: The fuzzy disk. *J. High Energy Phys.* 0308:057, 16pp. (electronic) (2003). hep-th/0306247.
6. V. Kac: *Infinite-dimensional Lie Algebras*, Third Edition, Cambridge University Press, Cambridge, 1990.
7. J. Madore: *Noncommutative Geometry and Applications*, Cambridge University Press, Cambridge, 1995.
8. J. Mickelsson: Kac-Moody groups, topology of the Dirac determinant bundle and fermionization. *Comm. Math. Phys.* **110**, 173-183 (1987).
9. J. Mickelsson: *Current Algebras and Groups*, Plenum Press, London and New York, 1989.
10. R. Melrose and F. Rochon: Periodicity of the determinant line bundle. Work in progress.
11. M. Murray: Another construction of the central extension of the loop group. *Comm. Math. Phys.* **116**, 73-80 (1988).

Received by the editors September 14, 2005; Revised December 2, 2005

Analysis, Geometry and Topology
of Elliptic Operators, pp. 411–426

AN ELEMENTARY PROOF OF THE HOMOTOPY EQUIVALENCE BETWEEN THE RESTRICTED GENERAL LINEAR GROUP AND THE SPACE OF FREDHOLM OPERATORS

TILMANN WURZBACHER
Laboratoire de Mathématiques
et Applications de Metz - UMR 7122
CNRS et Université Paul Verlaine - Metz,
Ile du Saulcy, F-57045 Metz, France
wurzbacher@math.univ-metz.fr

Dedicated to Krzysztof P. Wojciechowski on his 50th birthday

We complete a gap in the proof of a crucial result of Pressley and Segal's book on "Loop groups". It is well-known that given a polarized, separable complex Hilbert space of infinite dimension ($K = K_+ \oplus K_-$), the elements of the restricted linear group $GL_{\mathrm{res}}(K, K_+)$ can be written as two-by-two matrices of operators and that notably the upper left entry of these is a Fredholm operator from K_+ to itself. The resulting map from the restricted linear group to the space of Fredholm operators on K_+ is a homotopy equivalence. We complete the proof of this proposition (6.2.4 in "Loop groups") relying on a simple "Boardman-Vogt type lemma". We then remark on some applications of this result to classifying spaces of principal bundles, to geometric quantization in infinite dimensions and to string structures on loop spaces.

2000 *Mathematics Subject Classification.* Primary 55P10; Secondary 58B05, 22E65, 55R35

1. Introduction

The restricted Grassmannian of a polarized Hilbert space G_{res} (see the splendid fundamental reference "Loop groups" [15] of Pressley and Segal or the more recent [22] of the author) plays an important rôle in subjects as integrable partial differential equations, fermionic second quantization, infinite dimensional Lie groups and Lie algebras, and boundary problems of elliptic partial differential operators (to cite only some of the occurrences of this Grassmannian).

Though the fact that G_{res} is homotopy equivalent to the space of Fredholm operators on a separable complex Hilbert space is a folkloric fact since some time, the beautiful proof of it given in [15] (Proposition (6.2.4)), and quoted in several places in the literatur (see, e.g., the important [8] and [11] of Freed respectively of Mickelsson) seems to be incomplete.

We give here a direct, "elementary" argument for the crucial part of the proof relying on the (obvious) construction of a global section of a certain projection and on a "Boardman-Vogt lemma with parameters" (see Section 3 for the details).

The second section reviews the needed definitions and some basic facts, the third gives our proof, and the fourth section details the arguments of [15] on how to realize the universal bundle

$$E(U(\infty)) \longrightarrow B(U(\infty))$$

by a smooth homomorphism of Banach Lie groups, using the homotopy results of the third section.

Finally, in the fifth section we remark on applications to the theory of characteristic classes of $GL(\infty)$-bundles (compare [8]), to geometric quantization of the restricted Grassmannian (compare [22]) and to the geometry of free loop spaces (compare, e.g., a recent preprint [17] of Spera and the author).

Let us remark that related homotopy type calculations were made by various authors (e.g., by Booss-Bavnbek and Wojciechowski in [3] with a correction, together with Furutani, in [2], by Carey and Phillips in [6] and by Neeb in [12]). Nevertheless, we found it useful to complete the argument of Pressley and Segal (for its own sake and in view of other possible applications of it).

2. The restricted Grassmannian and the restricted general linear group

For convenience of the reader we follow as closely as possible the notations of [15], where most of the below mentioned facts are proven and detailed (see also [22]).

Let K be a complex separable Hilbert space and K_+ a closed complex subspace of infinite dimension and codimension. Denoting its orthogonal complement by K_- we have orthogonal projections p_+ and p_- onto K_+ respectively K_- and we refer to the orthogonal decomposition

$$K = K_+ \oplus K_-$$

as a *polarization of* K.

The *restricted Grassmannian (of the polarized Hilbert space $K = K_+ \oplus K_-$)* is then by definition the following set

$$\begin{aligned} G_{\text{res}} &= G_{\text{res}}(K, K_+) \\ &= \{W \subset K \mid W \text{ is a closed complex subspace s.t. } p_+|_W : W \to K_+ \\ &\qquad \text{is Fredholm and } p_-|_W : W \to K_- \text{ is Hilbert-Schmidt}\}. \end{aligned}$$

It is well-known that G_{res} is a complex Kähler manifold modelled on the separable Hilbert space $\mathcal{L}^2(K_+, K_-)$ of Hilbert-Schmidt operators from K_+ to K_-.

Remark 2.1.

(1) Up to isomorphy in the category of topological spaces, complex manifolds or Kähler manifolds, G_{res} is independent of the choice of a separabel complex Hilbert space K and its reference subspace K_+ of infinite dimension and codimension. This allows us the slight abuse of speaking of "the restricted Grassmannian" without mentioning the data K and K_+.

(2) Let us also mention that homotopy equivalent spaces can be defined by demanding that $p_-\Big|_W$ is in the p-th Schatten class $\mathcal{L}^p$ for $1 \leq p \leq \infty$ ($\mathcal{L}^\infty$ denotes the compact operators).

The natural automorphism groups of G_{res} are the *restricted general linear group (of the polarized Hilbert space $K = K_+ \oplus K_-$)*

$$\begin{aligned} GL_{\text{res}} &= GL_{\text{res}}(K, K_+) \\ &= \{A \in GL(K) \mid [A, p_+] \text{ is a Hilbert Schmidt operator}\} \end{aligned}$$

(where $GL(K)$ is the group of bounded linear isomorphisms of K) and the *restricted unitary group (of the polarized Hilbert space $K = K_+ \oplus K_-$)*

$$U_{\text{res}} = U_{\text{res}}(K, K_+) = GL_{\text{res}}(K, K_+) \cap U(K).$$

Writing an element A of $\mathcal{B}(K)$, the space of bounded linear maps from K to K, as a (2×2)-matrix of operators:

$$A = \begin{pmatrix} a & b \\ c & d \end{pmatrix}$$

with

$$a : K_+ \to K_+, c : K_+ \to K_- \text{ etc.},$$

it follows that an invertible element A of $\mathcal{B}(K)$ is in GL_{res} if and only if b and c are Hilbert-Schmidt operators. Furthermore, if A is in GL_{res}, then a and d are Fredholm operators.

Let us also recall that GL_{res} (respectively U_{res}) is a complex-analytic (respectively real-analytic) Banach Lie group with Lie algebra

$$\mathfrak{gl}_{\text{res}} = \left\{ \begin{pmatrix} \alpha & \beta \\ \gamma & \delta \end{pmatrix} \in \mathcal{B}(K) \middle| \beta \text{ and } \gamma \text{ are Hilbert-Schmidt} \right\}$$

$$(\text{resp. } \mathfrak{u}_{\text{res}} = \left\{ \begin{pmatrix} \alpha & \beta \\ \gamma & \delta \end{pmatrix} \in \mathfrak{gl}_{\text{res}} \middle| \alpha^* = -\alpha, \delta^* = -\delta, \beta = -\gamma^* \right\})$$

and with exponential map exp given by the exponential series.

We note the connected component of GL_{res} (respectively of U_{res}) that contains the neutral element by GL^0_{res} (respectively U^0_{res}), and write, furthermore G^0_{res} for the connected component of G_{res} that contains K_+.

Let us also recall the following by now well-known facts:

Lemma 2.1.

(i) *The action $\vartheta : GL_{\text{res}} \times G_{\text{res}} \to G_{\text{res}}, (A, W) \mapsto A(W)$ (respectively its restriction to U_{res}) is complex-analytic (respectively real-analytic) and transitive.*

(ii) *The isotropy group of $W = K_+$ under the GL_{res}-action is the connected and contractible Banach Lie group*

$$\mathcal{P} = \left\{ A = \begin{pmatrix} a & b \\ 0 & d \end{pmatrix} \in GL_{\text{res}} \right\}.$$

(iii) *The map $GL_{\text{res}} \xrightarrow{q} G_{\text{res}}, A \mapsto \vartheta(A, K_+) = A(K_+)$ is a holomorphic (locally trivial) $\mathcal{P}$-principal bundle, i.e. G_{res} is biholomorphically equivalent to the complex quotient manifold $GL_{\text{res}}/\mathcal{P}$.*

(iv) *The map q is a homotopy equivalence.*

Proof. The first statement and the determination of the stabilizer of K_+ are fairly obvious.

Since $\mathcal{P}$ is biholomorphic (as a complex manifold, not as a group) to $GL(K_+) \times GL(K_-) \times \mathcal{L}^2(K_-, K_+)$, it follows from Kuiper's theorem (saying that $GL(H)$ is contractible for H a separable complex Hilbert space, see Kuiper [9]) that $\mathcal{P}$ is connected and contractible.

The fibres of the holomorphic map q are the orbits of the right-action given by multiplication on the right of $\mathcal{P}$ on GL_{res}. In order to show local triviality we recall the graph coordinates near K_+ on G_{res}. Let

$$\mathcal{U}_{K_+} = \{W \in G_{\text{res}} \mid p_+|_W : W \to K_+ \text{ is an isomorphism}\}$$

and

$$\varphi : \mathcal{U}_{K_+} \to \mathcal{L}^2(K_+, K_-)$$

the inverse of the map

$$T \mapsto \Gamma_T = \{v + Tv \in K = K_+ \oplus K_- \mid v \in K_+\}$$

("the graph of T").

Observing that

$$\mathfrak{m} := \left\{ \begin{pmatrix} 0 & 0 \\ T & 0 \end{pmatrix} \middle| T \in \mathcal{L}^2(K_+, K_-) \right\} \cong \mathcal{L}^2(K_+, K_-)$$

is a topological complement of Lie $\mathcal{P} = \left\{ \begin{pmatrix} \alpha & \beta \\ 0 & \delta \end{pmatrix} \in \mathfrak{gl}_{\text{res}} \right\}$, the Lie algebra of $\mathcal{P}$, in $\mathfrak{gl}_{\text{res}} = \operatorname{Lie} GL_{\text{res}}$, there is an open neighborhood $V_\mathfrak{m}$ of 0 in $\mathfrak{m}$ such that $\exp|_{V_\mathfrak{m}}$ is a holomorphic diffeomorphism from $V_\mathfrak{m}$ onto its image $S := \exp(V_\mathfrak{m})$, a complex submanifold of GL_{res} having the property that $q|_S$ is injective. The map

$$\sigma := \exp \circ \varphi : \varphi^{-1}(V_\mathfrak{m}) \to GL_{\text{res}}$$

is a holomorphic local section of q since

$$(q \circ \exp) \begin{pmatrix} 0 & 0 \\ T & 0 \end{pmatrix} = \Gamma_T = \varphi^{-1}(T) \qquad \text{for all } T \in \mathcal{L}^2(K_+, K_-),$$

$$\text{i.e.} \qquad (q \circ \sigma)(W) = (q \circ \exp \circ \varphi)(W) = W \qquad \text{for all } W \in \varphi^{-1}(V_\mathfrak{m}).$$

Since q is obviously equivariant with respect to the natural left-actions of GL_{res}, it follows that q is a holomorphic locally trivial $\mathcal{P}$-principal bundle. It now easily follows that G_{res} is biholomorphic to $GL_{\text{res}}/\mathcal{P}$ with its natural manifold structure as a homogeneous space. Since q is GL_{res}-equivariant, all connected components of G_{res} (and of GL_{res}, of course) are diffeomorphically equivalent and homogeneous under GL^0_{res}. Let us thus concentrate on the restriction q^0 of q to GL^0_{res}, i.e. on

$$q^0 : GL^0_{\text{res}} \xrightarrow{\mathcal{P}} G^0_{\text{res}}.$$

Since G^0_{res} is a connected riemannian manifold (see [15] or [17] for more details on the riemannian structure), it is metrisable (and thus Hausdorff); furthermore this manifold, modelled on a separable Hilbert space, is paracompact and second countable. It follows notably that all locally trivial fibre bundles over G^0_{res} are *numerable* (i.e. allow for an open covering $\mathcal{U}$ of local trivializations such that there exists a locally finite partition of unity subordinate to it).

Since $\mathcal{P}$ is connected and contractible, we can take $\mathcal{P} \to (*)$ as the universal $\mathcal{P}$-principal bundle $E\mathcal{P} \to \mathcal{B}P$ and we conclude that all numerable $\mathcal{P}$-principal bundles are trivializable, i.e. allow for a global continuous section. Applying this to q^0 we find that it is isomorphic to the trivial principal bundle $G^0_{\text{res}} \times \mathcal{P} \to G^0_{\text{res}}$ with contractible fibre $\mathcal{P}$ and thus obviously a (fibre) homotopy equivalence.

Since the above argument holds for all connected compontents of G_{res}, we find that $q : GL_{\text{res}} \to G_{\text{res}}$ is a (fibre) homotopy equivalence. □

Remark 2.2. Using polar decomposition it is easy to show that U_{res} is a strong homotopy retract of GL_{res}, and thus notably homotopy-equivalent to it. A fortiori, U_{res} is then also homotopy-equivalent to G_{res}.

3. The fundamental homotopy equivalence

In order to complete the proof of Proposition (6.2.4) in [15] "in an elementary manner", we rely on the following *Boardman-Vogt type result* (compare, e.g., to Lemma 11.58 in Switzer's book [18]).

Lemma 3.1. *Let H_- be a separable complex Hilbert space and $\{e_n \mid n \in \mathbb{Z}$ and $n < 0\}$ be an orthonormal (Hilbert-)basis of H_-. Let furthermore $\psi : H_- \to H_- \oplus H_-$ be the unitary isomorphism determined by setting for $m < 0$:*

$$\psi(e_{2m}) = e_m \oplus 0 \quad \textit{and} \quad \psi(e_{2m+1}) = 0 \oplus e_m\,,$$

and let φ be the inverse of ψ, and let finally

$$i_1 : H_- \to H_- \oplus H_-$$

be the continuous linear injection onto the first factor, i.e., $i_1(v) = v \oplus 0$. Then there exists a continuous map

$$f : I \times H_- \to H_- \qquad (\textit{where } I = [0,1])$$

such that $f_0 = Id_{H_-}$, $f_1 = \varphi \circ i_1$ and for all $t \in I$, f_t is an injective linear continuous map from H_- to H_-.

Otherwise stated, the following diagram commutes up to homotopy:

$$\begin{array}{ccc} H_- & \overset{i_1}{\hookrightarrow} & H_- \oplus H_- \\ & & \\ Id_{H_-} & \searrow & \downarrow \varphi \\ & & \\ & & H_-. \end{array}$$

Proof. Let $b : H_- \to H_-$ be defined by $b(e_n) = e_{2n}$ for all $n < 0$. It follows that b is a linear isometry and thus notably injective. Furthermore, $\varphi \circ i_1 = b$ by direct calculation. Let now

$$f_t = (1-t) \cdot Id_{H_-} + t \cdot b \quad \text{for } t \in I.$$

Obviously, f_t is a continuous linear map for all t, $f_0 = Id_{H_-}$, $f_1 = b = \varphi \circ i_1$ and the map $f : I \times H_- \to H_-, (t, v) \mapsto f_t(v)$ is continuous. It remains only to show that f_t is injective for $t \in]0,1[$.

Let $v \in H_-$ and $t \in]0,1[$ such that $f_t(v) = 0$. Then $b(v) = \left(\frac{t-1}{t}\right) v$ and $\frac{t-1}{t} = -1$ since b is an isometry. Notably, it follows that $b^2(v) = v$, and for all $k \in \mathbb{N}, (b^2)^k(v) = v$.

On the other hand one has for all $n < 0$, $b^2(e_n) = e_{4n}$ and thus $(b^2)^k(e_n) = e_{4^k \cdot n}$. One easily deduces that for all w in H_-, the sequence $\left((b^2)^k(w)\right)_{k \in \mathbb{N}}$ converges weakly to zero.

One concludes that $f_t(v) = 0$ implies that $v = 0$, i.e. f_t is injective. □

Let now $\pi : GL_{\text{res}} \to \mathcal{B}(K_+) \times \mathcal{L}^2(K_+, K_-)$, $\pi \begin{pmatrix} a & b \\ c & d \end{pmatrix} = \binom{a}{c}$ and $\mathcal{F} := \pi(GL_{\text{res}})$. We now show that π is a homotopy equivalence. (We denote the open set of Fredholm operators in $\mathcal{B}$ by $\text{Fred}(K_+)$ in the sequel.)

Lemma 3.2.

(i) *$\mathcal{F} = \{\binom{a}{c} \in$ Fred$(K_+) \times \mathcal{L}^2(K_+, K_-) \mid c|_{\text{Ker } a}$ is injective $\}$ and thus $\mathcal{F}$ is open in $\mathcal{B}(K_+) \times \mathcal{L}^2(K_+, K_-)$.*

(ii) *The action $\theta : GL_{\text{res}} \times \mathcal{F} \to \mathcal{F}$,*

$$\theta\left(\begin{pmatrix} \alpha & \beta \\ \gamma & \delta \end{pmatrix}, \pi\begin{pmatrix} a & b \\ c & d \end{pmatrix}\right) = \pi\left(\begin{pmatrix} \alpha & \beta \\ \gamma & \delta \end{pmatrix} \cdot \begin{pmatrix} a & b \\ c & d \end{pmatrix}\right)$$

is well-defined, holomorphic and transitive.

(iii) *The stabilizer of* $\binom{Id_{K_+}}{0} = \binom{1}{0}$ *is the connected and contractible Banach Lie group* $\mathcal{B} := \left\{ \begin{pmatrix} 1 & b \\ 0 & d \end{pmatrix} \middle| \, b \in \mathcal{L}^2(K_-, K_+) \text{ and } d \in GL(K_-) \right\}$.

(iv) *The open set* $\mathcal{F}$ *is biholomorphic to the homogeneous space* $GL_{\text{res}}/\mathcal{B}$ *and the map* $\pi : GL_{\text{res}} \to \mathcal{F}$ *is a homotopy equivalence.*

Proof. Let

$$\check{\mathcal{F}} := \left\{ \begin{pmatrix} a \\ c \end{pmatrix} \in \text{Fred}(K_+) \times \mathcal{L}^2(K_+, K_-) \mid c\Big|_{\ker a} \right\}$$

is injective. It is clear that $\mathcal{F} \subset \check{\mathcal{F}}$. On the other hand, given $\binom{a}{c} \in \check{\mathcal{F}}$, one constructs easily an element A of GL_{res} such that the first column of A is $\binom{a}{c}$, thus proving the reversed inclusion. Openess of $\mathcal{F}$ is now elementary, as well as part (ii) of the lemma and the determination of the stabilizer of $\binom{1}{0}$. Since $\mathcal{B}$ is homeomorphic to $GL(K_-) \times \mathcal{L}^2(K_-, K_+)$, Kuiper's theorem shows that $\mathcal{B}$ is contractible.

Upon observing that $\text{Lie}\, \mathcal{B} = \left\{ \begin{pmatrix} 0 & B \\ 0 & D \end{pmatrix} \middle| B \in \mathcal{L}^2(K_-, K_+) \text{ and } D \in \mathcal{B}(K_-) \right\}$ allows a topological complement in $\mathfrak{gl}_{\text{res}}$, and using the contractibility of $\mathcal{B}$, the proof of the last part of the lemma is similar to the proof of Lemma 2.1, (iii) and (iv), and thus omitted. □

We can now prove the crucial homotopical result:

Lemma 3.3. *The map* $p := p_1\Big|_{\mathcal{F}} : \mathcal{F} \to \text{Fred}(K_+)$, $p\binom{a}{c} = a$ *is a homotopy equivalence.*

Proof.

Step 1: Construction of a continuous global section

Let $\{e_n \mid n \in \mathbb{Z}\}$ be a Hilbert basis of K such that K_- is generated by $\{e_n \mid n < 0\}$ and K_+ by $\{e_n \mid n \geq 0\}$. The map $c_0 : K_+ \to K_-$ defined by $c_0(v) = \sum_{n<0} \frac{1}{|n|} < e_{|n+1|}, v > e_n$ (with $\langle , \rangle$ the scalar product of K) is an injective, linear Hilbert-Schmidt operator. it follows that for $a \in \text{Fred}(K_+)$, the point $\binom{a}{c_0}$ is in $\mathcal{F}$, i.e. $s_0 : \text{Fred}(K_+) \to \mathcal{F}$, $s_0(a) = \binom{a}{c_0}$ is a continuous section of p. Note that this construction also shows that p is surjective.

Step 2: Construction of an auxiliary homotopy with values in a "doubled fibre"

Setting $\tilde{\mathcal{F}} = \left\{ \binom{a}{\tilde{c}} \in \mathrm{Fred}(K_+) \times \mathcal{L}^2(K_+, K_- \oplus K_-) \mid \tilde{c}\Big|_{\mathrm{Ker}\ a} \text{ is injective} \right\}$ and again $I = [0,1]$ we define

$$\tilde{G} : I \times \mathcal{F} \to \tilde{\mathcal{F}} \text{ by } \tilde{G}(t, \begin{pmatrix} a \\ c \end{pmatrix}) = \begin{pmatrix} a \\ (1-t)c \oplus tc_0 \end{pmatrix}.$$

(We define for $d', d'' \in \mathcal{L}^2(K_+, K_-)$ and $v \in K_+$, $(d' \oplus d'')(v) := d'(v) \oplus d''(v) \in K_- \oplus K_-$.)

The maps $(1-t)c \oplus tc_0$ are of course in $\mathcal{L}^2(K_+, K_- \oplus K_-)$, and $((1-t)c \oplus tc_0)\Big|_{\mathrm{Ker}\ a}$ is injective since $c\Big|_{\mathrm{Ker}\ a}$ is injective.

Setting, as usually, $\tilde{G}_t(\cdot) = \tilde{G}(t, \cdot)$, we have furthermore

$$\tilde{G}_0 \begin{pmatrix} a \\ c \end{pmatrix} = \begin{pmatrix} a \\ c \oplus 0 \end{pmatrix} \text{ and } \tilde{G}_1 \begin{pmatrix} a \\ c \end{pmatrix} = \begin{pmatrix} a \\ 0 \oplus c_0 \end{pmatrix}.$$

Step 3: Construction of a homotopy with values in $\mathcal{F}$

Let $\psi : K_- \to K_- \oplus K_-$ and $\varphi = \psi^{-1} : K_- \oplus K_- \to K_-$ be as in Lemma 3.1 upon taking $H_- = K_-$ and the above fixed Hilbert basis $\{e_n \mid n < 0\}$. We have an induced continuous map

$$\Phi : \tilde{\mathcal{F}} \to \mathcal{F},\ \Phi \begin{pmatrix} a \\ d' \oplus d'' \end{pmatrix} = \begin{pmatrix} a \\ \varphi \circ (d' \oplus d'') \end{pmatrix}$$

since $\varphi \circ (d' \oplus d'')$ is Hilbert-Schmidt and injective on the kernel of a (since $d' \oplus d''$ is injective on this kernel).

Putting now $G(t, \binom{a}{c}) = \Phi(\tilde{G}(t, \binom{a}{c}))$ we have a continuous homotopy $G : I \times \mathcal{F} \to \mathcal{F}$ such that $G_0 \binom{a}{c} = \binom{a}{\varphi \circ (c \oplus 0)}$ and $G_1 \binom{a}{c} = \binom{a}{\varphi \circ (0 \oplus c_0)}$. Note that

$$s : \mathrm{Fred}(K_+) \to \mathcal{F}, \qquad a \mapsto \begin{pmatrix} a \\ \varphi \circ (0 \oplus c_0) \end{pmatrix}$$

is a continuous section of p.

Step 4: Construction of a preliminary homotopy

Applying Lemma 3.1 to $H_- = K_-$ and the above fixed Hilbert basis we get a continuous family $\{f_t : K_- \to K_- \mid t \in I\}$ of injective continuous linear maps such that $f_0 = Id_{K_-}$ and $f_1 = \varphi \circ i_1$ (where i_1 denotes again the injection $K_- \to K_- \oplus K_-$ onto the first factor).

Let now $F = I \times \mathcal{F} \to \mathcal{F}$ be defined by $F(t, \binom{a}{c}) = \binom{a}{f_t \circ c}$ and observe that $f_t \circ c$ is Hilbert-Schmidt and injective on Ker a, i.e. $F_t \binom{a}{c}$ is in $\mathcal{F}$. Obviously, one has

$$F_0 \begin{pmatrix} a \\ c \end{pmatrix} = \begin{pmatrix} a \\ c \end{pmatrix} \text{ and } F_1 \begin{pmatrix} a \\ c \end{pmatrix} = \begin{pmatrix} a \\ f_1 \circ c \end{pmatrix} = \begin{pmatrix} a \\ \varphi \circ (c \oplus 0) \end{pmatrix} = G_0 \begin{pmatrix} a \\ c \end{pmatrix}.$$

Step 5: The homotopy

Since $p \circ s = Id_{\mathrm{Fred}(K_+)}$, it is enough to exhibit a continuous homotopy from $Id_{\mathcal{F}}$ to $s \circ p$ in order to show that p is a homotopy equivalence. Let $H : I \times \mathcal{F} \to \mathcal{F}$ be defined as $H_t = F_{2t}$ for $0 \leq t \leq 1/2$ and $H_t = G_{2t-1}$ for $1/2 \leq t \leq 1$. We then have that H is continuous, $H_0 = F_0 = Id_{\mathcal{F}}$, $H_{1/2} = F_1 = G_0$, and $H_1 = G_1 = s \circ p$. □

Remark 3.1. Observe that the homotopy $H : I \times \mathcal{F} \to \mathcal{F}$ preserves the fibration $p = \mathcal{F} \to \mathrm{Fred}(K_+)$, i.e. for all t in I, we have $p \circ H_t = p$.

We now arrive at

Corollary 3.1. (= Proposition (6.2.4) in "Loop groups" [15]) *The map* $GL_{\mathrm{res}} \xrightarrow{p \circ \pi} \mathit{Fred}(K_+), A = \begin{pmatrix} a & b \\ c & d \end{pmatrix} \mapsto a$, *is a homotopy equivalence.*

Proof. Factorizing the given map as follows

$$GL_{\mathrm{res}} \xrightarrow{\pi} \mathcal{F} \xrightarrow{p} \mathrm{Fred}(K_+)$$

and applying part (iv) of Lemma 3.2 and Lemma 3.3 immediately yields the claim. □

4. The restricted general linear group as the classifying space of $U(\infty)$

Following essentially the ideas of [15] but giving substantially more details, we will explain in this section how to realize the universal principal bundle $U(\infty) \to EU(\infty) \to BU(\infty)$ by smooth homomorphisms of Banach Lie groups:

$$GL^1(K_+) \longrightarrow \mathcal{E} \longrightarrow GL^0_{\mathrm{res}} \quad (*)$$

and how to deduce from $(*)$ — via Bott periodicity — all homotopy groups of the restricted general linear group land thus of the restricted Grassmannian as well.

Let us first, for H a separable complex Hilbert space and $1 \leq p \leq \infty$, define

$$GL^p(H) := \{T \in GL(H) \mid T - Id_H \in \mathcal{L}^p(H)\},$$

and for a polarized Hilbert space $K = K_+ \oplus K_-$ as in the preceding two sections,

$$\mathcal{E} := \{(A,q) \in GL^0_{\text{res}} \times GL(K_+) \mid a - q \in \mathcal{L}^1(K_+)\},$$

where, again, $A = \begin{pmatrix} a & b \\ c & d \end{pmatrix}$. We give $\mathcal{E}$ the topology induced from the set-theoretic inclusion $\mathcal{E} \hookrightarrow GL^0_{\text{res}} \times \mathcal{L}^1(K_+), (A,q) \mapsto (A, a-q)$.

Lemma 4.1.

(i) *For $1 \le p \le \infty$ and H a separable complex Hilbert space, $GL^p(H)$ is a complex Banach Lie group.*
(ii) *For a polarized complex Hilbert space $K = K_+ \oplus K_-$, $\mathcal{E}$ is a complex Banach Lie group.*
(iii) *The map $\beta : \mathcal{E} \to GL^0_{\text{res}}, (A,q) \mapsto A$ is a smooth (in fact, complex-analytic) surjective homomorphism of Banach Lie groups, and its kernel $\{(1,q) \in \mathcal{E} \mid 1-q \in \mathcal{L}^1(K_+)\}$ is canonically isomorphic to $GL^1(K_+)$.*

Proof. The set-theoretic inclusion

$$\kappa : GL^p(H) \hookrightarrow \mathcal{L}^p(H), T \mapsto T - Id_H$$

induces a topology on $GL^p(H)$ such that it becomes a topological group. Since the image of κ is open in the complex Banach space $\mathcal{L}^p(H)$, part (i) easily follows.

Parts (ii) and (iii) are from [15] and proven in detail in [22], Section II.3. □

We can now give the central property of $\mathcal{E}$.

Proposition 4.1. (= Proposition (6.6.2) in "Loop groups" [15]) *The Banach Lie group $\mathcal{E}$ is connected and contractible.*

Proof. Define

$$\delta : \mathcal{E} \to GL(K_+) \times \mathcal{L}^1(K_+), \delta(A,q) := (q, aq^{-1} - 1)$$

and

$$\begin{aligned}\gamma : GL(K_+) \times \mathcal{L}^1(K_+) &\to \text{Fred}^0(K_+) \\ &= \{a \in \text{Fred}(K_+) \mid a \text{ has index zero}\}, \quad \gamma(q,t) = (1+t)\cdot q.\end{aligned}$$

Since $p \circ \pi : GL^0_{\text{res}} \to \text{Fred}^0(K_+), A \mapsto a$ is surjective, both maps, δ and γ, are easily seen to be surjective as well. Furthermore, the γ-fibres are the orbits of the following (holomorphic) $GL^1(K_+)$-action:

$$GL^1(K_+) \times (GL(K_+) \times \mathcal{L}^1(K_+)) \to GL(K_+) \times \mathcal{L}^1(k_+),$$
$$((1+s),(q,t)) \mapsto (1+s)q, (1+t)(1+s)^{-1} - 1),$$

where an element of $GL^1(K_+)$ is written as $1+s$ with s in $\mathcal{L}^1(K_+)$ (and $1+s$ invertible!). Let us now consider the following diagram of (at least) continuous maps:

$$\begin{array}{ccc} \mathcal{E} & \xrightarrow{\delta} & GL(K_+) \times \mathcal{L}^1(K_+) \\ \beta \downarrow & & \downarrow \gamma \\ Gl^0_{\text{res}} & \xrightarrow{p\circ\pi} & \text{Fred}^0(K_+) \,. \end{array}$$

One directly checks that the diagram commutes and that the $GL^1(K_+)$-bundle β is isomorphic to the pullback of the bundle γ under the map $p \circ \pi$. Thus, since $p \circ \pi$ is a homotopy equivalence by the result of Corollary 3.1 (restricted to connected components) the same holds true for δ. We conclude that $\mathcal{E}$ is homotopically equivalent to the contractible space $GL(K_+) \times \mathcal{L}^1(K_+)$ and is thus itself contractible to a point. □

Les us recall that $GL(\infty, \mathbb{C}) = \cup_{N \geq 1} GL(N, \mathbb{C})$ and $U(\infty) = \cup_{N \geq 1} U(N)$ with their direct limit (or "colimit") topologies are homotopy equivalent by the usual polar decomposition argument. By results of Palais (Theorems (B) and (E) in [13]), $GL(\infty, \mathbb{C})$ is furthermore homotopy equivalent to $GL^p(H)$ for $1 \leq p < \infty$ and H any separable complex Hilbert space. We can now give the result announced in the title of this section.

Corollary 4.1. *The smooth sequence*

$$GL^1(K_+) \longrightarrow \mathcal{E} \xrightarrow{\beta} GL^0_{\text{res}} \quad (*)$$

realizes the universal principal bundle $U(\infty) \longrightarrow EU(\infty) \longrightarrow BU(\infty)$.

Proof. By the preceding discussion it is enough to show that $(*)$ realizes the universal bundle $GL^1(K_+) \longrightarrow E(GL^1(K_+) \longrightarrow B(GL^1(K_+))$. A numerable G-principal bundle $G \to P \to X$ is (up to homotopy equivalence, of course) the universal principal bundle $G \to EG \to BG$ if and only if the total space P is contractible (compare, e.g., Satz (4.9) in [19], together with Satz (8.11) in [20], in the books of tom Dieck respectively tom Dieck, Kamps and Puppe). Proposition 4.1 now implies the claim. □

As a first application we determine the homotopy groups of GL_{res}.

Proposition 4.2.

(i) As a complex manifold GL_{res} is isomorphic to $GL^0_{\text{res}} \times \mathbb{Z}$.
(ii) For all $k \geq 0$ we have

$$\pi_{2k+1}(GL^0_{\text{res}}) = \{0\} \quad \textit{and} \quad \pi_{2k+2}(GL^0_{\text{res}}) \cong \mathbb{Z}.$$

Proof. Since $p \circ \pi : GL_{\text{res}} \to \text{Fred}(K_+)$ is a homotopy equivalence by Corollary 3.1, and the connected components of $\text{Fred}(K_+)$ are $\text{Fred}^n(K_+) = \{a \in \text{Fred}(K_+) \mid a \text{ has index } n\}$ for $n \in \mathbb{Z}$ (see, e.g., Theorem 5.3.6 in Douglas' book [7]), the connected components of GL_{res} are given by $GL^n_{\text{res}} := (p \circ \pi)^{-1}(\text{Fred}^n(K_+))$ for $n \in \mathbb{Z}$. Since GL_{res} is a Lie group and $(p \circ \pi)^{-1}(\text{Fred}^0(K_+)) = GL^0_{\text{res}}$, the first claim follows easily.

Applying the long exact sequence of homotopy groups to the fibration β, we get for $j \geq 0$

$$\pi_{j+1}(GL^0_{\text{res}}) \cong \pi_j(GL^1(K_+))$$

since $\mathcal{E}$ is contractible. By Palais' results, we thus have that for $j \geq 0$ $\pi_{j+1}(GL^0_{\text{res}}) \cong \pi_j(U(\infty))$. Applying Bott periodicity (see Bott's original article [4], or, e.g., the recent book of Aguilar et al. [1]) to the homotopy groups of $U(\infty)$ we get part (ii) of the proposition. □

Remark 4.1. Since U_{res} and G_{res} are homotopy equivalent to GL_{res}, the preceding proposition determines obviously their homotopy groups as well. We denote in the sequel the connected components $\vartheta(GL^n_{\text{res}}, K_+)$ of G_{res} by G^n_{res} (the action ϑ being defined in Lemma 2.1).

5. Applications and remarks

5.1. *Characteristic classes of $GL(\infty)$-bundles*

Realizing the universal $GL(\infty)$-bundle $GL(\infty) \to BGL(\infty)$ as the short exact sequence of homomorphisms of Banach Lie groups

$$\{1\} \to GL^1(K_+) \to \mathcal{E} \xrightarrow{\beta} GL^0_{\text{res}} \to \{1\}$$

(compare Corollary 4.1) obviously helps to simplify and "geometrize" the theory of $GL(\infty)$- (or $U(\infty)$- or GL^1-) principal bundles. The space of universal characteristic classes of such bundles, $H^*(BGL(\infty), \mathbb{Z})$ (or, at least, $H^*(BGL(\infty), \mathbb{R})$) can then in fact be represented by differential forms coming from a connection form, à la Chern. See the work of Freed [8], extending the theory of finite-dimensional complex vector bundles and their Chern classes.

Intriguingly, Carey and Mickelsson show in [5] (Proposition 2) that $B(GL_{\text{res}})$ can be realized by $GL^1(H)$, with H a separable, complex Hilbert space. It would be interesting to have a conceptual proof for this "duality" and to know if there are other pairs (G, H) of topological groups such that

$$BG = H \quad \text{and} \quad BH = G.$$

5.2. *Geometric quantization of the restricted Grassmannian*

The homotopy result Proposition 4.2 easily yields certain unicity properties for the geometric quantization of G_{res}. First of all, by Hurewicz' theorem $H^2(G^0_{\text{res}}, \mathbb{Z}) \cong \mathbb{Z}$, and it is well-known that this group is generated by the Chern class $c_1(\text{Det})$ of the determinant bundle $\text{Det} \to G_{\text{res}}$, restricted to G^0_{res}. Furthermore, the negative of this class can be represented by ω, a natural Kähler form on G_{res} (see [15] and [22] for more details). Holomorphic geometric quantization thus leads unambigously to the holomorphic section space $\Gamma_{\mathcal{O}}(G_{\text{res}}, (\text{Det}^*)^{\otimes k})$ with $k > 0$, since negative powers of Det^* have no holomorphic sections. Up to the sacaling $\omega \mapsto k{\cdot}\omega\, (K \in \mathbb{N})$, holomorphic geometric quantization is thus unique on G^0_{res}.

Furthermore, looking from a not necessarily holomorphic point of view on geometric quantization (see, e.g., the book of Woodhouse [21] for more details on this theory), there is – up to gauge equivalence – only one connection on a complex line bundle on G^0_{res} since $H^1_{\text{deRham}}(G^0_{\text{res}}, \mathbb{R}) \cong H^1(G^0_{\text{res}}, \mathbb{R}) = \{0\}$. This implies unicity for Kostant's pre-quantization algorithm (compare again [21]).

Let us remark that (the dual of) the so-called "fermionic Fock space", arising in second quantization, is naturally (densely) injected into $\Gamma_{\mathcal{O}}(G_{\text{res}}, \text{Det}^*)$ and that the holomorphic sections of $\text{Det}^*\big|_{G^q_{\text{res}}}$ (for $q \in \mathbb{Z}$) contains densely the so-called "charge-q sector" of second quantization (see [15] and [22] for details). Work of Pickrell [14], and work in progress of the author together with Driver, singles out the Fock space as the "space of square-integrable holomorphic sections of the line bundle Det^* over G_{res}".

5.3. *String structures*

Given a real, separable Hilbert space (H, g) one associates to its complexification $H^{\mathbb{C}}$ the Hermitian extension $\langle , \rangle$ and the complex bilinear extension

$B = g^{\mathbb{C}}$ of g. Furthermore, one has the Clifford algebra $CL(H,g)$ defined via $[\gamma(u),\gamma(v)]_+ = g(u,v)$, its complexification $\mathbb{C}l(H,g) = Cl(H,g) \oplus \mathbb{C}$, and, given a maximal B-isotropic subspace W of $H^{\mathbb{C}}$, a so-called CAR-algebra. The latter algebra $CAR(W)$ is defined via $[a^*(w_1), a(w_2)]_+ = \langle w_1, w_2\rangle$, and is isomorphic to $\mathbb{C}l(H,g)$ as a C^*-algebra. The algebra $CAR(W)$ is naturally represented on the "spinor space" $S = S_W := \Lambda W$. Elements of the orthogonal group $O(H,g)$ act as automorphisms on $\mathbb{C}l(H,g)$ and are implemented on S if and only if they are in the "restricted orthogonal group" $O_{\mathrm{res}}(H,g;W) := O(H,g) \cap GL_{\mathrm{res}}(H^{\mathbb{C}},W)$ (where $H^{\mathbb{C}} = W \oplus \bar{W}$ is the polarization needed). This yields a central S^1-extension $O^{\sim}_{\mathrm{res}}(H,g;W)$ (see [15] and [17] for details).

The group $O_{\mathrm{res}}(H,g;W)$ is the orthogonal analog of U_{res}, and both play a prominent rôle as structure groups of infinite dimensional fibre bundles over loop spaces. For example, a "string structure" on the free loop space $\mathcal{L}M$ of a finite dimensional Spin manifold M (see McLaughlin's article [10] for this notion) exists if and only if a certain O_{res}-principal bundle over $\mathcal{L}M$ can be lifted to a $O^{\sim}_{\mathrm{res}}$-bundle (see [17] for more details). The homotopy types of U_{res} and O_{res} are thus obviously important for geometry on loop spaces.

Acknowledgement. This work was finalized during a sabbatical term partially supported by the C.N.R.S. Furthermore, I would like to thank Alan Carey, Karl-Hermann Neeb and Stephan Stolz for their interest in understanding the proof of Proposition (6.2.4) in "Loop groups" [15], and Hans-Werner Henn for insisting that I should look for an "elementary proof" thereof.

References

1. M. Aguilar, S. Gitler and C. Prieto: *Algebraic Topology from a Homotopical Viewpoint.* Springer-Verlag, New York, 2002.
2. B. Booss-Bavnbek and K. Furutani: Symplectic functional analysis and spectral invariants. In: *Geometric aspects of partial differential equations (Roskilde, 1998)*, 53–83, Contemp. Math., **242**, Amer. Math. Soc., Providence, RI, 1999.
3. B. Booss and K. Wojciechowski: Desuspension of splitting elliptic symbols. I. *Ann. Global Anal. Geom.* **3** (1985), no. 3, 337–383; and: Desuspension of splitting elliptic symbols. II. *Ann. Global Anal. Geom.* **4** (1986), no. 3, 349–400.
4. R. Bott: The stable homotopy of the classical groups. *Ann. of Math.* (2) **70** (1959), 313–337.
5. A. Carey and J. Mickelsson: A gerbe obstruction to quantization of fermions

on odd-dimensional manifolds with boundary. *Lett. Math. Phys.* **51** (2000), no. 2, 145–160.
6. A. Carey and J. Phillips: Algebras almost commuting with Clifford algebras in a II_∞ factor. *K-Theory* **4** (1991), no. 5, 445–478.
7. R.G. Douglas: *Banach Algebra Techniques in Operator Theory.* Academic Press, New York-London, 1972.
8. D. Freed: An index theorem for families of Fredholm operators parametrized by a group. *Topology* **27** (1988), no. 3, 279–300.
9. N.H. Kuiper: The homotopy type of the unitary group of Hilbert space. *Topology* **3** (1965), 19–30.
10. D.A. McLaughlin: Orientation and string structures on loop space. *Pacific J. Math.* **155** (1992), no. 1, 143–156.
11. J. Mickelsson: *Current Algebras and Groups.* Plenum Monographs in Nonlinear Physics. Plenum Press, New York, 1989.
12. K.-H. Neeb: Classical Hilbert-Lie groups, their extensions and their homotopy groups. In: *Geometry and Analysis on Finite- and Infinite-Dimensional Lie Groups (Bedlewo, 2000)*, 87–151, Banach Center Publ., 55, Polish Acad. Sci., Warsaw, 2002.
13. R.S. Palais: On the homotopy type of certain groups of operators. *Topology* **3** (1965), 271–279.
14. D. Pickrell: Measures on infinite-dimensional Grassmann manifolds. *J. Funct. Anal.* **70** (1987), no. 2, 323–356.
15. A. Pressley and G. Segal: *Loop Groups.* Oxford University Press, New York, 1986.
16. M. Spera and T. Wurzbacher: Differential geometry of Grassmannian embeddings of based loop groups. *Differential Geom. Appl.* **13** (2000), no. 1, 43–75.
17. M. Spera and T. Wurzbacher: Twistor spaces and spinors over loop spaces. Preprint 2005, see:
http://www.math.univ-metz.fr/~wurzbacher/publications.html
18. R.M. Switzer: *Algebraic Topology — Homotopy and Homology.* Classics in Mathematics. Springer-Verlag, Berlin, 2002.
19. T. tom Dieck: *Topologie.* de Gruyter Lehrbuch. Walter de Gruyter, Berlin, 1991.
20. T. tom Dieck, K.H. Kamps and D. Puppe: *Homotopietheorie.* Lecture Notes in Mathematics, Vol. 157. Springer-Verlag, Berlin-New York, 1970.
21. N.M.J. Woodhouse: *Geometric Quantization.* Second edition. Oxford University Press, New York, 1992.
22. T. Wurzbacher: Fermionic second quantization and the geometry of the restricted Grassmannian. *Infinite Dimensional Kähler Manifolds (Oberwolfach, 1995)*, 287–375, DMV Sem., **31**, Birkhäuser, Basel, 2001.

Received by the editors September 15, 2005; Revised October 14, 2005

Part V

Theoretical Particle, String and Membrane Physics, and Hamiltonian Dynamics

Analysis, Geometry and Topology
of Elliptic Operators, pp. 429–466

T-DUALITY FOR NON-FREE CIRCLE ACTIONS

ULRICH BUNKE
THOMAS SCHICK
Mathematisches Institut, Universität Göttingen
Bunsenstr. 3-5, 37073 Göttingen, Germany
bunke@uni-math.gwdg.de, schick@uni-math.gwdg.de

Dedicated to Krzysztof P. Wojciechowski on his 50th birthday

We study the topology of T-duality for pairs of $U(1)$-bundles and three-dimensional integral cohomology classes over orbispaces.

2000 *Mathematics Subject Classification.* Primary 55R15; Secondary 55R65

1. Introduction

1.1. *From spaces to orbispaces*

1.1.1. The concept of T-duality has its origin in string theory. Very roughly speaking, it relates one type of string theory on some target space with another type of string theory on a T-dual target space. Some topological aspects of T-duality in the presence of H-fields were studied in Bunke and Schick [2] (following earlier work by Bouwknegt, Mathai and Evslin [1], and others). In those preceeding investigations the main objects were pairs consisting of a $U(1)$-principal bundle and a three-dimensional integral cohomology class on its total space. Here we could replace the notion of an $U(1)$-principal bundle by the equivalent notion of a free $U(1)$-space satisfying some slice condition.

The main goal of the present paper is to extend the study of the topological aspects of T-duality to $U(1)$-spaces with finite stabilizers where we keep the slice condition. These spaces correspond to $U(1)$-bundles over orbispaces.

1.1.2. In order to deal properly with morphisms between orbispaces we will use the more general language of topological stacks. Orbispaces are particular topological stacks which admit an orbispace atlas. Morphisms between orbispaces are required to be representable maps. Our notion of an orbispace is a generalization of the notion of a topological space in the same spirit as the notion of an orbifold (see Moerdijk [7] for the definition of orbifolds which was motivating our definition of orbispaces) generalizes the notion of a smooth manifold.

Topological T-duality is now about pairs of $U(1)$-bundles in the category of orbispaces and three-dimensional cohomology classes in integral orbispace cohomology. We will explain these notions at the appropriate places.

1.1.3. Topological T-duality is the home for two different concepts. First it is a relation on the set $P(B)$ of isomorphism classes of pairs (E, h) over a base space B, where $E \to B$ is a $U(1)$-principal bundle and $h \in H^3(E, \mathbb{Z})$ is an integral cohomology class on the total space E of the bundle. Secondly, T-duality denotes a natural involution $T_B : P(B) \to P(B)$, which associates to each pair a canonical isomorphism class of T-dual pairs. In the present paper we generalize the definition of the T-duality relation as well as the construction of canonical T-dual pairs (see [2]). The main idea is to pass from orbispaces to spaces using a classifying space functor. Once this functor is established the extension of the results about the topology of T-duality of pairs from spaces to orbispaces is actually a formal matter.

1.1.4. Another aspect of T-duality is the T-duality transformation in twisted cohomology theories. It maps the twisted cohomology of the total space of one $U(1)$-bundle to the twisted cohomology of its T-dual, where the twists are classified by the corresponding three-dimensional cohomology classes. Of particular interest is the fact that under a T-admissibility assumption on the cohomology theory this transformation is an isomorphism. In the present paper we discuss the generalization of this aspect to the orbispace case. In general it is a non-trivial matter to extend a cohomology theory to the larger category of orbispaces. Of course, one could consider the Borel extension. In this case, where we again use the classifying space functor in order to pass from orbispaces to spaces, the generalization of the T-duality isomorphism is straight forward. On the other hand, having in mind the example of K-theory, the Borel extension might not be the most interesting extension of the given generalized cohomology theory from topological spaces to orbispaces.

At the moment we do not know if the correct extension of twisted K-theory to orbispaces is T-admissible.

1.1.5. It is an amusing fact that the topology of T-duality of $U(1)$-bundles over an orbispace as simple[a] as $[*/(\mathbb{Z}/n\mathbb{Z})]$ (a point with the isotropy group

[a] Actually the orbispaces $[*/\Gamma]$ are quite complicated. They are as complex as the classifying space $B\Gamma$.

$\mathbb{Z}/n\mathbb{Z}$) is already a non-trivial matter. We will develop this example in detail.
This example serves as a building block of the more general example of a Seifert bundle over a two-dimensional orbispace. As an illustration we will calculate the T-dual of a Seifert bundle equipped with a three-dimensional cohomology class in terms of topological invariants.
1.1.6. The problem of checking T-admissibility e.g. of twisted K-theory is equivalent to the verification that the T-duality transformations for all pairs over orbispaces of the form $[*/\Gamma]$ for all finite groups Γ are isomorphisms. Currently we do not have explicit general results about the topology of T-duality and the associated T-duality transformation in this large class of examples.

1.2. *A detailed description of the contents*

1.2.1. This paper is a continuation of [2]. In that paper we introduced a contravariant set-valued homotopy invariant functor P : *spaces* $\to$ *sets* which associates to each space B the set of isomorphism classes of pairs (E, h) over B. Here $E \to B$ is a $U(1)$-principal bundle and $h \in H^3(E, \mathbb{Z})$. We have shown that the functor can be represented by a space R carrying a universal pair. One of the main results was the determination of the homotopy type of R. Consider the map $K(\mathbb{Z}, 2) \times K(\mathbb{Z}, 2) \to K(\mathbb{Z}, 4)$ of Eilenberg-MacLane spaces given by the product of the canonical generators of the second cohomology of the two copies of $K(\mathbb{Z}, 2)$. Then R has homotopy type of the homotopy fibre of this map.
1.2.2. The notion of T-duality appeared first as a relation between isomorphism classes of pairs. We then have shown that the universal pair has a unique T-dual pair which determines and is determined by its classifying map $T : R \to R$. This map induces a natural transformation $T : P \to P$ which turns out to be two-periodic.
1.2.3. The following short reformulation of the results of [2] was suggested by the referee. It is close in spirit to the approach to T-duality for $U(1)^n$-principal bundles via T-duality triples Bunke, Rumpf and Schick [3]. For two $U(1)$-principal bundles $E \to B$ and $\hat{E} \to B$ let $E * \hat{E} \to B$ denote the fibrewise join. It is a bundle with fibre S^3. Let $\tilde{P}$: *spaces* $\to$ *sets* be the functor which associates to a space B the set of isomorphism classes of triples $(E, \hat{E}, Th)$, where $Th \in H^3(E * \hat{E}, \mathbb{Z})$ is a Thom class. Let $i : E \to E * \hat{E}$ be the natural inclusion map. Then $(E, \hat{E}, Th) \mapsto (E, i^*Th)$ defines a transformation $i : \tilde{P} \to P$. Using [2], Thm. 2.16 one can show that

this transformation is an isomorphism of functors. Under this isomorphism the T-duality transformation boils down to the involution $T : \tilde{P} \to \tilde{P}$ given by $(E, \hat{E}, Th) \mapsto (\hat{E}, E, Th)$. Note that this isomorphism $\tilde{P} \xrightarrow{\sim} P$ does not carry over to a corresponding result for $U(1)^n$-principal bundles if $n > 1$, see [3].

1.2.4. There are various pictures of twisted cohomology theories. In [2] we descided to axiomatize those properties of twists and twisted cohomology theories which are used in connection with T-duality.

In general, given a generalized cohomology theory represented by some spectrum E a twist of this cohomology theory over a space B is something like a bundle of spectra with fibre E, or a presheaf of spectra with stalk E, depending on the framework. The classification of twists is related to the classifying space $B\mathtt{Aut}(E)$ of the topological monoid of automorphisms of E. The twists considered in the present paper (as well as in the previous papers [2], [3]) are quite special and releated to the occurence of a map $K(\mathbb{Z}, 3) \to B\mathtt{Aut}(E)$ for cohomology theories like complex K-theory, $Spin^c$-cobordism theory, or periodized real cohomology. In connection with T-duality the restriction to this special sort of twists is crucial.

1.2.5. In this setting, twists should form a functor $\mathcal{T}$: *spaces* $\to$ *groupoids* such that the set of isomorphism classes of $\mathcal{T}(B)$ is in natural bijection with $H^3(B, \mathbb{Z})$, and such that the group of automorphisms of every $\mathcal{H} \in \mathcal{T}(B)$ is naturally isomorphic to $H^2(B, \mathbb{Z})$.

In order to have an explicit model choose a realization of the Eilenberg-MacLane space $K(\mathbb{Z}, 3)$. Then let $\mathcal{T}(B)$ be the set of maps $B \to K(\mathbb{Z}, 3)$. For two such maps $\mathcal{H}, \mathcal{H}'$ let $\mathtt{Hom}_{\mathcal{T}(B)}(\mathcal{H}, \mathcal{H}')$ be the set of homotopy classes of homotopies from $\mathcal{H}$ to $\mathcal{H}'$.

1.2.6. In [2] we have further introduced the notion of a T-admissible twisted cohomology theory. It associates to a space E and a twist $\mathcal{H} \in \mathcal{T}(E)$ the graded group $h(E, \mathcal{H})$. Twisted cohomology is functorial in both arguments. If $u : \mathcal{H} \to \mathcal{H}'$ is an isomorphism of twists, then we have an induced map $u^* : h(E, \mathcal{H}') \to h(E, \mathcal{H})$. If $f : B' \to B$ is a map of spaces, then we have a functorial map $f^* : h(B, \mathcal{H}) \to h(B', f^*\mathcal{H})$. It should furthermore admit an integration map for suitable oriented bundles. For details we refer to [2].

1.2.7. Given a pair (E, h) the class h determines an isomorphism class $[\mathcal{H}]$ of twists $\mathcal{H} \in \mathcal{T}(E)$. If $(\hat{E}, \hat{h})$ is dual to (E, h) and $[\hat{\mathcal{H}}] = \hat{h}$, then the T-duality transformation

$$T : h(E, \mathcal{H}) \to h(\hat{E}, \hat{\mathcal{H}})$$

is given by the following construction. Note that there is a unique class $(E, \hat{E}, Th) \in \tilde{P}(B)$ such that $(E, h) \cong i(E, \hat{E}, Th)$ and $(\hat{E}, \hat{h}) \cong i \circ T(E, \hat{E}, Th)$ (see 1.2.3 for the notation). Consider the fibre product

$$\begin{array}{ccccc} & & E \times_B \hat{E} & & \\ & p \swarrow & & \hat{p} \searrow & \\ E & & & & \hat{E} \ . \\ & \searrow & & \swarrow & \\ & & B & & \end{array}$$

As explained in [2] the Thom class Th determines an isomorphism $u : \hat{p}^*\hat{\mathcal{H}} \to p^*\mathcal{H}$. The T-duality transformation is defined as the composition

$$T := \hat{p}_! \circ u^* \circ p^* \ .$$

1.2.8. By definition, the twisted cohomology theory is T-admissible if the T-duality transformation is an isomorphism in the special case where B is a point. In [2] we have shown that T-admissibility implies, via a Mayer-Vietoris argument, that the T-duality transformation is an isomorphism in general.

1.2.9. With these results our contribution consisted in presenting an effective formalism and adding some precision and slight generalizations to the understanding of the topic as presented in [1] or Mathai, Rosenberg [5].

In the present paper we develop a formalism which allows a considerable generalization of T-duality. The spaces which were suitable for T-duality in [2] were total spaces E of principal $U(1)$-fibrations $E \to B$. In particular, the spaces E were free $U(1)$-spaces.

In the present paper we will relax this condition by admitting finite stabilizers. In order to keep track of all information it turns out to be necessary to consider the quotient $B := [E/U(1)]$ as a topological orbispace, i.e. as a proper topological stack on the category of topological spaces which admit an orbispace atlas. For the language we refer to Heinloth [4] and Noohi [8], but we will recall essential notions in Subsection 2.1. The brackets shall indicate that we consider the quotient as a stack and not just as a space. The map $E \to [E/U(1)]$ is an atlas which represents $[E/U(1)]$ as a topological stack. Since $U(1)$ is compact, this stack is proper. The requirement that $[E/U(1)]$ admits an orbispace atlas (note that $E \to [E/U(1)]$ is not an orbispace atlas) replaces the requirement of the existence of local trivializations in the case of principal bundles.

1.2.10. Consider the simple example of the $U(1)$-stack $[U(1)/(\mathbb{Z}/n\mathbb{Z})]$ (equipped with the trivial three-dimensional cohomology class) which is actually a space with a $U(1)$-action. It will turn out that its canonical T-dual is $U(1) \times [*/(\mathbb{Z}/n\mathbb{Z})]$ (equipped with a non-trivial three-dimensional cohomology class). This stack is not equivalent to a space. Therefore we are led to consider $U(1)$-bundles in the category of stacks as the domain and the target of the canonical T-duality from the beginning. By definition, a representable map $E \to B$ of topological stacks is a $U(1)$-principal bundle, if it admits a fibrewise action of $U(1)$, if in addition there is a $U(1)$-equivariant isomorphism

$$\begin{array}{ccc} E \times_B E & \cong & E \times U(1) \\ & \mathrm{pr}_1 \searrow \quad \mathrm{pr}_1 \swarrow & \\ & E & \end{array} ,$$

where $U(1)$ acts on the second factors (this means that $E \to B$ is a family of $U(1)$-torsors), and if for every map $T \to B$ with T a space the induced map $T \times_B E \to T$ has local sections. Note that $E \to [E/U(1)]$ is a $U(1)$-principal bundle in the category of stacks.

1.2.11. There are various equivalent ways to define the integral cohomology group $H^*(E, \mathbb{Z})$ of a topological stack E. One possibility is as the sheaf cohomology of the constant sheaf over E with fibre $\mathbb{Z}$. In the present paper we prefer to employ classifying spaces. An atlas $X \to E$ of the topological stack gives rise to a topological groupoid $X \times_E X \Rightarrow X$ and thus to a simplicial space $X^\cdot$. Let $|X^\cdot|$ denote its geometric realization. If E is an orbispace and X is an orbispace atlas, then (see Proposition 2.1) there is a natural isomorphism

$$H^*(E, \mathbb{Z}) \cong H^*(|X^\cdot|, \mathbb{Z}) .$$

1.2.12. A pair (E, h) over a stack B will be a $U(1)$-principal bundle $E \to B$ together with a class $h \in H^3(E, \mathbb{Z})$. Two pairs (E, h) and (E', h') over B are isomorphic if there exists an isomorphism of $U(1)$-bundles $\phi : E \to E'$ such that $\phi^* h' = h$.

If (E, h) is a pair over B, and $f : B' \to B$ is a representable map of topological stacks, then we can define the pull-back $f^*(E, h) := (f^*E, \tilde{f}^*h)$, where $f^*E := B' \times_B E \to B'$ is the induced $U(1)$-bundle, and $\tilde{f} : f^*E \to E$ is the induced map. This definition extends the functor P to a functor P : (*stacks, representable maps*) $\to$ *sets*. Note that stacks form a two-category, and P identifies two-isomorphic morphisms.

1.2.13. Assume that B is an orbispace, and let $Y \to B$ be an orbispace atlas of B. Let $Y^\cdot$ be the associated simplicial space, and $|Y^\cdot|$ be its geometric realization. It turns out (Proposition 2.1) that the homotopy type of $|Y^\cdot|$

is independent of the choice of Y in a natural way. In fact, if $i : Y' \to Y$ is a refinement of orbispace atlases, then $|i^\cdot| : |(Y')^\cdot| \to |Y^\cdot|$ is a homotopy equivalence, where $i^\cdot : (Y')^\cdot \to Y^\cdot$ is the induced map of simplicial spaces. Furthermore, if $Y_1 \to B$ is another orbispace atlas, then the common refinement $Y \leftarrow Y \times_B Y_1 \to Y_1$ is again an orbispace atlas.

1.2.14. A pair (E, h) over B gives rise to a pair $(|X^\cdot|, h) \in P(|Y^\cdot|)$ as follows. Note that $X := Y \times_B E \to E$ is an orbispace atlas of E. The natural map $X^\cdot \to Y^\cdot$ is a simplicial $U(1)$-bundle which induces an ordinary $U(1)$-bundle $|X^\cdot| \to |Y^\cdot|$. We can consider $h \in H^3(|X^\cdot|, \mathbb{Z})$. Therefore given an orbispace atlas $Y \to B$ we obtain a map

$$PA_Y : P(B) \to P(|Y^\cdot|) .$$

The map is natural in B and in the atlas Y as follows. Consider a representable map $f : B' \to B$. Then we have the equality

$$PA_{Y'} \circ f^* = |f^\cdot|^* \circ PA_Y ,$$

where $Y' := B' \times_B Y$ is the induced atlas of B', and $f^\cdot : (Y')^\cdot \to Y^\cdot$ is induced by the natural map $Y' \to Y$.

Consider now a refinement $i : Y' \to Y$ of the orbispace atlas $Y \to B$. Then we have the equality

$$|i^\cdot|^* \circ PA_Y = PA_{Y'} .$$

1.2.15. The following theorem is the key to our generalization from spaces to orbispaces of the results about T-duality of pairs.

Theorem 1.1. *If B is an orbispace with orbispace atlas $Y \to B$, then $PA_Y : P(B) \to P(|Y^\cdot|)$ is an isomorphism.*

This theorem will be proved in Section 4. The main intermediate result, Proposition 4.3, states that for a given orbispace atlas $Y \to B$ the construction above on the level of $U(1)$-principal bundles provides an equivalence between the categories of $U(1)$-principal bundles over B and $|Y^\cdot|$, where morphisms are homotopy classes of bundle isomorphisms.

1.2.16. We use Theorem 1.1 and the naturality properties of the transformation PA_Y in order to extend the transformation $T : P \to P$, which associates to an isomorphism class of pairs a natural isomorphism class of T-dual pairs, from spaces to orbispaces. Let B be an orbispace and $Y \to B$ be an orbispace atlas.

Definition 1.2. We define $T_B : P(B) \to P(B)$ by

$$T_B := PA_Y^{-1} \circ T_{|Y^\cdot|} \circ PA_Y .$$

By Theorem 1.1 the map T_B is well-defined. It follows from the functorial properties of PA_Y that T_B is independent of the choice of the orbispace atlas $Y \to B$. It furthermore follows that the maps T_B for all orbispaces assemble to an automorphism of the functor P.
If B is a space, then we can use the atlas $B \to B$. In this case T reduces to the original T on spaces. Therefore our construction provides an extension of T from spaces to orbispaces. Since the original T on spaces is involutive, the same is true for its extension to orbispaces.
1.2.17. The second topic of the present paper is the T-duality transformation in twisted cohomology. To this end we first introduce the notion of a twisted cohomology theory defined on orbispaces. Here we essentially repeat the axioms formulated in [2] and add an axiom dealings with two-isomorphisms. We show in Subsection 3.4 that every twisted cohomology defined on spaces has a Borel extension to orbispaces. But in general there might be different more interesting extensions (K-theory provides an example).
1.2.18. Let us fix a twisted cohomology theory h on orbispaces. Given two pairs (E_i, h_i), $i = 0, 1$, which are T-dual (this is the T-duality relation, see 3.1), we consider twists $\mathcal{H}_i$ on E_i classified by h_i. Then we define a T-duality transformation $T : h(E_0, \mathcal{H}_0) \to h(E_1, \mathcal{H}_1)$ of degree one which is natural in B. We extend the notion of T-admissibility of a twisted cohomology theory to the orbispace case (Definition 3.3). If h is T-admissible then the T-duality transformation is an isomorphism (Theorem 3.5).
Compared with the case of spaces, in the case of orbispaces T-admissibility is much more complicated to check. The reason is that an orbispace can have a complicated local structure. At the moment we are not able to show that in the orbispace case twisted K-theory is T-admissible. But we shall see in Subsection 3.4 that the Borel extension of a T-admissible twisted cohomology theory from spaces to orbispaces is again T-admissible.
1.2.19. The paper concludes with the computation of the canonical T-duals in some instructive examples in Section 5.

2. Some stack language

2.1. *Topological stacks and orbispaces*

2.1.1. In the present paper we consider stacks in topological spaces. A stack is a sheaf of groupoids on this category. The sheaf conditions are descend conditions for objects and morphisms with respect to open coverings of spaces. We refer to [4], [8] for details. Stacks form a two-category.

The category of topological spaces is embedded into stacks by mapping a space X to the sheaf of sets $Y \mapsto \mathtt{Hom}(Y, X)$, and we consider a set as a groupoid with only identity morphisms. We can and will consider spaces as stacks. This point of view is also reflected in our notation which uses the same type of letters for spaces and stacks.

2.1.2. We shall illustrate the stack notions in the example of quotient stacks. Let G be a topological group acting on a space B. Then we can form the quotient stack $[B/G]$. It associates to a space T the groupoid $[B/G](T)$ of pairs $(P \to T, \phi)$, where $P \to T$ is a G-principal bundle and $\phi : P \to B$ is a G-equivariant map. The morphisms $(P \to T, \phi) \to (P' \to T, \phi')$ are principal bundle isomorphisms $P \to P'$ which are compatible with the maps to B. If $f : T' \to T$ is a map of spaces, then $[B/G](f) : [B/G](T) \to [B/G](T')$ is given by pull-back.

A G-equivariant map $h : B \to B'$ induces a morphism of stacks $h_* : [B/G] \to [B'/G]$ by $(P \to T, \phi) \mapsto (P \to T, h \circ \phi)$.

2.1.3. A map $X \to Y$ between stacks is called representable if for each space T and map $T \to Y$ the stack $T \times_Y X$ is equivalent to a space.

2.1.4. Let us check that the map $h_* : [B/G] \to [B'/G]$ in 2.1.2 is representable. To this end we must calculate the fibre product $T \times_{[B'/G]} [B/G]$ for a map $f : T \to [B'/G]$ and show that it is equivalent to a space. Let f be given by $(P' \to T, \phi')$. We claim that $T \times_{[B'/G]} [B/G] \cong (P' \times_{B'} B)/G$. The map to $[B/G]$ is given by the pair $(P' \times_{B'} B \to (P \times_{B'} B)/G, \mathtt{pr}_2)$, and the map to T is given by the composition $(P' \times_{B'} B)/G \stackrel{\mathtt{pr}_1}{\to} P'/G \cong T$. Let S be a space. Then by definition of the fibre product of stacks an object in $(T \times_{[B'/G]} [B/G])(S)$ is a triple $(g, ((P \to S), \phi), u)$, where $g : S \to T$ is an object of $T(S)$, i.e. a map, $(P \to S, \phi)$ is an object of $[B/G](S)$, and $u : f(g) \to h(P \to S, \phi)$, i.e. an isomorphism $h : g^*P' \to P$ of principal bundles such that $\phi' \circ g^\sharp = \phi \circ h$, where $g^\sharp : g^*P' \to P'$ is the induced map of total spaces.

The equivalence $(T \times_{[B'/G]} [B/G])(S) \stackrel{\sim}{\to} ((P' \times_{B'} B)/G)(S)$ associates to $(g, ((P \to S), \phi), u)$ the map $S \to (P' \times_{B'} B)/G$ induced by the G-equivariant map $(g^\sharp \circ u^{-1}, \phi) : P \to P' \times_{B'} B$.

2.1.5. A topological stack is a stack which admits an atlas. An atlas of a stack B is a representable map $X \to B$ from a space X to B which admits local sections. Here we say that a map of stacks $X \to Y$ admits local sections if for each map $T \to Y$ from a space T to Y each point $y \in T$ has a neighborhood $U \subset T$ such that there exists a map $U \to X$ and a two-isomorphism from the composition $U \to X \to Y$ to the composition $U \to T \to Y$.

A refinement of an atlas $X \to B$ is given by an atlas $X' \to B$ and a diagram

$$\begin{array}{ccc} X' & \to & X \\ & \searrow \quad \swarrow & \\ & B & \end{array} .$$

2.1.6. Let us check that the quotient stack $[B/G]$ considered in 2.1.2 is topological. We claim that $B \to [B/G]$ is an atlas.
In order to see that this map is representable observe that $B \cong [G/G] \times B \cong [(G \times B)/G]$, where in the last term G acts on $G \times B$ by $h(g,b) := (gh^{-1}, hb)$. In order to see the first equivalence observe that $[G/G](S)$ is the groupoid of G-principal bundles with a section on S. This groupoid is connected and a set, hence equivalent to a one-point set. The second equivalence is induced by the G-equivariant map $G \times B \to G \times B$, $(g,b) \mapsto (g, g^{-1}b)$, where the action of G on the left $G \times B$ is given by $h(g,b) := (gh^{-1}, b)$. The map $B \cong [G \times B/G] \to [B/G]$ is now induced by the G-equivariant map $\mathrm{pr}_2 : G \times B \to B$. It is representable by 2.1.4.
Going through the definitions we see that the map $B \to [B/G]$ considered as an object of $[B/G](B)$ is given by $(G \times B \stackrel{\mathrm{pr}_2}{\to} B, \phi)$ with $\phi(g,b) := g^{-1}b$.
The existence of local sections can be seen as follows. Let $S \to [B/G]$ be a map given by a pair $(P \to S, \phi)$. Then we find a surjective map $f : A \to S$ such that f^*P is trivial, i.e. admits an isomorphism $f^*P \cong G \times A$. The composition $A \stackrel{a \mapsto (e,a)}{\longrightarrow} G \times A \cong f^*P \stackrel{f^\sharp}{\to} P \stackrel{\phi}{\to} B$ gives the required section.
2.1.7. Given an atlas $X \to B$ we can define a topological groupoid

$$X \times_B X \Rightarrow X .$$

If $X' \to X$ is a refinement, then we get an associated homomorphism of groupoids.
2.1.8. In the case of the quotient stack $[B/G]$ with the atlas $B \to [B/G]$ this groupoid is the action groupoid $G \times B \Rightarrow B$, where the range and source maps are given by $(g,b) \mapsto gb$ and $(g,b) \mapsto b$.
2.1.9. A topological stack B is called proper if the map of spaces

$$X \times_B X \to X \times X$$

is proper. This condition is independent of the choice of the atlas.
2.1.10. A topological groupoid $\mathcal{G}^1 \Rightarrow \mathcal{G}^0$ is called étale if the source and range maps $s, r : \mathcal{G}^1 \to \mathcal{G}^0$ are étale. An orbispace atlas of a proper topological stack is an atlas $X \to B$ such that $X \times_B X \Rightarrow X$ is an étale topological groupoid.

We define a topological orbispace to be a proper topological stack which admits an orbispace atlas. Our two-category of orbispaces (*orbispaces, representable morphisms*) has such orbispaces as objects and representable maps between orbispaces as one-morphisms.
2.1.11. We again consider quotient stack $[B/G]$ of 2.1.2. In view of 2.1.8 it is proper if and only if the action of G on B is proper, i.e. the map $G \times B \to B \times B$, $(g, b) \mapsto (gb, b)$, is proper. It is in addition étale if and only if G acts with finite stabilizers.
In particular, if G is a discrete group acting properly on B, then $[B/G]$ is an orbispace.
2.1.12. If G is a finite group acting on the one-point space, then $[*/G]$ is an orbispace. If $G \to H$ is a homomorphism of finite groups, then we obtain a map of stacks $[*/G] \to [*/H]$. It is a map of orbispaces (i.e. representable) if and only if the group homomorphism is injective. In fact, in this case we can factor this map as $[*/G] \cong [(G\backslash H)/H] \to [*/H]$, and the second map is prepresentable by 2.1.4.
2.1.13. More generally, let $\mathcal{G} : \mathcal{G}^1 \Rightarrow \mathcal{G}^0$ be a topological groupoid acting on a space B, i.e. there is a map $f : B \to \mathcal{G}^0$ and an action $B \times_{\mathcal{G}^0} \mathcal{G}^1 \to B$ (the fibre product employs the range map $r : \mathcal{G}^1 \to \mathcal{G}^0$). Then we have the quotient stack $[B/\mathcal{G}]$. Its value on a space X is given by the groupoid of pairs $(P \to X, \phi)$ of locally trivial $\mathcal{G}$-bundles $P \to X$ (see [4], Section. 3 for a definition) and maps $\phi : P \to B$ of $\mathcal{G}$-spaces, and the morphisms of the groupoid are the isomorphisms of such pairs. There is a canonical map $B \to [B/\mathcal{G}]$ which is an atlas. Thus $[B/\mathcal{G}]$ is a topological stack. If $\mathcal{G}$ is proper and étale then $[B/\mathcal{G}]$ is an orbispace. In particular, we can apply this construction to the $\mathcal{G}$-space $\mathcal{G}^0$. We obtain the orbispace $[\mathcal{G}^0/\mathcal{G}]$ which is the classifying stack for locally trivial $\mathcal{G}$-bundles.

2.2. *Cohomology of orbispaces*

2.2.1. Let $X \to B$ be an atlas of a topological stack and $X \times_B X \Rightarrow X$ be the associated groupoid. Then we obtain an associated simplicial space $X^\cdot$ such that $X^n := \underbrace{X \times_B \cdots \times_B X}_{n+1}$. By $|X^\cdot|$ we denote its geometric realization.
A refinement $u : X' \to X$ leads to a map of simplicial spaces $u^\cdot : (X')^\cdot \to X^\cdot$. It further induces a map $|u^\cdot| : |(X')^\cdot| \to |X^\cdot|$ of realizations.
2.2.2. In the present paper we heavily use the following fact (which we learned from I. Moerdijk).

Proposition 2.1. *If B is an orbispace, and $u : X' \to X$ is a refinement of orbispace atlases of B, then $|u^\cdot| : |(X')^\cdot| \to |X^\cdot|$ is a weak homotopy equivalence of spaces.*

Proof. The category of sheaves (of sets) on the groupoid $X \times_B X \Rightarrow X$ is equivalent to the category of sheaves on B. In particular, the homomorphism of groupoids

$$(X' \times_B X' \Rightarrow X') \to (X \times_B X \Rightarrow X)$$

induces an equivalence of categories of sheaves over groupoids. In Moerdijk [6] it is shown that the category of sheaves on $X \times_B X \Rightarrow X$ is equivalent to the category of sheaves on the space $|X^\cdot|$. If a map of spaces induces an equivalence of categories of sheaves, then it is a weak homotopy equivalence. This implies the result. □

2.2.3. If $h(\dots)$ is some generalized cohomology theory then we can extend this theory canonically to orbispaces. Given an orbispace B we choose an orbispace atlas $X \to B$. Then we define

$$h(B) := h(|X^\cdot|) .$$

This determines $h(B)$ up to natural isomorphisms (related to the various choices of the orbispace atlas).
If $f : B' \to B$ is a representable map, then $X' := B' \times_B X \to B'$ is again an orbispace atlas. We obtain an induced morphism of groupoids $(X' \times_{B'} X' \Rightarrow X') \to (X \times_B X \Rightarrow X)$, which induces a map of simplicial spaces $f^\cdot : (X')^\cdot \to X^\cdot$, and eventually a map $|f^\cdot| : |(X')^\cdot| \to |X^\cdot|$ of geometric realizations. The map $f^* : h(B) \to h(B')$ is now given by $|f^\cdot|^* : h(|(X')^\cdot|) \to h(|X^\cdot|)$.
2.2.4. Below we will apply this construction to integral cohomology $h(\dots) = H(\dots, \mathbb{Z})$. In order to distinguish the construction described above from other extensions of h to orbispaces it will be called the Borel extension and denoted by h_{Borel} (see also 3.4). This notation is justified by its close relationship with the Borel extension of a cohomology theory to an equivariant cohomology theory.

3. The T-duality relation

3.1. *Thom classes and T-duality*

3.1.1. Let B be a topological stack. We consider two $U(1)$-bundles $E_i \to B$, $i = 0, 1$ over B and let $L_i \to B$ be the associated Hermitian vector bundles.

Let $S := S(L_0 \oplus L_1) \to B$ denote the unit-sphere bundle in the sum of the two line bundles. Observe that the fibres of these bundles are spaces since the corresponding projection maps to B are representable. We will denote points in the fibre of S by (z_0, z_1), where $z_i \in L_i$ and $\|z_0\|^2 + \|z_1\|^2 = 1$. Then we have natural inclusions $s_i : E_i \to S$ which identify E_i with the subsets $\{\|z_i\| = 1\}$ for $i = 0, 1$, respectively.

3.1.2. A Thom class for a three-sphere bundle $S \to B$ is a class $Th \in H^3(S, \mathbb{Z})$ which specializes to a Thom class of the three-sphere bundle $|Y^{\cdot}| \to |X^{\cdot}|$ under the natural isomorphism $H^3(S, \mathbb{Z}) \cong H^3(|Y^{\cdot}|, \mathbb{Z})$ for some (and hence every) orbispace atlas $X \to B$, where $Y := S \times_B X \to S$ is the induced atlas of S.

3.1.3. Let $c_1(L_i) \in H^2(B, \mathbb{Z})$ denote the first Chern classes of L_i. As in the case of spaces the three-sphere bundle $S \to B$ admits a Thom class if and only if $c_1(L_0) \cup c_1(L_1) = 0$ in $H^4(B, \mathbb{Z})$.

3.1.4. We now introduce the T-duality relation between pairs. We consider classes $h_i \in H^3(E_i, \mathbb{Z})$ for $i = 0, 1$ and the pairs (E_0, h_0) and (E_1, h_1) over B.

Definition 3.1. We call the pairs (E_0, h_0) and (E_1, h_1) T-dual if there exists a Thom class $Th \in H^3(S, B)$ such that $h_i = s_i^* Th$ for $i = 0, 1$, respectively.

This is the direct generalization of [2], Definition 2.9.

3.2. *The T-duality transformation*

3.2.1. In this subsection we assume that we have a twisted cohomology theory defined on orbispaces. Thus given is a a functor of twists $\mathcal{T} : (orbispaces, representable\ maps) \to groupoids$ which satisfies the axioms listed in [2], Section 3.1 with spaces replaced by orbispaces. As an additional datum we require that a two-isomorphism $f \stackrel{\Phi}{\Rightarrow} f'$ between maps $f, f' : B' \to B$ induces an isomorphism of functors $f^* \stackrel{\Phi}{\Rightarrow} (f')^* : \mathcal{T}(B) \to \mathcal{T}(B')$ in a functorial way.

Furthermore, given is a bifunctor $h(\ldots, \ldots)$ which associates to each pair $(B, \mathcal{H})$ of an orbispace B and $\mathcal{H} \in \mathcal{T}(B)$ a graded group $h(B, \mathcal{H})$, and which satisfies the axioms listed again in [2], Section 3.1. In addition we assume that $f^* = \Phi^*_{\cdot} \circ (f')^* : h(B, \mathcal{H}) \to h(B', f^*\mathcal{H})$ for two-isomorphic morphisms using the notation above.

We require that the integration map $g_! : h(B', g^*\mathcal{H}) \to h(B, \mathcal{H})$ is defined for representable proper maps $g : B' \to B$ which are h-oriented. By definition, the datum of an h-orientation of g is equivalent to a compatible choice of h-orientations of the induced maps of spaces $T \times_B B' \to T$ for all maps $T \to B$, where T is a space.

3.2.2. We consider an orbispace B. Let (E_0, h_0) and (E_1, h_1) be pairs over B and $Th \in H^3(S, \mathbb{Z})$ be a Thom class such that $s_i^* Th = h_i$. We choose a twist $\mathcal{H} \in \mathcal{T}(S)$ such that $[\mathcal{H}] = Th$. Then we define the twists $\mathcal{H}_i := s_i^* \mathcal{H} \in \mathcal{T}(E_i)$ for $i = 0, 1$. In the present section we define the T-duality transformation

$$T_0 : h(E_0, \mathcal{H}_0) \to h(E_1, \mathcal{H}_1) .$$

3.2.3. We consider the two-torus bundle $F := E_0 \times_B E_1 \to B$. The map

$$F \ni (z_0, z_1) \to (\frac{1}{\sqrt{2}} z_0, \frac{1}{\sqrt{2}} z_1) \in S$$

defines embedding which gives rise to a decomposition

$$S \cong S_0 \cup_F S_1 ,$$

where

$$S_i := \{(z_0, z_1) \in S \mid \|z_i\| \geq \|z_{1-i}\|\} .$$

3.2.4. The composition $s_0 \circ \mathtt{pr}_0 : F \to S$ is homotopic to the inclusion by the homotopy

$$(z_0, z_1) \mapsto (\sqrt{1 - \frac{t}{2}} z_0, \sqrt{\frac{t}{2}} z_1) \ , t \in [0, 1] .$$

Similarly, $s_1 \circ \mathtt{pr}_1$ is homotopic to the inclusion. These homotopies give rise to isomorphism classes of isomorphisms of twists

$$v_i : \mathcal{H}_{|F} \xrightarrow{\sim} \mathtt{pr}_i^* \mathcal{H}_i .$$

3.2.5. **Definition 3.2.** We define the T-duality transformations

$$T_i : h(E_i, \mathcal{H}_i) \to h(E_{1-i}, \mathcal{H}_{1-i})$$

as the compositions

$$T_i := (\mathtt{pr}_{1-i})_! \circ (v_{1-i}^{-1})^* \circ v_i^* \circ \mathtt{pr}_i^* .$$

Here it is essential to use the transformation $(v_{1-i}^{-1})^* \circ v_i^* : \mathtt{pr}_{1-i}^* \mathcal{H}_{1-i} \to \mathtt{pr}_i^* \mathcal{H}_i$. With other choices we can not expect that the maps T_i become isomorphisms for T-admissible cohomology theories.

3.3. *T-admissible cohomology theories*

3.3.1. Let Γ be a finite group, and choose two characters $\chi_0, \chi_1 : \Gamma \to U(1)$. We consider the stack $B := [*/\Gamma]$ and the bundles $E_i := [U(1)/_{\chi_i}\Gamma] \to [*/\Gamma]$, where Γ acts on $U(1)$ by χ_i (this is indicated by the subscript), $i = 0, 1$. We further consider classes $h_i \in H^3(E_i, \mathbb{Z})$ such that (E_0, h_0) and (E_1, h_1) are T-dual according to Definition 3.1. This is a non-trivial condition as we shall see later in 5.1.

Definition 3.3. Following [2], Definition 3.1,2 we call a twisted cohomology theory $h(\dots,\dots)$ on orbispaces T-admissible if the T-duality transformations T_i are isomorphisms for all examples of the type described above (i.e. for all choices finite groups Γ, pairs of characters χ_0, χ_1, and choices of the classes h_i).

3.3.2. If the cohomology theory is T-admissible then the property that the T-duality transformation is an isomorphism can be extended to the large class of base orbispaces B which are build by glueing the local examples of the form $[*/\Gamma]$. The argument is based on the Mayer-Vietoris sequence. We call an orbispace B finite if it has a finite filtration

$$\bigsqcup_i^{finite} [*/\Gamma_{i,0}] = B^0 \subset B^1 \subset \cdots \subset B^r = B$$

such that there exists cartesian diagrams

$$\begin{array}{ccc} S^{n_\alpha - 1} \times [*/\Gamma_\alpha] & \to & B^{\alpha-1} \\ \downarrow & & \downarrow \\ D^{n_\alpha} \times [*/\Gamma_\alpha] & \stackrel{i_\alpha}{\to} & B^\alpha \end{array} \qquad (3.4)$$

for $n_\alpha \in \mathbb{N}$ and appropriate finite groups Γ_α, where the i_α are representable and induce inclusions of open substacks $(D^{n_\alpha} \setminus S^{n_\alpha - 1}) \times [*/\Gamma_\alpha] \to B^\alpha$ (see [4], Definition 2.8), and $D^{n_\alpha} \times [*/\Gamma_\alpha] \sqcup B^{\alpha-1} \to B^\alpha$ is surjective.
For example, if M is a compact smooth manifold on which a compact group G acts with finite stabilizers, then $[M/G]$ is a finite orbispace. In fact, M admits a G-equivariant triangulation (by G-simplices of the form $\Delta^k \times G/H$ with $H \subset G$ a finite subgroup). Using this triangulation we obtain the required filtration of $[M/G]$. We expect that compact orbifolds in the sense of [7] are finite orbispaces.

3.3.3. **Theorem 3.5.** *Assume that the twisted cohomology theory is T-admissible. Let B be a finite orbispace, and let (E_0, h_0) and (E_1, h_1) be pairs* *over B which are T-dual to each other. Then the T-duality transformations 3.2 are isomorphisms.*

Proof. This theorem is proved using induction over the number of cells of B and the Mayer-Vietoris sequence in the same way as [2], Thm. 3.13. □

Using the method of the proof of Proposition 3.10 we could weaken the finiteness condition.
3.3.4. It is natural to expect that an appropriate extension of twisted Atiyah-Segal K-theory to orbispaces is T-admissible. At the moment we do not have a proof. In the following Subsection 3.4 we provide examples of T-admissible cohomology theories.

3.4. *Borel-K-theory as an admissible cohomology theory on orbispaces*

3.4.1. The goal of the present subsection is to show that every twisted cohomology theory defined on spaces and satisfying the list of axioms stated in [2], Section 3.1, admits an extension to orbispaces by a Borel construction. For a demonstration we use K-theory. We shall see that the Borel extension of a T-admissible twisted cohomology theory is again T-admissible.
3.4.2. Note that in the case of K-theory the Borel construction is probably not the most interesting extension to orbispaces. A better extension is provided by the construction of Tu, Xu and Laurent [9].
3.4.3. An extension of a twisted cohomology theory from spaces to orbispaces consists of an extension of the notion of a twist from spaces to orbispaces, and then of the extension of the cohomology functor itself.
We start with the discussion of twists. In this subsection we will assume that we are given a functor $\mathcal{T}$ on spaces which associates to each space B the groupoid of twists $\mathcal{T}(B)$ (Note that in general twists form a two-category. Here we adjust the notion by identifying isomorphic isomorphisms.)
3.4.4. We now extend twists to orbispaces.

Definition 3.6. A twist of an orbispace B is given by an orbispace atlas $X \to B$ and a twist $\mathcal{H} \in \mathcal{T}(|X^\cdot|)$. A morphism of twists $\mathcal{H} \to \mathcal{H}'$, where $\mathcal{H} \in \mathcal{T}(|X^\cdot|)$ and $\mathcal{H}' \in \mathcal{T}(|(X')^\cdot|)$, is given by a common refinement $Y \to B$ of the orbispace atlases X and X' and a morphism $\phi : u^*\mathcal{H} \to (u')^*\mathcal{H}'$, where $u : |Y^\cdot| \to |X^\cdot|$ and $u' : |Y^\cdot| \to |(X')^\cdot|$ are the induced maps.

We identify morphisms which become equal on a common refinement of orbispace atlases. In this way we associate to each orbispace B a category of twists $\mathcal{T}(B)$.
3.4.5. Let $f : B' \to B$ be a morphism of orbispaces, i.e. a representable map of stacks. Then we define the pull-back $f^* : \mathcal{T}(B) \to \mathcal{T}(B')$ as follows. If $X \to B$ is an orbispace atlas then we get an orbispace atlas $X' := B' \times_B X$ and an induced map $\phi : |(X')^\cdot| \to |X^\cdot|$. If $\mathcal{H} \in \mathcal{T}(|X^\cdot|) \subset \mathcal{T}(B)$, then we define $f^*\mathcal{H} \in \mathcal{T}(B')$ as $\phi^*\mathcal{H} \in \mathcal{T}(|(X')^\cdot|)$. The pull-back of morphisms is defined similarly. In this way we obtain a functor $\mathcal{T}$: $(orbispaces, representable\ maps) \to groupoids$.
3.4.6. We consider a two-isomorphism $f \overset{\Phi}{\Rightarrow} f'$ between representable maps $f, f' : B' \to B$ of orbispaces. If $X \to B$ is an atlas, and $Y, Y' \to B'$ are the atlases obtained by pull-back via f, f', then Φ induces a map $\Phi : Y \to Y'$ which we consider as a refinement. Note that $\phi' \circ |\Phi^\cdot| = \phi : |Y^\cdot| \to |X^\cdot|$. For $\mathcal{H} \in \mathcal{T}(|X^\cdot|) \subset \mathcal{T}(B)$ we define $\Phi_.(\mathcal{H}) : \phi^*(\mathcal{H}) \to |\Phi^\cdot|^* \circ (\phi')^*(\mathcal{H})$ to be the associated canonical isomorphism, interpreted as an isomorphisms $f^*\mathcal{H} \to (f')^*\mathcal{H}$.
3.4.7. Now we extend the K-theory functor (or any other twisted cohomology theory) to orbispaces. Let $\mathcal{H} \in \mathcal{T}(|X^\cdot|)$ be a twist of B in the sense above.

Definition 3.7. We define

$$K_{Borel}(B, \mathcal{H}) := K(|X^\cdot|, \mathcal{H}) \ .$$

Let $f : B' \to B$ be a map of orbispaces. We use the notation of 3.4.5.

Definition 3.8. We define $f^* : K_{Borel}(B, \mathcal{H}) \to K_{Borel}(B', f^*\mathcal{H})$ to be the map $|\phi^\cdot|^* : K(|X^\cdot|, \mathcal{H}) \to K(|(X')^\cdot|, \phi^*\mathcal{H})$.

Let $\Phi : \mathcal{H} \to \mathcal{H}'$ be a morphism of twists given by $\phi : u^*\mathcal{H} \to (u')^*\mathcal{H}'$, where we use the notation of 3.6.

Definition 3.9. We define $\Phi^* : K_{Borel}(B, \mathcal{H}') \to K_{Borel}(B, \mathcal{H})$ to be the composition

$$\Phi^* := (u^*)^{-1} \circ \phi^* \circ (u')^* \ .$$

Here we us the fact that the refinement map $u : |Y^\cdot| \to |X^\cdot|$ is a homotopy equivalence (see Proposition 2.1), and therefore that u^* is invertible. We also see that Φ^* is an isomorphism.
It is straight forward to check that this bi-functor has the required properties of a twisted cohomology defined on orbispaces as explained in 3.2.1.

3.4.8. **Proposition 3.10.** *The twisted Borel K-theory $K_{Borel}(\ldots,\ldots)$ is T-admissible.*

Proof. We consider the orbispace chart $X := * \to [*/\Gamma]$. Then the corresponding classifying space $|X^{\cdot}|$ is a countable CW-complex of the homotopy type $B\Gamma$. The T-duality transformation in K_{Borel} for pairs over $[*/\Gamma]$ translates to the T-duality transformation for pairs over $|X^{\cdot}|$.
In [2] we have shown that the T-admissibility of K-theory implies that the T-duality transformation is an isomorphism for pairs over bases spaces which are equivalent to finite CW-complexes. In fact, this result can be extended to countable complexes as follows. Let

$$W_0 \subset W_1 \subset \cdots \subset W_i \subset \cdots \subset W$$

be a filtration of a countable CW-complex W by finite sub-complexes. Let (E_i, h_i), $i = 0, 1$, be T-dual pairs over W and consider twists $\mathcal{H}_i \in \mathcal{T}(E_i)$ such that $[\mathcal{H}_i] = h_i$. Let

$$T_0 : K^*(E_0, \mathcal{H}_0) \to K^{*-1}(E_1, \mathcal{H}_1)$$

be the associated T-duality transformation. We claim that T_0 is an isomorphism of groups.
Let $(E_i(k), h(k))$ be the pairs over W_k obtained by restriction. We have exact sequences

$$0 \to \lim_{\overleftarrow{k}}{}^{1} K^{*-1}(E_i(k), \mathcal{H}_i(k)) \to K(E_i, \mathcal{H}_i) \to \lim_{\overleftarrow{k}} K^*(E_i(k), \mathcal{H}_i(k)) \to 0$$

for $i = 0, 1$. The T-duality transformation T_0 is compatible with restriction and therefore induces a map of sequences $(K^*(E_0(k), \mathcal{H}_0(k)))_{k\geq 0} \stackrel{(T_0(k))_{k\geq 0}}{\to} (K^{*-1}(E_1(k), \mathcal{H}_1(k)))_{k\geq 0}$. Since the complexes W_k are finite, this map is an isomorphism. We thus obtain a map of short exact sequences

$$\begin{array}{ccc} 0 \to \lim^1_{\overleftarrow{k}} K^{*-1}(E_0(k), \mathcal{H}_0(k)) & \longrightarrow & K(E_0, \mathcal{H}_0) \\ (T_0(k))_{k\geq 0} \downarrow & & T_0 \downarrow \\ 0 \to \lim^1_{\overleftarrow{k}} K^{*-2}(E_1(k), \mathcal{H}_1(k)) & \longrightarrow & K(E_1, \mathcal{H}_1) \end{array}$$

$$\begin{array}{cccc} \longrightarrow & \lim_{\overleftarrow{k}} K^*(E_0(k), \mathcal{H}_0(k)) & \to & 0 \\ & (T_0(k))_{k\geq 0} \downarrow & & \\ \longrightarrow & \lim_{\overleftarrow{k}} K^{*-1}(E_1(k), \mathcal{H}_1(k)) & \to & 0 \end{array}$$

By the five lemma we see that T_0 is an isomorphism. This proves the claim.

We can now apply the claim in order to show that K_{Borel} is T-admissible since the CW-complexes $|X^{\cdot}|$ obtained from $* \to [*/\Gamma]$ for finite groups Γ are countable. □

4. Groupoids and classifying spaces

4.1. *Continuous cohomology*

4.1.1. We consider a topological groupoid $\mathcal{G} : \mathcal{G}^1 \Rightarrow \mathcal{G}^0$ and a topological abelian group A. Then we define a cochain complex of abelian groups

$$\cdots \to C^p_{cont}(\mathcal{G}, A) \xrightarrow{\delta} C^{p+1}_{cont}(\mathcal{G}, A) \to \ldots ,$$

where

$$C^0(\mathcal{G}, A) = C(\mathcal{G}^0, A) , \quad C^p_{cont}(\mathcal{G}, A) := C(\underbrace{\mathcal{G}^1 \times_{\mathcal{G}^0} \cdots \times_{\mathcal{G}^0} \mathcal{G}^1}_{p}, A)$$

and

$$(\delta a)(\gamma_1, \ldots, \gamma_{p+1}) := a(\gamma_2, \ldots, \gamma_{p+1})$$
$$+ \sum_{i=1}^{p} (-1)^i a(\gamma_1, \ldots, \gamma_i \circ \gamma_{i+1}, \ldots, \gamma_{p+1}) + (-1)^{p+1} a(\gamma_1, \ldots, \gamma_p) .$$

Definition 4.1. The continuous cohomology $H_{cont}(\mathcal{G}, A)$ of $\mathcal{G}$ with values in A is the cohomology of the complex $(C^*_{cont}(\mathcal{G}, A), \delta)$.

This definition is an immediate extension of the definition of the continuous cohomology of a topological group.
4.1.2. We now assume that $\mathcal{G}$ is proper and étale, and that A admits the structure of a $\mathbb{Q}$-vector space. The following Lemma generalizes the well-known fact that the higher cohomology of a finite group with coefficients in a $\mathbb{Q}$-vector space is trivial.

Lemma 4.2. *We have $H^p(\mathcal{G}, A) = 0$ for $p \geq 1$.*

Proof. Let $a \in C^{p+1}_{cont}(\mathcal{G}, A)$ be a cocycle. We define the continuous cochain $b \in C^p_{cont}(\mathcal{G}, A)$ by

$$b(\gamma_1, \ldots, \gamma_p) := (-1)^{p+1} \int_{\mathcal{G}^{s(\gamma_p)}_{s(\gamma_p)}} a(\gamma_1, \ldots, \gamma_p, \gamma) d\gamma ,$$

where $d\gamma$ is the normalized counting measure on the finite group $\mathcal{G}^{s(\gamma_p)}_{s(\gamma_p)}$. Then by a straight forward computation we have $\delta b = a$. □

4.2. *The Borel construction and $U(1)$-bundles*

4.2.1. We consider a $U(1)$-bundle $E \to B$ over an orbispace B. We choose an orbispace atlas $X \to B$ and get an induced orbispace atlas $Y := X \times_B E \to E$ of E. Then we have the groupoids $\mathcal{G} : X \times_B X \Rightarrow X$ and $\mathcal{E} : Y \times_E Y \Rightarrow Y$ together with a homomorphism $\mathcal{E} \to \mathcal{G}$. The latter can be considered as a $U(1)$-bundle over $\mathcal{G}$.
It gives rise to a simplicial $U(1)$-bundle $Y^\cdot \to X^\cdot$ (using the notation 1.2.11), and thus to an ordinary $U(1)$-bundle $|Y^\cdot| \to |X^\cdot|$.
This construction extends in an obvious manner to a functor A_X from the category of $U(1)$-bundles over B to $U(1)$-bundles over $|X^\cdot|$. The morphisms in these categories here are homotopy classes of bundle isomorphisms. The main step in the proof of 1.1 is the following proposition.

Proposition 4.3. *A_X is an equivalence of categories.*

The remainder of the present subsection is devoted to the proof. It consists of three steps. In the first step we show that A_X is surjective on the level of sets of isomorphisms classes. Then we show that it is full. In the last step we show that it is faithful.
4.2.2. We have an equivalence of stacks $B \cong [\mathcal{G}^0/\mathcal{G}]$. Moreover the category of $U(1)$-bundles over B is equivalent to the category of $U(1)$-bundles over $\mathcal{G}$. In fact, given a $U(1)$-bundle $E \to B$ in stacks we obtain by the construction above a $U(1)$-bundle $\mathcal{E} \to \mathcal{G}$ in a functorial manner. In the other direction we funtorially associate to a $U(1)$-bundle $\mathcal{E} \to \mathcal{G}$ of groupoids a $U(1)$-bundle $[\mathcal{E}^0/\mathcal{E}] \to [\mathcal{G}^0/\mathcal{G}]$ of stacks.
A $U(1)$-bundle $\mathcal{E} \to \mathcal{G}$ in groupoids can equivalently be considered as a $\mathcal{G}$-equivariant $U(1)$-bundle, i.e. a $U(1)$-bundle $\mathcal{E}^0 \to \mathcal{G}^0$ together with an action $\mathcal{G}^1 \times_{\mathcal{G}^0} \mathcal{E}^0 \to \mathcal{E}^0$. Below we will freely switch between these two points of view.
4.2.3. If $\mathcal{G}$ is a topological groupoid then we let $B(\mathcal{G})$ denote the associated simplicial space, and we let $|B(\mathcal{G})|$ denote its geometric realization.
In order to prove Proposition 4.3 it suffices to show that the functor which associates $|B(\mathcal{E})| \to |B(\mathcal{G})|$ to $\mathcal{E} \to \mathcal{G}$ is an equivalence of categories. We will denote it by A.
We first show that A induces a surjection on the level of sets of isomorphisms classes of objects.
4.2.4. For the following discussion we employ the smooth bundle $U \to P\mathbb{C}^\infty$ as a model for the universal $U(1)$-principal bundle. To be precise we consider this bundle in the category of *ind*-manifolds such that $U :=$

$\lim_{\overrightarrow{n}} S^{2n+1}$ and $P\mathbb{C}^\infty := \lim_{\overrightarrow{n}} P\mathbb{C}^n$, and the connecting maps are in both cases induced by the canonical embeddings $\mathbb{C}^n \to \mathbb{C}^{n+1}$.
We choose a connection on this $U(1)$ bundle which induces a parallel transport and a curvature two-form $\omega \in \Omega^2(P\mathbb{C}^\infty)$. In detail this amounts to choose a compatible family of connections on the bundles $S^{2n+1} \to P\mathbb{C}^n$ (e.g. the one induced by the round metric on the spheres), and the curvature form is interpreted as a compatible family of two-forms on the family of complex projective spaces, i.e. $\omega \in \lim_{\overleftarrow{n}} \Omega^2(P\mathbb{C}^n)$.
A map $c : |B(\mathcal{G})| \to P\mathbb{C}^\infty$ determines a $U(1)$-bundle $c^*U \to |B(\mathcal{G})|$. Homotopic maps give isomorphic $U(1)$-bundles. We want to show that the isomorphism class of $c^*U \to |B(\mathcal{G})|$ is in the image of A. Let $\mathbf{c}$ denote the homotopy class of c.
4.2.5. For all $n \geq 0$ we have a natural map

$$i_n : \Delta^n \times \underbrace{\mathcal{G}^1 \times_{\mathcal{G}^0} \cdots \times_{\mathcal{G}^0} \mathcal{G}^1}_{n} \to |B(\mathcal{G})| \ .$$

If $(\gamma_1, \ldots, \gamma_n) \in \underbrace{\mathcal{G}^1 \times_{\mathcal{G}^0} \cdots \times_{\mathcal{G}^0} \mathcal{G}^1}_{n}$, then we let

$$\begin{aligned} i_n(\gamma_1, \ldots, \gamma_n) : \Delta^n &\cong \Delta^n \times \{(\gamma_1, \ldots, \gamma_n)\} \\ &\subset \Delta^n \times \underbrace{\mathcal{G}^1 \times_{\mathcal{G}^0} \cdots \times_{\mathcal{G}^0} \mathcal{G}^1}_{n} \xrightarrow{i_n} |B(\mathcal{G})| \ . \end{aligned}$$

4.2.6. We plan to use the parallel transport along one-simplices. Furthermore we want to apply Stokes theorem to the curvature form on three-simplices. Therefore we need a representative of $\mathbf{c}$ which is smooth in the interior of each simplex. Let $\Delta^n_{int} \subset \Delta^n$ denote the interior of the standard simplex.

Lemma 4.4. *The class* $\mathbf{c}$ *has a representative* c *such that for all* $n \geq 1$ *the composition* $c \circ i_n$ *induces a continuous map*

$$\underbrace{\mathcal{G}^1 \times_{\mathcal{G}^0} \cdots \times_{\mathcal{G}^0} \mathcal{G}^1}_{n} \to C^\infty(\Delta^n_{int}, P\mathbb{C}^\infty) \ .$$

Proof. For all $n \geq 1$ we set up one of the usual procedures to smooth out maps $\Delta^n \to P\mathbb{C}^\infty$ in the interior $\Delta^n_{int} \subset \Delta^n$ without changing the restriction to the boundary. In this way we obtain a family of continuous maps $C(\Delta^n, P\mathbb{C}^\infty) \to C^\infty(\Delta^n_{int}, P\mathbb{C}^\infty) \cap C(\Delta^n, P\mathbb{C}^\infty)$. We apply these procedures to the maps $i_n(\gamma_1, \ldots, \gamma_n)$ for all $(\gamma_1, \ldots, \gamma_n) \in \underbrace{\mathcal{G}^1 \times_{\mathcal{G}^0} \cdots \times_{\mathcal{G}^0} \mathcal{G}^1}_{n}$, increasing n from 1 to ∞ inductively. The resulting maps assemble to a representative of $\mathbf{c}$ with the required properties. $\square$

4.2.7. We define a $U(1)$-bundle $E \to \mathcal{G}^0$ by the iterated pull-back

$$\begin{array}{ccccc} E & \to & c^*U & \to & U \\ \downarrow & & \downarrow & & \downarrow \\ \mathcal{G}^0 & \subset & |B(\mathcal{G})| & \stackrel{c}{\to} & P\mathbb{C}^\infty \end{array} .$$

The idea is to define an action of $\mathcal{G}$ on E so that if we apply A to the resulting bundle $\mathcal{E} \to \mathcal{G}$ we get back the isomorphism class of $c^*U \to |B(\mathcal{G})|$.

4.2.8. For $\gamma \in \mathcal{G}^1$ we have a path $c \circ i_1(\gamma) : \Delta^1 \to P\mathbb{C}^\infty$ from $c(s(\gamma))$ to $c(r(\gamma))$. We let $\phi(\gamma) : E_{s(\gamma)} \to E_{r(\gamma)}$ denote the isomorphism such that

$$\begin{array}{ccc} E_{s(\gamma)} & \stackrel{\phi(\gamma)}{\to} & E_{r(\gamma)} \\ \| & & \| \\ U_{c(s(\gamma))} & \to & U_{c(r(\gamma))} \end{array} ,$$

where the lower horizontal arrow is the parallel transport along the path. The maps $\phi(\gamma)$, $\gamma \in \mathcal{G}^1$, combine to a map $\phi : \mathcal{G}^1 \times_{\mathcal{G}^0} E \to E$. This is not yet an action. In the following we modify this map to make it associative. In fact, the non-associativity will be measured by a continuous groupoid cocycle a with coefficients in $U(1)$, and the crucial fact will be that it represents the trivial cohomology class.

4.2.9. Consider a pair $(\gamma_1, \gamma_2) \in \mathcal{G}^1 \times_{\mathcal{G}^0} \mathcal{G}^1$. We define

$$a(\gamma_1, \gamma_2) := \phi(\gamma_1 \circ \gamma_2)^{-1} \circ \phi(\gamma_1) \circ \phi(\gamma_2) \in \mathtt{Aut}(E_{s(\gamma_2)}) \cong U(1) .$$

Note that $a \in C^2_{cont}(\mathcal{G}, U(1))$ is a cocyle which represents a class $[a] \in H^2_{cont}(\mathcal{G}, U(1))$.

Lemma 4.5. *We have* $[a] = 0$.

Proof. We consider the continuous homomorphism $e : \mathbb{R} \to U(1)$ given by $t \mapsto \exp(2\pi i t)$. In induces a map of complexes $e_* : C^2_{cont}(\mathcal{G}, \mathbb{R}) \to C^2_{cont}(\mathcal{G}, U(1))$. The key to the proof is the observation that the cocycle a can be lifted to a cocycle $\tilde{a} \in C^2_{cont}(\mathcal{G}, \mathbb{R})$ such that $e_*\tilde{a} = a$. By Lemma 4.2 we have $[\tilde{a}] = 0$ so that $[a] = e_*[\tilde{a}] = 0$, too.

Note that (γ_1, γ_2) determines a smooth map $c \circ i_2(\gamma_1, \gamma_2) : \Delta^2 \to P\mathbb{C}^\infty$. The restriction of this map to the boundary of the simplex determines a piecewise differentiable loop in $P\mathbb{C}^\infty$, and $a(\gamma_1, \gamma_2)$ is exactly the holonomy of the parallel transport along this loop. We thus get

$$a(\gamma_1, \gamma_2) = e\left(\int_{\Delta^2} (c \circ i_2(\gamma_1, \gamma_2))^*\omega\right) .$$

We now define the continuous $\mathbb{R}$-valued groupoid-cochain

$$\tilde{a}(\gamma_1,\gamma_2) := \int_{\Delta^2} (c \circ i_2(\gamma_1,\gamma_2))^*\omega \ . \tag{4.6}$$

We claim that $\tilde{a}$ is a cocycle. In fact, for $(\gamma_1,\gamma_2,\gamma_3) \in \mathcal{G}^1 \times_{\mathcal{G}^0} \mathcal{G}^1 \times_{\mathcal{G}^0} \mathcal{G}^1$ the number

$$(\delta\tilde{a})(\gamma_1,\gamma_2,\gamma_3) = \tilde{a}(\gamma_2,\gamma_3) - \tilde{a}(\gamma_1 \circ \gamma_2,\gamma_3) + \tilde{a}(\gamma_1,\gamma_2 \circ \gamma_3) - \tilde{a}(\gamma_1,\gamma_2)$$

is the integral over the boundary of Δ^3 of $i_3(\gamma_1,\gamma_2,\gamma_3)^*\omega$. Since ω is closed, this integral vanishes by Stokes theorem. □

4.2.10. By Lemma 4.5 we can choose $b \in C^1_{cont}(\mathcal{G}, U(1))$ such that

$$\delta b = a \ . \tag{4.7}$$

We now define

$$m(\gamma) := \phi(\gamma)b(\gamma)^{-1}$$

Then it is easy to check that $m : \mathcal{G}^1 \times_{\mathcal{G}^0} E \to E$ is an action. Let $\mathcal{E} \to \mathcal{G}$ denote the corresponding equivariant $U(1)$-bundle.
4.2.11. Let $F := |B(\mathcal{E})| \to |B(\mathcal{G})|$.

Lemma 4.8. *We have an isomorphism of $U(1)$-bundles $F \cong c^*U$.*

Proof. We will prove the assertion by explicitly defining an isomorphism $\psi : F \to c^*(U)$.
If $(a_0,\ldots,a_n)$ are the labels of the vertices of Δ^n, then let t_{a_i} denote the linear coordinate on Δ^n which vanishes at the vertex labeled by a_i, and which is equal to 1 on the opposite face.
First note that we can find a cochain $\tilde{b} \in C^1_{Cont}(\mathcal{G},\mathbb{R})$ such that $\delta\tilde{b} = \tilde{a}$ and $e(\tilde{b}) = b$ (using the notation of 4.2.10). Let Δ^n denote the copy of the standard simplex in $|B(\mathcal{G})|$ corresponding to

$$(\gamma_1,\ldots,\gamma_n) \in \underbrace{\mathcal{G}^1 \times_{\mathcal{G}^0} \cdots \times_{\mathcal{G}^0} \mathcal{G}^1}_{n} \ .$$

The vertices of Δ^n are naturally labeled by the ordered set $\{r(\gamma_1),\ldots,r(\gamma_n),s(\gamma_n)\}$. Let $\Delta^n_\circ := \Delta^n \setminus \partial_{s(\gamma_n)}\Delta^n$, where $\partial_{s(\gamma_n)}\Delta^n$ is the unique face not containing the vertex labeled by $s(\gamma_n)$. We define ψ over the subset $\Delta^n_\circ \times (\gamma_1,\ldots,\gamma_n) \subset |B(\mathcal{G})|$ as follows. By construction the fiber of $F_{|\Delta^n_\circ \times (\gamma_1,\ldots,\gamma_n)}$ is canonically isomorphic to $E_{s(\gamma_n)} = U_{c(s(\gamma_n))}$. Each point $s \in \Delta^n_\circ$ can be joined by a linear path with the vertex with label

$s(\gamma_n)$. Let $\psi(s,(\gamma_1,\ldots,\gamma_n)) : F_{(s,(\gamma_1,\ldots,\gamma_n))} \cong U_{c(s(\gamma_n))} \to U_{c(s,(\gamma_1,\ldots,\gamma_n))}$ be given by the parallel transport along this path multiplied by

$$e(-t_{s(\gamma_n)}\tilde{b}(\gamma_n))e(-t_{s(\gamma_n)}t_{s(\gamma_{n-1})}\tilde{b}(\gamma_{n-1}))\ldots e(-t_{s(\gamma_n)}\cdots t_{s(\gamma_1)}\tilde{b}(\gamma_1)) .$$

We use the construction for all $n \geq 1$ and points $(\gamma_1,\ldots,\gamma_n) \in \underbrace{\mathcal{G}^1 \times_{\mathcal{G}^0} \cdots \times_{\mathcal{G}^0} \mathcal{G}^1}_{n}$. It is now easy to check that ψ is an everywhere defined continuous bundle isomorphism. □

This finishes the proof of the fact that A is surjective on the level of sets of isomorphism classes of objects.

4.2.12. Our next task is to show that A is full. We consider the following intermediate construction. Let $\mathcal{E} \to \mathcal{G}$ be a $U(1)$-bundle. Then we have a cartesian diagram

$$\begin{array}{ccccc} |B(\mathcal{E})| & \stackrel{\cong}{\to} & c^*U & \to & U \\ \downarrow & & \downarrow & & \downarrow \\ |B(\mathcal{G})| & \stackrel{=}{\to} & |B(\mathcal{G})| & \stackrel{c}{\to} & P\mathbb{C}^\infty \end{array} , \qquad (4.9)$$

where c is uniquely determined up to homotopy. After a further homotopy we can assume that c satisfies the condition of Lemma 4.4. We apply to this map c the construction of the first part of the proof and obtain a $U(1)$-bundle $\tilde{\mathcal{E}} \to \mathcal{G}$.

4.2.13. **Lemma 4.10.** *We have $\tilde{\mathcal{E}} \cong \mathcal{E}$ as $U(1)$-bundles over $\mathcal{G}$.*

Proof. Let $E, \tilde{E} \to \mathcal{G}^0$ be the underlying $U(1)$-bundles. Note that (4.9) induces a canonical isomorphism $\Psi : \tilde{E} \stackrel{\sim}{\to} E$ as $U(1)$-principal bundles over $\mathcal{G}^0$. We must compare the action $\tilde{m}$ of $\mathcal{G}$ on $\tilde{E}$ with the original action m on E. The difference between these two actions is measured by the continuous cocycle $h \in C^1_{cont}(\mathcal{G}, U(1))$ defined by

$$h(\gamma) = \Psi^{-1} \circ m(\gamma)^{-1} \circ \Psi \circ \tilde{m}(\gamma) \in \mathtt{Aut}(\tilde{E}_{s(\gamma)}) \cong U(1) .$$

The cohomology class of this cocycle is the obstruction against making Ψ equivariant by multiplying it by a $U(1)$-valued function on $\mathcal{G}^0$. □

4.2.14. **Lemma 4.11.** *We have $[h] = 0$.*

Proof. The key is again the construction of a lift of h to a cocycle $\tilde{h} \in C^1_{cont}(\mathcal{G}, \mathbb{R})$ such that $e_*(\tilde{h}) = h$. By Lemma 4.2 we then have $[h] = e_*([\tilde{h}]) = 0$.

We consider $\gamma \in \mathcal{G}^1$. It induces a smooth path $c \circ i_1(\gamma) : \Delta^1 \to P\mathbb{C}^\infty$ and therefore a parallel transport $\phi(\gamma) : U_{c(s(\gamma))} \to U_{c(r(\gamma))}$. We have $\tilde{m}(\gamma) = \phi(\gamma)b(\gamma)^{-1}$, where b is as in (4.7). As in the proof of Lemma 4.8 will again use the cochain $\tilde{b} \in C_{cont}(\mathcal{G}, \mathbb{R})$ such that $\delta\tilde{b} = \tilde{a}$ and $b = e_*(\tilde{b})$. The identification $|B(\mathcal{E})| \cong c^*U$ induces a trivialization $i_1(\gamma)^*U \cong \Delta^1 \times E_{s(\gamma)}$. If $\alpha(\gamma)$ denotes the connection-one form in this trivialization, then we can write

$$\phi(\gamma) = e\left(\int_{\Delta^1} \alpha(\gamma)\right) .$$

By construction we have $h(\gamma) = e\left(\int_{\Delta^1} \alpha(\gamma)\right) b(\gamma)^{-1}$. We define the cochain $\tilde{h} \in C^1_{cont}(\mathcal{G}, \mathbb{R})$

$$\tilde{h}(\gamma) := \int_{\Delta^1} \alpha(\gamma) - \tilde{b}(\gamma) .$$

It satisfies $e_*(\tilde{h}) = h$. We claim that $\tilde{h}$ is in fact a cocycle. Let $(\gamma_1, \gamma_2) \in \mathcal{G}^1 \times_{\mathcal{G}^0} \mathcal{G}^1$. The identification $|B(\mathcal{E})| \cong c^*U$ induces a trivialization $(c \circ i_2(\gamma_1, \gamma_2))^*U \cong \Delta^2 \times U_{E_{s(\gamma_2)}}$. Let $\alpha(\gamma_1, \gamma_2)$ denote the connection one-form in this trivialization. Then we have

$$\delta\tilde{h}(\gamma_1, \gamma_2) = \int_{\partial\Delta^2} \alpha(\gamma_1, \gamma_2) - \delta\tilde{b}(\gamma_1, \gamma_2) .$$

By Stoke's theorem the first term of the right-hand side is equal to

$$\int_{\Delta^2} d\alpha(\gamma_1, \gamma_2) .$$

Now the claim follows in view of $d\alpha(\gamma_1, \gamma_2) = (c \circ i_2(\gamma_1, \gamma_2))^*\omega$, $\delta\tilde{b} = \tilde{a}$, and (4.6). □

4.2.15. By Lemma 4.11 we can choose a cochain $f \in C^0_{cont}(\mathcal{G}, U(1))$ such that $\delta f = h$. If we define the isomorphism $\tilde{\Psi} : \tilde{E} \to E$ by $\tilde{\Psi}(x) = \Psi(x)f^{-1}(x)$ then $\tilde{\Psi}$ is $\mathcal{G}$-equivariant.

4.2.16. We now finish the proof of the fact that A is full. To this end we consider $U(1)$-bundles $\mathcal{E}, \mathcal{E}' \to \mathcal{G}$ and an isomorphism of $U(1)$-bundles $\Lambda : |B(\mathcal{E}')| \to |B(\mathcal{E})|$ over $|B(\mathcal{G})|$. We must show that Λ can be written as $A(\lambda)$ for some $\lambda : \mathcal{E}' \to \mathcal{E}$ over $\mathcal{G}$. We apply to $\mathcal{E}$ and $\mathcal{E}'$ the intermediate construction started in 4.2.12, where we use the same map $c : |B(\mathcal{G})| \to P\mathbb{C}^\infty$ in both cases. We obtain a chain of isomorphisms

$$\mathcal{E} \stackrel{\tilde{\Psi}}{\cong} \tilde{\mathcal{E}} = \tilde{\mathcal{E}}' \stackrel{\tilde{\Psi}'}{\cong} \mathcal{E}' .$$

Let $\mathcal{E} \stackrel{\lambda}{\cong} \mathcal{E}'$ be the composition.
In general $A(\lambda)$ is not equal to Λ (recall that we consider homotopy classes). But the following result shows that we can find an automorphism ϕ of $\mathcal{E}$ such that $A(\lambda \circ \phi) = \Lambda$.
4.2.17. Let $\phi : \mathcal{G}^0 \to U(1)$ be a $\mathcal{G}^1$-invariant function. We can interpret ϕ as an automorphism of the $U(1)$-bundle $\mathcal{E} \to \mathcal{G}$. Applying the classifying space functor we get an automorphism $|B(\phi)|$ of the $U(1)$-bundle $|B(\mathcal{E})| \to |B(\mathcal{G})|$, i.e. a function $|B(\phi)| : |B(\mathcal{G})| \to U(1)$.

Lemma 4.12. *Every homotopy class of maps* $[|B(\mathcal{G})|, U(1)]$ *has a representative of the form* $|B(\phi)|$ *for a* $\mathcal{G}^1$*-invariant function* $\phi : \mathcal{G}^0 \to U(1)$.

Proof. We consider a homotopy class of maps $|B(\mathcal{G})| \to U(1)$ and choose a representative $\tilde{f}$. The restriction of $\tilde{f} : |B(\mathcal{G})| \to U(1)$ to $\mathcal{G}^0 \subset |B(\mathcal{G})|$ gives a function $\tilde{\phi} : \mathcal{G}^0 \to U(1)$. In general it is not $\mathcal{G}^1$-invariant.
We consider $\tilde{\phi} \in C^0_{cont}(\mathcal{G}, U(1))$. Then the non-invariance is measured by $h := \delta\tilde{\phi} \in C^1_{cont}(\mathcal{G}, U(1))$.
We have $h(\gamma) = \phi(r(\gamma))\phi(s(\gamma))^{-1}$. We now construct a lift $\tilde{h} \in C^1_{cont}(\mathcal{G}, \mathbb{R})$ as follows. Let $\gamma \in \mathcal{G}^1$. It gives rise to a path $i_1(\gamma) : \Delta^1 \to |B(\mathcal{G})|$. The restriction $i_1(\gamma)^*\tilde{f}$ has a lift to an $\mathbb{R}$-valued function $\kappa(\gamma) : \Delta^1 \to \mathbb{R}$. The difference $\tilde{h}(\gamma) := \kappa(\gamma)(1) - \kappa(\gamma)(0)$ is independent of the choice of the lift. We claim that $\delta\tilde{h} = 0$. This follows from the fact that $\tilde{f}$ is defined on the image of $i_2(\gamma_1, \gamma_2) : \Delta^2 \to |B(\mathcal{G})|$ for all composeable $\gamma_1, \gamma_2 \in \mathcal{G}^1$. By Lemma 4.2 we can find a function $a \in C^0_{cont}(\mathcal{G}, \mathbb{R})$ such that $\delta a = \tilde{h}$. We now define the $\mathcal{G}^1$-invariant $U(1)$-valued function

$$\phi = \tilde{\phi}\exp(-2\pi i a) .$$

We can consider a as an $\mathbb{R}$-valued continuous function defined on the closed subset $\mathcal{G}^0 \subset |B(\mathcal{G})|$. Let $\tilde{a} : |B(\mathcal{G})| \to \mathbb{R}$ be any continuous extension, and set $f := \tilde{f}\exp(-2\pi i\tilde{a})$. Then clearly $[f] = [\tilde{f}]$. It remains to show that $[f] = [|B(\phi)|]$.
Note that $i_n(\gamma_1, \ldots, \gamma_n)^*B(\phi) = \phi(s(\gamma_n)) = \phi(r(\gamma_i))$ for all $i = 1, \ldots, n$. We now consider the function $g : |B(\mathcal{G})| \to U(1)$ defined by $g = fB(\phi)^{-1}$. It has the property that $g_{|\mathcal{G}^0} = 1$. We must show that g is homotopic to the constant function, or equivalently, that it admits a lift to an $\mathbb{R}$-valued function. In fact, in this case $[f] = [|B(\phi)|]$.
We have a natural map $p : |B(\mathcal{G})| \to \mathcal{G}^0/\mathcal{G}^1$ (the target is the quotient space of $\mathcal{G}^0$ with respect to the equivalence relation generated by $\mathcal{G}^1$) given by $p(\sigma, (\gamma_1, \ldots, \gamma_n)) := s(\gamma_n)$, where $\sigma \in \Delta^n$. The fibre of p over the

class $[x] \in \mathcal{G}^0/\mathcal{G}^1$ is homotopy equivalent to the classifying space $|B(\mathcal{G}^x_x)|$. Since $\mathcal{G}^x_x$ is a finite group we have $H^1(|B(\mathcal{G}^x_x)|, \mathbb{Z}) = 0$. This shows that the restriction of the $U(1)$-valued function g to $p^{-1}([x])$ admits a lift to an $\mathbb{R}$-valued function which is unique up to an additive integer.
Let $[x] \in \mathcal{G}^0/\mathcal{G}^1$ and $\gamma \in \mathcal{G}^1$ such that $s(\gamma) \in [x]$. Let $\tilde{g}_{[x]}$ be a lift of $g_{|p^{-1}([x])}$. Then we have $\tilde{g}_{[x]}(r(\gamma)) - \tilde{g}_{[x]}(s(\gamma)) = \kappa(\gamma)(1) - \kappa(\gamma)(0) - \tilde{a}(r(\gamma)) + \tilde{a}(s(\gamma)) = \tilde{h}(\gamma) - \tilde{a}(r(\gamma)) + \tilde{a}(s(\gamma)) = 0$. This allows us to normalize the lift $\tilde{g}_{[x]}$ such that $(\tilde{g}_{[x]})_{|[x]} = 0$. These normalized lifts fit together to a lift $\tilde{g} : |B(\mathcal{G})| \to \mathbb{R}$ of g. □

This finishes the proof of the fact that A is full. Note that this implies that A is injective on the level of sets of isomorphism classes of objects.
4.2.18. In the final step of the proof of Proposition 4.3 we show that A is faithful. It suffices to show that A is injective on the group of automorphisms of a $U(1)$-bundle $\mathcal{E} \to \mathcal{G}$. Via a mapping torus construction we can translate this assertion to the injectivity of A on the set of isomorphism classes of $U(1)$-bundles over $S^1 \times \mathcal{G}$. Therefore faithfulness is implied by the preceeding results. This finishes the proof of Proposition 4.3. □

4.3. *The Borel construction for pairs*

4.3.1. In this subsection we finish the proof of Theorem 1.1. Let $Y \to B$ be an atlas of an orbispace B. Recall that $PA_Y : P(B) \to P(|Y^\cdot|)$ maps the pair (E, h) to $(|X^\cdot|, h)$, where $X := E \times_B Y \to E$ is the induced atlas of E, $|X^\cdot| \to |Y^\cdot|$ is the induced $U(1)$-principal bundle, and $h \in H^3(|X^\cdot|, \mathbb{Z}) \cong H^3(E, \mathbb{Z})$.
We must show that PA_Y induces an isomorphism on the level of isomorphism classes pairs. Since the construction is functorial it is clear that PA_Y descends to isomorphism classes.
We first show that it is surjective. Consider a pair (F, h) over $|Y^\cdot|$. Then by Proposition 4.3 we find a $U(1)$-bundle $E \to B$ such that $|X^\cdot| \cong F$ as $U(1)$-bundles over $|Y^\cdot|$. Using this isomorphism we consider $h \in H^3(E, \mathbb{Z})$. It follows that A_Y maps (E, h) to (F, h). Hence, PA_Y hits all isomorphism classes.
We now consider two pairs (E_i, h_i), $i = 0, 1$ over B. We assume that they become isomorphic under PA_Y, i.e. we have an isomorphism of $U(1)$-bundles $\phi : |X_0^\cdot| \to |X_1^\cdot|$ such that $\phi^* h_1 = h_0$. We apply again Proposition 4.3 in order to find an isomorphism $\Phi : E_0 \to E_1$ such that $PA_Y(\Phi)$ is homotopic to ϕ. It therefore gives an isomorphism of pairs $(E_0, h_0) \cong$

(E_1, h_1). This shows that PA_Y is injective. □

5. Examples

5.1. Γ-*Points - cyclic groups*

5.1.1. *L*et Γ be a finite group. Let Γ act on the one point space $*$ and consider the orbispace $B := [*/\Gamma]$. The map $* \mapsto [*/\Gamma]$ is an atlas. The associated groupoid is $\mathcal{G} : \Gamma \Rightarrow *$, and $B(\mathcal{G})$ is the usual bar construction on Γ. We have $|B(\mathcal{G})| \cong B\Gamma$.
5.1.2. The group of characters of Γ can be identified with the group cohomology $H^1(\Gamma, U(1))$. Let $\chi \in H^1(\Gamma, U(1))$ be a character. It induces an action of Γ on $U(1)$. We obtain a $U(1)$-principal bundle $E := [U(1)/\Gamma] \to B$. In order to extend E to a pair over B we must choose a class $h \in H^3(E, \mathbb{Z})$. We use the Gysin sequence in order to get some information about this cohomology group.
5.1.3. The topology of the bundle $E \to B$ enters into the Gysin sequence through its first Chern class. In order to describe this class in terms of the character χ we consider the boundary operator of the long exact sequence in group cohomology associated to the sequence of coefficients

$$0 \to \mathbb{Z} \to \mathbb{R} \to U(1) \to 0 .$$

It provides an isomorphism

$$\delta : H^1(\Gamma, U(1)) \xrightarrow{\sim} H^2(\Gamma, \mathbb{Z}) \cong H^2(B\Gamma, \mathbb{Z}) \cong H^2(B, \mathbb{Z}) .$$

Let $c_1(E) \in H^2(B, \mathbb{Z})$ denote the first Chern class of E. We then have

$$c_1(E) = \delta(\chi) .$$

5.1.4. Since Γ is finite we have $H^1(B\Gamma, \mathbb{Z}) = H^1(B, \mathbb{Z}) = 0$. The relevant part of the Gysin sequence has the form

$$0 \to H^3(B, \mathbb{Z}) \xrightarrow{\pi^*} H^3(E, \mathbb{Z}) \xrightarrow{\pi_!} H^2(B, \mathbb{Z}) \xrightarrow{\dots \cup c_1(E)} H^4(B, \mathbb{Z}) \to \dots .$$

5.1.5. Let us from now on assume that Γ is the cyclic group $\mathbb{Z}/n\mathbb{Z}$. We identify $\hat{\Gamma} \cong \mathbb{Z}/n\mathbb{Z}$ such that the character corresponding to $[q] \in \mathbb{Z}/n\mathbb{Z}$ is given by

$$\chi([p]) = \exp(\frac{2\pi i p q}{n}) .$$

The cohomology of $B\Gamma$ is given by

i	$H^i(B\Gamma,\mathbb{Z})$
0	$\mathbb{Z}$
$2l-1$	0
$2l$	$\mathbb{Z}/n\mathbb{Z}$

,

where $l \geq 1$.
Under this identification we have $c_1(E) = [q]$. The Gysin sequence specializes to

$$0 \to H^3(E,\mathbb{Z}) \stackrel{\pi_!}{\to} \mathbb{Z}/n\mathbb{Z} \stackrel{[q]}{\to} \mathbb{Z}/n\mathbb{Z} \to \dots$$

so that

$$H^3(E,\mathbb{Z}) \cong \{[s] \in \mathbb{Z}/n\mathbb{Z} \mid n|sq\} \subset \mathbb{Z}/n\mathbb{Z} .$$

We fix a class $h = [s]$ in this group.
5.1.6. We can now calculate the T-dual pair $(\hat{E},\hat{h})$. Note that by [2], Lemma 2.12, we have $c_1(\hat{E}) = -\pi_!(h)$. Therefore, we have $c_1(\hat{E}) = [-s] \in \mathbb{Z}/n\mathbb{Z} \cong H^2(B,\mathbb{Z})$. We can determine $\hat{h}$ by the condition $\hat{\pi}_!(\hat{h}) = -c_1(E)$. The relevant part of the Gysin sequence for $\hat{E}$ has the form

$$0 \to H^3(\hat{E},\mathbb{Z}) \stackrel{\hat{\pi}_!}{\to} \mathbb{Z}/n\mathbb{Z} \stackrel{[-s]}{\to} \mathbb{Z}/n\mathbb{Z} \to \dots ,$$

so that

$$H^3(\hat{E},\mathbb{Z}) = \{[r] \in \mathbb{Z}/n\mathbb{Z} \mid n|sr\} \subset \mathbb{Z}/n\mathbb{Z} ,$$

and we have $\hat{h} = [-q]$.
5.1.7. Note that the stack $E = [U(1)/\mathbb{Z}/n\mathbb{Z}]$ is equivalent to a space which is homeomorphic to $U(1)$. But the action of $U(1)$ on this space is not free. Let us assume that $(q,n) = 1$. Then we have $H^3(E,\mathbb{Z}) = 0$ and thus $h = 0$. The dual bundle is then given by the orbispace $\hat{E} = [U(1)/\mathbb{Z}/n\mathbb{Z}]$, where the group $\mathbb{Z}/n\mathbb{Z}$ now acts trivially. This orbispace is not equivalent to a space. We have $H^3(\hat{E},\mathbb{Z}) \cong \mathbb{Z}/n\mathbb{Z}$, and $h = [-q]$. This example shows that in general the T-dual of a space with a non-free $U(1)$-action is an orbispace which is not equivalent to a space anymore.
5.1.8. We now calculate the twisted Borel K-groups for E and $\hat{E}$. As predicted by the general theory they turn out to be isomorphic (up to degree-shift). We keep the assumption $(n,q) = 1$.
Since $h = 0$ and $E \cong U(1)$ we have

i	$K^i_{Borel}(E,\mathcal{H})$
$2l-1$	$\mathbb{Z}$
$2l$	$\mathbb{Z}$

,

where $l \in \mathbb{Z}$ and $\mathcal{H}$ is a trivializable twist.

5.1.9. We employ the Mayer-Vietoris sequence in order to calculate $K^*_{Borel}(\hat{E}, \hat{\mathcal{H}})$, where $\hat{\mathcal{H}}$ is a twist of $\hat{E} \cong U(1) \times [*/\mathbb{Z}/n\mathbb{Z}]$ classified by $\hat{h}$. We fix the atlas $* \to [*/\mathbb{Z}/n\mathbb{Z}]$. Then $X := U(1) \times * \to U(1) \times [*/\mathbb{Z}/n\mathbb{Z}]$ is an atlas of $\hat{E}$. We get $|X^\cdot| \cong U(1) \times B\mathbb{Z}/n\mathbb{Z}$. We have $\hat{h} = \mathtt{or}_{U(1)} \times [-q]$, where $\mathtt{or}_{U(1)} \in H^1(U(1), \mathbb{Z})$ is the positive generator, and $[-q] \in H^2(B\mathbb{Z}/n\mathbb{Z}, \mathbb{Z}) \cong \mathbb{Z}/n\mathbb{Z}$. We can assume that $\hat{\mathcal{H}}$ is a twist on $|X^\cdot|$. We decompose $U(1)$ into the union of an upper and a lower hemisphere $I^\pm$. The restriction of $\hat{\mathcal{H}}$ to $I^\pm \times B\mathbb{Z}/n\mathbb{Z}$ is trivializable.

5.1.10. We have a ring isomorphism $K(B\mathbb{Z}/n\mathbb{Z}) \cong R(\mathbb{Z}/n\mathbb{Z})_{(I)}$, where $I \subset R(\mathbb{Z}/n\mathbb{Z})$ is the dimension ideal in the representation ring of $\mathbb{Z}/n\mathbb{Z}$, and $(\dots)_{(I)}$ denotes the I-adic completion. In particular we have $K^1(B\mathbb{Z}/n\mathbb{Z}) \cong \{0\}$. We have a natural map $\mathbb{Z}/n\mathbb{Z} \to K(B\mathbb{Z}/n\mathbb{Z})$ which associates to $[q]$ the class of the line bundle over $B\mathbb{Z}/n\mathbb{Z}$ associated to the character $[s] \mapsto \exp(2\pi i \frac{sq}{n})$.

5.1.11. We can now write out the Mayer-Vietoris sequence in twisted K-theory associated to the decomposition

$$|X^\cdot| \cong (I^+ \times B\mathbb{Z}/n\mathbb{Z}) \cup (I^- \times B\mathbb{Z}/n\mathbb{Z}) \ .$$

$$0 \to K^0_{Borel}(\hat{E}, \hat{\mathcal{H}}) \to$$

$$K^0(B\mathbb{Z}/n\mathbb{Z}) \oplus K^0(B\mathbb{Z}/n\mathbb{Z}) \stackrel{\begin{pmatrix} 1 & 1 \\ -[-q] & -1 \end{pmatrix}}{\to} K^0(B\mathbb{Z}/n\mathbb{Z}) \oplus K^0(B\mathbb{Z}/n\mathbb{Z})$$

$$\to K^1_{Borel}(\hat{E}, \hat{\mathcal{H}}) \to 0 \ .$$

Here, since $I^\pm$ is contractible and the restriction of the twist is trivializable, we identify $K(I^\pm \times B\mathbb{Z}/n\mathbb{Z}, \hat{\mathcal{H}}_{|I^\pm \times B\mathbb{Z}/n\mathbb{Z}})$ with $K(B\mathbb{Z}/n\mathbb{Z})$. The appearance of $[-q]$ instead of -1 in the lower left corner of the matrix is due to the presence of twists. We now use the isomorphism $K(B\mathbb{Z}/n\mathbb{Z}) \cong R(\mathbb{Z}/n\mathbb{Z})_{(I)}$ and calculate that

$$K^0_{Borel}(\hat{E}, \hat{\mathcal{H}}) \cong \ker(([-q] - 1) : R(\mathbb{Z}/n\mathbb{Z})_{(I)} \to R(\mathbb{Z}/n\mathbb{Z})_{(I)}) \cong \mathbb{Z}$$

and

$$K^1_{Borel}(\hat{E}, \hat{\mathcal{H}}) \cong \mathtt{coker}(([-q] - 1) : R(\mathbb{Z}/n\mathbb{Z})_{(I)} \to R(\mathbb{Z}/n\mathbb{Z})_{(I)}) \cong \mathbb{Z} \ .$$

Therefore we get

i	$K^i_{Borel}(\hat{E}, \hat{\mathcal{H}})$
$2l-1$	$\mathbb{Z}$
$2l$	$\mathbb{Z}$

as predicted by the T-duality isomorphism.

5.2. *Seifert fibrations*

5.2.1. In this subsection we consider T-duality of $U(1)$-bundles over certain two-dimensional orbispaces. In order to describe such an orbispace B we fix numbers $r, g \in \mathbb{N}_0$, and an element $(n_1, \ldots, n_r) \in (\mathbb{Z}\backslash\{0\})^r$. We set $n_0 := 1$. We consider $\Gamma_i := \mathbb{Z}/n_i\mathbb{Z}$ as a subgroup of $U(1)$ via $[q] \mapsto \exp(2\pi i \frac{q}{n_i})$.
Let Σ be an oriented surface of genus g. We fix pairwise distinct points $p_0, p_1, \ldots, p_r \in \Sigma$. We further choose orientation preserving identifications $(\bar{U}_i, p_i) \cong (D^2, 0)$ of suitable pairwise disjoint closed pointed neighborhoods $\bar{U}_i$ of p_i for all $i = 0, \ldots r$. The group Γ_i acts naturally on the disk $\tilde{D} \subset \mathbb{C}$. We consider the associated branched covering $\tilde{D} \to D$, $z \mapsto z^{|n_i|}$, and let $\tilde{\bar{U}}_i \to \bar{U}_i$ be the branched covering induced via our identification $\bar{U}_i \cong D$.
5.2.2. This data determines a topological groupoid $\mathcal{G}$ which represents the orbispace $B := [\mathcal{G}^1/\mathcal{G}^0]$. Let $\Sigma^0 := \Sigma \setminus \bigcup_{i=0}^r U_i$, where $U_i \subset \bar{U}_i$ denotes the interior. We define

$$\mathcal{G}^0 := \Sigma^0 \sqcup \bigsqcup_{i=0}^{r} \tilde{\bar{U}}_i \ .$$

The set of morphisms is defined as follows. First of all the restriction of $\mathcal{G}$ to Σ^0 is the trivial groupoid. The restriction of $\mathcal{G}$ to $\tilde{\bar{U}}_i$ is the action groupoid of the Γ_i-action on $\tilde{\bar{U}}_i$, i.e. $\Gamma_i \times \tilde{\bar{U}}_i \Rightarrow \tilde{\bar{U}}_i$. It remains to describe the morphisms over the overlaps. A point $s^\Sigma \in \partial\Sigma^0$ determines an index i and a point $\bar{s} \in \bar{U}_i$. For any lift $\tilde{s} \in \tilde{\bar{U}}_i$ of $\bar{s}$ we require that there is exactly one morphism $s^\Sigma \to \tilde{s}$ in $\mathcal{G}^1$. As a topological space $\mathcal{G}^1$ is fixed by the requirement that $s : s^{-1}(\partial\Sigma^0) \to \partial\Sigma^0$ is a connected covering over each connected component of $\partial\Sigma^0$, where $s : \mathcal{G}^1 \to \mathcal{G}^0$ is the source map.
In fact, this groupoid describes an orbispace structure on Σ with singular points $p_1, \ldots, p_r$ of multiplicity $n_1, \ldots, n_r$. The point p_0 will be used later in order to introduce a non-trivial topology on $U(1)$-bundles over B in the case $r = 0$.
5.2.3. We now describe $U(1)$-bundles over B. To this end we choose a number $c \in \mathbb{Z}$ and an element $(\chi_1, \ldots, \chi_r) \in \hat{\Gamma}_1 \times \cdots \times \hat{\Gamma}_r$. This data together with additional choices (the ϕ_i introduced below) determines a $U(1)$-bundle $E \to B$ as follows. We will describe it as a quotient $E := [\mathcal{E}/\mathcal{G}^1]$, where $\mathcal{E} \to \mathcal{G}$ is an equivariant $U(1)$-bundle. It is given by a $U(1)$-bundle $\mathcal{E} \to \mathcal{G}^0$ together with an action $\mathcal{G}^1 \times_{\mathcal{G}^0} \mathcal{E} \to \mathcal{E}$. We set $\mathcal{E} := U(1) \times \mathcal{G}^0$. The data fixed above determines the action of $\mathcal{G}$. On $\mathcal{E}_{|\tilde{\bar{U}}_i}$ we

let Γ_i act on the fibre with character χ_i.
For all $i = 1,\ldots,r$ we choose a map $\phi_i : \partial\tilde{\bar{U}}_i \to U(1)$ such that $\phi_i(\gamma\tilde{s}) = \chi_i(\gamma)\phi_i(\tilde{s})$, $\gamma \in \Gamma_i$. We identify $\hat{\Gamma}_i \cong \mathbb{Z}/n_i\mathbb{Z}$ such that $[q] \in \mathbb{Z}/n_i\mathbb{Z}$ corresponds to the character $[p] \mapsto \exp(2\pi i \frac{pq}{n_i})$. Note that in $\hat{\Gamma}_i \cong \mathbb{Z}/n_i\mathbb{Z}$ we have $[\deg(\phi_i)] = \chi_i$. Here in order to define the degree $\deg(\phi_i) \in \mathbb{Z}$, we choose the orientation of $\partial\tilde{\bar{U}}_i$ as the boundary of the oriented disk $\tilde{\bar{U}}_i$. Furthermore note that two choices of ϕ_i differ by a function $\partial\bar{U}_i \to U(1)$. Thus we can realize all elements of the residue class of χ as $\deg(\phi_i)$ for an appropriate choice of ϕ_i.
We let the morphism $s^\Sigma \to \tilde{s}$ act as multiplication by $\phi_i(\tilde{s})$, if s^Σ is in the ith component of $\partial\Sigma^0$, $i = 1,\ldots,r$.
Finally, we take a function $u : \partial\bar{U}_0 \to U(1)$ of degree c and let the morphism $s^\Sigma \to s$ act by multiplication by $u(s)$, if s^Σ is in the zero-component of $\partial\Sigma^0$.
5.2.4. If χ_i are generators of $\hat{\Gamma}_i$ for all $i = 1,\ldots r$, then E is a space. Otherwise E is an orbispace which is not equivalent to a space.
5.2.5. We first compute $H^*(B,\mathbb{Z})$ using a Mayer-Vietoris sequence. We obtain

$$\cdots \to \bigoplus_{i=0}^{r} H^{*-1}(\partial\bar{U}_i,\mathbb{Z}) \to H^*(B,\mathbb{Z}) \to H^*(\Sigma^0,\mathbb{Z}) \oplus \bigoplus_{i=0}^{r} H^*(B\Gamma_i,\mathbb{Z})$$
$$\to \bigoplus_{i=0}^{r} H^*(\partial\bar{U}_i,\mathbb{Z}) \to \ldots \quad .$$

We have a canonical identification $H^2(B\Gamma_i,\mathbb{Z}) \cong \hat{\Gamma}_i$. The fixed embedding $\Gamma_i \hookrightarrow U(1)$ induces a map

$$B\Gamma_i \to BU(1) \cong K(\mathbb{Z},2)$$

and therefore a generator $c_i \in H^2(B\Gamma_i,\mathbb{Z})$. The multiplication with the powers of c_i provides the isomorphisms $\hat{\Gamma}_i \cong H^{2l}(B\Gamma_i,\mathbb{Z})$. Furthermore, $H^{2l-1}(B\Gamma_i,\mathbb{Z}) \cong \{0\}$.
5.2.6. The Mayer-Vietoris sequence now gives the following information.

l	$H^l(B,\mathbb{Z})$
0	$\mathbb{Z}$
1	$\mathbb{Z}^{2g}$
2	$0 \to \mathbb{Z} \xrightarrow{\delta} H^2(B,\mathbb{Z}) \to \bigoplus_{i=1}^{r} \hat{\Gamma}_i \to 0$
$2l+1, l \geq 1$	0
$2l, l \geq 2$	$\bigoplus_{i=1}^{r} \hat{\Gamma}_i$

The data chosen in the construction 5.2.3 provides a split s of the exact sequence for $H^2(B,\mathbb{Z})$. In fact, given $(\chi_1,\ldots,\chi_r) \in \bigoplus_{i=1}^{r} \hat{\Gamma}_i$ we construct the line bundle $E \to B$ associated to this tuple and $c = 0$. Then we set $s(\chi_1, \ldots, \chi_r) := c_1(E)$. It will follow from the calculations in 5.2.7 that this gives a split. Since there is no non-trivial homomorphism from $\bigoplus_{i=1}^r \hat{\Gamma}_i$ to $\mathbb{Z}$ the split s is independent of the choices. Therefore we can unambiguously write

$$H^2(B, \mathbb{Z}) \cong \mathbb{Z} \oplus \bigoplus_{i=1}^{r} \hat{\Gamma}_i \ .$$

We will write elements in the form $(e, (\kappa_1, \ldots, \kappa_r))$.

5.2.7. By Proposition 4.3 the topological type of the $U(1)$-bundle $E \to B$ is classified by its first Chern class $c_1(E)$. In the following paragraph we calculate this invariant. To this end we consider the following part of the Gysin sequence of $\pi : E \to B$:

$$\mathbb{Z} \cong H^0(B, \mathbb{Z}) \stackrel{c_1(E)}{\to} H^2(B, \mathbb{Z}) \stackrel{\pi^*}{\to} H^2(E, \mathbb{Z}) \ .$$

We see that we can calculate $c_1(E)$ by determining the corresponding generator of the kernel of $\pi^* : H^2(B, \mathbb{Z}) \to H^2(E, \mathbb{Z})$.

We obtain information on $H^2(E, \mathbb{Z})$ using the Mayer-Vietoris sequence. The relevant part has the form

$$\begin{aligned} H^1(U(1) \times \Sigma^0, \mathbb{Z}) \oplus \bigoplus_{i=0}^{r} H^1([U(1)/_{\chi_i}\Gamma_i], \mathbb{Z}) &\stackrel{\alpha}{\to} \bigoplus_{i=0}^{r} H^1(U(1) \times \partial\bar{U}_i, \mathbb{Z}) \\ &\to H^2(E, \mathbb{Z}) \to \\ H^2(U(1) \times \Sigma^0, \mathbb{Z}) \oplus \bigoplus_{i=0}^{r} H^2([U(1)/_{\chi_i}\Gamma_i], \mathbb{Z}) &\stackrel{\beta}{\to} \bigoplus_{i=0}^{r} H^2(U(1) \times \partial\bar{U}_i, \mathbb{Z}) \ . \end{aligned}$$

The known cohomology groups are

$$\begin{aligned} H^1(U(1) \times \Sigma^0, \mathbb{Z}) &\cong 1_{U(1)} \times H^1(\Sigma^0, \mathbb{Z}) \oplus \mathtt{or}_{U(1)} \times (1_{\Sigma^0})\mathbb{Z} \\ H^1(\Sigma_0, \mathbb{Z}) &\cong \mathbb{Z}^{2g+r} \\ H^1(U(1) \times \partial\bar{U}_i, \mathbb{Z}) &\cong (1_{U(1)} \times \mathtt{or}_{\partial\bar{U}_i})\mathbb{Z} \oplus (\mathtt{or}_{U(1)} \times 1_{\partial\bar{U}_i})\mathbb{Z} \\ H^1([U(1)/_{\chi_i}\Gamma_i], \mathbb{Z}) &\cong \mathbb{Z} \\ H^2(U(1) \times \Sigma^0, \mathbb{Z}) &\cong \mathtt{or}_{U(1)} \times H^1(\Sigma^0, \mathbb{Z}) \\ H^2(U(1) \times \partial\bar{U}_i, \mathbb{Z}) &\cong (\mathtt{or}_{U(1)} \times \mathtt{or}_{\partial\bar{U}_i})\mathbb{Z} \\ H^2([U(1)/_{\chi_i}\Gamma_i], \mathbb{Z}) &\cong \hat{\Gamma}_i/\chi_i\hat{\Gamma}_i \ , \end{aligned}$$

where the definition of $\hat{\Gamma}_i/\chi\hat{\Gamma}_i$ uses the ring structure on $\hat{\Gamma}_i \cong \mathbb{Z}/n\mathbb{Z}$.

The map β vanishes on the torsion subgroups $H^2([U(1)/_{\chi_i}\Gamma_i],\mathbb{Z})$. The range of the restriction of β to $H^2(U(1)\times\Sigma^0,\mathbb{Z})$ has rank r. We see that

$$\ker(\beta)\cong\mathbb{Z}^{2g}\oplus\bigoplus_{r=1}^{r}\hat{\Gamma}_i/_{\chi_i}\hat{\Gamma}_i\ .$$

We now determine the cokernel of α. We proceed in stages. We first determine the cokernel of the restriction of α to $1_{U(1)}\times H^1(\Sigma^0,\mathbb{Z})$. It is given by

$$\bigoplus_{i=0}^{r}(1_{U(1)}\times\mathtt{or}_{\partial\bar{U}_i})\mathbb{Z}\oplus\bigoplus_{i=0}^{r}(\mathtt{or}_{U(1)}\times 1_{\partial\bar{U}_i})\mathbb{Z}\to\mathbb{Z}\oplus\bigoplus_{i=0}^{r}(\mathtt{or}_{U(1)}\times 1_{\partial\bar{U}_i})\mathbb{Z}\ ,$$

where the first component maps $\sum_{i=0}^r a_i(1_{U(1)}\times\mathtt{or}_{\partial\bar{U}_i})$ to $\sum_{i=0}^r a_i$, and the second component is the identity. Let

$$\alpha_1:(\mathtt{or}_{U(1)}\times 1_{\Sigma^0})\mathbb{Z}\oplus\bigoplus_{i=0}^{r}H^1([U(1)/_{\chi_i}\Gamma_i],\mathbb{Z})\to\mathbb{Z}\oplus\bigoplus_{i=0}^{r}(\mathtt{or}_{U(1)}\times 1_{\partial\bar{U}_i})\mathbb{Z}$$

be the induced map. We have

$$\alpha_1(\mathtt{or}_{U(1)}\times 1_{\Sigma^0})=0\oplus\oplus_{i=0}^r(\mathtt{or}_{U(1)}\times 1_{\partial\bar{U}_i})\ .$$

We now describe the restriction of α_1 to the summand $H^1([U(1)/_{\chi_i}\Gamma_i],\mathbb{Z})$. It is given by the composition of pull-backs along the following sequence of maps:

$$U(1)\times\partial\bar{U}_i\cong[U(1)\times\partial\tilde{\bar{U}}_i/_1\Gamma_i]\stackrel{I_{\phi_i}}{\to}[U(1)\times\partial\tilde{\bar{U}}_i/_{\chi_i}\Gamma_i]\to[U(1)\times\tilde{\bar{U}}_i/_{\chi_i}\Gamma_i]\\ \to[U(1)/_{\chi_i}\Gamma_i]\quad ,$$

where I_{ϕ_i} is induced by the map ϕ_i (see 5.2.3) $I_{\phi_i}(z,\tilde{s}):=(\phi_i(\tilde{s})z,\tilde{s})$, and the remaining maps are the obvious inclusions and projections. In the case $i=0$ we set $\phi_0:=u$.

We have $H^1([U(1)/_{\chi_i}\Gamma_i],\mathbb{Z})\cong H^1(U(1)\times_{\Gamma_i,\chi_i}E\Gamma_i,\mathbb{Z})$. We consider the $U(1)$-bundle $U(1)\times_{\Gamma_i,\chi_i}E\Gamma_i\to B\Gamma_i$. Using the Serre spectral sequence we see that the restriction to the fibre r^* fits into an exact sequence

$$0\to H^1([U(1)/_{\chi_i}\Gamma_i],\mathbb{Z})\stackrel{r^*}{\to}\mathbb{Z}\stackrel{\chi_i}{\to}\mathbb{Z}/n_i\mathbb{Z}\ .$$

Similarly, restriction to the fibre of the bundle $(U(1)\times\partial\tilde{\bar{U}}_i)_{\Gamma_i,1}\times E\Gamma_i\to B\Gamma_i$ gives an exact sequence

$$0\to H^1([U(1)\times\partial\tilde{\bar{U}}_i/_1\Gamma_i],\mathbb{Z})\to\mathbb{Z}\oplus\mathbb{Z}\stackrel{\mathtt{pr}_2}{\to}\mathbb{Z}/n_i\mathbb{Z}\ ,$$

where we use the basis $\mathbb{Z}\oplus\mathbb{Z}\cong(\mathtt{or}_{U(1)}\times 1_{\partial\tilde{\bar{U}}_i})\mathbb{Z}\oplus(1_{U(1)}\times\mathtt{or}_{\partial\tilde{\bar{U}}_i})\mathbb{Z}$.

Let $a \in \mathbb{Z}$ represent an element of $H^1([U(1)/_{\chi_i}\Gamma_i], \mathbb{Z})$, i.e. $\chi_i[a] = 0 \in \mathbb{Z}/n_i\mathbb{Z}$. Then one can check that $\alpha_1(a) = (a, \deg(\phi_i)a)$. Fortunately, as observed in 5.2.3, $[\deg(\phi_i)] = \chi$ in $\hat{\Gamma}_i \cong \mathbb{Z}/n_i\mathbb{Z}$ so that $n_i | \deg(\phi_i)a$, and thus $(a, \deg(\phi_i)a) \in H^1([U(1) \times \partial\tilde{\bar{U}}_i/_1\Gamma_i], \mathbb{Z})$. Combining these calculations we obtain the following explicit description of

$$\alpha_1 : \mathbb{Z} \oplus \bigoplus_{i=0}^{r} \ker(\chi_i : \mathbb{Z} \to \mathbb{Z}/n_i\mathbb{Z}) \to \mathbb{Z} \oplus \mathbb{Z}^{r+1} ,$$

$$\alpha_1(x, (a_0, \ldots, a_r)) = (\sum_{i=0}^{r} \frac{\deg(\phi_i)a_i}{n_i}, (a_0 + x, \ldots, a_r + x)) ,$$

where on the right-hand side we identify $\mathbb{Z}^{r+1} \cong \bigoplus_{i=0}^{r}(\mathtt{or}_{U(1)} \times 1_{\partial\bar{U}_i})\mathbb{Z}$. We now have collected sufficient information on $H^2(E, \mathbb{Z})$ in order to calculate $c_1(E)$. By the compatibility of the Mayer-Vietoris sequences with the pull-back

$$\pi^* : H^2(B, \mathbb{Z}) \to H^2(E, \mathbb{Z})$$

we get the diagram

$$\begin{array}{ccccccc} 0 \to & \mathtt{coker}(\alpha_1) & \to H^2(E,\mathbb{Z}) & \to & \mathbb{Z}^{2g} \oplus \bigoplus_{i=1}^{r} \hat{\Gamma}_i/\chi_i\hat{\Gamma}_i & \to 0 \\ & f \uparrow & \uparrow & & g \uparrow & \\ 0 \to & \mathbb{Z} & \to H^2(B,\mathbb{Z}) & \stackrel{\oplus_{i=0}^{r} t_i^*}{\to} & \bigoplus_{i=0}^{r} \hat{\Gamma}_i & \to 0 \end{array} ,$$

where $t_i : [p_i/\Gamma_i] \to B$ is the canonical embedding. We must determine generators of $\ker(f)$ and $\ker(g)$. We have a factorization of f as $\mathbb{Z} \stackrel{(\mathrm{id},0)}{\to} \mathbb{Z} \oplus \mathbb{Z}^{r+1} \to \mathtt{coker}(\alpha_1)$. We see that $f(b) = 0$ is equivalent to the condition that the system

$$b = x \sum_{i=0}^{r} \frac{\deg(\phi_i)}{n_i}$$
$$\chi_i[x] = 0 \in \mathbb{Z}/n_i\mathbb{Z} , \quad i = 0, \ldots r$$

has a solution $x \in \mathbb{Z}$. We see that $\ker(f) \subset \mathbb{Z}$ is a non-trivial subgroup, and we fix the generator $e \in \mathbb{Z}$ which is given by the component of $c_1(E)$. It is determined by the subgroup up to sign.
The kernel of g is the sum of the kernels of the projections $\hat{\Gamma}_i \to \hat{\Gamma}_i/\chi_i\hat{\Gamma}_i$. In order to find the generators which correspond to the Chern character of E we use the fact that the Chern character is compatible with restriction.

We consider the pull-back

$$\begin{array}{ccc} E_{p_i} & \rightarrow & E \\ \downarrow & & \downarrow . \\ [p_i/\Gamma_i] & \stackrel{t_i}{\rightarrow} & B \end{array}$$

Since we know that $c_1(E_{p_i}) = \chi_i \in \hat{\Gamma}_i$ we see that $t_i^* c_1(E) = \chi_i$ is the correct generator of the kernel of the corresponding component of g. Combining these calculations we get

$$c_1(E) = (e, (\chi_1, \ldots, \chi_1)) \in \mathbb{Z} \oplus \bigoplus_{i=1}^{r} \hat{\Gamma}_i \ ,$$

where e was described above.

5.2.8. We now compute $H^3(E, \mathbb{Z})$, again using a Mayer-Vietoris sequence. Let $[U(1)/_{\chi_i}\Gamma_i]$ denote the orbispace given by the action of Γ_i on $U(1)$ via χ_i. The relevant part of the Mayer-Vietoris sequence has the form

$$H^2(U(1) \times \Sigma^0, \mathbb{Z}) \oplus \bigoplus_{i=0}^{r} H^2([U(1)/_{\chi_i}\Gamma_i], \mathbb{Z}) \to \bigoplus_{i=0}^{r} H^2(U(1) \times \partial\bar{U}_i, \mathbb{Z})$$
$$\to H^3(E, \mathbb{Z}) \to \bigoplus_{i=0}^{r} H^3([U(1)/_{\chi_i}\Gamma_i], \mathbb{Z}) \to 0 \ .$$

We now use the facts that the restriction

$$H^2([U(1)/_{\chi_i}\Gamma_i], \mathbb{Z}) \to H^2(U(1) \times \partial\bar{U}_i, \mathbb{Z})$$

is trivial, that the cokernel of

$$H^2(U(1) \times \Sigma^0, \mathbb{Z}) \to \bigoplus_{i=0}^{r} H^2(U(1) \times \partial\bar{U}_i, \mathbb{Z})$$

is isomorphic to $\mathbb{Z}$, and that

$$H^3([U(1)/_{\chi_i}\Gamma_i], \mathbb{Z}) \cong \mathtt{Ann}(\chi_i),$$

where the definition of $\mathtt{Ann}(\chi_i) \subset \hat{\Gamma}_i$ uses the ring structure of $\hat{\Gamma}_i$ (see 5.1.5 for the computation of $H^3([U(1)/_{\chi_i}\Gamma_i], \mathbb{Z})$). The sequence thus simplifies to

$$0 \to \mathbb{Z} \stackrel{\delta_E}{\to} H^3(E, \mathbb{Z}) \to \bigoplus_{i=1}^{r} \mathtt{Ann}(\chi_i) \to 0 \ .$$

Let $\pi : E \to B$ be the projection. Then the following diagram commutes:

$$\begin{array}{ccc} \mathbb{Z} & \stackrel{\delta_E}{\to} & H^3(E,\mathbb{Z}) \\ \| & & \pi_! \downarrow \\ \mathbb{Z} & \stackrel{\delta}{\to} & H^2(B,\mathbb{Z}) \end{array} .$$

Therefore the decomposition

$$H^2(B,\mathbb{Z}) = \mathbb{Z} \oplus \bigoplus_{i=1}^{r} \hat{\Gamma}_i$$

induces a split $s_E : H^3(E,\mathbb{Z}) \to \mathbb{Z}$, so that we obtain an identification

$$H^3(E,\mathbb{Z}) \cong \mathbb{Z} \oplus \bigoplus_{i=1}^{r} \mathtt{Ann}(\chi_i) .$$

Note that this decomposition is again canonical. A cohomology class $h \in H^3(E,\mathbb{Z})$ is thus identified with an element

$$(f,(a_1,\ldots,a_r)) \in \mathbb{Z} \oplus \mathtt{Ann}(\chi_1) \oplus \cdots \oplus \mathtt{Ann}(\chi_r) .$$

5.2.9. It follows from Proposition 4.3 that the topological type of E is classified by $c_1(E)$.
We further observe that $\pi_! : H^3(E,\mathbb{Z}) \to H^2(B,\mathbb{Z})$ is injective. Therefore we can characterize a class in $H^3(E,\mathbb{Z})$ by its image under $\pi_!$. It follows that automorphisms of the $U(1)$-bundle E act trivially on $H^3(E,\mathbb{Z})$. We see that the isomorphism class of the pair (E,h) is determined by

$$(c_1(E),\pi_!(h)) = (e,(\chi_1,\ldots,\chi_r),f,(a_1,\ldots,a_r)) \in H^2(B,\mathbb{Z}) \oplus H^2(B,\mathbb{Z})$$

(see 5.2.6 for the notation). It therefore makes sense to calculate the T-dual pair $(\hat{E},\hat{h})$ in terms of its topological invariants $(c_1(\hat{E}),\hat{\pi}_!(\hat{h}))$. We get

$$(c_1(\hat{E}),\hat{\pi}_!(\hat{h})) = (-f,(-a_1,\ldots,-a_r),-e,(-\chi_1,\ldots,-\chi_r)) .$$

References

1. P. Bouwknegt and J. Evslin and V. Mathai. T-Duality: Topology change from H-flux. *Comm. Math. Phys.*, 249 (2004), 383-415, arXiv:hep-th/0306062.
2. U. Bunke and Th. Schick. On the topology of T-duality. *Rev. Math. Phys.*, 17 (2005), 77–112, arXiv:math.GT/0405132.
3. U. Bunke, Ph. Rumpf and Th. Schick. On the topology of T-duality for T^n-bundles. arXiv:math.GT/0501487.
4. J. Heinloth. Survey on topological and smooth stacks. *Mathematisches Institut Göttingen, WS04-05 (Y. Tschinkel, ed.)*, 1–31 (2005).

5. V. Mathai and J. Rosenberg. T-Duality for torus bundles with H-fluxes via noncommutative topology. *Comm. Math. Phys.*, 253 (2005), 707–721, arXiv:hep-th/0401168.
6. I. Moerdijk. Classifying Toposes and Foliations. *Ann. Inst. Fourier, Grenoble*, 41 (1991), 189–209.
7. I. Moerdijk. Orbispaces as groupoids: An introduction. *Adem, A. et al (eds.), Orbifolds in mathematics and physics (Madison, WI 2001) Amer. Math. Soc., Contemp. Math.* 310 (2002), 205–222, arXiv:math.DG/0203100.
8. B. Noohi. Foundations of topological stacks I. arXiv:math.AG/0503247.
9. J.L. Tu, P. Xu and C. Laurent. Twisted K-theory of differentiable stacks. *Ann. Sci. École Norm. Sup.*, 37 (2004), 841–910, arXiv:math.KT/0306318.

Received by the editors August 29, 2005; Revised December 31, 2005

Analysis, Geometry and Topology
of Elliptic Operators, pp. 467–492

A NEW SPECTRAL CANCELLATION IN QUANTUM GRAVITY

GIAMPIERO ESPOSITO

INFN, Sezione di Napoli, Complesso Universitario di Monte S. Angelo, Via Cintia, Edificio N', 80126 Napoli, Italy
Università degli Studi di Napoli Federico II, Dipartimento di Scienze Fisiche, Complesso Universitario di Monte S. Angelo, Via Cintia, Edificio N', 80126 Napoli, Italy
giampiero.esposito@na.infn.it

GUGLIELMO FUCCI

Department of Physics, New Mexico Institute of Mining and Technology, Leroy Place 801, Socorro, NM 87801, USA
gfucci@nmt.edu

ALEXANDER KAMENSHCHIK

Dipartimento di Fisica and INFN, Via Irnerio 46, 40126 Bologna, Italy
L.D. Landau Institute for Theoretical Physics, Kosygin str. 2, 119334 Moscow, Russia
alexander.kamenshchik@bo.infn.it

KLAUS KIRSTEN

Department of Mathematics, Baylor University, Waco TX 76798, USA
Klaus_Kirsten@baylor.edu

Dedicated to Krzysztof P. Wojciechowski on his 50th birthday

A general method exists for studying Abelian and non-Abelian gauge theories, as well as Euclidean quantum gravity, at one-loop level on manifolds with boundary. In the latter case, boundary conditions on metric perturbations h can be chosen to be completely invariant under infinitesimal diffeomorphisms, to preserve the invariance group of the theory and BRST symmetry. In the de Donder gauge, however, the resulting boundary-value problem for the Laplace type operator acting on h is known to be self-adjoint but not strongly elliptic.

The present paper shows that, on the Euclidean four-ball, only the scalar part of perturbative modes for quantum gravity is affected by the lack of strong ellipticity. Interestingly, three sectors of the scalar-perturbation problem remain elliptic, while lack of strong ellipticity is "confined" to the remaining fourth sector. The integral representation of the resulting ζ-function asymptotics on the Euclidean four-ball is also obtained; this remains regular at the origin by virtue of a peculiar spectral identity obtained by the authors. There is therefore encouraging evidence in favour of the $\zeta(0)$ value with fully diff-invariant boundary conditions remaining well defined, at least on the four-ball, although severe technical obstructions remain in general.

2000 *Mathematics Subject Classification.* Primary 58J35; Secondary 83C45,81S40, 81T20.

1. Introduction

This paper is motivated by the authors' struggle over many years with an important problem in quantum field theory and spectral geometry, i.e. the functional determinant in Euclidean quantum gravity on manifolds with non-empty boundary. The related open issues are not yet settled, but there is a sufficient amount of new calculations to justify further efforts, as we are going to see shortly.

The subject of boundary effects in quantum field theory (Deutsch and Candelas [1]) has always received a careful consideration in the literature by virtue of very important physical and mathematical motivations, that can be summarized as follows.

(i) Boundary data play a crucial role in the functional-integral approach (DeWitt [2]), in the quantum theory of the early universe (Hartle and Hawking, Hawking [3]) in supergravity (Hawking [4]) and even in string theory (Abouelsaood et al. [5]).

(ii) The way in which quantum fields react to the presence of boundaries is responsible for remarkable physical effects, e.g. the attractive Casimir force among perfectly conducting parallel plates (Bordag et al., Milton, Nesterenko et al. [6]), which can be viewed as arising from differences of zero-point energies of the quantized electromagnetic field.

(iii) The spectral geometry of a Riemannian manifold (Gilkey [7]) with boundary is a fascinating problem where many new results have been derived over the last few years (Kirsten [8], Vassilevich [9]).

(iv) Boundary terms (Moss [10]) in heat-kernel expansions have become a major subject of investigation in quantum gravity (Avramidi [11]), since they shed new light on one-loop conformal anomalies (Esposito et al., Moss and Poletti [12], Tsoupros [13]) and one-loop divergences (Esposito [14],

Esposito et al. [15]).

In our paper we are interested in boundary conditions for metric perturbations that are completely invariant under infinitesimal diffeomorphisms, since they are part of the general scheme according to which the boundary conditions are preserved under the action of the symmetry group of the theory (Barvinsky [16], Moss and Silva [17], Avramidi and Esposito [18]). In field-theoretical language, this means setting to zero at the boundary that part πA of the gauge field A that lives on the boundary $\mathcal{B}$ (π being a projection operator):

$$\left[\pi A\right]_{\mathcal{B}} = 0, \tag{1}$$

as well as the gauge-fixing functional,

$$\left[\Phi(A)\right]_{\mathcal{B}} = 0, \tag{2}$$

and the whole ghost field

$$[\varphi]_{\mathcal{B}} = 0. \tag{3}$$

For Euclidean quantum gravity, Eq. (1) reads as

$$[h_{ij}]_{\mathcal{B}} = 0, \tag{4}$$

where h_{ij} are perturbations of the induced three-metric. To arrive at the gravitational counterpart of Eqs. (2) and (3), note first that, under infinitesimal diffeomorphisms, metric perturbations $h_{\mu\nu}$ transform according to

$$\widehat{h}_{\mu\nu} \equiv h_{\mu\nu} + \nabla_{(\mu}\,\varphi_{\nu)}, \tag{5}$$

where ∇ is the Levi–Civita connection on the background four-geometry with metric g, and $\varphi_\nu dx^\nu$ is the ghost one-form (strictly, our presentation is simplified: there are two independent ghost fields obeying Fermi statistics, and we will eventually multiply by -2 the effect of φ_ν to take this into account). In geometric language, the infinitesimal variation $\delta h_{\mu\nu} \equiv \widehat{h}_{\mu\nu} - h_{\mu\nu}$ is given by the Lie derivative along φ of the four-metric g. For manifolds with boundary, Eq. (5) implies that (Esposito et al. [19], Avramidi et al. [20])

$$\widehat{h}_{ij} = h_{ij} + \varphi_{(i|j)} + K_{ij}\varphi_0, \tag{6}$$

where the stroke denotes three-dimensional covariant differentiation tangentially with respect to the intrinsic Levi–Civita connection of the boundary, while K_{ij} is the extrinsic-curvature tensor of the boundary. Of course, φ_0

and φ_i are the normal and tangential components of the ghost, respectively. By virtue of Eq. (6), the boundary conditions (4) are "gauge invariant", i.e.

$$\left[\widehat{h}_{ij}\right]_{\mathcal{B}} = 0, \tag{7}$$

if and only if the whole ghost field obeys homogeneous Dirichlet conditions, so that

$$[\varphi_0]_{\mathcal{B}} = 0, \tag{8}$$

$$[\varphi_i]_{\mathcal{B}} = 0. \tag{9}$$

The conditions (8) and (9) are necessary and sufficient since φ_0 and φ_i are independent, and three-dimensional covariant differentiation commutes with the operation of restriction to the boundary. We are indeed assuming that the boundary $\mathcal{B}$ is smooth and not totally geodesic, i.e. $K_{ij} \neq 0$. However, for totally geodesic boundaries, having $K_{ij} = 0$, the condition (8) is no longer necessary.

On imposing boundary conditions on the remaining set of metric perturbations, the key point is to make sure that *the invariance of such boundary conditions under the infinitesimal transformations* (5) *is again guaranteed by* (8) *and* (9), since otherwise one would obtain incompatible sets of boundary conditions on the ghost field. Indeed, on using the DeWitt–Faddeev–Popov formalism for the $\langle \text{out}|\text{in}\rangle$ amplitudes of quantum gravity, it is necessary to use a gauge-averaging term in the Euclidean action, of the form[2]

$$I_{g.a.} = \frac{1}{16\pi G}\int_{\mathcal{M}} \frac{\Phi_\nu \Phi^\nu}{2\alpha}\sqrt{\det g}\, d^4x, \tag{10}$$

where Φ_ν is any functional which leads to self-adjoint (elliptic) operators on metric and ghost perturbations. One then finds that

$$\delta\Phi_\mu(h) \equiv \Phi_\mu(h) - \Phi_\mu(\widehat{h}) = \mathcal{F}_\mu^{\ \nu}\, \varphi_\nu, \tag{11}$$

where $\mathcal{F}_\mu^{\ \nu}$ is an elliptic operator that acts linearly on the ghost field. Thus, if one imposes the boundary conditions

$$\left[\Phi_\mu(h)\right]_{\mathcal{B}} = 0, \tag{12}$$

and if one assumes that the ghost field can be expanded in a complete orthonormal set of eigenfunctions $u_\nu^{(\lambda)}$ of $\mathcal{F}_\mu^{\ \nu}$ which vanish at the boundary, i.e.

$$\mathcal{F}_{\mu}^{\ \nu} u_{\nu}^{(\lambda)} = \lambda u_{\mu}^{(\lambda)}, \tag{13}$$

$$\varphi_{\nu} = \sum_{\lambda} C_{\lambda} u_{\nu}^{(\lambda)}, \tag{14}$$

$$\left[u_{\mu}^{(\lambda)} \right]_{\mathcal{B}} = 0, \tag{15}$$

the boundary conditions (12) are automatically gauge-invariant under the Dirichlet conditions (8) and (9) on the ghost.

Having obtained the general recipe expressed by Eqs. (4) and (12), we can recall what they imply on the Euclidean four-ball. This background is relevant for one-loop quantum cosmology in the limit of small three-geometry on the one hand (Schleich [21]), and for spectral geometry and spectral asymptotics on the other hand [8, 9]. As shown in [19], if one chooses the de Donder gauge-fixing functional

$$\Phi_{\mu}(h) = \nabla^{\nu} \left(h_{\mu\nu} - \frac{1}{2} g_{\mu\nu} g^{\rho\sigma} h_{\rho\sigma} \right), \tag{16}$$

which has the virtue of leading to an operator of Laplace type on $h_{\mu\nu}$ in the one-loop functional integral, Eq. (12) yields the mixed boundary conditions

$$\left[\frac{\partial h_{00}}{\partial \tau} + \frac{6}{\tau} h_{00} - \frac{\partial}{\partial \tau} (g^{ij} h_{ij}) + \frac{2}{\tau^2} h_{0i}^{\ |i} \right]_{\mathcal{B}} = 0, \tag{17}$$

$$\left[\frac{\partial h_{0i}}{\partial \tau} + \frac{3}{\tau} h_{0i} - \frac{1}{2} \frac{\partial h_{00}}{\partial x^i} \right]_{\mathcal{B}} = 0. \tag{18}$$

In [15], [19], the boundary conditions (4), (17) and (18) were used to evaluate the full one-loop divergence of quantized general relativity on the Euclidean four-ball, including all $h_{\mu\nu}$ and all ghost modes. However, the meaning of such a calculation became unclear after the discovery in [18] that the boundary-value problem for the Laplacian P acting on metric perturbations is not strongly elliptic by virtue of tangential derivatives in the boundary conditions (17) and (18). Moreover, the work by Dowker and Kirsten [22] had proved even earlier, in a simpler case, that the boundary-value problem with tangential derivatives is, in general, not strongly elliptic. Strong ellipticity [8, 18] is a technical requirement ensuring that a unique smooth solution of the boundary-value problem exists which vanishes at infinite geodesic distance from the boundary. If it is fulfilled, this ensures that the L^2 trace of the heat semigroup e^{-tP} exists, with the associated global heat-kernel asymptotics that yields one-loop divergence and one-loop effective action. However, when strong ellipticity does not hold, the L^2 trace of e^{-tP} acquires a singular part [18] and hence ζ-function calculations may become ill-defined.

All of this has motivated our analysis, which therefore derives in Sec. 2 the eigenvalue conditions for scalar modes. Section 3 obtains the first pair of resulting scalar-mode ζ-functions and Sec. 4 studies the remaining elliptic and non-elliptic parts of spectral asymptotics. Results and open problems are described in Sec. 5.

2. Eigenvalue conditions for scalar modes on the four-ball

On the Euclidean four-ball, which can be viewed as the portion of flat Euclidean four-space bounded by a three-sphere of radius q, metric perturbations $h_{\mu\nu}$ can be expanded in terms of hyperspherical harmonics as (Lifshitz and Khalatnikov [23], Esposito et al. [24])

$$h_{00}(x,\tau) = \sum_{n=1}^{\infty} a_n(\tau) Q^{(n)}(x), \tag{19}$$

$$h_{0i}(x,\tau) = \sum_{n=2}^{\infty} \left[b_n(\tau) \frac{Q^{(n)}_{|i}(x)}{(n^2-1)} + c_n(\tau) S_i^{(n)}(x) \right], \tag{20}$$

$$\begin{aligned} h_{ij}(x,\tau) &= \sum_{n=3}^{\infty} d_n(\tau) \left[\frac{Q^{(n)}_{|ij}(x)}{(n^2-1)} + \frac{c_{ij}}{3} Q^{(n)}(x) \right] + \sum_{n=1}^{\infty} \frac{e_n(\tau)}{3} c_{ij} Q^{(n)}(x) \\ &+ \sum_{n=3}^{\infty} \left[f_n(\tau) \left(S^{(n)}_{i|j}(x) + S^{(n)}_{j|i}(x) \right) + k_n(\tau) G^{(n)}_{ij}(x) \right], \end{aligned} \tag{21}$$

where $\tau \in [0,q]$ and $Q^{(n)}(x)$, $S_i^{(n)}(x)$ and $G_{ij}^{(n)}(x)$ are scalar, transverse vector and transverse-traceless tensor hyperspherical harmonics, respectively, on a unit three-sphere with metric c_{ij}. By insertion of the expansions (19)-(21) into the eigenvalue equation for the Laplacian acting on $h_{\mu\nu}$, and by setting $\sqrt{E} \rightarrow iM$, which corresponds to a rotation of contour in the ζ-function analysis (Barvinsky et al. [25]) one finds the modes as linear combinations of modified Bessel functions of first kind. Modified Bessel functions of the second kind are not included to ensure regularity at the origin $\tau = 0$. For details, we refer the reader to the work by Esposito et al. [26].

The boundary conditions (4), (17), (18), (8), (9), jointly with the mode-expansions on the four-ball, can be used to obtain homogeneous linear systems that yield, implicitly, the eigenvalues of our problem. The conditions for finding non-trivial solutions of such linear systems are given by the vanishing of the associated determinants; these yield the eigenvalue conditions $\delta(E) = 0$, i.e. the equations obeyed by the eigenvalues by virtue of the boundary conditions. For the purpose of a rigorous analysis, we need the full expression of such eigenvalue conditions for each set of coupled modes. Upon setting $\sqrt{E} \rightarrow iM$, we denote by $D(Mq)$ the counterpart of $\delta(E)$, bearing in mind that, strictly, only $\delta(E)$ yields implicitly the eigenvalues, while $D(Mq)$ is more convenient for ζ-function calculations [25].

In particular, we here focus on scalar modes (for the whole set of modes, see again the work in [26]). For all $n \geq 3$, coupled scalar modes a_n, b_n, d_n, e_n are ruled by a determinant reading as

$$D_n(Mq) = \det \rho_{ij}(Mq), \tag{22}$$

with degeneracy n^2, where ρ_{ij} is a 4×4 matrix with entries (hereafter, I_n are modified Bessel functions of first kind)

$$\rho_{11} = I_n(Mq) - MqI'_n(Mq), \ \rho_{12} = MqI'_n(Mq), \tag{23}$$

$$\begin{aligned} \rho_{13} &= (2-n)I_{n-2}(Mq) + MqI'_{n-2}(Mq), \\ \rho_{14} &= (2+n)I_{n+2}(Mq) + MqI'_{n+2}(Mq), \end{aligned} \tag{24}$$

$$\rho_{21} = -(n^2-1)I_n(Mq), \ \rho_{22} = 2MqI'_n(Mq) + 6I_n(Mq), \tag{25}$$

$$\rho_{23} = 2(n+1)MqI'_{n-2}(Mq) - (n^2-6n-7)I_{n-2}(Mq), \tag{26}$$

$$\rho_{24} = -2(n-1)MqI'_{n+2}(Mq) - (n^2+6n-7)I_{n+2}(Mq), \tag{27}$$

$$\rho_{31} = 0, \ \rho_{32} = -I_n(Mq), \tag{28}$$

$$\rho_{33} = \frac{(n+1)}{(n-2)}I_{n-2}(Mq), \ \rho_{34} = \frac{(n-1)}{(n+2)}I_{n+2}(Mq), \tag{29}$$

$$\begin{aligned} \rho_{41} &= 3I_n(Mq), \ \rho_{42} = -2I_n(Mq), \\ \rho_{43} &= -I_{n-2}(Mq), \ \rho_{44} = -I_{n+2}(Mq). \end{aligned} \tag{30}$$

The hardest part of our analysis is the investigation of the equation obtained by setting to zero the determinant (22). For this purpose, we first exploit the recurrence relations among I_n, I_{n+1} and I'_n to find (from now on, $w \equiv Mq$)

$$\begin{aligned} \rho_{11} &= I_n(w) - wI'_n(w), \ \rho_{12} = wI'_n(w), \\ \rho_{13} &= wI'_n(w) + nI_n(w), \ \rho_{14} = wI'_n(w) - nI_n(w), \end{aligned} \tag{31}$$

$$\rho_{21} = -(n^2 - 1)I_n(w), \ \rho_{22} = 2(wI'_n(w) + 3I_n(w)), \tag{32}$$

$$\begin{aligned}\rho_{23} =&(n+1)\Biggl\{ \left[3(n+1) + \frac{2n(n-1)(n+3)}{w^2}\right] I_n(w) \\ &+ 2\left[w + \frac{(n-1)(n+3)}{w}\right] I'_n(w)\Biggr\},\end{aligned} \tag{33}$$

$$\begin{aligned}\rho_{24} =&(n-1)\Biggl\{ \left[3(n-1) + \frac{2n(n+1)(n-3)}{w^2}\right] I_n(w) \\ &- 2\left[w + \frac{(n+1)(n-3)}{w}\right] I'_n(w)\Biggr\},\end{aligned} \tag{34}$$

$$\rho_{31} = 0, \ \rho_{32} = -I_n(w), \tag{35}$$

$$\rho_{33} = \frac{(n+1)}{(n-2)}\left[\left(1 + \frac{2n(n-1)}{w^2}\right) I_n(w) + \frac{2(n-1)}{w} I'_n(w)\right], \tag{36}$$

$$\rho_{34} = \frac{(n-1)}{(n+2)}\left[\left(1 + \frac{2n(n+1)}{w^2}\right) I_n(w) - \frac{2(n+1)}{w} I'_n(w)\right], \tag{37}$$

$$\rho_{41} = 3I_n(w), \ \rho_{42} = -2I_n(w), \tag{38}$$

$$\rho_{43} = -\left(1 + \frac{2n(n-1)}{w^2}\right) I_n(w) - \frac{2(n-1)}{w} I'_n(w), \tag{39}$$

$$\rho_{44} = -\left(1 + \frac{2n(n+1)}{w^2}\right) I_n(w) + \frac{2(n+1)}{w} I'_n(w). \tag{40}$$

The resulting determinant, despite its cumbersome expression, can be studied by introducing the variable

$$y \equiv \frac{I'_n(w)}{I_n(w)}, \tag{41}$$

which leads to

$$D_n(w) = \frac{48n(1-n^2)}{(n^2-4)} I_n^4(w)(y-y_1)(y-y_2)(y-y_3)(y-y_4), \tag{42}$$

where

$$y_1 \equiv -\frac{n}{w}, \ y_2 \equiv \frac{n}{w}, \ y_3 \equiv -\frac{n}{w} - \frac{w}{2}, \ y_4 \equiv \frac{n}{w} - \frac{w}{2}, \tag{43}$$

and hence

$$
\begin{aligned}
&\frac{(n^2-4)}{48n(1-n^2)} D_n(w) \\
&= \left(I_n'(w) + \frac{n}{w} I_n(w)\right)\left(I_n'(w) - \frac{n}{w} I_n(w)\right) \\
&\cdot \left(I_n'(w) + \left(\frac{w}{2} + \frac{n}{w}\right) I_n(w)\right)\left(I_n'(w) + \left(\frac{w}{2} - \frac{n}{w}\right) I_n(w)\right).
\end{aligned}
\tag{44}
$$

3. First pair of scalar-mode ζ-functions

In our problem, the differential operator under investigation is the Laplacian on the Euclidean four-ball acting on metric perturbations. The boundary conditions for vector, tensor and ghost modes correspond to a familiar mixture of Dirichlet and Robin boundary conditions for which an integral representation of the ζ-function and heat-kernel coefficients are immediately obtained. New features arise instead from Eq. (44), that gives rise to four different ζ-functions. On studying the first line of Eq. (44), we exploit the Cauchy integral formula to express the power $-s$ of the eigenvalues and hence turn the ζ-function

$$
\zeta_A^{\pm}(s) \equiv \sum_{n=3}^{\infty} n^2 \lambda_{A\pm}^{-s}
$$

into an integral, i.e. we use

$$
\sum_{l=1}^{\infty} x_l^{-s} = \int_{\gamma} dx\, x^{-s} \frac{d}{dx} \log H_n(x),
$$

where γ encloses the zeros $x_1, x_2, ..., x_\infty$ of the function H_n, which here equals $J_n'(x) \pm \frac{n}{x} J_n(x)$. Such a combination of J_n and J_n' is proportional to the power of degree $(\beta_\pm - 1)$ of the independent variable multiplied by an infinite product, with $\beta_+(n) \equiv n, \beta_-(n) \equiv n+2$. Only the infinite product encodes information on the countable infinity of non-vanishing zeros, and hence one should divide $xJ_n'(x) \pm nJ_n(x)$ by $x^{\beta_\pm}$. Last, rotation of contour to the imaginary axis (Dowker and Kirsten [22], Bordag et al. [27]), which brings in modified Bessel functions I_n, jointly with setting $w = zn$, leads to the following integral formula:

$$
\zeta_A^{\pm}(s) \equiv \frac{(\sin \pi s)}{\pi} \sum_{n=3}^{\infty} n^{-(2s-2)} \int_0^{\infty} dz\, z^{-2s} \frac{\partial}{\partial z} \log \left[\frac{\left(znI_n'(zn) \pm nI_n(zn)\right)}{z^{\beta_\pm(n)}} \right].
\tag{45}
$$

The uniform asymptotic expansion of modified Bessel functions and their first derivatives (see Appendix) can be used to find (hereafter $\tau = \tau(z) \equiv (1+z^2)^{-\frac{1}{2}}$)

$$znI_n'(zn) \pm nI_n(zn) \sim \frac{n}{\sqrt{2\pi n}} \frac{e^{n\eta}}{\sqrt{\tau}}(1\pm\tau)\left(1+\sum_{k=1}^{\infty}\frac{p_{k,\pm}(\tau)}{n^k}\right), \tag{46}$$

where (see Eqs. (139) and (141) in the Appendix for the functions u_k and v_k)

$$p_{k,\pm}(\tau) \equiv (1\pm\tau)^{-1}\Big(v_k(\tau) \pm \tau u_k(\tau)\Big), \tag{47}$$

for all $k \geq 1$, and

$$\log\left(1+\sum_{k=1}^{\infty}\frac{p_{k,\pm}(\tau)}{n^k}\right) \sim \sum_{k=1}^{\infty}\frac{T_{k,\pm}(\tau)}{n^k}. \tag{48}$$

Thus, the ζ-functions (45) obtain, from the first pair of round brackets in Eq. (46), the contributions (cf. [22])

$$A_+(s) \equiv \sum_{n=3}^{\infty} n^{-(2s-2)}\frac{(\sin\pi s)}{\pi}\int_0^{\infty} dz\; z^{-2s}\frac{\partial}{\partial z}\log\Big(1+(1+z^2)^{-\frac{1}{2}}\Big), \tag{49}$$

$$A_-(s) \equiv \sum_{n=3}^{\infty} n^{-(2s-2)}\frac{(\sin\pi s)}{\pi}\int_0^{\infty} dz\; z^{-2s}\frac{\partial}{\partial z}\log\left(\frac{1-(1+z^2)^{-\frac{1}{2}}}{z^2}\right), \tag{50}$$

where z^2 in the denominator of the argument of the log arises, in Eq. (50), from the extra z^{-2} in the prefactor $z^{-\beta_-(n)}$ in the definition (45). Moreover, the second pair of round brackets in Eq. (46) contributes $\sum_{j=1}^{\infty} A_{j,\pm}(s)$, having defined

$$A_{j,\pm}(s) \equiv \sum_{n=3}^{\infty} n^{-(2s+j-2)}\frac{(\sin\pi s)}{\pi}\int_0^{\infty} dz\; z^{-2s}\frac{\partial}{\partial z}T_{j,\pm}(\tau(z)), \tag{51}$$

where, from the formulae

$$T_{1,\pm} = p_{1,\pm}, \tag{52}$$

$$T_{2,\pm} = p_{2,\pm} - \frac{1}{2}p_{1,\pm}^2, \tag{53}$$

$$T_{3,\pm} = p_{3,\pm} - p_{1,\pm}p_{2,\pm} + \frac{1}{3}p_{1,\pm}^3, \tag{54}$$

we find

$$T_{1,\pm} = -\frac{3}{8}\tau \pm \frac{1}{2}\tau^2 - \frac{5}{24}\tau^3, \tag{55}$$

$$T_{2,\pm} = -\frac{3}{16}\tau^2 \pm \frac{3}{8}\tau^3 + \frac{1}{8}\tau^4 \mp \frac{5}{8}\tau^5 + \frac{5}{16}\tau^6, \tag{56}$$

$$T_{3,\pm} = -\frac{21}{128}\tau^3 \pm \frac{3}{8}\tau^4 + \frac{509}{640}\tau^5 \mp \frac{25}{12}\tau^6 + \frac{21}{128}\tau^7 \pm \frac{15}{8}\tau^8 - \frac{1105}{1152}\tau^9, \tag{57}$$

and hence, in general,

$$T_{j,\pm}(\tau) = \sum_{a=j}^{3j} f_a^{(j,\pm)} \tau^a. \tag{58}$$

We therefore find, from the first line of Eq. (44), contributions to the generalized ζ-function, from terms in round brackets in Eq. (46), equal to

$$\chi_A^{\pm}(s) = \omega_0(s) F_0^{\pm}(s) + \sum_{j=1}^{\infty} \omega_j(s) F_j^{\pm}(s), \tag{59}$$

where, for all $\lambda = 0, j$ (ζ_R and ζ_H being the Riemann and Hurwitz ζ-functions, respectively),

$$\begin{aligned} \omega_\lambda(s) &\equiv \sum_{n=3}^{\infty} n^{-(2s+\lambda-2)} = \zeta_H(2s+\lambda-2;3) \\ &= \zeta_R(2s+\lambda-2) - 1 - 2^{-(2s+\lambda-2)}, \end{aligned} \tag{60}$$

while, from Eqs. (49)–(51),

$$F_0^+(s) \equiv \frac{(\sin \pi s)}{\pi} \int_0^\infty dz\, z^{-2s} \frac{\partial}{\partial z} \log\Bigl(1 + (1+z^2)^{-\frac{1}{2}}\Bigr), \tag{61}$$

$$F_0^-(s) \equiv -2\frac{(\sin \pi s)}{\pi} \int_0^\infty dz\, \frac{z^{-(2s-1)}}{(1+z^2)} - F_0^+(s) = -1 - F_0^+(s), \tag{62}$$

$$F_j^{\pm}(s) \equiv \frac{(\sin \pi s)}{\pi} \sum_{a=j}^{3j} L^{\pm}(s,a,0) f_a^{(j,\pm)}, \tag{63}$$

having set (this general definition will prove useful later, and arises from a more general case, where τ^a is divided by the b-th power of $(1 \pm \tau)$ in Eq. (58))

$$L^{\pm}(s,a,b) \equiv \int_0^1 \tau^{2s+a}(1-\tau)^{-s}(1+\tau)^{-s}\Bigl(\pm b(1\pm\tau)^{-b-1} - a\tau^{-1}(1\pm\tau)^{-b}\Bigr) d\tau. \tag{64}$$

Moreover, on considering

$$L_0^+(s) \equiv \frac{\pi}{\sin \pi s} F_0^+(s), \tag{65}$$

and changing variable from z to τ therein, all L-type integrals above can be obtained from

$$Q(\alpha, \beta, \gamma) \equiv \int_0^1 \tau^\alpha (1-\tau)^\beta (1+\tau)^\gamma d\tau. \tag{66}$$

In particular, we will need

$$L_0^+(s) = -Q(2s, -s, -s-1), \tag{67}$$

$$\begin{aligned} L^+(s, a, b) = & \; bQ(2s+a, -s, -s-b-1) \\ & - aQ(2s+a-1, -s, -s-b), \end{aligned} \tag{68}$$

where, from the integral representation of the hypergeometric function, one has (Gradshteyn and Ryzhik [28])

$$Q(\alpha, \beta, \gamma) = \frac{\Gamma(\alpha+1)\Gamma(\beta+1)}{\Gamma(\alpha+\beta+2)} F(-\gamma, \alpha+1; \alpha+\beta+2; -1). \tag{69}$$

For example, explicitly,

$$L_0^+(s) = -\frac{\Gamma(2s+1)\Gamma(1-s)}{\Gamma(s+2)} F(s+1, 2s+1; s+2; -1). \tag{70}$$

Now we exploit Eqs. (45), (46) and (59) to write

$$\begin{aligned} \zeta_A^+(s) = \chi_A^+(s) + \frac{(\sin \pi s)}{\pi} \sum_{n=3}^{\infty} n^{-(2s-2)} \int_0^\infty dz \Bigg[& \frac{z^{-(2s-1)}}{2(1+z^2)} \\ & + n z^{-(2s+1)} \left(\sqrt{1+z^2} - 1 \right) \Bigg]. \end{aligned} \tag{71}$$

Hence we find

$$\zeta_A^+(0) = \lim_{s \to 0} \left[\omega_0(s) F_0^+(s) + \sum_{j=1}^{\infty} \omega_j(s) F_j^+(s) + \left(\zeta_A^+(s) - \chi_A^+(s) \right) \right]. \tag{72}$$

The first limit in Eq. (72) is immediately obtained by noting that

$$\lim_{s \to 0} L_0^+(s) = -\log(2), \tag{73}$$

and hence

$$\lim_{s \to 0} \omega_0(s) F_0^+(s) = \lim_{s \to 0} \left[\zeta_H(2s-2; 3) \frac{(\sin \pi s)}{\pi} L_0^+(s) \right] = 0. \tag{74}$$

To evaluate the second limit in Eq. (72), we use

$$\lim_{s \to 0} L^+(s, a, 0) = -1, \tag{75}$$

and bear in mind that $\omega_j(s)$ is a meromorphic function with first-order pole, as $s \to 0$, only at $j = 3$ by virtue of the limit

$$\lim_{y \to 1} \left[\zeta_R(y) - \frac{1}{(y-1)} \right] = \gamma. \tag{76}$$

Hence we find (see coefficients in Eq. (57))

$$\begin{aligned} \lim_{s \to 0} \sum_{j=1}^{\infty} \omega_j(s) F_j^+(s) &= \lim_{s \to 0} \frac{(\sin \pi s)}{\pi} \sum_{j=1}^{\infty} \omega_j(s) \left[\sum_{a=j}^{3j} L^+(s, a, 0) f_a^{(j,+)} \right] \\ &= -\frac{1}{2} \sum_{a=3}^{9} f_a^{(3,+)} = -\frac{1}{720}, \end{aligned} \tag{77}$$

while, from Eqs. (71) and (69),

$$\begin{aligned} &\lim_{s \to 0} \left(\zeta_A^+(s) - \chi_A^+(s) \right) \\ &= \lim_{s \to 0} \left(\frac{1}{4} \zeta_H(2s - 2; 3) + \frac{1}{4\sqrt{\pi}} \frac{\Gamma\left(s - \frac{1}{2}\right)}{\Gamma(s+1)} \zeta_H(2s - 3; 3) \right) \\ &= -\frac{5}{4} + \frac{1079}{240}. \end{aligned} \tag{78}$$

We therefore find, with the same algorithms as in [27],

$$\zeta_A^+(0) = -\frac{5}{4} + \frac{1079}{240} - \frac{1}{2} \sum_{a=3}^{9} f_a^{(3,+)} = \frac{146}{45}, \tag{79}$$

$$\zeta_A^-(0) = -\frac{5}{4} + \frac{1079}{240} + 5 - \frac{1}{2} \sum_{a=3}^{9} f_a^{(3,-)} = \frac{757}{90}. \tag{80}$$

These results have been double-checked by using also the powerful analytic technique in [25].

4. Further spectral asymptotics: elliptic and non-elliptic parts

As a next step, the second line of Eq. (44) suggests considering ζ-functions having the integral representation (using again the Cauchy theorem and rotation of contour as in Eq. (45))

$$\zeta_B^{\pm}(s) \equiv \frac{(\sin \pi s)}{\pi} \sum_{n=3}^{\infty} n^{-(2s-2)} \int_0^{\infty} dz \; z^{-2s} \cdot \frac{\partial}{\partial z} \log \left[z^{-\beta_{\pm}(n)} \left(z n I_n'(zn) + \left(\frac{z^2 n^2}{2} \pm n \right) I_n(zn) \right) \right]. \tag{81}$$

To begin, we exploit again the uniform asymptotic expansion of modified Bessel functions and their first derivatives to find (cf. Eq. (46))

$$zn I_n'(zn) + \left(\frac{z^2 n^2}{2} \pm n \right) I_n(zn) \sim \frac{n^2}{2\sqrt{2\pi n}} \frac{e^{n\eta}}{\sqrt{\tau}} \left(\frac{1}{\tau} - \tau \right) \left(1 + \sum_{k=1}^{\infty} \frac{r_{k,\pm}(\tau)}{n^k} \right), \tag{82}$$

where we have (bearing in mind that $u_0 = v_0 = 1$)

$$r_{k,\pm}(\tau) \equiv u_k(\tau) + \frac{2\tau}{(1-\tau^2)} \Big((v_{k-1}(\tau) \pm \tau u_{k-1}(\tau) \Big), \tag{83}$$

for all $k \geq 1$. Hereafter we set

$$\Omega \equiv \sum_{k=1}^{\infty} \frac{r_{k,\pm}(\tau(z))}{n^k}, \tag{84}$$

and rely upon the formula

$$\log(1+\Omega) \sim \sum_{k=1}^{\infty} (-1)^{k+1} \frac{\Omega^k}{k} \tag{85}$$

to evaluate the uniform asymptotic expansion (cf. Eq. (48))

$$\log \left(1 + \sum_{k=1}^{\infty} \frac{r_{k,\pm}(\tau(z))}{n^k} \right) \sim \sum_{k=1}^{\infty} \frac{R_{k,\pm}(\tau(z))}{n^k}. \tag{86}$$

The formulae yielding $R_{k,\pm}$ from $r_{k,\pm}$ are exactly as in Eqs. (52)–(54), with T replaced by R and p replaced by r (see, however, comments below Eq.

(90)). Hence we find, bearing in mind Eq. (83),

$$R_{1,\pm} =(1 \mp \tau)^{-1} \left(\frac{17}{8}\tau \mp \frac{1}{8}\tau^2 - \frac{5}{24}\tau^3 \pm \frac{5}{24}\tau^4 \right), \tag{87}$$

$$\begin{aligned} R_{2,\pm} =&(1 \mp \tau)^{-2} \left(-\frac{47}{16}\tau^2 \pm \frac{15}{8}\tau^3 - \frac{21}{16}\tau^4 \pm \frac{3}{4}\tau^5 \right. \\ &\left. - \frac{1}{16}\tau^6 \mp \frac{5}{8}\tau^7 + \frac{5}{16}\tau^8 \right), \end{aligned} \tag{88}$$

$$\begin{aligned} R_{3,\pm} =&(1 \mp \tau)^{-3} \left(\frac{1721}{384}\tau^3 \mp \frac{441}{128}\tau^4 + \frac{597}{320}\tau^5 \mp \frac{1033}{960}\tau^6 + \frac{239}{80}\tau^7 \right. \\ &\left. \mp \frac{28}{5}\tau^8 + \frac{2431}{576}\tau^9 \pm \frac{221}{192}\tau^{10} - \frac{1105}{384}\tau^{11} \pm \frac{1105}{1152}\tau^{12} \right), \end{aligned} \tag{89}$$

and therefore

$$R_{j,\pm}(\tau(z)) = (1 \mp \tau)^{-j} \sum_{a=j}^{4j} C_a^{(j,\pm)} \tau^a, \tag{90}$$

where, unlike what happens for the $T_{j,\pm}$ polynomials, the exponent of $(1 \mp \tau)$ never vanishes. Note that, at $\tau = 1$ (i.e. $z = 0$), our $r_{k,+}(\tau)$ and $R_{k,+}(\tau)$ are singular. Such a behaviour is not seen for any of the strongly elliptic boundary-value problems [8]. This technical difficulty motivates our efforts below and is interpreted by us as a clear indication of the lack of strong ellipticity proved, on general ground, in [18].

The $\zeta_B^-(s)$ function is more easily dealt with. It indeed receives contributions from terms in round brackets in Eq. (82) equal to (cf. Eq. (50) and bear in mind that $\beta_- - \beta_+ = 2$ in Eq. (81))

$$\begin{aligned} B_-(s) &\equiv \sum_{n=3}^{\infty} n^{-(2s-2)} \frac{(\sin \pi s)}{\pi} \int_0^\infty dz\ z^{-2s} \frac{\partial}{\partial z} \log \left(\frac{\frac{1}{\tau(z)} - \tau(z)}{z^2} \right) \\ &= \omega_0(s) \frac{(\sin \pi s)}{\pi} \int_0^\infty dz\ z^{-2s} \frac{\partial}{\partial z} \log \frac{1}{\sqrt{1+z^2}} = -\frac{1}{2}\omega_0(s), \end{aligned} \tag{91}$$

and $\sum_{j=1}^{\infty} B_{j,-}(s)$, having defined, with $\lambda = 0, j$ (cf. Eq. (51))

$$\omega_\lambda(s) \equiv \sum_{n=3}^{\infty} n^{-(2s+\lambda-2)} = \zeta_H(2s+\lambda-2; 3), \tag{92}$$

$$B_{j,-}(s) \equiv \omega_j(s) \frac{(\sin \pi s)}{\pi} \int_0^\infty dz\ z^{-2s} \frac{\partial}{\partial z} R_{j,-}(\tau(z)). \tag{93}$$

On using the same method as in Sec. 3, the formulae (81)–(93) lead to

$$\zeta_B^-(0) = -\frac{5}{4} + \frac{1079}{240} + \frac{5}{2} - \frac{1}{16}\sum_{a=3}^{12} C_a^{(3,-)} = \frac{206}{45}, \tag{94}$$

a result which agrees with a derivation of $\zeta_B^-(0)$ relying upon the method of [25].

Although we have stressed after Eq. (90) the problems with the $\zeta_B^+(s)$ part, for the moment let us proceed formally in the same way as above. Thus we define, in analogy to Eq. (91),

$$B_+(s) \equiv \omega_0(s)\frac{(\sin \pi s)}{\pi}\int_0^\infty dz\, z^{-2s}\frac{\partial}{\partial z}\log\left(\frac{1}{\tau(z)} - \tau(z)\right), \tag{95}$$

and, in analogy to Eq. (93),

$$B_{j,+}(s) \equiv \omega_j(s)\frac{(\sin \pi s)}{\pi}\int_0^\infty dz\, z^{-2s}\frac{\partial}{\partial z}R_{j,+}(\tau(z)). \tag{96}$$

In order to make the presentation as transparent as possible, we write out the derivatives of $R_{j,+}$. On changing integration variable from z to τ we define

$$C_j(\tau) \equiv \frac{\partial}{\partial \tau}R_{j,+}(\tau), \tag{97}$$

and we find the following results:

$$C_1(\tau) =(1-\tau)^{-2}\left(\frac{17}{8} - \frac{1}{4}\tau - \frac{1}{2}\tau^2 + \frac{5}{4}\tau^3 - \frac{5}{8}\tau^4\right), \tag{98}$$

$$\begin{aligned} C_2(\tau) =&(1-\tau)^{-3}\Big(-\frac{47}{8}\tau + \frac{45}{8}\tau^2 - \frac{57}{8}\tau^3 + \frac{51}{8}\tau^4 \\ &- \frac{21}{8}\tau^5 - \frac{33}{8}\tau^6 + \frac{45}{8}\tau^7 - \frac{15}{8}\tau^8\Big), \end{aligned} \tag{99}$$

$$\begin{aligned} C_3(\tau) =&(1-\tau)^{-4}\Big(\frac{1721}{128}\tau^2 - \frac{441}{32}\tau^3 + \frac{1635}{128}\tau^4 - \frac{163}{16}\tau^5 + \frac{1545}{64}\tau^6 - \frac{227}{4}\tau^7 \\ &+ \frac{4223}{64}\tau^8 - \frac{221}{16}\tau^9 - \frac{5083}{128}\tau^{10} + \frac{1105}{32}\tau^{11} - \frac{1105}{128}\tau^{12}\Big), \end{aligned} \tag{100}$$

so that the general expression of $C_j(\tau)$ reads as

$$C_j(\tau) = (1-\tau)^{-j-1}\sum_{a=j-1}^{4j} K_a^{(j)}\tau^a,\ \forall j = 1,\ldots,\infty\,. \tag{101}$$

These formulae engender a $\zeta_B^+(0)$ which can be defined, after change of variable from z to τ, by splitting the integral with respect to τ, in the integral representation of $\zeta_B^+(s)$, according to the identity

$$\int_0^1 d\tau = \int_0^\mu d\tau + \int_\mu^1 d\tau,$$

and taking the limit as $\mu \to 1$ *after having evaluated the integral.* More precisely, since the integral on the left-hand side is independent of μ, we can choose μ small on the right-hand side so that, in the interval $[0, \mu]$ (and only there!), we can use the uniform asymptotic expansion of the integrand where the negative powers of $(1-\tau)$ are harmless. Moreover, independence of μ also implies that, after having evaluated the integrals on the right-hand side, we can take the $\mu \to 1$ limit. Within this framework, the limit as $\mu \to 1$ of the second integral on the right-hand side yields vanishing contribution to the asymptotic expansion of $\zeta_B^+(s)$.

With this *caveat*, on defining (cf. (66))

$$Q_\mu(\alpha, \beta, \gamma) \equiv \int_0^\mu \tau^\alpha (1-\tau)^\beta (1+\tau)^\gamma d\tau, \tag{102}$$

we obtain the representations

$$\begin{aligned} B_+(s) = &- \omega_0(s) \frac{(\sin \pi s)}{\pi} \Big[- Q_\mu(2s, -s-1, -s) \\ &+ Q_\mu(2s, -s, -s-1) - Q_\mu(2s-1, -s, -s) \Big], \end{aligned} \tag{103}$$

$$B_{j,+}(s) = - \omega_j(s) \frac{(\sin \pi s)}{\pi} \sum_{a=j-1}^{4j} K_a^{(j)} Q_\mu(2s+a, -s-j-1, -s). \tag{104}$$

The relevant properties of $Q_\mu(\alpha, \beta, \gamma)$ can be obtained by observing that this function is nothing but a hypergeometric function of two variables [28], i.e.

$$Q_\mu(\alpha, \beta, \gamma) = \frac{\mu^{\alpha+1}}{\alpha+1} F_1(\alpha+1, -\beta, -\gamma, \alpha+2; \mu, -\mu). \tag{105}$$

In detail, a summary of results needed to consider the limiting behaviour of $\zeta_B^+(s)$ as $s \to 0$ is

$$\omega_0(s)\frac{(\sin \pi s)}{\pi} \sim -5s + O(s^2), \tag{106}$$

$$\omega_j(s)\frac{(\sin \pi s)}{\pi} \sim \frac{1}{2}\delta_{j,3} + \tilde{b}_{j,1}s + O(s^2), \tag{107}$$

$$\lim_{\mu \to 1} Q_\mu(2s, -s-1, -s) \sim -\frac{1}{s} + O(s^0), \tag{108}$$

$$\lim_{\mu \to 1} Q_\mu(2s, -s, -s-1) \sim \log(2) + O(s), \tag{109}$$

$$\lim_{\mu \to 1} Q_\mu(2s-1, -s, -s) \sim \frac{1}{2s} + O(s), \tag{110}$$

$$\begin{aligned} \lim_{\mu \to 1} Q_\mu(2s+a, -s-j-1, -s) &= \frac{\Gamma(-j-s)\Gamma(a+2s+1)}{\Gamma(a-j+s+1)} \\ &\quad \cdot {}_2F_1(a+2s+1, s, a-j+s+1; -1) \\ &\sim \frac{b_{j,-1}(a)}{s} + b_{j,0}(a) + O(s), \end{aligned} \tag{111}$$

where

$$\tilde{b}_{j,1} = -1 - 2^{2-j} + \zeta_R(j-2)(1-\delta_{j,3}) + \gamma\delta_{j,3}, \tag{112}$$

$$b_{j,-1}(a) = \frac{(-1)^{j+1}}{j!}\frac{\Gamma(a+1)}{\Gamma(a-j+1)}(1-\delta_{a,j-1}), \tag{113}$$

and we only strictly need $b_{3,0}(a)$ which, unlike the elliptic cases studied earlier, now depends explicitly on a and is given by (ψ being the standard notation for the logarithmic derivative of the Γ-function)

$$\begin{aligned} b_{3,0}(a) = \frac{1}{6}\frac{\Gamma(a+1)}{\Gamma(a-2)}\bigg[&-\log(2) - \frac{1}{4}(6a^2 - 9a + 1)\frac{\Gamma(a-2)}{\Gamma(a+1)} \\ &+ 2\psi(a+1) - \psi(a-2) - \psi(4)\bigg]. \end{aligned} \tag{114}$$

Remarkably, the coefficient of $\frac{1}{s}$ in the small-s behaviour of the generalized ζ-function $\zeta_B^+(s)$ is zero because it is equal to

$$\lim_{s \to 0} s\zeta_B^+(s) = \sum_{a=2}^{12} b_{3,-1}(a)K_a^{(3)} = \frac{1}{6}\sum_{a=3}^{12} a(a-1)(a-2)K_a^{(3)}, \tag{115}$$

which vanishes by virtue of the rather peculiar general property

$$\sum_{a=j}^{4j} \frac{\Gamma(a+1)}{\Gamma(a-j+1)}K_a^{(j)} = \sum_{a=j}^{4j}\prod_{l=0}^{j-1}(a-l)K_a^{(j)} = 0, \ \forall j = 1, \ldots, \infty, \tag{116}$$

and hence we find eventually

$$
\begin{aligned}
\zeta_B^{\pm}(0) &= -\frac{5}{4} + \frac{1079}{240} + \frac{5}{2} - \frac{1}{2}\sum_{a=2}^{12} b_{3,0}(a)K_a^{(3)} - \sum_{j=1}^{\infty} \tilde{b}_{j,1} \sum_{a=j-1}^{4j} b_{j,-1}(a)K_a^{(j)} \\
&= \frac{5}{4} + \frac{1079}{240} + \frac{599}{720} = \frac{296}{45},
\end{aligned} \tag{117}
$$

because the infinite sum on the first line of Eq. (117) vanishes by virtue of Eqs. (113) and (116), and exact cancellation of log(2) terms is found to occur by virtue of Eq. (116).

To cross-check our analysis, we use Eq. (83) to evaluate

$$
r_{k,+}(\tau) - r_{k,-}(\tau) = \frac{4\tau^2}{(1-\tau^2)} u_{k-1}(\tau), \tag{118}
$$

and hence we find

$$
R_{1,+} = R_{1,-} + \frac{4\tau^2}{(1-\tau^2)}, \tag{119}
$$

$$
R_{2,+} = R_{2,-} + \frac{4\tau^2}{(1-\tau^2)}\left(u_1 - \frac{2\tau^2}{(1-\tau^2)} - R_{1,-}\right), \tag{120}
$$

$$
\begin{aligned}
R_{3,+} = R_{3,-} &+ \frac{4\tau^2}{(1-\tau^2)}\left(u_2 - \frac{4\tau^2}{(1-\tau^2)}u_1 - u_1 R_{1,-} - R_{2,-}\right. \\
&\left. + \frac{4\tau^2}{(1-\tau^2)}R_{1,-}\right) + \frac{64}{3}\frac{\tau^6}{(1-\tau^2)^3} + \frac{2\tau^2}{(1-\tau^2)}R_{1,-}^2,
\end{aligned} \tag{121}
$$

and so on. This makes it possible to evaluate $B_{j,+}(s) - B_{j,-}(s)$ for all $j = 1, 2, ...\infty$. Only $j = 3$ contributes to $\zeta_B^{\pm}(0)$ (see below) and we find

$$
\begin{aligned}
B_{3,+}(s) - B_{3,-}(s) = &- \omega_3(s)\frac{(\sin \pi s)}{\pi} \lim_{\mu \to 1} \int_0^{\mu} d\tau \\
&\cdot \tau^{2s}(1-\tau)^{-s}(1+\tau)^{-s}\frac{\partial}{\partial \tau}(R_{3,+} - R_{3,-}).
\end{aligned} \tag{122}
$$

The derivative in the integrand on the right-hand side of Eq. (122) reads as

$$
\frac{\partial}{\partial \tau}(R_{3,+} - R_{3,-}) = (1-\tau)^{-4}(1+\tau)^{-4}\left(80\tau^3 - 24\tau^5 + 32\tau^7 - 8\tau^9\right), \tag{123}
$$

and hence we can use again the definition (102) and the formula (105) to express (122) through the functions $Q_\mu(2s + a, -s - 4, -s - 4)$, with

$a = 3, 5, 7, 9$. This leads to

$$\begin{aligned}\zeta_B^+(0) &= \zeta_B^-(0) + B_{3,+}(0) - B_{3,-}(0) \\ &= \zeta_B^-(0) - \frac{1}{24}\sum_{l=1}^{4}\frac{\Gamma(l+1)}{\Gamma(l-2)}\left[\psi(l+2) - \frac{1}{(l+1)}\right]\kappa_{2l+1}^{(3)} \\ &= \frac{206}{45} + 2 = \frac{296}{45},\end{aligned} \tag{124}$$

where $\kappa_{2l+1}^{(3)}$ are the four coefficients on the right-hand side of (123). Regularity of $\zeta_B^+(s)$ at the origin is guaranteed because $\lim_{s\to 0} s\zeta_B^+(s)$ is proportional to

$$\sum_{l=1}^{4}\frac{\Gamma(l+1)}{\Gamma(l-2)}\kappa_{2l+1}^{(3)} = 0,$$

which is a particular case of the peculiar spectral cancellation (cf. (116))

$$\sum_{a=a_{\min}(j)}^{a_{\max}(j)} \frac{\Gamma\left(\frac{(a+1)}{2}\right)}{\Gamma\left(\frac{(a+1)}{2} - j\right)}\kappa_a^{(j)} = 0, \tag{125}$$

where a takes both odd and even values. The case $j = 3$ is simpler because then only $\kappa_a^{(j)}$ coefficients with odd a are non-vanishing.

Remaining contributions to $\zeta(0)$, being obtained from strongly elliptic sectors of the boundary-value problem, are easily found to agree with the results in [19], i.e.

$$\zeta(0)[\text{transverse traceless modes}] = -\frac{278}{45}, \tag{126}$$

$$\zeta(0)[\text{coupled vector modes}] = \frac{494}{45}, \tag{127}$$

$$\zeta(0)[\text{decoupled vector mode}] = -\frac{15}{2}, \tag{128}$$

$$\zeta(0)[\text{scalar modes}(a_1, e_1; a_2, b_2, e_2)] = -17, \tag{129}$$

$$\zeta(0)[\text{scalar ghost modes}] = -\frac{149}{45}, \tag{130}$$

$$\zeta(0)[\text{vector ghost modes}] = \frac{77}{90}, \tag{131}$$

$$\zeta(0)[\text{decoupled ghost mode}] = \frac{5}{2}. \tag{132}$$

Our full $\zeta(0)$ is therefore, from (79), (80), (94), (117), (126)-(132), $\zeta(0) = \frac{142}{45}$.

5. Concluding remarks

We have studied the analytically continued eigenvalue conditions for metric perturbations on the Euclidean four-ball, in the presence of boundary conditions completely invariant under infinitesimal diffeomorphisms in the de Donder gauge and with the α parameter set to 1 in Eq. (10). This has made it possible to prove that only one sector of the scalar-mode determinant is responsible for lack of strong ellipticity of the boundary-value problem (see second line of Eq. (44) and the analysis in Secs. 3 and 4). The first novelty with respect to the work in [18] is a clear separation of the elliptic and non-elliptic sectors of spectral asymptotics for Euclidean quantum gravity. We have also shown that one can indeed obtain a regular ζ-function asymptotics at small s in the non-elliptic case by virtue of the remarkable identity (116). Our prescription for the $\zeta(0)$ value differs from the result first obtained in [19], where, however, neither the strong ellipticity issue [18] nor the non-standard spectral asymptotics of our Sec. 4 had been considered.

As far as we can see, the issues raised by our results are as follows.

(i) The integral representation (81) is legitimate because the second line of Eq. (44) corresponds to the eigenvalue conditions, for $n \geq 3$,

$$F_B^{\pm}(n,x) \equiv J_n'(x) + \left(-\frac{x}{2} \pm \frac{n}{x}\right) J_n(x) = 0. \tag{133}$$

For both choices of sign in front of $\frac{n}{x}$, if x_i is a root, then so is $-x_i$, with positive eigenvalue $E_i = x_i^2$ (having set the 3-sphere radius $q = 1$ for simplicity). For any fixed n, there is a countable infinity of roots x_i and they grow approximately linearly with the integer i counting such roots. The function $F_B^{\pm}$ admits therefore a canonical-product representation (Ahlfors [29]) which ensures that the integral representation (81) reproduces the standard definition of generalized ζ-function, i.e.

$$\zeta(s) \equiv \sum_{E_k>0} d(E_k) E_k^{-s},$$

where $d(E_k)$ is the degeneracy of the eigenvalue E_k.

(ii) Even though the lack of strong ellipticity implies that the functional trace of the heat semigroup no longer exists, and hence the Mellin transform relating ζ-function to integrated heat kernel cannot be exploited, it remains possible to define the functional determinant of the operator P acting on metric perturbations. For this purpose, a weaker assumption provides a sufficient condition, i.e. the existence of a sector in the complex plane free of eigenvalues of the leading symbol of P (Seeley [30]). Note also that, if one looks at the A_1 heat-kernel coefficient for boundary conditions involving tangential derivatives [8], it is exactly for the ball that the potentially divergent pieces involving the extrinsic curvature in A_1 cancel. Thus, on the Euclidean ball cancellations take place that maybe could explain why $\zeta(0)$ is finite. This might be therefore *a very particular result for the ball.*

(iii) By virtue of standard recurrence relations among Bessel functions, the eigenvalue conditions (133) are equivalent to studying the eigenvalue conditions

$$\widetilde{F}_B^{\pm}(n,x) = J_n(x) \mp \frac{2}{x} J_{n-1}(x) = 0, \tag{134}$$

where the eigenvalues $E(i,n,\pm)$ are obtained by squaring up the roots $x(i,n,\pm)$. The equation for $\widetilde{F}_B^{-}(n,x)$ can be further re-expressed in the form

$$\left(1 + \frac{4n}{x^2}\right) J_n(x) - \frac{2}{x} J_{n-1}(x) = 0. \tag{135}$$

The functions $\widetilde{F}_B^{\pm}$ differ therefore by one term only, and this term gets small as x gets larger. The numerical analysis confirms indeed that a $\rho(i,n)$ positive and much smaller than 1 exists such that one can write (Esposito et al. [31])

$$E(i,n,+) = E(1,n,+)\delta_{i,1} + E(i-1,n,-)(1+\rho(i,n))(1-\delta_{i,1}), \tag{136}$$

for all $n \geq 3$ and for all $i \geq 1$.

(iv) The remarkable factorization of eigenvalue conditions, with resulting isolation of elliptic part of spectral asymptotics (transverse-traceless, vector and ghost modes, all modes in finite-dimensional sub-spaces and three of the four equations for scalar modes), suggests trying to re-assess functional integrals on manifolds with boundary, with the hope of being able to obtain unique results from the non-elliptic contribution. If this cannot be achieved, the two alternatives below should be considered again.

(v) Luckock boundary conditions (Luckock [32]), which engender BRST-invariant amplitudes but are not diffeomorphism invariant [15]. They have already been applied by Moss and Poletti [12, 33].

(vi) Non-local boundary conditions that lead to surface states in quantum cosmology and pseudo-differential operators on metric and ghost modes

(Marachevsky and Vassilevich, Esposito [34]). Surface states are particularly interesting since they describe a transition from quantum to classical regime in cosmology entirely ruled by the strong ellipticity requirement, while pseudo-differential operators are a source of technical complications.

There is therefore encouraging evidence in favour of Euclidean quantum gravity being able to drive further developments in quantum field theory, quantum cosmology and spectral asymptotics (see early mathematical papers by Grubb [35], Gilkey and Smith [36]) in the years to come.

Appendix: Olver expansions

In Secs. 3 and 4 we use the uniform asymptotic expansion of modified Bessel functions I_ν first found by Olver [37]:

$$I_\nu(z\nu) \sim \frac{e^{\nu\eta}}{\sqrt{2\pi\nu}(1+z^2)^{\frac{1}{4}}}\left(1+\sum_{k=1}^{\infty}\frac{u_k(\tau)}{\nu^k}\right), \tag{137}$$

where

$$\tau \equiv (1+z^2)^{-\frac{1}{2}},\ \eta \equiv (1+z^2)^{\frac{1}{2}} + \log\left(\frac{z}{1+\sqrt{1+z^2}}\right). \tag{138}$$

This holds for $\nu \to \infty$ at fixed z. The polynomials $u_k(\tau)$ can be found from the recurrence relation [27]

$$u_{k+1}(\tau) = \frac{1}{2}\tau^2(1-\tau^2)u_k'(\tau) + \frac{1}{8}\int_0^\tau d\rho\,(1-5\rho^2)u_k(\rho), \tag{139}$$

starting with $u_0(\tau) = 1$. Moreover, the first derivative of I_ν has the following uniform asymptotic expansion at large ν and fixed z:

$$I_\nu'(z\nu) \sim \frac{e^{\nu\eta}}{\sqrt{2\pi\nu}}\frac{(1+z^2)^{\frac{1}{4}}}{z}\left(1+\sum_{k=1}^{\infty}\frac{v_k(\tau)}{\nu^k}\right), \tag{140}$$

with the v_k polynomials determined from the u_k according to [27]

$$v_k(\tau) = u_k(\tau) + \tau(\tau^2-1)\left[\frac{1}{2}u_{k-1}(\tau) + \tau u_{k-1}'(\tau)\right], \tag{141}$$

starting with $v_0(\tau) = u_0(\tau) = 1$.

Acknowledgement. We are grateful to Bernhelm Booß–Bavnbek and Krysztof Wojciechowski for their kind invitation and encouragement. Moreover, we are indebted to Ivan Avramidi for inspiration provided by previous collaboration with some of us and by continuous correspondence, and to Gerd Grubb for enlightening correspondence. K. Kirsten is grateful to the

Baylor University Research Committee, to the Max-Planck-Institute for Mathematics in the Sciences (Leipzig, Germany) and to the INFN for financial support. The work of G. Esposito and K. Kirsten has been partially supported also by PRIN *SINTESI*. The work of A.Yu. Kamenshchik was partially supported by the Russian Foundation for Basic Research under the Grant No. 02-02-16817 and by the Scientific School Grant No. 2338.2003.2

References

1. D. Deutsch and P. Candelas, Boundary effects in quantum field theory, *Phys. Rev.* D **20** (1979), 3063–3080.
2. B. S. DeWitt, "The Global Approach to Quantum Field Theory", Oxford University Press, Oxford, 2003.
3. J. B. Hartle and S. W. Hawking, Wave function of the universe, *Phys. Rev.* D **28** (1983), 2960–2975; S. W. Hawking, The quantum state of the universe, *Nuclear Phys.* B **239** (1984), 257–276.
4. S. W. Hawking, The boundary conditions for gauged supergravity, *Phys. Lett.* B **126** (1983), 175–177.
5. A. Abouelsaood, C. G. Callan, C. R. Nappi and S. A. Yost, Open strings in background gauge fields, *Nuclear Phys.* B **280** (1987), 599–624.
6. M. Bordag, U. Mohideen and V. M. Mostepanenko, New developments in the Casimir effect, *Phys. Rep.* **353** (2001), 1–205 (quant-ph/0106045); K. Milton, "The Casimir Effect: Physical Manifestations of Zero-Point Energy", World Scientific, River Edge, 2001; V. V. Nesterenko, G. Lambiase and G. Scarpetta, Calculation of the Casimir energy at zero and finite temperature: some recent results, *Riv. Nuovo Cim.* **27**, Ser. 4, Num. 6 (2004), 1–74 (hep-th/0503100).
7. P. Gilkey, The spectral geometry of a Riemannian manifold, *J. Differential Geom.* **10** (1975), 601–618.
8. K. Kirsten, "Spectral Functions in Mathematics and Physics", CRC Press, Boca Raton, 2001.
9. D. V. Vassilevich, Heat kernel expansion: user's manual, *Phys. Rep.* **388** (2003), 279–360 (hep-th/0306138).
10. I. G. Moss, Boundary terms in the heat kernel expansion, *Classical Quantum Gravity* **6** (1989), 759–765.
11. I. G. Avramidi, "Heat Kernel and Quantum Gravity", Lecture Notes in Physics m**64**, Springer–Verlag, Berlin, 2000.
12. G. Esposito, A. Yu. Kamenshchik, I. V. Mishakov and G. Pollifrone, Euclidean Maxwell theory in the presence of boundaries. II, *Classical Quantum Gravity* **11** (1994), 2939–2950 (gr-qc/9506061); I. G. Moss and S. Poletti, Conformal anomalies on Einstein spaces with boundary, *Phys. Lett.* B **333** (1994), 326–330 (gr-qc/9405044).
13. G. Tsoupros, Conformal anomalies for interacting scalar fields on curved manifolds with boundary, *Internat. J. Modern Phys.* A **20**

(2005), 1027–1064 (hep-th/0311100); Conformal anomaly for free scalar propagation on curved bounded manifolds, *Gen. Relativity Gravitation* **37** (2005), 399–406 (hep-th/0409163).
14. G. Esposito, "Quantum Gravity, Quantum Cosmology and Lorentzian Geometries", Lecture Notes in Physics Vol. m12, Springer–Verlag, Berlin, 1994.
15. G. Esposito, A. Yu. Kamenshchik and G. Pollifrone, "Euclidean Quantum Gravity on Manifolds with Boundary", Fundamental Theories of Physics Vol. 85, Kluwer, Dordrecht, 1997.
16. A. O. Barvinsky, The wave function and the effective action in quantum cosmology: covariant loop expansion, *Phys. Lett.* B **195** (1987), 344–348.
17. I. G. Moss and P. Silva, BRST invariant boundary conditions for gauge theories, *Phys. Rev.* D **55** (1997), 1072–1078 (gr-qc/9610023).
18. I. G. Avramidi and G. Esposito, Gauge theories on manifolds with boundary, *Comm. Math. Phys.* **200** (1999), 495–543 (hep-th/9710048); Lack of strong ellipticity in Euclidean quantum gravity, *Classical Quantum Gravity* **15** (1998), 1141–1152 (hep-th/9708163).
19. G. Esposito, A. Yu. Kamenshchik, I. V. Mishakov and G. Pollifrone, One-loop amplitudes in Euclidean quantum gravity, *Phys. Rev.* D **52** (1995), 3457–3465 (gr-qc/9504016).
20. I. G. Avramidi, G. Esposito and A. Yu. Kamenshchik, Boundary operators in Euclidean quantum gravity, *Classical Quantum Gravity* **13** (1996), 2361–2373 (hep-th/9603021).
21. K. Schleich, Semiclassical wave function of the universe at small three geometries, *Phys. Rev.* D **32** (1985), 1889–1898.
22. J. S. Dowker and K. Kirsten, Heat kernel coefficients for oblique boundary conditions, *Classical Quantum Gravity* **14** (1997), L169–L175 (hep-th/9706129).
23. E. M. Lifshitz and I. M. Khalatnikov, Investigations in relativistic cosmology, *Adv. in Phys.* **12** (1963), 185–249.
24. G. Esposito, A. Yu. Kamenshchik, I. V. Mishakov and G. Pollifrone, Gravitons in one-loop quantum cosmology. Correspondence between covariant and non-covariant formalisms, *Phys. Rev.* D **50** (1994), 6329–6337 (gr-qc/9506066).
25. A. O. Barvinsky, A. Yu. Kamenshchik and I. P. Karmazin, One-loop quantum cosmology: ζ-function technique for the Hartle–Hawking wave function of the universe, *Ann. Physics* **219** (1992), 201–242.
26. G. Esposito, G. Fucci, A. Yu. Kamenshchik and K. Kirsten, Spectral asymptotics of Euclidean quantum gravity with diff-invariant boundary conditions, *Classical Quantum Gravity* **22** (2005), 957–974 (hep-th/0412269).
27. M. Bordag, E. Elizalde and K. Kirsten, Heat kernel coefficients of the Laplace operator on the D-dimensional ball, *J. Math. Phys.* **37** (1996), 895–916 (hep-th/9503023); M. Bordag, J. S. Dowker and K. Kirsten, Heat kernels and functional determinants on the generalized

cone, *Comm. Math. Phys.* **182** (1996), 371–394 (hep-th/9602089).

28. I. S. Gradshteyn and I. M. Ryzhik, "Table of Integrals, Series and Products", Academic Press, New York, 1965.
29. L. V. Ahlfors, "Complex Analysis", McGraw–Hill, New York, 1966.
30. R. T. Seeley, Complex powers of an elliptic operator, *Amer. Math. Soc. Proc. Sympos. Pure Math.* **10** (1967), 288–307.
31. G. Esposito, G. Fucci, A. Yu. Kamenshchik and K. Kirsten, A non-singular one-loop wave function of the universe from a new eigenvalue asymptotics in quantum gravity, *J. High Energy Phys.* **0509** (2005), 063, 1–16 (hep-th/0507264).
32. H. C. Luckock, Mixed boundary conditions in quantum field theory, *J. Math. Phys.* **32** (1991), 1755–1766.
33. I. G. Moss and S. Poletti, Boundary conditions for quantum cosmology, *Nucl. Phys.* B **341** (1990), 155–161.
34. V. N. Marachevsky and D. V. Vassilevich, A diffeomorphism-invariant eigenvalue problem for metric perturbations in a bounded region, *Classical Quantum Gravity* **13** (1996), 645–652 (gr-qc/9509051); G. Esposito, Non-local boundary conditions in Euclidean quantum gravity, *Classical Quantum Gravity* **16** (1999), 1113–1126 (gr-qc/9806057); New kernels in quantum gravity, *Classical Quantum Gravity* **16** (1999), 3999–4010 (hep-th/9906169); Boundary operators in quantum field theory, *International J. Mod. Phys.* A **15** (2000), 4539–4555 (hep-th/0001086).
35. G. Grubb, Properties of normal boundary problems for elliptic even-order systems, *Ann. Scuola Norm. Sup. Pisa Cl. Sci.* (4) **1** (1974), 1–61.
36. P. B. Gilkey and L. Smith, The twisted index problem for manifolds with boundary, *J. Differential Geom.* **18** (1983), 393–444.
37. F. W. J. Olver, On Bessel functions of large order, *Philos. Trans. Roy. Soc. London* Ser. A **247** (1954), 328–368.

Received by the editors July 22, 2005; Revised October 10, 2005

Analysis, Geometry and Topology
of Elliptic Operators, pp. 493–540

A GENERALIZED MORSE INDEX THEOREM

CHAOFENG ZHU
Chern Institute of Mathematics
Key Lab of Pure Mathematics and
Combinatorics of Ministry of Education
Nankai University, Tianjin 300071
The People's Republic of China
zhucf@nankai.edu.cn

Dedicated to Krzysztof P. Wojciechowski on his 50th birthday

In this paper, we prove a Morse index theorem for the index form of even order linear Hamiltonian systems on the closed interval with self-adjoint boundary conditions. The highest order term is assumed to be nondegenerate.

2000 *Mathematics Subject Classification.* Primary 58E10; Secondary 53D12

1. Introduction

1.1. *History*

Let (M, g) be an n-dimensional Riemannian manifold. The classical Morse Index Theorem states that the number of conjugate points along a geodesic $\gamma : [a, b] \to M$ counted with multiplicities is equal to the Morse index of the second variation of the Riemannian action functional $E(c) = \frac{1}{2}\int_a^b g(\dot{c}, \dot{c})dt$ at the critical point γ, where $\dot{c}$ denotes $\frac{d}{dt}c$. Such second variation is called the **index form** for E at γ. The theorem has later been extended in several directions (see Agrachev and Sarychev, Ambrose, Duistermaat, Piccione and Tausk, Smale, Uhlenbeck [1, 2, 15, 29, 30, 34, 35] for versions of this theorem in different contexts). In [15] of 1976, J. J. Duistermaat proved his general Morse index theorem for Lagrangian system with positive definite second order term and self-adjoint boundary conditions. In [1] of 1996, A. A. Agrachev and A. V. Sarychev studied the Morse index and rigidity of the abnormal sub-Riemannian geodesics. In [5, 6] of 1979, J. K. Beem and

Work partially supported by FANEDD 200215, 973 Program of MOST, Fok Ying Tung Edu. Funds 91002, LPMC of MOE of China, and Nankai University.

P. E. Ehrlich considered the semi-Riemannian case. Later in [20] of 1994, A. D. Helfer gave a generalization. In [29, 30] of 2000, P. Piccione and D. V. Tausk proved a version of the Morse index theorem for geodesics in semi-Riemannian geometry with both endpoints varying on two submanifolds of M under some nondegenerate conditions (cf. Theorem 6.4 in [30]). In the electronic preprint [38] of 2001, the author proved the Morse index theorem for regular Lagrangian systems with general boundary conditions.

The present paper can be considered as a revised and extended version of [38], generalizing [16] of 1964, where H. Edwards considered the higher even order case. He proved a version of the Morse index theorem for even order Hamiltonian systems on the closed interval with positive definite highest order term and boundary condition $\beta = \gamma \oplus \infty$ (cf. Theorem 3.1 in [16]). He then reduced the general case to the case of $\beta = \infty$ by Morse concavity (cf. Propositions 2.6 and 8.3 in [16]).

1.2. *Setup for regular Lagrangian systems*

Let M be a smooth manifold of dimension n. Points in its tangent bundle TM will be denoted by (m, v), with $m \in M$, $v \in T_m M$. Let f be a real-valued C^3 function on an open subset Z of $\mathbf{R} \times TM$. Then for $T > 0$,

$$E(c) := \int_0^T f\left(t, c(t), \dot{c}(t)\right) dt \tag{1}$$

defines a real-valued C^2 function E on the space of curves

$$\mathcal{C} = \left\{c \in C^1([0,T], M); (t, c(t), \dot{c}(t)) \in Z \text{ for all } t \in [0,T]\right\}. \tag{2}$$

Equipped with the usual topology of uniform convergence of curves and their derivatives, the set $\mathcal{C}$ has a C^2 Banach manifold structure modelled on the Banach space $C^1([0,T], \mathbf{R}^n)$.

Boundary conditions will be introduced by restricting E to the set of curves

$$\mathcal{C}_N = \{c \in \mathcal{C}; (c(0), c(T)) \in N\}, \tag{3}$$

where N is a given smooth submanifold of $M \times M$. The most familiar examples are $N = \{m(0), m(T)\}$ and $N = \{(m_1, m_2) \in M \times M; m_1 = m_2\}$. In the general case, $\mathcal{C}_N$ is a smooth submanifold of $\mathcal{C}$ with its tangential space $T_c\mathcal{C}_N$ consisting of all C^1 sections δc of the pull-back bundle c^*TM satisfying

$$(\delta c(0), \delta c(T)) \in T_{(c(0),c(T))}N. \tag{4}$$

Generalizing the concept of a geodesic, a curve $c \in \mathcal{C}_N$ of class C^2 is called a **stationary curve** (or **extremal** or **critical**) for the boundary condition N if the restriction of E to $\mathcal{C}_N$ has a stationary point at c, i.e., if $DE(c)(\delta c) = 0$ for all $\delta c \in T_c\mathcal{C}_N$.

Let $c \in \mathcal{C}_N$ be a stationary curve for the boundary condition N. Then the second order differential $D^2E(c)$ of E at c is a symmetric bilinear form on $T_c\mathcal{C}_N$, which is called the **index form** of E at c with respect to the boundary condition N. We want to understand the **Morse index** of this form, i.e., the dimension of the maximal negative definite subspace of the space $T_c\mathcal{C}_N$ for the form $D^2E(c)$. In general the Morse index of the form $D^2E(c)$ on $T_c\mathcal{C}_N$ will be infinite. In order to get a well-defined integer, we introduce the following concept.

Assume that the function f is **regular Lagrangian**, that is,

$$D_v^2 f(t,m,v) \text{ is nondegenerate for all } (t,m,v) \in Z. \tag{5}$$

Here D_v denotes the differential of functions on Z with respect to $v \in T_mM$, keeping t and m fixed. The condition (5) is called the **Legendre condition.**

Let $H = H^1(T_c\mathcal{C}_N)$ denote the H^1 completion of $T_c\mathcal{C}_N$. By the Sobolev embedding theorem, $H \subset C([0,T]; c^*TM)$, the space of all C^0 sections of the pull-back c^*TM. Then $D^2E(c)$ is well-defined on H. In local coordinates, we have

$$\begin{aligned} D^2E(c)(X,Y) = \int_0^T \Big(& D_v^2 f(\tilde{c}(t))(\dot{\alpha},\dot{\beta}) + D_m D_v f(\tilde{c}(t))(\alpha,\dot{\beta}) \\ & + D_v D_m f(\tilde{c}(t))(\dot{\alpha},\beta) + D_m^2 f(\tilde{c}(t))(\alpha,\beta)\Big)\, dt, \end{aligned} \tag{6}$$

where $X, Y \in H$, α, β are the local coordinate expressions of X, Y defined by $X = (\alpha, \partial m)$, $Y = (\beta, \partial m)$, ∂m is the natural frame of T_mM, and we use the abbreviation

$$\tilde{c}(t) = (t, c(t), \dot{c}(t)).$$

In general ∂m and α are not globally well-defined along the curve c. Choose a C^1 frame e of $T_c\mathcal{C}_N$. Such a frame can be obtained by the parallel transformation of the induced connection on c^*TM of a connection on TM (for example, the Levi-Civita connection with respect to the semi-Riemannian metric on M). Then in local coordinates, there is a C^1 path $a(t) \in \mathrm{GL}(n, \mathbf{R})$ such that ∂m at $c(t)$ is the pairing $(a(t), e(t))$. Note that $a(t)$ is only locally defined in general. Then the vector fields $X, Y \in H$ along

c can be written as $X = (x, e)$, $Y = (y, e)$, where $x, y \in H^1([0,T]; \mathbf{R}^n)$ and $(x(0), x(T)), (y(0), y(T) \in R$, R is defined by

$$R = \left\{(\xi, \eta) \in \mathbf{R}^{2n}; ((\xi, e(0)), (\eta, e(T))) \in T_{(c(0),c(T))}N\right\}.$$

So we have

$$x = a\alpha, \quad \dot{x} = a\dot{\alpha} + \dot{a}\alpha, \quad y = a\beta, \quad \dot{y} = a\dot{\beta} + \dot{a}\beta. \tag{7}$$

Substituting (7) to (6), we get the following form of the index form:

$$D^2E(c)(X, Y) = \int_0^T (\langle p\dot{x} + qx, \dot{y}\rangle + \langle q^*\dot{x}, y\rangle + \langle rx, y\rangle)\, dt, \tag{8}$$

where $p, q, r \in C([0,T], \mathrm{gl}(n, \mathbf{R}))$, p is of class C^1, $p(t) = p^*(t)$, $r(t) = r^*(t)$, $p(t)$ are invertible for all $t \in [0,T]$, and $*$ denotes taking the adjoint matrix.

Now define

$$\mathcal{I}_{s,R}(x, y) = \int_0^T (\langle p\dot{x} + sqx, \dot{y}\rangle + \langle sq^*\dot{x}, y\rangle + \langle srx, y\rangle)\, dt, \quad s \in [0,1], \tag{9}$$

where $x, y \in H^1([0,T]; \mathbf{R}^n)$ and $(x(0), x(T)), ((y(0), y(T)) \in R$. Since p is of class C^1 and all $p(t)$ are nondegenerate, we can associate the path $\mathcal{I}_{s,R}$ with a well-defined finite integer, the spectral flow $\mathrm{sf}\{\mathcal{I}_{s,R}\}$. Then we can define the **relative Morse index** $I(\mathcal{I}_{0,R}, \mathcal{I}_{1,R})$ as $-\mathrm{sf}\{\mathcal{I}_{s,R}\}$. When p is positive definite, $I(\mathcal{I}_{0,R}, \mathcal{I}_{1,R})$ is the Morse index of $D^2E(c)$. Note that the forms $\mathcal{I}_{s,R}$ will depend on the choice of the frame e.

1.3. *The highlights of the paper*

As mentioned before, this paper can be viewed as a revised version of [38]. In this paper, we shall prove a general version of the Morse index theorem for the index form of even order linear Hamiltonian systems on the closed interval with reasonable self-adjoint boundary conditions (Theorem 2.1). The highest order term is assumed to be nondegenerate. As a special case, we prove the Morse index theorem for regular Lagrangian systems with self-adjoint boundary conditions (Corollary 2.1). Note that the index form (see (9) above) will take different forms under different choices of the frames e. Then we show how the indices vary under such choices (Theorem 2.3).

Our approach is inspired by the recent papers [10, 11] of B. Booss-Bavnbek and the author. We do not use the perturbation method. Our index theorem does not contain any assumption on nondegeneracy for the index form. Moreover, we consider the spectral flow of a path connecting two given index forms. The index forms along such a path are in general not compact perturbations of a given index form. Such phenomena occur when we consider connected trajectories between two geodesics on a manifold. These achievements make it easy to apply our Theorem 2.1 below to the variational problems.

Our paper is arranged as follows. In §1, we give the background of the problem. In §2, we state our main results. In §3, we discuss the properties of the spectral flow. In §4, we discuss the properties of the Maslov indices. In §5, we prove our main results. In this paper, dim denotes the complex dimension if not otherwise specified.

2. Main results

We shall consider the general case of even order linear Hamiltonian systems. We will restrict us to the complex case. The real case is an obvious consequence of the complex case (cf. Proposition 8.2 in [16]).

Let $m, n \in \mathbf{Z}^+$ be positive integers, and $T \in \mathbf{R}^+$ be a positive real number. Let $p_{k,l}(s,t) \in \mathrm{gl}(n, \mathbf{C})$, $(s,t) \in [0,1]\times[0,T]$ be $(m+1)^2$ continuous families of matrices, where $k, l = 0, \ldots, m$. Assume that for all $(s,t) \in [0,1] \times [0,T]$, $p_s(t) = (p_{m-k,m-l}(s,t))_{k,l=0,\ldots,m} \in \mathrm{gl}((m+1)n, \mathbf{C})$ are self-adjoint, and $p_{m,n}(s,t)$ are nondegenerate. Assume further that for all $s \in [0,1]$ and $k, l = 0 \ldots, m$, $p_{k,l}(s,t) \in C^{\max\{k,l\}}([0,T], \mathrm{gl}(n, \mathbf{C}))$. Then we have a continuous family of quadratic forms for all $x, y \in H^m([0,T]; \mathbf{R}^n)$

$$\mathcal{I}_{p_s}(x,y) = \int_0^T \left(\sum_{k,l=0}^{m} \left\langle p_{k,l}(s,t) \frac{d^\ell}{dt^\ell} x, \frac{d^k}{dt^k} y \right\rangle \right) dt.$$

Here $\langle \cdot, \cdot \rangle$ denotes the standard hermitian inner product in $\mathbf{C}^n$, and the norm of the Sobolev space $H^m([0,T]; \mathbf{C}^n)$ is defined by

$$\langle x, y \rangle_m = \int_0^T \left(\sum_{k=0}^{m} \left\langle \frac{d^k}{dt^k} x, \frac{d^k}{dt^k} y \right\rangle \right) dt, \quad \text{for all } x, y \in H^m([0,T]; \mathbf{C}^n).$$

Now we define the boundary condition. Let $R \subset \mathbf{C}^{2mn}$ be a given linear subspace. Let H_R denote the space which consists of all $x \in H^m([0,T]; \mathbf{C}^n)$ with

$$\left(\frac{d^{m-1}}{dt^{m-1}} x(0), \ldots, x(0), \frac{d^{m-1}}{dt^{m-1}} x(T), \ldots, x(T)\right) \in R. \tag{10}$$

Let Q_s, $0 \le s \le 1$ be a continuous family of quadratic forms on R. Then each Q_s defines a quadratic form $\tilde{Q}_s$ on H_R by

$$\tilde{Q}_s(x,y) = Q_s\Big(\big(\frac{d^{m-1}}{dt^{m-1}}x(0),\dots,x(0),\frac{d^{m-1}}{dt^{m-1}}x(T),\dots,x(T)\big),\\ \big(\frac{d^{m-1}}{dt^{m-1}}y(0),\dots,y(0),\frac{d^{m-1}}{dt^{m-1}}y(T),\dots,y(T)\big)\Big),$$

where $x, y \in H_R$.

We define (cf. §2.2 in [16])

$$\mathcal{I}_{p_s,Q_s}(x,y) = \mathcal{I}_{p_s}(x,y) - \tilde{Q}_s(x,y). \tag{11}$$

When we consider the index form deduced from the calculus of variations, the boundary condition is defined by (10) (see §8 of [16] for details). In this case Q_s is zero on R, and we denote $\mathcal{I}_{p_s,Q_s}$ by $\mathcal{I}_{p_s,R}$. The central problem in this paper is to understand the Morse index of the form $\mathcal{I}_{p_1,Q_1}$, i.e., the maximal dimension of negative definite subspaces of the form $\mathcal{I}_{p_1,Q_1}$. As in §1.2, we shall use the negative spectral flow $-\mathrm{sf}\{\mathcal{I}_{p_s,Q_s}\}$ as the "difference" between the "Morse indices" of the forms $\mathcal{I}_{p_1,Q_1}$ and $\mathcal{I}_{p_0,Q_0}$.

Let L_{p_s} denote the unbounded operator on $L^2([0,T];\mathbf{C}^n)$ with domain $H^{2m}([0,T];\mathbf{C}^n)$ defined for all $x \in H^{2m}([0,T];\mathbf{C}^n)$ by

$$(L_{p_s}x)(t) = \sum_{k,l=0}^{m}(-1)^k\frac{d^k}{dt^k}\left(p_{k,l}(s,t)\frac{d^l}{dt^l}x(t)\right).$$

Define R^{2m,Q_s} and $W_{2m}(Q_s)$ by

$$\begin{aligned} R^{2m,Q_s} = \Big\{(x_1,\dots,x_{2m}) \in \mathbf{C}^{2mn};\ &\sum_{k=1}^{m}(-1)^{k-1}\langle x_k, y_{m-k+1}\rangle \\ &+ \sum_{k=m+1}^{2m}(-1)^{k-m}\langle x_k, y_{3m-k+1}\rangle \\ &+ Q_s((x_1,\dots,x_{2m}),(y_1,\dots,y_{2m})) = 0 \\ &\quad \text{for all } (y_1,\dots,y_{2m}) \in R\Big\}, \end{aligned} \tag{12}$$

$$\begin{aligned} W_{2m}(Q_s) = \{&(x_1,x_2,x_3,x_4) \in \mathbf{C}^{4mn}; x_1,x_2,x_3,x_4 \in \mathbf{C}^{mn}, \\ &(x_1,x_3) \in R^{2m,Q_s}, (x_2,x_4) \in R\}. \end{aligned} \tag{13}$$

If Q_s is zero on R, we write

$$R^{2m,b} = R^{2m,Q_s} \text{ and } W_{2m}(R) = W_{2m}(Q_s).$$

For each $x \in H^{2m}([0,T];\mathbf{C}^n)$, let $u_{p_s,x} \in H^1([0,T];\mathbf{C}^{2mn})$ and $u^k_{p_s,x}$, $k = 0,\dots,2m$ be defined by

$$\begin{aligned} u_{p_s,x}(t) &= (u^{2m-1}_{p_s,x}(t),\dots,u^0_{p_s,x}(t)), \\ u^k_{p_s,x}(t) &= \frac{d^k}{dt^k}x(t), \qquad k = 0,\dots,m-1, \\ u^k_{p_s,x}(t) &= \sum_{\substack{2m-k\le\alpha\le m \\ 0\le\beta\le m}} (-1)^{\alpha-m}\frac{d^{\alpha+k-2m}}{dt^{\alpha+k-2m}}\left(p_{\alpha,\beta}(s,t)\frac{d^\beta}{dt^\beta}x(t)\right), \\ &\qquad k = m,\dots,2m. \end{aligned} \tag{14}$$

Let $L_{p_s,W_{2m}(Q_s)}$ denote the restriction of L_{p_s} on the domain

$$\{x \in H^{2m}([0,T];\mathbf{C}^n); (u_{p_s,x}(0),u_{p_s,x}(T)) \in W_{2m}(Q_s)\}.$$

If Q_s is zero on R, we denote $L_{p_s,W_{2m}(Q_s)}$ by $L_{p_s,W_{2m}(R)}$. By Proposition 6.1 in [16], all self-adjoint boundary conditions of L_{p_s} arise in this way. By Lemma 3.5 in [11], $L_{p_s,W_{2m}(Q_s)}$, $0 \le s \le 1$ is a continuous family (in the gap norm sense) of unbounded self-adjoint Fredholm operators. We will consider the negative spectral flow $-\mathrm{sf}\{L_{p_s,W_{2m}(Q_s)}\}$ of the path $\{L_{p_s,W_{2m}(Q_s)}\}$.

Let $J_{2m,n} \in \mathrm{GL}(2mn,\mathbf{C})$ denote the matrix $(j_{k,l})_{k,l=0,\dots,2m-1}$, where $j_{k,l} = 0_n$ for $k+l \ne 2m-1$, $j_{k,l} = (-1)^{k+m}I_n$ for $k+l = 2m-1$, and we denote by I_n and 0_n the identity matrix and the zero matrix on $\mathbf{C}^n$ respectively. When there is no confusion, we will omit the subindex n of I_n and 0_n. Set

$$\bar u_{p_s,x} = (u^m_{p_s,x},\dots,u^0_{p_s,x}), \quad \bar u_{0,x} = (\frac{d^m}{dt^m}x,\dots,x).$$

From (14), we can define the matrices $U(p_s(t))$ and $V(p_s(t))$ for each $(s,t) \in [0,1]\times[0,T]$ by

$$\bar u_{p_s,x}(t) = U(p_s(t))\bar u_{0,x}(t), \quad \bar u_{0,x}(t) = V(p_s(t))\bar u_{p_s,x}(t). \tag{15}$$

Let $\Theta_{2m,n} \in \mathrm{gl}(2mn,\mathbf{C})$ denote the matrix $(\theta_{k,l})_{k,l=0,\dots,2m-1}$, where $\theta_{k,l} = 0_n$ for $k+l \ne 2m-2$ or one of $k = l = m-1$, $\theta_{k,l} = (-1)^{k+m+1}I_n$ for $k+l = 2m-2$ and $k,l \ne m-1$. For each $(s,t) \in [0,1]\times[0,T]$, define the matrices $P(p_s(t))$ and $b(p_s(t))$ in $\mathrm{gl}((m+1)n,\mathbf{C})$ by

$$P(p_s(t)) = (P_{k,l}(s,t))_{k,l=0,\dots,m} \tag{16}$$

$$b(p_s(t)) = \Theta_{2m,n} + \mathrm{diag}(0_{(m-1)n}, P(p_s(t))), \tag{17}$$

where

$$\begin{aligned}
P_{0,0}(s,t) &= p_{m,m}(s,t)^{-1},\\
P_{0,l}(s,t) &= -p_{m,m}(s,t)^{-1}p_{m,m-l}(s,t),\\
P_{k,0}(s,t) &= -p_{m-k,m}(s,t)p_{m,m}(s,t)^{-1},\\
P_{k,l}(s,t) &= p_{m-k,m-l}(s,t) - p_{m-k,m}(s,t)p_{m,m}(s,t)^{-1}p_{m,m-l}(s,t)
\end{aligned}$$

for $k,l = 1,\ldots,m$. For each $s \in [0,1]$, let $\gamma_{p_s}(t)$ denote the fundamental solution of the linear Hamiltonian system

$$\dot{u} = J_{2m,n}b(p_s)u. \tag{18}$$

Then $\gamma_{p_s}(t)$ are symplectic matrices and we can associate the symplectic path $\gamma_{p_s}(t)$, $0 \le t \le T$ with the Maslov-type index $i_{W_{2m}(Q_s)}\{\gamma_{p_s}(t)\}$ for each fixed $s \in [0,1]$ (see Definition 4.6 below).

In this paper we want to address the following problems for the even order case:

(I) Give the relationship between the integers $i_{W_{2m}(Q_s)}\{\gamma_{p_s}(t)\}$, $-\mathrm{sf}\{\mathcal{I}_{p_s,Q_s}\}$ and $-\mathrm{sf}\{L_{p_s,W_{2m}(Q_s)}\}$ for $0 \le s \le 1$.

(II) Calculate $i_{W_{2m}(R)}\{\gamma_{p_0}\}$ for $p_0(t) = \mathrm{diag}(p_{0,0}(0,t), 0_{mn})$.

(III) For two different choices of the frame e, the resulting index forms $\mathcal{I}_{p_s,R}$ defined by (11) will look differently. In this case, calculate the difference between the resulting integers $i_{W_2(R)}\{\gamma_{p_1}\}$.

The following three theorems solve the above problems.

Theorem 2.1. *Let* $\mathrm{sf}\{\mathcal{I}_{p_s,Q_s}, 0 \le s \le 1\}$ *denote the spectral flow of* $\mathcal{I}_{p_s,Q_s}$, $\mathrm{sf}\{L_{p_s,W_{2m}(Q_s)}, 0 \le s \le 1\}$ *denote the spectral flow of* $L_{p_s,W_{2m}(Q_s)}$, *and* $i_{W_{2m}(Q_s)}\{\gamma_{p_s}\}$ *denote the Maslov-type index of* γ_{p_s} *defined below. Then we have*

$$\begin{aligned}
-\mathrm{sf}\{\mathcal{I}_{p_s,Q_s}, 0 \le s \le 1\} &= -\mathrm{sf}\{L_{p_s,W_{2m}(Q_s)}, 0 \le s \le 1\}\\
&= i_{W_{2m}(Q_1)}(\gamma_{p_1}) - i_{W_{2m}(Q_0)}(\gamma_{p_0}).
\end{aligned} \tag{19}$$

Assume that $p_0(t) = \mathrm{diag}(p_{m,m}(0,t), 0_{mn})$ for all $t \in [0,T]$. Then we have $(P(p_0))(t) = (p_0(t))^{-1}$, $b(p_0)(t) = (b_{k,l}(t))_{k,l=0,\ldots,2m-1}$, and $\gamma_{p_0}(t) = (\gamma_{k,l}(t))_{k,l=0,\ldots,2m-1}$, where $b_{k,l}(t) = 0_n$ for $k-l \ne 1$, $b_{k,l}(t) = I_n$ for $k-l=1$ and $k \ne m$, $b_{m,m-1}(t) = (p_{m,m}(0,t))^{-1}$, $\gamma_{k,l}(t) = 0$ for $k < l$,

$\gamma_{k,l}(t) = \frac{t^{k-l}}{(k-l)!} I_n$ for $k \geq l$ and $k \leq m-1$, or $k \geq l$ and $l \geq m$, and

$$\gamma_{k,l}(t)$$
$$= \frac{1}{(m-l-1)!} \int_0^t dt_{k-m} \int_0^{t_{k-m}} dt_{k-m-1} \cdots \int_0^{t_1} t_0^{m-l-1} (p_{m,m}(0,t_0))^{-1} dt_0$$
$$= \frac{1}{(k-m)!(m-l-1)!} \int_0^t s^{m-l-1} (t-s)^{k-m} (p_{m,m}(0,s))^{-1} ds$$

for $k \geq m$ and $l \leq m-1$.

The form of our symplectic path $\gamma_{p_0}(t)$ looks rather complicated. We will consider the following more general situation to simplify our problem.

Let $K \in \mathrm{GL}(n,\mathbf{C})$. Set $J_K = \begin{pmatrix} 0 & -K^* \\ K & 0 \end{pmatrix}$. Then $(\mathbf{C}^{2n}, \langle J_K \cdot, \cdot \rangle)$ is a symplectic space. Let $\gamma(t) = \begin{pmatrix} M_{1,1}(t) & 0 \\ M_{2,1}(t) & M_{2,2}(t) \end{pmatrix}$, $0 \leq t \leq T$ be a path in $\mathrm{GL}(2n,\mathbf{C})$ with $M_{2,2}(t)^* K M_{1,1}(t) = K$ and $M_{1,1}(t)^* K^* M_{2,1}(t)$ self-adjoint for each $t \in [0,T]$. Then $\gamma(t)$ is a symplectic path, i.e., $\gamma(t)^* J_K \gamma(t) = J_K$. Let $R \subset \mathbf{C}^{2n}$ be a given linear subspace. Define R^K and $W_K(R)$ by

$$R^K = \{(x_1,x_2) \in \mathbf{C}^{2n}; \langle K x_1, y_1 \rangle - \langle K x_2, y_2 \rangle = 0 \text{ for all } (y_1,y_2) \in R\}, \tag{20}$$

$$W_K(R) = \{(x_1,x_2,x_3,x_4) \in \mathbf{C}^{4n}; x_1,x_2,x_3,x_4 \in \mathbf{C}^n, (x_1,x_3) \in R^K, (x_2,x_4) \in R\}. \tag{21}$$

Theorem 2.2. *For the symplectic path γ and the Lagrangian space $W_K(R)$ defined above, we have (denoting graphs by* $\mathrm{Gr}(\cdot)$*)*

$$\begin{aligned} \dim(\mathrm{Gr}(\gamma(t)) \cap W_K(R)) = {} & \dim\ker\left((M_{1,1}(T)^* K^* M_{2,1}(t))|_{S(t)}\right) \\ & + \dim S(t) + \dim(\mathrm{Gr}(I_n) \cap R) \\ & - \dim(\mathrm{Gr}(I_n) \cap R^K), \end{aligned} \tag{22}$$

$$\begin{aligned} i_{W_K(R)}(\gamma) = {} & m^+\left((M_{1,1}(T)^* K^* M_{2,1}(T))|_{S(T)}\right) \\ & - m^+\left((M_{1,1}(0)^* K^* M_{2,1}(0))|_{S(0)}\right) \\ & + \dim S(0) - \dim S(T). \end{aligned} \tag{23}$$

Here m^+ denotes the Morse positive index defined below in (30), *and*

$$S(t) = \{x \in \mathbf{C}^n; (x, M_{1,1}(t)x) \in R^K\}.$$

In our case, set $K_{m,n} = (k_{k,l})_{k,l=0,\ldots,m-1}$, where $k_{k,l} = 0_n$ for $k+l \neq m-1$, $k_{k,l} = (-1)^l I_n$ for $k+l = m-1$. Then we have $R^{K_{m,n}} = R^{2m,b}$

and $W_{2m}(R) = W_{K_{m,n}}(R)$. Moreover for the symplectic path $\gamma = \gamma_{p(0)}$, the block $M_{1,1}(T)^* K_{m,n}^* M_{2,1}(T)$ is given by

$$\left(\frac{1}{(m-k-1)!(m-l-1)!} \int_0^T t^{2m-k-l-2} (p_{m,m}(0,t))^{-1} dt \right)_{k,l=0,\dots,m-1}. \tag{24}$$

As a special case, we get the following higher order generalization of Theorem 4.3 in J. J. Duistermaat [15].

Corollary 2.1. *Assume that $p_{m,m}(1,t)$ is positive definite for each $t \in [0,T]$. Then we have*

$$m^-(\mathcal{I}_{p_1,Q_1}) = m^-(L_{p_1,W_{2m}(Q_1)}) = i_{W_{2m}(Q_1)}(\gamma_{p_1}) - \dim S, \tag{25}$$

where m^- denotes the Morse (negative) index defined below in (30), and

$$S = \{x \in \mathbf{C}^{mn}; (x,x) \in R^{2m,b}\}.$$

Now we consider the third problem. Then $m = 1$ and everything is real. Let $a(t)$ be a C^1 path in $\mathrm{GL}(n,\mathbf{R})$, and

$$R' = \{(x,y) \in \mathbf{R}^{2n}; (a(0)x, a(T)y) \in R\}.$$

After the change of the frame $e \mapsto a^{*-1}e$, we have $x \mapsto ax$ and the quadratic form $\mathcal{I}_{p_1,R}$ is changed to the restriction of the form $\mathcal{I}_1(ax,ay)$ on $H_{R'}$. Then, like in (8) we get the corresponding p', q' and r'. Set $p_1 = \begin{pmatrix} p & q \\ q^* & r \end{pmatrix}$ and $p_1' = \begin{pmatrix} p' & q' \\ (q')^* & r' \end{pmatrix}$. Let γ_{p_1} and $\gamma_{p_1'}$ be defined by (18). Then we can prove

$$\gamma_{p_1'} = \mathrm{diag}(a^*, a^{-1})\gamma_{p_1}\mathrm{diag}(a(0)^{*-1}, a(0)). \tag{26}$$

Theorem 2.3. *Let $a(t)$, $0 \le t \le T$ be a path in $\mathrm{GL}(n,\mathbf{C})$, and*

$$R' = \{(x,y) \in \mathbf{C}^{2n}; (a(0)x, a(T)y) \in R\}.$$

Let γ be a symplectic path, i.e., $\gamma(t)^* J_{2,n}\gamma(t) = J_{2,n}$ for all $0 \le t \le T$. Define the symplectic path γ' by

$$\gamma' = \mathrm{diag}(a^*, a^{-1})\gamma_{p_1}\mathrm{diag}(a(0)^{*-1}, a(0)). \tag{27}$$

Then we have

$$i_{W_2(R')}(\gamma') - i_{W_2(R)}(\gamma) = \dim(\mathrm{Gr}(I_n) \cap (R')^{2,b})) - \dim(\mathrm{Gr}(I_n) \cap R^{2,b}). \tag{28}$$

The proof of the preceding three theorems and of Corollary 2.1 will be postponed to Section 5. It depends on the subtle relationship between two concepts, namely the spectral flow and various types of the Maslov index which we shall discuss in the following two sections to some detail.

3. Spectral flow

3.1. *Definition of the spectral flow*

Roughly speaking, the spectral flow counts the net number of eigenvalues changing from the negative real half axis to the non-negative one. The definition goes back to a famous paper by M. Atiyah, V. Patodi, and I. Singer [4], and was made rigorous by J. Phillips [28] for continuous paths of bounded self-adjoint Fredholm operators, by K. P. Wojciechowski [36] and C. Zhu and Y. Long [39] in various non-self-adjoint cases, and by B. Booss-Bavnbek, M. Lesch, and J. Phillips [8] in the unbounded self-adjoint case.

Let X be a complex Hilbert space. For a self-adjoint Fredholm operator A on X, there exists a unique orthogonal decomposition

$$X = X^+(A) \oplus X^0(A) \oplus X^-(A) \tag{29}$$

such that $X^+(A)$, $X^0(A)$ and $X^-(A)$ are invariant subspaces associated to A, and $A|_{X^+(A)}$, $A|_{X^0(A)}$ and $A|_{X^-(A)}$ are positive definite, zero and negative definite respectively. We introduce vanishing, natural, or infinite numbers

$$m^+(A) := \dim X^+(A),\ m^0(A) := \dim X^0(A),\ m^-(A) := \dim X^-(A), \tag{30}$$

and call them **Morse positive index**, **nullity** and **Morse index** of A respectively. For finite-dimensional X, the **signature** of A is defined by $\operatorname{sign}(A) = m^+(A) - m^-(A)$ which yields an integer. The **APS projection** Q_A (where APS stands for Atiyah-Patodi-Singer) is defined by

$$Q_A(x^+ + x^0 + x^-) := x^+ + x^0,$$

for all $x^+ \in X^+(A), x_0 \in X^0(A), x^- \in X^-(A)$.

Let $\{A_s\}$, $0 \le s \le 1$ be a continuous family of self-adjoint Fredholm operators. The spectral flow $\operatorname{sf}\{A_s\}$ of the family should be equal to $m^-(A_0) - m^-(A_1)$ if $\dim X < +\infty$. We will generalize this definition to general Banach space X and general continuous families of admissible operators to be defined below.

Our goal is to minimize the assumptions for defining a spectral flow. We pursue two aims: (i) We provide the same frame for continuous paths of (not necessarily bounded) self-adjoint Fredholm operators and continuous paths of unitary operators which are Fredholm perturbations of the identity. (ii) We neglect completely any global picture of the spectra and restrict our special attention to a bounded region of **C**.

We shall admit that we could do the applications in this paper independently of (i) and (ii). We could exploit the exponential transformation for relating Fredholm operators and unitary operators, as in Booss-Bavnbek and Wojciechowski [9], Chapter 16 and Definition 17.9 and Kirk and Lesch [23], Section 6; and we could exploit global properties of the spectrum, in particular, that the spectrum is discrete and rapidly growing as for all elliptic operators, and corresponding nice properties of the Lagrangians popping up in our applications. We, however, prefer to do it our way.

Let X be a complex Banach space. We denote the set of closed, bounded, and compact operators on X by $\mathcal{C}(X)$, $\mathcal{B}(X)$ and $\mathcal{K}(X)$ respectively. We will denote the spectrum, the resolvent set and the domain of an operator $A \in \mathcal{C}(X)$ by $\sigma(A)$, $\rho(A)$ and $\mathrm{dom}(A)$ respectively. Let N be a bounded open subset of $\mathbf{C}$ and $A \in \mathcal{C}(X)$. If there exists a bounded open subset $\tilde{N} \subset N$ with C^1 boundary $\partial\tilde{N}$ such that $\partial\tilde{N} \cap \sigma(A) = \emptyset$ and $N \cap \sigma(A) \subset \tilde{N}$, we define the spectral projection $P(A, N)$ by

$$P(A, N) := -\frac{1}{2\pi\sqrt{-1}} \int_{\partial\tilde{N}} (A - \zeta I)^{-1} d\zeta.$$

The orientation of $\partial\tilde{N}$ is chosen to make $\tilde{N}$ stay on the left side of $\partial\tilde{N}$.

Inspired by [28], we find that the necessary data for defining the spectral flow are the following:

- a co-oriented bounded real 1-dimensional regular C^1 submanifold ℓ of $\mathbf{C}$, closed or with 2-point boundary $\bar{\ell} \setminus \ell$ where $\bar{\ell}$ denotes the closure of ℓ in $\mathbf{C}$ (we call such an ℓ **admissible**, and write $\ell \in \mathcal{A}(\mathbf{C})$);
- a complex Banach space X (for real X, we consider $X \bigotimes \mathbf{C}$);
- and a continuous family (in the gap norm sense) of admissible operators A_s, $0 \le s \le 1$ in $\mathcal{A}_\ell(X)$.

Here we define $A \in \mathcal{C}(X)$ to be **admissible** with respect to ℓ, if there exists a bounded open neighbourhood N of $\bar{\ell}$ in $\mathbf{C}$ with C^1 boundary ∂N such that (i) $\partial N \cap \sigma(A) = \emptyset$; (ii) $N \cap \sigma(A) \subset \ell$ is a finite set; and (iii) $P(A, N)$ is a finite rank projection.

Then $P(A, N)$ does not depend on the choice of such N. We set

$P_\ell^0(A) := P(A, N)$ and call $\nu_{h,\ell}(A) := \dim \operatorname{im} P_\ell^0(A)$ the **hyperbolic nullity** of A with respect to ℓ. We denote by $\mathcal{A}_\ell(X)$ the set of closed admissible operators with respect to ℓ. It is an open subset of $\mathcal{C}(X)$.

Example 3.1. a) In the self-adjoint case, $\ell = \sqrt{-1}(-\epsilon, \epsilon)$ $(\epsilon > 0)$ with co-orientation from left to right. Then a self-adjoint operator A is admissible with respect to ℓ if and only if A is Fredholm.

b) Another important case is that $\ell = (1-\epsilon, 1+\epsilon)$ $(\epsilon \in (0,1))$ with co-orientation from downward to upward, and all A_s unitary. A unitary operator A is admissible with respect to ℓ if and only if $A - I$ is Fredholm.

Similarly as the definition in [28, 39], we can define the spectral flow $\mathrm{sf}_\ell\{A_s\}$ as follows. It counts the number of spectral lines of A_s coming from the negative side of ℓ to the non-negative side of ℓ.

For each $t \in [0,1]$, there exist bounded open subsets N_t, $N_t^\pm$ of $\mathbf{C}$ such that $\sigma(A_t) \cap \partial N_t = \emptyset$, $\sigma(A_t) \cap \bar{\ell} \subset N_t \cap \ell$, $N_t = N_t^+ \cup (N_t \cap \ell) \cup N_t^-$ is a disjoint union, $N_t^\pm$ stays in the positive (negative) side of ℓ near $N_t \cap \ell$, and $P(A_t, N_t)$ is a finite rank projection. Then $\sigma(A_t) \cap (\partial N_t \cup (\bar{\ell} \setminus (N_t \cap \ell))) = \emptyset$. The set $(\partial N_t \cup (\bar{\ell} \setminus (N_t \cap \ell)))$ is compact since it is a bounded closed subset of $\mathbf{C}$. Since the family $\{A_s\}$, $0 \le s \le 1$ is continuous, there exists a $\delta(t) > 0$ for each $t \in [0,1]$ such that

$$\sigma(A_s) \cap (\partial N_t \cup (\bar{\ell} \setminus (N_t \cap \ell))) = \emptyset \quad \text{for all } s \in (t - \delta(t), t + \delta(t)) \cap [0,1].$$

Then $\sigma(A_s) \cap \bar{\ell} \subset N_t \cap \ell$, and

$$\{P(A_s, N_t)\}_{s \in (t-\delta(t), t+\delta(t)) \cap [0,1]} \quad \text{for fixed } t \in [0,1],$$

is a continuous family of projections. By Lemma I.4.10 in Kato [22], the operators in the family have the same rank. Since $[0,1]$ is compact, there exist a partition $0 = s_0 < \ldots < s_n = 1$ and $t_k \in [s_k, s_{k+1}]$, $k = 0, \ldots, n-1$ such that $[s_k, s_{k+1}] \subset (t_k - \delta(t_k), t_k + \delta(t_k))$ for each $k = 0, \ldots, n-1$.

Definition 3.1. Let $\ell \in \mathcal{A}(\mathbf{C})$ be admissible and let $\{A_s\}$, $0 \le s \le 1$ be a curve in $\mathcal{A}_\ell(X)$. The **spectral flow** $\mathrm{sf}_\ell\{A_s\}$ of the family $\{A_s\}$, $0 \le s \le 1$ with respect to the curve ℓ is defined by

$$\mathrm{sf}_\ell\{A_s\} = \sum_{k=0}^{n-1} \left(\dim \operatorname{im} P(A_{s_k}, N_{t_k}^-) - \dim \operatorname{im} P(A_{s_{k+1}}, N_{t_k}^-)\right). \tag{31}$$

The spectral flow has the following properties (cf. [28] and Lemma 2.6 and Proposition 2.2 in [39]).

Proposition 3.1. *Let $\ell \in \mathcal{A}(\mathbf{C})$ be admissible and let $\{A_s\}$, $0 \le s \le 1$ be a curve in $\mathcal{A}_\ell(X)$. Then the spectral flow $\mathrm{sf}_\ell\{A_s\}$ is well-defined, and the following properties hold:*
(i) ***Catenation.*** *Assume $t \in [0,1]$. Then we have*

$$\mathrm{sf}_\ell\{A_s; 0 \le s \le t\} + \mathrm{sf}_\ell\{A_s; t \le s \le 1\} = \mathrm{sf}_\ell\{A_s; 0 \le s \le 1\}.$$

(ii) ***Homotopy invariance.*** *Let $A(s,t)$, $(s,t) \in [0,1] \times [0,1]$ be a continuous family in $\mathcal{A}_\ell(X)$. Then we have*

$$\mathrm{sf}_\ell\{A(s,t); (s,t) \in \partial([0,1] \times [0,1])\} = 0.$$

(iii) ***Endpoint dependence for Riesz continuity.*** *Let $\mathcal{B}^{\mathrm{sa}}(X)$, respectively $\mathcal{C}^{\mathrm{sa}}(X)$ denote the spaces of bounded, respectively closed self-adjoint operators in X. Let*

$$R : \mathcal{C}^{\mathrm{sa}} \to \mathcal{B}^{\mathrm{sa}}(X),\ A \mapsto A(A^2 + I)^{-\frac{1}{2}}$$

denote the ***Riesz transformation****. Let $A_s \in \mathcal{C}^{\mathrm{sa}}(X)$ for $s \in [0,1]$. Assume that $\{R(A_s)\}$ is a continuous family. If $m^-(A_0) < +\infty$, then $m^-(A_1) < +\infty$ and we have*

$$\mathrm{sf}\{A_s\} = m^-(A_0) - m^-(A_1).$$

(iv) ***Product.*** *Let $\{P_s\}$ be a curve of projections on X such that $P_s A_s \subset A_s P_s$ for all $s \in [0,1]$. Set $Q_s = I - P_s$. Then we have $P_s A_s P_s \in \mathcal{A}_\ell(\mathrm{im}\ P_s) \subset \mathcal{C}(\mathrm{im}\ P_s)$, $Q_s A_s Q_s \in \mathcal{A}_\ell(\mathrm{im}\ Q_s) \subset \mathcal{C}(\mathrm{im}\ Q_s)$, and*

$$\mathrm{sf}_\ell\{A_s\} = \mathrm{sf}_\ell\{P_s A_s P_s\} + \mathrm{sf}_\ell\{Q_s A_s Q_s\}.$$

(v) ***Bound.*** *For $A \in \mathcal{A}_\ell(X)$, there exists a neighbourhood $\mathcal{N}$ of A in $\mathcal{C}(X)$ such that $\mathcal{N} \subset \mathcal{A}_\ell(X)$, and for curves $\{A_s\}$ in $\mathcal{N}$ with endpoints $A_0 =: A$ and $A_1 =: B$, the* ***relative Morse index*** *$I_\ell(A,B) := -\mathrm{sf}_\ell\{A_s, 0; \le s \le 1\}$ is well defined and satisfies*

$$0 \le I_\ell(A,B) \le \nu_{h,\ell}(A) - \nu_{h,\ell}(B).$$

(vi) ***Reverse orientation.*** *Let $\hat{\ell}$ denote the curve ℓ with opposite co-orientation. Then we have*

$$\mathrm{sf}_\ell\{A_s\} + \mathrm{sf}_{\hat{\ell}}\{A_s\} = \nu_{h,\ell}(A_1) - \nu_{h,\ell}(A_0).$$

(vii) ***Zero.*** *Suppose that the hyperbolic nullities $\nu_{h,\ell}(A_s)$ are constant for $s \in [0,1]$. Then $\mathrm{sf}_\ell\{A_s\} = 0$.*
(viii) ***Invariance.*** *Let $\{T_s\}_{s\in[0,1]}$ be a curve of bounded invertible operators. Then we have*

$$\mathrm{sf}_\ell\{T_s^{-1} A_s T_s\} = \mathrm{sf}_\ell\{A_s\}.$$

Proof. We shall only prove that the spectral flow is well-defined. The proof for the rest of the proposition is the same as that in [28] and Lemma 2.6 and Proposition 2.2 in [39] and is omitted.

Since two different partitions of $[0,1]$ have a common refinement, we only need to prove the following local result:

Claim. Let N_l, $N_l^{\pm}$, $l=1,2$ be open subsets in $\mathbf{C}$. Assume that for all $s\in[0,1]$ and $l=1,2$, we have $\sigma(A_s)\cap\partial N_l=\emptyset$, $\sigma(A_s)\cap\bar{\ell}\subset N_l\cap\ell$, $N_l=N_l^+\cup(N_l\cap\ell)\cup N_l^-$, $N_l^{\pm}$ stays on the positive (negative) side of ℓ near $N_l\cap\ell$, and $P(A_s,N_l)$ is a finite rank projection. Then we have

$$\begin{aligned}&\dim\operatorname{im}P(A_0,N_1^-)-\dim\operatorname{im}P(A_1,N_1^-)\\&\quad=\dim\operatorname{im}P(A_0,N_2^-)-\dim\operatorname{im}P(A_1,N_2^-).\end{aligned}$$

In fact, our assumptions imply

$$\sigma(A_s)\cap\partial(N_1^-\setminus N_2^-)=\sigma(A_s)\cap\partial(N_2^-\setminus N_1^-)=\emptyset.$$

Then $P(A_s,N_1^-\setminus N_2^-)$ and $P(A_s,N_2^-\setminus N_1^-)$, $s\in[0,1]$ are continuous families of projections. By Lemma I.4.10 in Kato [22], im $P(A_t,N_1^-\setminus N_2^-)$ and im $P(A_t,N_2^-\setminus N_1^-)$ are constants. So we have

$$\begin{aligned}&\big(\dim\operatorname{im}P(A_0,N_1^-)-\dim\operatorname{im}P(A_1,N_1^-)\big)\\&\quad-\big(\dim\operatorname{im}P(A_0,N_2^-)-\dim\operatorname{im}P(A_1,N_2^-)\big)\\&=\big(\dim\operatorname{im}P(A_0,N_1^-)-\dim\operatorname{im}P(A_0,N_2^-)\big)\\&\quad-\big(\dim\operatorname{im}P(A_1,N_1^-)-\dim\operatorname{im}P(A_1,N_2^-)\big)\\&=\big(\dim\operatorname{im}P(A_0,N_1^-\setminus N_2^-)-\dim\operatorname{im}P(A_0,N_2^-\setminus N_1^-)\big)\\&\quad-\big(\dim\operatorname{im}P(A_1,N_1^-\setminus N_2^-)-\dim\operatorname{im}P(A_1,N_2^-\setminus N_1^-)\big)\\&=0.\end{aligned}$$

Thus our claim is proved. □

Remark 3.1. In (iv) of the above proposition, we allow the Banach space im P_s to vary continuously. By Lemma I.4.10 in [22], for $t\in[0,1]$ being close enough to s, there is a continuous family of invertible operators $U_{s,t}\in\mathcal{B}(X)$ such that

$$P_tU_{s,t}=U_{s,t}P_s,\qquad U_{s,t}\to I,\ \text{as}\ t\to s.$$

So locally we can define the spectral flow of $B_t\in\mathcal{C}(\operatorname{im}P_t)$ as that of $U_{s,t}^{-1}B_tU_{s,t}:\operatorname{im}P_s\to\operatorname{im}P_s$ (s fixed), and globally patch them together.

3.2. *Calculation of the spectral flow*

In this subsection we shall give a method of calculating the spectral flow of differentiable curves, inspired among others by J.J. Duistermaat [15] and J. Robbin and D. Salamon [32].

Let X be a complex Banach space, $\tilde{N} \subset N$ be bounded open subsets of $\mathbf{C}$, and γ be a closed C^1 curve in $\mathbf{C}$ which bounds $\tilde{N}$. Let A_s, $s \in (-\epsilon, \epsilon)$, where $\epsilon > 0$, be a curve in $\mathcal{C}(X)$. Assume that $\gamma \cap \sigma(A_s) = \emptyset$ and $N \cap \sigma(A_s) \subset \tilde{N}$ for all $s \in (-\epsilon, \epsilon)$. Set $A := A_0$, $P_s := P(A_s, N)$, and $P := P_0$. Assume that im $P \subset \operatorname{dom}(A_s)$ for all $s \in (-\epsilon, \epsilon)$, im P is a finite dimensional subspace of X, and $\frac{d}{ds}|_{s=0}(A_s P) = B$ (in the bounded operator sense). Let f be a polynomial. Then $P_s f(A_s) P_s$, $s \in (-\epsilon, \epsilon)$ is a continuous family of bounded operators, and

$$P_s f(A_s) P_s = -\frac{1}{2\pi\sqrt{-1}} \int_\gamma f(\zeta)(A - \zeta I)^{-1} d\zeta. \tag{32}$$

Since P_s, $s \in (-\epsilon, \epsilon)$ is a continuous family, we have $\|P_s - P\| < 1$ if $|s|$ is small. For such s, set $R_s = (I - (P_s - P)^2)^{-\frac{1}{2}}$. Since $P(P_s - P)^2 = (P_s - P)^2 P$ and $P_s(P_s - P)^2 = (P_s - P)^2 P_s$, we have $R_s P = P R_s$ and $R_s P_s = P_s R_s$. Set

$$\begin{aligned} U_s' &= P_s P + (I - P_s)(I - P), \qquad & U_s &= U_s' R_s, \\ V_s' &= P P_s + (I - P)(I - P_s), \qquad & V_s &= V_s' R_s. \end{aligned}$$

Then we have

$$\begin{aligned} U_s V_s &= V_s U_s = I, \\ U_s P &= P_s U_s = P_s R_s P, \\ P V_s &= V_s P_s = P R_s P_s. \end{aligned}$$

Lemma 3.1. *We have*

$$\frac{d}{ds}|_{s=0}(U_s^{-1} P_s A_s P_s U_s) = \frac{1}{2\pi\sqrt{-1}} \int_\gamma \zeta (A - \zeta I)^{-1} P B (A - \zeta I)^{-1} d\zeta. \tag{33}$$

If $(PAP)(PB) = (PB)(PAP)$, *then we have*

$$\frac{d}{ds}|_{s=0}(P_s P) = 0, \text{ and } \frac{d}{ds}|_{s=0}(U_s^{-1} P_s A_s P_s U_s) = PB. \tag{34}$$

Proof. By the definition of U_s and V_s we have

$$U_s^{-1} P_s A_s P_s U_s = V_s P_s A_s P_s U_s = P R_s P_s A_s P_s R_s P.$$

By (33) we have

$$
\begin{aligned}
&(P_s f(A_s)P_s - Pf(A)P)P \\
&= \frac{1}{2\pi\sqrt{-1}}\int_\gamma f(\zeta)(A_s-\zeta I)^{-1}(A_sP - AP)(A-\zeta I)^{-1}d\zeta. \qquad (35)
\end{aligned}
$$

Since A_s, $s\in(-\epsilon,\epsilon)$ is a curve in $\mathcal{C}(X)$ and im P has finite dimension, we have

$$
\frac{d}{ds}|_{s=0}(P_sf(A_s)P_sP) = \frac{1}{2\pi\sqrt{-1}}\int_\gamma f(\zeta)(A-\zeta I)^{-1}B(A-\zeta I)^{-1}d\zeta. \quad (36)
$$

Take $f=1$, we have that $\frac{d}{ds}|_{s=0}(P_sP)$ exists. By the definition of R_s we have $\frac{d}{ds}|_{s=0}(R_sP)=0$. Hence we have

$$
\begin{aligned}
\frac{d}{ds}|_{s=0}(U_s^{-1}P_sA_sP_sU_s) &= \frac{d}{ds}|_{s=0}(PR_sP_sA_sP_sR_sP) \\
&= \frac{d}{ds}|_{s=0}((R_sP)(P_sA_sP_sP)(R_sP)) \\
&= \frac{1}{2\pi\sqrt{-1}}\int_\gamma \zeta P(A-\zeta I)^{-1}B(A-\zeta I)^{-1}Pd\zeta \\
&= \frac{1}{2\pi\sqrt{-1}}\int_\gamma \zeta(A-\zeta I)^{-1}PB(A-\zeta I)^{-1}d\zeta.
\end{aligned}
$$

In the case of $(PAP)(PB)=(PB)(PAP)$, we have

$$
\frac{d}{ds}|_{s=0}(P_sP) = \frac{1}{2\pi\sqrt{-1}}\int_\gamma (A-\zeta I)^{-2}Bd\zeta = 0,
$$

and

$$
\begin{aligned}
\frac{d}{ds}|_{s=0}(U_s^{-1}P_sA_sP_sU_s) &= \frac{1}{2\pi\sqrt{-1}}\int_\gamma \zeta P(A-\zeta I)^{-2}Bd\zeta \\
&= \frac{1}{2\pi\sqrt{-1}}\int_\gamma \left(PA(A-\zeta I)^{-2} - P(A-\zeta I)^{-1}\right)Bd\zeta \\
&= P^2B = PB.
\end{aligned}
$$

□

Proposition 3.2. (cf. Theorem 4.1 in [39]) *Let X be a Banach space and ℓ be a bounded open submanifold of $\sqrt{-1}\mathbf{R}$ with co-orientation from left to right. Let A_s, $-\epsilon\le s\le\epsilon$ ($\epsilon>0$), be a curve in $\mathcal{A}_\ell(X)$. Set $P=P_\ell^0(A_0)$, $A=A_0$. Assume that* im $P\subset$ dom(A_s) *and $B:=\frac{d}{ds}|_{s=0}(A_sP)$ exists. Assume that*

$$
(PAP)(PB) = (PB)(PAP), \quad (37)
$$

where $PAP, PB \in \mathcal{B}(\text{im } P)$, and $PB : \text{im } P \to \text{im } P$ is hyperbolic, i.e., $\sigma(PB) \cap (\sqrt{-1}\mathbf{R}) = \emptyset$. Then there is a $\delta \in (0, \epsilon)$ such that $\nu_{h,\ell}(A_s) = 0$ for all $s \in [-\delta, 0) \cup (0, \delta]$ and

$$\text{sf}_\ell\{A_s; 0 \le s \le \delta\} = -m^-(PB), \tag{38}$$

$$\text{sf}_\ell\{A_s; -\delta \le s \le 0\} = m^+(PB). \tag{39}$$

Here we denote by $m^+(PB)$ ($m^-(PB)$) the total algebraic multiplicity of eigenvalues of PB with positive (negative) imaginary part respectively.

Proof. We follow the proof of Theorem 4.1 in [39]. Since $A \in \mathcal{A}_\ell(X)$, there exist bounded open subsets N and $N^\pm$ of $\mathbf{C}$ such that $N = N^+ \cup (N \cap \ell) \cup N^-$, $N^\pm$ stays on the right (left) side of the imaginary axis, $\sigma(A) \cap \bar{\ell} \subset N \cap \ell$, $\sigma(A) \cap \partial N = \emptyset$, and $P(A, N) = P$. Since A_s, $s \in (-\epsilon, \epsilon)$ is a continuous family in $\mathcal{C}(X)$, $\sigma(A_s) \cap (\partial N \cup (\bar{\ell} \setminus (N \cap \ell))) = \emptyset$ for $|s|$ small. For such s, let P_s be defined in Lemma 3.1. Then $\|P_s - P\| < 1$ for $|s|$ small, and R_s and U_s in Lemma 3.1 are well-defined for such s. Then we have

$$\sigma(A_s) \cap \ell \subset \sigma(A_s) \cap N = \sigma(U_s^{-1} P_s A_s P_s U_s).$$

Now we work in the finite dimensional vector space im P. Since PB commutes with PAP, we can assume that they are both in Jordan normal form. Then $P(A+sB)P$ is also in Jordan norm form for each s. By Lemma 3.1, we have $\frac{d}{ds}|_{s=0}(U_s^{-1} P_s A_s P_s U_s) = PB$. Then there exists a $\delta \in (0, \epsilon)$ such that $U_s^{-1} P_s A_s P_s U_s$ are hyperbolic for all $s \in [-\delta, 0) \cup (0, \delta]$, and

$$m^-(U_s^{-1} P_s A_s P_s U_s) = m^-(PB) \quad \text{for all } s \in (0, \delta],$$

$$m^-(U_s^{-1} P_s A_s P_s U_s) = m^+(PB) \quad \text{for all } s \in [-\delta, 0).$$

Then our results follow from the definition of the spectral flow and the fact that

$$\dim \text{im } P(A_s, N^-) = m^-(U_s^{-1} P_s A_s P_s U_s) \quad \text{for all } s \in [-\delta, 0) \cup (0, \delta]. \quad \square$$

3.3. *Spectral flow for curves of quadratic forms*

Let X be a complex Hilbert space and $\ell = \sqrt{-1}(-\epsilon, \epsilon)$ ($\epsilon > 0$) with co-orientation from left to right. Let A_s, $0 \le s \le 1$ be a curve of closed self-adjoint Fredholm operators. We will denote $\text{sf}_\ell\{A_s\}$ by $\text{sf}\{A_s\}$.

Lemma 3.2. *Let X be a Hilbert space. Let A_s, $0 \le s \le 1$ be a curve of (not necessarily bounded) self-adjoint Fredholm operators. Then for any curve $P_s \in \mathcal{B}(X)$ of invertible operators, we have*

$$\text{sf}\{P_s P_s^* A_s\} = \text{sf}\{P_s^* A_s P_s\} = \text{sf}\{A_s\}. \tag{40}$$

Proof. Since A_s is a curve of (closed) self-adjoint Fredholm operators and P_s is a curve of bounded invertible operators, the families $P_s^* A_s P_s$ and $P_s P_s^* A_s$, $0 \le s \le 1$ are curves of closed Fredholm operators. By (viii) of Proposition 3.1 we have

$$\mathrm{sf}\{P_s P_s^* A_s\} = \mathrm{sf}\{P_s(P_s^* A_s P_s)P_s^{-1}\} = \mathrm{sf}\{P_s^* A_s P_s\}. \tag{41}$$

Since $P_s^* A_t P_s$ are self-adjoint Fredholm operators and $\dim\ker(P_s^* A_t P_s) = \dim\ker A_t$, we have

$$\begin{aligned}\mathrm{sf}\{P_s^* A_s P_s\} &= \mathrm{sf}\{P_0^* A_s P_0\} + \mathrm{sf}\{P_s^* A_1 P_s\}\\ &= \mathrm{sf}\{P_0^* A_s P_0\} = \mathrm{sf}\{P_1^* A_s P_1\}.\end{aligned} \tag{42}$$

Let Q_s, $0 \le s \le 1$ be a curve of bounded positive definite operators on X with $Q_0 = I$, $Q_1^2 = P_0 P_0^*$. By (41) and (42) we have

$$\begin{aligned}\mathrm{sf}\{P_s^* A_s P_s\} &= \mathrm{sf}\{P_0^* A_s P_0\} = \mathrm{sf}\{P_0 P_0^* A_s\} = \mathrm{sf}\{Q_1 A_s Q_1\}\\ &= \mathrm{sf}\{Q_0 A_s Q_0\} = \mathrm{sf}\{A_s\}.\end{aligned}$$

□

The above lemma leads to the following definition.

Definition 3.2. Let X be a Hilbert space. Let $\mathcal{I}_s$, $0 \le s \le 1$ be a curve of **bounded Fredholm quadratic forms**, i.e., $\mathcal{I}_s(x,y) = \langle A_s x, y\rangle_X$ for all $x, y \in X$, where A_s, $0 \le s \le 1$ is a curve of bounded self-adjoint Fredholm operators, and $\langle\cdot,\cdot\rangle_X$ denotes the inner product in X.

(a) The **spectral flow** $\mathrm{sf}\{\mathcal{I}_s\}$ of $\mathcal{I}_s$ is defined to be the spectral flow $\mathrm{sf}\{A_s\}$.

(b) If $A_1 - A_0$ is compact, the **relative Morse index** $I(\mathcal{I}_0, \mathcal{I}_1)$ is defined to be the relative Morse index $I(A_0, A_1) := -\mathrm{sf}\{A_0 + s(A_1 - A_0); 0 \le s \le 1\}$.

Based on this notation we have the following lemma.

Lemma 3.3. *Let X be a Hilbert space. Let $A_s \in \mathcal{B}(X)$, $0 \le s \le 1$ be a curve of self-adjoint Fredholm operators and $\mathcal{I}_s$ be quadratic forms defined by $\mathcal{I}_s(x,y) = \langle A_s x, y\rangle$ for all $x, y \in X$. Assume that $P_s \in \mathcal{B}(X)$, $0 \le s \le 1$ is a curve of operators such that $P_s^2 = P_s$ and $\mathcal{I}_s(x,y) = 0$ for all $x \in \mathrm{im}\, P_s$, $y \in \mathrm{im}\, Q_s$, where $Q_s = I - P_s$. Then we have*

$$\mathrm{sf}\{\mathcal{I}_s\} = \mathrm{sf}\{\mathcal{I}_s|_{\mathrm{im}\, P_s}\} + \mathrm{sf}\{\mathcal{I}_s|_{\mathrm{im}\, Q_s}\}. \tag{43}$$

Proof. Set $R_s := P_s^* P_s + Q_s^* Q_s$, $s \in [0,1]$. Since $P_s + Q_s = I$ and $P_s^2 = I$, we have

$$R_s = \frac{I}{2} + 2(\frac{I}{2} - P_s^*)(\frac{I}{2} - P_s) > 0.$$

Consider the new inner product $\langle R_s x, y\rangle$, $x, y \in X$ on X. For this inner product P_s is an orthogonal projection, i.e. $R_s P_s = P_s^* R_s$.

Now we work in the Hilbert space X with the new inner product. So we can assume that P_s is orthogonal. By the fact that im P_s and im Q_s are $\mathcal{I}_{p_s}$ orthogonal, we have $P_s A_s Q_s = Q_s A_s P_s = 0$. Then we have

$$A_s = (P_s + Q_s)A_s(P_s + Q_s) = P_s A_s P_s + Q_s A_s Q_s.$$

So $P_s A_s = A_s P_s$. By (iv) of Proposition 3.1, $P_s A_s P_s$ is a Fredholm operator on im P_s, $Q_s A_s Q_s$ is a Fredholm operator on im Q_s, and we have

$$\begin{aligned} \mathrm{sf}\{\mathcal{I}_s\} &= \mathrm{sf}\{A_s\} \\ &= \mathrm{sf}\{P_s A_s P_s : \mathrm{im}\ P_s \to \mathrm{im}\ P_s\} + \mathrm{sf}\{Q_s A_s Q_s : \mathrm{im}\ Q_s \to \mathrm{im}\ Q_s\} \\ &= \mathrm{sf}\{\mathcal{I}_s|_{\mathrm{im}\ P_s}\} + \mathrm{sf}\{\mathcal{I}_s|_{\mathrm{im}\ Q_s}\}. \end{aligned}$$

□

Lemma 3.4. *Let X be a Hilbert space and M a closed subspace of finite codimension. Let $A \in \mathcal{B}(M)$ be a self-adjoint Fredholm operator and $\mathcal{I}(x,y) = \langle Ax, y\rangle$ for all $x, y \in M$. Let N_1 and N_2 be subspaces of H such that $X = M \oplus N_1 = M \oplus N_2$. Define $\mathcal{I}_k$ on H, $k = 1,2$ by*

$$\mathcal{I}_k(x+u, y+v) = \langle Ax, y\rangle, \qquad \textit{for all } x, y \in M \textit{ and } u, v \in N_k.$$

Then we have $I(\mathcal{I}_1, \mathcal{I}_2) = 0$.

Proof. Let N_0 denote the orthogonal complement of M. Set $A_0 = \mathrm{diag}(A, 0)$ under the direct sum decomposition $X = M \oplus N_0$. Define $\mathcal{I}_0$ and A_1, A_2 by $\mathcal{I}_k(x,y) = \langle A_k x, y\rangle$, for all $x, y \in H$, where $k = 0,1,2$. Let $B : N_1 \to N_0$ be a linear isomorphism. Define $P_1 \in \mathcal{B}(X)$ by $P_1(x+y) = x + By$ for all $x \in M$, $y \in N_1$. Then P_1 is invertible, $P_1 - I$ is compact, and $A_1 = P_1^* A_0 P_1$. So $A_1 - A_0$ is compact. Let $P_s \in \mathcal{B}(X)$, $0 \le s \le 1$ be a curve of invertible operators such that $P_0 = I$ and $P_s - I$ are compact. By the definition of the relative Morse index and Lemma 3.2, we have

$$\begin{aligned} I(\mathcal{I}_0, \mathcal{I}_1) = I(A_0, A_1) = I(A_0, A_1) &= -\mathrm{sf}\{P_s^* A_0 P_s\} \\ &= -\mathrm{sf}\{A_0\} = 0. \end{aligned}$$

Similarly we have that $A_2 - A_0$ is compact and $\mathcal{I}(A_0, A_2) = 0$. So $A_2 - A_1$ is compact, and

$$I(\mathcal{I}_1, \mathcal{I}_2) = I(\mathcal{I}_0, \mathcal{I}_2) - I(\mathcal{I}_0, \mathcal{I}_1) = 0. \qquad \square$$

The following proposition gives a generalization of a formula of M. Morse, more specifically, of Proposition 5.3 in [1].

Proposition 3.3. *Let X be a Hilbert space and $A \in \mathcal{B}(X)$ be a self-adjoint Fredholm operator. Let P be an orthogonal projection with* $\ker P$ *of finite dimension. Let $\mathcal{I}$ denote the quadratic form on X defined by $\mathcal{I}(x, y) = \langle Ax, y\rangle$, $x, y \in X$. Set $M =$ im P and let N denote the $\mathcal{I}$-orthogonal complement of M, i.e., $N = \{x \in X; \mathcal{I}(x, y) = 0$ for all $y \in M\}$. Then we have*

$$I(PAP, A) = m^-(\mathcal{I}|_N) + \dim\ker \mathcal{I}|_N - \dim\ker \mathcal{I}. \qquad (44)$$

Proof. Since $PAP - A$ is of finite rank, $sPAP + (1 - s)A$, $0 \le s \le 1$ is a curve of self-adjoint Fredholm operators. We divide our proof into four steps.

Step1. Assume that $\ker A = \{0\}$. Let $M_0 = \ker \mathcal{I}|_M$, M_1 denote the orthogonal complement of M_0 in M, and P_0, P_1 denote the orthogonal projection onto M_0, M_1 respectively. Then $P = P_0 + P_1$. Since AM is of finite codimension and $M_0 = (AM)^\perp \cap M$, P_0 is of finite rank. Let N_1 denote the $\mathcal{I}$-orthogonal complement of M_1. Since $M = M_0 + M_1$, we have $M_1 \cap N_1 \subset M_0$. So $M_1 \cap N_1 = \{0\}$. Moreover we have

$$\begin{aligned}\dim N_1 &= \dim\ker(AP_1) - \operatorname{ind}(AP_1) = \dim\ker P_1 - \operatorname{ind} A - \operatorname{ind} P_1\\ &= \dim\ker P_1 < +\infty,\end{aligned}$$

where we denote the index of a Fredholm operator A by ind$A$. So $X = M_1 \oplus N_1$. Since $\mathcal{I}$ is nondegenerate, $\mathcal{I}|_{N_1}$ is nondegenerate.

Let $\mathcal{I}_1$ be defined by $\mathcal{I}_1(x+u, y+v) = \mathcal{I}(x, y)$ for all $x, y \in M_1$, $u, v \in N_1$. By Lemma 3.3 and Lemma 3.4 we have

$$\begin{aligned}I(PAP, A) &= I(PAP, P_1AP_1) + I(P_1AP_1, A) = I(P_1AP_1, A)\\ &= I(\mathcal{I}_1, \mathcal{I}) = I(\mathcal{I}_1|_{M_1}, \mathcal{I}|_{M_1}) + I(\mathcal{I}_1|_{N_1}, \mathcal{I}|_{N_1}) = m^-(\mathcal{I}|_{N_1}).\end{aligned}$$

Step 2. Equation (44) holds if $\ker A = \{0\}$ and $N \subset M$. In this case, $M_0 = N \subset N_1$, $m^-(\mathcal{I}|_N) = 0$ and $\ker \mathcal{I}|N = N$. For each $x \in N_1$ such that $\mathcal{I}(x, y) = 0$ for all $y \in N$, we have $\mathcal{I}(x, y) = 0$ for all $y \in M_1$ and hence for all $y \in M$. Then $x \in N$. Thus N is the $\mathcal{I}|_{N_1}$-orthogonal complement of N. N_1 has an orthogonal decomposition $N_1 = N^+ \oplus N^-$ such that N^+ and N^- are $\mathcal{I}$-orthogonal, $\mathcal{I}|_{N^+} > 0$ and $\mathcal{I}|_{N^-} < 0$. Let $P^\pm$ denote the orthogonal projections onto $N^\pm$. Then $P^\pm|_{M_0}$ are isomorphisms. So we have $\dim N_1 = 2\dim N = 2m^-(\mathcal{I}|_{N_1})$. By Step 1 we have

$$I(PAP, A) = m^-(\mathcal{I}|_{N_1}) = m^-(\mathcal{I}|_N) + \dim\ker\mathcal{I}|_N - \dim\ker\mathcal{I}.$$

Step 3. Equation (44) holds if $M + N = X$. In this case we have

$$\ker\mathcal{I}|_N = \ker\mathcal{I} = M \cap N.$$

Firstly we assume that $\ker A = \{0\}$. Then $M_0 = \{0\}$, $N_1 = N$ and $\ker\mathcal{I}|_N = \ker\mathcal{I} = \{0\}$. By Step 1, equation (44) holds.

In the general case, we apply the above special case by taking the quotient space with $\ker A$ and get $I(PAP, A) = m^-(\mathcal{I}|_N)$.

Step 4. Equation (44) holds.

Firstly we assume that $\ker A = \{0\}$. Let Q denote the orthogonal projection onto $M + N$. Then the $\mathcal{I}$-orthogonal complement of $M + N$ is $\ker\mathcal{I}|_N$. By Step 2 and Step 3 we have

$$I(PAP, A) = I(PAP, QAQ) + I(QAQ, A) = m^-(\mathcal{I}|_N) + \dim\ker\mathcal{I}|_N.$$

In the general case, we apply the above special case by taking the quotient space with $\ker A$ and get equation (44). □

3.4. *A spectral flow formula*

The aim of this subsection is to prove Proposition 3.4 below. It can be deduced by Proposition 3.3 in the finite dimensional case.

Let X_k, Y_k, $k = 1, 2$ be Banach spaces. Define the map $D : \mathcal{B}(X_1, Y_1) \times \mathcal{C}(X_2, Y_1) \times \mathcal{C}(X_1, Y_2) \times \mathcal{B}(X_2, Y_2) \to \mathcal{C}(X_1 \oplus X_2, Y_1 \oplus Y_2)$ by

$$D(A_{11}, A_{12}, A_{21}, A_{22}) = \begin{pmatrix} A_{11} & A_{12} \\ A_{21} & A_{22} \end{pmatrix}.$$

Lemma 3.5. *The map D is a well-defined continuous map.*

Proof. Our result follows from the fact that

$$\operatorname{Gr}\begin{pmatrix} A_{11} & A_{12} \\ A_{21} & A_{22} \end{pmatrix} = \begin{pmatrix} I_{X_1 \oplus X_2} & 0 \\ \operatorname{diag}(A_{11}, A_{22}) & I_{Y_1 \oplus Y_2} \end{pmatrix} \operatorname{Gr}\begin{pmatrix} 0 & A_{12} \\ A_{21} & 0 \end{pmatrix}. \qquad \square$$

Lemma 3.6. *Assume that $A_{11} \in \mathcal{B}(X_1, Y_1)$, $A_{12} \in \mathcal{C}(X_2, Y_1)$ and $A_{21} \in \mathcal{C}(X_1, Y_2)$. If A_{12} and A_{21} are Fredholm operators, then the operator $D(A_{11}, A_{12}, A_{21}, 0)$ is Fredholm.*

Proof. We divide the proof into two steps.

Step1. $\dim X_1 < +\infty$ and $Y_2 = \{0\}$.

In this case we have $\ker(A_{11}, A_{12}) \subset X_1 \oplus \ker A_{12}$ and $\mathrm{im}\,(A_{11}, A_{12}) \supset \mathrm{im}\, A_{12}$. Since A_{12} is Fredholm, (A_{11}, A_{12}) is Fredholm.

Step 2. The general case.

Set $X_{10} := \ker A_{21}$ and $Y_{21} := \mathrm{im}\, A_{21}$. Since $A_{21} \in \mathcal{C}(X_1, Y_2)$ is Fredholm, X_{10} is finite dimensional and Y_{21} is closed with finite codimension. Then there exist a closed subspace X_{11} of X_1 and a closed subspace Y_{20} of Y_2 such that

$$X_1 = X_{10} \oplus X_{11}, \qquad Y_2 = Y_{20} \oplus Y_{22}.$$

Set $B_{21} := A_{21}|_{X_{21}}$. Then $B_{21} \in \mathcal{C}(X_{21}, Y_{21})$ is injective and surjective. By the closed graph theorem, B_{21}^{-1} is bounded. Define the operator $\tilde{A}_{21} \in \mathcal{B}(Y_2, X_1)$ by $\tilde{A}_{21} := \mathrm{diag}(0, B_{21}^{-1})$. Define

$$\begin{aligned} D : &= \begin{pmatrix} A_{11}(I_{X_1} - \tilde{A}_{21}A_{21}) & A_{12} \\ A_{21} & 0 \end{pmatrix} \\ &= \left(\begin{pmatrix} A_{11}(I_{X_1} - \tilde{A}_{21}A_{21}) \\ 0 \end{pmatrix}, \begin{pmatrix} 0 & A_{12} \\ A_{21} & 0 \end{pmatrix} \right). \end{aligned}$$

Then $D \in \mathcal{C}(X_{10} \oplus (X_{11} \oplus X_2), Y_1 \oplus Y_2)$. By Step 1, D is Fredholm. Then our statement follows from the fact that

$$\begin{pmatrix} A_{11} & A_{12} \\ A_{21} & 0 \end{pmatrix} = \begin{pmatrix} I_{Y_1} & A_{11}\tilde{A}_{21} \\ 0 & I_{Y_2} \end{pmatrix} \begin{pmatrix} A_{11}(I_{X_1} - \tilde{A}_{21}A_{21}) & A_{12} \\ A_{21} & 0 \end{pmatrix}. \qquad \square$$

Lemma 3.7. *Let X, Y be Hilbert spaces and $H = X \oplus Y$. Let $B_s \in \mathcal{C}(X, Y)$, $0 \le s \le 1$ be a curve of Fredholm operators. Let the operators $D_s \in \mathcal{C}(H)$ be defined by $D_s = \begin{pmatrix} 0 & B_s^* \\ B_s & 0 \end{pmatrix}$. Then we have*

$$\mathrm{sf}\{D_s\} = \dim\ker B_1 - \dim\ker B_0. \tag{45}$$

Proof. By Theorem IV.2.23 in [22], B_s^*, $0 \le s \le 1$ is a curve of closed operators.

Note that $\lambda \in \sigma(D_s)$ if and only if $\lambda^2 \in \sigma(B_s^* B_s)$, and the algebraic multiplicities of them are the same if $|\lambda| \ne 0$ is small. Moreover we have

$$\begin{aligned} \dim\ker D_s &= \dim\ker B_s + \dim\ker B_s^*, \\ \mathrm{ind} B_s = \mathrm{ind} B_0 &= \dim\ker B_s - \dim\ker B_s^*. \end{aligned}$$

By the definition of the spectral flow we have

$$\mathrm{sf}\{D_s\} = \frac{1}{2}(\dim\ker D_1 - \dim\ker D_0) = \dim\ker B_1 - \dim\ker B_0. \qquad \square$$

Lemma 3.8. *Let X, Y be Hilbert spaces and $H = X \oplus Y$. Let $B \in \mathcal{C}(X,Y)$ be a Fredholm operator and $A \in \mathcal{B}(X)$ be a self-adjoint operator. Define linear operators $D_s \in \mathcal{C}(H)$ by $D_s = \begin{pmatrix} sA & B^* \\ B & 0 \end{pmatrix}$. Then D_s, $0 \le s \le 1$ is a curve of self-adjoint Fredholm operators, and we have*

$$\dim\ker D_s = \dim\ker A|_{\ker B} + \dim\ker B^* \qquad \textit{for all } s \in (0,1], \quad (46)$$

$$\mathrm{sf}\{D_s\} = -m^-(A|_{\ker B}). \quad (47)$$

Proof. By Lemmas 3.5 and 3.6, D_s, $0 \le s \le 1$ is a curve of closed Fredholm operators.

For each $s \in (0,1]$ we have

$$\begin{aligned}\ker D_s &= \{(x,y) \in H; sAx + B^*y = 0, Bx = 0\} \\ &= \{(x,y) \in H; x \in \ker B, sAx = -B^*y \in \mathrm{im}\, B^* = (\ker B)^\perp\} \\ &= \{(x,y) \in H; x \in \ker A|_{\ker B}, sAx = -B^*y\}.\end{aligned}$$

Define $\varphi : \ker D_s \to \ker A|_{\ker B}$ by $\varphi(x,y) = x$ for $(x,y) \in \ker D_s$. Then φ is a linear surjective map, and $\ker\varphi = \{0\} \times \ker B^*$. Then we get (46).

Let $\lambda_t \in \sigma(D_t)$ be a spectral point of D_t near 0 for $t \ne 0$ small. Then there exists $(x_t, y_t) \in H \setminus \{0\}$ such that $D_t(x_t,y_t) = \lambda_t(x_t,y_t)$. Then one of the following cases holds.

Case 1. $\lambda_t = 0$. In this case we have $(x_t,y_t) \in \ker D_t$. The algebraic multiplicity of the eigenvalue 0 of D_t is $\dim\ker D_t$.

Case 2. $\lambda_t \ne 0$ and $Bx_t = 0$. In this case we have $y_t = 0$ and $tAx_t = \lambda_t x_t$. Let P denote the orthogonal projection of X onto $\ker B$. Then $tPAPx_t = \lambda_t x_t$. So the total algebraic multiplicity of these eigenvalues λ_t of D_t with such eigenvectors is

$$m^+(tPAP) + m^-(tPAP) = m^+(A|_{\ker B}) + m^-(A|_{\ker B}).$$

Case 3. $\lambda_t \ne 0$ and $Bx_t \ne 0$. We denote by k_t the total algebraic multiplicity of these eigenvalues of D_t with such eigenvectors. Since D_s, $0 \le s \le 1$ is continuously varying, we have for $t \ne 0$ small

$$\dim\ker D_0 = \dim\ker D_t + m^+(A|_{\ker B}) + m^-(A|_{\ker B}) + k_t. \quad (48)$$

By (46) we have

$$\begin{aligned}\dim\ker D_0 &= \dim\ker D_t + \dim\ker B - \dim\ker A|_{\ker B} \\ &= \dim\ker D_t + m^+(A|_{\ker B}) + m^-(A|_{\ker B}). \quad (49)\end{aligned}$$

Then we have $k_t = 0$.

By the definition of the spectral flow and the analysis of the three cases above we have

$$\mathrm{sf}\{D_s\} = -m^-(A|_{\ker B}). \qquad \square$$

Proposition 3.4. *Let X, Y be Hilbert spaces and $H = X \oplus Y$. Let $B_s \in \mathcal{C}(X,Y)$, $0 \le s \le 1$ be a curve of Fredholm operators, and $A_s \in \mathcal{B}(X)$, $0 \le s \le 1$ be a curve of self-adjoint operators. Define linear operators D_s on H by $D_s = \begin{pmatrix} A_s & B_s^* \\ B_s & 0 \end{pmatrix}$. Then $D_s \in \mathcal{C}(H)$, $0 \le s \le 1$ is a curve of (self-adjoint) Fredholm operators, and we have*

$$\dim\ker D_s = \dim\ker A_s|_{\ker B_s} + \dim\ker B_s^* \qquad \textit{for all } s \in [0,1], \tag{50}$$

$$\mathrm{sf}\{D_s\} = m^-(A_0|_{\ker B_0}) - m^-(A_1|_{\ker B_1}) + \dim\ker B_1 - \dim\ker B_0.$$

Proof. (50) follows from (46). Set $D_{s,t} = \begin{pmatrix} tA_s & B_s^* \\ B_s & 0 \end{pmatrix}$ for $s,t \in [0,1]$. By Lemmas 3.5 and 3.6, $D_{s,t}$, $0 \le s,t \le 1$ is a continuous family of closed Fredholm operators.

By Proposition 3.1, Lemmas 3.7 and 3.8 we have

$$\begin{aligned}\mathrm{sf}\{D_s\} &= -\mathrm{sf}\{D_{0,t}; 0 \le t \le 1\} + \mathrm{sf}\{D_{s,0}; 0 \le s \le 1\} + \mathrm{sf}\{D_{1,t}; 0 \le t \le 1\} \\ &= m^-(A_0|_{\ker B_0}) + (\dim\ker B_1 - \dim\ker B_0) - m^-(A_1|_{\ker B_1}) \\ &= m^-(A_0|_{\ker B_0}) - m^-(A_1|_{\ker B_1}) + \dim\ker B_1 - \dim\ker B_0. \qquad \square\end{aligned}$$

4. Maslov-type index theory

4.1. *Symplectic functional analysis and Maslov index*

A main feature of symplectic analysis is the study of the *Maslov index*. It is an intersection index between a path of Lagrangian subspaces with the *Maslov cycle*, or, more generally, with another path of Lagrangian subspaces. The Maslov index assigns an integer to each continuous path of Fredholm pairs of Lagrangian subspaces of a fixed Hilbert space with continuously varying symplectic structures.

Firstly we define symplectic Hilbert spaces and Lagrangian subspaces.

Definition 4.1. Let H be a complex vector space. A mapping

$$\omega : H \times H \to \mathbf{C}$$

is called a (weak) **symplectic form** on H, if it is sesquilinear, skew-symmetric, and non-degenerate, i.e.,

(i) $\omega(x,y)$ is linear in x and conjugate linear in y;
(ii) $\omega(y,x) = -\overline{\omega(y,x)}$;
(iii) $H^\omega := \{x \in H; \omega(x,y) = 0 \text{ for all } y \in H\} = \{0\}$.

Then we call (H,ω) a **complex symplectic vector space.**

Definition 4.2. Let (H,ω) be a complex symplectic vector space.

(a) The **annihilator** of a subspace λ of H is defined by

$$\lambda^\omega := \{y \in H; \omega(x,y) = 0 \text{ for all } x \in \lambda\}.$$

(b) A subspace λ is called **isotropic**, **co-isotropic**, or **Lagrangian** if

$$\lambda \subset \lambda^\omega, \quad \lambda \supset \lambda^\omega, \quad \lambda = \lambda^\omega$$

respectively.

(c) The **Lagrangian Grassmannian** $\mathcal{L}(H,\omega)$ consists of all Lagrangian subspaces of (H,ω).

Definition 4.3. Let H be a complex Hilbert space. A mapping $\omega : H \times H \to \mathbf{C}$ is called a (strong) **symplectic form** on H, if $\omega(x,y) = \langle Jx, y\rangle_H$ for some bounded invertible skew-symmetric operator J. (H,ω) is called a (strong) **symplectic Hilbert space.**

For the ease of application we have dropped the common additional assumption of unitary J yielding $J^2 = -I$ in the preceding definition. Clearly, $J^2 = -I$ can always be obtained by smooth deformation of the inner product of H, see [10], Lemma 1.6.

Before giving a rigorous definition of the Maslov index, we fix the terminology and give a simple lemma.

We recall:

Definition 4.4.

(a) The space of (algebraic) **Fredholm pairs** of linear subspaces of a vector space H is defined by

$$\mathcal{F}^2_{\text{alg}}(H) := \{(\lambda,\mu) \mid \dim(\lambda \cap \mu) < +\infty \text{ and } \dim\big(H/(\lambda+\mu)\big) < +\infty\}$$

with

$$\operatorname{ind}(\lambda,\mu) := \dim(\lambda \cap \mu) - \dim(H/(\lambda+\mu)).$$

(b) In a Banach space H, the space of (topological) **Fredholm pairs** is defined by

$$\mathcal{F}^2(H) := \{(\lambda, \mu) \in \mathcal{F}^2_{\mathrm{alg}}(H) \mid \lambda, \mu, \text{ and } \lambda + \mu \subset H \ \text{ closed}\}.$$

We need the following well-known lemma (see, e.g., Lemma 1.7 in [10]).

Lemma 4.1. *Let (H, ω) be a (strong) symplectic Hilbert space. Then*

(i) *there exists an ω-orthogonal splitting*

$$H = H^+ \oplus H^-$$

such that $-\sqrt{-1}\omega$ is positive (negative) definite on $H^\pm$, and we call it a ***canonical symplectic splitting****;*

(ii) *there is a 1-1 correspondence between the space $\mathcal{U}(H^+, H^-, \omega)$ of all $U \in \mathcal{B}(H^+, H^-)$ with*

$$\omega(Ux, Uy) = -\omega(x, y), \text{ for all } x, y \in H^+$$

and $\mathcal{L}(H, \omega)$ under the mapping $U \to L := \mathrm{Gr}(U)$;

(iii) *if $U, V \in \mathcal{U}(H^+, H^-, \omega)$ and $\lambda := \mathrm{Gr}(U)$, $\mu := \mathrm{Gr}(V)$, then (λ, μ) is a Fredholm pair if and only if $U - V$, or, equivalently, $UV^{-1} - I$ is Fredholm. Moreover, we have a natural isomorphism*

$$\ker(UV^{-1} - I) \simeq \lambda \cap \mu\,.$$

Definition 4.5. Let $(H, \langle\cdot,\cdot\rangle_s)$, $s \in [0, 1]$ be a continuous family of Hilbert spaces, and $\omega_s(x, y) = \langle J_s x, y\rangle_s$ be a continuous family of symplectic forms on H, i.e., $\{A_{s,0}\}$ and $\{J_s\}$ are two continuous families of bounded invertible operators, where $A_{s,0}$ is defined by

$$\langle x, y\rangle_s = \langle A_{s,0}x, y\rangle_0 \quad \text{for all } x, y \in H.$$

Let $\{(\lambda_s, \mu_s)\}$ be a continuous family of Fredholm pairs of Lagrangian subspaces of $(H, \langle\cdot,\cdot\rangle_s, \omega_s)$. Then there is a continuous family of canonical symplectic splitting

$$H = H_s^+ \oplus H_s^- \tag{51}$$

for all $s \in [0, 1]$. Such $H_s^\pm$ can be chosen to be the positive (negative) space associated to the self-adjoint operator $-\sqrt{-1}J_s \in \mathcal{B}(H, \langle\cdot,\cdot\rangle_s)$. By Lemma 4.1, $\lambda_s = \mathrm{Gr}_s(U_s)$ and $\mu_s = \mathrm{Gr}_s(V_s)$ with U_s, $V_s \in \mathcal{U}(H_s^+, H_s^-, \omega_s)$, where Gr_s denotes the graph associated to the splitting (51). We define the **Maslov index** $\mathrm{Mas}\{\lambda_s, \mu_s\}$ by

$$\mathrm{Mas}\{\lambda_s, \mu_s\} = -\mathrm{sf}_\ell\{U_s V_s^{-1}\}, \tag{52}$$

where $\ell := (1-\epsilon, 1+\epsilon)$ with $\epsilon \in (0,1)$ and with upward co-orientation.

Remark 4.1. For finite dimensional (H,ω), constant $\mu_s = \mu_0$, and a loop $\{\lambda_s\}$, i.e., for $\lambda_0 = \lambda_1$, we notice that $\mathrm{Mas}\{\lambda_s, \mu_s\}$ is the winding number of the closed curve $\{\det(U_s^{-1}V_0)\}_{s\in[0,1]}$. This is the original definition of the Maslov index as explained in Arnol'd [3].

Lemma 4.2. *The Maslov index is independent of the choice of the canonical symplectic splitting of H.*

Proof. Let $H = H_{s,k}^+ \oplus H_{s,k}^-$, $s \in [0,1]$ with $k = 0,1$ be two continuous families of canonical symplectic splitting. For each $s \in [0,1]$ and $k = 0,1$, set

$$\langle \cdot,\cdot \rangle_{s,k} = (-\sqrt{-1}\omega|_{H_{s,k}^+}) \oplus (\sqrt{-1}\omega|_{H_{s,k}^-}).$$

Then $(H, \langle \cdot,\cdot \rangle_{s,k})$ is a Hilbert space for each $s \in [0,1]$ and $k = 0,1$. Set

$$\langle \cdot,\cdot \rangle_{s,t} = (1-t)\langle \cdot,\cdot \rangle_{s,0} + t\langle \cdot,\cdot \rangle_{s,1}$$

for each $(s,t) \in [0,1] \times [0,1]$. For each $(s,t) \in [0,1] \times [0,1]$, define $J_{s,t} \in \mathcal{B}(H)$ by

$$\omega(x,y)_s = \langle J_{s,t}x, y \rangle_{s,t} \quad \text{for all } x,y \in H.$$

Then $H_{s,k}^{\pm}$ is the positive (negative) space associated with the self-adjoint operator $-\sqrt{-1}J_{s,k}$ for each $s \in [0,1]$ and $k = 0,1$. Let $H_{s,t}^{\pm}$ denote the positive (negative) space associated with the self-adjoint operator $-\sqrt{-1}J_{s,t}$ for each $s \in [0,1]$ and $t \in [0,1]$.

Let (λ_s, μ_s) be a continuous family of Fredholm pairs of Lagrangian subspaces of (H, ω_s). For each canonical symplectic splitting $H = H_{s,t}^+ \oplus H_{s,t}^-$, we denote by $U_{s,t}$ and $V_{s,t}$ the associated generated "unitary" operators of λ_s and μ_s respectively. We also denote by Mas_t the Maslov index defined with $\langle \cdot,\cdot \rangle_{s,t}$ for each $t \in [0,1]$. By Proposition 3.1 we have

$$\begin{aligned}\mathrm{Mas}_0\{\lambda_s,\mu_s\} - \mathrm{Mas}_1\{\lambda_s,\mu_s\} &= -\mathrm{sf}_\ell\{U_{s,0}V_{s,0}^{-1}\} + \mathrm{sf}_\ell\{U_{s,1}V_{s,1}^{-1}\} \\ &= -\mathrm{sf}_\ell\{U_{s,t}V_{s,t}^{-1}; (s,t) \in \partial([0,1]\times[0,1])\} = 0. \qquad \square\end{aligned}$$

Corollary 4.1. (Symplectic invariance) *Let $(H_k, \omega_{s,k})$, $k = 1,2$ be two continuous families of symplectic Hilbert spaces. Let $M(s) \in \mathcal{B}(H_1, H_2)$, $0 \le s \le 1$ be a curve of invertible operators such that*

$$\omega_{s,2}(M_s x, M_s y) = \omega_{s,1}(x,y) \quad \textit{for all } x,y \in H_1 \textit{ and } s \in [0,1].$$

Then for any curve $(\lambda(s), \mu(s))$, $0 \le s \le 1$ *of Fredholm pairs of Lagrangian subspaces of* H_1,

$$\mathrm{Mas}\{M\lambda, M\mu\} = \mathrm{Mas}\{\lambda, \mu\}. \tag{53}$$

Proof. Let $H_1 = H^+_{s,1} \oplus H^-_{s,1}$ be a continuous family of canonical symplectic splittings of the family $(H_1, \omega_{s,1})$, $0 \le s \le 1$. Then $H_2 = H^+_{s,2} \oplus H^-_{s,2}$ is a continuous family of canonical symplectic splittings of the family $(H_2, \omega_{s,2})$, $0 \le s \le 1$, where $H^+_{s,2} = M_s H^+_{s,1}$ and $H^-_{s,2} = M_s H^-_{s,1}$. For each $s \in [0,1]$ and $k = 1, 2$, we denote by $U_{s,k}$ and $V_{s,k}$ the generating "unitary" operators of λ_s and μ_s, associated to the canonical symplectic splittings $H_k = H^+_{s,k} \oplus H^-_{s,k}$ respectively. Then we have

$$U_{s,2} = M_s U_{s,1} M_s^{-1}, \qquad V_{s,2} = M_s V_{s,1} M_s^{-1}.$$

By the definition of the Maslov index we have

$$\begin{aligned} \mathrm{Mas}\{M\lambda, M\mu\} &= -\mathrm{sf}_\ell\{(M_s U_{s,1} M_s^{-1})(M_s V_{s,1} M_s^{-1})^{-1}; 0 \le s \le 1\} \\ &= -\mathrm{sf}_\ell\{U_{s,1} V_{s,1}^{-1}; 0 \le s \le 1\} = \mathrm{Mas}\{\lambda, \mu\}. \end{aligned}$$ □

Now we give a method of using the crossing form to calculate Maslov indices (cf. [15], [32] and Theorem 2.1 in [7]).

Let $\lambda = \{\lambda_s\}_{s \in [0,1]}$ be a C^1 curve of Lagrangian subspaces of (H, ω). Let $t \in [0,1]$ and W be a fixed Lagrangian complement of λ_t. For $v \in \lambda_t$ and $|s - t|$ small, define $w(s) \in W$ by $v + w(s) \in \lambda_s$. The form

$$Q(\lambda, t)(u, v) := Q(\lambda, W, t)(u, v) = \frac{d}{ds}|_{s=t}\omega(u, w(s)), \quad \text{for all } u, v \in \lambda_t$$

is independent of the choice of W. Let $\{(\lambda_s, \mu_s)\}$, $0 \le s \le 1$ be a curve of Fredholm pairs of Lagrangian subspaces of H. For $t \in [0,1]$, the **crossing form** $\Gamma(\lambda, \mu, t)$ is a quadratic form on $\lambda_t \cap \mu_t$ defined by

$$\Gamma(\lambda, \mu, t)(u, v) = Q(\lambda, t)(u, v) - Q(\mu, t)(u, v), \quad \text{for all } u, v \in \lambda_t \cap \mu_t.$$

A **crossing** is a time $t \in [0,1]$ such that $\lambda_t \cap \mu_t \ne \{0\}$. A crossing is called **regular** if $\Gamma(\lambda, \mu, t)$ is nondegenerate. It is called **simple** if it is regular and $\lambda_t \cap \mu_t$ is one-dimensional.

Now let (H, ω) be a symplectic Hilbert space with $\omega(x, y) = \langle Jx, y \rangle$, for all $x, y \in H$, where $J \in \mathcal{B}(H)$ is an invertible skew self-adjoint operator. Then we have a symplectic Hilbert space $X = (H \oplus H, (-\omega) \oplus \omega)$. For each $M \in \mathrm{Sp}(H, \omega)$, i.e., $M \in \mathcal{B}(H)$ invertible and ω-invariant, its graph $\mathrm{Gr}(M)$ is a Lagrangian subspace of X. The following lemma is Lemma 3.1 in [15].

Lemma 4.3. *Let $M(s) \in \mathrm{Sp}(H,\omega)$, $0 \le s \le 1$ be a curve of linear symplectic maps. Assume that $M(s)$ is differentiable at $t \in [a,b]$. Set $B_1(t) = -J\dot{M}(t)M(t)^{-1}$ and $B_2(t) = -JM(t)^{-1}\dot{M}(t)$. Then $B_1(t)$, $B_2(t)$ are self-adjoint, $B_2(t) = M(t)^*B_1(t)M(t)$ and we have*

$$Q(\mathrm{Gr}(M),t)((x,M(t)x),(y,M(t)y)) = \langle B_2(t)x,y\rangle. \qquad \square$$

Proposition 4.1. *Let (H,ω) be a symplectic Hilbert space and $\{(\lambda_s,\mu_s)\}$, $0 \le s \le 1$ be a C^1 curve of Fredholm pairs of Lagrangian subspaces of H with only regular crossings. Then we have*

$$\mathrm{Mas}\{\lambda,\mu\} = m^+(\Gamma(\lambda,\mu,0)) - m^-(\Gamma(\lambda,\mu,1)) + \sum_{0<t<1} \mathrm{sign}(\Gamma(\lambda,\mu,t)).$$

Proof. Pick an invertible skew self-adjoint operator $J \in \mathcal{B}(H)$ such that $J^2 = -I$ and $\omega(x,y) = \langle Jx,y\rangle$. Let $H_1 = \ker(J - \sqrt{-1}I)$ and $H_2 = \ker(J + \sqrt{-1}I)$. By Lemma 4.1 there are curves of isometric $U(t)$, $V(t)$ in $\mathcal{U}(H_1,H_2,\omega)$ such that $\lambda(t) = \mathrm{Gr}(U(t))$ and $\mu(t) = \mathrm{Gr}(V(t))$. Applying Lemma 4.3 for $(H_1, \langle -\sqrt{-1}x,y\rangle)$, for any $x,y \in \ker(U(t)-V(t))$ and $t \in [a,b]$ we have

$$\begin{aligned}
&\tfrac{d}{ds}\big|_{s=t}\langle -\sqrt{-1}V^{-1}Ux,y\rangle = \langle \sqrt{-1}V^{-1}\dot{V}V^{-1}Ux,y\rangle - \langle \sqrt{-1}V^{-1}\dot{U}x,y\rangle \\
&= \langle \sqrt{-1}V^{-1}\dot{V}x,y\rangle - \langle \sqrt{-1}U^{-1}VV^{-1}\dot{U}x,U^{-1}Vy\rangle \\
&= \langle \sqrt{-1}V^{-1}\dot{V}x,y\rangle - \langle \sqrt{-1}U^{-1}\dot{U}x,y\rangle = -\Gamma(\lambda,\mu,t)((x,Ux),(y,Uy)).
\end{aligned}$$

By Proposition 3.2 we obtain (53). $\square$

4.2. *Spectral flow formula for fixed maximal domain*

Let $D_m \hookrightarrow D_M \hookrightarrow X$ be three Hilbert spaces. We assume that D_m is a closed subspace of D_M and a dense subspace of X. Let $\{A_s\}_{s\in[0,1]}$ be a family of symmetric densely defined operators in $\mathcal{C}(X)$ with domain $\mathrm{dom}(A_s) = D_m$. Assume that $\mathrm{dom}(A_s^*) = D_M$, i.e., the domain of the extension A_s^* of A_s is independent of s.

We recall from [7] (partly reproduced in Everitt and Markus [17], Theorem 1.14) for each $s \in [0,1]$:

(I) The space D_M is a Hilbert space with the graph inner product

$$\langle x,y\rangle_{\mathrm{Gr}_s} := \langle x,y\rangle_X + \langle A_s^*x, A_s^*y\rangle_X \quad \text{for } x,y \in D_M\,. \qquad (54)$$

(II) The space D_m is a closed subspace in the graph norm and the quotient space D_M/D_m is a strong symplectic Hilbert space with the (bounded) symplectic form induced by Green's form

$$\omega_s(x+D_m, y+D_m) := \langle A_s^* x, y\rangle_X - \langle x, A_s^* y\rangle_X \quad \text{for } x, y \in D_M\,. \tag{55}$$

(III) If A_s admits a self-adjoint Fredholm extension $A_{s,D_s} := A_s^*|_{D_s}$ with domain $D_s \subset X$, then the **natural Cauchy data space** $(\ker A_s^* + D_m)/D_m$ is a Lagrangian subspace of $(D_M/D_m, \omega_s)$.

(IV) Moreover, self-adjoint Fredholm extensions are characterized by the property of the domain D_s that $(D_s + D_m)/D_m$ is a Lagrangian subspace of $(D_M/D_m, \omega_s)$ and forms a Fredholm pair with $(\ker A_s^* + D_m)/D_m$.

(V) We denote the natural projection (which is independent of s) by

$$\gamma : D_M \to D_M/D_m.$$

We call γ the **abstract trace map**.

We have the following spectral flow formula (cf. Theorem 5.1 in [7], Corollary 2.14 in [10] and Theorem 1.5 [11]).

Proposition 4.2. *We assume that on D_M the graph norms induced by A_s^* and the original norm are equivalent. Assume that $\{A_s^* : D_M \to X\}$ is a continuous family of bounded operators and each A_s is injective. Let $\{D_s/D_m\}$ be a continuous family of Lagrangian subspaces of $(D_M/D_m, \omega_s)$, such that each A_{s,D_s} is a Fredholm operator. Then:*

(a) *Each $\left(D_s/D_m, \gamma(\ker(A_s^*))\right)$ is a Fredholm pair in D_M/D_m.*
(b) *Each Cauchy data space $\gamma(\ker A_s^*)$ is a Lagrangian subspace of $(D_M/D_m, \omega_s)$.*
(c) *The family $\{\gamma(\ker A_s^*)\}$ is a continuous family in D_M/D_m.*
(d) *The family $\{A_{s,D_s}\}$ is a continuous family of self-adjoint Fredholm operators in $\mathcal{C}(X)$.*
(e) *Finally, we have*

$$\mathrm{sf}\{A_{s,D_s}\} = -\mathrm{Mas}\{\gamma(D_s), \gamma(\ker A_s^*)\}. \qquad \square$$

4.3. *The Maslov-type indices*

Definition 4.6. Let (X_l, ω_l) be symplectic Hilbert spaces with $\omega_l(x,y) = (J_l x, y)$, $x, y \in X_l$, $J_l \in \mathcal{B}(X)$ are invertible, and $J_l^* = -J_l$, where $l = 1, 2$. Then we have a symplectic Hilbert space $(H = X_1 \oplus X_2, (-\omega_1) \oplus \omega_2)$. Let $W \in \mathcal{L}(H)$. Let $M(t)$, $0 \le t \le T$ be a curve in $\mathrm{Sp}(X_1, X_2)$ such that $\mathrm{Gr}(M(t))$ forms a Fredholm pair of Lagrangian subspaces with W for all $t \in [0,T]$. The **Maslov-type index** $i_W\{M(t)\}$ is defined to be $\mathrm{Mas}\{\mathrm{Gr}(M(t)), W\}$. If $a = 0$, $b = T$, $(X_1, \omega_1) = (X_2, \omega_2)$ and $M(0) = I$, we denote $\dim(\mathrm{Gr}(M(T)) \cap W)$ by $\nu_{T,W}(M(t))$.

The Maslov-type indices have the following property.

Lemma 4.4. *Let (X_l, ω_l) be symplectic Hilbert spaces with $\omega_l(x,y) = (J_l x, y)$, where $x, y \in X_l$, $J_l \in \mathcal{B}(X_l)$ are invertible, and $J_l^* = -J_l$, $l = 1,2,3,4$. Let W be a Lagrangian subspace of $(X_1 \oplus X_4, (-\omega_1) \oplus \omega_4)$. Let $\gamma_l \in C([0,1], \mathrm{Sp}(X_l, X_{l+1}))$, $l = 1,2,3$ be symplectic paths such that $\mathrm{Gr}(\gamma_3(s)\gamma_2(t)\gamma_1(s))$ forms a Fredholm pair of Lagrangian subspaces with W for all $(s,t) \in [0,1] \times [0,1]$. Then we have*

$$i_W(\gamma_3\gamma_2\gamma_1) = i_{W'}(\gamma_2) + i_W(\gamma_3\gamma_2(0)\gamma_1), \tag{56}$$

where $W' = \mathrm{diag}(\gamma_1(1), \gamma_3(1)^{-1})W$.

Proof. Let $M = \mathrm{diag}(\gamma_1(1), \gamma_3(1)^{-1})$. By the homotopy invariance relative endpoints of the Maslov-type indices and Corollary 4.1, we have

$$\begin{aligned} i_W(\gamma_3\gamma_2\gamma_1) &= i_W(\gamma_3(1)\gamma_2\gamma_1(1)) + i_W(\gamma_3\gamma_2(0)\gamma_1) \\ &= \mathrm{Mas}(M\mathrm{Gr}(\gamma_3(1)\gamma_2\gamma_1(1)), MW) + i_W(\gamma_3\gamma_2(0)\gamma_1) \\ &= i_{W'}(\gamma_2) + i_W(\gamma_3\gamma_2(0)\gamma_1). \end{aligned}$$ □

The following properties of fundamental solutions for linear ODE will be used later.

Lemma 4.5. *Let $J \in C^1([0,+\infty), \mathrm{GL}(m, \mathbf{C}))$ be a curve of skew self-adjoint matrices, and $b \in C([0,+\infty), \mathrm{gl}(m, \mathbf{C}))$ be a curve of self-adjoint matrices. Let $\gamma \in C^1([0,+\infty), \mathrm{GL}(m, \mathbf{C}))$ denote the fundamental solution of*

$$-J\dot{x} - \frac{1}{2}\dot{J}x = bx. \tag{57}$$

Then we have $\gamma(t)^* J(t)\gamma(t) = J(0)$ for all t.

Proof. By the definition of the fundamental solution, we have

$\gamma(0)^*J(0)\gamma(0) = J(0)$. Since $J^* = -J$ and $b^* = b$, we have

$$\begin{aligned}\frac{d}{dt}(\gamma(t)^*J(t)\gamma(t)) &= \dot{\gamma}^*J\gamma + \gamma^*\dot{J}\gamma + \gamma^*J\dot{\gamma}\\ &= (-b\gamma - \frac{1}{2}\dot{J}\gamma)^*J^{*-1}J\gamma + \gamma^*\dot{J}\gamma + \gamma^*JJ^{-1}(-b\gamma - \frac{1}{2}\dot{J}\gamma)\\ &= \gamma^*(b - \frac{1}{2}\dot{J} + \dot{J} - b - \frac{1}{2}\dot{J})\gamma = 0.\end{aligned}$$

So we have $\gamma(t)^*J(t)\gamma(t) = J(0)$. □

Lemma 4.6. *For two curves of matrices $B \in C([0, +\infty), \mathrm{gl}(m, \mathbf{C}))$ and $P \in C^1([0, +\infty), \mathrm{GL}(m, \mathbf{C}))$, let $\gamma \in C^1([0, +\infty), \mathrm{GL}(m, \mathbf{C}))$ denote the fundamental solution of*

$$\dot{x} = Bx, \tag{58}$$

and $\gamma' \in C^1([0, +\infty), \mathrm{GL}(m, \mathbf{C}))$ denote the fundamental solution of

$$\dot{y} = (PBP^{-1} + \dot{P}P^{-1})y. \tag{59}$$

Then we have

$$\gamma' = P\gamma P(0)^{-1}. \tag{60}$$

Proof. Direct calculation shows

$$\frac{d}{dt}(P\gamma P(0)^{-1}) = (PBP^{-1} + \dot{P}P^{-1})P\gamma P(0)^{-1}$$

and $P(0)\gamma P(0)^{-1} = I$. By definition, $P\gamma P(0)^{-1}$ is the fundamental solution of (59). □

Corollary 4.2. *Let $J_1, J_2 \in C^1([0, +\infty), \mathrm{GL}(m, \mathbf{C}))$ be two curves of skew self-adjoint matrices. Let $P \in C^1([0, +\infty), \mathrm{GL}(m, \mathbf{C}))$ be a curve of matrices such that $P^*J_2P = J_1$, and $b \in C([0, +\infty), \mathrm{GL}(m, \mathbf{C}))$ be a curve of self-adjoint matrices. Let $\gamma \in C^1([0, +\infty), \mathrm{GL}(m, \mathbf{C}))$ denote the fundamental solution of*

$$-J_1\dot{x} - \frac{1}{2}\dot{J}_1x = bx, \tag{61}$$

and $\gamma' \in C^1([0, +\infty), \mathrm{GL}(m, \mathbf{C}))$ denote the fundamental solution of

$$J_2\dot{y} - \frac{1}{2}\dot{J}_2y = (P^{*-1}bP^{-1} + Q)y, \tag{62}$$

where $Q = \frac{1}{2}(P^{*-1}\dot{P}^* J_2 - J_2\dot{P}P^{-1})$. *Then we have*

$$\gamma' = P\gamma P(0)^{-1}. \tag{63}$$

In particular, when J_1 *and* J_2 *are constant matrices, we have*

$$Q = P^{*-1}\dot{P}^* J_2 = -J_2\dot{P}P^{-1}.$$

Proof. Taking $B = -J_1^{-1}(b + \frac{1}{2}\dot{J}_1)$ in Lemma 4.6, we obtain

$$\begin{aligned}
&-J_2(PBP^{-1} + \dot{P}P^{-1}) - \frac{1}{2}\dot{J}_2 \\
&\quad = -J_2(P(-J_1)^{-1}(b + \frac{1}{2}\dot{J}_1)P^{-1} + \dot{P}P^{-1}) - \frac{1}{2}\dot{J}_2 \\
&\quad = P^{*-1}(b + \frac{1}{2}\dot{J}_1)P^{-1} - J_2\dot{P}P^{-1} - \frac{1}{2}\dot{J}_2 \\
&\quad = P^{*-1}bP^{-1} - J_2\dot{P}P^{-1} + \frac{1}{2}(P^{*-1}\dot{J}_1P^{-1} - J_2) \\
&\quad = P^{*-1}bP^{-1} - J_2\dot{P}P^{-1} + \frac{1}{2}(P^{*-1}\frac{d}{dt}(P^*J_2P)P^{-1} - J_2) \\
&\quad = P^{*-1}bP^{-1} + Q.
\end{aligned}$$

By Lemma 4.6, our results hold. □

The following is a special case of the spectral flow formula.

Let $J \in C^1([0,T], \mathrm{GL}(m, \mathbf{C}))$ be a curve of skew self-adjoint matrices. Then we have symplectic Hilbert spaces $(\mathbf{C}^m, \omega(t))$ with standard quadratic inner product and $\omega(t)(x,y) = \langle J(t)x, y\rangle$, for all $x, y \in \mathbf{C}^m$ and $t \in [0,T]$. Then we have a symplectic Hilbert space $(V = \mathbf{C}^m \oplus \mathbf{C}^m, (-\omega(0)) \oplus \omega(T))$. Let $W \in \mathcal{L}(V)$. Let $b_s(t) \in \mathcal{B}(\mathbf{C}^m)$, $0 \le s \le 1$, $0 \le t \le T$ be a continuous family of self-adjoint matrices such that $b_0(t) = 0$. By Lemma 4.5, there are continuous families of matrices $M_s(t) \in \mathrm{GL}(m, \mathbf{C})$ such that $M_s(0) = I$, $M_s(t)^* J(t) M_s(t) = J(0)$ and

$$-J\frac{d}{dt}M_s(t) - \frac{1}{2}(\frac{d}{dt}J)M_s(t) = b_s(t)M_s(t).$$

Set

$$\begin{aligned}
X &= L^2([0,T], \mathbf{C}^m), \quad D_m = H_0^1([0,T], \mathbf{C}^m), \\
D_M &= H^1([0,T], \mathbf{C}^m), \quad D_W = \{x \in D_M; (x(0), x(T)) \in W\}.
\end{aligned}$$

Let $A_M \in \mathcal{C}(X)$ with domain D_M be defined by

$$A_M x = -J\frac{d}{dt}x - \frac{1}{2}(\frac{d}{dt}J)x.$$

Set $x \in D_M$, $A = A_M|_{D_m}$, $A_W = A_M|_{D_W}$. Let $C_s \in \mathcal{B}(X)$ be defined by $(C_s x)(t) = b_s(t)x(t)$, $x \in X$, $t \in [0,T]$.

Proposition 4.3. *Set* $W' = \mathrm{diag}(I, M_0(T)^{-1})W$. *Then we have*

$$I(A_W, A_W - C_1) = i_{W'}(M_0^{-1}M_1). \tag{64}$$

Proof. The Sobolev embedding theorem shows that $D_M \subset C([0,T], \mathbf{C}^m)$. For any $x \in D_M$, define $\gamma(x) = (x(0), x(T))$. Direct calculation shows that $D_M/D_m = \mathbf{C}^m \oplus \mathbf{C}^m$ with symplectic structure $(\mathrm{diag}(J(0), -J(T))\gamma(x), \gamma(y))$, $x, y \in D_M$, and γ is the abstract trace map. Moreover, $A^* = A_M$, $\gamma(A^* - C_s) = \mathrm{Gr}(M_s(T))$, and $\gamma(D_W) = W$. By Proposition 4.2 and Lemma 4.4, we have

$$\begin{aligned}
I(A_W, A_W - C_1) &= -\mathrm{sf}\{A_W - C_s\} \\
&= \mathrm{Mas}(\{\mathrm{Gr}(M_s(T)); 0 \le s \le 1\}, W) \\
&= i_W(M_0(T)(M_0(T)^{-1}M_s(T))I; 0 \le s \le 1) \\
&= i_{W'}(M_0(T)^{-1}M_s(T); 0 \le s \le 1) \\
&= -i_{W'}(M_0(t)^{-1}M_0(t); 0 \le t \le T) \\
&\quad + i_{W'}(M_0(0)^{-1}M_s(0); 0 \le s \le 1) \\
&\quad + i_{W'}(M_0(t)^{-1}M_1(t); 0 \le t \le T) \\
&= i_{W'}(M_0^{-1}M_1).
\end{aligned}$$

□

5. Proof of the main results

In this section we shall use the notations of §2.

5.1. *Proof of Theorem 2.1*

Lemma 5.1. *The index forms* $\mathcal{I}_{p_s,Q_s}$, $0 \le s \le 1$ *form a curve of bounded Fredholm quadratic forms on* H_R.

Proof. Since all $\mathcal{I}_{p_s,Q_s}$ are bounded symmetric quadratic forms on H_R, by Riesz representation theorem, they form a continuous curve. By Sobolev embedding theorem, each $\tilde{Q}_s$ defines a compact operator on H_R. Thus we need only consider the case when Q_s is zero on R, i.e. the forms $\mathcal{I}_{p_s,R}$.

For each $k, l = 0, \ldots, m$ and $s \in [0,1]$, we define the bounded operators $P_{k,l}(s) \in \mathcal{B}(H_R)$ by

$$\langle P_{k,l}(s)x, y\rangle_m = \int_0^T \langle p_{k,l}(s,t)\frac{d^k x}{dt^k}, \frac{d^l y}{dt^l}\rangle dt \quad \text{for all } x, y \in H_R.$$

Claim. $P_{k,l}(s)$ is compact for either $k \neq m$ or $l \neq m$.

Since $P_{k,l}(s) = P^*_{l,k}(s)$, without loss of generality we can assume that $k \neq m$. Pick a bounded sequence $\{x_\alpha; \alpha \in \mathbf{N}\}$ in H_R. By Sobolev embedding theorem, the sequence $\{p_{k,l}(s,t)\frac{d^k x_\alpha}{dt^k}\}$ has a convergent subsequence, which is denoted by the original sequence. Since $P_{k,l}(s)$ is bounded, we have

$$\begin{aligned}
&\lim_{\alpha,\beta\to+\infty} \|P_{k,l}(s)(x_\alpha - x_\beta)\|_m^2 \\
&= \lim_{\alpha,\beta\to+\infty} \int_0^T \langle p_{k,l}(s,t)\frac{d^k(x_\alpha - x_\beta)}{dt^k}, \frac{d^l(P_{k,l}(s)(x_\alpha - x_\beta))}{dt^l}\rangle dt \\
&= 0.
\end{aligned}$$

So the sequence $\{P_{k,l}(s)(x_\alpha)\}$ converges and $P_{k,l}(s)$ is a compact operator.

Now we prove that $P_{m,m}(s)$ is Fredholm and then our lemma is proved. If $p_{m,m}(s,t)$ is positive definite for each $s, t \in [0,1]$, we can choose $p_{k,l}(s,t)$ such that $\mathcal{I}_{p_s,R}$ is positive definite for each s. So $P_{m,m}(s)$ is a compact perturbation of a Fredholm operator and is Fredholm. Here it is only required that $p_{m,m}(s,t)$ is continuous in t. In the general case, we have to assume that $p_{m,m}(s,t)$ is C^m in t. Consider the operator $p_{m,m}(s,\cdot) : H \to H$. Let $j : H_R \to H$ denote the inclusion. Then $p_{m,m}(s,\cdot)$ is invertible and $p_{m,m}(s,\cdot)j$ is Fredholm. For any $x \in H_R$ and $y \in H$, the inner product $\langle (P_{m,m}(s) - p_{m,m}(s,\cdot))x, y\rangle_m$ consists only of the lower-order terms (i.e., no second-order differential involved) and some boundary terms. As in the above proof, we can conclude that the lower-order terms correspond to compact operators. The boundary terms correspond to finite rank operators. So $jP_{m,m}(s) - p_{m,m}(s,\cdot)j$ is compact. Since $p_{m,m}(s)j$ and j are Fredholm, $jP_{m,m}(s)$ and $P_{m,m}(s)$ are Fredholm. □

The following lemma is the key to the proof of Theorem 2.1.

Lemma 5.2. (i) *Any solution $u \in H^1([0,T]; \mathbf{C}^{2mn})$ of (18) can be expressed by $u = u_{p_s,x}$ for some $x \in H^m([0,T]; \mathbf{C}^n)$, and the following three conditions are equivalent:*

(a) $x \in \ker \mathcal{I}_{s,Q_s}$;
(b) $x \in \ker L_{p_s,W_{2m}(Q_s)}$;
(c) $u_{p_s,x}$ *is a solution of* (18) *and* $(u_{p_s,x}(0), u_{p_s,x}(T)) \in W_{2m}(Q_s)$.

(ii) *If p_s is C^1 in s, then for any $x, y \in H^m([0,T]; \mathbf{C}^n)$, we have*

$$\left\langle \left(\frac{d}{ds}p_s\right) \bar{u}_{0,x}, \bar{u}_{0,y} \right\rangle = -\left\langle \left(\frac{d}{ds}b(p_s)\right) u_{p_s,x}, u_{p_s,y} \right\rangle. \tag{65}$$

(iii) *Let $J \in \mathrm{GL}(\mathbf{C}^m)$ be skew self-adjoint, and $b_s(t) \in \mathrm{gl}(\mathbf{C}^m)$, $0 \le s \le 1$, $0 \le t \le T$ is a continuous family of self-adjoint matrices. Let γ_s denote the fundamental solutions of the linear Hamiltonian system*

$$-J\dot{u} = b_s u. \tag{66}$$

If b_s is C^1 in s, we have

$$\frac{\partial}{\partial t}(-J\gamma_s^{-1}\frac{\partial \gamma_s}{\partial s}) = \gamma_s^* \frac{\partial b_s}{\partial s}\gamma_s. \tag{67}$$

(iv) *If p_s is C^1 in s, then for any $x, y \in \ker L_{p_s}$, we have*

$$\begin{aligned}
&\left\langle -J_{2m,n}\gamma_{p_s}(T)^{-1}\frac{d\gamma_{p_s}(T)}{ds}u_{p_s,x}(0), u_{p_s,y}(0)\right\rangle \\
&= -\int_0^T \left\langle \left(\frac{d}{ds}p_s\right)\bar{u}_{0,x}, \bar{u}_{0,y}\right\rangle dt.
\end{aligned} \tag{68}$$

Proof. (i) The proof for the solution u of (18) can be expressed by $u = u_{p_s,x}$ and (a)⇔(b) is standard and we omit it. Now we prove (b)⇔(c). By (14), we have $\frac{d}{dt}u^k_{p_s,x}(t) = u^{k+1}_{p_s,x}(t)$ for $k = 0, \dots, m-2$,

$$\begin{aligned}
\frac{d}{dt}u^{m-1}_{p_s,x}(t) &= \frac{d^m}{dt^m}x(t) \\
&= p_{m,m}(s,t)^{-1}u^m_{p_s,x}(t) - \sum_{0\le\beta\le m-1} p_{m,m}(s,t)^{-1}p_{m,\beta}(s,t)u^\beta_{p_s,x}(t)
\end{aligned}$$

and

$$\begin{aligned}
\frac{d}{dt}u^k_{p_s,x}(t) &= \sum_{2m-k\le\alpha\le m, 0\le\beta\le m} (-1)^{\alpha-m}\frac{d^{\alpha+k+1-2m}}{dt^{\alpha+k+1-2m}}\left(p_{\alpha,\beta}(s,t)\frac{d^\beta}{dt^\beta}x(t)\right) \\
&= u^{k+1}_{p_s,x}(t) - \sum_{0\le\beta\le m}(-1)^{m+k+1}\left(p_{2m-k-1,\beta}(s,t)\frac{d^\beta}{dt^\beta}x(t)\right) \\
&= u^{k+1}_{p_s,x}(t) + (-1)^{m+k}p_{2m-k-1,m}(s,t)p_{m,m}(s,t)^{-1}u^m_{p_s,x}(t) \\
&\quad + \sum_{0\le\beta\le m-1}(-1)^{m+k}(p_{2m-k-1,\beta}(s,t) \\
&\qquad\qquad - p_{2m-k-1,m}(s,t)p_{m,m}(s,t)^{-1}p_{m,\beta}(s,t))u^\beta_{p_s,x}(t)
\end{aligned}$$

for $k = m, \dots, 2m-1$. Combine the above equations and we get

$$\frac{d}{dt}u_{p_s,x}(t) = J_{2m,n}b(p_s)u_{p_s,x}(t) + (u^{2m}_{p_s,x}(t), 0, \dots, 0). \tag{69}$$

By the fact that $L_{p_s}x = (-1)^m u^{2m}_{p_s,x}(t)$, we get (b)⇔(c).

(ii) By the definition of $U(p_s)$, $V(p_s)$, $\bar{u}_{p_s,x}$ and $\bar{u}_{0,x}$ in §2, direct computation shows

$$V(p_s)^*\left(\frac{d}{ds}p_s\right)V(p_s) = -\frac{d}{ds}P(p_s).$$

Thus for all $x, y \in H_R$, we have

$$\begin{aligned}\left\langle\left(\frac{d}{ds}p_s\right)\bar{u}_{0,x}, \bar{u}_{0,y}\right\rangle &= -\left\langle U(p_s)^*\left(\frac{d}{ds}P(p_s)\right)U(p_s)\bar{u}_{0,x}, \bar{u}_{0,y}\right\rangle \\ &= -\left\langle\left(\frac{d}{ds}P(p_s)\right)\bar{u}_{p_s,x}, \bar{u}_{p_s,y}\right\rangle \\ &= -\left\langle\left(\frac{d}{ds}b(p_s)\right)u_{p_s,x}, u_{p_s,y}\right\rangle.\end{aligned}$$

(iii) By the definition of γ_s, we have $\gamma_s^* J\gamma_s = J$, and

$$\begin{aligned}\frac{\partial}{\partial t}(-J\gamma_s^{-1}\frac{\partial\gamma_s}{\partial s}) &= J\gamma_s^{-1}\dot{\gamma}_s\gamma_s^{-1}\frac{\partial\gamma_s}{\partial s} - J\gamma_s^{-1}\frac{\partial^2\gamma_s}{\partial s\partial t} \\ &= J\gamma_s^{-1}(-J^{-1}b_s)\frac{\partial\gamma_s}{\partial s} - J\gamma_s^{-1}\frac{\partial}{\partial s}(-J^{-1}b_s\gamma_s) \\ &= J\gamma_s^{-1}J^{-1}\frac{\partial b_s}{\partial s}\gamma_s \\ &= \gamma_s^*\frac{\partial b_s}{\partial s}\gamma_s.\end{aligned}$$

(iv) follows from (ii), (iii) and the fact that $\gamma_{p_s}u_{p_s,x}(0) = u_{p_s,x}$ for all $x \in \ker L_{p_s}$. □

Now we can prove Theorem 2.1.

We begin with a simple case.

Lemma 5.3. *Let $\mathcal{I}_{Id,R}$ denote the inner product on H_R. If $\epsilon > 0$ satisfies $[-\epsilon, 0] \cap \sigma(p_{m,m}(0,t)) = \emptyset$ for all $t \in [0,T]$, we have*

$$-\mathrm{sf}\{\mathcal{I}_{p_0,Q_0} + a\mathcal{I}_{Id,R}; a \in [0,\epsilon]\} = i_{W_{2m}(Q_0)}(\{\gamma_{p_s+aI_{(m+1)n}}(T); 0 \le a \le T\}). \tag{70}$$

Proof. By Lemma 5.1, $\mathcal{I}_{p_0,Q_0} + a\mathcal{I}_{Id,R}$, $a \in [0,\epsilon]$ is a continuous family of Fredholm quadratic forms. By the definition of the spectral flow we have

$$\mathrm{sf}\{\mathcal{I}_{p_0,Q_0} + a\mathcal{I}_{Id,R}; a \in [0,\epsilon]\} = \sum_{a\in(0,\epsilon]} \dim\ker(\mathcal{I}_{p_0,Q_0} + a\mathcal{I}_{Id,R}). \tag{71}$$

Set

$$Z_a = -J_{2m,n}\left(\gamma_{p_s+aI_{(m+1)n}}(T)\right)^{-1}\frac{d\gamma_{p_s+aI_{(m+1)n}}(T)}{da}$$

for $a \in [0, \epsilon]$. By (iv) of Lemma 5.2, the matrix $-Z_a$ is non negatively definite. Let $v \in \mathbf{C}^{2mn}$ be a vector such that $\langle Z_a v, v\rangle = 0$. By (i) of Lemma 5.2, there exists $x \in \ker L_{p_s}$ such that $v = u_{p_s + aI_{(m+1)n}, x}(0)$. By (iv) of Lemma 5.2, we have $\bar{u}_{0,x}(t) = 0$ for all $t \in [0, T]$. Thus $x = 0$, $u_{p_s + aI_{(m+1)n}, x} = 0$ and $v = 0$. So Z_a is negative definite. By Lemma 4.3, Proposition 4.1, (i) of Lemma 5.2 and the definition of Maslov-type index we have

$$\begin{aligned}
&i_{W_{2m}(Q_0)}(\{\gamma_{p_s + aI_{(m+1)n}}(T); 0 \le a \le T\}) \\
&= - \sum_{a \in (0,\epsilon]} \dim \mathrm{Gr}\left((\gamma_{p_s + aI_{(m+1)n}}(T)) \cap W_{2m}(Q_0)\right) \\
&= - \sum_{a \in (0,\epsilon]} \dim \ker(\mathcal{I}_{p_0, Q_0} + a\mathcal{I}_{Id,R}). \qquad (72)
\end{aligned}$$

Combine (71) and (72), we get (70). □

Proof of Theorem 2.1. We divide the proof into two steps.

Step 1. We apply Proposition 4.2. Set

$$A_s = L^*_{p_s}, \quad D_m = H_0^{2m}([0, T]; \mathbf{C}^n), \quad D_M = H^{2m}([0, T]; \mathbf{C}^n).$$

Then A_s is injective for each s and $L_{p_s, W_{2m}(R)}$, $0 \le s \le 1$ is a continuous family of self-adjoint operators. Define the trace map $\hat{\gamma} : D_M \to \mathbf{C}^{4mn}$ by $\hat{\gamma}(x) = (u_{p_s,x}(0), u_{p_s,x}(T))$ for $x \in D_M$. Then $\hat{\gamma}$ induce an isomorphism $\mathrm{D}_M/D_m \to \mathbf{C}^{4mn}$. After identifying the two spaces D_M/D_m and $\mathbf{C}^{4mn}$, we have $\hat{\gamma} = \gamma$. Direct computation shows

$$\omega_s(x + D_m, y + D_m) = \langle J_{2m,n} u_{p_s,x}(0), u_{p_s,y}(0)\rangle - \langle J_{2m,n} u_{p_s,x}(T), u_{p_s,y}(T)\rangle.$$

Let D_s denote the domain of $L_{p_s, W_{2m}(Q_s)}$. Then $\gamma(D_s) = W_{2m}(Q_s)$ and $\gamma(\ker A_s^*) = \mathrm{Gr}(\gamma_{p_s}(T))$. By Proposition 4.2 we have

$$\begin{aligned}
-\mathrm{sf}\{L_{p_s, W_{2m}(Q_s)}; 0 \le s \le 1\} &= \mathrm{Mas}\{W_{2m}(Q_s), \mathrm{Gr}(\gamma_{p_s}(T)); 0 \le s \le 1; \omega_s\} \\
&= \mathrm{Mas}\{\mathrm{Gr}(\gamma_{p_s}(T)), W_{2m}(Q_s); 0 \le s \le 1; -\omega_s\} \\
&= i_{W_{2m}(Q_s)}(\{\gamma_{p_s}(T); 0 \le s \le 1\}).
\end{aligned}$$

Step2. We claim that

$$-\mathrm{sf}\{\mathcal{I}_{p_s, Q_s}; 0 \le s \le 1\} = i_{W_{2m}(Q_s)}(\{\gamma_{p_s}(T); 0 \le s \le 1\}). \qquad (73)$$

Let $\mathcal{I}_{Id,R}$ denote the inner product on H_R. Let $\epsilon > 0$ be small enough such that $[-\epsilon, 0] \cap \sigma(p_{m,m}(s,t)) = \emptyset$ for all $(s,t) \in [0,1] \times [0,T]$. By Lemma 5.1, $\mathrm{sf}\{\mathcal{I}_{p_s,Q_s} + a\mathcal{I}_{Id,R}\}$ is well-defined. For each $c \in [0, 1]$, there exist $\delta_c > 0$ and $\epsilon_c \in (0, \epsilon]$ such that $\ker(\mathcal{I}_{p_s,Q_s} + \epsilon_c \mathcal{I}_{Id,R}) = \{0\}$ for all $s \in (c - \delta_c, c + \delta_c) \cap [0, 1]$.

Let $[s_0, s_1]$ be a subinterval of $(c - \delta_c, c + \delta_c) \cap [0, 1]$. Consider the spectral flow $\mathrm{sf}\{\mathcal{I}_{p_s,Q_s} + a\mathcal{I}_{Id,R}\}$ and the Maslov-type index $i_{W_{2m}(Q_s)}(\gamma_{p_s + aI_{(m+1)n}}(T))$. Because of the homotopy invariance of spectral flow and Maslov-type index, both integers must vanish for the boundary loop going counter clockwise around the rectangular domain from the corner point $(s_0, 0)$ via the corner points $(s_1, 0)$, (s_1, ϵ_c), and (s_0, ϵ_c) back to $(s_0, 0)$. The spectral flow and Maslov index vanish on the top segment of our box. By the preceding lemma, the left and the right side segments of our curves yield vanishing sum of spectral flow and Maslov index. So, by the additivity under catenation, we have

$$-\mathrm{sf}\{\mathcal{I}_{p_s,Q_s}; s_0 \le s \le s_1\} = i_{W_{2m}(Q_s)}(\{\gamma_{p_s}(T); s_0 \le s \le s_1\}).$$

Since $[0, 1]$ is compact, there exist $c_0, \ldots, c_{N-1} \in [0, 1]$ and a partition of $0 = s_0 < s_1 < \ldots < s_N = 1$ of $[0, 1]$ such that $[s_j, s_{j+1}] \subset (c_j - \delta_{c_j}, c_j + \delta_{c_j}]$ for $j = 0, \ldots N - 1$. Then (73) follows from additivity under catenation of spectral flow and Maslov-type index.

Step 3. Since $\gamma_{p_s}(0) = I_{2mn}$, by the homotopy invariance of the Maslov-type index we have

$$i_{W_{2m}(Q_s)}(\{\gamma_{p_s}(T); 0 \le s \le 1\}) = i_{W_{2m}(Q_1)}(\gamma_{p_1}) - i_{W_{2m}(Q_0)}(\gamma_{p_0}). \qquad \square$$

5.2. *Proof of Theorem 2.2*

We divide the proof into three steps.

Step 1. (22), (23) holds for the C^1 path γ with $\gamma_0 = I_{2n}$.

Set $H = L^2([0, T]; \mathbf{C}^n)$ and $H_R = \{x \in H; (x(0), x(T)) \in R\}$. Let F_R denote the closed operator on H with domain H_{R^K} defined by $F_R x = -K\dot{x}$ for all $x \in H_R$. Set $X = L^2([0, T], \mathbf{C}^{2n})$ and

$$D_{W_K(R)} = \{x \in H^1([0, T]; \mathbf{C}^{2n}); (x(0), x(t)) \in W(R)\}.$$

Let $A_{W_K(R)} \in \mathcal{C}(X)$ with domain $D_{W_K(R)}$ be defined by $A_{W_K(R)} x = -J_K \dot{x}$ for $x \in D_{W_K(R)}$. Let $b(t) \in \mathrm{gl}(\mathbf{C}^{2n})$ and $C \in \mathcal{B}(X)$ be defined by $b(t) = -J_K \dot{\gamma}(t)\gamma(t)^{-1}$, $t \in [0, T]$ and $(Cx)(t) = b(t)x(t)$ for $x \in X$, $t \in [0, T]$. Then we have $F_R^* = -F_{R^K}$.

Consider the standard orthogonal decomposition

$$\mathbf{C}^{2n} = (\mathbf{C}^n \times \{0\}) \oplus (\{0\} \times \mathbf{C}^n).$$

It induces orthogonal decompositions $X = H \oplus H$ and $D_{W_K(R)} = H_{R^K} \oplus H_R$. Under such orthogonal decompositions, $A_{W_K(R)}$ is in block form $A_{W_K(R)} = \begin{pmatrix} 0 & F_R^* \\ F_R & 0 \end{pmatrix}$. Let C be in block form

$$C = \begin{pmatrix} C_{1,1} & C_{1,2} \\ C_{2,1} & C_{2,2} \end{pmatrix}.$$

By the definition of $b(t)$ and the symplectic path γ we have

$$b(t) = \begin{pmatrix} K^*(\dot{M}_{2,1}M_{1,1}^{-1} - \dot{M}_{2,2}M_{2,2}^{-1}M_{2,1}M_{1,1}^{-1}) & K^*\dot{M}_{2,2}M_{2,2}^{-1} \\ -K\dot{M}_{1,1}M_{1,1}^{-1} & 0 \end{pmatrix}.$$

Since $M_{2,2}^*KM_{1,1} = K$, we have $K^*\dot{M}_{2,2}M_{2,2}^{-1} = -(M_{1,1}^*)^{-1}\dot{M}_{1,1}^*K^*$. So there holds

$$\begin{aligned} &K^*(\dot{M}_{2,1}M_{1,1}^{-1} - \dot{M}_{2,2}M_{2,2}^{-1}M_{2,1}M_{1,1}^{-1}) \\ &= K^*\dot{M}_{2,1}M_{1,1}^{-1} + (M_{1,1}^*)^{-1}\dot{M}_{1,1}^*K^*M_{2,1}M_{1,1}^{-1} \\ &= (M_{1,1}^*)^{-1}\left(\frac{d}{dt}(M_{1,1}^*K^*M_{2,1})\right)M_{1,1}^{-1}. \end{aligned}$$

Clearly we have

$$\ker(F_R - C_{2,1}) = \{M_{1,1}x(0); (x(0), M_{1,1}(T)x(0)) \in R^K\}.$$

Since $\operatorname{ind}(F_R - C_{2,1}) = \operatorname{ind}F_R = \dim(\operatorname{Gr}(I_n)\cap R^K) - \dim(\operatorname{Gr}(I_n)\cap R)$, we have

$$\dim\ker(F_R - C_{2,1})^* = \dim S(T) + \dim(\operatorname{Gr}(I_n)\cap R) - \dim(\operatorname{Gr}(I_n)\cap R^K).$$

Let $x, y \in \ker(F_R - C_{2,1})$. Then we have

$$\begin{aligned} &\langle C_{1,1}x, y\rangle \\ &= \int_0^T \left\langle (M_{1,1}^*)^{-1}\left(\frac{d}{dt}(M_{1,1}^*K^*M_{2,1})\right)M_{1,1}^{-1}x, y\right\rangle dt \\ &= \int_0^T \left\langle (M_{1,1}^*)^{-1}\left(\frac{d}{dt}(M_{1,1}^*K^*M_{2,1})\right)M_{1,1}^{-1}M_{1,1}x(0), M_{1,1}y(0)\right\rangle dt \\ &= \int_0^T \left\langle \left(\frac{d}{dt}(M_{1,1}^*K^*M_{2,1})\right)x(0), y(0)\right\rangle dt \\ &= \langle M_{1,1}(T)^*K^*M_{2,1}(T)x(0), y(0)\rangle. \end{aligned}$$

By Proposition 3.4, Proposition 4.3 and the definition of $S(t)$, we have (22) and

$$\begin{aligned} i_{W_K(R)}(\gamma) &= -\mathrm{sf}\{A_{W_K(R)} - sC; 0 \le s \le 1\} \\ &= m^+((M_{1,1}(T)^* K^* M_{2,1}(T))|_{S(T)}) \\ &\quad + \dim(\mathrm{Gr}(I_n) \cap R^K) - \dim S(T). \end{aligned}$$

Step 2. Define the set

$$Y = \{M \in \mathrm{GL}(\mathbf{C}^{2n}); M = \begin{pmatrix} M_{1,1} & 0 \\ M_{2,1} & M_{2,2} \end{pmatrix}, M^* J_K M = J_K\}.$$

Note that any symplectic loop γ in Y is homotopic to the loop in Y starting from I_{2n}. By the homotopy invariance of the Maslov-type index and Step 1, we have $i_{W_K(R)}(\gamma) = 0$ for any loop in γ in Y. For a general γ in Y, we can connect I_{2n} and the endpoints $\gamma(0)$ and $\gamma(T)$ in Y by C^1 paths. Then (22) follows from Step 1 and the path additivity of Maslov-type index under catenation. □

5.3. *The positive definite leading term case*

Firstly we give an alternative proof of the following part (A) of Theorem 3.1 in [16].

Proposition 5.1. *Assume that $p_{m,m}(1,t)$ is positive definite for each $t \in [0,T]$. Then $m^-(\mathcal{I}_{p_1,Q_1}) < +\infty$. Let $\tilde{p}(t) = (\tilde{p}_{k,l}(t)) \in \mathrm{gl}((\tilde{m}+1)n, \mathbf{C})$ be a continuous family of self-adjoint matrix such that $\tilde{p}_{k,l} \in C^{\max\{k,l\}}([0,T], \mathrm{gl}(n,\mathbf{C}))$. Assume that $\mathcal{I}_{\tilde{p}}$ is positive definite. Then we have*

$$m^-(\mathcal{I}_{p_1,Q_1}) = \sum_{\lambda>0} m^0(\mathcal{I}_{p_1+\lambda\tilde{p},Q_1}). \tag{74}$$

Remark 5.1. Here $p_1 + \lambda\tilde{p}$ denotes

$$\mathrm{diag}(0_{(\hat{m}-m)n}, p_1) + \lambda \mathrm{diag}(0_{(\hat{m}-\tilde{m})n}, \tilde{p}),$$

where $\hat{m} = \max\{m, \tilde{m}\}$.

The proof of the following two lemmas is standard and is omitted.

Lemma 5.4. *Let H be a complex vector space and V be a linear subspace of H. Let $\mathcal{I}$ be a quadratic form on H and Λ be an inner product on H. If $m^-(\mathcal{I}) < +\infty$, then there holds*

$$m^-(\mathcal{I}) \ge m^-(\mathcal{I}|_V) \ge \sum_{\lambda>0} m^0(\mathcal{I}|_V + \lambda\Lambda).$$ □

Lemma 5.5. *Let H be a normed vector space and V a dense subspace. Let $\mathcal{I}$ be a quadratic form on H. Assume that $\mathcal{I}$ is bounded, i.e., there exists a constant $M > 0$ such that $\mathcal{I}(x,y) \le M\|x\|\|y\|$ for all $x, y \in H$. If $m^-(\mathcal{I}) < +\infty$, then there holds*

$$m^-(\mathcal{I}) = m^-(\mathcal{I}|_V). \qquad \square$$

Proof of Proposition 5.1. We divide the proof into four steps.

Step 1. Set $\tilde{p}_0 = I_{(m+1)n}$. Then (74) holds for $\tilde{p} = \tilde{p}_0$.

For $\tilde{p}_0$, there exists a $\lambda^+(0) > 0$ such that $\mathcal{I}_{p_1+\lambda^+\tilde{p}_0,Q_1}$ is positive definite. By Lemma 5.1, $\mathcal{I}_{p_1+\lambda\tilde{p}_0,Q_1}$ is Fredholm for each $\lambda \ge 0$. By the spectral properties of self-adjoint Fredholm operators we have

$$m^-(\mathcal{I}_{p_1,Q_1}) = \sum_{\lambda>0} m^0(\mathcal{I}_{p_1+\lambda\tilde{p}_0,Q_1}) < +\infty.$$

Step 2. Assume that $\tilde{m} \le m$. Set $\tilde{p}_s = (1-s)\tilde{p}_0 + s\mathrm{diag}(0_{(m-\tilde{m})n}, \tilde{p})$ for $0 \le s \le 1$ Then there exists $\lambda^+ > 0$ such that $\mathcal{I}_{p_1+\lambda^+\tilde{p}_s,Q_1}$ is positive definite.

Set $\tilde{p}_2 = I_n$. By Lemma 5.4 and Step 1 we have

$$m^-(\mathcal{I}_{p_1,Q_1}) \ge \sum_{\lambda>0} m^0(\mathcal{I}_{p_1+\lambda\tilde{p}_2,Q_1}). \tag{75}$$

Since $L_{p_1,W_{2m}(Q_1)}$ is a self-adjoint operator with compact resolvent, by (75), the operator is bounded from below. Thus there exists $\lambda^+(2) > 0$ such that $\mathcal{I}_{p_1+\lambda^+(2)\tilde{p}_2,Q_1}$ is positive definite. Since $\mathcal{I}_{\tilde{p}}$ is positive definite, there exists $M > 0$ such that $\mathcal{I}_{\tilde{p}_s} - M\mathcal{I}_{\tilde{p}_2}$ is positive definite for each $s \in [0,1]$. Then our claim holds for $\lambda^+ = \frac{\lambda^+(2)}{M}$.

Step 3. Equation (74) holds for $\tilde{m} \le m$.

Since $\mathcal{I}_{\tilde{p}}$ is positive definite, by Lemmas 5.2 and 4.3, the crossing form $\Gamma\left(\mathrm{Gr}(\gamma_{p_1+\lambda\tilde{p}}(T)), W_{2m}(Q_1), \lambda\right)$ is negative definite if $\tilde{m} \le m$ and $\lambda \le 0$, or $\tilde{m} > m$ and $\lambda < 0$. By Step 2, Proposition 4.1 and the definition of the Maslov-type index, for $s \in [0,1]$ we have

$$-i_{W_{2m}(Q_1)}\left(\gamma_{p_1+\lambda\tilde{p}_s}(T); 0 \le \lambda \le \lambda^+\right) = \sum_{\lambda>0} m^0(\mathcal{I}_{p_1+\lambda\tilde{p}_s,Q_1}). \tag{76}$$

Since $m^0(\mathcal{I}_{p_1+\lambda^+\tilde{p}_s,Q_1})=0$, by (76), Steps 1, 2 and the homotopy invariance of the Maslov-type index, we have

$$\begin{aligned} m^-(\mathcal{I}_{p_1,Q_1}) &= \sum_{\lambda>0} m^0(\mathcal{I}_{p_1+\lambda\tilde{p}_0,Q_1}) \\ &= -i_{W_{2m}(Q_1)}\left(\gamma_{p_1+\lambda\tilde{p}_0}(T); 0\le\lambda\le\lambda^+\right) \\ &= -i_{W_{2m}(Q_1)}\left(\gamma_{p_1+\lambda\tilde{p}}(T); 0\le\lambda\le\lambda^+\right) \\ &= \sum_{\lambda>0} m^0(\mathcal{I}_{p_1+\lambda\tilde{p},Q_1}). \end{aligned}$$

Step 4. Equation (74) holds for $\tilde{m}>m$. Since $H^{\tilde{m}}([0,T];\mathbf{C}^n)\cap H_R$ is dense in H_R, by Steps 1,3, Lemma 5.5 and Proposition 4.1, for sufficiently small $\epsilon>0$ and sufficiently large λ^+, we have

$$\begin{aligned} m^-(\mathcal{I}_{p_1,Q_1}) &= m^-(\mathcal{I}_{p_1,Q_1}|_{H^{\tilde{m}}([0,T];\mathbf{C}^n)\cap H_R}) \\ &= \sum_{\lambda>0} m^0(\mathcal{I}_{p_1+\lambda\tilde{p},Q_1}) \\ &= -i_{W_{2m}(Q_1)}\left(\gamma_{p_1+\lambda\tilde{p}}(T); \epsilon\le\lambda\le\lambda^+\right). \end{aligned}$$

□

Proof of Corollary 2.1. Set $p_0(t)=\operatorname{diag}(p_{m,m}(1,t),0_{mn})$, and

$$\gamma_{p_0}(t)=\begin{pmatrix} M_{1,1}(t) & 0 \\ M_{2,1}(t) & M_{2,2}(t)\end{pmatrix}.$$

Let $x=(x_0,\ldots,x_{m-1})$ and $y=(y_0,\ldots,y_{m-1})$ be two vectors in $\mathbf{C}^{mn}$. By direct calculation we get our form of $\gamma_{p_0}=(\gamma_{k,l}(t))_{k,l=0,\ldots,2m-1}$ and (24) with $p_{m,m}(0,t)=p_{m,m}(1,t)$. Then we have

$$\begin{aligned} &\langle M_{1,1}(T)^* K_{m,n}^* M_{2,1}(T)x, y\rangle \\ &= \sum_{k,l=0,\ldots,m-1} \left\langle \left(\frac{1}{(m-k-1)!(m-l-1)!} \right.\right. \\ &\qquad \left.\left. \int_0^T t^{2m-k-l-2}(p_{m,m}(1,t))^{-1}dt\right) x_l, y_k \right\rangle \\ &= \int_0^T \left\langle (p_{m,m}(1,t))^{-1} \sum_{l=0,\ldots,m-1} \frac{t^{m-l-1}}{(m-l-1)!}x_l, \right. \\ &\qquad \left. \sum_{k=0,\ldots,m-1} \frac{t^{m-k-1}}{(m-k-1)!}y_k \right\rangle dt\,. \end{aligned}$$

Since $p_{m,m}(1,t)$ is positive definite for each $t \in [0,T]$, we have $\langle M_{1,1}(T)^* K_{m,n}^* M_{2,1}(T)x, x\rangle \geq 0$. If $\langle M_{1,1}(T)^* K_{m,n}^* M_{2,1}(T)x, y\rangle = 0$, we have $\sum_{k=0,\dots,m-1} \frac{t^{m-k-1}}{(m-k-1)!} x_k = 0$ for all $t \in [0,T]$. By taking derivatives with t, we have $\sum_{l=0,\dots,k} \frac{t^{k-l-1}}{(k-l-1)!} x_l = 0$ for all $k = 0,\dots,m-1$ and $t \in [0,T]$. Then we get $x_k = 0$ for $k = 0,\dots,m-1$ and $x = 0$. Thus $M_{1,1}(T)^* K_{m,n}^* M_{2,1}(T)$ is positive definite.

Let $p_s = (1-s)p_0 + sp_1$ and $Q_s = sQ_1$. Clearly $\mathcal{I}_{p_0,R}$ and $L_{p_0,W_{2m}(R)}$ is non-negative definite. Note that $M_{1,1}(0) = I_{mn}$ and $S(0) = S$. By the definition of the spectral flow, Theorem 2.1 and Theorem 2.2, we have

$$
\begin{aligned}
m^-(\mathcal{I}_{p_1,Q_1}) &= -\mathrm{sf}\{\mathcal{I}_{p_s,Q_s}; 0 \leq s \leq 1\} \\
&= i_{W_{2m}(Q_1)}(\gamma_{p_1}) - i_{W_{2m}(R)}(\gamma_{p_0}) \\
&= i_{W_{2m}(Q_1)}(\gamma_{p_1}) - (\dim S(T) + \dim S(0) - \dim S(T)) \\
&= i_{W_{2m}(Q_1)}(\gamma_{p_1}) - \dim S.
\end{aligned}
$$

Applying Proposition 5.1 for $\tilde{p} = I_n$, we obtain

$$
m^-(\mathcal{I}_{p_1,Q_1}) = m^-(L_{p_1,W_{2m}(Q_1)}). \qquad \square
$$

5.4. *Proof of Theorem 2.3*

Let a, p_1, p_1' and R' be as in §2. Firstly we prove (26). The following lemma follows from direct calculation.

Lemma 5.6. *We have*

$$
p_1' = \begin{pmatrix} a^* & 0 \\ \dot{a}^* & a^* \end{pmatrix} p_1 \begin{pmatrix} a & \dot{a} \\ 0 & a \end{pmatrix},
$$

$$
b(p_1') = \mathrm{diag}(a^{-1}, a^*) b(p_1) \mathrm{diag}(a^{*-1}, a) + \begin{pmatrix} 0 & -a^{-1}\dot{a} \\ -\dot{a}^* a^{*-1} & 0 \end{pmatrix}. \qquad \square
$$

By Corollary 4.2 we have

Corollary 5.1. *We have*

$$
\gamma_1' = \mathrm{diag}(a^*, a^{-1}) \gamma_1 \mathrm{diag}(a(0)^{*-1}, a(0)). \tag{77}
$$

Proof of Theorem 2.3. By the definition of R' we have

$$
(R')^{2,b} = \{(x,y) \in \mathbf{C}^{2n}; (a(0)^* x, a(T)^* y) \in R^{2,b}\}.
$$

By Theorem 2.2 and Lemma 4.4, we have

$$\begin{aligned}
&i_{W_2(R')}(\gamma_1') \\
&= i_{W_2(R')}(\mathrm{diag}(a^*, a^{-1})\gamma_1 \mathrm{diag}(a(0)^{*-1}, a(0))) \\
&= i_{W_2(R)}(\gamma_1) + i_{W_2(R')}(\mathrm{diag}(a^*, a^{-1})\mathrm{diag}(a(0)^{*-1}, a(0))) \\
&= i_{W_2(R)}(\gamma_1) + \dim(\mathrm{Gr}(I_n) \cap (R')^{2,b})) - \dim(\mathrm{Gr}(a(T)^* a(0)^{*-1}) \cap (R')^{2,b}) \\
&= i_{W_2(R)}(\gamma_1) + \dim(\mathrm{Gr}(I_n) \cap (R')^{2,b})) - \dim(\mathrm{Gr}(I_n) \cap R^{2,b}). \qquad \square
\end{aligned}$$

Acknowledgement. This work was partially done when the author visited MIT in 2001 and MPI, Leipzig in 2002-2003. The research atmosphere is very nice there. The author sincerely thanks Professor Gang Tian and Professor Chun-gen Liu for helpful discussions and valuable suggestions, and the referees for their careful reading, pointing out mistakes and typos, and valuable comments on the earlier version of this paper.

References

1. A. A. Agrachev, A. V. Sarychev, *Abnormal sub-Riemannian geodesics: Morse index and rigidity*, Ann. Inst. H. Poincaré Anal. Non Linéaire **13** (1996), 635-690.
2. W. Ambrose, *The index theorem in Riemannian geometry*, Ann. of Math. **73** (1961), 49-86.
3. V. I. Arnol'd, *A characteristic class entering into quantization conditions*, Funkcional Anal. i Priložen. **1** (1967), 1-14 (Russian); Funct. Anal. Appl. **1** (1967), 1-13 (English translation).
4. M. F. Atiyah, V. K. Patodi, and I. M. Singer, *Spectral asymmetry and Riemannian geometry. III*, Math. Proc. Cambridge Phil. Soc. **79** (1976), 71-99.
5. J. K. Beem and P. E. Ehrlich, *Cut points, conjugate points and Lorentzian comparison theorems*, Math. Proc. Cambridge Phil. Soc. **86** 1979, 365-384.
6. J. K. Beem and P. E. Ehrlich, *A Morse index theorem for null geodesics*, Duke Math. J., **46** (1979), 561-569.
7. B. Booss-Bavnbek and K. Furutani, *The Maslov index – a functional analytical definition and the spectral flow formula*, Tokyo J. Math. **21** (1998), 1-34.
8. B. Booss-Bavnbek, M. Lesch, and J. Phillips, *Unbounded Fredholm operators and spectral flow*, Canad. J. Math. **57** (2005), 225-250 (arXiv: math.FA/0108014).
9. B. Booss-Bavnbek and K. P. Wojciechowski, Elliptic Boundary Problems for Dirac Operators. Birkhäuser, Boston, 1993.
10. B. Booss-Bavnbek and C. Zhu, *Weak symplectic functional analysis and general spectral flow formula*, Preprint, Roskilde 2004 (arXiv: math.DG/0406139).
11. B. Booss-Bavnbek and C. Zhu, *General spectral flow formula for*

fixed maximal domain, Cent. Eur. J. Math. **3** (2005), 558-577 (arXiv: math.DG/0504125).

12. S. E. Cappell, R. Lee, and E. Y. Miller, *On the Maslov index*, Comm. Pure Appl. Math. **47** (1994), 121-186.
13. X. Dai and W. Zhang, *Splitting of the family index*, Comm. Math. Phys. **182** (1996), 303-317.
14. X. Dai and W. Zhang, *Higher spectral flow*, J. Funct. Anal. **157** (1998), 432-469.
15. J. J. Duistermaat, *On the Morse index in variational calculus*, Adv. Math. **21** (1976), 173-195.
16. H. Edwards, *A generalized Sturm Theorem*, Ann. of Math. **80** (1964), 2-57.
17. W. N. Everitt and L. Markus, *Complex symplectic spaces and boundary value problems*, Bull. Amer. Math. Soc. **42**/4 (2005), 461-500.
18. P. M. Fitzpatrick, J. Pejsachowicz and L. Recht, *Spectral flow and bifurcation of critical points of strongly-indefinite functionals. I. General theory*, J. Funct. Anal. **162** (1999), 52-95.
19. A. Floer, *A relative Morse index for the symplectic action*, Comm. Pure Appl. Math., **41** (1988), 393-407.
20. A. D. Helfer, *Conjugate points on spacelike geodesics or pseudo-self-adjoint Morse-Sturm-Liouville systems*, Pacific J. Math. **164** (1994), 321-340.
21. L. Hörmander, *Fourier integral operators I*, Acta Math., **127** (1971), 79-183.
22. T. Kato, Pertubation Theory for Linear Operators. Springer-Verlag, Berlin, 1966, 3d ed., 1980.
23. P. Kirk and M. Lesch, *The η–invariant, Maslov index, and spectral flow for Dirac–type operators on manifolds with boundary*, Forum Math. **16** (2004), 553-629 (arXiv: math.DG/0012123).
24. Y. Long, *Bott formula of the Maslov-type index theory*, Pacific J. Math. **187** (1999), 113-149.
25. R. B. Melrose and P. Piazza, *Families of Dirac operators, boundaries and the b-calculus*, J. Differential Geom. **46** (1997), 99-180.
26. Y. Long and C. Zhu, *Maslov-type index theory for symplectic paths and spectral flow (II)*, Chinese Ann. of Math. **21B**/1 (2000), 89-108.
27. M. Morse, The Calculus of Variations in the Large. A.M.S. Coll. Publ., Vol.18, Amer. Math. Soc., New York, 1934.
28. J. Phillips, *Self–adjoint Fredholm operators and spectral flow*, Canad. Math. Bull. **39** (1996), 460-467.
29. P. Piccione and D. V. Tausk, *The Maslov index and a generalized Morse index theorem for non-positive definite metrics*, C. R. Acad. Sci. Paris Sér. I Math. **331** (2000), 385-389.
30. P. Piccione and D. V. Tausk, *The Morse index theorem in semi-Riemannian Geometry*, Topology **41** (2002), 1123-1159 (arXiv: math.DG/0011090).
31. P. Piccione and D. V. Tausk, *An index theory for paths that are solutions of a class of strongly indefinite variational problems*, Calc. Var. Partial Differential Equations, **15**/4 (2002), 529-551 (arXiv: math.DG/0108044v1).
32. J. Robbin and D. Salamon, *The Maslov index for paths*, Topology **32** (1993), 827-844.

33. J. Robbin and D. Salamon, *The spectral flow and the Maslov index*, Bull. London Math. Soc. **27** (1995), 1-33.
34. S. Smale, *On the Morse index theorem*, J. Math. Mech. **14** (1965), 1049-1056.
35. K. Uhlenbeck, *The Morse index theorem in Hilbert space*, J. Differential Geom. **8** (1973), 555-564.
36. K. P. Wojciechowski, *Spectral flow and the general linear conjugation problem*, Simon Stevin **59** (1985), 59-91.
37. C. Zhu, Maslov-type index theory and closed characteristic on compact convex hypersurfaces in $\mathbf{R}^{2n}$. Ph. D. Thesis. Nankai Institute of Mathematics, Tianjin, 1999.
38. C. Zhu, *The Morse index theorem for regular Lagrangian systems* (arXiv: math.DG/0109117).
39. C. Zhu and Y. Long, *Maslov-type index theory for symplectic paths and spectral flow (I)*, Chinese Ann. of Math. **20B**/4 (1999), 413-424.

Received by the editors September 15, 2005; Revised January 3, 2006